Illustrating Statistical Procedures: Finding Meaning in Quantitative Data

Ray W. Cooksey

Illustrating Statistical Procedures: Finding Meaning in Quantitative Data

Third Edition

Ray W. Cooksey
UNE Business School
University of New England
Armidale, NSW, Australia

ISBN 978-981-15-2539-1   ISBN 978-981-15-2537-7 (eBook)
https://doi.org/10.1007/978-981-15-2537-7

1st edition: © Tilde Publishing and Distribution 2007
2nd edition: © Tilde Publishing and Distribution 2014
© Springer Nature Singapore Pte Ltd. 2020

This work is subject to copyright. All rights are reserved by the Publisher, whether the whole or part of the material is concerned, specifically the rights of translation, reprinting, reuse of illustrations, recitation, broadcasting, reproduction on microfilms or in any other physical way, and transmission or information storage and retrieval, electronic adaptation, computer software, or by similar or dissimilar methodology now known or hereafter developed.

The use of general descriptive names, registered names, trademarks, service marks, etc. in this publication does not imply, even in the absence of a specific statement, that such names are exempt from the relevant protective laws and regulations and therefore free for general use.

The publisher, the authors, and the editors are safe to assume that the advice and information in this book are believed to be true and accurate at the date of publication. Neither the publisher nor the authors or the editors give a warranty, expressed or implied, with respect to the material contained herein or for any errors or omissions that may have been made. The publisher remains neutral with regard to jurisdictional claims in published maps and institutional affiliations.

This Springer imprint is published by the registered company Springer Nature Singapore Pte Ltd.
The registered company address is: 152 Beach Road, #21-01/04 Gateway East, Singapore 189721, Singapore

Preface

This third edition provides a further updated and extended set of descriptions and illustrations for a comprehensive range of statistical procedures. I have retained the broad focus on statistical analysis in business, behavioural and social sciences, encompassing areas such as general business, management, marketing, human resource management, organisational behaviour, some aspects of economics, finance and accounting, education, psychology, sociology, criminology, some aspects of public policy, health and biomedical research. During your journey, I introduce you to statistical analysis as a research activity as well as provide an overview and illustration of each statistical procedure – what it is, what it does, what it gives you, what its benefits and drawbacks are and where you can go to find out more about it. Along the way, I explain a number of fundamental statistical principles and concepts and review a number of useful computer packages, indicating how they might be used to accomplish each type of analysis.

There are still very few formulas within the covers of this book; the few that are included appear in an Appendix right at the end. Instead, I want to help you attain a high level of conceptual understanding, unobscured by mathematical and computational details.

In the larger scheme of things, statistical analysis of quantitative data comes rather late in the execution of any research investigation. A great deal of literature review, preliminary planning, research design and measurement construction work must be done before any data are collected. Then, observations/measurements must be systematically obtained in appropriate ways and represented in quantified (numerical) form prior to any analyses. If there are problems with the logic or execution of these early stages of the research process, no amount of sophisticated statistical analysis will save the day. However, if all goes well in these preliminary stages to produce good quantitative data, then statistics provides coherent pathways for making sense of those data.

Statistical analysis is one important means of getting to defensible and convincing research stories, although it is certainly not the only means. An important thing to remember is that statistical procedures are nothing more than research tools. This book helps you explore the diverse statistical analysis toolkit by demystifying and illustrating each of the tools. Many people find that statistical analysis employs highly specialised language and concepts that impose a major barrier to learning. I recognised this problem and have written the book in a way that hopefully makes the specialised language and concepts more accessible, thereby setting the stage for you to achieve a better understanding of statistical analysis processes and purposes to which they might be put.

Statistical analysis can be challenging but can be fun as well. The idea of discovering meaning in data you have personally worked hard to collect can be very rewarding. If you have a research project in which you must use statistics to help address your research questions, I hope you find this book provides a useful foundation for understanding the varied possibilities for analysis. If you are primarily a consumer of the published work of other researchers, then I hope this book provides the foundations you need to come to grips with the concepts and procedures that authors employ to help make their research conclusions and generalisations convincing to you as a reader.

In this third edition, I have worked with a new publisher to produce a refined text that can reach a more diversified audience. My philosophy remains the same, but there is more polish on the product!

Enjoy your journey!

Armidale, NSW, Australia Ray W. Cooksey
March 2020

Acknowledgements

I would like to acknowledge the contributions played by my students and colleagues over many years, whose influence is reflected in the content of this book. Ever since I wrote the first and slimmest version of this book over 36 years ago, resulting in what was affectionately called the "little green book", I have been encouraged to keep this book updated without losing its fundamental premise: give my readers the essential concepts, with lots of illustrations, without obscuring their learning with formulas and theorems. I hope I have continued to succeed in this endeavour. Finally, I wish to acknowledge the support of my wife, Christie, who has accompanied me on every step of the journey taken to produce the many versions of this book over the years.

Armidale, NSW, Australia   Ray W. Cooksey

Contents

1 The Logic & Language of Statistical Analysis 1
    The Language of Statistics 4
        Bivariate Statistics 4
        Cause and Effect 4
        Correlation 6
        Correlational Statistics 7
        Descriptive Statistics 7
        Dependent Variable 8
        Distribution 8
        Extraneous Variable 11
        Independent Variable 11
        Inferential Statistics 12
        Mean 13
        Mediating Variable 13
        Moderating Variable 14
        Multivariate Statistics 14
        Nonparametric Statistics 15
        Normal Distribution 15
        Parameter 16
        Parametric Statistics 16
        Population 17
        Probability 17
        Random Sampling 18
        Sample 19
        Statistic 19
        Statistical Inference 19
        Test of Significance 20
        Univariate Statistics 21

	Variability	21
	Variable	21
	Variance Explained	22
	Reference	22
2	**Measurement Issues in Quantitative Research**	**23**
	Hypothetical Constructs, Operational Definitions and Construct Validity	23
	Measurement Scales	26
	References	31
3	**Computer Programs for Analysing Quantitative Data**	**33**
	SPSS	33
	SYSTAT	35
	NCSS	37
	STATGRAPHICS Centurion	38
	R	39
	Comparing the Software Packages	41
	Other Available Statistical Packages	42
	Stata, STATISTICA, SAS, eViews & Mplus	43
	UNISTAT, XLSTAT & Excel	44
	AMOS	45
	Considerations When Using Computers for Quantitative Analysis	45
	Preparing Data for Analysis by Computer	46
	References	49
4	**Example Research Context & Quantitative Data Set**	**51**
	The Quality Control Inspectors (QCI) Database	52
	Research Context	52
	Quantitative Stories That Could Be Pursued in the QCI Database	56
	References	60
5	**Descriptive Statistics for Summarising Data**	**61**
	Procedure 5.1: Frequency Tabulation, Distributions & Crosstabulation	62
	Frequency Tabulation and Distributions	62
	Crosstabulation	64
	Advantages	66
	Disadvantages	67
	Where Is This Procedure Useful?	67
	Software Procedures	68
	Procedure 5.2: Graphical Methods for Displaying Data	68
	Bar and Pie Charts	69
	Histograms and Frequency Polygons	70
	Line Graphs	73
	Scatterplots	75

Contents

- Advantages 78
- Disadvantages 78
- Where Is This Procedure Useful? 78
- Software Procedures 79
- Procedure 5.3: Multivariate Graphs & Displays 79
  - Scatterplot Matrices 80
  - Radar Plots 82
  - Multiplots 84
  - Parallel Coordinate Displays 86
  - Icon Plots 87
  - Advantages 92
  - Disadvantages 92
  - Where Is This Procedure Useful? 93
  - Software Procedures 93
- Procedure 5.4: Assessing Central Tendency 94
  - Mean 94
  - Median 96
  - Mode 97
  - Advantages 98
  - Disadvantages 98
  - Where Is This Procedure Useful? 99
  - Software Procedures 99
- Procedure 5.5: Assessing Variability 100
  - Range 100
  - Interquartile Range 101
  - Variance 103
  - Standard Deviation 105
  - Advantages 106
  - Disadvantages 106
  - Where Is This Procedure Useful? 107
  - Software Procedures 107
- Fundamental Concept I: Basic Concepts in Probability 108
  - The Concept of Simple Probability 108
  - The Concept of Conditional Probability 110
- Procedure 5.6: Exploratory Data Analysis 112
  - Stem & Leaf Displays 113
  - Boxplots 114
  - Violin Plots 118
  - Advantages 119
  - Disadvantages 119
  - Where Is This Procedure Useful? 120
  - Software Procedures 120

Procedure 5.7: Standard (z) Scores	121
Advantages	126
Disadvantages	126
Where Is This Procedure Useful?	127
Software Procedures	127
Fundamental Concept II: The Normal Distribution	128
Checking for Normality	130
References	134
6 Correlational Statistics for Characterising Relationships	**141**
Fundamental Concept III: Correlation and Association	142
Interpretive Guidelines for a Correlation Coefficient	145
Data Patterns that Can Influence or Distort Correlations	146
Procedure 6.1: Assessing Correlation	150
Pearson Correlation	150
Spearman's Rho Correlation	152
Point-Biserial Correlation	154
Biserial Correlation	155
Tetrachoric Correlation	156
Advantages	157
Disadvantages	158
Where Is This Procedure Useful?	158
Software Procedures	159
Procedure 6.2: Assessing Association in Contingency Tables	160
Phi Coefficient and Cramer's V	161
Proportional Reduction in Error (PRE) Statistics	163
Gamma Statistic, Kendall's Tau Statistic, and Somer's D Statistic	164
Advantages	166
Disadvantages	167
Where Is This Procedure Useful?	167
Software Procedures	168
Procedure 6.3: Simple Linear Regression & Prediction	168
Advantages	173
Disadvantages	174
Where Is This Procedure Useful?	174
Software Procedures	175
Procedure 6.4: Multiple Regression & Prediction	175
Advantages	182
Disadvantages	183
Where Is This Procedure Useful?	184
Software Procedures	184
Fundamental Concept IV: Partial & Semi-Partial Correlation (Hierarchical Regression)	185
Semi-Partial Correlation	186

Partial Correlation	189
What Can Be Partialled?	190
Procedure 6.5: Exploratory Factor Analysis	191
Advantages	199
Disadvantages	200
Where Is This Procedure Useful?	201
Software Procedures	202
Procedure 6.6: Cluster Analysis	203
Advantages	211
Disadvantages	211
Where Is This Procedure Useful?	213
Software Procedures	213
Procedure 6.7: Multidimensional Scaling and Correspondence Analysis	215
Multidimensional Scaling	216
Correspondence Analysis	219
Advantages	221
Disadvantages	221
Where Is This Procedure Useful?	222
Software Procedures	222
Procedure 6.8: Canonical & Set Correlation	224
Canonical Correlation	224
Set Correlation	228
Advantages	230
Disadvantages	231
Where Is This Procedure Useful?	232
Software Procedures	232
References	233
7 Inferential Statistics for Hypothesis Testing	**241**
Fundamental Concept V: The Logic and Rules of Statistical Inference for Hypothesis Testing	243
The Role of Sampling Distributions in Hypothesis Testing	246
The Statistical Inference Process in Practice	251
The Risk of Inflating the Alpha or Type I Error	253
Assumptions About the Data Required by Hypothesis Testing	254
Fundamental Concepts VI: The General Linear Model	256
Building Models	256
Configuring Predictors	258
How General Is the General Linear Model?	260
Limits to a General Linear MODEL	260
Fundamental Concept VII: Principles of Experimental & Quasi-Experimental Design	264
The Concept of Control Over Extraneous Variables	265
Common Types of Experimental and Quasi-Experimental Designs	270

Fundamental Concept VIII: Principles of Sampling 274
  Types of Sampling Schemes 274
  The Concept of Standard Error 278
Procedure 7.1: Contingency Tables: Test of Significance 280
  Advantages 283
  Disadvantages 284
  Where Is This Procedure Useful? 284
  Software Procedures 285
Procedure 7.2: t-Test for Two Independent Groups 286
  Advantages 289
  Disadvantages 290
  Where Is This Procedure Useful? 290
  Software Procedures 291
Procerdure 7.3: Mann-Whitney U-Test for Two Independent Groups ... 292
  Advantages 295
  Disadvantages 295
  Where Is This Procedure Useful? 296
  Software Procedures 296
Procedure 7.4: t-Test for Two Related Groups 297
  Advantages 303
  Disadvantages 303
  Where Is This Procedure Useful? 304
  Software Procedures 305
Procedure 7.5: Wilcoxon Signed Ranks Text for Two Related
Groups 305
  Advantages 310
  Disadvantages 310
  Where Is This Procedure Useful? 310
  Software Procedures 311
Procedure 7.6: One-Way Analysis of Variance (ANOVA) 312
  Advantages 316
  Disadvantages 316
  Where Is This Procedure Useful? 317
  Software Procedures 317
Procedure 7.7: Assessing Effect Sizes 318
  Advantages 321
  Disadvantages 321
  Where Is This Procedure Useful? 322
  Software Procedures 322
Procedure 7.8: Planned Comparisons and Posthoc Multiple
Comparisons 323
  Planned Comparison Tests 324
  Posthoc Multiple Comparison Tests 328
  Advantages 332

Disadvantages	332
Where Is This Procedure Useful?	333
Software Procedures	333
Procedure 7.9: Kruskal-Wallis Rank Test for Several Independent Groups	335
Advantages	339
Disadvantages	339
Where Is This Procedure Useful?	340
Software Procedures	340
Procedure 7.10: Factorial Between-Groups ANOVA	341
Advantages	346
Disadvantages	347
Where Is This Procedure Useful?	349
Software Procedures	349
Fundamental Concept IX: The Concept of Interaction	350
Interaction in ANOVA Designs	350
Interactions as Moderator Effects	356
Procedure 7.11: Repeated Measures ANOVA	357
Advantages	363
Disadvantages	363
Where Is This Procedure Useful?	365
Software Procedures	365
Procedure 7.12: Friedman Rank Test for Repeated Measures	366
Advantages	370
Disadvantages	371
Where Is This Procedure Useful?	371
Software Procedures	371
Procedure 7.13: Multiple Regression: Tests of Significance	372
Testing R^2	375
Testing Regression Coefficients	376
Checking Residual Patterns & Assumptions	380
Testing Squared Semi-partial Correlations Hierarchically	383
Advantages	388
Disadvantages	389
Where Is This Procedure Useful?	390
Software Procedures	391
Procedure 7.14: Logistic Regression	393
Advantages	399
Disadvantages	400
Where Is This Procedure Useful?	400
Software Procedures	401
Procedure 7.15: Analysis of Covariance (ANCOVA)	402
Advantages	407
Disadvantages	407

	Where Is This Procedure Useful?	407
	Software Procedures	408
	Procedure 7.16: Multivariate Analysis of Variance & Covariance (MANOVA & MANCOVA)	409
	Advantages	416
	Disadvantages	416
	Where Is This Procedure Useful?	417
	Software Procedures	418
	Procedure 7.17: Discriminant Analysis	420
	Advantages	426
	Disadvantages	427
	Where Is This Procedure Useful?	428
	Software Procedures	428
	Procedure 7.18: Log-Linear Models for Contingency Tables	429
	General Log-Linear Models	430
	Logit Log-Linear Models	432
	Advantages	435
	Disadvantages	436
	Where Is This Procedure Useful?	437
	Software Procedures	437
	References	438
8	**Other Commonly Used Statistical Procedures**	453
	Procedure 8.1: Reliability & Classical Item Analysis	455
	Internal Consistency Reliability	456
	Classical Item Analysis	457
	Inter-Observer Reliability (Inter-Rater Agreement)	464
	Advantages	466
	Disadvantages	466
	Where Is This Procedure Useful?	467
	Software Procedures	468
	Procedure 8.2: Data Screening & Missing Value Analysis	469
	Data Screening	470
	Missing Value Analysis	476
	Advantages	484
	Disadvantages	484
	Where Is This Procedure Useful?	485
	Software Procedures	485
	Procedure 8.3: Confidence Intervals	487
	Advantages	490
	Disadvantages	490
	Where Is This Procedure Useful?	491
	Software Procedures	491
	Procedure 8.4: Bootstrapping & Jackknifing	492
	Bootstrapping	493

Contents   xvii

    Jackknifing 498
    Advantages 498
    Disadvantages 499
    Where Is This Procedure Useful? 499
    Software Procedures 500
  Procedure: 8.5 Time Series Analysis 500
    Time Series Analysis Without an Intervention 502
    Time Series Analysis with an Intervention 506
    Advantages 512
    Disadvantages 512
    Where Is This Procedure Useful? 513
    Software Procedures 513
  Procedure 8.6: Confirmatory Factor Analysis 515
    Advantages 521
    Disadvantages 522
    Where Is This Procedure Useful? 523
    Software Procedures 524
  Procedure 8.7: Structural Equation (Causal Path) Models (SEM) 525
    The Covariance-Based Approach to SEM 526
    The Partial Least Squares (PLS) Approach to SEM 532
    Advantages 536
    Disadvantages 536
    Where Is This Procedure Useful? 538
    Software Procedures 538
  Procedure 8.8: Meta-analysis 540
    Advantages 546
    Disadvantages 547
    Where Is This Procedure Useful? 548
    Software Procedures 549
  References 549

9 Specialised Statistical Procedures 557
  Fundamental Concept X: Bayesian Statistical Inference – An
  Alternative Logic 560
  Procedure 9.1: Rasch Models & Item Response Theory 564
    Rasch Models 564
    More General Item Response Theory Models 568
    Advantages 571
    Disadvantages 571
    Where Is This Procedure Useful? 572
    Software Procedures 573
  Procedure 9.2: Survival/Failure Analysis 574
    Advantages 578
    Disadvantages 578

	Where Is This Procedure Useful?	579
	Software Procedures	579
Procedure 9.3: Quality Control Charts		580
	Pareto Charts	581
Control X-Bar Charts		582
	R charts	584
	Advantages	586
	Disadvantages	586
	Where Is This Procedure Useful?	587
	Software Procedures	588
Procedure 9.4: Conjoint Measurement & Choice Modelling		589
	Conjoint Measurement or Preference Modelling	590
	Choice Modelling	596
	Best-Worst Scaling Models	596
	Advantages	599
	Disadvantages	600
	Where Is This Procedure Useful?	600
	Software Procedures	601
Procedure 9.5: Multilevel (Hierarchical Linear/Mixed) Models		603
	Advantages	611
	Disadvantages	611
	Where Is This Procedure Useful?	612
	Software Procedures	613
Procedure 9.6: Classification & Regression Trees		614
	Advantages	619
	Disadvantages	620
	Where Is This Procedure Useful?	621
	Software Procedures	621
Procedure 9.7: Social Network Analysis		622
	Advantages	631
	Disadvantages	632
	Where Is This Procedure Useful?	632
	Software Procedures	632
Procedure 9.8: Specialised Forms of Regression Analysis		633
	Other Forms Of Regression Analysis	634
	Alternative Estimation Approaches for Regression Modelling	640
	More Sophisticated Regression Modelling Approaches	640
	Software Procedures	651
Procedure 9.9: Data Mining & Text Mining		652
	Data Mining Approaches	653
	Text Mining Approaches	661
	Advantages	664
	Disadvantages	664

Where Is This Procedure Useful?	666
Software Procedures	666
Procedure 9.10: Simulation & Computational Modelling	668
Monte Carlo Simulation Methods	670
Computational Modelling	673
Advantages	681
Disadvantages	681
Where Is This Procedure Useful?	682
Software Procedures	683
References	685

Appendices 695

Index 729

Introduction

Abstract This book focuses on a wide range of concepts and procedures collectively considered to constitute the domain of statistical analysis. The approach taken is to describe, in plain English, various statistical concepts and procedures without readers having to worry about the mathematical formulas behind them. Conceptual understanding of statistics is separated from understanding the mathematical basis of statistics; thus, readers will not find a single formula in any of the descriptions and illustrations. The discussion of statistics is divided into three broad areas: *descriptive statistics*, *correlational statistics* and *inferential statistics*, each reflecting an emphasis on supporting different types of stories about data. Additionally, a range of commonly used but more advanced statistical procedures are discussed, as is a range of rather more specialised approaches. Within each of these broad areas, various topics are categorised as either a *fundamental concept* or a *procedure*. The illustrations of procedures provided throughout the book focus on a single general research context and its associated quantitative data. Each procedure is discussed within a common format: general description and illustration(s), discussion of advantages and disadvantages, discussion of where the procedure might be useful and review of software approaches for implementing the procedure.

The purpose of this book is to focus on a wide range of concepts and procedures which are available for analysing quantitative research data. These concepts and procedures are collectively considered to constitute the domain of statistical analysis. It is an inescapable fact that in order to either read or conduct good business, behavioural and social science research, you must become acquainted with how such data are analysed.

- **Quantitative data** necessarily require the application of specific statistical formulas, analytical sequences and computations which transform and condense raw numerical data (e.g. people's marks on a survey questionnaire or readings from an instrument) into forms that can be interpreted as telling particular stories or as

supporting or not supporting a particular research question, hypothesis or prediction.
- **Qualitative data**, on the other hand, confront the researcher in non-numerical (often textual) form, and this necessitates quite different analytical approaches to extracting meaning from the data and generating appropriate interpretations.

However, in some circumstances, it may be appropriate to transform qualitative data into quantitative form, which means that there may be instances where qualitative research data can be usefully analysed using quantitative methods. This book should prove very useful to the reader in either case.

The Philosophy Behind This Book

My task in this book is to make your introduction to the domain of statistics as painless as possible. It has been my experience, over many years as a teacher of statistics to undergraduate students, postgraduate students, academic colleagues and professionals, e.g. psychologists, teachers, marketers, managers, accountants and nurses, that many students are averse to the topic of statistics because it relies on mathematics and many people are either anxious about their competence in mathematics or see little utility in their particular discipline for learning statistics, or both. I must agree that learning statistics is somewhat like learning a foreign language, but it need not be traumatic and impenetrable.

There are at least two ways in which knowledge of statistics is useful. Firstly, if you ever plan to conduct your own research investigation(s) involving quantitative data, the need for statistics is obvious. However, even if you never do a research study of your own, knowledge of statistics is of great benefit (I would go so far as to say absolutely essential) in helping you read and understand quantitative research published by others in your field. Furthermore, such knowledge can help you detect abuses of statistical methods which are very prevalent in the media and popular literature. These abuses may lead the unenlightened to draw unwarranted conclusions about some issue of importance, perhaps to the point of inappropriately influencing public policy or organisational strategy. It is easy to mislead with statistics, but you need some training in order to detect it; the people who abuse statistics rely on an ignorant public who don't know any better. This book offers an important steppingstone in such training.

This book provides a solid foundation on which to build more formal learning of statistics (i.e. how to do/compute them). Furthermore, it provides you with the conceptual awareness you need to become an enlightened consumer of quantitative research findings. *No matter what your initial state of knowledge and anxiety, you can learn statistics – anyone can – and be competent at using and interpreting them.* My job is to show you that the big mystery in statistics is really no mystery at all once you understand the logic, the language and some essential concepts.

The approach I took in this book is to describe, in plain English, various statistical concepts and procedures without you having to worry about the mathematical formulas behind them. That part of your learning of statistics can come later, if necessary. Hence, you will not find a single formula in any of the descriptions and illustrations in this book. I try to separate the conceptual understanding of statistics from the understanding of the mathematical basis of statistics. However, I also realise that if you go on to study statistics in a formal course, you will need to deal with the mathematical side of statistics; therefore, I have included Appendix A to ease you into some of the more important mathematical considerations you should know when you begin to deal with statistical formulas. Also, there may be rare occasions where you may need to compute some statistical quantity yourself, because your statistical package doesn't provide it. Appendix A will also show you some of these formulas.

This book will teach you about statistics – what they mean and what they can tell you about certain types of research questions – but it will not teach you how to compute statistics. I have left that task to any of the very good texts explicitly written for that purpose. Many of these good texts are cited in this book so you can acquire a clear roadmap to more in-depth learning in the area.

How This Book Is Organised

Much of the discussion of statistics in this book is divided into three broad areas: **descriptive statistics**, **correlational statistics** and **inferential statistics**, each reflecting an emphasis on supporting different types of stories about data. Additionally, a range of commonly used but more advanced statistical procedures is discussed, as is a range of rather more specialised approaches. Within each of these broad areas, various topics are categorised as either a **fundamental concept** or a **procedure**.

- **Fundamental concepts** are aspects of the domain of statistics or research design which provide crucial foundations for the understanding of material to follow later in the book. Each concept is located in the book at the point where it will provide maximum benefit to your learning. Therefore, each fundamental concept section should be read prior to undertaking any reading of the material that immediately follows it. A list of suggested additional readings accompanies the discussion of each fundamental concept.
- The discussion format for a topic classified as a **procedure** has a simple structure to facilitate your understanding. For each procedure, its central purpose(s) is identified, and appropriate measurement considerations are summarised. The procedure is then more fully described and illustrated, using figures and tables produced using a relevant statistical package, where appropriate. The advantages and disadvantages of each procedure are highlighted as are the situations/research contexts where the procedure could be most useful. Finally, relevant computer packages for conducting the procedure are listed, and a set of references for finding more detailed information on the procedure is provided.

The illustrations of procedures provided throughout the book focus on a single general research context and its associated quantitative data. This context is described more fully in Chap. 4 – *Example Research Context and Quantitative Data Set*. Illustrations of most procedures are based on analyses produced by one of the five major statistical programs/packages (*SPSS*, *SYSTAT*, *NCSS*, *STATGRAPHICS* or **R**) discussed in Chap. 3 – *Computer Programs for Analysing Quantitative Data*.

Most of the procedures discussed in this book are classified according to two distinct sets of categories – one category from each of the first two dimensions listed below; the third dimension regarding *assumptions* comes into play when the inferential category is invoked:

Classification dimension		
Number of dependent variables analysed	**Basic purpose of the statistical procedure**	**Assumptions required?**
Univariate	Descriptive	Parametric
Bivariate	Correlational	Nonparametric
Multivariate	Inferential	

Software Website References

Throughout Chap. 3, website addresses are included which provide summaries of the main features of each program, show how to acquire the program and, in some cases, allow you to download a trial or student version of the program for evaluation. Some websites (e.g. for *SPSS*) also have downloadable papers that have been written to illustrate the use of specific features of the program. While the web addresses are current at the time of publication, they may be subject to change out of the control of the publisher.

Software Procedures

Where software procedures are listed, the arrow symbol → indicates following a drop-down menu path to the next named option.

Supplementary Video Lecturettes

I have uploaded a series of short (less than 25 min each) video lecturettes that I have used in my teaching, both for statistical analysis and for general research methods (quantitative and qualitative), to YouTube. To access the playlist containing the 11 lecturettes on statistical methods, go to https://www.youtube.com/ playlist?list=PLiDmwXFBUqUv-x9MnJQULJDBgbZl8idcv. To access the playlist containing the 14 lecturettes on research methods, go to https://www.youtube.com/playlist?list=PLiDmwXFBUqUsIVrFkRJTifkW8vua1TPFp.

Chapter 1
The Logic & Language of Statistical Analysis

If someone talks to you about doing quantitative research in organisational, social, educational, nursing or psychological contexts, many topics and issues associated with this specific human activity may surface during the course of your conversation. These issues could include: reviewing the literature, shaping a specific research problem and research questions, framing, configuring and shaping the overall research plan, identifying and sampling research participants, designing of instruments and testing procedures, data analysis and interpretation, hypothesis testing, and reporting of results (Cooksey and McDonald 2019, review all of these issues in the context of postgraduate research). All of these issues as well as many others are of critical importance to the conduct and reporting of 'good' or 'convincing' research as judged by other professionals in a particular discipline as well as, perhaps, by other stakeholders such as members of the general public, government officials, managers and executives, practitioners and clinicians, teachers and students.

As you can see, data analysis is only one aspect of the much broader activity of 'doing quantitative behavioural research', but it is an indispensable aspect. Throughout this book, the term 'behavioural' is used as a catchall term to refer to researchers, approaches and contexts which, in other books, might be characterised as dealing with psychological research, educational research, social research, business research, marketing research, nursing research and any other type of research where the main focus is gathering quantitative data on or about the behaviour of people, singly or in groups, institutions, cultures or societies. This approach is taken to avoid unnecessary awkwardness in writing while still signalling a very broad focus in applicability.

Inevitably, when behavioural researchers configure and conduct a research investigation, they must consider empirical (observable) data, in whatever forms the study yields, e.g. numerical ratings, checklists, rankings, categories, instrument readings, subjective opinions, comments and interviews.

When considering such data, those same researchers must constantly reflect on why the data were collected in the first place: to provide information that would help them to make a judgment or decision (some of which may be theoretical and others quite practical) about some research question or hypothesis of interest. That is a major reason to collect data – to inform judgments about research questions.

While the nature of the judgments and decisions may vary according to the particular research question, we can illustrate them with a few concrete examples:

> - Is an improvement in job satisfaction observed when employees participate in making company decisions compared with when they do not participate?
>   **Judgment**: Yes or no? **Potential decision**: Should company culture and management approaches be changed to encourage employee participation in decision making?
> - How different are two groups of consumers in their preferences for a new brand of cola drink if each group tastes a different formula of that cola compared with another competitor's drink?
>   **Judgment**: Are groups different – yes or no – and by how much and in favour of which cola? **Potential decision**: Should we launch the new brand into the marketplace?
> - How effective is a particular approach to counselling patients diagnosed with post-traumatic stress disorder?
>   **Judgment**: Is the approach effective – yes or no – and do the effects last? **Potential decision**: Is it worth retaining this approach to counselling patients?
> - Why does lower self-concept seem to lead to lower achievement in school (or vice-versa)?
>   **Judgment**: Which explanation for the outcomes of a study appears to best account for the results? Do changes in self-concept cause changes in achievement or do changes in achievement cause changes in self-concept?
>   **Potential decisions**: Should we revise a particular theory relating self-concept and achievement in light of this new evidence? Do we implement a program to improve students' self-concept in hopes of improving their achievement or to raise student achievement levels in order to enhance students' self-concept?

In order to make judgments and decisions, researchers need information; hence the need to collect data. However, it is seldom the case that the data present themselves in a form which will directly answer a research question. Something needs to be done to the raw data before we can understand the general story which the data seem to be conveying. That something is data analysis. When we analyse data, we boil them down, summarise them, transform them, condense them, graph them, paraphrase them, and so on. In short, we reduce the raw data to forms which are more useful for addressing our research questions.

Raw data, typically, come to us in one of two forms: **quantitative** – taking numerical form as in counts, attitude ratings, test scores, and instrument readings;

and **qualitative** – taking non-numerical form as in categorical observations, images, interview transcripts, company documents, and field observation notes.

In some cases, we can do some switching back and forth between these two forms of data depending upon our needs.

> We would have qualitative data if we record the different words that the North American Inuit use to refer to the concept of 'snow'. On the other hand, we could count the number of times particular words for the concept of 'snow' are used within a particular geographical region or within a specified time period – these counts constituting quantitative data. Which form we use depends on which would be the most useful for addressing our research question. If we want to describe or characterise the Inuit language in terms of its richness, the qualitative form would perhaps serve us better. However, if we want to look for relationships between usage of particular words and geographical regions, the quantitative form would serve us better.

The important factor in all of this is that social and behavioural researchers have control over the forms their data should take, and those forms should be selected by considering which will provide more appropriate and accurate information with respect to their research questions.

Assuming the data are in the form the researcher desires, the conclusions which those data will support are still unlikely to be immediately obvious. More needs to be done. Data analysis entails the summarisation and interrelating of many separate bits of data; each bit of data on its own would not carry enough weight to inform a clear judgment or conclusion regarding a research question. But, collectively considered, these individual bits of data may provide sufficient evidence for a clear judgment to be made.

> A marketing questionnaire generally consists of a number of questions and each questionnaire is normally administered to several hundred people. Focusing on one person's rating for a single question would be unlikely to lead a marketing researcher to conclude that consumers in general have a strong preference for this or that product. However, when the responses to a particular question from each consumer who completed the questionnaire are collectively considered, we may obtain a sufficiently strong indication of product preference.

The procedures of statistical analysis give researchers the tools they need to perform this collective consideration step. In order to analyse data, we need some concepts to use, some rules to follow and, generally, a computer program to do the math. The analytical concepts and procedures needed to analyse qualitative data are very different from those used to analyse quantitative data (or qualitative data converted to quantitative form) and it would require a massive volume indeed to do justice to the whole range of possibilities.

The Language of Statistics

Throughout this book you will be exposed to various concepts specific to the field of statistics. When you read the published research of other researchers or other types of statistics texts, you will invariably encounter one or more of these concepts. It is well-known that unclear, confusing or ambiguous language can interfere with learning, making it much harder than it needs to be. My goal is to help negate such interference.

Accordingly, this section describes and clarifies the meaning of a number of important concepts in the realm of statistical analysis. For some concepts, other important related concepts in the section are identified for easy cross-referencing purposes.

Bivariate Statistics

Bivariate statistics encompass statistical procedures that deal simultaneously with **two dependent variables** obtained from the same sample of research participants. Most introductory statistics courses spend a smaller but significant portion of their time exploring bivariate statistics.

For example ... *Bivariate statistical procedures include the Pearson correlation, contingency table analysis and scatterplots.*

See also: dependent variable; sample.

Cause and Effect

The issue of **cause** and **effect** is often important in experimental, quasi-experimental and, frequently, observational and survey behavioural research. Independent variables are either manipulated or measured as putative causes for observable changes in one or more dependent variables. Behavioural researchers attempt to establish as clear a link as possible between putative causes and their effects by neutralising, to the extent possible, the influence of extraneous or confounding variables.

Such neutralisation can be done through (i) alterations in experimental designs to explicitly account for the influence of an extraneous variable or (ii) through procedures such as 'randomisation' (where the intervention or treatment each research participant receives is randomly determined) or (iii) through 'statistical computations' which mathematically 'remove 'the influence of an extraneous variable. All of these procedures come under the general heading of 'control procedures'. In order to unambiguously demonstrate a link between a causal factor and its effect, three conditions must be met.

- **Temporal ordering:** The cause must *precede* the effect in time. The arrow indicates the direction of the suspected causal influence. Temporal precedence is typically accomplished by explicit experimental arrangements.

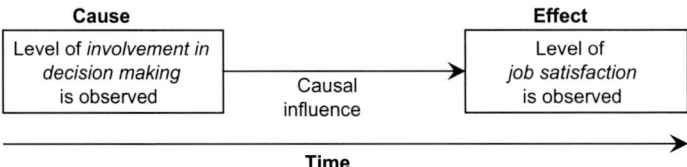

- **Covariation:** When changes, e.g. an increase or decrease, are made to the proposed causal factor, an associated change, i.e. increase or decrease, in the effect can be observed. That is, the cause and the effect can be seen to co-vary or correlate with each other. Statistical procedures are most useful in helping to demonstrate that this condition has been met.

- **Rule out alternative plausible causes:** This condition explicitly refers to the nullifying of possible extraneous variables as possible causal reasons for why changes in the effect (the dependent variable) were observed. Meeting this condition is facilitated by control procedures. Such procedures may be implemented by good research design practices (control by design) or by using more sophisticated statistical procedures (statistical control).

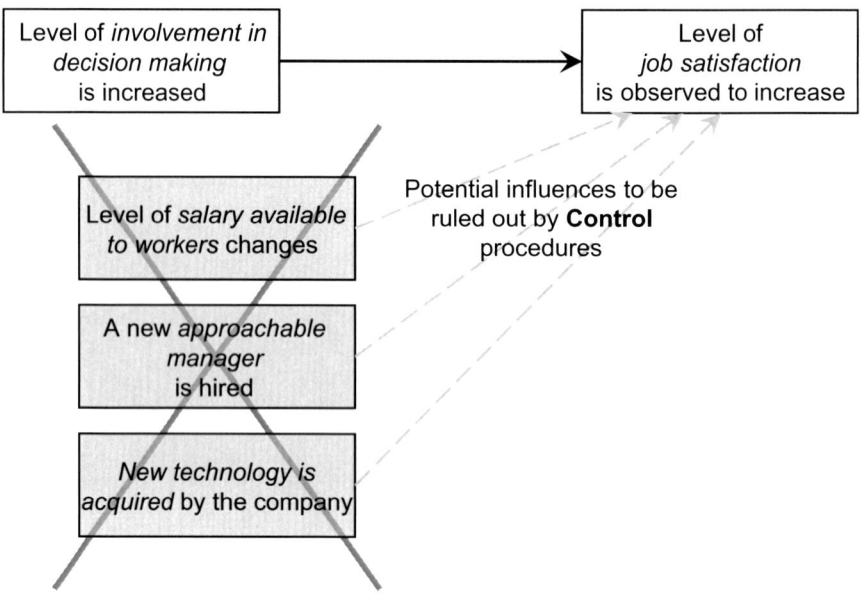

See also: correlation; dependent variable; extraneous variable; independent variable.

Correlation

A correlation is a numerical expression of the strength and direction of the relationship between two **variables** or measures. Correlation (see also *Fundamental concept III*) is another important way that behavioural researchers have of quantifying uncertainty, only this time what they are quantifying is uncertainty in relationship (or, equivalently, 'association'). It is important to note that correlation does not equal causation; finding two variables to be correlated does not necessarily mean that one variable is causing the behaviour of the other variable. Thus, correlation only addresses the covariation requirement for establishing a cause and effect relationship.

Correlation is measured using a number which can range between -1.0 and $+1.0$. The sign of the number indicates the direction of relationship. A negative correlation indicates that values of one variable tend to get larger while the values of the other variable tend to get smaller; a positive correlation indicates that values of one variable tend to get larger while the values of the other variable also tend to get larger.

The closeness of the value of a correlation to either zero or 1 gives an indication of the strength of the relationship. Correlations near zero indicate very weak relationships (thus, high relationship uncertainty) and correlations near +1.0 indicate very strong relationships (thus, high relationship certainty).

A correlation of exactly 0.0 means no relationship (complete uncertainty; knowing something about one variable tells us nothing about the other variable) between the two variables; a correlation of exactly −1.0 or +1.0 indicates a perfect relationship (complete certainty) between the two variables. As you might imagine, in the behavioural sciences where the measures we construct are fallible, correlations near zero would be rather common, whereas correlations near −1.0 or +1.0 would be very rare.

> A health researcher investigates the relationships between daily dietary intakes of aspartame artificial sweetener (e.g. Equal), monosodium glutamate (MSG) and frequency of occurrence of headaches. His results showed that daily intake of aspartame and frequency of headaches were correlated +.50 and daily intake of MSG and frequency of headache were correlated -.10.
>
> Thus, there was a moderate tendency for increasing intake of aspartame in the diet to be associated with increasing frequency of headaches. Conversely, there was only a very slight tendency for increasing intake of MSG in the diet to be associated with decreasing frequency of headaches.

Correlational Statistics

Correlational statistics encompass statistical procedures which examine the relationships or associations between two or more variables (bivariate or multivariate situations). Examples include the Pearson and Spearman correlations, factor and cluster analysis, canonical correlation and scatterplots.

See: *Procedure 6.1, Procedure 6.5, Procedure 6.6* and *Procedure 6.8*.

See also: bivariate statistics; correlation; multivariate statistics; variable.

Descriptive Statistics

Descriptive statistics comprise statistical procedures for summarising, organising and communicating characteristics of samples of data.

Examples of descriptive statistical procedures include the mean, measures of variability, the *z*-score, and many graphical display techniques including histograms, scatterplots, stem-and-leaf displays and boxplots.

See: *Procedure 5.1, Procedure 5.2, Procedure 5.3, Procedure 5.4* and *Procedure 5.6*.

See also: mean; sample; variability.

Dependent Variable

A dependent variable is a set of measurements obtained from a sample whose values the researcher hopes to be able to predict or explain from knowledge of relevant independent variables. Therefore, a dependent variable constitutes the putative effect in a research study; it is the measure for which the independent variables are expected to cause or predict changes. Often, behavioural researchers are interested in more than one dependent variable in a study, which means that statistical analyses must be more sophisticated. This then becomes the province of multivariate statistics. Synonyms for dependent variable include **dependent measure** or **criterion measure** (in the case of regression analysis).

> A market researcher explicitly presents different brands of the same product to consumers and measures their degree of preference for each brand. The theory is that exposing consumers to different brands is expected to lead to or cause different degrees of product preference.
>
> An Equity Officer at a university, who does a study to examine gender differences in work expectations, is testing the theory that being male or female will be predictive of differential expectations for work outcomes (e.g. salary, hours). In these two scenarios, a measure of degree of product preference and measure(s) of work outcomes would constitute the dependent variables in each case.

See also: cause and effect; independent variable; multivariate statistics; sample.

Distribution

A distribution is a display, often graphically presented, of the possible values of a variable paired with the frequency (i.e. the number of times observed) or probability of occurrence in a sample or the population associated with each value.

A popular graphical method for displaying a distribution is the **histogram**, examples of which are shown in Figs. 1.1, 1.2, 1.3 and 1.4. These histograms summarise the distributions obtained from simulated measurements on a sample of 1000 people. In the style of histogram shown below, each bar tallies the number of times that values within a specific range were observed in a sample. The entire graphical picture then represents the distribution of possible values for the variable being displayed.

Four important characteristics of a distribution are its **mean**, **variability** (measured by the variance or the standard deviation), **kurtosis** and **skewness**. Kurtosis reflects how flat or peaked the distribution is (see Figs. 1.1 and 1.2). Skewness

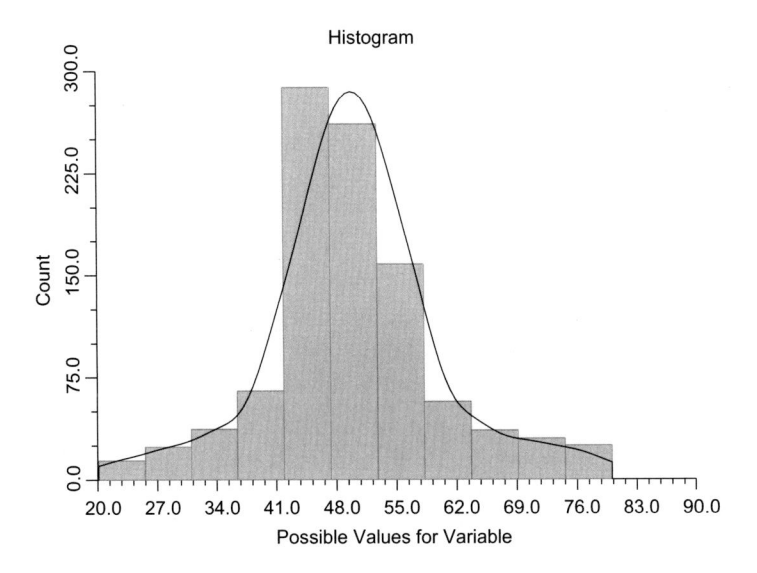

Fig. 1.1 Leptokurtosis (tall centre peak; thin tails)

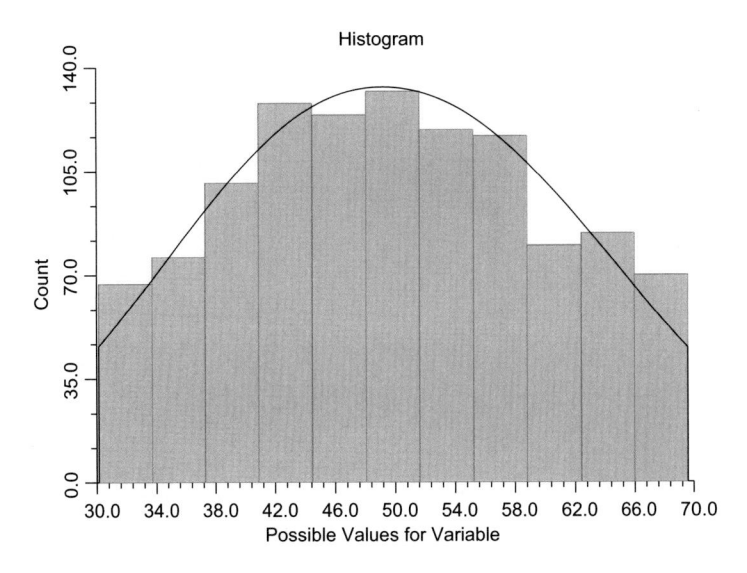

Fig. 1.2 Platykurtosis (short centre peak; thick tails)

reflects the degree to which there are more extreme values present at one end of the distribution than at the other (see Figs. 1.3 and 1.4).

See also: distribution; mean; sample; population; probability; variability; variable.

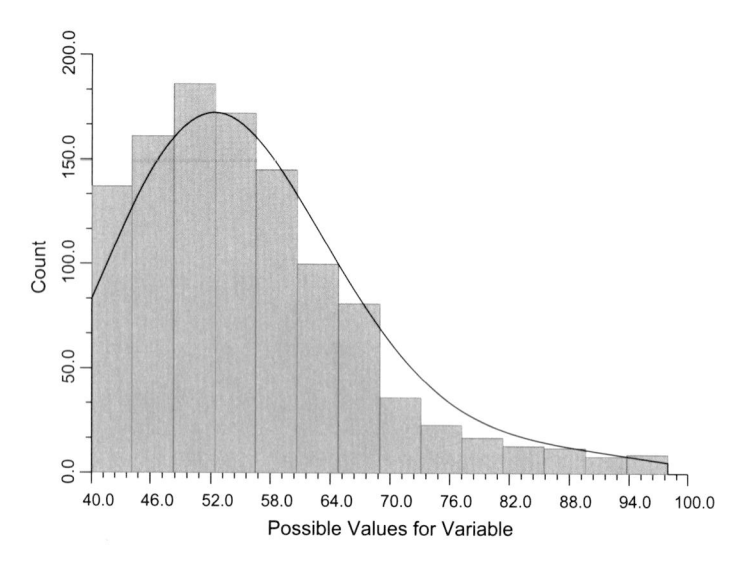

Fig. 1.3 Positive skew (extreme values prevalent in right tail)

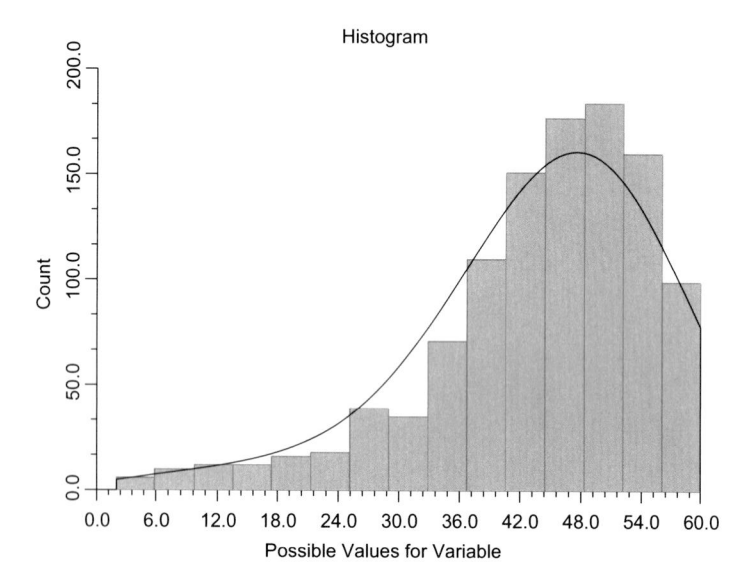

Fig. 1.4 Negative skew (extreme values prevalent in left tail)

Extraneous Variable

Extraneous variables are variables whose existence may or may not be known when research is conducted but whose influence potentially interferes with the clear linking of changes in independent variables to changes in a dependent variable. Extraneous variables thus constitute rival possible causes for any effect which might be observed in the study. Synonyms for extraneous variable include **confounding variable** or **nuisance variable**.

The primary goal of research planning and design is to eliminate the influence of as many extraneous variables as possible so that an unambiguous cause and effect story may be told. Behavioural researchers must therefore spend a great deal of time thinking about such variables and how to control for, minimise or remove their influence. This is one important function of experimental and quasi-experimental designs (see *Fundamental concept VII*).

See also: cause and effect; dependent variable; independent variable; sample, variable.

> A market researcher explicitly presents different brands of the same product to consumers and measures their degree of preference for each brand. The theory is that exposing consumers to different brands is expected to lead to or cause different degrees of product preference. If while presenting different brands of a product to consumers, the researcher inadvertently presented the products in the same order to all consumers, then any changes in brand preference could be due to the order of presentation (an extraneous variable) rather than to brand differences (the desired independent variable). An Equity Officer at a university, who does a study to examine gender differences in work expectations, is testing the theory that being male or female will be predictive of differential expectations for work outcomes (e.g. salary, hours). A possible extraneous variable which could influence results in this scenario is the cultural background of the people in the sample. People from an Islamic background might show differences in work expectations because of the way Islamic culture views the roles of men and women as compared to people from Christian or Buddhist societies. In each case, extraneous variables cloud the desired interpretation and causal inferences.

Independent Variable

This is a variable whose values are explicitly controlled, manipulated, or measured by the researcher. Independent variables frequently serve as the hypothesised causes of any observed changes (effects) in the values of a dependent variable. Synonyms for independent variable include **predictor** in the case of regression analysis, or **treatment** in the case of experimental or quasi-experimental designs (see *Fundamental concept VII*).

Independent variables that are explicitly manipulated or controlled by the researcher (e.g. where the researcher determines which participants are exposed to or experience which value of an independent variable) offer the strongest opportunity to support an inference of causation. Independent variables that are simply measured by the researcher (e.g. via questions or observations on a questionnaire) generally cannot support strong inferences of causation.

> A market researcher explicitly presents different brands of the same product to consumers and measures their degree of preference for each brand. The different product brands constitute the different values of a 'manipulated' independent variable and exposing consumers to different brands is expected to lead to or cause different degrees of product preference. An Equity Officer at a university, who does a study to examine gender differences in work expectations, is using gender as a 'measured' independent variable. Being male or female is expected to be predictive of differential expectations for work outcomes, e.g. salary, hours. Comparing the two scenarios, the market researcher would be on stronger ground, other things being equal, in claiming that brand operates as a causal factor for product preferences.

See also: cause and effect; dependent variable; variable.

Inferential Statistics

Inferential statistical procedures are explicitly designed to permit inferences from observed sample statistics to corresponding population parameters through the use of the rules for statistical inference associated with a test of significance. Examples include *t*-tests, univariate and multivariate analysis of variance, structural equation models, discriminant analysis, and multiple regression analysis (see *Procedure 7.2*, *Procedure 7.6*, *Procedure 7.13*, *Procedure 7.16*, *Procedure 7.17* and *Procedure 8.7*).

See also: bivariate statistics; multivariate statistics; parameter; population; sample; statistic; statistical inference; test of significance; univariate statistics.

Mean

The mean is defined as the average score in the distribution of scores for a variable, found by adding up the scores for every participant on a particular variable in a sample and dividing the resulting total by the number of participants in that sample. It summarises, using a single number, what is happening in the centre of a distribution. The mean is the most commonly used measure of the 'typical' value for a variable in a distribution and is one of several measures of central tendency. If you had to guess what score any randomly chosen participant would show on a variable, your best guess, in the absence of any other information, would be the mean.

See also: distribution; sample; variable.

Mediating Variable

A mediating variable is a specific type of independent variable that operates in between another independent variable and a dependent variable of interest. By 'in between' is meant that the relationship (e.g. correlation) between an independent variable and a dependent variable may be partially or totally attributable to the operation of a third variable, the mediating variable. Without knowledge of the effects of a mediating variable, a researcher could draw false conclusions about the existence of a relationship between an independent variable and a dependent variable (called a **spurious correlation**).

> A researcher conducts an investigation looking at whether or not high school students' gender influences their decision to pursue further studies in mathematics. He finds that male students are much more likely to decide to pursue further studies in mathematics than female students and therefore concludes that gender differences do influence the decision. However, a colleague points out that gender might not be the causative factor here; instead the attitudes toward mathematics that exist in the students' home situations may be the actual determining factor in a student's decision. Female students may simply live in homes where women are not 'supposed' to like mathematics. If the colleague is right, then attitude toward mathematics in the home may be operating as a mediating variable between gender and the decision to pursue further studies in mathematics. This means that gender is not directly causally linked to the decision but only indirectly (and maybe very weakly) linked through the actions of attitudes toward mathematics in the home.

See also: correlation; dependent variable; independent variable.

Moderating Variable

A moderating variable is a specific type of independent variable that conditionally influences or changes the effect of a second independent variable on some dependent variable of interest. Statisticians refer to such a conditional influence as an **interaction** (see *Fundamental concept IX*). Without knowledge of the effects of a moderating variable, a researcher could falsely conclude that an independent variable has consistent effects on a dependent variable under all conditions (called a **main effect**).

> A researcher is interested in measuring the time it takes a driver to react to an obstruction on the road by braking as a function of the amount of headlight glare through the windscreen (bright or dim headlights coming toward the driver). The researcher concluded, on the basis of her analysis, that bright headlight glare (one value of the independent variable) leads to slower braking reaction times (the dependent variable) compared to dim headlight glare (the other value of the independent variable). However, on reflection, she began to suspect that age may have had an impact on drivers' braking reaction times. The researcher redoes her study using older (60+ years old) and younger (20–30 years old) drivers as well as manipulating the level of headlight glare. She finds that younger drivers show a less strong effect of headlight glare on their braking reaction times that do older drivers. This means that driver age operated as a moderating variable for the relationship between headlight glare and braking reaction time.

See also: dependent variable; independent variable.

Multivariate Statistics

Multivariate statistics encompass statistical procedures that deal simultaneously with two or more dependent variables obtained from the same sample of research participants. Most introductory statistics courses spend virtually no time exploring multivariate statistics, leaving such exploration to more advanced courses. Examples of multivariate statistical procedures include factor analysis, cluster analysis, multivariate analysis of variance, and structural equation models.

See also: dependent variable; sample.

Nonparametric Statistics

This category comprises statistical procedures which make very few if any assumptions about the nature of the distribution of dependent variable values in a population when conducting a test of significance.

Nonparametric tests are most useful under conditions where the researcher is not convinced that his or her data satisfy the assumptions required by parametric statistics. Some simpler parametric statistics such as the *t*-test and the analysis of variance have a direct nonparametric counterpart, namely the Mann-Whitney *U*-test and the Kruskal-Wallis *H*-test, respectively (see *Procedure 7.3* and *Procedure 7.9*). Such tests are called analogue tests. Most analogue nonparametric statistical tests focus on data either collected as, or transformed into, ranks or counts rather than explicit numerical measurements.

A newer conceptualisation of nonparametric testing has emerged which focuses on the use of intensive computer resampling methods to construct a picture of what the distribution of values probably looks like in the population. This resampling approach is called **bootstrapping** (see *Procedure 8.4*) and it provides a way to do nonparametric testing on data that do not satisfy parametric assumptions, but which also have not been transformed into ranks or counts.

See also: dependent variable; distribution; means; normal distribution; parametric statistics; population; statistical inference; test of significance; variability.

Normal Distribution

The normal (or bell-shaped or Gaussian) distribution (see *Fundamental concept II*) is the most commonly used theoretical distribution and is the one most often assumed to be true in the population for data analysed by parametric statistics. A normal distribution based on 1000 simulated observations looks as shown in Fig. 1.5. The theoretical smooth shape of the normal distribution is shown by the black curve; this would be the shape of the normal distribution if we assumed a population of infinite size.

Notice where the average score or mean of this distribution falls – the dotted vertical line in Fig. 1.5. The mean of a normal distribution cuts the distribution exactly in half and each half is the mirror image of the other. If you folded the distribution along the line for the mean, the two halves would exactly match – a property of distributions called **symmetry**. A normal distribution has zero skewness (i.e. is not skewed toward either tail) and its kurtosis is referred to as **mesokurtic**. The fact that the normal distribution is symmetric around its mean is very important when a test of significance is performed using inferential statistics.

See also: distribution; inferential statistics; means; parametric statistics; population; test of significance.

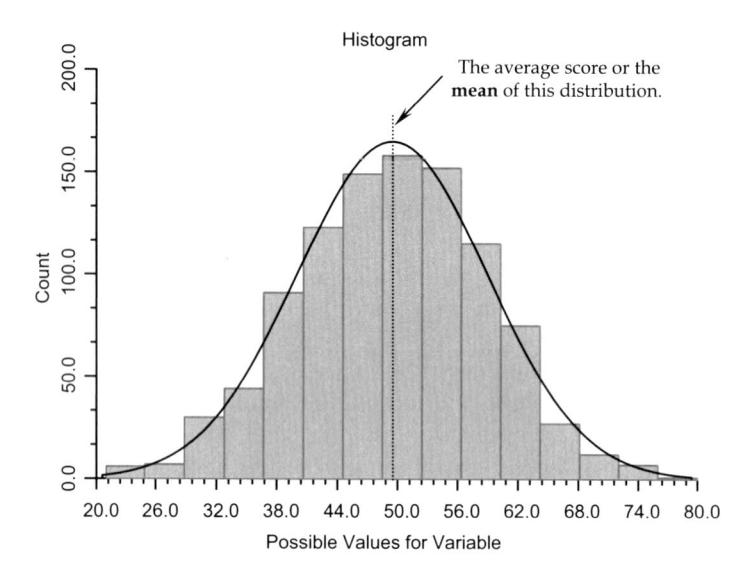

Fig. 1.5 The normal distribution histogram

Parameter

A parameter is a numerical quantity that summarises a specific characteristic of a population. In all but the most mundane and specialised of instances, the precise value of a parameter is never known by a researcher and it must therefore be estimated from sample data using a statistic.

> For example, the **population mean** is a parameter which describes the average of scores for a specific variable reflected by all of the members of a population.

See also: population; sample; statistic.

Parametric Statistics

This category consists of statistical procedures which make specific assumptions about the nature of the distribution of dependent variable values in a population. Most parametric tests of significance assume, for example, that in the population, values for a particular dependent variable follow a normal distribution. Many parametric statistical tests focus on means and the variance measure of variability as the starting points for the statistical inference process.

Examples of parametric statistical test procedures include the *t*-test, analysis of variance, discriminant analysis, and structural equation models (see *Procedure 7.2*, *Procedure 7.6*, *Procedure 7.17* and *Procedure 8.7*).

See also: dependent variable; distribution; means; normal distribution; population; statistical inference; test of significance; variability.

Population

A population contains the complete set of objects, events, or people of interest from which we select a sample or a subset to be observed/measured. It is to this population that we wish to generalise our behavioural research findings. We use statistics computed on sample data to estimate and draw statistical inferences about some characteristics or parameters of the population.

See also: parameter; sample; statistic; statistical inference.

Probability

A **probability** is a numerical expression of the chances that a specific event (or class of events) will be observed. Probability is perhaps the most important way behavioural researchers have of quantifying uncertainty (see *Fundamental concept I*). Probability is measured using a number which can range between 0.0 and 1.0.

Note: Any probability can be multiplied by 100% to give a number which is expressed as a percentage. Thus, saying that the probability of rain tomorrow is .20 is the same as saying there is a 20% chance of rain tomorrow.

It is only at the two end points of this 0–1 continuum that we are certain in our knowledge of events. A probability of 0.0 indicates that a specific event has absolutely no chance of being observed whereas a probability of 1.0 indicates that we are certain that a specific event will occur. Probabilities between 0 and 1 indicate more or less uncertainty with respect to the occurrence of an event and may be estimated either subjectively or objectively. Subjectively, if we feel fairly sure an event will occur, a probability near 1.0 might be appropriate; if we feel fairly sure an event won't occur a probability near 0.0 might be appropriate. And if we really have no idea whether an event will occur or not, then a probability near .50 (which is the chance of a coin landing on heads when it is flipped) may be appropriate. Another more formal and statistically appropriate way of estimating and interpreting a probability is to count how many occasions one observes an event of interest out of N observation instances – this objective approach to estimating a probability is called the frequentist approach.

> The probability of throwing a six on a roll of a fair 6-faced die is known to be 1/6 or .167 (this number represents the application of the classical approach to probability and is a precise calculation of probability rather than an estimate).
> However, one could also throw a fair die 100 times, and count the number of 'six' outcomes observed. If 19 'six' outcomes were observed out of 100 throws, this would yield a probability estimate of 19/100 or .19. Researchers often use probabilities to quantify risk, such as the risk of a particular part failing in a machine, the risk of congestive heart failure if one is a smoker, the risk of failing a unit of study and, importantly, the risks of making certain types of errors when drawing conclusions from a particular analysis of data. In a sample of 1000 smokers, for example, if one were to observe that 452 smokers exhibited symptoms of congestive heart failure, one might conclude that the probability of a smoker exhibiting signs of congestive heart failure was .452. This, of course, assumes the sample has been randomly drawn and is representative of the population of smokers to which generalisation is to be made.

There are very few events in the realm of human behaviour which can be characterised with perfect certainty and the rules of statistical inference are explicitly set up to acknowledge that any conclusion from a test of significance will always have some probability of error. The researcher's task, therefore, is to attempt to keep this probability low, usually to somewhere near .05 or 5% (meaning that the researcher would be wrong 5 times out of 100 times they did the study). However, the probability of statistical inference error can never be reduced to 0.0.

See also: random sampling; statistical inference; test of significance.

Random Sampling

Random sampling describes a procedure used to systematically obtain representative samples from a population. Most commonly, it is a method whereby each member of the population has an equal chance or probability of being selected for the sample and each sample of a given size from that population has an equal chance of being assembled.

There are various types of random sampling schemes including:

- **simple random sampling** – sampling randomly from the entire population; and
- **stratified random sampling** – sampling randomly within defined categories of a population, such as gender or ethnic background categories (see *Fundamental concept VIII*).

See also: population; probability; sample.

Sample

A sample is a subset or group of objects, people or events, usually selected from a population of interest, from which specific observations or measurements are to be obtained. Statistically speaking, the preferred sampling method is some type of random sampling scheme which places the researcher's ability to generalise from sample to population on its strongest footing. However, for a variety of reasons, samples in behavioural research are often not random (e.g. **convenience, quota** or **volunteer** samples – see *Fundamental concept VIII*) and, if this is the case, generalisations from such samples are on statistically weaker ground.

See also: population; random sampling.

Statistic

A statistic is a numerical measure, computed using data obtained from a sample, which can be used to:

- describe or summarise some variable measured in that sample; and/or
- estimate the value of the corresponding parameter in the population from which the sample was drawn.

> For example, the **sample mean** is a statistic which describes the average of all observed values for a specific variable within the sample. This sample mean or average can be used by itself as an index for describing the sample and/or it can be used as an estimate of the parameter value for the mean of the population.

See also: parameter; population; sample.

Statistical Inference

Statistical inference refers to systematic process through which a behavioural researcher comes to make an explicit judgment about specific parameter estimates, relationships or differences which are observed in sample data. Statistical inference is guided by a specific set of rules but, in the end, is a cognitive process engaged in by human researchers (see *Fundamental concept V*). This inevitably means that there are some elements of subjectivity in the inferential judgments that behavioural researchers make. What counts as a meaningful or significant result for one researcher may not necessarily count as meaningful or significant to another, which leads to the evolution of different standards or conventions for an acceptable degree of uncertainty, depending upon the intentions of the researcher.

Therefore, the rules of statistical inference, as embodied in the logic of a test of significance, provide guidance for making judgments but do not determine those judgments. An example of statistical inference would be a researcher's decision whether or not to conclude that consumer preferences for brand A of a product were significantly higher than for brands B or C. The rules of statistical inference operate in a similar fashion to the rules in a court of law: differences or relationships are assumed not to exist ('innocence') unless there is sufficient evidence to warrant rejection of this assumption, given an acceptably small allowance for doubt and potential error ('guilty beyond a reasonable doubt'). This allowance for doubt and error is quantified by statisticians using the concept of probability, as for example, in expressions such as the "probability of inference error is .05 or 5%". What counts as an 'acceptable' and 'reasonable' amount of error, of course, is where subjectivity and societal or academic expectations enter the frame.

See also: inferential statistics; parameter; probability; sample; test of significance.

Test of Significance

A test of significance refers to any result from an inferential statistics procedure which evaluates a statistic or set of statistics from a sample relative to some comparative standard expectation (parameter) in the population. This evaluation permits conclusions (inferences) to be drawn regarding the validity of a researcher's hypothesis.

> A researcher may employ an inferential statistical procedure such as the t-test in order to assess whether or not two groups of managers (a group of senior managers and a group of front-line managers) differ significantly from one another in terms of their mean job satisfaction. The comparative standard in this case, as in most cases, is the presumption at the outset that the groups do not differ. Specifically, this translates to an expectation that, in the population from which these managers were selected, the mean difference in degree of job satisfaction expressed by senior and front-line managers would be zero. The rules of statistical inference would help this researcher decide whether or not the observed mean difference between the two groups of managers in his/her sample is large enough to warrant the rejection of the "no difference" assumption in the population.

See also: inferential statistics; mean; parameter; population; sample; statistic; statistical inference.

Univariate Statistics

Univariate statistics encompass statistical procedures that deal with one dependent variable at a time. Most introductory statistics courses spend the bulk of their time exploring univariate statistics. Examples of univariate statistical procedures include the histogram, *t*-tests, analysis of variance, and *z*-scores.

See also: dependent variable.

Variability

Variability refers to a statistical measure of the extent to which values for a variable in a distribution differ from a statistical measure of typical value in the centre of the distribution. Variability thus measures the extent of the 'spread' in scores of a distribution around a typical value. The most commonly used measure of variability is the 'standard deviation' which provides a descriptive measure of how values differ, on the average, from the mean of a distribution. The standard deviation is a direct function of another measure of variability, the 'variance', which forms the statistical platform for estimating effects and errors in tests of significance (giving rise to the class of tests known as 'analysis of variance', for example).

See also: distribution; mean; sample; variable, test of significance; variance explained.

Variable

A variable represents the measurement of a specific characteristic or observation, obtainable from either a sample or a population, which has a range of possible values. The term **measure** is often used as a synonym for variable.

Job satisfaction scores obtained using a 7-point Likert-type rating scale on a questionnaire, from a sample of workers in a factory, define a variable we could call 'Job Satisfaction'. This measure could take on one of a range of possible values from 1 (the lowest possible value) to 7 (the highest possible value) for any particular person. An indication of the gender of persons in a sample defines a variable we could call 'Gender' and this variable would typically take on one of two possible values for any person: male or female.

See also: population; sample.

Variance Explained

When behavioural researchers talk about **variance explained** (or **variability explained** or **variance accounted for**), they are referring to how observed or manipulated changes in the values of an independent variable relate to observed changes in the values of the dependent variable in a sample. Variance explained is an important measure of effect size (see *Procedure 7.7*) and is related to the concept of correlation (see *Fundamental concept III*). It is often reported as a proportion or a percentage.

In behavioural research, especially when testing theories and making predictions, we aspire to explain 100% of the variability of a dependent variable. Of course, this is never realistically achievable since virtually all measures employed in behavioural research have random and unexplainable measurement errors associated with them and we could never measure all the possible things that might systematically influence a dependent variable. However, if we can conclude that an independent variable of interest can explain a substantive portion (or percentage) of the variance in a dependent variable, we have learned something important.

Other things being equal, i.e. controlling for relevant extraneous variables, the stronger the relationship between an independent variable and a dependent variable, the greater the amount of variance we can explain.

> An educational researcher might report that children's self-concept explains 5% of the variance in school achievement and parent's attitude toward schooling explains 14%. This means that both self-concept and parents' attitude toward schooling are predictive of school achievement and that, of the two independent variables, parents' attitude toward schooling is more important.

See also: correlation; dependent variable; extraneous variable; independent variable; sample, variability.

Reference

Cooksey, R. W., & McDonald, G. (2019). *Surviving and thriving in postgraduate research* (2nd ed.). Singapore: Springer.

Chapter 2
Measurement Issues in Quantitative Research

Collecting quantitative data necessarily entails making decisions about how various phenomena of interest to the researcher will be defined and quantified. This is the domain of **measurement** and there are several important considerations. [Note that more detailed discussion of measurement issues in social and behavioural research appears in Chapter 18 of Cooksey and McDonald 2019.]

Hypothetical Constructs, Operational Definitions and Construct Validity

Research in the behavioural and social sciences depends upon having coherent measures available for analysis. Where do these measures come from? In most cases, the measures used by behavioural researchers come from theory. Without getting into the philosophical issues associated with theorising, it will suffice to describe a theory as a researcher's idea about what determines or influences how people behave in specific contexts. The components of a theory are:

- hypothetical constructs; and
- hypothesised links – some or all of which may be causal in nature – connecting those constructs.

We call constructs 'hypothetical' (a synonym, often used by sociologists and psychologists, is 'latent') because they are hypothesised to exist yet cannot be directly observed.

> A researcher has a theory that argues that job satisfaction (JS) and organisational commitment (OC) causally influence a person's intention to leave their job (ILJ). This simple theory, diagrammed below, proposes two causes or independent variables (JS and OC) each contributing to the same effect or dependent variable (ILJ).
>
> JS, OC and ILJ each represent a hypothetical construct, since there is no way to directly observe these constructs in operation. That is, the researcher cannot place a probe in the person's brain to directly read out their level of JS, OC and ILJ.

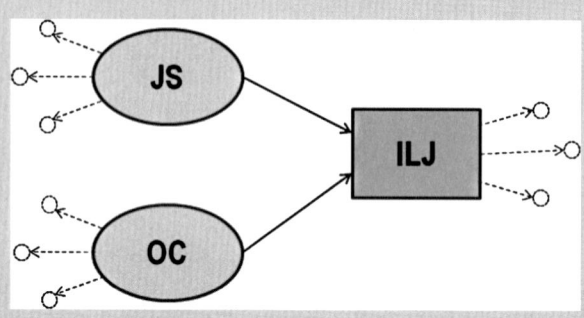

In order to conduct quantitative behavioural research, we must first translate each hypothetical construct into a process that will produce the numerical measurements that will be analysed. This translation process is called **operational definition** and the meaning is fairly clear: operationally defining a construct means spelling out, very clearly, the steps the researcher will go through to obtain concrete measures of that construct. Such steps might involve constructing an instrument to measure a specific physical manifestation of the construct of interest, writing survey questions to measure the construct or spelling out what to look for when observing and scoring particular behaviours reflecting that construct. The measurements resulting from the process of operational definition are then taken to reflect the intended hypothetical construct.

> In the previous example, there are three hypothetical constructs, JS, OC and ILJ. In order to test her theory, the behavioural researcher must first operationally define each construct in terms of what she will do to obtain measures for them. She could write a series of questionnaire items asking about issues she thinks relate to JS, OC and ILJ. The scores on those items would then be taken to reflect the participant's level of JS, OC and ILJ. This process is symbolised in the earlier diagram of the theory using dashed circles to represent specific measurement items and dashed arrows moving from the construct to the specific measurement item to represent the construct reflection relationship.

If you have been thinking carefully about the above process, you may have spotted a potential flaw: how do we know that the process of operational definition has provided a good translation of the hypothetical construct into a measurement system? In other words, how do we know the measures reflect the construct we intended them to?

The answers to these questions implicate an important concept in behavioural research, namely **construct validity**. We say a construct is valid to the extent that the measure(s) used to operationally define it behave in the ways that we would expect the construct to behave (if we could only directly measure it). If you think this sounds a bit circular, you are right. In the end, establishing construct validity is about building an argument strong enough to convince others that your measures are working in the way you theorised that they would.

If you cannot argue that your measures have construct validity, then no amount of statistically sophisticated analysis will provide you with coherent interpretations – a classic case of 'garbage in, garbage out'. Statistical procedures such as correlation and factor analysis (see *Procedure 6.1*, *Procedure 6.5* and *Procedure 8.4*) can help you demonstrate construct validity. They do this by showing how multiple measurement items (such as attitude items on a questionnaire), written to reflect specific constructs, relate to one another in a pattern that makes it clear they are reflecting the intended constructs.

Another statistical approach to demonstrating construct validity is called the 'known groups' method (see Hattie and Cooksey 1984). 'Known groups' validation works by identifying types of people that theoretically should differ, in a predictable manner, on the construct being measured. The researcher then builds an instrument to measure that construct and samples a number of those types of people. If the sampled types of participants are shown to differ in their scores on the new instrument in ways predicted by theory, then 'known groups' validity would be demonstrated. For this type of construct validation, group comparison procedures like ANOVA (see *Procedures 7.6* and *7.8*) and MANOVA (see *Procedure 7.16*) would be most useful.

> Continuing the previous example, let's focus on the JS construct. Suppose the researcher wrote 3 different questionnaire items, each designed to assess what she thought was a specific aspect of job satisfaction. Her goal is to add the scores from the 3 items together to get a score for JS. If she could conduct an analysis showing the 3 items were all highly correlated with each other (meaning they all tended to 'point' in the same direction), she would have the beginnings of an argument for the construct validity of the JS measure.
>
> The procedure of factor analysis (see *Procedure 6.5*) is designed to produce just such evidence. If her analysis showed that some of her items were uncorrelated (meaning they did not measure the same thing as other items that were correlated), but she added all 3 items together to get her JS scores

(continued)

anyway, then any analysis of the JS scores would be suspect because the measure would not be adequately reflecting the construct it was intended to. This could happen if some of the items were poorly conceived, poorly written or ambiguous.

Suppose the researcher, after a thorough review of the literature, theorised that employees who have been with their company for more than 10 years and who are not union members would be more satisfied in their jobs than either employees who had joined a union or former employees who had recently left the company and employees who had joined a union would be more satisfied than former employees. She could then sample each type of person and administer her JS scale to them. Using a one-way ANOVA (*Procedure 7.6* and *7.8*), if she showed that mean JS scores differed in a way that matched her theoretical prediction, she would have evidence for 'known groups' validity.

Measurement Scales

In any behavioural research application where observations must be quantified through a process of operational definition, there is a hierarchy of four scales of measurement or quantification from which to choose. A scale merely refers to a set of numbers which can potentially be assigned to objects, people, or events to reflect something about a characteristic which they possess. Each scale possesses certain properties that determine the relationships between the numbers along the scale.

- **Nominal scale** – permits the classification of objects, people, or events into one of a set of mutually exclusive categories representing some characteristic of interest. The numbers assigned are entirely arbitrary and have no meaning other than as symbolic names for the categories.

The characteristic of gender is commonly defined as having two mutually exclusive (dichotomous) categories—male and female—and we can arbitrarily decide that males will be scored as 1s and females will be scored as 2s (note, however, this is not the only way one could define the characteristic of gender). A particular test item may be scored as right (1) or wrong (0). In both cases, the numbers chosen merely serve as convenient labels for naming categories.

The nominal scale represents the simplest measurement scale available for quantifying observations and constitutes the base level of the scale hierarchy.

- **Ordinal scale** – permits a researcher to rank objects, people, or events along a single continuum representing the relative degree of possession of a characteristic of interest. The numbers assigned reflect this rank ordering but nothing more.

> We could rank 40 students in a class from highest to lowest in terms of academic performance by assigning a 1 to the best student, 2 to the next best, and so on down through 40 for the worst performing student. The student ranked 1 is interpreted as being a better performer than the student ranked 2 but we cannot say much better he/she is.

Thus, the numbers on an ordinal scale reflect the ordering of observational units (e.g. object, event, or person) but they cannot tell us how much more of the characteristic of interest one observational unit possesses compared to other observational units. The actual numbers chosen to represent the ordering are arbitrary but for convenience, an ordinal ranking scale usually starts at 1 and continues up to the number of observational units or objects being ranked.

The choice of whether or not to use 1 to represent the best or the worst observational unit to commence the ranking process is somewhat arbitrary, but once a choice has been made, the direction of the ranks is entirely determined (if 1 is best, the largest number represents the worst; if 1 is worst, the largest number represents the best). This choice will only affect the sign (not the strength) of any correlations between the rankings and other variables of interest.

- **Interval scale** – permits the scoring of objects, persons, or events along a single quantitative dimension which indicates the degree to which they possess some characteristic of interest. The numbers assigned and the intervals between them have rational meaning. However, the zero point for an interval scale is arbitrary, meaningless or non-existent, which means that making ratio comparisons between numbers on the scale is also meaningless. An important assumption regarding interval scales is that the intervals between successive values on the scale are at least approximately 'equal'.

> Celsius temperature is a good example of an interval scale where the difference between 10° and 11° represents the same amount of change in temperature as the difference between 21° and 22°. However, it does *not* make sense to make ratio-type comparisons between measurements (such as 20 °C being twice as hot as 10 °C) on an interval scale.

In the social, management, psychological, and educational domains, a commonly used interval scale (although some researchers dispute this), particularly in questionnaires, is the **Likert-type scale**:

> The colonisation of space via space stations and bases on other worlds is important for making progress toward world peace (circle one number only):
>
> | Strongly Disagree | Disagree | Neutral | Agree | Strongly Agree |
> |---|---|---|---|---|
> | 1 | 2 | 3 | 4 | 5 |
> | −2 | −1 | 0 | 1 | 2 |
>
> Note: On the above scale, it would make little sense to speak of someone having zero amount of an attitude (even if the Neutral point was scored as a 0 as shown by the italicised numbers which represent another completely acceptable way to quantify this scale). Neither would it make sense to say that a person who scores 4 has twice as much of the attitude as a person who scores 2.

We should note here that some behavioural researchers disagree as to the 'interval' nature of a Likert-type scale since, for example, the difference between an attitude of 1 and 2 in the above example may not be the same from person to person nor may it be the same amount of attitude difference as between an attitude rating of 4 and 5. However, it appears that violation of this assumption of equal intervals does not greatly influence statistical procedures—most research journals in the behavioural sciences do not seriously query the treatment of Likert-type scales as interval scales.

Likert-type scales are useful for measuring attitudes, beliefs, opinions, and similar psychological constructs and sentiments. Other types of scales and tests also yield interval measures, such as intelligence and ability tests. Even in the most competently designed test of mathematics comprehension, for example, it would make little conceptual sense to say that someone who scores 40% on the test knows half as much or is half as smart as someone who scores 80%. Most of the measures used in the behavioural sciences are measured at the interval level or lower.

- **Ratio scale** – permits the scoring of objects, people, or events along a single quantitative dimension, where equal intervals exist between successive scale values, reflecting the degree to which they possess some characteristic. What differentiates the ratio scale from the interval scale is the existence of a defensible absolute zero point in the scale (representing zero amount of the characteristic of interest) which makes ratio-type comparisons between two measurements possible and conceptually meaningful.

> Weight is a good example of a ratio scale. It has an absolute zero point (0 kg achieved in the absence of gravity), the difference between 100 kg and 101 kg is the same amount of weight change as the difference between 209 kg and 210 kg, and it makes conceptual sense to say that an object weighing 50 kg is twice as heavy as an object weighing 25 kg.

The ratio scale of measurement is the most complex and precise system for quantifying observations. However, in most domains of behavioural research, unlike the physical or biological sciences, a good example of a ratio scale is hard to come by because most of the measurements of hypothetical constructs (e.g. intelligence, job satisfaction, morality) we might be interested in cannot be said to have a rational absolute zero point or to permit meaningful ratio comparisons. Therefore, you cannot meaningfully and defensibly say that a person has zero intelligence/job satisfaction/morality or that person A is twice as smart/ satisfied/moral as person B.

Any measurement based on counts can be considered to produce a ratio scale (e.g., pulse rate, achievement test scores (number of items correct), number of defective parts, number of traffic accidents and so on). However, the interpretation of such measures needs to be carefully thought through because counts have unknown relationships with underlying constructs they may be hypothesised to reflect (unless you do additional validation work to connect the count measure to the construct you are interested in). Thus, counts are usually interpreted as just that – counts of observed things.

> On a simplistic level, you might think that counting the number of items correct on a test could be considered to be a ratio scale (where no items correct is the absolute zero point) and it would thus make sense to say that getting ten items correct is twice as many correct as five items. But this simplistic scale has limited utility as a measure of human behaviour because it does not readily translate into a ratio scale for achievement (the hypothetical or theoretical construct regarding human knowledge that the test would have been designed to reflect) in that having ten items correct on the test wouldn't necessarily mean that you were twice as good or knowledgeable as a person scoring only five items correct.

There is, however, one natural class of measures which are available to behavioural researchers in the form of a ratio scale, namely time-based measures.

> Psychologists frequently study people's reaction times to stimulus events as a way of gaining insights into information processing demands and motor skills. Organisational analysts may use time-and-motion studies to assess how much time workers spend doing particular components of a work task. Educational researchers may measure the amount of time children spend reading portions of printed texts.

In each of these instances, the time-based measures do constitute precise ratio scale measures. However, one must be careful what inferences are drawn about unobservable psychological processes based on time measures since other factors could potentially contribute to them, particularly in longer response time intervals (e.g. fatigue, distraction, boredom).

Note that in economics and finance, measuring worth or value in terms of dollars and cents, for example, produces a ratio-scale, but one must be careful what inferences are made from such measures. As objective measures of value (e.g. price paid for a commodity in a market) they work fine; as subjective measures of value (e.g. psychological value of a house), they cannot be said to work as a ratio-scale.

In terms of increasing complexity, rigor, and degree of precision, the four scales of measurement may be hierarchically ranked as nominal → ordinal → interval → ratio where the more complex and rigorous scales have all the properties of the scales of lesser complexity. For example, interval scales have all the properties of nominal and ordinal scales as well as the interval scale properties.

We can sometimes convert data from one scale of measurement to another. This conversion, formally known as a transformation, can be to another form of scale at the same level of measurement. For example, when reaction times are transformed by taking their logarithms to reduce the influence of extremely large times produced by very slow people, this transformation merely changes the original ratio scale of reaction time to another ratio scale of the logarithm of reaction time.

It also is possible to transform a measurement scale from one level of complexity to another, but the direction of the transformation is almost always in the direction of a scale of lesser complexity. Thus, it is possible to transform an interval scale down to an ordinal scale (as is often done in nonparametric statistical procedures where original measurements are converted to ranks), but it is generally not possible, except in the context of some highly specialised statistical procedures (covered in Chap. 9), to convert an ordinal scale to an interval or a ratio scale. Thus, in converting downward between scales of measurement we inevitably lose information as well as precision.

> In converting an interval scale to an ordinal scale, we sacrifice the information represented by the intervals between successive numbers on the scale.
>
> Suppose we convert the interval responses to the 'space colonisation' Likert-type statement shown earlier to a dichotomous measure: didn't agree (1); or did agree (2). To do this, we could recode the scores for all people in the sample who gave a response of 1, 2, or 3 on the scale into the "didn't agree" category (all get a score of 1) and all people in the sample who gave a response of 4 or 5 on the scale into the "did agree" category (all get a score of 2).
>
> We will then have transformed the attitude item from an interval scale of measurement to an ordinal scale where people scoring a 2 have a more positive attitude toward space colonisation than people scoring 1. In this new scale, however, we cannot now talk about how much agreement there was with the statement, only that there was or was not agreement.

Converting down in scale complexity is no problem since each scale of measurement has all the properties of those scales below it in complexity. Often the downward transformation of a variable measured on a more complex scale to a simpler measurement scale is performed by behavioural researchers because:

- they are unsure of the fidelity of the data in their original form;
- they wish to tell a simpler story to a particular audience; or
- the distribution of scores on the original scale is such that the key assumptions, required by parametric statistical analyses, are substantially violated.

Different types of statistical procedures are suited for analysing data measured using specific scales.

> ***For example*** *... Parametric statistical procedures generally deal with ratio or interval scales of measurement for dependent variables whereas nonparametric statistical procedures focus on ordinal or nominal scales of measurement for dependent variables.*

With each statistical procedure described in this book, the prerequisite, preferred, or assumed measurement levels for variables are clearly indicated.

References

Cooksey, R. W., & McDonald, G. (2019). *Surviving and thriving in postgraduate research* (2nd ed.). Singapore: Springer.

Hattie, J. A., & Cooksey, R. W. (1984). Procedures for assessing the validity of tests using the "known groups" method. *Applied Psychological Measurement, 8*, 295–305.

Chapter 3
Computer Programs for Analysing Quantitative Data

There are numerous statistical analysis software systems of generally high quality currently available. We will review five such software systems – statistical packages – with a view to what they can accomplish and why they might be useful to business, social and behavioural researchers. These five packages will serve as primary focal points when reviewing available computer programs for conducting specific statistical procedures throughout this book. A rough indication of the ease of use and expense of obtaining each will be given during this discussion.

SPSS

Probably the best known of all statistical computer packages is the IBM **Statistical Package for the Social Sciences**[1] (IBM SPSS; see http://www-01.ibm.com/support/docview.wss?uid=swg27049428; see also George and Mallery 2019). SPSS first emerged as a major software innovation for large mainframe computers in the 1960s. From that point, it built up a dominant position in the statistical package market. In 2009, IBM acquired SPSS Inc. and, as a consequence, the package is now formally known as IBM SPSS Statistics. SPSS now offers regularly maintained and updated versions for the Windows and Mac OS desktop and laptop computer environments as well as a range of student versions (called 'gradpacks', see www-01.ibm.com/software/analytics/spss/products/statistics/gradpack/). It would be fair to say that most universities worldwide would offer access to SPSS in some form. SPSS offers a very comprehensive suite of procedures for the manipulation and transformation of files and of data as well as procedures for nearly every commonly used (and a number of less commonly used) statistical analysis procedures, many of which are reviewed and illustrated in this book. Figure 3.1 shows a sample screen shot from the

---

[1]For details on SPSS, see https://www.ibm.com/products/spss-statistics

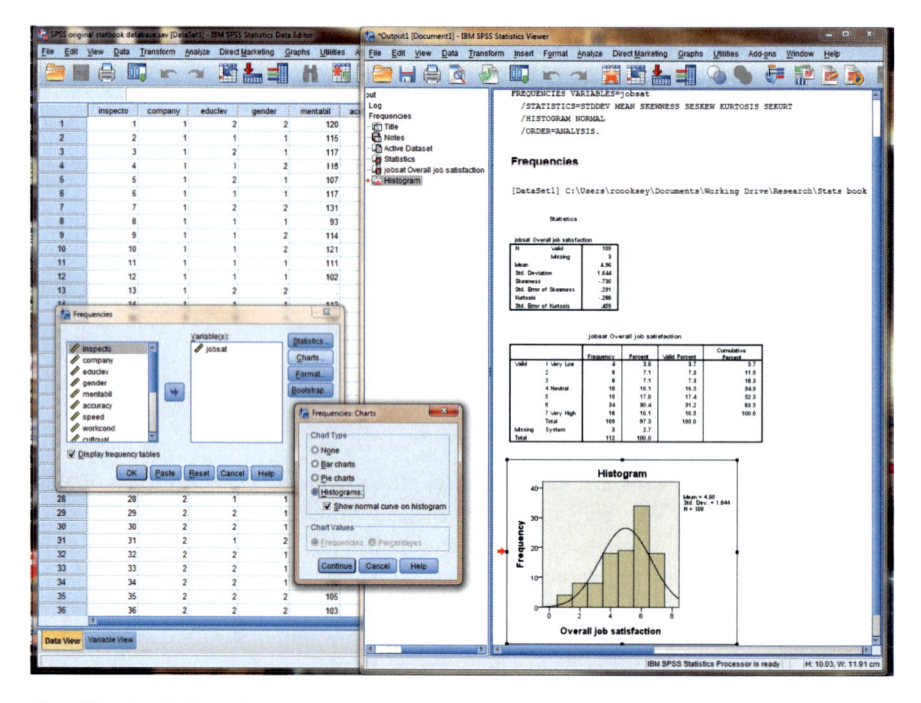

Fig. 3.1 The SPSS environment

Windows platform (version 20) of the SPSS software package showing its spreadsheet data entry system (in the left-side open window) and a sample of the analytical output and high resolution histogram plot produced by the Frequencies procedure (in the right-side open window). Note that the Output window, at the top, shows some of the SPSS language syntax, specifically the SPSS command structure, needed to produce the Frequency analysis output and Histogram chart shown in Fig. 3.1. The Output window also shows the analysis outline window (top centre of Fig. 3.1) which provides a running log of all analyses conducted in the current session.

For most types of analyses, the Dialog boxes (see open small windows on lower left-side of Fig. 3.1) provided by SPSS offer appropriate choices to select and control most analysis options. However, more advanced options for specific procedures may require the use of actual SPSS commands in the programming language so that some familiarity with the SPSS syntax would be highly beneficial for more sophisticated use of the package. A number of good books exist which can provide a decent and fairly well-rounded background in the use of SPSS (see, for example, Allen et al. 2019; Field 2018; George and Mallery 2019; Pallant 2020; Wagner 2013). SPSS is regularly updated in a new version by IBM, so keeping pace with current versions is a dynamically changing proposition. For this book, version 20 of SPSS was used to generate most of the SPSS-related output and figures.

Note the menu bars at the top of each open window as well as the various functional icons directly under the menu bars. These features are characteristic of nearly every good statistical package now available for computers today: drop-down menus provide access to dialog boxes where specific procedures and the analysis and output options for those particular procedures can be easily selected without the user needing to resort to a programming language to make things run.

These window-oriented operating environments thus facilitate ease of user access to sophisticated statistical techniques without sacrificing the statistical rigour and computational accuracy needed for the analyses being conducted. All graphic images produced by SPSS can be individually edited and cosmetically altered. SPSS can be used to conduct most of the statistical *Procedures* described in this book, although certain add-on modules, at extra expense, may be required for certain procedures like classification and regression trees. One major drawback to SPSS is its cost—a single user could expect to pay several thousand dollars for a single user's licence to run a fairly complete version of the package. The best bet for gaining access to SPSS is to have access to a major university or business which has purchased a software site licence that allows individual users to either use the package for free or to purchase a copy of the system for a few hundred dollars. Alternatively, a more affordable 'gradpack' can be purchased which has an extensive, but not full range, of capabilities and a time-limited license.

SYSTAT

SYSTAT[2] (SYSTAT Software Inc 2009) is another comprehensive statistical package that has the reputation of being one of the more statistically and numerically accurate packages on the market. It has a simple yet powerful language which can be used to accomplish more advanced types of analyses and data manipulations not accommodated in the drop-down menus and procedure dialog boxes. Additionally, SYSTAT is very well-known for its graphical capabilities and flexibility as well as for its attention to addressing issues regarding how human perception influences graphical interpretation.

Figure 3.2 shows a sample screen shot of the SYSTAT (version 13.1) Output Window interface illustrating a correlation matrix and a high-resolution Scatterplot Matrix (or SPLOM). The Output window shows some details of the programming language used by SYSTAT; choosing options in specific dialog boxes (see smaller open windows on the right side of Fig. 3.2) automatically causes SYSTAT to

---

[2]For details on SYSTAT, see https://systatsoftware.com/products/systat/

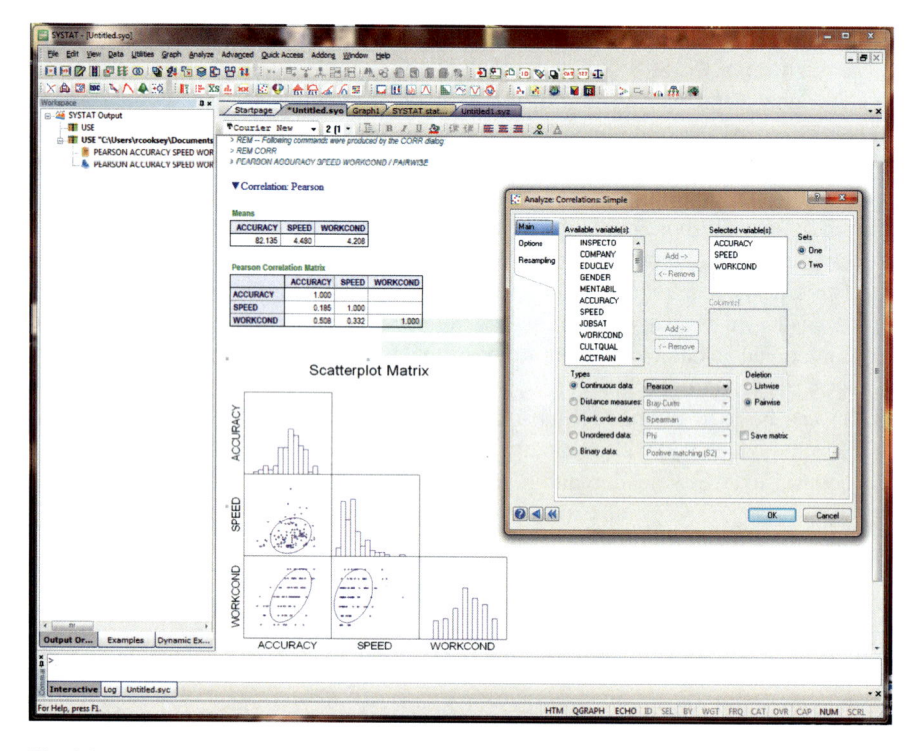

Fig. 3.2 The SYSTAT environment

generate the appropriate commands. Alternatively, the user can input the command language directly. Drop-down menus again feature largely in this package. A unique feature of the SYSTAT program is the availability of push-button icons for accomplishing simple yet standard types of data analysis (note the row of icon buttons underneath the menu bar in Fig. 3.2).

SYSTAT offers some capabilities for editing graphs to suit user requirements, but these features are generally more limited than SPSS. Much of the control over the format of graphs must be accomplished before the graph is produced. This may mean having several tries at producing a graph before it ends up with the look you want. SYSTAT can be used to conduct most of the statistical procedures discussed in this book. As a package, SYSTAT is relatively expensive, but not prohibitively so and a substantially reduced academic pricing is generally available. There is also a reduced capability version, called MYSTAT, which students can download for free from https://systatsoftware.com/products/systat/mystat-statistical-analysis-product-for-student-use/. For this book, version 13.1 of SYSTAT was used to generate many of the SYSTAT-related output and figures.

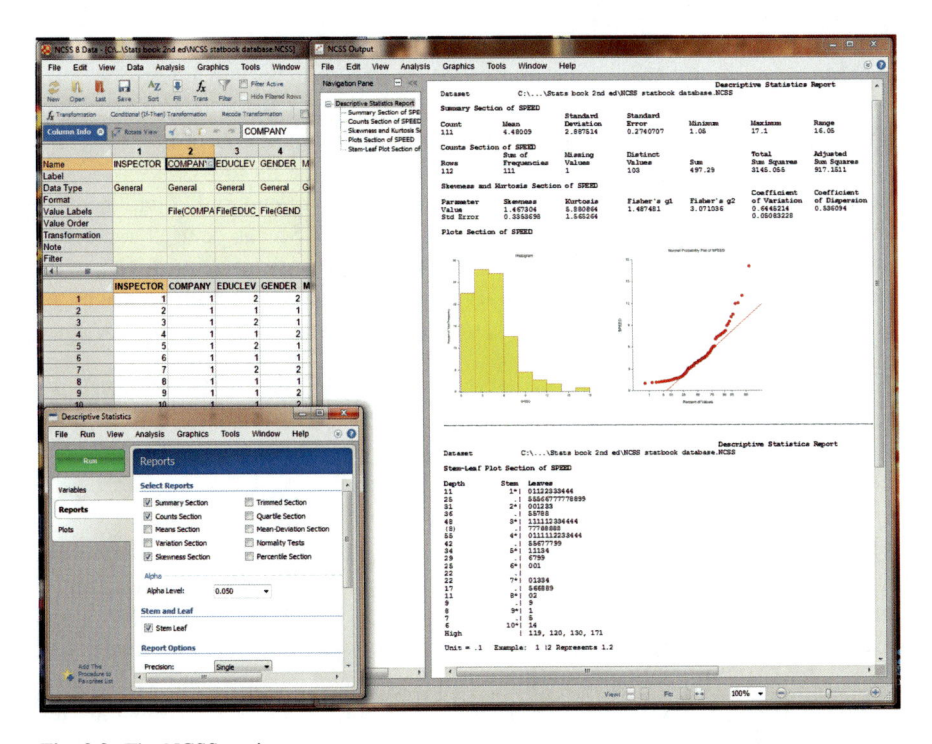

Fig. 3.3 The NCSS environment

NCSS

The **Number Cruncher Statistical System**[3] (NCSS—Hintze 2012) is a comprehensive statistical package developed exclusively for the desktop computing environment. NCSS has the virtue of being relatively easy to use and it offers some types of analysis and graphical procedures that are not available in many other packages (including SPSS and SYSTAT). It is a very useful package for data screening (looking for abnormalities and unusual observations in a set of data) and exploratory data analysis (implementing useful and innovative graphs such as 'violin' plots which very few other packages offer).

Another virtue of the NCSS package is that it is relatively inexpensive to buy; it is certainly cheaper than SPSS or SYSTAT. The user interface has a lot in common with SPSS with the exception that the Task windows (see bottom left-side small open window in Fig. 3.3) for conducting statistical procedures and graphs offer detailed choices for controlling all aspects of analysis. Thus, all aspects of analysis are controlled through selection of options from extensive lists—there is no programming language at all to learn for NCSS, even for very advanced applications.

---

[3]For details on NCSS, see https://www.ncss.com/software/ncss/

Figure 3.3 shows a sample screen shot from the NCSS (version 8) software package showing its spreadsheet data entry system (in the upper left-side open window) and the Output window (in the right-side open window). Like SPSS and SYSTAT, NCSS operates using a drop-down menu bar to select specific tasks and push-button icons to carry out specific functions. Despite its low cost, NCSS has been shown to be a highly accurate, statistically sophisticated and comprehensive alternative to the more expensive statistical packages. NCSS can be used to conduct many of the procedures discussed in this book, except those associated with structural equation models and confirmatory factor analysis for which it makes no provision. For this book, NCSS 8 was used to generate a number of NCSS-related output and figures.

STATGRAPHICS Centurion

STATGRAPHICS Centurion[4] (StatPoint Technologies, Inc. 2010) is a comprehensive statistical package developed for the desktop computing environment but is available only for the Windows platform. However, STATGRAPHICS also offers scaled-down versions for mobile devices as well as a subscriber-based STATGRAPHICS Online environment for doing data analyses online. STATGRAPHICS offers a range of analysis and graphical procedures that are not available in many other packages (including SPSS and SYSTAT). It is a very useful package for a variety of social and behavioural research purposes (including quality control and other business and engineering management statistics). STATGRAPHICS is especially innovative in its data visualisation and graphical capabilities.

STATGRAPHICS is an intuitive and comprehensive all-in-one package requiring no add-on modules to obtain additional functionality (it is like SYSTAT in this respect). The interface is very straightforward to learn and work with and there is extensive online and pdf documentation help associated with this program. Publication quality graphs can easily be produced and edited using STATGRAPHICS. Uniquely, STATGRAPHICS offers a StatAdvisor as an integrated aspect of the program. The StatAdvisor provides annotations on all output which aids interpretation and even suggests possible additional analyses for further exploration of the data. In this respect, STATGRAPHICS is especially novice friendly. The program also offers a StatReport function for assembling a data analysis report within the program's working environment.

Figure 3.4 shows a sample screen shot from the STATGRAPHICS Centurion (version 16) software package showing its spreadsheet data entry system (in the upper left-side open window) and the Output window (see the right-most two-thirds of Fig. 3.4). The Output window is split into two sections (left for numerical output,

---

[4]For details on STATGRAPHICS Centurion, see http://www.statgraphics.com/

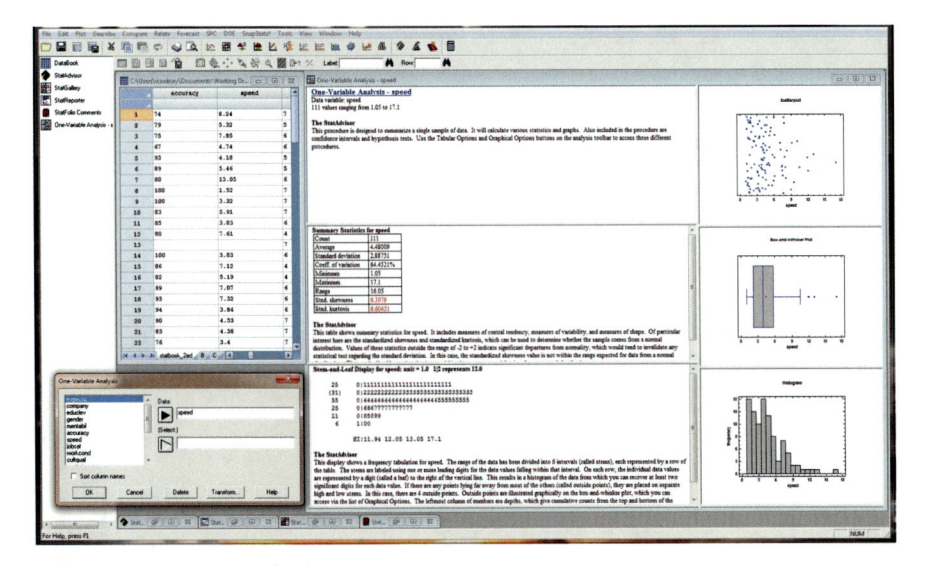

Fig. 3.4 The STATGRAPHICS Centurion environment

annotated by the StatAdvisor; the right for graphical output). The far-left side of Fig. 3.4 shows the navigation window where the user can switch between different work environments within STATGRAPHICS (e.g. StatAdvisor, StatReport and StatGallery for collating graphs). STATGRAPHICS Centurion operates using drop-down menus and dialog boxes to select specific tasks (see bottom left open window in Fig. 3.4) as well as offering push-button icons to carry out specific statistical and graphical functions. STATGRAPHICS is totally menu-driven, there is no programming language to learn; an entire analysis session can be saved via the StatFolio function. For this book, STATGRAPHICS Centurion XVI was used to generate STATGRAPHICS-related output and figures.

R

R[5] (see, for example, Field et al. 2012; Hogan 2010) is an open source (freely downloadable) statistical package developed in the United Kingdom. It is rapidly acquiring the status of a highly flexible and programmable approach to statistical computing and graphics (in addition to being free). **R** is more properly considered as a programming environment, where users execute a series of commands within a very simple interface. This generally means that users need to be both comfortable

---

[5]For details on **R**, see https://www.r-project.org/. For details on **R** Commander (the *Rcmdr* package), see http://www.rcommander.com/. For details on the *psych* package, see https://cran.r-project.org/web/packages/psych/index.html. For details on **R** Studio, see https://www.rstudio.com/

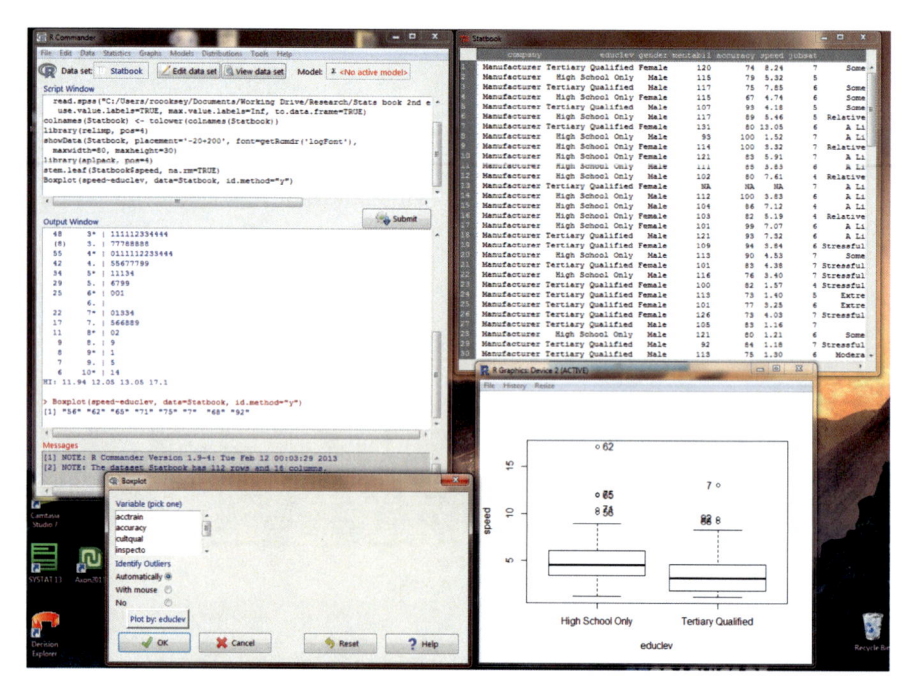

Fig. 3.5 The R Commander environment

working in such a command-driven environment (as opposed to the SPSS-type drop-down menu/dialog box Windows environment) and more statistically savvy.

As an open source system, **R** users can write their own add-on programs and distribute them as 'packages' for use with **R**. One popular package is *Rcmdr* (known as **R** Commander), originally developed by John Fox in the UK (see, for example, Hogan 2010, chapter 2 and Karp 2010). **R** Commander is intended to provide a graphical interface for **R**, in a style similar to a simplified SPSS interface. This has obvious appeal to those users who are not so comfortable working in a command-driven environment. **R** Commander cannot be considered a comprehensive statistical package; its goal is to offer a data entry/editing window along with a range of commonly used data manipulation, statistical and graphing procedures to the user. Another very useful **R** package is *psych* (Revelle 2019a, b) which offers a range of psychometric statistical procedures, including the ICLUST item clustering package referred to in *Procedure 6.6*. More recently, **R** Studio has emerged as a free open source analysis development and display environment for conducting analyses using **R** (see Love, 'The Advantages of R Studio', https://www.theanalysisfactor.com/the-advantages-of-rstudio/, accessed 19 August 2019).

Figure 3.5 shows a sample screen shot from **R** Commander (using version 2.15.2 of **R**) package showing its basic spreadsheet data entry system (in the upper right-side open window) and the Output window (in the upper left-side open window; graphics appear in a separate window as shown in the lower right corner of Fig. 3.5). Like SPSS and SYSTAT, **R** Commander operates using a basic drop-down menu bar to select specific tasks; it does not offer push-button icons for carrying out specific functions. Despite its being open source, **R** has been shown to be a highly accurate, statistically sophisticated and comprehensive alternative to the more expensive statistical packages. Many statistics departments now routinely teach **R** as do many psychology and other social/behavioural science departments. **R** can be used to conduct versions of many of the procedures discussed in this book and the list of its capabilities is increasing daily as more and more users contribute their 'packages' to the system. For this book, **R** Commander, implemented via **R** (version 2.15.2) was used to generate most of **R**-related output and figures.

Comparing the Software Packages

For a novice user, it would probably take about the same amount of time to learn how to use either SYSTAT, STATGRAPHICS or SPSS, whereas NCSS is a bit quicker to learn, but a bit more fiddly to get graphs and analyses to come out just right because of having to twiddle many options in the Task window. Editing graphs in SYSTAT, SPSS and STATGRAPHICS is rather easier because you can operate directly on the graphs themselves. However, SPSS has a broader selection of editing tools than SYSTAT or STATGRAPHICS, its editing process is easier, and more features of the graph can be manipulated after the graph has been produced by the program. SYSTAT is flexible regarding the design of graphs, but most graphical features must be preset before the graph is produced. STATGRAPHICS offers a large suite of pre-configured types of graphs, including a diverse range of options for exploratory data analysis (see *Procedure 5.6*) and producing these graphs and tweaking many of their features is relatively straightforward. With NCSS, you need to configure the features you want in your graph before you produce them. The same is true of **R**, but the process is more complicated because the user needs to manipulate figures and tables at the command level.

In terms of comprehensiveness of analytical capabilities, SPSS clearly has the edge in many ways: it is widely available in the university sector being very popular with business, social and behavioural scientists, it is relatively easy to learn, a student version is available, a lot of books on how to use it have been published and it is regularly maintained and updated (SYSTAT is not that regularly updated). SYSTAT's main claim to fame is its unique and human-centred approach to graphics and it offers some capabilities not available in any of the other four packages (for example, set correlation, see *Procedure 6.8*, perceptual mapping, see *Procedure 6.7*, and logistic test item analysis, see *Procedure 9.1*). STATGRAPHICS is superior in data visualisation methods and has a good user support system. Each package therefore reflects different idiosyncratic choices and preferences on the part of the system designers. However, for most down-to-earth social and behavioural researchers, this should not pose any real problem. Cost is often a limiting factor, in which case, if you can't get access to SPSS somewhere, NCSS or STATGRAPHICS are two of the better packages around for the money, if ease of use is also factored in (**R** may be free but it is not that easy to learn or use for the average person). Most of the illustrations in this book will be produced using SPSS; illustrations produced by the other packages will be used selectively, where they best showcase the procedure being illustrated.

For any of the packages described above, except for **R**, the amount of sophistication the user must possess with respect to their knowledge of statistics is relatively small (to use SYSTAT, you probably have to be a bit more statistically sophisticated compared to SPSS, STATGRAPHICS or NCSS). Obviously, some sound knowledge is important in order to be able to correctly interpret the output and to be able to make rational choices among possible analysis options. However, an in-depth level of statistical or mathematical knowledge is generally not required. If you want to use **R**, but prefer the ease of use of an SPSS-like interface (technically called a GUI or Graphical User Interface), then **R** Commander is the package you have to use within **R** and this offers a limited collection of procedures (it is particularly light in the area of multivariate statistics).

Virtually all the statistical packages described above, except for **R** (outside of **R** Commander), require the same general process in terms of getting data analysed. The first step in each case is to prepare a data file for processing. All of the above packages (including **R** Commander) provide a spreadsheet-like data entry system where the rows of the spreadsheet correspond to separate individuals (or whatever observational units are of interest in the research) in the sample and the columns of the spreadsheet correspond to distinct measures or variables obtained on each entity in the sample. In each statistical package, the columns of the spreadsheet can be labelled with names to facilitate remembering what data each column holds. Each package has its own variable naming conventions and restrictions. For example, in SPSS, a variable name must follow these rules: the name must start with a letter; can be a maximum of 64 characters long; and can only use only letters, numbers, the period and the underscore (but the name cannot end with such punctuation characters). Once the data have been edited and checked for entry errors, analyses can commence. Analyses and graphs can be generated one at a time, inspected, edited, printed and then decisions as to what to do next may be made conditional upon what has been learned in earlier analyses. Consistent with the modern emphasis on exploratory data analysis, this interactive style of analysis is highly appropriate.

Other Available Statistical Packages

There are a number of other reasonably comprehensive statistical packages available for computers that have not been discussed in any detail here, but which are worth a brief mention.

Other Available Statistical Packages   43

Stata, STATISTICA, SAS, eViews & Mplus

Stata[6] (Acock 2018; Hamilton 2013) is a comprehensive statistical package developed exclusively for the desktop computing environment and is available for both Windows and Mac OS platforms (a UNIX version is also available). Stata has the virtue of offering some types of analysis and graphical procedures that are not available in many other packages (including SPSS and SYSTAT). It is a very useful package for econometric as well as epidemiological/biomedical research but serves a variety of purposes for social and behavioural researchers as well. Stata is particularly powerful in terms of the diversity and depth of its time series and regression modelling capabilities.

STATISTICA[7] is a very well-rounded and sophisticated statistical package, offering very similar capabilities to SYSTAT as well as a number of add-on modules. However, STATISTICA is not cheap as a package. One nice feature of STATISTICA is that, for a number of years, it has been associated with an electronic statistics textbook (https://statisticasoftware.wordpress.com/2012/04/12/electronic-statistics-textbook/), which provides sound introductions to many statistical concepts and procedures.

SAS[8] (Statistical Analysis System) refers to a range of very comprehensive statistical packages. SAS is especially powerful and flexible, which is why many statisticians prefer using it. However, users need a fair bit of statistical sophistication to effectively use SAS. SAS is available for the Windows environment with redesigned interfaces, but in terms of cost and knowledge required to make effective use of them, it is expensive and generally are less widely available than the ones described in previous sections.

eViews[9] is a comprehensive statistical package that is especially useful for econometric research. Its look and feel are similar to packages like SPSS, but with strengths, especially, in time series analysis and forecasting. It can also handle a wide range of standard statistical analyses, has excellent graphical and data management capabilities. In some areas, such as time series analysis, eViews has more diverse and up-to-date algorithms for dealing with a wide variety of designs and data anomalies, compared to packages like SPSS, SYSTAT and NCSS. You need a reasonable exposure to econometric methods to get maximum value from using this package and a range of multivariate procedures that are used by social and behavioural researchers (e.g. cluster analysis, MANOVA, multidimensional scaling) are not offered in eViews.

---

[6] For details on Stata, see https://www.stata.com/
[7] For details on STATISTICA, see https://www.statsoft.de/en/statistica
[8] For details on SAS, see https://www.sas.com/en_au/home.html
[9] For details on eViews, see http://www.eviews.com/home.html

MPlus[10] (see, for example, Geiser 2013) is a statistical package devoted to building and testing statistical models for cross-sectional, longitudinal and multi-level research designs. It employs up-to-date algorithms to estimate a wide variety of data models, especially those employing what are called 'latent' variables or constructs. Latent variables or constructs are inferred from sets of directly measured variables rather than being directly observed (for example, 'intelligence' is a technically a latent construct that is indirectly measured through patterns of observed responses to items on an intelligence test). MPlus is ideally configured to effectively test models involving such constructs. While packages like SPSS (via the AMOS package described below) and SYSTAT can deal with such models as well, MPlus has more potent and more flexible modelling capabilities. The drawbacks are a somewhat steeper learning curve and a less intuitive researcher-program interface for MPlus. In addition, building the types of models that MPlus can test requires some degree of statistical knowledge and sophistication.

UNISTAT, XLSTAT & Excel

UNISTAT[11] and XLSTAT[12] are two packages that have been designed to be purchased and used as add-in programs to supplement Microsoft Excel[13].

While Excel is supplied with a basic **Data Analysis TookPak** add-in package, this package is rather limited in analytical capabilities, offering only selected types of descriptive, correlational and inferential procedures and virtually nothing in the multivariate arena. However, Excel does have rather extensive graphical capabilities.

Both UNISTAT and XLSTAT offer a wide-range of capabilities, including many popular multivariate procedures, but neither is as comprehensive as the five packages reviewed in the previous section. Both packages take advantage of Excel's flexibility in graphical production. With either add-in package, one can do a relatively complete job of analysing behavioural data from within the Excel environment. One important observation to make about Excel is that many researchers use it as the major spreadsheet program for entering their research data. Since virtually every statistical package worth its salt today can read Excel data files, this means that Excel provides a data entry platform independent of the idiosyncrasies of the spreadsheet interfaces that come with each different statistical package.

---

[10] For details on MPlus, see https://www.statmodel.com/
[11] For details on UNISTAT, see https://www.unistat.com/
[12] For details on XLSTAT, see https://www.xlstat.com/en/
[13] For details on Excel, see https://products.office.com/en-au/excel

AMOS

There is one final statistical package worthy of mention here, not because it is a comprehensive package, but because it has made some very complex statistical analyses relatively painless to accomplish. This package is AMOS[14] (Analysis of Moment Structures) for Windows. AMOS (Arbuckle 2011; Byrne 2016) is explicitly designed to facilitate analysis of Common Factor and Structural Equation Models (see *Procedure 8.6* and *Procedure 8.7*) as well as a range of other types of multivariate analyses. AMOS was originally written as a teaching program to help students learn how to design and analyse complex structural equation models without having to resort to a complicated matrix specification language. It has since been incorporated as an add-on module for SPSS.

AMOS offers users the capacity to draw a picture of the hypothesised causal model for their data, then use that picture as input for the program to conduct the analysis (AMOS generates all the commands required to translate the drawing into equations without the user having to know how this is done). This graphical method of generating input to guide the analysis of complex causal modelling data is an important and user-friendly strength of AMOS. Other specialised packages for accomplishing the same tasks do exist (e.g. MPlus, LISREL, EQS, RAMONA included in the SYSTAT package, EZPath), but their input and output interpretation requirements tend to be more complex and, in some cases, equation-oriented rather than graphically oriented. [In *Procedure 8.7*, we will discuss a more specialised software package for implementing the partial least squares approach to structural equation modelling called SmartPLS, see Hair et al. 2016; Ringle et al. 2014.]

Considerations When Using Computers for Quantitative Analysis

A few points are worth emphasising with respect to the use of computers in the analysis of business, social and behavioural research data. Learning how to use a specific statistical computing package in no way relieves the researcher of the responsibility of learning about statistics, how and when they should be used, and how they should be interpreted. Virtually all statistical packages are 'mindless' and very flexible. They will do precisely what they are instructed to do even if those instructions are inappropriate (such as when an analytical or graphing procedure is applied to data measured on an inappropriate scale of measurement or to data where many data entry/typing errors have been made). There is a truism, conveyed by the acronym GIGO, which aptly describes this problem: 'Garbage In, Garbage Out'. No

---

[14]For details on AMOS, see https://www.ibm.com/au-en/marketplace/structural-equation-modeling-sem

amount of sophisticated data analysis can save a researcher if the data being analysed are of poor quality in the first place or if poor analytical choices have been made.

The key and perhaps more difficult part of data analysis is learning to instruct the package in the correct things to do. This translates into possessing at least a working knowledge of statistics and statistical procedures, not to the level of mathematical detail in most cases, but enough to know the when, where, and why for applying specific techniques. It also means that the researcher must know how to make sense of the output that is produced during the course of a computer-based analysis. He/she must know what numbers are critical to interpret, what numbers are supplementary, what numbers can be safely ignored and what graphs or tables are important to focus on – in short, the researcher needs to know what to extract from the analysis and what to display in order to tell a convincing story.

It is an unfortunate consequence of programming very general and flexible statistical packages that they tend to produce far too much output. Even when the appearance of certain aspects of that output can be user-controlled; the user must still be able to glean the most relevant information from that output to provide the evidence from which conclusions will be drawn.

This book sets you on the road to achieving this level of understanding, but it is only a beginning. Computational knowledge may not be a necessity, although I am convinced that such knowledge helps understanding in the end, but a working knowledge is absolutely essential. *Chapter 4 – Example research context & quantitative data set* begins your quest toward building this knowledge.

Preparing Data for Analysis by Computer

There are several things to consider when preparing your data for analysis by computer. The first thing to understand is the structure of the database itself – the spreadsheet containing all the data you have collected. Nowadays, the prevalence in the use of questionnaires and other measurement methods means that quantitative data collection is likely to yield many measurements per participant. These data need to be properly organised for analyses to proceed.

A database spreadsheet, such as you might create using SPSS, NCSS or any other statistical package, has a typical structure: the rows represent distinct participants, entities or cases and the columns represent the measurements obtained on each participant or case. Table 3.1 illustrates a typical database structure for a program like SPSS; showing 10 cases by 11 variables (ignore the dashed line columns for the moment). The first column typically contains a participant or case ID, which can link a participant's data in the database to their original data gathering instrument(s), such as a questionnaire. Many analyses will be able to use this ID to identify specific cases for particular purposes.

Each column represents one variable or measurement obtained from a participant and each has a name for the variable. Every statistical package will have slightly

Table 3.1 Typical structure for a statistical package database

case_id	gender	industry	like_boss	like_coworkers	like_job	like_company	Like_salary	llike_benefits	years_w_company	want_to_leave	computed_var_1	computed_var_2
1	1	2	6	5	7	5	6	8	15	90		
2	2	1	5	4	5	5	6	7	66	55		
3	2	3	9	3	5	4	5	5	42	46		
4	1	1	2	.	3	7	6	7	47	72		
5	1	1	4	5	7	5	5	4	28	69		
6	1	3	5	6	6	4	4	5	23	30		
7	2	1	5	4	5	.	5	4	37	59		
8	2	2	2	3	6	6	6	8	.	65		
9	1	1	6	3	6	7	8	9	.	75		
10	2	2	3	4	4	5	5	6	29	54		
.	.	.	.	.	.	.	.	.	.	.	.	.

different variable naming rules and conventions, so you will need to learn what these are before you plan your data entry spreadsheet. Over and above the package-specific variable naming rules, it is good practice to use variable names that have mnemonic value, helping you to remember what each one measures. Table 3.1 reflects this practice where it is easy to see, at a glance, what each variable measures. If the package you are using offers a way to attach an extended label to the variable, this is also useful – I use such labels to remind myself of how the measurement scale worked (for example, in Table 3.1, a label for the variable *like_boss* might be 'How well do you like your boss? 1 = not at all; 7 = extremely well').

Prior to data entry, you will need to make decisions about how to numerically represent each of the categorical, typically nominal scale, variables you have. To provide a record of such decisions, most statistical packages offer the capability of attaching labels to each value for a categorical variable. For example, in Table 3.1, the *gender* variable has two categories, numbered '1' and '2'. To facilitate interpretation, good practice dictates that each number be given a label to reflect the nature of the category, e.g. '1' might represent 'Females' and '2' might represent 'Males'. Most statistical packages offer data management functions that will allow you to recode these values to different numbers (e.g. changing '1' and '2' values to '0' and '1' values) or to combine several category values under a single value, if that is required for certain analyses (e.g. combining three categories, say '1', '2', and '3' into one category with the value of '1').

When preparing your spreadsheet for data entry, you need to decide how you will handle missing data (e.g. where participants do not give a response or where participants do not use a measurement scale in the way you intended). For a package like SPSS, missing values can easily be represented by leaving the relevant variable for that participant blank in the spreadsheet (see, for example, case 4 in the *like_coworkers* variable and cases 8 and 9 in the *years_w_company* variable in Table 3.1). However, you might wish to distinguish between a truly omitted response and a response where the participant did not use the scale in the intended way (like circling two numbers for a Likert-type item on a questionnaire). In such cases, you could use a blank entry to represent truly missing data (no response given) and a number outside the measurement scale (like '9' for a 1 to 7 scale) to represent the situation where the participant gave a response but it was not appropriate for the scale. This will allow you to differentiate the two types of missing data in your analyses. If you then want to combine the types for certain analyses, a package like SPSS can easily accommodate this.

I would offer some final bits of advice on data entry using a statistical package or Excel:

- Firstly, be sure to save your work often. Some packages, like SPSS, can occasionally 'hang up' your computer for some reason and you may lose whatever you have entered since the last time you saved. Frequent saving will avoid big losses. If your package offers the option to save automatically, say every 15 min, activate that option.
- Secondly, it is a good practice to make periodic back-up copies, not only of your spreadsheet database, but also of all analyses you conduct, and store these safely in a different location from your usual work area.
- Thirdly, when you have completed your data entry for all participants or cases, go back and double check the data to ensure that what is in the database is what the original questionnaire or other data gathering record showed. Ideally, it is best to check every row in the database. However, if you have a very large database, it may be more feasible to systematically sample and check, say, every 5th row. Also remember that some early data screening analyses (like frequency tabulation and cross-tabulation in *Procedure 5.1* or correlation in *Procedure 6.1*) can reveal certain types of data entry errors (e.g. numbers out of range; variables not being correlated when you know they should be), so such analyses are worth running as check procedures.

You may wonder what the final two dashed-line columns in Table 3.1, labelled *computed_var_1* and *computed_var_2*, were intended to show. I included these columns to show that, during the process of analysing data, a specific analysis in the statistical package may produce, or you may deliberately create, new variables based on the information in existing variables. These new variables will be added to the database as new columns and saved as part of it. Thus, for many researchers, the database you start with will not be the database you end up with – it will be a dynamic, growing and changing thing.

References

Acock, A. C. (2018). *A gentle introduction to Stata* (6th ed.). College Station: Stata Press.
Allen, P., Bennett, K., & Heritage, B. (2019). *SPSS statistics: A practical guide* (4th ed.). South Melbourne: Cengage Learning Australia Pty.
Arbuckle, J. L. (2011). *IBM SPSS AMOS 20 user's guide*. ftp://public.dhe.ibm.com/software/analytics/spss/documentation/amos/20.0/en/Manuals/IBM_SPSS_Amos_User_Guide.pdf
Byrne, B. M. (2016). *Structural equation modelling with AMOS: Basic concepts, applications, and programming* (3rd ed.). New York: Routledge.
Field, A. (2018). *Discovering statistics using SPSS* (5th ed.). Los Angeles: Sage Publications.
Field, A., Miles, J., & Field, Z. (2012). *Discovering statistics using R*. London: Sage Publications.
Geiser, C. (2013). *Data analysis with Mplus*. New York: The Guilford Press.
George, D., & Mallery, P. (2019). *IBM SPSS statistics 25 step by step: A simple guide and reference* (15th ed.). New York: Routledge.
Hair, J. F., Jr., Hult, G. T. M., Ringle, C., & Sarstedt, M. (2016). *A primer on partial least squares structural equation modeling (PLS-SEM)*. Los Angeles: Sage Publications.
Hamilton, L. C. (2013). *Statistics with stata: Version 12*. Boston: Brooks/Cole.
Hintze, J. L. (2012). *NCSS 8 help system: Introduction*. Kaysville: Number Cruncher Statistical System.
Hogan, T. P. (2010). *Bare-bones R: A brief introductory guide*. Los Angeles: Sage Publications.
Karp, N. A. (2010). *R commander: An introduction*. Cambridge: Wellcome Trust Sanger Institute. https://cran.r-project.org/doc/contrib/Karp-Rcommander-intro.pdf
Pallant, J. (2020). *SPSS survival manual: A step-by-step guide to data analysis using IBM SPSS* (7th ed.). Crows Nest, NSW: Allen & Unwin.
Revelle, W. R. (2019a). *An introduction to the psych package: Part I: Data entry and data description*. Evanston: Northwestern University. https://cran.r-project.org/web/packages/psych/vignettes/intro.pdf
Revelle, W. R. (2019b). *An introduction to the psych package: Part II: Scale construction and psychometrics*. Evanston: Northwestern University. https://cran.r-project.org/web/packages/psych/vignettes/overview.pdf
Ringle, C., da Silva, D., & Bido, D. (2014). Structural equation modeling with the SmartPLS. *Brazil J Market, 13*(2), 57–73.
StatPoint Technologies, Inc. (2010). *STATGRAPHICS Centurion XVI user manual*. Warrenton: StatPoint Technologies Inc.
SYSTAT Software Inc. (2009). *SYSTAT 13: Getting started*. Chicago: SYSTAT Software Inc.
Wagner, W. E., III. (2013). *Using IBM SPSS statistics for research methods and social science statistics*. Los Angeles: Sage Publications.

Chapter 4
Example Research Context & Quantitative Data Set

An important aspect of the learning value for this book is that nearly all the statistical procedures will be illustrated with reference to data emerging from a single coherent research context. This means that, as you read, the variables used in the illustrations will remain familiar to you. The research context has been constructed to resonate not only with readers from a business background, but also with readers from social science, psychological, educational and health-related backgrounds.

The research context and database are entirely fictitious and were created so that specific relationships and comparisons could be illustrated. Throughout the book, we will refer to the database as the **Quality Control Inspectors** (**QCI**) database. Most of the procedures to be illustrated will be performed on variables from this database using SPSS, with some periodic use of SYSTAT, NCSS, STATGRAPHICS or **R**, as appropriate, for specific analytical and illustrative purposes. For certain more sophisticated procedures (such as time series analysis and Rasch models for item analysis) and procedures involving different research designs (such as repeated measures designs), we will need to extend the QCI research context to encompass new design aspects and additional data for the illustrations and to demonstrate the use of more specialised computer software programs. However, as far as possible, illustrations will be framed within the general QCI research context.

The Quality Control Inspectors (QCI) Database[1]

Research Context

A behavioural researcher, Maree Lakota, undertook a consultancy for a consortium of manufacturing industries. Her primary task was to conduct an in-depth study of the performance of Quality Control Inspectors in electronics manufacturing companies. Such companies ranged from those involved in the assembly of microcomputers and small electronic control devices and appliances to those involved in the manufacture of major electrical appliances and transportation systems.

As part of her research, Maree created a new test designed to help electronics manufacturing companies evaluate the performance of quality control inspectors on their product assembly lines. This test measured both the speed and the accuracy with which an inspector could visually detect assembly errors in electronic circuit boards and other complex wiring arrangements. She had also been asked to explore the relationships between inspector performance and other psychological and behavioural variables such as job satisfaction, general mental ability, perceived stressfulness of work conditions, and inspectors' attitudes toward management, toward quality control as an organisational process, and toward provisions for training in quality control. Certain demographic characteristics of inspectors, such as the type of company the inspector worked for as well as their education level and gender, were also of interest. The goal of Maree's research endeavour was to try to understand the factors which influenced quality control inspection accuracy and speed. She also conducted more specialised investigations to address specific research questions.

For her primary survey investigation, Maree sampled Quality Control Inspectors who work on assembly lines in five different electronics manufacturing industries/companies. She employed random sampling to obtain at least 20 inspectors from each company yielding a total sample size of 112 quality control inspectors. Maree obtained measures on 17 variables from each of the sampled Inspectors. The *QCI Database Variable Coding Guide* below describes the nature and interpretation of each of these 17 variables.

---

[1] Versions of the QCI database, formatted for use within specific statistical packages, can be obtained, on request by email, from Ray Cooksey at rcooksey@une.edu.au. Formatted versions that can be requested include: Excel, SPSS, SYSTAT, NCSS, STATGRAPHICS Centurion, Stata and R.

The Quality Control Inspectors (QCI) Database

QCI database variable coding guide

Variable	Variable name[a]	Variable details
Company	**Company**	A nominal measure recording the type of company the inspector works for: 1. = PC manufacturer 2. = large electrical appliance manufacturer 3. = small electrical appliance manufacturer 4. = large business computer manufacturer 5. = automobile manufacturer
Education level achieved	**Educlev**	A dichotomous ordinal measure recording the highest level of education attained by the inspector: 1. = High school level (to Year 12) education 2. = Tertiary qualification (University or vocational (TAFE))
Gender	**Gender**	A dichotomous nominal measure recording the gender of the inspector: 1. = Male 2. = Female
General mental ability	**Mentabil**	The inspector's score on a standardised test of general mental ability (having a mean of 100 in the population and a standard deviation of 10). For example, a score of 110 would be interpreted as a level of general mental ability that was one standard deviation above the general population mean. Given that general mental ability is typically normally distributed in the population, this implies that roughly 68% of all people who take the test will score between 90 and 110 and about 95% of all people who take the test will score between 80 and 120.
		As a further guide to interpreting this score for all tested people, consider the following: a score of 80 is better than scores obtained by 2.28% of people a score of 90 is better than scores obtained by 15.87% of people a score of 100 is better than scores obtained by 50.0% of people a score of 110 is better than scores obtained by 84.13% of people a score of 120 is better than scores obtained by 97.72% of people a score of 130 is better than scores obtained by 99.86% of people
Inspection accuracy	**Accuracy**	The inspector's score on the accuracy component of Maree's new test. This score is expressed as a percentage between 0% and 100% indicating the percentage of defects correctly detected in her test.
Inspection speed	**Speed**	The inspector's score on the speed component of Maree's new test. This score is expressed in the average or mean number of seconds that the inspector took to judge the quality ('defective' or 'not defective') of each test assembly.

(continued)

Variable	Variable name[a]	Variable details
Overall job satisfaction	**Jobsat**	An indication of the overall level of satisfaction the inspector felt toward his/her quality control job. The survey question was: 'All things being considered, please rate your general level of satisfaction with your job as Quality Control Inspector in this company?' This general attitude was measured using a 7-point scale where $1 =$ Very Low, $4 =$ Neutral (or 'So-So'), and $7 =$ Very High.
Work conditions	**Workcond**	An indication of the inspector's perception of the general stressfulness of the conditions under which he/she works on a daily basis. The survey question was: 'All things being considered, please rate how stressful you find the conditions under which you have to work each day?' This general attitude was measured using a 7-point scale where $1 =$ Highly Stressful, $4 =$ Moderately Stressful but Tolerable, and $7 =$ Relatively Stress-free. Thus, **higher** scores indicated a less stressful working environment.

The remaining 9 variables were all specific attitude items measured using a 7-point scale where $1 =$ Strongly Disagree, $4 =$ Neither Agree nor Disagree, and $7 =$ Strongly Agree

[a]Variable names used for the illustrative analyses performed using the SPSS, SYSTAT, NCSS, STATGRAPHICS and **R** statistical packages

Variable	Variable name	Variable details
Culture of quality	**Cultqual**	This item measured the extent to which the inspector perceived that achieving high product quality was continually reinforced in the culture of the organisation; the specific question was: 'The importance of high product quality is clearly evident in the culture of my organisation'.
Access to training	**Acctrain**	This item measured the extent to which the inspector perceived that their organisation was providing adequate access to training programs relevant to his/her job as Inspector. The specific question was: 'My company provides me with adequate opportunities to access training programs that could help me in my job as inspector'.
Training quality	**Trainqua**	This item measured the extent to which the inspector perceived that they had access to good quality training. The specific question was: 'The training which I have undertaken in order to improve my skills as Inspector has been of generally high quality'.
Quality important for marketing	**Qualmktg**	This item measured the extent to which the inspector believed that maintaining high product quality was important for successful marketing of those products. The specific question was: 'I believe that maintaining high product quality is important for the successful marketing of my company's products'.

(continued)

The Quality Control Inspectors (QCI) Database

Variable	Variable name	Variable details
Management communication adequacy	**Mgmtcomm**	This item measured the extent to which the inspector perceived that senior management in the company had adequate communication links (both upward and downward) with the Quality Control Inspectors. The specific question was: 'The senior managers in my company have established generally good communication linkages with us as Quality Control Inspectors in terms of providing us with information as well as listening to our ideas'.
Quality important for own satisfaction	**Qualsat**	This item measured the extent to which the inspector believed that maintaining high product quality was important in helping to attain a good level of satisfaction in their job. The specific question was: 'I believe that part of what helps me to feel satisfied in my job is my role in maintaining high product quality for this company'.
Training appropriate	**Trainapp**	This item measured the extent to which the inspector perceived that the training which they had access to was appropriately targeted to the work that they did. The specific question was: 'The training which I have undertaken in order to improve my specific skills as an Inspector has appropriately targeted those skills'.
Satisfaction with supervisor	**Satsupvr**	This item measured the extent to which the inspector perceived that they had a generally satisfactory working relationship with their immediate supervisor or manager. The specific question was: 'My working relationship with my immediate supervisor or manager is generally satisfactory'.
Satisfaction with management policies	**Polsatis**	This item measured the extent to which the inspector perceived that they were generally satisfied with those policies, established by senior management, concerning quality control inspection practices. The specific question was: 'I am generally satisfied with the policies that senior management in my company has established concerning quality control inspection practices'.

Figure 4.1 shows what the data for the first 36 quality control inspectors looks like in the SPSS version of the database. Where an inspector did not provide data for a specific variable, the relevant space in each version of database is left blank. This could happen through the inspector not answering or forgetting to answer a specific questionnaire item or not successfully completing Maree's quality inspection test; e.g. note that inspector 13 in Fig. 4.1 did not provide data for the **mentabil**, **accuracy**, **speed** or **cultqual** variables. The SPSS, SYSTAT, STATGRAPHICS and NCSS programs are all designed to read a blank space for a variable as what is called a 'missing value'. **R** can be told to process blanks as missing values, but then converts them to 'NA' when it saves the datafile. Missing values are assigned a special number by each software program so that they can be identified and treated differently for analyses.

Most often, missing values are omitted when analysing data from a specific variable. However, as will be illustrated in *Procedure 8.2*, there are occasions where patterns of missing values are specifically analysed.

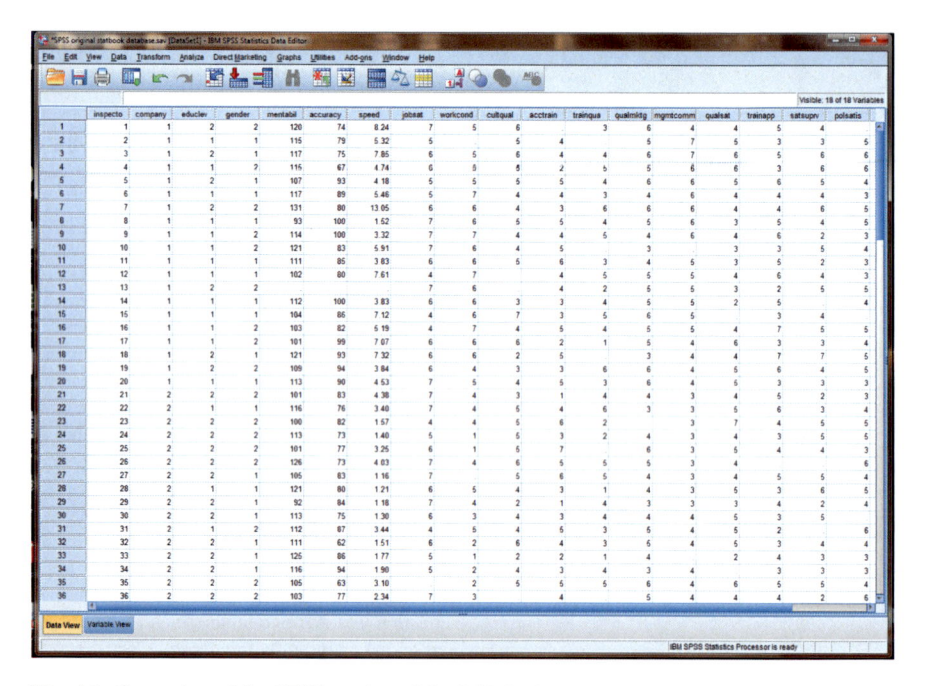

Fig. 4.1 Screenshot of the SPSS version of the QCI database

Quantitative Stories That Could Be Pursued in the QCI Database

It is important to realise that in analysing quantitative data, there are many more stories to learn about and convey than just addressing research questions. This is true whether you are writing a postgraduate thesis, dissertation or portfolio or writing a journal article or conference paper – writing up and interpreting results really is a story-telling exercise. Often you will need to tell a range of preliminary stories before you ever get down to testing a hypothesis.

Below are just some of the stories that need to be conveyed with respect to research based on quantitative data, such as the QCI database. Maree Lakota, in the QCI research context, would certainly need to unfold each of these stories in order to produce a convincing research report.

- *Choice of approaches to analysis and reasoning for those choices*: This story clues the reader into the researcher's reasoning behind the choices of analytical procedures that will be used. These choices must suit the nature of the data themselves as well as being appropriate for addressing the research problem, question or hypothesis at hand.

The Quality Control Inspectors (QCI) Database

> With respect to the QCI research context Maree would need to convey a story about the approaches she will take to check for construct validity of her measures, test the predictions in her hypotheses and/or evaluate group differences and other hypothesised relationships. This story may intersect another one identified below with respect to checking of assumptions because if her data do not meet assumptions for an intended procedure she would need to discuss and give reasons for either transforming her data to meet those assumptions or choosing and implementing an alternative analytical approach.

- *Data preparation and coding*: This story informs the reader about how the data were prepared and checked for analysis, including decisions made about how anomalous outcomes (e.g. missing data, incorrect or inappropriate responses) would be handled, how categorical measures would be coded, how any qualitative data (e.g. from open-ended survey items) would be handled and so on. Chapter 20 in Cooksey and McDonald (2019) talks more about issues associated with data preparation.

> With respect to the QCI context, Maree would want to summarise how the QCI database was prepared, how the **company**, **educlev** and **gender** categorical variables were coded and signal that there were no anomalous questionnaire responses beyond the occasional missing value.

- *Data screening and checking for anomalous data patterns*: The next two stories are most likely closely interwoven, but it is useful to distinguish them. A story about data screening and anomalous data patterns is a more general story about the nature of the data themselves. Part of this story could be indications of any extreme data values ('outliers') and discussion of any notable patterns in missing data (see *Procedure 8.2*). Missing data patterns are especially important to explore (although many researchers neglect this) because the last thing a researcher wants is for something in their methodology (e.g. the way a question is asked) to cause some participants to not give a legitimate response. Missing data should be random and sparse and there are statistical procedures to help the researcher provide evidence for this. In most cases, this will be a narrative story requiring few if any actual statistics or graphs to be presented, unless the nature of an anomaly needs to be displayed.

> In the QCI context, Maree would want to unfold a story about any outlier values in the variables in the database and about any patterns of missing data (e.g. do high school educated participants tend to show more missing values than tertiary educated participants). Figure 4.1, in the first 36 cases, shows evidence of missing data; the question is does it occur randomly or is it related to something about the participants or the survey construction? Unless she is transparent in this story, her generalisations about any hypotheses will be weakened.

- *Checking that assumptions are satisfied for specific procedures*: This story, while related to the previous story, is much more focused on specific procedures. It addresses the issue of whether the assumptions (e.g. normality, homogeneity of variance) required by the analytical procedure to be used (e.g. multiple regression, analysis of variance, correlation) can be assumed to be satisfied. Statistical packages offer many diagnostic statistics and graphical techniques to help the researcher make such judgments. If assumptions are satisfied, the story to be told is straightforward. If, however, assumptions are not satisfied, the researcher is obligated to discuss/defend what they will do about the problem (which may impact on the analytical choices story highlighted earlier).

> In the QCI context, Maree should report checking the assumptions for the inferential procedures she plans to use. Her story could include reference to both statistics (e.g. tests of normality) and graphs (e.g. plots of distributions and/or residuals), but normally, she would not actually include any of these in the write up. She could simply unfold a narrative about what she looked at, what she learned and what she did about any problems identified.

- *Describing the demographics/essential characteristics of the sample*: This is the first substantive analytical story a researcher needs to convey and judicious use of statistics, tables and/or figures will help to make the story convincing. This is the story about who the sources of data were, how they were chosen, the nature and number of participants/companies/schools of different types included and, where possible, how the sample compares to the population. Any questions about possible sampling bias would be addressed in this story. This is largely a descriptive analysis story, so any of the procedures discussed in Chap. 5 could potentially be useful.

The Quality Control Inspectors (QCI) Database

> Maree, in the QCI context, would want to tell the story of who the 112 participants in her survey were, by company, education level and gender, at the very least. If she had access to some sort of industry level employment data, she could compare her returns for various categories of participant to those data to make judgments about sampling bias or representativeness.

- *Describing procedures for checking for measurement validity and reliability, including any procedures for combining measures*: This is the measurement story; one that is very important to convey as it goes to the heart of the quality of the data for addressing the research questions. For a good deal of behavioural research, this story is really a preliminary one, rather than one that explicitly targets research questions or hypotheses. It sets the foundation for the analyses to come. The story requires a narrative that addresses the operational definition steps undertaken to produce any measures. This would be followed by a judicious reporting of appropriate statistics (e.g. a factor analysis followed by reliability analysis) to provide the evidence in the narrative. Remember, the goal is to convince the reader of the quality of your measures (e.g. their construct validity or other types of validity, as appropriate).

> In the QCI database, Maree would want to unfold a narrative that discusses how she constructed and pilot tested her survey (including where she sourced the items, if relevant), and shows how her attitude items, for example, worked together to measure focal constructs.

- *Describing the results of analyses that directly address hypotheses and/or research questions*: This is the story that many researchers think is all there is to tell. However, hopefully you can see that this is definitely not the case. What this story requires is a close focus on the research questions or hypotheses, with judicious use of statistics, tables and figures, as appropriate, to provide the evidence. The goal is **not** to overwhelm the reader with numbers and figures. The story must be told strategically, for maximum impact, by choosing precisely what analytical outcomes need to be reported to make the story convincing, no more, no less (Appendix C provides concrete examples for many analytical procedures). If a hypothesis is not supported or the outcomes are unexpected, the researcher needs to devote time to unpacking what may have happened. In fact, how a researcher handles unexpected findings can make the difference between research that is convincing and research that is not convincing.

> In the QCI context, this story is where Maree would get down to addressing her concrete research questions (e.g. do mental ability, job satisfaction, working conditions and speed of decision making predict the accuracy of a QCI inspector's decisions? Do speed and accuracy of quality decision differ for participants from different industries, with different levels of education and of different genders?). Each research question or hypothesis would get its own mini-story, which reported what analysis was done, what was learned (with evidence) and what the outcomes meant (surface level interpretation at this point; deeper interpretations and broader conclusions and connections to the literature would be a story for a later section in her report or paper).

- *Describing the results of any supplemental analyses which explore trends in the data that may not have been anticipated*: While analysing quantitative data, certain patterns and relationships may emerge that the researcher did not anticipate or about which no hypothesis was proposed. Communicating these can provide value for learning and fertilisation of ideas for other researchers so they are worth a bit of attention, not as main events, but as interesting and unexpected findings.

> In the QCI context, if Maree discovered a pattern of relationships between attitudes and speed of decision making that neither she nor the literature had anticipated, it would be worth her telling that story laced judiciously with her evidence to set the stage for other researchers to build on her initial findings.

To help you gain additional insights into the range of stories that a quantitative researcher may need to tell, Appendix C provides concrete examples and broader discussion. You may also find it useful to read chapters 21 and 22 in Cooksey and McDonald (2019) for more insights into writing a convincing research story.

References

Cooksey, R. W., & McDonald, G. (2019). *Surviving and thriving in postgraduate research* (2nd ed.). Singapore: Springer Nature.

Chapter 5
Descriptive Statistics for Summarising Data

The first broad category of statistics we discuss concerns *descriptive statistics*. The purpose of the procedures and fundamental concepts in this category is quite straightforward: to facilitate the description and summarisation of data. By 'describe' we generally mean either the use of some pictorial or graphical representation of the data or the computation of an index or number designed to summarise a specific characteristic of a variable or measurement.

We seldom interpret individual data points or observations primarily because it is too difficult for the human brain to extract or identify the essential nature, patterns, or trends evident in the data, particularly if the sample is large. Rather we utilise procedures and measures which provide a general depiction of how the data are behaving. These statistical procedures are designed to identify or display specific patterns or trends in the data. What remains after their application is simply for us to interpret and tell the story.

> Reflect on the QCI research scenario and the associated data set discussed in *Chap. 4*. Consider the following questions that Maree might wish to address with respect to decision accuracy and speed scores:
>
> - What was the typical level of accuracy and decision speed for inspectors in the sample? [see *Procedure 5.4* – Assessing central tendency.]
> - What was the most common accuracy and speed score amongst the inspectors? [see *Procedure 5.4* – Assessing central tendency.]
> - What was the range of accuracy and speed scores; the lowest and the highest scores? [see *Procedure 5.5* – Assessing variability.]
> - How frequently were different levels of inspection accuracy and speed observed? What was the shape of the distribution of inspection accuracy and speed scores? [see *Procedure 5.1* – Frequency tabulation, distributions & crosstabulation.]
>
> (continued)

© Springer Nature Singapore Pte Ltd. 2020
R. W. Cooksey, *Illustrating Statistical Procedures: Finding Meaning in Quantitative Data*, https://doi.org/10.1007/978-981-15-2537-7_5

- What percentage of inspectors would have 'failed' to 'make the cut' assuming the industry standard for acceptable inspection accuracy and speed combined was set at 95%? [see *Procedure 5.7* – Standard (z) scores.]
- How variable were the inspectors in their accuracy and speed scores? Were all the accuracy and speed levels relatively close to each other in magnitude or were the scores widely spread out over the range of possible test outcomes? [see *Procedure 5.5* – Assessing variability.]
- What patterns might be visually detected when looking at various QCI variables singly and together as a set? [see *Procedure 5.2* – Graphical methods for dispaying data, *Procedure 5.3* – Multivariate graphs & displays, and *Procedure 5.6* – Exploratory data analysis.]

This chapter includes discussions and illustrations of a number of procedures available for answering questions about data like those posed above. In addition, you will find discussions of two fundamental concepts, namely *probability* and the *normal distribution*; concepts that provide building blocks for *Chaps. 6* and *7*.

Procedure 5.1: Frequency Tabulation, Distributions & Crosstabulation

Classification	Univariate (crosstabulations are bivariate); descriptive.
Purpose	To produce an efficient counting summary of a sample of data points for ease of interpretation.
Measurement level	Any level of measurement can be used for a variable summarised in frequency tabulations and crosstabulations.

Frequency Tabulation and Distributions

Frequency tabulation serves to provide a convenient counting summary for a set of data that facilitates interpretation of various aspects of those data. Basically, frequency tabulation occurs in two stages:

- First, the scores in a set of data are rank ordered from the lowest value to the highest value.
- Second, the number of times each specific score occurs in the sample is counted. This count records the *frequency* of occurrence for that specific data value.

Procedure 5.1: Frequency Tabulation, Distributions & Crosstabulation

Consider the overall job satisfaction variable, **jobsat**, from the QCI data scenario. Performing frequency tabulation across the 112 Quality Control Inspectors on this variable using the *SPSS Frequencies* procedure (Allen et al. 2019, ch. 3; George and Mallery 2019, ch. 6) produces the frequency tabulation shown in Table 5.1. Note that three of the inspectors in the sample did not provide a rating for **jobsat** thereby producing three missing values (= 2.7% of the sample of 112) and leaving 109 inspectors with valid data for the analysis.

Table 5.1 Frequency tabulation of overall job satisfaction scores

jobsat

		Frequency	Percent	Valid Percent	Cumulative Percent
Valid	1 Very Low	4	3.6	3.7	3.7
	2	8	7.1	7.3	11.0
	3	8	7.1	7.3	18.3
	4 Neutral	18	16.1	16.5	34.9
	5	19	17.0	17.4	52.3
	6	34	30.4	31.2	83.5
	7 Very High	18	16.1	16.5	100.0
	Total	109	97.3	100.0	
Missing	System	3	2.7		
Total		112	100.0		

The display of frequency tabulation is often referred to as the *frequency distribution* for the sample of scores. For each value of a variable, the frequency of its occurrence in the sample of data is reported. It is possible to compute various percentages and percentile values from a frequency distribution.

Table 5.1 shows the 'Percent' or *relative frequency* of each score (the percentage of the 112 inspectors obtaining each score, including those inspectors who were missing scores, which SPSS labels as 'System' missing). Table 5.1 also shows the 'Valid Percent' which is computed only for those inspectors in the sample who gave a valid or non-missing response.

Finally, it is possible to add up the 'Valid Percent' values, starting at the low score end of the distribution, to form the *cumulative distribution* or 'Cumulative Percent'. A cumulative distribution is useful for finding *percentiles* which reflect what percentage of the sample scored at a specific value or below.

We can see in Table 5.1 that 4 of the 109 valid inspectors (a 'Valid Percent' of 3.7%) indicated the lowest possible level of job satisfaction—a value of 1 (Very Low) – whereas 18 of the 109 valid inspectors (a 'Valid Percent' of 16.5%) indicated the highest possible level of job satisfaction—a value of 7 (Very High). The 'Cumulative Percent' number of 18.3 in the row for the job satisfaction score of 3 can be interpreted as "roughly 18% of the sample of inspectors reported a job satisfaction score of 3 or less"; that is, nearly a fifth of the sample expressed some degree of negative satisfaction with their job as a quality control inspector in their particular company.

If you have a large data set having many different scores for a particular variable, it may be more useful to tabulate frequencies on the basis of intervals of scores.

For the **accuracy** scores in the QCI database, you could count scores occurring in intervals such as 'less than 75% accuracy', 'between 75% but less than 85% accuracy', 'between 85% but less than 95% accuracy', and '95% accuracy or greater', rather than counting the individual scores themselves. This would yield what is termed a 'grouped' frequency distribution since the data have been grouped into intervals or score classes. Producing such an analysis using *SPSS* would involve extra steps to create the new category or 'grouping' system for scores prior to conducting the frequency tabulation.

Crosstabulation

In a *frequency crosstabulation*, we count frequencies on the basis of two variables simultaneously rather than one; thus we have a bivariate situation.

For example, Maree might be interested in the number of male and female inspectors in the sample of 112 who obtained each **jobsat** score. Here there are two variables to consider: inspector's **gender** and inspector's **jobsat** score. Table 5.2 shows such a crosstabulation as compiled by the SPSS *Crosstabs* procedure (George and Mallery 2019, ch. 8). Note that inspectors who did not report a score for **jobsat** and/or **gender** have been omitted as missing values, leaving 106 valid inspectors for the analysis.

(continued)

Table 5.2 Frequency crosstabulation of **jobsat** scores by **gender** category for the QCI data

jobsat * gender Crosstabulation

			gender		
			1 Male	2 Female	Total
jobsat	1 Very Low	Count	2	1	3
		% within jobsat	66.7%	33.3%	100.0%
		% within gender	3.5%	2.0%	2.8%
	2	Count	3	5	8
		% within jobsat	37.5%	62.5%	100.0%
		% within gender	5.3%	10.2%	7.5%
	3	Count	2	6	8
		% within jobsat	25.0%	75.0%	100.0%
		% within gender	3.5%	12.2%	7.5%
	4 Neutral	Count	11	7	18
		% within jobsat	61.1%	38.9%	100.0%
		% within gender	19.3%	14.3%	17.0%
	5	Count	14	5	19
		% within jobsat	73.7%	26.3%	100.0%
		% within gender	24.6%	10.2%	17.9%
	6	Count	17	15	32
		% within jobsat	53.1%	46.9%	100.0%
		% within gender	29.8%	30.6%	30.2%
	7 Very High	Count	8	10	18
		% within jobsat	44.4%	55.6%	100.0%
		% within gender	14.0%	20.4%	17.0%
Total		Count	57	49	106
		% within jobsat	53.8%	46.2%	100.0%
		% within gender	100.0%	100.0%	100.0%

The crosstabulation shown in Table 5.2 gives a composite picture of the distribution of satisfaction levels for male inspectors and for female inspectors. If frequencies or 'Counts' are added across the **gender** categories, we obtain the numbers in the 'Total' column (the percentages or relative frequencies are also shown immediately below each count) for each discrete value of **jobsat** (note this column of statistics differs from that in Table 5.1 because the **gender** variable was missing for certain inspectors). By adding down each **gender** column, we obtain, in the bottom row labelled 'Total', the number of males and the number of females that comprised the sample of 106 valid inspectors.

The totals, either across the rows or down the columns of the crosstabulation, are termed the *marginal distributions* of the table. These

(continued)

marginal distributions are equivalent to frequency tabulations for each of the variables **jobsat** and **gender**. As with frequency tabulation, various percentage measures can be computed in a crosstabulation, including the percentage of the sample associated with a specific count within either a row ('% within **jobsat**') or a column ('% within **gender**'). You can see in Table 5.2 that 18 inspectors indicated a job satisfaction level of 7 (Very High); of these 18 inspectors reported in the 'Total' column, 8 (44.4%) were male and 10 (55.6%) were female. The marginal distribution for **gender** in the 'Total' row shows that 57 inspectors (53.8% of the 106 valid inspectors) were male and 49 inspectors (46.2%) were female. Of the 57 male inspectors in the sample, 8 (14.0%) indicated a job satisfaction level of 7 (Very High). Furthermore, we could generate some additional interpretive information of value by adding the '% within gender' values for job satisfaction levels of 5, 6 and 7 (i.e. differing degrees of positive job satisfaction). Here we would find that 68.4% (= 24.6% + 29.8% + 14.0%) of male inspectors indicated some degree of positive job satisfaction compared to 61.2% (= 10.2% + 30.6% + 20.4%) of female inspectors.

This helps to build a picture of the possible relationship between an inspector's gender and their level of job satisfaction (a relationship that, as we will see later, can be quantified and tested using *Procedure 6.2* and *Procedure 7.1*).

It should be noted that a crosstabulation table such as that shown in Table 5.2 is often referred to as a *contingency table* about which more will be said later (see *Procedure 7.1* and *Procedure 7.18*).

Advantages

Frequency tabulation is useful for providing convenient data summaries which can aid in interpreting trends in a sample, particularly where the number of discrete values for a variable is relatively small. A cumulative percent distribution provides additional interpretive information about the relative positioning of specific scores within the overall distribution for the sample.

Crosstabulation permits the simultaneous examination of the distributions of values for two variables obtained from the same sample of observations. This examination can yield some useful information about the possible relationship between the two variables. More complex crosstabulations can be also done where the values of three or more variables are tracked in a single systematic summary. The use of frequency tabulation or cross-tabulation in conjunction with various other statistical measures, such as measures of central tendency (see *Procedure 5.4*) and measures of variability (see *Procedure 5.5*), can provide a relatively complete descriptive summary of any data set.

Disadvantages

Frequency tabulations can get messy if interval or ratio-level measures are tabulated simply because of the large number of possible data values. Grouped frequency distributions really should be used in such cases. However, certain choices, such as the size of the score interval (group size), must be made, often arbitrarily, and such choices can affect the nature of the final frequency distribution.

Additionally, percentage measures have certain problems associated with them, most notably, the potential for their misinterpretation in small samples. One should be sure to know the sample size on which percentage measures are based in order to obtain an interpretive reference point for the actual percentage values.

For example In a sample of 10 individuals, 20% represents only two individuals whereas in a sample of 300 individuals, 20% represents 60 individuals. If all that is reported is the 20%, then the mental inference drawn by readers is likely to be that a sizeable number of individuals had a score or scores of a particular value—but what is 'sizeable' depends upon the total number of observations on which the percentage is based.

Where Is This Procedure Useful?

Frequency tabulation and crosstabulation are very commonly applied procedures used to summarise information from questionnaires, both in terms of tabulating various demographic characteristics (e.g. gender, age, education level, occupation) and in terms of actual responses to questions (e.g. numbers responding 'yes' or 'no' to a particular question). They can be particularly useful in helping to build up the data screening and demographic stories discussed in *Chap. 4*. Categorical data from observational studies can also be analysed with this technique (e.g. the number of times Suzy talks to Frank, to Billy, and to John in a study of children's social interactions).

Certain types of experimental research designs may also be amenable to analysis by crosstabulation with a view to drawing inferences about distribution differences across the sets of categories for the two variables being tracked.

> You could employ crosstabulation in conjunction with the tests described in *Procedure 7.1* to see if two different styles of advertising campaign differentially affect the product purchasing patterns of male and female consumers.
>
> In the QCI database, Maree could employ crosstabulation to help her answer the question "do different types of electronic manufacturing firms (**company**) differ in terms of their tendency to employ male versus female quality control inspectors (**gender**)?"

Software Procedures

Application	Procedures
SPSS	*Analyze → Descriptive Statistics → Frequencies...* or *Crosstabs...* and select the variable(s) you wish to analyse; for the *Crosstabs* procedure, hitting the '*Cells*' button will allow you to choose various types of statistics and percentages to show in each cell of the table.
NCSS	*Analysis → Descriptive Statistics → Frequency Tables* or *Cross Tabulation* and select the variable(s) you wish to analyse.
SYSTAT	*Analyze → One-Way Frequency Tables...* or *Tables → Two-Way...* and select the variable(s) you wish to analyse and choose the optional statistics you wish to see.
STATGRAPHICS	*Describe → Categorical Data → Tabulation* or *→ Crosstabulation* and select the variable(s) you wish to analyse; hit '*OK*' and when the 'Tables and Graphs' window opens, choose the Tables and Graphs you wish to see.
R Commander	*Statistics → Summaries → Frequency Tables...* or *Crosstabulation → Two-way table...* and select the variable(s) you wish to analyse and choose the optional statistics you wish to see.

Procedure 5.2: Graphical Methods for Displaying Data

Classification	Univariate (scatterplots are bivariate); descriptive.
Purpose	To visually summarise characteristics of a data sample for ease of interpretation.
Measurement level	Any level of measurement can be accommodated by these graphical methods. Scatterplots are generally used for interval or ratio-level data.

Graphical methods for displaying data include bar and pie charts, histograms and frequency polygons, line graphs and scatterplots. It is important to note that what is presented here is a small but representative sampling of the types of simple graphs one can produce to summarise and display trends in data. Generally speaking, SPSS offers the easiest facility for producing and editing graphs, but with a rather limited range of styles and types. SYSTAT, STATGRAPHICS and NCSS offer a much wider range of graphs (including graphs unique to each package), but with the drawback that it takes somewhat more effort to get the graphs in exactly the form you want.

Procedure 5.2: Graphical Methods for Displaying Data

Bar and Pie Charts

These two types of graphs are useful for summarising the frequency of occurrence of various values (or ranges of values) where the data are categorical (nominal or ordinal level of measurement).

- A *bar chart* uses vertical and horizontal axes to summarise the data. The vertical axis is used to represent frequency (number) of occurrence or the relative frequency (percentage) of occurrence; the horizontal axis is used to indicate the data categories of interest.
- A *pie chart* gives a simpler visual representation of category frequencies by cutting a circular plot into wedges or slices whose sizes are proportional to the relative frequency (percentage) of occurrence of specific data categories. Some pie charts can have a one or more slices emphasised by 'exploding' them out from the rest of the pie.

Consider the **company** variable from the QCI database. This variable depicts the types of manufacturing firms that the quality control inspectors worked for. Figure 5.1 illustrates a bar chart summarising the percentage of female inspectors in the sample coming from each type of firm. Figure 5.2 shows a pie chart representation of the same data, with an 'exploded slice' highlighting the percentage of female inspectors in the sample who worked for large business computer manufacturers – the lowest percentage of the five types of companies. Both graphs were produced using SPSS.

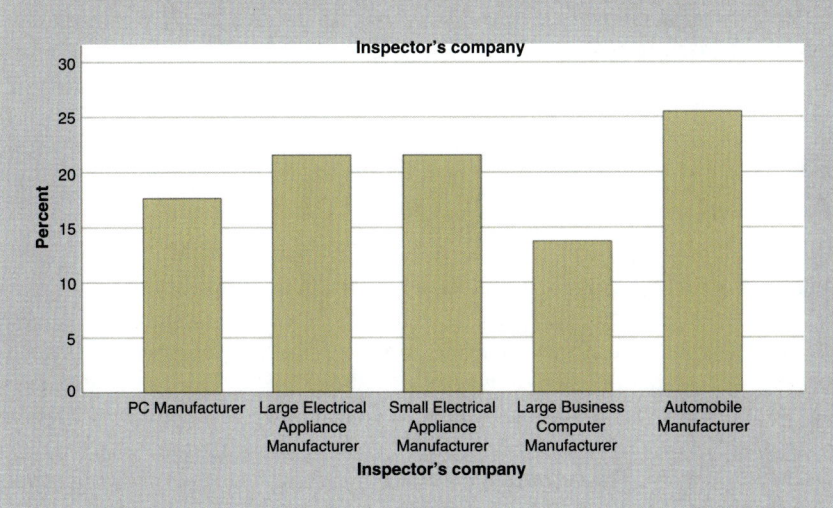

Fig. 5.1 Bar chart: Percentage of female inspectors

(continued)

Fig. 5.2 Pie chart: Percentage of female inspectors

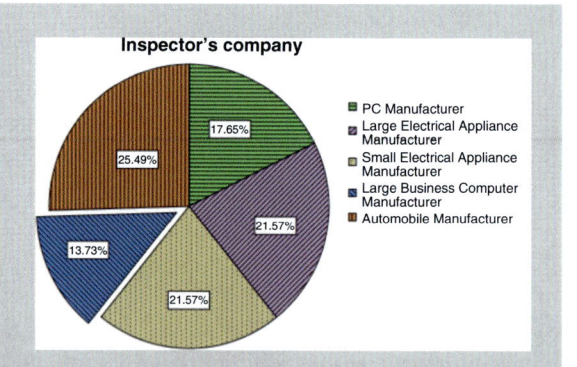

Inspector's company

- PC Manufacturer
- Large Electrical Appliance Manufacturer
- Small Electrical Appliance Manufacturer
- Large Business Computer Manufacturer
- Automobile Manufacturer

17.65%, 25.49%, 13.73%, 21.57%, 21.57%

The pie chart was modified with an option to show the actual percentage along with the label for each category. The bar chart shows that computer manufacturing firms have relatively fewer female inspectors compared to the automotive and electrical appliance (large and small) firms. This trend is less clear from the pie chart which suggests that pie charts may be less visually interpretable when the data categories occur with rather similar frequencies. However, the 'exploded slice' option can help interpretation in some circumstances.

Certain software programs, such as SPSS, STATGRAPHICS, NCSS and Microsoft Excel, offer the option of generating 3-dimensional bar charts and pie charts and incorporating other 'bells and whistles' that can potentially add visual richness to the graphic representation of the data. However, you should generally be careful with these fancier options as they can produce distortions and create ambiguities in interpretation (e.g. see discussions in Jacoby 1997; Smithson 2000; Wilkinson 2009). Such distortions and ambiguities could ultimately end up providing misinformation to researchers as well as to those who read their research.

Histograms and Frequency Polygons

These two types of graphs are useful for summarising the frequency of occurrence of various values (or ranges of values) where the data are essentially continuous (interval or ratio level of measurement) in nature. Both histograms and frequency polygons use vertical and horizontal axes to summarise the data. The vertical axis is used to represent the frequency (number) of occurrence or the relative frequency (percentage) of occurrences; the horizontal axis is used for the data values or ranges of values of interest. The *histogram* uses bars of varying heights to depict frequency; the *frequency polygon* uses lines and points.

There is a visual difference between a histogram and a bar chart: the bar chart uses bars that do not physically touch, signifying the discrete and categorical nature of the data, whereas the bars in a histogram physically touch to signal the potentially continuous nature of the data.

Procedure 5.2: Graphical Methods for Displaying Data

Suppose Maree wanted to graphically summarise the distribution of **speed** scores for the 112 inspectors in the QCI database. Figure 5.3 (produced using NCSS) illustrates a histogram representation of this variable. Figure 5.3 also illustrates another representational device called the 'density plot' (the solid tracing line overlaying the histogram) which gives a smoothed impression of the overall shape of the distribution of **speed** scores. Figure 5.4 (produced using STATGRAPHICS) illustrates the frequency polygon representation for the same data.

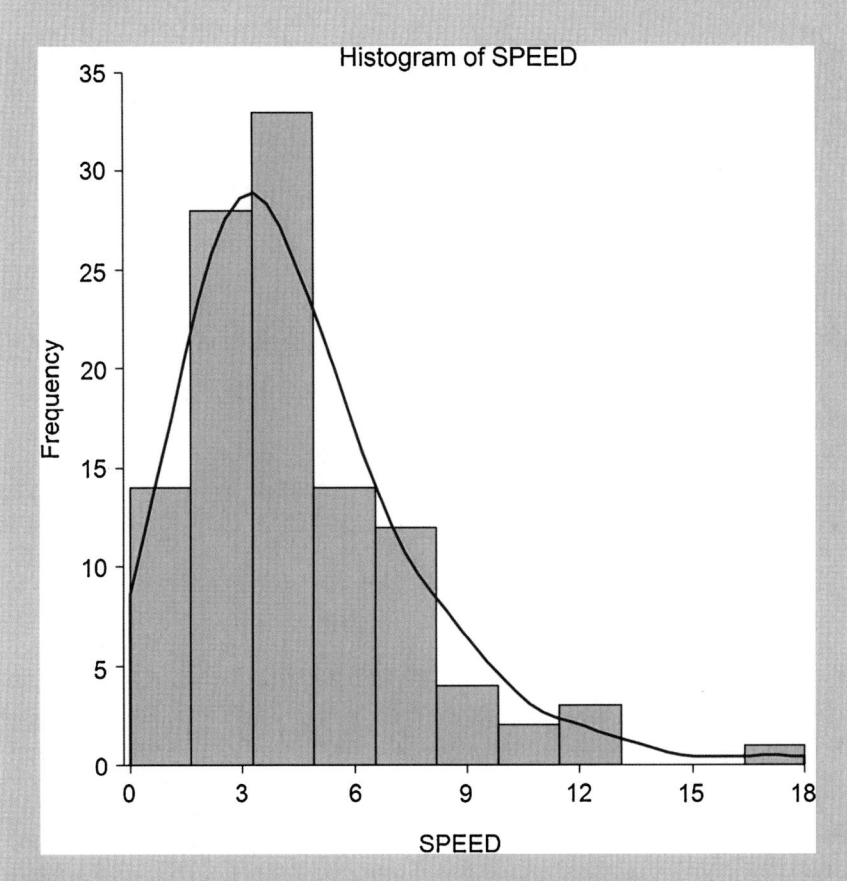

Fig. 5.3 Histogram of the **speed** variable (with density plot overlaid)

These graphs employ a grouped format where **speed** scores which fall within specific intervals are counted as being essentially the same score. The shape of the data distribution is reflected in these plots. Each graph tells us that the inspection **speed** scores are positively skewed with only a few inspectors taking very long times to make their inspection judgments and the majority of inspectors taking rather shorter amounts of time to make their decisions.

(continued)

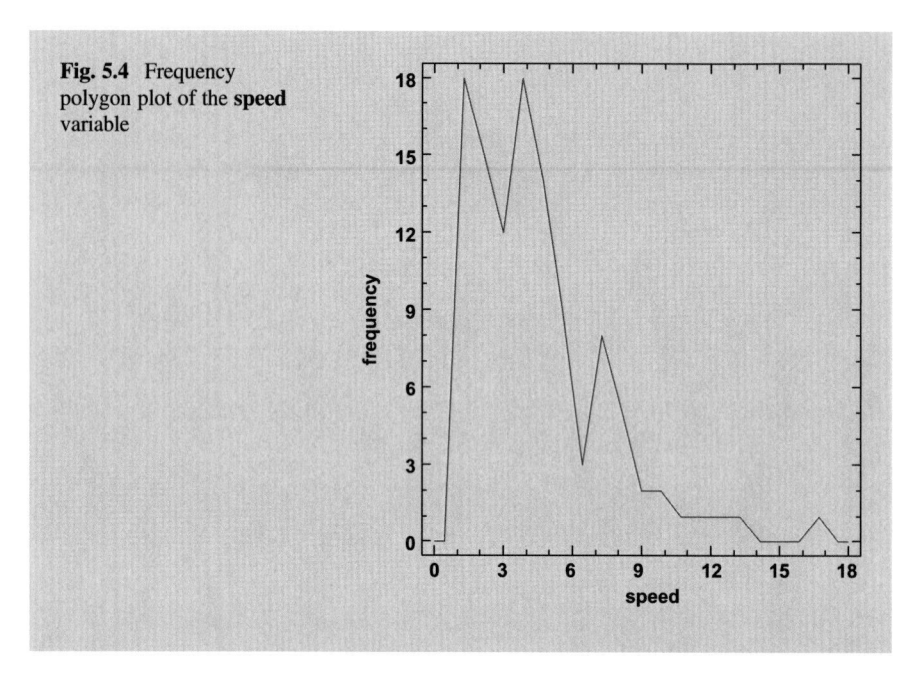

Fig. 5.4 Frequency polygon plot of the **speed** variable

Both representations tell a similar story; the choice between them is largely a matter of personal preference. However, if the number of bars to be plotted in a histogram is potentially very large (and this is usually directly controllable in most statistical software packages), then a frequency polygon would be the preferred representation simply because the amount of visual clutter in the graph will be much reduced.

It is somewhat of an art to choose an appropriate definition for the width of the score grouping intervals (or 'bins' as they are often termed) to be used in the plot: choose too many and the plot may look too lumpy and the overall distributional trend may not be obvious; choose too few and the plot will be too coarse to give a useful depiction. Programs like SPSS, SYSTAT, STATGRAPHICS and NCSS are designed to choose an 'appropriate' number of bins to be used, but the analyst's eye is often a better judge than any statistical rule that a software package would use.

There are several interesting variations of the histogram which can highlight key data features or facilitate interpretation of certain trends in the data. One such variation is a graph is called a **dual histogram** (available in SYSTAT; a variation called a 'comparative histogram' can be created in NCSS) – a graph that facilitates visual comparison of the frequency distributions for a specific variable for participants from two distinct groups.

> Suppose Maree wanted to graphically compare the distributions of **speed** scores for inspectors in the two categories of education level (**educlev**) in the QCI database. Figure 5.5 shows a dual histogram (produced using SYSTAT) that accomplishes this goal. This graph still employs the grouped

(continued)

Procedure 5.2: Graphical Methods for Displaying Data

format where **speed** scores falling within particular intervals are counted as being essentially the same score. The shape of the data distribution within each group is also clearly reflected in this plot. However, the story conveyed by the dual histogram is that, while the inspection **speed** scores are positively skewed for inspectors in both categories of **educlev,** the comparison suggests that inspectors with a high school level of education (= 1) tend to take slightly longer to make their inspection decisions than do their colleagues who have a tertiary qualification (= 2).

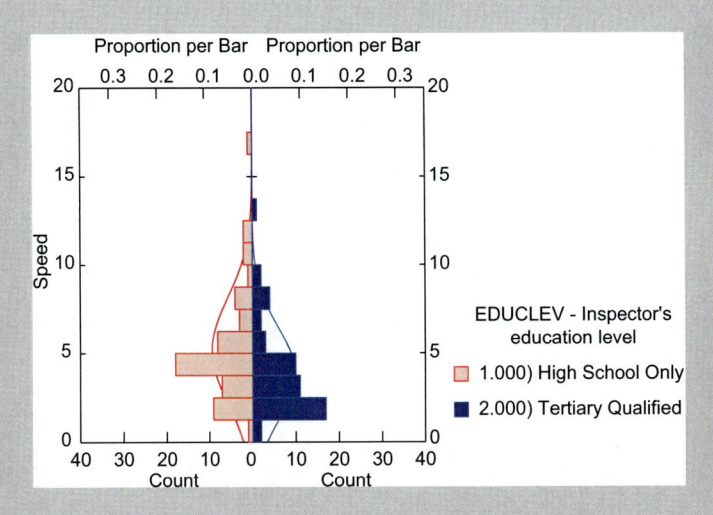

Fig. 5.5 Dual histogram of *speed* for the two categories of **educlev**

Line Graphs

The *line graph* is similar in style to the frequency polygon but is much more general in its potential for summarising data. In a line graph, we seldom deal with percentage or frequency data. Instead we can summarise other types of information about data such as averages or means (see *Procedure 5.4* for a discussion of this measure), often for different groups of participants. Thus, one important use of the line graph is to break down scores on a specific variable according to membership in the categories of a second variable.

In the context of the QCI database, Maree might wish to summarise the average inspection **accuracy** scores for the inspectors from different types of manufacturing companies. Figure 5.6 was produced using SPSS and shows such a line graph.

(continued)

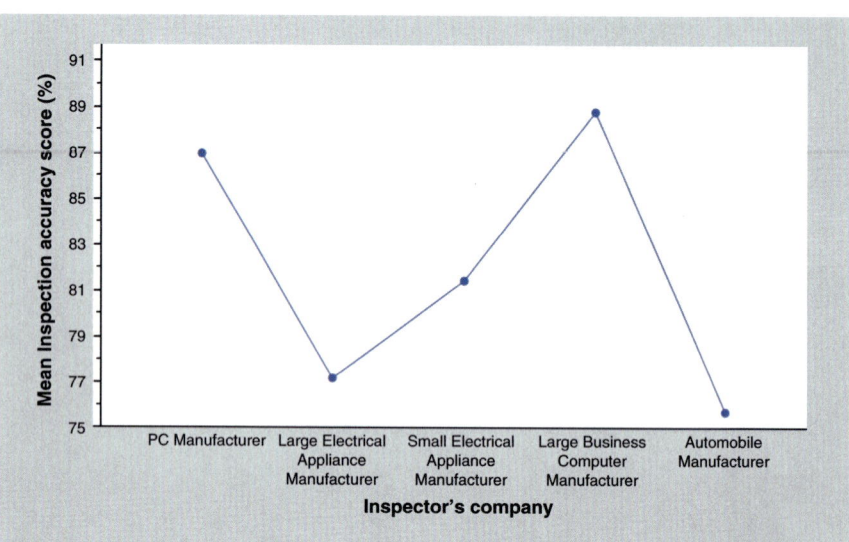

Fig. 5.6 Line graph comparison of companies in terms of average inspection **accuracy**

Note how the trend in performance across the different companies becomes clearer with such a visual representation. It appears that the inspectors from the Large Business Computer and PC manufacturing companies have better average inspection accuracy compared to the inspectors from the remaining three industries.

With many software packages, it is possible to further elaborate a line graph by including error or confidence intervals bars (see *Procedure 8.3*). These give some indication of the precision with which the average level for each category in the population has been estimated (narrow bars signal a more precise estimate; wide bars signal a less precise estimate).

Figure 5.7 shows such an elaborated line graph, using 95% confidence interval bars, which can be used to help make more defensible judgments (compared to Fig. 5.6) about whether the companies are substantively different from each other in average inspection performance. Companies whose confidence interval bars do not overlap each other can be inferred to be substantively different in performance characteristics.

The accuracy confidence interval bars for participants from the Large Business Computer manufacturing firms do not overlap those from the Large or Small Electrical Appliance manufacturers or the Automobile manufacturers.

(continued)

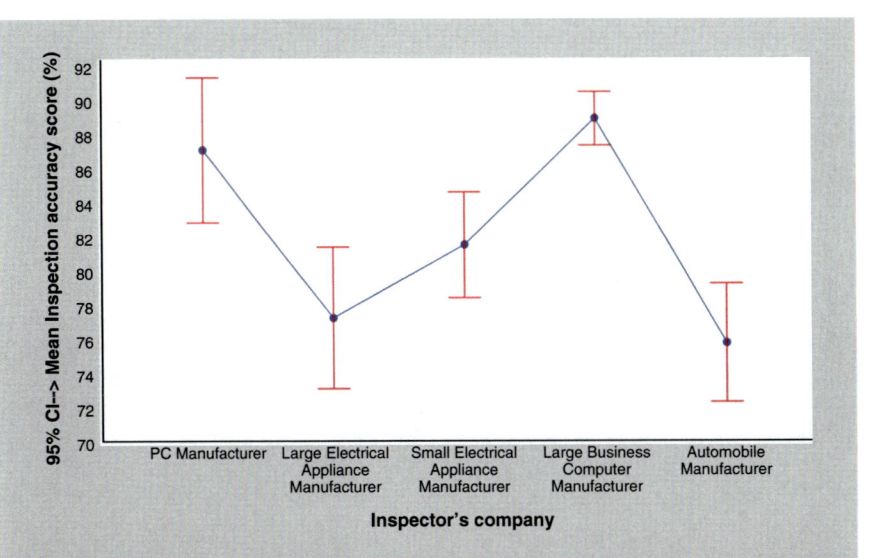

Procedure 5.2: Graphical Methods for Displaying Data

Fig. 5.7 Line graph using confidence interval bars to compare **accuracy** across companies

We might conclude that quality control inspection accuracy is substantially better in the Large Business Computer manufacturing companies than in these other industries but is not substantially better than the PC manufacturing companies. We might also conclude that inspection accuracy in PC manufacturing companies is not substantially different from Small Electrical Appliance manufacturers.

Scatterplots

Scatterplots are useful in displaying the relationship between two interval- or ratio-scaled variables or measures of interest obtained on the same individuals, particularly in correlational research (see *Fundamental Concept III* and *Procedure 6.1*).

In a scatterplot, one variable is chosen to be represented on the horizontal axis; the second variable is represented on the vertical axis. In this type of plot, all data point pairs in the sample are graphed. The shape and tilt of the cloud of points in a scatterplot provide visual information about the strength and direction of the relationship between the two variables. A very compact elliptical cloud of points signals a strong relationship; a very loose or nearly circular cloud signals a weak or non-existent relationship. A cloud of points generally tilted upward toward the right side of the graph signals a positive relationship (higher scores on one variable associated with higher scores on the other and vice-versa). A cloud of points generally tilted downward toward the right side of the graph signals a negative relationship (higher scores on one variable associated with lower scores on the other and vice-versa).

Maree might be interested in displaying the relationship between inspection **accuracy** and inspection **speed** in the QCI database. Figure 5.8, produced using SPSS, shows what such a scatterplot might look like. Several characteristics of the data for these two variables can be noted in Fig. 5.8. The shape of the distribution of data points is evident. The plot has a fan-shaped characteristic to it which indicates that accuracy scores are highly variable (exhibit a very wide range of possible scores) at very fast inspection speeds but get much less variable and tend to be somewhat higher as inspection speed increases (where inspectors take longer to make their quality control decisions). Thus, there does appear to be some relationship between inspection **accuracy** and inspection **speed** (a weak positive relationship since the cloud of points tends to be very loose but tilted generally upward toward the right side of the graph – slower speeds tend to be slightly associated with higher accuracy.

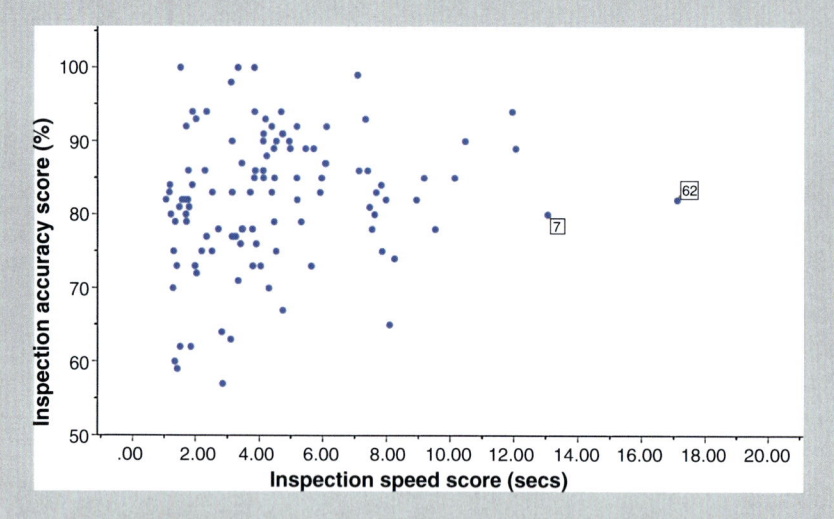

Fig. 5.8 Scatterplot relating inspection **accuracy** to inspection **speed**

However, it is not the case that the inspection decisions which take longest to make are necessarily the most accurate (see the labelled points for inspectors 7 and 62 in Fig. 5.8). Thus, Fig. 5.8 does not show a simple relationship that can be unambiguously summarised by a statement like "the longer an inspector takes to make a quality control decision, the more accurate that decision is likely to be". The story is more complicated.

Some software packages, such as SPSS, STATGRAPHICS and SYSTAT, offer the option of using different plotting symbols or *markers* to represent the members of different groups so that the relationship between the two focal variables (the ones anchoring the X and Y axes) can be clarified with reference to a third categorical measure.

Procedure 5.2: Graphical Methods for Displaying Data

Maree might want to see if the relationship depicted in Fig. 5.8 changes depending upon whether the inspector was tertiary-qualified or not (this information is represented in the **educlev** variable of the QCI database).

Figure 5.9 shows what such a modified scatterplot might look like; the legend in the upper corner of the figure defines the marker symbols for each category of the **educlev** variable. Note that for both High School only-educated inspectors and Tertiary-qualified inspectors, the general fan-shaped relationship between **accuracy** and **speed** is the same. However, it appears that the distribution of points for the High School only-educated inspectors is shifted somewhat upward and toward the right of the plot suggesting that these inspectors tend to be somewhat more accurate as well as slower in their decision processes.

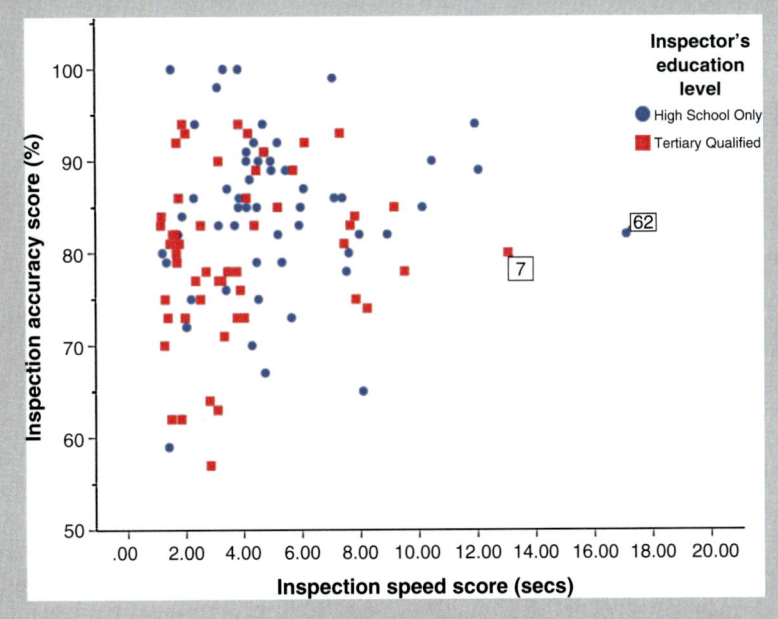

Fig. 5.9 Scatterplot displaying **accuracy** vs **speed** conditional on **educlev** group

There are many other styles of graphs available, often dependent upon the specific statistical package you are using. Interestingly, NCSS and, particularly, SYSTAT and STATGRAPHICS, appear to offer the most variety in terms of types of graphs available for visually representing data. A reading of the user's manuals for these programs (see the Useful additional readings) would expose you to the great diversity of plotting techniques available to researchers. Many of these techniques go by rather interesting names such as: Chernoff's faces, radar plots, sunflower plots, violin plots, star plots, Fourier blobs, and dot plots.

Advantages

These graphical methods provide summary techniques for visually presenting certain characteristics of a set of data. Visual representations are generally easier to understand than a tabular representation and when these plots are combined with available numerical statistics, they can give a very complete picture of a sample of data. Newer methods have become available which permit more complex representations to be depicted, opening possibilities for creatively visually representing more aspects and features of the data (leading to a style of visual data storytelling called *infographics*; see, for example, McCandless 2014; Toseland and Toseland 2012). Many of these newer methods can display data patterns from multiple variables in the same graph (several of these newer graphical methods are illustrated and discussed in *Procedure 5.3*).

Disadvantages

Graphs tend to be cumbersome and space consuming if a great many variables need to be summarised. In such cases, using numerical summary statistics (such as means or correlations) in tabular form alone will provide a more economical and efficient summary. Also, it can be very easy to give a misleading picture of data trends using graphical methods by simply choosing the 'correct' scaling for maximum effect or choosing a display option (such as a 3-D effect) that 'looks' presentable but which actually obscures a clear interpretation (see Smithson 2000; Wilkinson 2009).

Thus, you must be careful in creating and interpreting visual representations so that the influence of aesthetic choices for sake of appearance do not become more important than obtaining a faithful and valid representation of the data—a very real danger with many of today's statistical packages where 'default' drawing options have been pre-programmed in. No single plot can completely summarise all possible characteristics of a sample of data. Thus, choosing a specific method of graphical display may, of necessity, force a behavioural researcher to represent certain data characteristics (such as frequency) at the expense of others (such as averages).

Where Is This Procedure Useful?

Virtually any research design which produces quantitative data and statistics (even to the extent of just counting the number of occurrences of several events) provides opportunities for graphical data display which may help to clarify or illustrate important data characteristics or relationships. Remember, graphical displays are communication tools just like numbers—which tool to choose depends upon the message to be conveyed. Visual representations of data are generally more useful in communicating to lay persons who are unfamiliar with statistics. Care must be taken though as these same lay people are precisely the people most likely to misinterpret a graph if it has been incorrectly drawn or scaled.

Software Procedures

Application	Procedures
SPSS	*Graphs → Chart Builder...* and choose from a range of gallery chart types: *Bar, Pie/Polar, Histogram, Line, Scatter/Dot*; drag the chart type into the working area and customise the chart with desired variables, labels, etc. many elements of a chart, including error bars, can be controlled.
NCSS	*Graphics → Bar Charts → Bar Charts* or *Graphics → Pie Charts* or *Graphics → Histograms → Histograms* or *Histograms – Comparative* or *Graphics → Error Bar Charts → Error Bar Charts* or *Graphics → Scatter Plots → Scatter Plots;* whichever type of chart you choose, you can control many features of the chart from the dialog box that pops open upon selection.
STATGRAPHICS	*Plot → Business Charts → Barchart...* or *Piechart...* or *Plot → Exploratory Plots → Frequency Histogram...* or *Plot → Scatterplots → X – Y Plot...;* whichever type of chart you choose, you can control a number of features of the chart from the series of dialog boxes that pops open upon selection.
SYSTAT	*Graph → Bar...* or *Pie...* or *Histogram...* or *Line...* or *Scatterplot...* or *Graph Gallery...* (which offers a range of other more novel graphical displays, including the dual histogram). For each choice, a dialog box opens which allows you to control almost every characteristic of the graph you want.
R Commander	*Graphs → Bar graph* or *Pie chart* or *Histogram* or *Scatterplot* or *Plot of Means*; for some graphs (*Scatterplot* being the exception), there is minimal control offered by **R** Commander over the appearance of the graph (you need to use full **R** commands to control more aspects; e.g. see Chang 2019).

Procedure 5.3: Multivariate Graphs & Displays

Classification	Multivariate; descriptive.
Purpose	To simultaneously and visually summarise characteristics of many variables obtained on the same entities for ease of interpretation.
Measurement level	Multivariate graphs and displays are generally produced using interval or ratio-level data. However, such graphs may be grouped according to a nominal or ordinal categorical variable for comparison purposes.

Graphical methods for displaying multivariate data (i.e. many variables at once) include scatterplot matrices, radar (or spider) plots, multiplots, parallel coordinate displays, and icon plots. Multivariate graphs are useful for visualising broad trends and patterns across many variables (Cleveland 1995; Jacoby 1998). Such graphs typically sacrifice precision in representation in favour of a snapshot pictorial summary that can help you form general impressions of data patterns.

It is important to note that what is presented here is a small but reasonably representative sampling of the types of graphs one can produce to summarise and display trends in multivariate data. Generally speaking, SYSTAT offers the best facilities for producing multivariate graphs, followed by STATGRAPHICS, but with the drawback that it is somewhat tricky to get the graphs in exactly the form you want. SYSTAT also has excellent facilities for creating new forms and combinations of graphs – essentially allowing graphs to be tailor-made for a specific communication purpose. Both SPSS and NCSS offer a more limited range of multivariate graphs, generally restricted to scatterplot matrices and variations of multiplots. Microsoft Excel or STATGRAPHICS are the packages to use if radar or spider plots are desired.

Scatterplot Matrices

A *scatterplot matrix* is a useful multivariate graph designed to show relationships between pairs of many variables in the same display.

> Figure 5.10 illustrates a scatterplot matrix, produced using SYSTAT, for the **mentabil, accuracy, speed, jobsat** and **workcond** variables in the QCI database. It is easy to see that all the scatterplot matrix does is stack all pairs of scatterplots into a format where it is easy to pick out the graph for any 'row' variable that intersects a column 'variable'.
>
> In those plots where a 'row' variable intersects itself in a column of the matrix (along the so-called 'diagonal'), SYSTAT permits a range of univariate displays to be shown. Figure 5.10 shows univariate histograms for each variable (recall *Procedure 5.2*). One obvious drawback of the scatterplot matrix is that, if many variables are to be displayed (say ten or more); the graph gets very crowded and becomes very hard to visually appreciate.
>
> Looking at the first column of graphs in Fig. 5.10, we can see the scatterplot relationships between **mentabil** and each of the other variables. We can get a

(continued)

visual impression that **mentabil** seems to be slightly negatively related to **accuracy** (the cloud of scatter points tends to angle downward to the right, suggesting, very slightly, that higher **mentabil** scores are associated with lower levels of **accuracy**).

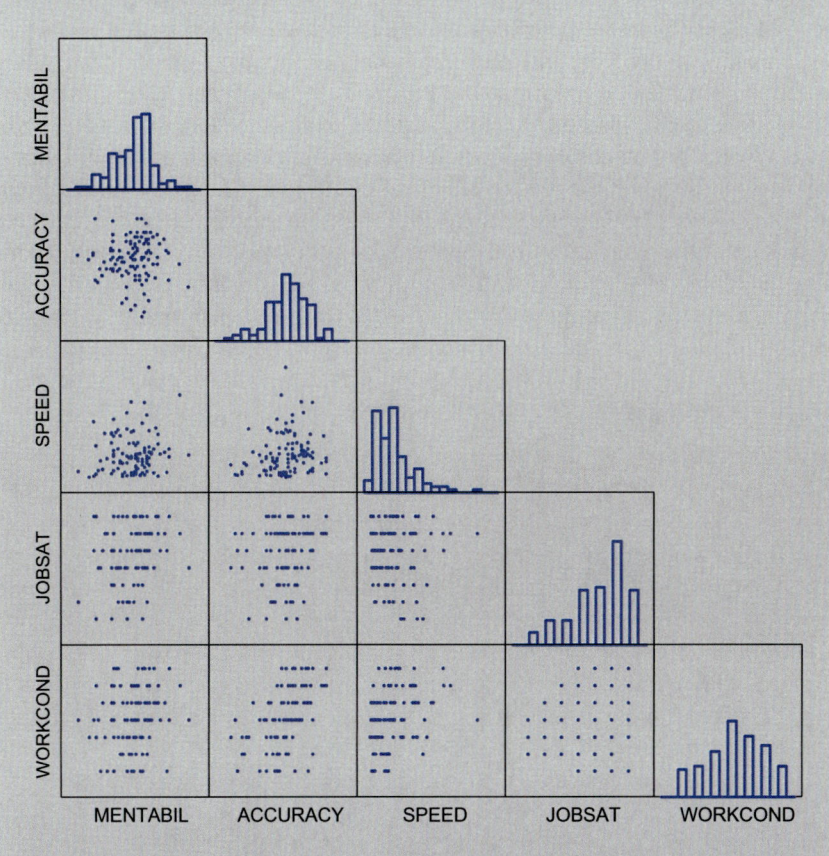

Fig. 5.10 Scatterplot matrix relating **mentabil, accuracy, speed, jobsat** & **workcond**

Conversely, the visual impression of the relationship between **mentabil** and **speed** is that the relationship is slightly positive (higher **mentabil** scores tend to be associated with higher **speed** scores = longer inspection times). Similar types of visual impressions can be formed for other parts of Fig. 5.10. Notice that the histogram plots along the diagonal give a clear impression of the shape of the distribution for each variable.

Radar Plots

The *radar plot* (also known as a *spider graph* for obvious reasons) is a simple and effective device for displaying scores on many variables. Microsoft Excel offers a range of options and capabilities for producing radar plots, such as the plot shown in Fig. 5.11. Radar plots are generally easy to interpret and provide a good visual basis for comparing plots from different individuals or groups, even if a fairly large number of variables (say, up to about 25) are being displayed. Like a clock face, variables are evenly spaced around the centre of the plot in clockwise order starting at the 12 o'clock position. Visual interpretation of a radar plot primarily relies on shape comparisons, i.e. the rise and fall of peaks and valleys along the spokes around the plot. Valleys near the centre display low scores on specific variables, peaks near the outside of the plot display high scores on specific variables. [Note that, technically, radar plots employ polar coordinates.] SYSTAT can draw graphs using polar coordinates but not as easily as Excel can, from the user's perspective. Radar plots work best if all the variables represented are measured on the same scale (e.g. a 1 to 7 Likert-type scale or 0% to 100% scale). Individuals who are missing any scores on the variables being plotted are typically omitted.

> The radar plot in Fig. 5.11, produced using Excel, compares two specific inspectors, 66 and 104, on the nine attitude rating scales. Inspector 66 gave the highest rating (= 7) on the **cultqual** variable and inspector 104 gave the lowest rating (= 1). The plot shows that inspector 104 tended to provide very low ratings on all nine attitude variables, whereas inspector 66 tended to give very high ratings on all variables except **acctrain** and **trainapp**, where the scores were similar to those for inspector 104. Thus, in general, inspector 66 tended to show much more positive attitudes toward their workplace compared to inspector 104.
>
> While Fig. 5.11 was generated to compare the scores for two individuals in the QCI database, it would be just as easy to produce a radar plot that compared the five types of companies in terms of their average ratings on the nine variables, as shown in Fig. 5.12.
>
> Here we can form the visual impression that the five types of companies differ most in their average ratings of **mgmtcomm** and least in the average ratings of **polsatis**. Overall, the average ratings from inspectors from PC manufacturers (black diamonds with solid lines) seem to be generally the most positive as their scores lie on or near the outer ring of scores and those

(continued)

Procedure 5.3: Multivariate Graphs & Displays

from Automobile manufacturers tend to be least positive on many variables (except the training-related variables).

Extrapolating from Fig. 5.12, you may rightly conclude that including too many groups and/or too many variables in a radar plot comparison can lead to so much clutter that any visual comparison would be severely degraded. You may have to experiment with using colour-coded lines to represent different groups versus line and marker shape variations (as used in Fig. 5.12), because choice of coding method for groups can influence the interpretability of a radar plot.

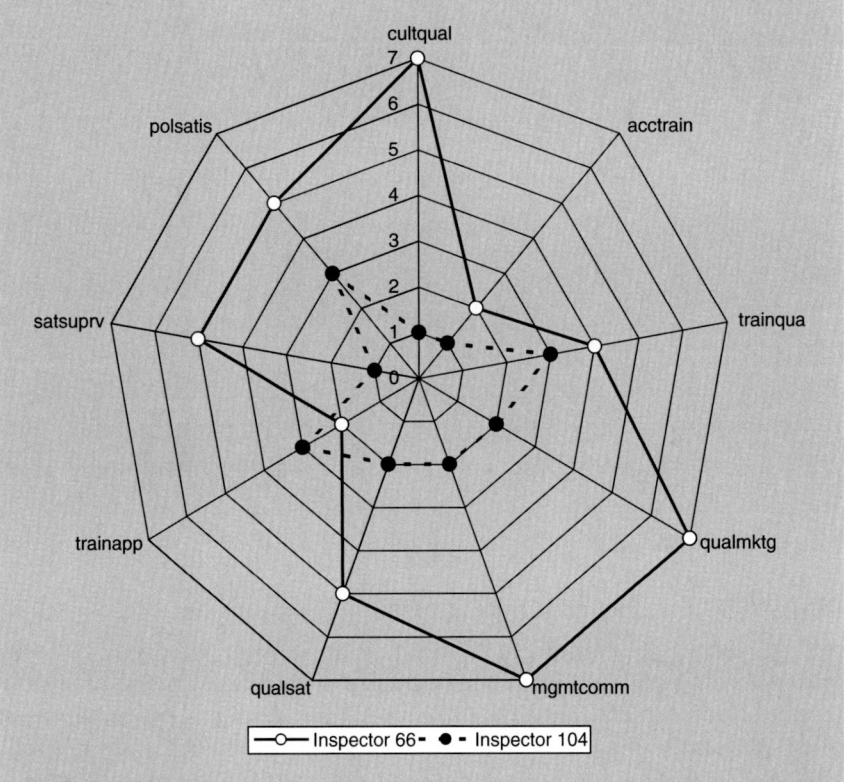

Fig. 5.11 Radar plot comparing attitude ratings for inspectors 66 and 104

(continued)

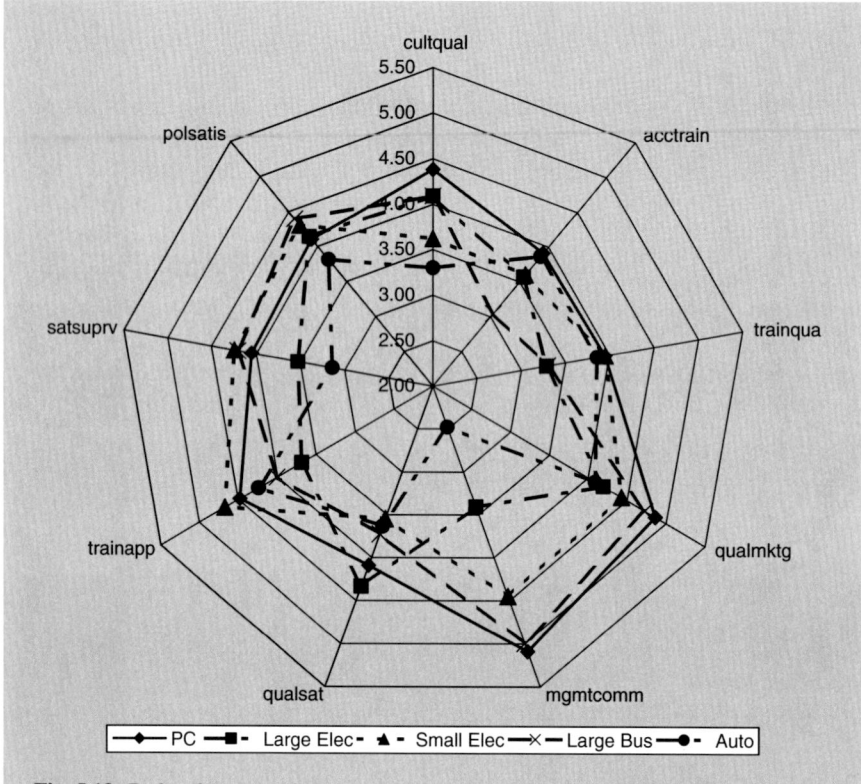

Fig. 5.12 Radar plot comparing average attitude ratings for five types of company

Multiplots

A *multiplot* is simply a hybrid style of graph that can display group comparisons across a number of variables. There are a wide variety of possible multiplots one could potentially design (SYSTAT offers great capabilities with respect to multiplots). Figure 5.13 shows a multiplot comprising a side-by-side series of profile-based line graphs – one graph for each type of company in the QCI database.

> The multiplot in Fig. 5.13, produced using SYSTAT, graphs the profile of average attitude ratings for all inspectors within a specific type of company. This multiplot shows the same story as the radar plot in Fig. 5.12, but in a different graphical format. It is still fairly clear that the average ratings from inspectors from PC manufacturers tend to be higher than for the other types of companies and the profile for inspectors from automobile manufacturers tends to be lower than for the other types of companies.

(continued)

Procedure 5.3: Multivariate Graphs & Displays

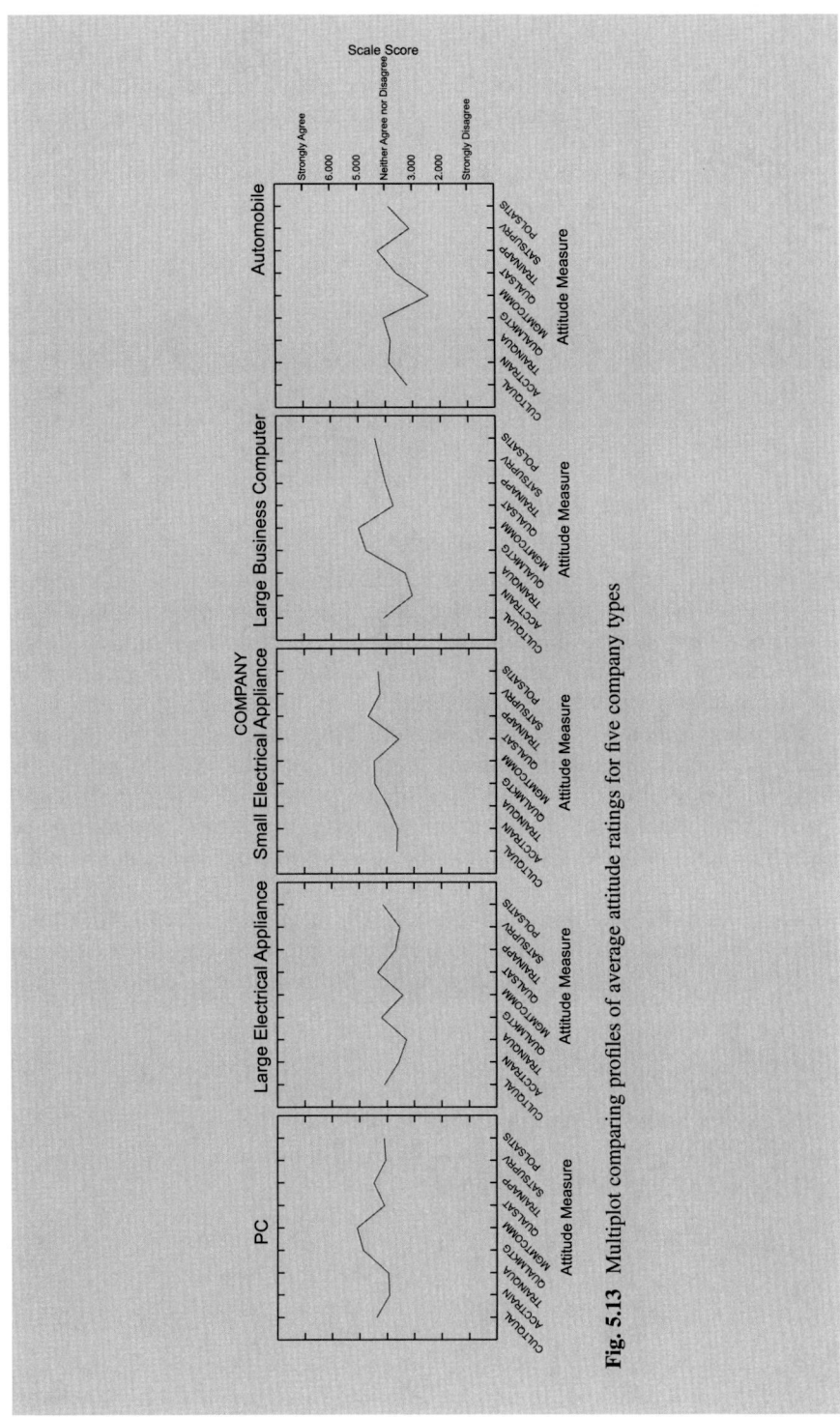

Fig. 5.13 Multiplot comparing profiles of average attitude ratings for five company types

(continued)

> The profile for inspectors from large electrical appliance manufacturers is the flattest, meaning that their average attitude ratings were less variable than for other types of companies. Comparing the ease with which you can glean the visual impressions from Figs. 5.12 and 5.13 may lead you to prefer one style of graph over another. If you have such preferences, chances are others will also, which may mean you need to carefully consider your options when deciding how best to display data for effect.
>
> Frequently, choice of graph is less a matter of which style is right or wrong, but more a matter of which style will suit specific purposes or convey a specific story, i.e. the choice is often strategic.

Parallel Coordinate Displays

A *parallel coordinate display* is useful for displaying individual scores on a range of variables, all measured using the same scale. Furthermore, such graphs can be combined side-by-side to facilitate very broad visual comparisons among groups, while retaining individual profile variability in scores. Each line in a parallel coordinate display represents one individual, e.g. an inspector.

The interpretation of a parallel coordinate display, such as the two shown in Fig. 5.14, depends on visual impressions of the peaks and valleys (highs and lows) in the profiles as well as on the density of similar profile lines. The graph is called 'parallel coordinate' simply because it assumes that all variables are measured on the same scale and that scores for each variable can therefore be located along vertical axes that are parallel to each other (imagine vertical lines on Fig. 5.14 running from bottom to top for each variable on the X-axis). The main drawback of this method of data display is that only those individuals in the sample who provided legitimate scores on all of the variables being plotted (i.e. who have no missing scores) can be displayed.

> The parallel coordinate display in Fig. 5.14, produced using SYSTAT, graphs the profile of average attitude ratings for all inspectors within two specific types of company: the left graph for inspectors from PC manufacturers and the right graph for automobile manufacturers.
>
> There are fewer lines in each display than the number of inspectors from each type of company simply because several inspectors from each type of company were missing a rating on at least one of the nine attitude variables. The graphs show great variability in scores amongst inspectors within a company type, but there are some overall patterns evident.

(continued)

Procedure 5.3: Multivariate Graphs & Displays

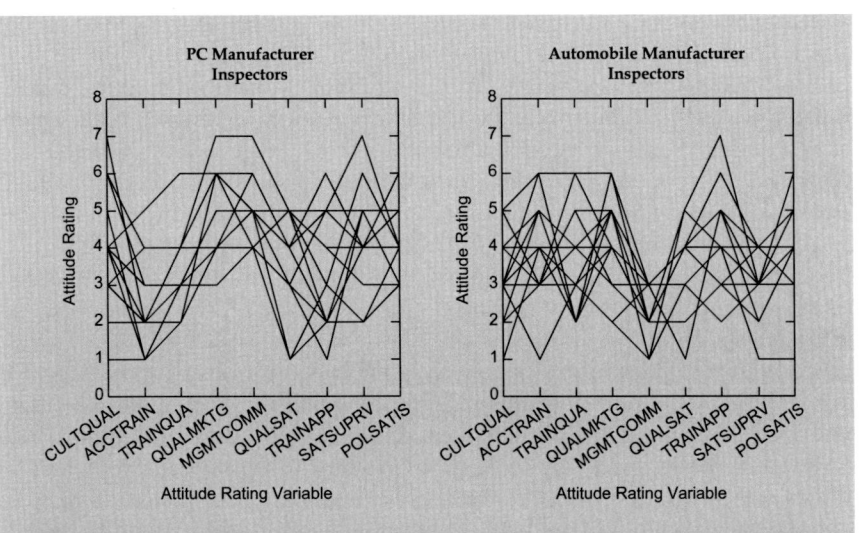

Fig. 5.14 Parallel coordinate displays comparing profiles of average attitude ratings for five company types

For example, inspectors from automobile companies clearly and fairly uniformly rated **mgmtcomm** toward the low end of the scale, whereas the reverse was generally true for that variable for inspectors from PC manufacturers. Conversely, inspectors from automobile companies tend to rate **acctrain** and **trainapp** more toward the middle to high end of the scale, whereas the reverse is generally true for those variables for inspectors from PC manufacturers.

Icon Plots

Perhaps the most creative types of multivariate displays are the so-called *icon plots*. SYSTAT and STATGRAPHICS offer an impressive array of different types of icon plots, including, amongst others, Chernoff's faces, profile plots, histogram plots, star glyphs and sunray plots (Jacoby 1998 provides a detailed discussion of icon plots).

Icon plots generally use a specific visual construction to represent variables scores obtained by each individual within a sample or group. All icon plots are thus methods for displaying the response patterns for individual members of a sample, as long as those individuals are not missing any scores on the variables to be displayed (note that this is the same limitation as for radar plots and parallel coordinate displays). To illustrate icon plots, without generating too many icons to focus on, Figs. 5.15, 5.16, 5.17 and 5.18 present four different icon plots for QCI inspectors classified, using a new variable called **BEST_WORST**, as either the worst performers (= 1 where their accuracy scores were less than 70%) or the best performers (= 2 where their accuracy scores were 90% or greater).

The *Chernoff's faces plot* gets its name from the visual icon used to represent variable scores – a cartoon-type face. This icon tries to capitalise on our natural human ability to recognise and differentiate faces. Each feature of the face is controlled by the scores on a single variable. In SYSTAT, up to 20 facial features are controllable; the first five being curvature of mouth, angle of brow, width of nose, length of nose and length of mouth (SYSTAT Software Inc., 2009, p. 259). The theory behind Chernoff's faces is that similar patterns of variable scores will produce similar looking faces, thereby making similarities and differences between individuals more apparent.

The *profile plot* and *histogram plot* are actually two variants of the same type of icon plot. A profile plot represents individuals' scores for a set of variables using simplified line graphs, one per individual. The profile is scaled so that the vertical height of the peaks and valleys correspond to actual values for variables where the variables anchor the X-axis in a fashion similar to the parallel coordinate display. So, as you examine a profile from left to right across the X-axis of each graph, you are looking across the set of variables. A histogram plot represents the same information in the same way as for the profile plot but using histogram bars instead.

> Figure 5.15, produced using SYSTAT, shows a Chernoff's faces plot for the best and worst performing inspectors using their ratings of job satisfaction, working conditions and the nine general attitude statements.
>
> Each face is labelled with the inspector number it represents. The gaps indicate where an inspector had missing data on at least one of the variables, meaning a face could not be generated for them. The worst performers are drawn using red lines; the best using blue lines. The first variable is **jobsat** and this variable controls mouth curvature; the second variable is **workcond** and this controls angle of brow, and so on. It seems clear that there are differences in the faces between the best and worst performers with, for example, best performers tending to be more satisfied (smiling) and with higher ratings for working conditions (brow angle).
>
> Beyond a broad visual impression, there is little in terms of precise inferences you can draw from a Chernoff's faces plot. It really provides a visual sketch, nothing more. The fact that there is no obvious link between facial features, variables and score levels means that the Chernoff's faces icon plot is difficult to interpret at the level of individual variables – a holistic impression of similarity and difference is what this type of plot facilitates.
>
> Figure 5.16 produced using SYSTAT, shows a profile plot for the best and worst performing inspectors using their ratings of job satisfaction, working conditions and the nine attitude variables.
>
> (continued)

Procedure 5.3: Multivariate Graphs & Displays

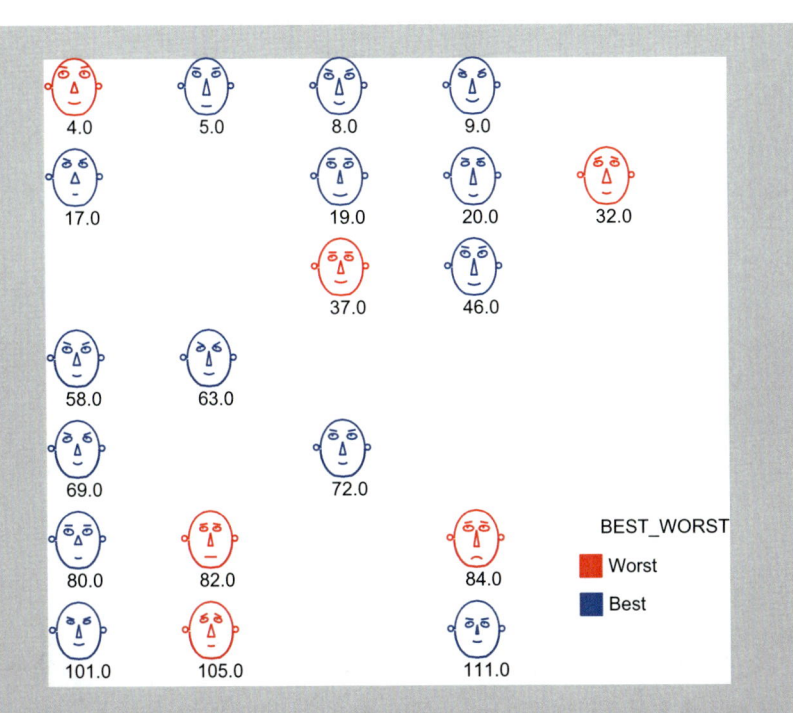

Fig. 5.15 Chernoff's faces icon plot comparing individual attitude ratings for best and worst performing inspectors

Like the Chernoff's faces plot (Fig. 5.15), as you read across the rows of the plot from left to right, each plot corresponds respectively to a inspector in the sample who was either in the worst performer (red) or best performer (blue) category. The first attitude variable is **jobsat** and anchors the left end of each line graph; the last variable is **polsatis** and anchors the right end of the line graph. The remaining variables are represented in order from left to right across the X-axis of each graph. Figure 5.16 shows that these inspectors are rather different in their attitude profiles, with best performers tending to show taller profiles on the first two variables, for example.

Figure 5.17 produced using SYSTAT, shows a histogram plot for the best and worst performing inspectors based on their ratings of job satisfaction, working conditions and the nine attitude variables. This plot tells the same story as the profile plot, only using histogram bars. Some people would prefer the histogram icon plot to the profile plot because each histogram bar corresponds to one variable, making the visual linking of a specific bar to a specific variable much easier than visually linking a specific position along the profile line to a specific variable.

(continued)

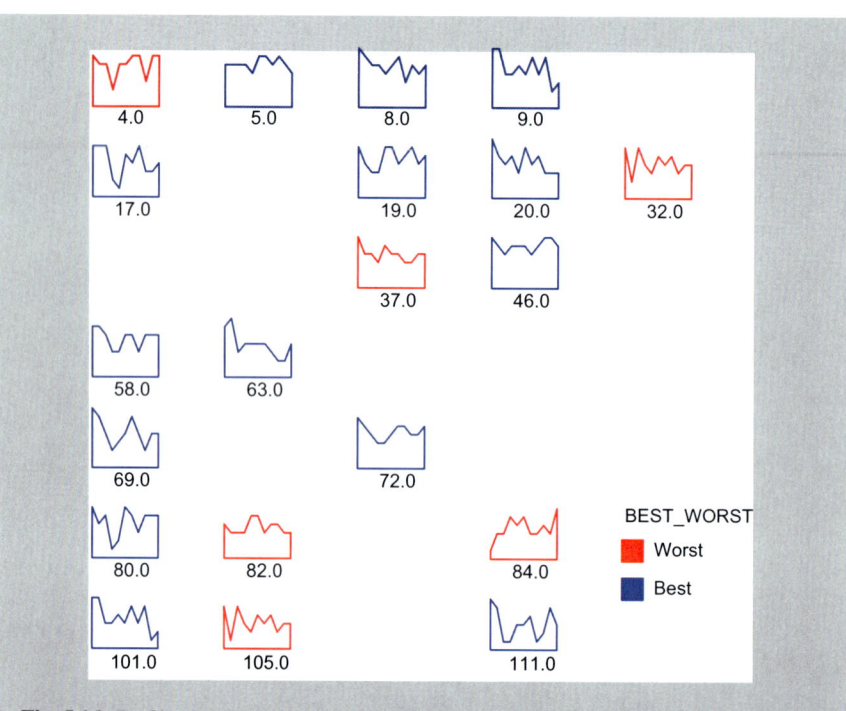

Fig. 5.16 Profile plot comparing individual attitude ratings for best and worst performing inspectors

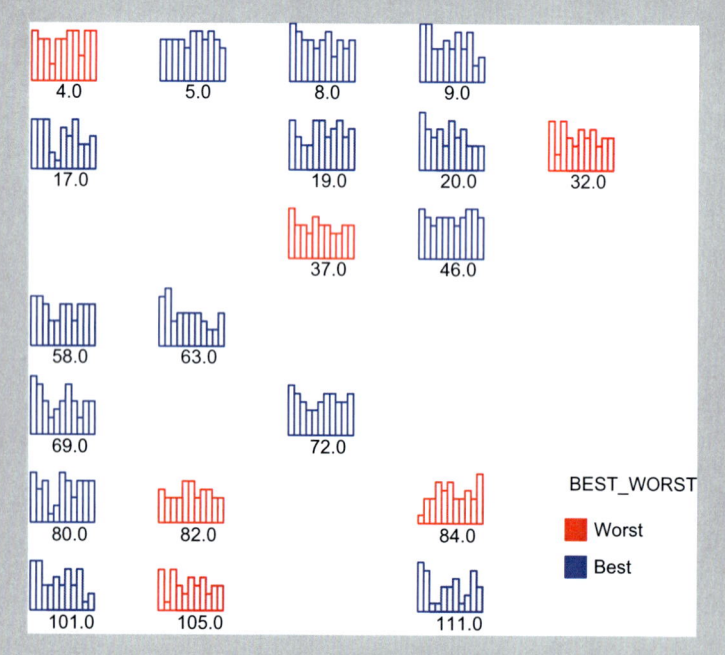

Fig. 5.17 Histogram plot comparing individual attitude ratings for best and worst performing inspectors

Procedure 5.3: Multivariate Graphs & Displays

The *sunray plot* is actually a simplified adaptation of the radar plot (called a "star glyph") used to represent scores on a set of variables for each individual within a sample or group. Remember that a radar plot basically arranges the variables around a central point like a clock face; the first variable is represented at the 12 o'clock position and the remaining variables follow around the plot in a clockwise direction.

Unlike a radar plot, while the spokes (the actual 'star' of the glyph's name) of the plot are visible, no interpretive scale is evident. A variable's score is visually represented by its distance from the central point. Thus, the star glyphs in a sunray plot are designed, like Chernoff's faces, to provide a general visual impression, based on icon shape. A wide diameter well-rounded plot indicates an individual with high scores on all variables and a small diameter well-rounded plot vice-versa. Jagged plots represent individuals with highly variable scores across the variables. 'Stars' of similar size, shape and orientation represent similar individuals.

Figure 5.18, produced using STATGRAPHICS, shows a sunray plot for the best and worst performing inspectors. An interpretation glyph is also shown in the lower right corner of Fig. 5.18, where variables are aligned with the spokes of a star (e.g. **jobsat** is at the 12 o'clock position). This sunray plot could lead you to form the visual impression that the worst performing inspectors (group 1) have rather less rounded rating profiles than do the best performing inspectors (group 2) and that the **jobsat** and **workcond** spokes are generally lower for the worst performing inspectors.

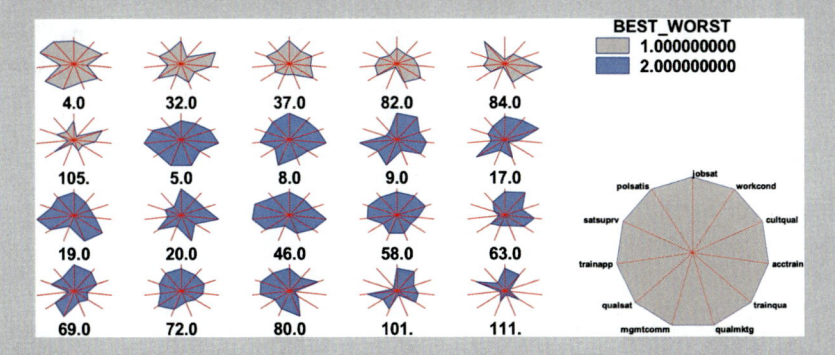

Fig. 5.18 Sunray plot comparing individual attitude ratings for best and worst performing inspectors

Comparatively speaking, the sunray plot makes identifying similar individuals a bit easier (perhaps even easier than Chernoff's faces) and, when ordered as STATGRAPHICS showed in Fig. 5.18, permits easier visual comparisons between groups of individuals, but at the expense of precise knowledge about variable scores. Remember, a holistic impression is the goal pursued using a sunray plot.

Advantages

Multivariate graphical methods provide summary techniques for visually presenting certain characteristics of a complex array of data on variables. Such visual representations are generally better at helping us to form holistic impressions of multivariate data rather than any sort of tabular representation or numerical index. They also allow us to compress many numerical measures into a finite representation that is generally easy to understand. Multivariate graphical displays can add interest to an otherwise dry statistical reporting of numerical data. They are designed to appeal to our pattern recognition skills, focusing our attention on features of the data such as shape, level, variability and orientation. Some multivariate graphs (e.g. radar plots, sunray plots and multiplots) are useful not only for representing score patterns for individuals but also providing summaries of score patterns across groups of individuals.

Disadvantages

Multivariate graphs tend to get very busy-looking and are hard to interpret if a great many variables or a large number of individuals need to be displayed (imagine any of the icon plots, for a sample of 200 questionnaire participants, displayed on a A4 page – each icon would be so small that its features could not be easily distinguished, thereby defeating the purpose of the display). In such cases, using numerical summary statistics (such as averages or correlations) in tabular form alone will provide a more economical and efficient summary. Also, some multivariate displays will work better for conveying certain types of information than others.

For example *Information about variable relationships may be better displayed using a scatterplot matrix. Information about individual similarities and difference on a set of variables may be better conveyed using a histogram or sunray plot. Multiplots may be better suited to displaying information about group differences across a set of variables. Information about the overall similarity of individual entities in a sample might best be displayed using Chernoff's faces.*

Because people differ greatly in their visual capacities and preferences, certain types of multivariate displays will work for some people and not others. Sometimes, people will not see what you see in the plots. Some plots, such as Chernoff's faces, may not strike a reader as a serious statistical procedure and this could adversely influence how convinced they will be by the story the plot conveys. None of the multivariate displays described here provide sufficiently precise information for solid inferences or interpretations; all are designed to simply facilitate the formation of holistic visual impressions. In fact, you may have noticed that some displays (scatterplot matrices and the icon plots, for example) provide no numerical scaling information that would help make precise interpretations. If precision in summary information is desired, the types of multivariate displays discussed here would not be the best strategic choices.

Where Is This Procedure Useful?

Virtually any research design which produces quantitative data/statistics for multiple variables provides opportunities for multivariate graphical data display which may help to clarify or illustrate important data characteristics or relationships. Thus, for survey research involving many identically-scaled attitudinal questions, a multivariate display may be just the device needed to communicate something about patterns in the data. Multivariate graphical displays are simply specialised communication tools designed to compress a lot of information into a meaningful and efficient format for interpretation—which tool to choose depends upon the message to be conveyed.

Generally speaking, visual representations of multivariate data could prove more useful in communicating to lay persons who are unfamiliar with statistics or who prefer visual as opposed to numerical information. However, these displays would probably require some interpretive discussion so that the reader clearly understands their intent.

Software Procedures

Application	Procedures
SPSS	*Graphs* → *Chart Builder*... and choose *Scatter/Dot* from the gallery; drag the *Scatterplot Matrix* chart type into the working area and customise the chart with desired variables, labels, etc. Only a few elements of each chart can be configured and altered.
NCSS	*Graphics* → *Scatter Plots* → *Scatter Plot Matrix*. Only a few elements of this plot are customisable in NCSS.
SYSTAT	*Graph* → *Scatterplot Matrix (SPLOM)*... (and you can select what type of plot you want to appear in the diagonal boxes) or *Graph* → *Line Chart*... (*Multiplots* can be selected by choosing a *Grouping* variable. e.g. **company**) or *Graph* → *Multivariate Display* → *Parallel Coordinate Display*... or *Icon Plot*... (for icon plots, you can choose from a range of icons including Chernoff's faces, histogram, star, sun or profile amongst others). A large number of elements of each type of plot are easily customisable, although it may take some trial and error to get exactly the look you want.
STATGRAPHICS	*Plot* → *Multivariate Visualization* → *Scatterplot Matrix*... or *Parallel Coordinates Plot*... or *Chernoff's Faces* or *Star Glyphs and Sunray Plots*... Several elements of each type of plot are easily customisable, although it may take some trial and error to get exactly the look you want.
R commander	*Graphs* → *Scatterplot Matrix*... You can select what type of plot you want to appear in the diagonal boxes, and you can control some other features of the plot. Other multivariate data displays are available via various **R** packages (e.g. the *lattice* or *car* package), but not through **R** commander.

Procedure 5.4: Assessing Central Tendency

Classification	Univariate; descriptive.
Purpose	To provide numerical summary measures that give an indication of the central, average or typical score in a distribution of scores for a variable.
Measurement level	*Mean* – variables should be measured at the interval or ratio-level.
	Median – variables should be measured at least at the ordinal-level.
	Mode – variables can be measured at any of the four levels.

The three most commonly reported measures of central tendency are the mean, median and mode. Each measure reflects a specific way of defining central tendency in a distribution of scores on a variable and each has its own advantages and disadvantages.

Mean

The *mean* is the most widely used measure of central tendency (also called the arithmetic average). Very simply, a mean is the sum of all the scores for a specific variable in a sample divided by the number of scores used in obtaining the sum. The resulting number reflects the average score for the sample of individuals on which the scores were obtained. If one were asked to predict the score that any single individual in the sample would obtain, the best prediction, in the absence of any other relevant information, would be the sample mean. Many parametric statistical methods (such as *Procedures 7.2, 7.4, 7.6* and *7.10*) deal with sample means in one way or another. For any sample of data, there is one and only one possible value for the mean in a specific distribution. For most purposes, the mean is the preferred measure of central tendency because it utilises all the available information in a sample.

> In the context of the QCI database, Maree could quite reasonably ask what inspectors scored on the average in terms of mental ability (**mentabil**), inspection accuracy (**accuracy**), inspection speed (**speed**), overall job satisfaction (**jobsat**), and perceived quality of their working conditions (**workcond**). Table 5.3 shows the mean scores for the sample of 112 quality control inspectors on each of these variables. The statistics shown in Table 5.3 were computed using the SPSS *Frequencies...* procedure. Notice that the table indicates how many of the 112 inspectors had a valid score for each variable

(continued)

Procedure 5.4: Assessing Central Tendency

and how many were missing a score (e.g. 109 inspectors provided a valid rating for **jobsat;** 3 inspectors did not).

Each mean needs to be interpreted in terms of the original units of measurement for each variable. Thus, the inspectors in the sample showed an average mental ability score of 109.84 (higher than the general population mean of 100 for the test), an average inspection **accuracy** of 82.14%, and an average **speed** for making quality control decisions of 4.48 s. Furthermore, in terms of their work context, inspectors reported an average overall job satisfaction of 4.96 (on the 7-point scale, or a level of satisfaction nearly one full scale point above the Neutral point of 4—indicating a generally positive but not strong level of job satisfaction, and an average perceived quality of work conditions of 4.21 (on the 7-point scale which is just about at the level of Stressful but Tolerable.

Table 5.3 Measures of central tendency for specific QCI variables

		mentabil	accuracy	speed	jobsat	workcond
N	Valid	111	111	111	109	106
	Missing	1	1	1	3	6
Mean		109.84	82.14	4.4801	4.96	4.21
Median		111.00	83.00	3.8900	5.00	4.00
Mode		111	82	3.14	6	4
Percentiles	25	104.00	77.00	2.1900	4.00	3.00
	50	111.00	83.00	3.8900	5.00	4.00
	75	116.00	89.00	5.7100	6.00	6.00

The mean is sensitive to the presence of extreme values, which can distort its value, giving a biased indication of central tendency. As we will see below, the median is an alternative statistic to use in such circumstances. However, it is also possible to compute what is called a *trimmed mean* where the mean is calculated after a certain percentage (say, 5% or 10%) of the lowest and highest scores in a distribution have been ignored (a process called 'trimming'; see, for example, the discussion in Field 2018, pp. 262–264). This yields a statistic less influenced by extreme scores. The drawbacks are that the decision as to what percentage to trim can be somewhat subjective and trimming necessarily sacrifices information (i.e. the extreme scores) in order to achieve a less biased measure. Some software packages, such as SPSS, SYSTAT or NCSS, can report a specific percentage trimmed mean, if that option is selected for descriptive statistics or exploratory data analysis (see *Procedure 5.6*) procedures. Comparing the original mean with a trimmed mean can provide an indication of the degree to which the original mean has been biased by extreme values.

Median

Very simply, the *median* is the centre or middle score of a set of scores. By 'centre' or 'middle' is meant that 50% of the data values are smaller than or equal to the median and 50% of the data values are larger when the entire distribution of scores is rank ordered from the lowest to highest value. Thus, we can say that the median is that score in the sample which occurs at the 50th percentile. [Note that a 'percentile' is attached to a specific score that a specific percentage of the sample scored at or below. Thus, a score at the 25th percentile means that 25% of the sample achieved this score or a lower score.] Table 5.3 shows the 25th, 50th and 75th percentile scores for each variable – *note how the 50th percentile score is exactly equal to the median in each case.*

The median is reported somewhat less frequently than the mean but does have some advantages over the mean in certain circumstances. One such circumstance is when the sample of data has a few extreme values in one direction (either very large or very small relative to all other scores). In this case, the mean would be influenced (biased) to a much greater degree than would the median since all of the data are used to calculate the mean (including the extreme scores) whereas only the single centre score is needed for the median. For this reason, many nonparametric statistical procedures (such as *Procedures 7.3, 7.5* and *7.9*) focus on the median as the comparison statistic rather than on the mean.

A discrepancy between the values for the mean and median of a variable provides some insight to the degree to which the mean is being influenced by the presence of extreme data values. In a distribution where there are no extreme values on either side of the distribution (or where extreme values balance each other out on either side of the distribution, as happens in a normal distribution – see *Fundamental Concept II*), the mean and the median will coincide at the same value and the mean will not be biased.

For highly skewed distributions, however, the value of the mean will be pulled toward the long tail of the distribution because that is where the extreme values lie. However, in such skewed distributions, the median will be insensitive (statisticians call this property 'robustness') to extreme values in the long tail. For this reason, the direction of the discrepancy between the mean and median can give a very rough indication of the direction of skew in a distribution ('mean larger than median' signals possible positive skewness; 'mean smaller than median' signals possible negative skewness). Like the mean, there is one and only one possible value for the median in a specific distribution.

> In Fig. 5.19, the left graph shows the distribution of **speed** scores and the right-hand graph shows the distribution of **accuracy** scores. The **speed** distribution clearly shows the mean being pulled toward the right tail of the distribution whereas the **accuracy** distribution shows the mean being just slightly pulled toward the left tail. The effect on the mean is stronger in the **speed** distribution indicating a greater biasing effect due to some very long inspection decision times.

(continued)

Procedure 5.4: Assessing Central Tendency

If we refer to Table 5.3, we can see that the median score for each of the five variables has also been computed. Like the mean, the median must be interpreted in the original units of measurement for the variable. We can see that for **mentabil**, **accuracy**, and **workcond**, the value of the median is very close to the value of the mean, suggesting that these distributions are not strongly influenced by extreme data values in either the high or low direction. However, note that the median **speed** was 3.89 s compared to the mean of 4.48 s, suggesting that the distribution of **speed** scores is positively skewed (the mean is larger than the median—refer to Fig. 5.19). Conversely, the median **jobsat** score was 5.00 whereas the mean score was 4.96 suggesting very little substantive skewness in the distribution (mean and median are nearly equal).

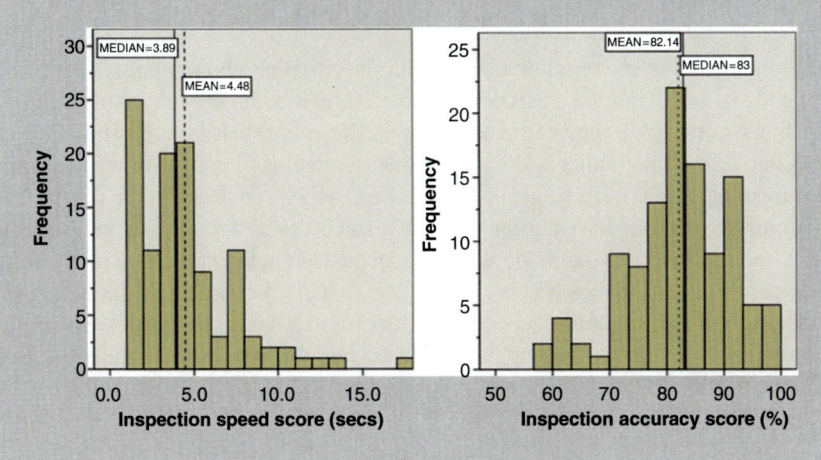

Fig. 5.19 Effects of skewness in a distribution on the values for the mean and median

Mode

The *mode* is the simplest measure of central tendency. It is defined as the most frequently occurring score in a distribution. Put another way, it is the score that more individuals in the sample obtain than any other score. An interesting problem associated with the mode is that there may be more than one in a specific distribution. In the case where multiple modes exist, the issue becomes which value do you report? The answer is that you must report all of them. In a 'normal' bell-shaped distribution, there is only one mode and it is indeed at the centre of the distribution, coinciding with both the mean and the median.

> Table 5.3 also shows the mode for each of the five variables. For example, more inspectors achieved a **mentabil** score of 111 more often than any other score and inspectors reported a **jobsat** rating of 6 more often than any other rating. SPSS only ever reports one mode even if several are present, so one must be careful and look at a histogram plot for each variable to make a final determination of the mode(s) for that variable.

Advantages

All three measures of central tendency yield information about what is going on in the centre of a distribution of scores. The mean and median provide a single number which can summarise the central tendency in the entire distribution. The mode can yield one or multiple indices. With many measurements on individuals in a sample, it is advantageous to have single number indices which can describe the distributions in summary fashion. In a normal or near-normal distribution of sample data, the mean, the median, and the mode will all generally coincide at the one point. In this instance, all three statistics will provide approximately the same indication of central tendency. Note however that it is seldom the case that all three statistics would yield exactly the same number for any particular distribution. The mean is the most useful statistic, unless the data distribution is skewed by extreme scores, in which case the median should be reported.

Disadvantages

While measures of central tendency are useful descriptors of distributions, summarising data using a single numerical index necessarily reduces the amount of information available about the sample. Not only do we need to know what is going on in the centre of a distribution, we also need to know what is going on around the centre of the distribution. For this reason, most social and behavioural researchers report not only measures of central tendency, but also measures of variability (see *Procedure 5.5*). The mode is the least informative of the three statistics because of its potential for producing multiple values.

Where Is This Procedure Useful?

Measures of central tendency are useful in almost any type of experimental design, survey or interview study, and in any observational studies where quantitative data are available and must be summarised. The decision as to whether the mean or median should be reported depends upon the nature of the data which should ideally be ascertained by visual inspection of the data distribution. Some researchers opt to report both measures routinely. Computation of means is a prelude to many parametric statistical methods (see, for example, *Procedure 7.2, 7.4, 7.6, 7.8, 7.10, 7.11* and *7.16*); comparison of medians is associated with many nonparametric statistical methods (see, for example, *Procedure 7.3, 7.5, 7.9* and *7.12*).

Software Procedures

Application	Procedures
SPSS	*Analyze → Descriptive Statistics → Frequencies* ... then press the '*Statistics*' button and choose mean, median and mode. To see trimmed means, you must use the *Analyze → Descriptive Statistics → Explore* ... Exploratory Data Analysis procedure; see *Procedure 5.6*.
NCSS	*Analysis → Descriptive Statistics → Descriptive Statistics* then select the reports and plots that you want to see; make sure you indicate that you want to see the 'Means Section' of the Report. If you want to see trimmed means, tick the 'Trimmed Section' of the Report.
SYSTAT	*Analyze → Basic Statistics* ... then select the mean, median and mode (as well as any other statistics you might wish to see). If you want to see trimmed means, tick the 'Trimmed mean' section of the dialog box and set the percentage to trim in the box labelled 'Two-sided'.
STATGRAPHICS	*Describe → Numerical Data → One-Variable Analysis...* or *Multiple-Variable Analysis...* then choose the variable(s) you want to describe and select Summary Statistics (you don't get any options for statistics to report – measures of central tendency and variability are automatically produced). STATGRAPHICS will not report modes and you will need to use *One-Variable Analysis...* and request 'Percentiles' in order to see the 50%ile score which will be the median; however, it won't be labelled as the median.
R Commander	*Statistics → Summaries → Numerical summaries...* then select the central tendency statistics you want to see. **R** Commander will not produce modes and to see the median, make sure that the 'Quantiles' box is ticked – the .5 quantile score (= 50%ile) score is the median; however, it won't be labelled as the median.

Procedure 5.5: Assessing Variability

Classification	Univariate; descriptive.
Purpose	To give an indication of the degree of spread in a sample of scores; that is, how different the scores tend to be from each other with respect to a specific measure of central tendency.
Measurement level	For the *variance* and *standard deviation*, interval or ratio-level measures are needed if these measures of variability are to have any interpretable meaning. At least an ordinal-level of measurement is required for the *range* and *interquartile range* to be meaningful.

There are a variety of measures of variability to choose from including the range, interquartile range, variance and standard deviation. Each measure reflects a specific way of defining variability in a distribution of scores on a variable and each has its own advantages and disadvantages. Most measures of variability are associated with a specific measure of central tendency so that researchers are now commonly expected to report both a measure of central tendency and its associated measure of variability whenever they display numerical descriptive statistics on continuous or ranked-ordered variables.

Range

This is the simplest measure of variability for a sample of data scores. The *range* is merely the largest score in the sample minus the smallest score in the sample. The range is the one measure of variability not explicitly associated with any measure of central tendency. It gives a very rough indication as to the extent of spread in the scores. However, since the range uses only two of the total available scores in the sample, the rest of the scores are ignored, which means that a lot of potentially useful information is being sacrificed. There are also problems if either the highest or lowest (or both) scores are atypical or too extreme in their value (as in highly skewed distributions). When this happens, the range gives a very inflated picture of the typical variability in the scores. Thus, the range tends not be a frequently reported measure of variability.

> Table 5.4 shows a set of descriptive statistics, produced by the SPSS *Frequencies* procedure, for the **mentabil, accuracy, speed, jobsat** and **workcond** measures in the QCI database. In the table, you will find three rows labelled 'Range', 'Minimum' and 'Maximum'.

(continued)

Procedure 5.5: Assessing Variability

Table 5.4 Measures of central tendency and variability for specific QCI variables

Statistics		mentabil	accuracy	speed	jobsat	workcond
N	Valid	111	111	111	109	106
	Missing	1	1	1	3	6
Mean		109.84	82.14	4.4801	4.96	4.21
Median		111.00	83.00	3.8900	5.00	4.00
Mode		111	82	3.14	6	4
Std. Deviation		8.764	9.172	2.88751	1.644	1.717
Variance		76.810	84.118	8.338	2.702	2.947
Range		50	43	16.05	6	6
Minimum		85	57	1.05	1	1
Maximum		135	100	17.10	7	7
Percentiles	25	104.00	77.00	2.1900	4.00	3.00
	50	111.00	83.00	3.8900	5.00	4.00
	75	116.00	89.00	5.7100	6.00	6.00

Using the data from these three rows, we can draw the following descriptive picture. **Mentabil** scores spanned a range of 50 (from a minimum score of 85 to a maximum score of 135). **Speed** scores had a range of 16.05 s (from 1.05 s – the fastest quality decision to 17.10 – the slowest quality decision). **Accuracy** scores had a range of 43 (from 57% – the least accurate inspector to 100% – the most accurate inspector). Both work context measures (**jobsat** and **workcond**) exhibited a range of 6 – the largest possible range given the 1 to 7 scale of measurement for these two variables.

Interquartile Range

The *Interquartile Range* (*IQR*) is a measure of variability that is specifically designed to be used in conjunction with the median. The IQR also takes care of the extreme data problem which typically plagues the range measure. The IQR is defined as the range that is covered by the middle 50% of scores in a distribution once the scores have been ranked in order from lowest value to highest value. It is found by locating the value in the distribution at or below which 25% of the sample scored and subtracting this number from the value in the distribution at or below which 75% of the sample scored. The IQR can also be thought of as the range one would compute after the bottom 25% of scores and the top 25% of scores in the distribution have been 'chopped off' (or 'trimmed' as statisticians call it).

The IQR gives a much more stable picture of the variability of scores and, like the median, is relatively insensitive to the biasing effects of extreme data values. Some behavioural researchers prefer to divide the IQR in half which gives a measure called the *Semi-Interquartile Range (S-IQR)*. The S-IQR can be interpreted as the distance one must travel away from the median, in either direction, to reach the value which separates the top (or bottom) 25% of scores in the distribution from the remaining 75%.

The IQR or S-IQR is typically not produced by descriptive statistics procedures by default in many computer software packages; however, it can usually be requested as an optional statistic to report or it can easily be computed by hand using percentile scores. Both the median and the IQR figure prominently in Exploratory Data Analysis, particularly in the production of *boxplots* (see *Procedure 5.6*).

Figure 5.20 illustrates the conceptual nature of the IQR and S-IQR compared to that of the range. Assume that 100% of data values are covered by the distribution curve in the figure. It is clear that these three measures would provide very different values for a measure of variability. Your choice would depend on your purpose. If you simply want to signal the overall span of scores between the minimum and maximum, the range is the measure of choice. But if you want to signal the variability around the median, the IQR or S-IQR would be the measure of choice.

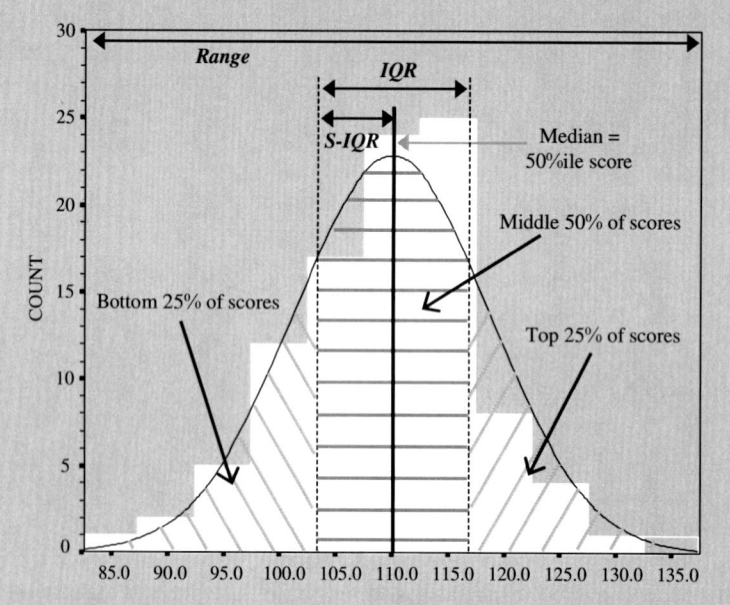

Fig. 5.20 How the range, IQR and S-IQR measures of variability conceptually differ

---

Note: Some behavioural researchers refer to the IQR as the *hinge-spread* (or *H-spread*) because of its use in the production of boxplots:

Procedure 5.5: Assessing Variability

- the 25th percentile data value is referred to as the 'lower hinge';
- the 75th percentile data value is referred to as the 'upper hinge'; and
- their difference gives the H-spread.

Midspread is another term you may see used as a synonym for interquartile range.

> Referring back to Table 5.4, we can find statistics reported for the median and for the 'quartiles' (25th, 50th and 75th percentile scores) for each of the five variables of interest. The 'quartile' values are useful for finding the IQR or S-IQR because SPSS does not report these measures directly. The median clearly equals the 50th percentile data value in the table.
>
> If we focus, for example, on the **speed** variable, we could find its IQR by subtracting the 25th percentile score of 2.19 s from the 75th percentile score of 5.71 s to give a value for the IQR of 3.52 s (the S-IQR would simply be 3.52 divided by 2 or 1.76 s). Thus, we could report that the median decision speed for inspectors was 3.89 s and that the middle 50% of inspectors showed scores spanning a range of 3.52 s. Alternatively, we could report that the median decision speed for inspectors was 3.89 s and that the middle 50% of inspectors showed scores which ranged 1.76 s either side of the median value.
>
> Note: We could compare the 'Minimum' or 'Maximum' *scores* to the 25th percentile score and 75th percentile score respectively to get a feeling for whether the minimum or maximum might be considered extreme or uncharacteristic data values.

Variance

The *variance* uses information from every individual in the sample to assess the variability of scores relative to the sample mean. Variance assesses the average squared deviation of each score from the mean of the sample. *Deviation* refers to the difference between an observed score value and the mean of the sample—they are squared simply because adding them up in their naturally occurring unsquared form (where some differences are positive and others are negative) always gives a total of zero, which is useless for an index purporting to measure something.

If many scores are quite different from the mean, we would expect the variance to be large. If all the scores lie fairly close to the sample mean, we would expect a small variance. If all scores exactly equal the mean (i.e. all the scores in the sample have the same value), then we would expect the variance to be zero.

Figure 5.21 illustrates some possibilities regarding variance of a distribution of scores having a mean of 100. The very tall curve illustrates a distribution with small variance. The distribution of medium height illustrates a distribution with medium variance and the flattest distribution ia a distribution with large variance.

If we had a distribution with no variance, the curve would simply be a vertical line at a score of 100 (meaning that all scores were equal to the mean). You can see that as variance increases, the tails of the distribution extend further outward and the concentration of scores around the mean decreases. You may have noticed that variance and range (as well as the IQR) will be related, since the range focuses on the difference between the ends of the two tails in the distribution and larger variances extend the tails. So, a larger variance will generally be associated with a larger range and IQR compared to a smaller variance.

Fig. 5.21 The concept of variance

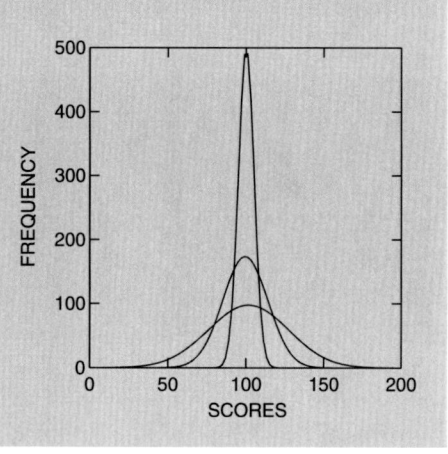

It is generally difficult to descriptively interpret the variance measure in a meaningful fashion since it involves squared deviations around the sample mean. [Note: If you look back at Table 5.4, you will see the variance listed for each of the variables (e.g. the variance of **accuracy** scores is 84.118), but the numbers themselves make little sense and do not relate to the original measurement scale for the variables (which, for the **accuracy** variable, went from 0% to 100% accuracy).] Instead, we use the variance as a steppingstone for obtaining a measure of variability that we can clearly interpret, namely the *standard deviation*. However, you should know that variance is an important concept in its own right simply because it provides the statistical foundation for many of the correlational procedures and statistical inference procedures described in *Chaps. 6, 7* and *8*.

For example When considering either correlations or tests of statistical hypotheses, we frequently speak of one variable explaining or sharing variance with another (see Procedure 6.4 and 7.7). In doing so, we are invoking the concept of variance as set out here—what we are saying is that variability in the behaviour of scores on one

Procedure 5.5: Assessing Variability

particular variable may be associated with or predictive of variability in scores on another variable of interest (e.g. it could explain why those scores have a non-zero variance).

Standard Deviation

The *standard deviation* (often abbreviated as SD, sd or Std. Dev.) is the most commonly reported measure of variability because it has a meaningful interpretation and is used in conjunction with reports of sample means. Variance and standard deviation are closely related measures in that the standard deviation is found by taking the square root of the variance. The standard deviation, very simply, is a summary number that reflects the 'average distance of each score from the mean of the sample'. In many parametric statistical methods, both the sample mean and sample standard deviation are employed in some form. Thus, the standard deviation is a very important measure, not only for data description, but also for hypothesis testing and the establishment of relationships as well.

Referring again back to Table 5.4, we'll focus on the results for the **speed** variable for discussion purposes. Table 5.4 shows that the mean inspection **speed** for the QCI sample was 4.48 s. We can also see that the standard deviation (in the row labelled 'Std Deviation') for **speed** was 2.89 s.

This standard deviation has a straightforward interpretation: we would say that 'on the average, an inspector's quality inspection decision speed differed from the mean of the sample by about 2.89 s in either direction'. In a normal distribution of scores (see *Fundamental Concept II*), we would expect to see about 68% of all inspectors having decision speeds between 1.59 s (the mean minus one amount of the standard deviation) and 7.37 s (the mean plus one amount of the standard deviation).

We noted earlier that the range of the **speed** scores was 16.05 s. However, the fact that the maximum **speed** score was 17.1 s compared to the 75th percentile score of just 5.71 s seems to suggest that this maximum speed might be rather atypically large compared to the bulk of **speed** scores. This means that the range is likely to be giving us a false impression of the overall variability of the inspectors' decision speeds.

Furthermore, given that the mean **speed** score was higher than the median **speed** score, suggesting that **speed** scores were positively skewed (this was confirmed by the histogram for **speed** shown in Fig. 5.19 in *Procedure 5.4*), we might consider emphasising the median and its associated IQR or S-IQR rather than the mean and standard deviation. Of course, similar diagnostic and interpretive work could be done for each of the other four variables in Table 5.4.

Advantages

Measures of variability (particularly the standard deviation) provide a summary measure that gives an indication of how variable (spread out) a particular sample of scores is. When used in conjunction with a relevant measure of central tendency (particularly the mean), a reasonable yet economical description of a set of data emerges. When there are extreme data values or severe skewness is present in the data, the IQR (or S-IQR) becomes the preferred measure of variability to be reported in conjunction with the sample median (or 50th percentile value). These latter measures are much more resistant ('robust') to influence by data anomalies than are the mean and standard deviation.

Disadvantages

As mentioned above, the range is a very cursory index of variability, thus, it is not as useful as variance or standard deviation. Variance has little meaningful interpretation as a descriptive index; hence, standard deviation is most often reported. However, the standard deviation (or IQR) has little meaning if the sample mean (or median) is not reported along with it.

For example *Knowing that the standard deviation for **accuracy** is 9.17 tells you little unless you know the mean **accuracy** (82.14) that it is the standard deviation from.*

Like the sample mean, the standard deviation can be strongly biased by the presence of extreme data values or severe skewness in a distribution in which case the median and IQR (or S-IQR) become the preferred measures. The biasing effect will be most noticeable in samples which are small in size (say, less than 30 individuals) and far less noticeable in large samples (say, in excess of 200 or 300 individuals). [Note that, in a manner similar to a trimmed mean, it is possible to compute a **trimmed standard deviation** to reduce the biasing effect of extreme data values, see Field 2018, p. 263.]

It is important to realise that the resistance of the median and IQR (or S-IQR) to extreme values is only gained by deliberately sacrificing a good deal of the information available in the sample (nothing is obtained without a cost in statistics). What is sacrificed is information from all other members of the sample other than those members who scored at the median and 25th and 75th percentile points on a variable of interest; information from all members of the sample would automatically be incorporated in mean and standard deviation for that variable.

Where Is This Procedure Useful?

Any investigation where you might report on or read about measures of central tendency on certain variables should also report measures of variability. This is particularly true for data from experiments, quasi-experiments, observational studies and questionnaires. It is important to consider measures of central tendency and measures of variability to be inextricably linked—one should never report one without the other if an adequate descriptive summary of a variable is to be communicated.

Other descriptive measures, such as those for skewness and kurtosis[1] may also be of interest if a more complete description of any variable is desired. Most good statistical packages can be instructed to report these additional descriptive measures as well.

Of all the statistics you are likely to encounter in the business, behavioural and social science research literature, means and standard deviations will dominate as measures for describing data. Additionally, these statistics will usually be reported when any parametric tests of statistical hypotheses are presented as the mean and standard deviation provide an appropriate basis for summarising and evaluating group differences.

Software Procedures

Application	Procedures
SPSS	*Analyze → Descriptive Statistics → Frequencies* ... then press the 'Statistics' button and choose Std. Deviation, Variance, Range, Minimum and/or Maximum as appropriate. SPSS does not produce or have an option to produce either the IQR or S-IQR, however, if your request 'Quantiles' you will see the 25th and 75th %ile scores, which can then be used to quickly compute either variability measure. Remember to select appropriate central tendency measures as well.
NCSS	*Analysis → Descriptive Statistics → Descriptive Statistics* then select the reports and plots that you want to see; make sure you indicate that you want to see the Variance Section of the Report. Remember to select appropriate central tendency measures as well (by opting to see the Means Section of the Report).
SYSTAT	*Analyze → Basic Statistics* ... then select SD, Variance, Range, Interquartile range, Minimum and/or Maximum as appropriate. Remember to select appropriate central tendency measures as well.

(continued)

---

[1]For more information, see *Chap. 1 – The language of statistics*.

Application	Procedures
STATGRAPHICS	*Describe* → *Numerical Data* → *One-Variable Analysis...* or *Multiple-Variable Analysis...* then choose the variable(s) you want to describe and select Summary Statistics (you don't get any options for statistics to report – measures of central tendency and variability are automatically produced). STATGRAPHICS does not produce either the IQR or S-IQR, however, if you use *One-Variable Analysis...* 'Percentiles' can be requested in order to see the 25th and 75th %ile scores, which can then be used to quickly compute either variability measure.
R Commander	*Statistics* → *Summaries* → *Numerical summaries...* then select either the Standard Deviation or Interquartile Range as appropriate. **R** Commander will not produce the range statistic or report minimum or maximum scores. Remember to select appropriate central tendency measures as well.

Fundamental Concept I: Basic Concepts in Probability

The Concept of Simple Probability

In *Procedures 5.1* and *5.2*, you encountered the idea of the frequency of occurrence of specific events such as particular scores within a sample distribution. Furthermore, it is a simple operation to convert the frequency of occurrence of a specific event into a number representing the relative frequency of that event. The relative frequency of an observed event is merely the number of times the event is observed divided by the total number of times one makes an observation. The resulting number ranges between 0 and 1 but we typically re-express this number as a percentage by multiplying it by 100%.

> In the QCI database, Maree Lakota observed data from 112 quality control inspectors of which 58 were male and 51 were female (gender indications were missing for three inspectors). The statistics 58 and 51 are thus the frequencies of occurrence for two specific types of research participant, a male inspector or a female inspector.
>
> If she divided each frequency by the total number of observations (i.e. 112), whe would obtain .52 for males and .46 for females (leaving .02 of observations with unknown gender). These statistics are relative frequencies which indicate the proportion of times that Maree obtained data from a male or female inspector. Multiplying each relative frequency by 100% would yield 52% and 46% which she could interpret as indicating that 52% of her sample was male and 46% was female (leaving 2% of the sample with unknown gender).

Fundamental Concept I: Basic Concepts in Probability

It does not take much of a leap in logic to move from the concept of 'relative frequency' to the concept of 'probability'. In our discussion above, we focused on relative frequency as indicating the proportion or percentage of times a specific category of participant was obtained in a sample. The emphasis here is on data from a sample.

> Imagine now that Maree had infinite resources and research time and was able to obtain ever larger samples of quality control inspectors for her study. She could still compute the relative frequencies for obtaining data from males and females in her sample but as her sample size grew larger and larger, she would notice these relative frequencies converging toward some fixed values.
>
> If, by some miracle, Maree could observe all of the quality control inspectors on the planet today, she would have measured the entire population and her computations of relative frequency for males and females would yield two precise numbers, each indicating the proportion of the population of inspectors that was male and the proportion that was female.
>
> If Maree were then to list all of these inspectors and randomly choose one from the list, the chances that she would choose a male inspector would be equal to the proportion of the population of inspectors that was male and this logic extends to choosing a female inspector. The number used to quantify this notion of 'chances' is called a probability. Maree would therefore have established the probability of randomly observing a male or a female inspector in the population on any specific occasion.

Probability is expressed on a 0.0 (the observation or event will certainly *not* be seen) to 1.0 (the observation or event will certainly be seen) scale where values close to 0.0 indicate observations that are less certain to be seen and values close to 1.0 indicate observations that are more certain to be seen (a value of .5 indicates an even chance that an observation or event will or will not be seen – a state of maximum uncertainty). Statisticians often interpret a probability as the *likelihood* of observing an event or type of individual in the population.

> In the QCI database, we noted that the relative frequency of observing males was .52 and for females was .46. If we take these relative frequencies as estimates of the proportions of each gender in the population of inspectors, then .52 and .46 represent the probability of observing a male or female inspector, respectively.
>
> Statisticians would state this as "the probability of observing a male quality control inspector is .52" or in a more commonly used shorthand code, the likelihood of observing a male quality control inspector is $p = .52$ (*p* for probability). For some, probabilities make more sense if they are converted to percentages (by multiplying by 100%). Thus, $p = .52$ can also understood as a 52% chance of observing a male quality control inspector.

We have seen that relative frequency is a sample statistic that can be used to estimate the population probability. Our estimate will get more precise as we use larger and larger samples (technically, as the size of our samples more closely approximates the size of our population). In most behavioural research, we never have access to entire populations so we must always estimate our probabilities.

In some very special populations, having a known number of fixed possible outcomes, such as results of coin tosses or rolls of a die, we can analytically establish event probabilities without doing an infinite number of observations; all we must do is assume that we have a fair coin or die. Thus, with a fair coin, the probability of observing a H or a T on any single coin toss is ½ or .5 or 50%; the probability of observing a 6 on any single throw of a die is 1/6 or .16667 or 16.667%. With behavioural data, though, we can never measure all possible behavioural outcomes, which thereby forces researchers to depend on samples of observations in order to make estimates of population values.

The concept of probability is central to much of what is done in the statistical analysis of behavioural data. Whenever a behavioural scientist wishes to establish whether a particular relationship exists between variables or whether two groups, treated differently, actually show different behaviours, he/she is playing a probability game. Given a sample of observations, the behavioural scientist must decide whether what he/she has observed is providing sufficient information to conclude something about the population from which the sample was drawn.

This decision always has a non-zero probability of being in error simply because in samples that are much smaller than the population, there is always the chance or probability that we are observing something rare and atypical instead of something which is indicative of a consistent population trend. Thus, the concept of probability forms the cornerstone for statistical inference about which we will have more to say later (see *Fundamental Concept VI*). Probability also plays an important role in helping us to understand theoretical statistical distributions (e.g. the normal distribution) and what they can tell us about our observations. We will explore this idea further in *Fundamental Concept II*.

The Concept of Conditional Probability

It is important to understand that the concept of probability as described above focuses upon the likelihood or chances of observing a specific event or type of observation for a specific variable relative to a population or sample of observations. However, many important behavioural research issues may focus on the question of the probability of observing a specific event given that the researcher has knowledge that some other event has occurred or been observed (this latter event is usually measured by a second variable). Here, the focus is on the potential relationship or link between two variables or two events.

Fundamental Concept I: Basic Concepts in Probability 111

With respect to the QCI database, Maree could ask the quite reasonable question "what is the probability (estimated in the QCI sample by a relative frequency) of observing an inspector being female *given* that she knows that an inspector works for a Large Business Computer manufacturer.

To address this question, all she needs to know is:

- how many inspectors from Large Business Computer manufacturers are in the sample (**22**); and
- how many of those inspectors were female (**7**) (inspectors who were missing a score for either **company** or **gender** have been ignored here).

If she divides 7 by 22, she would obtain the probability that an inspector is female *given* that they work for a Large Business Computer manufacturer – that is, $p = .32$.

This type of question points to the important concept of *conditional probability* ('conditional' because we are asking "what is the probability of observing one event conditional upon our knowledge of some other event").

Continuing with the previous example, Maree would say that the conditional probability of observing a female inspector working for a Large Business Computer manufacturer is .32 or, equivalently, a 32% chance. Compare this conditional probability of $p = .32$ to the overall probability of observing a female inspector in the entire sample ($p = .46$ as shown above).

This means that there is evidence for a connection or relationship between gender and the type of company an inspector works for. That is, the chances are lower for observing a female inspector from a Large Business Computer manufacturer than they are for simply observing a female inspector at all.

Maree therefore has evidence suggesting that females may be relatively underrepresented in Large Business Computer manufacturing companies compared to the overall population. Knowing something about the company an inspector works for therefore can help us make a better prediction about their likely gender.

Suppose, however, that Maree's conditional probability had been exactly equal to $p = .46$. This would mean that there was exactly the same chance of observing a female inspector working for a Large Business Computer manufacturer as there was of observing a female inspector in the general population. Here, knowing something about the company an inspector works doesn't help Maree make any better prediction about their likely gender. This would mean that the two variables are statistically independent of each other.

A classic case of events that are statistically independent is two successive throws of a fair die: rolling a six on the first throw gives us no information for predicting how likely it will be that we would roll a six on the second throw. The conditional probability of observing a six on the second throw given that I have observed a six on the first throw is 0.16667 (= 1 divided by 6) which is the same as the simple probability of observing a six on any specific throw. This statistical independence also means that if we wanted to know what the probability of throwing two sixes on two successive throws of a fair die, we would just multiply the probabilities for each independent event (i.e., throw) together; that is, $.16667 \times .16667 = .02789$ (this is known as the *multiplication rule* of probability, see, for example, Smithson 2000, p. 114).

Finally, you should know that conditional probabilities are often asymmetric. This means that for many types of behavioural variables, reversing the conditional arrangement will change the story about the relationship. Bayesian statistics (see *Fundamental Concept IX*) relies heavily upon this asymmetric relationship between conditional probabilities.

> Maree has already learned that the conditional probability that an inspector is female *given* that they worked for a Large Business Computer manufacturer is $p = .32$. She could easily turn the conditional relationship around and ask what is the conditional probability that an inspector works for a Large Business Computer manufacturer *given* that the inspector is female?
>
> From the QCI database, she can find that 51 inspectors in her total sample were female and of those 51, 7 worked for a Large Business Computer manufacturer. If she divided 7 by 51, she would get $p = .14$ (did you notice that all that changed was the number she divided by?). Thus, there is only a 14% chance of observing an inspector working for a Large Business Computer manufacturer *given* that the inspector is female – a rather different probability from $p = .32$, which tells a different story.

As you will see in *Procedures 6.2* and *7.1*, conditional relationships between categorical variables are precisely what crosstabulation contingency tables are designed to reveal.

Procedure 5.6: Exploratory Data Analysis

Classification	Univariate; descriptive.
Purpose	To visually summarise data, displaying some key characteristics of their distribution, while maintaining as much of their original integrity as possible.
Measurement level	Exploratory Data Analysis (EDA) procedures are most usefully employed to explore data measured at the ordinal, interval or ratio-level.

There are a variety of visual display methods for EDA, including stem & leaf displays, boxplots and violin plots. Each method reflects a specific way of displaying features of a distribution of scores or measurements and, of course, each has its own advantages and disadvantages. In addition, EDA displays are surprisingly flexible and can combine features in various ways to enhance the story conveyed by the plot.

Procedure 5.6: Exploratory Data Analysis

Stem & Leaf Displays

The *stem & leaf display* is a simple data summary technique which not only rank orders the data points in a sample but presents them visually so that the shape of the data distribution is reflected. Stem & leaf displays are formed from data scores by splitting each score into two parts: the first part of each score serving as the 'stem', the second part as the 'leaf' (e.g. for 2-digit data values, the 'stem' is the number in the tens position; the 'leaf' is the number in the ones position). Each stem is then listed vertically, in ascending order, followed horizontally by all the leaves in ascending order associated with it. The resulting display thus shows all of the scores in the sample, but reorganised so that a rough idea of the shape of the distribution emerges. As well, extreme scores can be easily identified in a stem & leaf display.

Consider the **accuracy** and **speed** scores for the 112 quality control inspectors in the QCI sample. Figure 5.22 (produced by the **R** Commander *Stem-and-leaf display . . .* procedure) shows the stem & leaf displays for inspection **accuracy** (left display) and **speed** (right display) data.

[The first six lines reflect information from **R** Commander about each display: lines 1 and 2 show the actual **R** command used to produce the plot (the variable name has been highlighted in bold); line 3 gives a warning indicating that inspectors with missing values (= NA in **R**) on the variable have been omitted from the display; line 4 shows how the stems and leaves have been defined; line 5 indicates what a leaf unit represents in value; and line 6 indicates the total number (n) of inspectors included in the display).] In Fig. 5.22, for the **accuracy** display on the left-hand side, the 'stems' have been split into 'half-stems'—one (which is starred) associated with the 'leaves' 0 through 4 and the other associated with the 'leaves' 5 through 9—a strategy that gives the display better balance and visual appeal.

```
> stem.leaf(statbook_2ed_RS$accuracy, trim.outliers=FALSE,
  na.rm=TRUE)
[1] "Warning: NA elements have been removed!!"
1 | 2: represents 12
 leaf unit: 1
              n: 111
    2      5. | 79
    7      6* | 02234
    9      6. | 57
   19      7* | 0012333334
   39      7. | 5555566777888888889999
  (25)     8* | 0000111222222222333333444
   47      8. | 5555556666666777899999
   25      9* | 000001112222333344444
    5      9. | 89
    3     10* | 000
```

```
> stem.leaf(statbook_2ed_RS$speed, trim.outliers=FALSE,
  na.rm=TRUE)
[1] "Warning: NA elements have been removed!!"
1 | 2: represents 1.2
 leaf unit: 0.1
              n: 111
   25      1 | 0112233344455566777778899
   36      2 | 00123355788
  (20)     3 | 11111233444477788888
   55      4 | 011111223344455677799
   34      5 | 111346799
   25      6 | 001
   22      7 | 01334566889
   11      8 | 029
    8      9 | 15
    6     10 | 14
    4     11 | 9
    3     12 | 0
    2     13 | 0
          14 |
          15 |
          16 |
    1     17 | 1
```

Fig. 5.22 Stem & leaf displays produced by **R** Commander

(continued)

Fig. 5.23 Stem & leaf display, produced by SYSTAT, of the **accuracy** QCI variable

```
Stem and Leaf Plot of Variable: ACCURACY, N = 111
Minimum       :            57.000
Lower Hinge   :            77.000
Median        :            83.000
Upper Hinge   :            89.000
Maximum       :           100.000

     5   7
     
     5   9
     6   02234
     6   57
     7   0012333334
     7 H 55555667778888889999
     8 M 000011122222223333333444
     8 H 555555566666677789999
     9   00000111222233344444
     9   89
    10   000

 1 Cases with missing values excluded from plot.
```

Notice how the left stem & leaf display conveys a fairly clear (yet sideways) picture of the shape of the distribution of **accuracy** scores. It has a rather symmetrical bell-shape to it with only a slight suggestion of negative skewness (toward the extreme score at the top). The right stem & leaf display clearly depicts the highly positively skewed nature of the distribution of **speed** scores. Importantly, we could reconstruct the entire sample of scores for each variable using its display, which means that unlike most other graphical procedures, we didn't have to sacrifice any information to produce the visual summary.

Some programs, such as SYSTAT, embellish their stem & leaf displays by indicating in which stem or half-stem the 'median' (50th percentile), the 'upper hinge score' (75th percentile), and 'lower hinge score' (25th percentile) occur in the distribution (recall the discussion of *interquartile range* in Procedure 5.5). This is shown in Fig. 5.23, produced by SYSTAT, where M and H indicate the stem locations for the median and hinge points, respectively. This stem & leaf display labels a single extreme **accuracy** score as an 'outside value' and clearly shows that this actual score was 57.

Boxplots

Another important EDA technique is the *boxplot* or, as it is sometimes known, the *box-and-whisker plot*. This plot provides a symbolic representation that preserves less of the original nature of the data (compared to a stem & leaf display) but typically gives a better picture of the distributional characteristics. The basic boxplot, shown in Fig. 5.24, utilises information about the median (50th percentile score) and

Procedure 5.6: Exploratory Data Analysis

the upper (75th percentile score) and lower (25th percentile score) hinge points in the construction of the 'box' portion of the graph (the 'median' defines the centre line in the box; the 'upper' and 'lower hinge values' define the end boundaries of the box—thus the box encompasses the middle 50% of data values).

Additionally, the boxplot utilises the IQR (recall *Procedure 5.5*) as a way of defining what are called 'fences' which are used to indicate score boundaries beyond which we would consider a score in a distribution to be an 'outlier' (or an extreme or unusual value). In SPSS, the inner fence is typically defined as 1.5 times the IQR in each direction and a 'far' outlier or extreme case is typically defined as 3 times the IQR in either direction (Field 2018, p. 193). The 'whiskers' in a boxplot extend out to the data values which are closest to the upper and lower inner fences (in most cases, the vast majority of data values will be contained within the fences). Outliers beyond these 'whiskers' are then individually listed. 'Near' outliers are those lying just beyond the inner fences and 'far' outliers lie well beyond the inner fences.

> Figure 5.24 shows two simple boxplots (produced using SPSS), one for the **accuracy** QCI variable and one for the **speed** QCI variable. The **accuracy** plot shows a median value of about 83, roughly 50% of the data fall between about 77 and 89 and there is one outlier, inspector 83, in the lower 'tail' of the distribution. The **accuracy** boxplot illustrates data that are relatively symmetrically distributed without substantial skewness. Such data will tend to have their median in the middle of the box, whiskers of roughly equal length extending out from the box and few or no outliers.
>
> **Fig. 5.24** Boxplots for the **accuracy** and **speed** QCI variables
>
> The **speed** plot shows a median value of about 4 s, roughly 50% of the data fall between 2 s and 6 s and there are four outliers, inspectors 7, 62, 65 and 75 (although inspectors 65 and 75 fall at the same place and are rather difficult

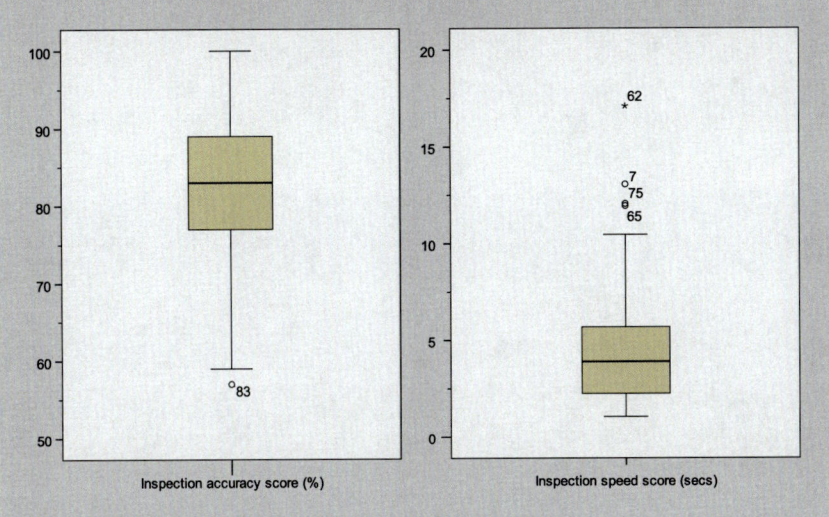

(continued)

to read), all falling in the slow speed 'tail' of the distribution. Inspectors 65, 75 and 7 are shown as 'near' outliers (open circles) whereas inspector 62 is shown as a 'far' outlier (asterisk). The **speed** boxplot illustrates data which are asymmetrically distributed because of skewness in one direction. Such data may have their median offset from the middle of the box and/or whiskers of unequal length extending out from the box and outliers in the direction of the longer whisker. In the **speed** boxplot, the data are clearly positively skewed (the longer whisker and extreme values are in the slow speed 'tail').

Boxplots are very versatile representations in that side-by-side displays for sub-groups of data within a sample can permit easy visual comparisons of groups with respect to central tendency and variability. Boxplots can also be modified to incorporate information about error bands associated with the median producing what is called a 'notched boxplot'. This helps in the visual detection of meaningful subgroup differences, where boxplot 'notches' don't overlap.

Figure 5.25 (produced using NCSS), compares the distributions of **accuracy** and **speed** scores for QCI inspectors from the five types of companies, plotted side-by-side

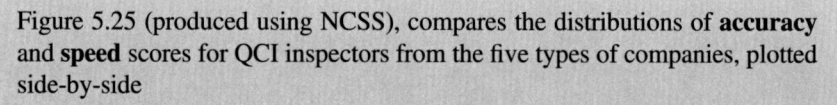

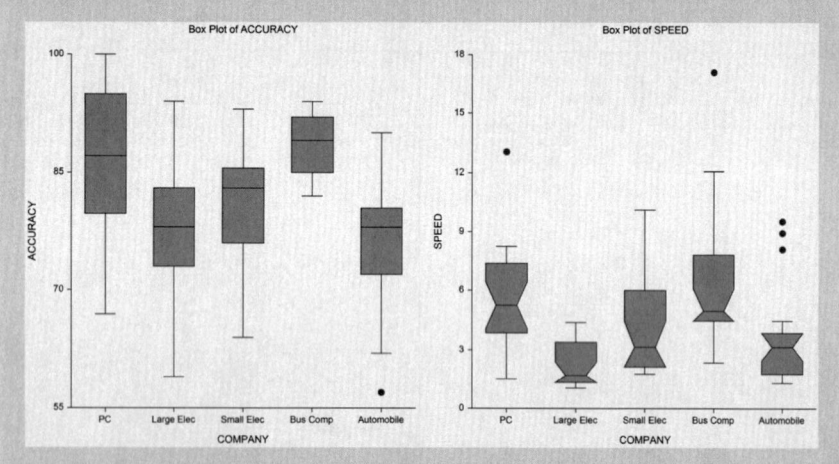

Fig. 5.25 Comparisons of the accuracy (regular boxplots) and speed (notched boxplots) QCI variables for different types of companies

Focus first on the left graph in Fig. 5.25 which plots the distribution of **accuracy** scores broken down by **company** using regular boxplots. This plot clearly shows the differing degree of skewness in each type of company (indicated by one or more outliers in one 'tail', whiskers which are not the same length and/or the median line being offset from the centre of a box), the differing variability of scores within each type of company (indicated by the

(continued)

Procedure 5.6: Exploratory Data Analysis 117

overall length of each plot—box and whiskers), and the differing central tendency in each type of company (the median lines do not all fall at the same level of **accuracy** score). From the left graph in Fig. 5.25, we could conclude that: inspection **accuracy** scores are most variable in PC and Large Electrical Appliance manufacturing companies and least variable in the Large Business Computer manufacturing companies; Large Business Computer and PC manufacturing companies have the highest median level of inspection accuracy; and inspection accuracy scores tend to be negatively skewed (many inspectors toward higher levels, relatively fewer who are poorer in inspection performance) in the Automotive manufacturing companies. One inspector, working for an Automotive manufacturing company, shows extremely poor inspection accuracy performance.

The right display compares types of companies in terms of their inspection **speed** scores, using' notched' boxplots. The notches define upper and lower error limits around each median. Aside from the very obvious positive skewness for **speed** scores (with a number of slow speed outliers) in every type of company (least so for Large Electrical Appliance manufacturing companies), the story conveyed by this comparison is that inspectors from Large Electrical Appliance and Automotive manufacturing companies have substantially faster median decision speeds compared to inspectors from Large Business Computer and PC manufacturing companies (i.e. their 'notches' do not overlap, in terms of **speed** scores, on the display).

Boxplots can also add interpretive value to other graphical display methods through the creation of hybrid displays. Such displays might combine a standard histogram with a boxplot along the X-axis to provide an enhanced picture of the data distribution as illustrated for the **mentabil** variable in Fig. 5.26 (produced using NCSS). This hybrid

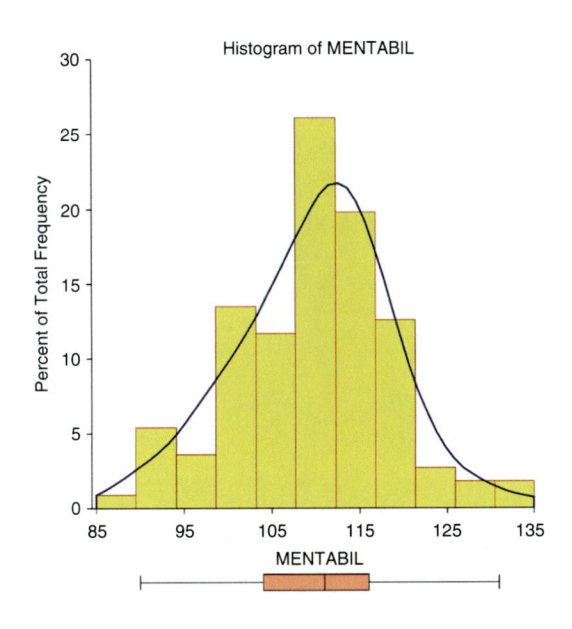

Fig. 5.26 A hybrid histogram-density-boxplot of the **mentabil** QCI variable

plot also employs a data 'smoothing' method called a **density trace** to outline an approximate overall shape for the data distribution. Any one graphical method would tell some of the story, but combined in the hybrid display, the story of a relatively symmetrical set of **mentabil** scores becomes quite visually compelling.

Violin Plots

Violin plots are a more recent and interesting EDA innovation, implemented in the NCSS software package (Hintze 2012). The violin plot gets its name from the rough shape that the plots tend to take on. Violin plots are another type of hybrid plot, this time combining density traces (mirror-imaged right and left so that the plots have a sense of symmetry and visual balance) with boxplot-type information (median, IQR and upper and lower inner 'fences', but not outliers). The goal of the violin plot is to provide a quick visual impression of the shape, central tendency and variability of a distribution (the length of the violin conveys a sense of the overall variability whereas the width of the violin conveys a sense of the frequency of scores occurring in a specific region).

> Figure 5.27 (produced using NCSS), compares the distributions of **speed** scores for QCI inspectors across the five types of companies, plotted side-by-side. The violin plot conveys a similar story to the boxplot comparison for **speed** in the right graph of Fig. 5.25. However, notice that with the violin plot, unlike with a boxplot, you also get a sense of distributions that have 'clumps' of scores in specific areas. Some violin plots, like that for Automobile manufacturing companies in Fig. 5.27, have a shape suggesting a multi-modal distribution (recall *Procedure 5.4* and the discussion of the fact that a distribution may have multiple modes). The violin plot in Fig. 5.27 has also been produced to show where the median (solid line) and mean (dashed line) would fall within each violin. This facilitates two interpretations: (1) a relative comparison of central tendency across the five companies and (2) relative degree of skewness in the distribution for each company (indicated by the separation of the two lines within a violin; skewness is particularly bad for the Large Business Computer manufacturing companies).
>
> (continued)

Procedure 5.6: Exploratory Data Analysis

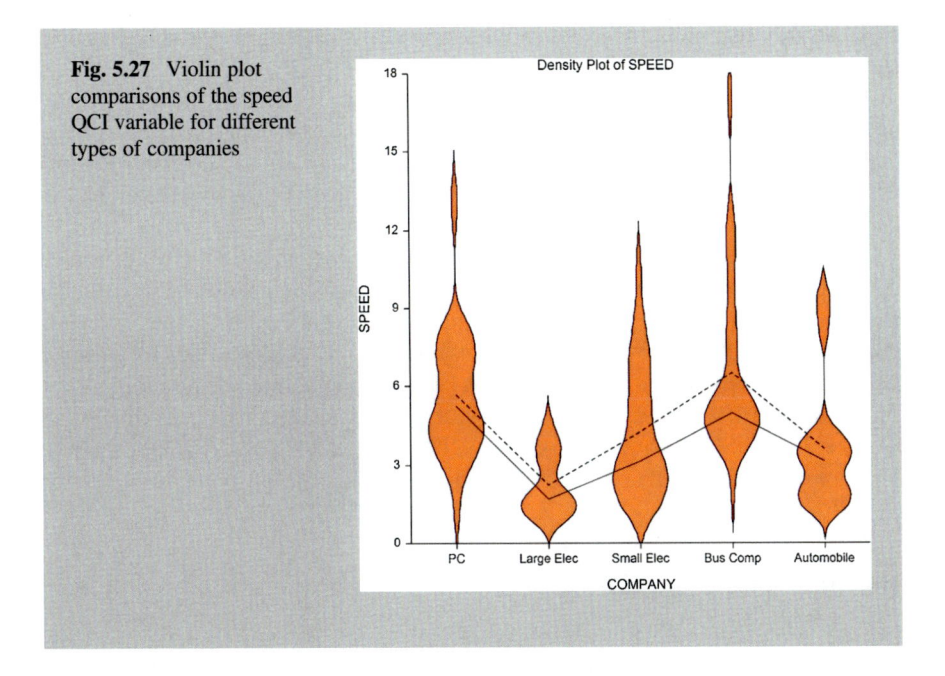

Fig. 5.27 Violin plot comparisons of the speed QCI variable for different types of companies

Advantages

EDA methods (of which we have illustrated only a small subset; we have not reviewed dot density diagrams, for example) provide summary techniques for visually displaying certain characteristics of a set of data. The advantage of the EDA methods over more traditional graphing techniques such as those described in *Procedure 5.2* is that as much of the original integrity of the data is maintained as possible while maximising the amount of summary information available about distributional characteristics.

Stem & leaf displays maintain the data in as close to their original form as possible whereas boxplots and violin plots provide more symbolic and flexible representations. EDA methods are best thought of as communication devices designed to facilitate quick visual impressions and they can add interest to any statistical story being conveyed about a sample of data. NCSS, SYSTAT, STATGRAPHICS and **R** Commander generally offer more options and flexibility in the generation of EDA displays than SPSS.

Disadvantages

EDA methods tend to get cumbersome if a great many variables or groups need to be summarised. In such cases, using numerical summary statistics (such as means and standard deviations) will provide a more economical and efficient summary.

Boxplots or violin plots are generally more space efficient summary techniques than stem & leaf displays.

Often, EDA techniques are used as data screening devices, which are typically not reported in actual write-ups of research (we will discuss data screening in more detail in *Procedure 8.2*). This is a perfectly legitimate use for the methods although there is an argument for researchers to put these techniques to greater use in published literature.

Software packages may use different rules for constructing EDA plots which means that you might get rather different looking plots and different information from different programs (you saw some evidence of this in Figs. 5.22 and 5.23). It is important to understand what the programs are using as decision rules for locating fences and outliers so that you are clear on how best to interpret the resulting plot—such information is generally contained in the user's guides or manuals for NCSS (Hintze 2012), SYSTAT (SYSTAT Inc. 2009a, b), STATGRAPHICS (StatPoint Technologies Inc. 2010) and SPSS (Norušis 2012).

Where Is This Procedure Useful?

Virtually any research design which produces numerical measures (even to the extent of just counting the number of occurrences of several events) provides opportunities for employing EDA displays which may help to clarify data characteristics or relationships. One extremely important use of EDA methods is as data screening devices for detecting outliers and other data anomalies, such as non-normality and skewness, before proceeding to parametric statistical analyses. In some cases, EDA methods can help the researcher to decide whether parametric or nonparametric statistical tests would be best to apply to his or her data because critical data characteristics such as distributional shape and spread are directly reflected.

Software Procedures

Application	Procedures
SPSS	*Analyze → Descriptive Statistics → Explore* ... produces stem-and-leaf displays and boxplots by default; variables may be explored on a whole-of-sample basis or broken down by the categories of a specific variable (called a 'factor' in the procedure). Cases can also be labelled with a variable (like **inspector** in the QCI database), so that outlier points in the boxplot are identifiable. *Graphs → Chart Builder*... can also be used to custom build different types of boxplots.
NCSS	*Analysis → Descriptive Statistics → Descriptive Statistics* produces a stem-and-leaf display by default. *Graphics → Box Plots → Box Plots* can be used to produce box plots with different features (such as 'notches' and connecting lines). *Graphics → Density Plots → Density Plots* can be configured to produce violin plots (by selecting the plot shape as 'density with reflection').

(continued)

Application	Procedures
SYSTAT	*Analyze → Stem-and-Leaf...* can be used to produce stem-and-leaf displays for variables; however, you cannot really control any features of these displays. *Graph → Box Plot...* can be used to produce boxplots of many types, with a number of features being controllable.
STATGRAPHICS	*Describe → Numerical Data → One-Variable Analysis...* allows you to do a complete exploration of a single variable, including stem-and-leaf display (you need to select this option) and boxplot (produced by default). Some features of the boxplot can be controlled, but not features of the stem-and-leaf diagram. *Plot → Exploratory Plots → Box-and-Whisker Plots...* and select either *One Sample...* or *Multiple Samples...* which can produce not only descriptive statistics but also boxplots with some controllable features.
R Commander	*Graphs → Stem-and-leaf display...* or *Boxplots...* the dialog box for each procedure offers some features of the display or plot that can be controlled; whole-of-sample boxplots or boxplots by groups are possible.

Procedure 5.7: Standard (*z*) Scores

Classification   Univariate; descriptive.
Purpose   To transform raw scores from a sample of data to a standardised form which permits comparisons with other scores within the same sample or with scores from other samples of data.
Measurement level   Generally, standard scores are computed from interval or ratio-level data.

In certain practical situations in behavioural research, it may be desirable to know where a specific individual's score lies relative to all other scores in a distribution. A convenient measure is to observe how many standard deviations (see *Procedure 5.5*) above or below the sample mean a specific score lies. This measure is called a *standard score* or *z*-score. Very simply, any raw score can be converted to a *z*-score by subtracting the sample mean from the raw score and dividing that result by the sample's standard deviation. *z*-scores can be positive or negative and their sign simply indicates whether the score lies above (+) or below (−) the mean in value. A *z*-score has a very simple interpretation: it measures the number of standard deviations above or below the sample mean a specific raw score lies.

In the QCI database, we have a sample mean for **speed** scores of 4.48 s, a standard deviation for **speed** scores of 2.89 s (recall Table 5.4 in *Procedure 5.5*). If we are interested in the z-score for Inspector 65's raw **speed** score of 11.94 s, we would obtain a z-score of +2.58 using the method described above (subtract 4.48 from 11.94 and divide the result by 2.89). The interpretation of this number is that a raw decision **speed** score of 11.94 s lies about 2.9 standard deviations above the mean decision **speed** for the sample.

z-scores have some interesting properties. First, if one converts (statisticians would say 'transforms') every available raw score in a sample to z-scores, the mean of these z-scores will always be zero and the standard deviation of these z-scores will always be 1.0. These two facts about z-scores (mean = 0; standard deviation = 1) will be true no matter what sample you are dealing with and no matter what the original units of measurement are (e.g. seconds, percentages, number of widgets assembled, amount of preference for a product, attitude rating, amount of money spent). This is because transforming raw scores to z-scores automatically changes the measurement units from whatever they originally were to a new system of measurements expressed in standard deviation units.

Suppose Maree was interested in the performance statistics for the top 25% most accurate quality control inspectors in the sample. Given a sample size of 112, this would mean finding the top 28 inspectors in terms of their **accuracy** scores. Since Maree is interested in performance statistics, **speed** scores would also be of interest. Table 5.5 (generated using the SPSS *Descriptives* ... procedure, listed using the *Case Summaries* ... procedure and formatted for presentation using Excel) shows **accuracy** and **speed** scores for the top 28 inspectors in descending order of **accuracy** scores. The z-score transformation for each of these scores is also shown (last two columns) as are the type of company, education level and gender for each inspector.

There are three inspectors (8, 9 and 14) who scored maximum accuracy of 100%. Such accuracy converts to a z-score of +1.95. Thus 100% accuracy is 1.95 standard deviations above the sample's mean accuracy level. Interestingly, all three inspectors worked for PC manufacturers and all three had only high school-level education. The least accurate inspector in the top 25% had a z-score for **accuracy** that was .75 standard deviations above the sample mean.

Interestingly, the top three inspectors in terms of accuracy had decision speeds that fell below the sample's mean speed; inspector 8 was the fastest inspector of the three with a speed just over 1 standard deviation ($z = -1.03$) below the sample mean. The slowest inspector in the top 25% was inspector 75 (case #28 in the list) with a **speed** z-score of +2.62; i.e., he was over two and a half standard deviations slower in making inspection decisions relative to the sample's mean speed.

(continued)

Procedure 5.7: Standard (z) Scores

Table 5.5 Listing of the 28 (top 25%) most accurate QCI inspectors' **accuracy** and **speed** scores as well as standard (z) score transformations for each score

Case number	Inspector	company	educlev	gender	accuracy	speed	Zaccuracy	Zspeed
1	8	PC Manufacturer	High School Only	Male	100	1.52	1.95	−1.03
2	9	PC Manufacturer	High School Only	Female	100	3.32	1.95	−0.40
3	14	PC Manufacturer	High School Only	Male	100	3.83	1.95	−0.23
4	17	PC Manufacturer	High School Only	Female	99	7.07	1.84	0.90
5	101	PC Manufacturer	High School Only	Female	98	3.11	1.73	−0.47
6	19	PC Manufacturer	Tertiary Qualified	Female	94	3.84	1.29	−0.22
7	34	Large Electrical Appliance Manufacturer	Tertiary Qualified	Male	94	1.90	1.29	−0.89
8	63	Large Business Computer Manufacturer	High School Only	Male	94	11.94	1.29	2.58
9	67	Large Business Computer Manufacturer	High School Only	Male	94	2.34	1.29	−0.74
10	80	Large Business Computer Manufacturer	High School Only	Female	94	4.68	1.29	0.07
11	5	PC Manufacturer	Tertiary Qualified	Male	93	4.18	1.18	−0.10
12	18	PC Manufacturer	Tertiary Qualified	Male	93	7.32	1.18	0.98
13	46	Small Electrical Appliance Manufacturer	Tertiary Qualified	Female	93	2.01	1.18	−0.86
14	64	Large Business Computer Manufacturer	High School Only	Female	92	5.18	1.08	0.24
15	77	Large Business Computer Manufacturer	Tertiary Qualified	Female	92	6.11	1.08	0.56
16	79	Large Business Computer Manufacturer	High School Only	Male	92	4.38	1.08	−0.03
17	106	Large Electrical Appliance Manufacturer	Tertiary Qualified	Male	92	1.70	1.08	−0.96
18	58	Small Electrical Appliance Manufacturer	High School Only	Male	91	4.12	0.97	−0.12
19	63	Large Business Computer Manufacturer	High School Only	Male	91	4.73	0.97	0.09

(continued)

Table 5.5 (continued)

Case number	Inspector	company	educlev	gender	accuracy	speed	Zaccuracy	Zspeed
20	72	Large Business Computer Manufacturer	Tertiary Qualified	Male	91	4.72	0.97	0.08
21	20	PC Manufacturer	High School Only	Male	90	4.53	0.86	0.02
22	69	Large Business Computer Manufacturer	High School Only	Male	90	4.94	0.86	0.16
23	71	Large Business Computer Manufacturer	High School Only	Female	90	10.46	0.86	2.07
24	85	Automobile Manufacturer	Tertiary Qualified	Female	90	3.14	0.86	−0.46
25	111	Large Business Computer Manufacturer	High School Only	Male	90	4.11	0.86	−0.13
26	6	PC Manufacturer	High School Only	Male	89	5.46	0.75	0.34
27	61	Large Business Computer Manufacturer	Tertiary Qualified	Male	89	5.71	0.75	0.43
28	75	Large Business Computer Manufacturer	High School Only	Male	89	12.05	0.75	2.62

(continued)

Procedure 5.7: Standard (z) Scores

> The fact that z-scores always have a common measurement scale having a mean of 0 and a standard deviation of 1.0 leads to an interesting application of standard scores. Suppose we focus on inspector number 65 (case #8 in the list) in Table 5.5. It might be of interest to compare this inspector's quality control performance in terms of both his decision accuracy and decision speed. Such a comparison is impossible using raw scores since the inspector's **accuracy** score and **speed** scores are different measures which have differing means and standard deviations expressed in fundamentally different units of measurement (percentages and seconds). However, if we are willing to assume that the score distributions for both variables are approximately the same shape and that both **accuracy** and **speed** are measured with about the same level of reliability or consistency (see *Procedure 8.1*), we can compare the inspector's two scores by first converting them to z-scores within their own respective distributions as shown in Table 5.5.
>
> Inspector 65 looks rather anomalous in that he demonstrated a relatively high level of **accuracy** (raw score = 94%; $z = +1.29$) but took a very long time to make those accurate decisions (raw score = 11.94 s; $z = +2.58$). Contrast this with inspector 106 (case #17 in the list) who demonstrated a similar level of **accuracy** (raw score = 92%; $z = +1.08$) but took a much shorter time to make those accurate decisions (raw score = 1.70 s; $z = -.96$). In terms of evaluating performance, from a company perspective, we might conclude that inspector 106 is performing at an overall higher level than inspector 65 because he can achieve a very high level of accuracy but much more quickly; accurate and fast is more cost effective and efficient than accurate and slow.
>
> Note: We should be cautious here since we know from our previous explorations of the **speed** variable in *Procedure 5.6*, that **accuracy** scores look fairly symmetrical and **speed** scores are positively skewed, so assuming that the two variables have the same distribution shape, so that z-score comparisons are permitted, would be problematic.

You might have noticed that as you scanned down the two columns of z-scores in Table 5.5, there was a suggestion of a pattern between the signs attached to the respective z-scores for each person. There seems to be a very slight preponderance of pairs of z-scores where the signs are reversed (12 out of 22 pairs). This observation provides some very preliminary evidence to suggest that there may be a relationship between inspection accuracy and decision speed, namely that a more accurate decision tends to be associated with a faster decision speed. Of course, this pattern would be better verified using the entire sample rather than the top 25% of inspectors. However, you may find it interesting to learn that it is precisely this sort of suggestive evidence (about agreement or disagreement between z-score signs for pairs of variable scores throughout a sample) that is captured and summarised by a single statistical indicator called a 'correlation coefficient' (see *Fundamental Concept III* and *Procedure 6.1*).

z-scores are not the only type of standard score that is commonly used. Three other types of standard scores are: *stanines* (standard nines), *IQ scores* and *T-scores* (not to be confused with the *t*-test described in *Procedure 7.2*). These other types of scores have the advantage of producing only positive integer scores rather than positive and negative decimal scores. This makes interpretation somewhat easier for certain applications. However, you should know that almost all other types of standard scores come from a specific transformation of z-scores. This is because once you have converted raw scores into z-scores, they can then be quite readily transformed into any other system of measurement by simply multiplying a person's z-score by the new desired standard deviation for the measure and adding to that product the new desired mean for the measure.

For example *T-scores are simply z-scores transformed to have a mean of 50.0 and a standard deviation of 10.0; IQ scores are simply z-scores transformed to have a mean of 100 and a standard deviation of 15 (or 16 in some systems). For more information, see* Fundamental Concept II.

Advantages

Standard scores are useful for representing the position of each raw score within a sample distribution relative to the mean of that distribution. The unit of measurement becomes the number of standard deviations a specific score is away from the sample mean. As such, z-scores can permit cautious comparisons across samples or across different variables having vastly differing means and standard deviations within the constraints of the comparison samples having similarly shaped distributions and roughly equivalent levels of measurement reliability. z-scores also form the basis for establishing the degree of correlation between two variables. Transforming raw scores into z-scores does not change the shape of a distribution or rank ordering of individuals within that distribution. For this reason, a z-score is referred to as a *linear transformation* of a raw score. Interestingly, z-scores provide an important foundational element for more complex analytical procedures such as factor analysis (*Procedure 6.5*), cluster analysis (*Procedure 6.6*) and multiple regression analysis (see, for example, *Procedure 6.4 and 7.13*).

Disadvantages

While standard scores are useful indices, they are subject to restrictions if used to compare scores across samples or across different variables. The samples must have similar distribution shapes for the comparisons to be meaningful and the measures must have similar levels of reliability in each sample. The groups used to generate the z-scores should also be similar in composition (with respect to age, gender distribution, and so on). Because *z*-scores are not an intuitively meaningful way of presenting scores to lay-persons, many other types of standard score schemes have been devised to improve interpretability. However, most of these schemes produce scores that run a greater risk of facilitating lay-person misinterpretations simply because their connection with *z*-scores is hidden or because the resulting numbers 'look' like a more familiar type of score which people do intuitively understand.

For example *It is extremely rare for a T-score to exceed 100 or go below 0 because this would mean that the raw score was in excess of 5 standard deviations away from the sample mean. This unfortunately means that T-scores are often misinterpreted as percentages because they typically range between 0 and 100 and therefore 'look' like percentages. However, T-scores are definitely not percentages.*

Finally, a common misunderstanding of *z*-scores is that transforming raw scores into *z*-scores makes them follow a normal distribution (see *Fundamental Concept II*). This is not the case. The distribution of *z*-scores will have exactly the same shape as that for the raw scores; if the raw scores are positively skewed, then the corresponding *z*-scores will also be positively skewed.

Where Is This Procedure Useful?

z-scores are particularly useful in evaluative studies where relative performance indices are of interest. Whenever you compute a correlation coefficient (*Procedure 6.1*), you are implicitly transforming the two variables involved into *z*-scores (which equates the variables in terms of mean and standard deviation), so that only the patterning in the relationship between the variables is represented. *z*-scores are also useful as a preliminary step to more advanced parametric statistical methods when variables differing in scale, range and/or measurement units must be equated for means and standard deviations prior to analysis.

Software Procedures

Application	Procedures
SPSS	*Analyze* → *Descriptive Statistics* → *Descriptives*... and tick the box labelled 'Save standardized values as variables'. *z*-scores are saved as new variables (labelled as Z followed by the original variable name as shown in Table 5.5) which can then be listed or analysed further.
NCSS	*Data* → *Transformations* → *Transformation* and select a new variable to hold the *z*-scores, then select the 'STANDARDIZE' transformation from the list of available functions. *z*-scores are saved as new variables which can then be listed or analysed further.

(continued)

Application	Procedures
SYSTAT	*Data → Standardize* ... where z-scores are saved as new variables which can then be listed or analysed further.
STATGRAPHICS	Open the *Databook* window, and select an empty column in the database, then *Edit → Generate Data...* and choose the 'STANDARDIZE' transformation, choose the variable you want to transform and give the new variable a name.
R Commander	*Data → Manage variables in active data set → Standardize variables...* and select the variables you want to standardize; R Commander automatically saves the transformed variable to the data base, appending Z. to the front of each variable's name.

Fundamental Concept II: The Normal Distribution

Arguably the most fundamental distribution used in the statistical analysis of quantitative data in the behavioural and social sciences is the *normal distribution* (also known as the *Gaussian* or *bell-shaped distribution*). Many behavioural phenomena, if measured on a large enough sample of people, tend to produce 'normally distributed' variable scores. This includes most measures of ability, performance and productivity, personality characteristics and attitudes. The normal distribution is important because it is the one form of distribution that you must assume describes the scores of a variable in the population when parametric tests of statistical inference are undertaken. The standard normal distribution is defined as having a population mean of 0.0 and a population standard deviation of 1.0. The normal distribution is also important as a means of interpreting various types of scoring systems.

> Figure 5.28 displays the standard normal distribution (mean = 0; standard deviation = 1.0) and shows that there is a clear link between z-scores and the normal distribution. Statisticians have analytically calculated the probability (also expressed as percentages or percentiles) that observations will fall above or below any specific z-score in the theoretical standard normal distribution. Thus, a z-score of +1.0 in the standard normal distribution will have 84.13% (equals a probability of .8413) of observations in the population falling at or below one standard deviation above the mean and 15.87% falling above that point. A z-score of −2.0 will have 2.28% of observations falling at that point or below and 97.72% of observations falling above that point. It is clear then that, in a standard normal distribution, z-scores have a direct relationship with percentiles.

(continued)

Fundamental Concept II: The Normal Distribution

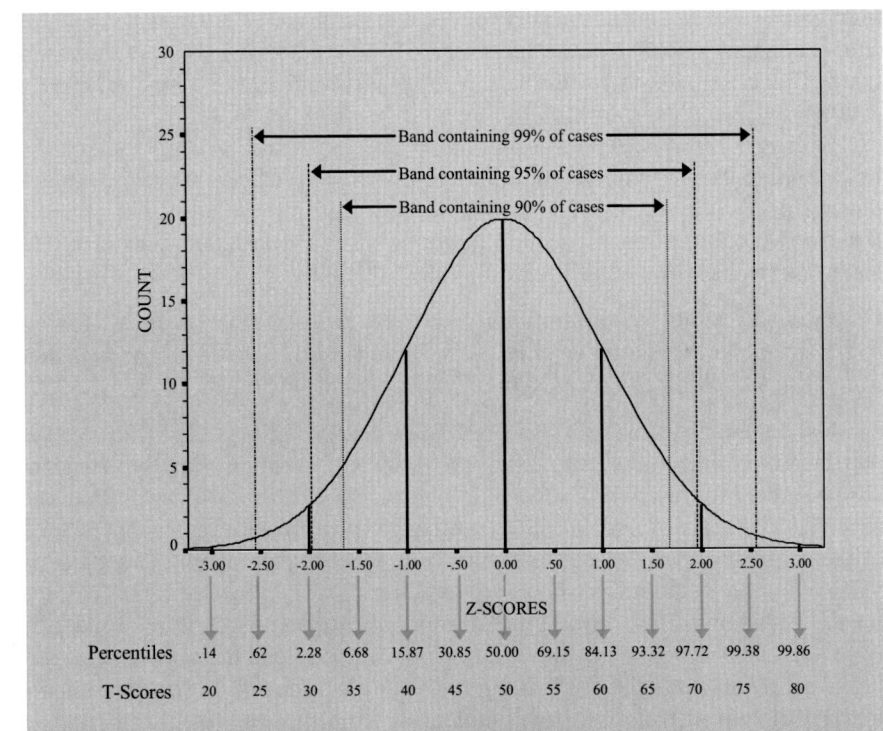

Fig. 5.28 The normal (bell-shaped or Gaussian) distribution

Figure 5.28 also shows how T-scores relate to the standard normal distribution and to z-scores. The mean T-score falls at 50 and each increment or decrement of 10 T-score units means a movement of another standard deviation away from this mean of 50. Thus, a T-score of 80 corresponds to a z-score of +3.0—a score 3 standard deviations higher than the mean of 50.

Of special interest to behavioural researchers are the values for z-scores in a standard normal distribution that encompass 90% of observations ($z = \pm 1.645$—isolating 5% of the distribution in each tail), 95% of observations ($z = \pm 1.96$—isolating 2.5% of the distribution in each tail), and 99% of observations ($z = \pm 2.58$—isolating 0.5% of the distribution in each tail).

Depending upon the degree of certainty required by the researcher, these bands describe regions outside of which one might define an observation as being atypical or as perhaps not belonging to a distribution being centred at a mean of 0.0. Most often, what is taken as atypical or rare in the standard normal distribution is a score at least two standard deviations away from the mean, in either direction. Why choose two standard deviations? Since in the standard normal distribution, only about 5% of observations will fall outside a band defined by z-scores of ± 1.96 (rounded to 2 for simplicity), this equates to data values that are 2 standard deviations away from their mean. This can give us a defensible way to identify outliers or extreme values in a distribution.

Thinking ahead to what you will encounter in *Chap. 7*, this 'banding' logic can be extended into the world of statistics (like means and percentages) as opposed to just the world of observations. You will frequently hear researchers speak of some statistic estimating a specific value (a *parameter*) in a population, plus or minus some other value.

For example *A survey organisation might report political polling results in terms of a percentage and an error band, e.g. 59% of Australians indicated that they would vote Labour at the next federal election, plus or minus 2%.*

Most commonly, this error band ($\pm 2\%$) is defined by possible values for the population parameter that are about two standard deviations (or two standard errors—a concept discussed further in *Fundamental Concept VIII*) away from the reported or estimated statistical value. In effect, the researcher is saying that on 95% of the occasions he/she would theoretically conduct his/her study, the population value estimated by the statistic being reported would fall between the limits imposed by the endpoints of the error band (the official name for this error band is a *confidence interval*; see *Procedure 8.3*). The well-understood mathematical properties of the standard normal distribution are what make such precise statements about levels of error in statistical estimates possible.

Checking for Normality

It is important to understand that transforming the raw scores for a variable to z-scores (recall *Procedure 5.7*) does not produce z-scores which follow a normal distribution; rather they will have the same distributional shape as the original scores. However, if you are willing to assume that the normal distribution is the correct reference distribution in the population, then you are justified is interpreting z-scores in light of the known characteristics of the normal distribution.

In order to justify this assumption, not only to enhance the interpretability of z-scores but more generally to enhance the integrity of parametric statistical analyses, it is helpful to actually look at the sample frequency distributions for variables (using a *histogram* (illustrated in *Procedure 5.2*) or a *boxplot* (illustrated in *Procedure 5.6*), for example), since non-normality can often be visually detected. It is important to note that in the social and behavioural sciences as well as in economics and finance, certain variables tend to be non-normal by their very nature. This includes variables that measure time taken to complete a task, achieve a goal or make decisions and variables that measure, for example, income, occurrence of rare or extreme events or organisational size. Such variables tend to be positively skewed in the population, a pattern that can often be confirmed by graphing the distribution.

Fundamental Concept II: The Normal Distribution

If you cannot justify an assumption of 'normality', you may be able to force the data to be normally distributed by using what is called a 'normalising transformation'. Such transformations will usually involve a nonlinear mathematical conversion (such as computing the logarithm, square root or reciprocal) of the raw scores. Such transformations will force the data to take on a more normal appearance so that the assumption of 'normality' can be reasonably justified, but at the cost of creating a new variable whose units of measurement and interpretation are more complicated. [For some non-normal variables, such as the occurrence of rare, extreme or catastrophic events (e.g. a 100-year flood or forest fire, coronavirus pandemic, the Global Financial Crisis or other type of financial crisis, man-made or natural disaster), the distributions cannot be 'normalised'. In such cases, the researcher needs to model the distribution as it stands. For such events, *extreme value theory* (e.g. see Diebold et al. 2000) has proven very useful in recent years. This theory uses a variation of the Pareto or Weibull distribution as a reference, rather than the normal distribution, when making predictions.]

> Figure 5.29 displays before and after pictures of the effects of a logarithmic transformation on the positively skewed **speed** variable from the QCI database. Each graph, produced using NCSS, is of the hybrid histogram-density trace-boxplot type first illustrated in *Procedure 5.6*. The left graph clearly shows the strong positive skew in the **speed** scores and the right graph shows the result of taking the $\log_{10}$ of each raw score.
>
> **Fig. 5.29** Combined histogram-density trace-boxplot graphs displaying the before and after effects of a 'normalising' $\log_{10}$ transformation of the **speed** variable
>
> Notice how the long tail toward slow **speed** scores is pulled in toward the mean and the very short tail toward fast **speed** scores is extended away from the mean. The result is a more 'normal' appearing distribution. The assumption would then be that we could assume normality of **speed** scores, but only in a $\log_{10}$ format (i.e. it is the log of **speed** scores that we assume is normally distributed in the population). In general, taking the logarithm of raw scores provides a satisfactory remedy for positively skewed distributions (but not for negatively skewed ones). Furthermore, anything we do with the transformed **speed** scores now has to be interpreted in units of $\log_{10}$ (seconds) which is a more complex interpretation to make.

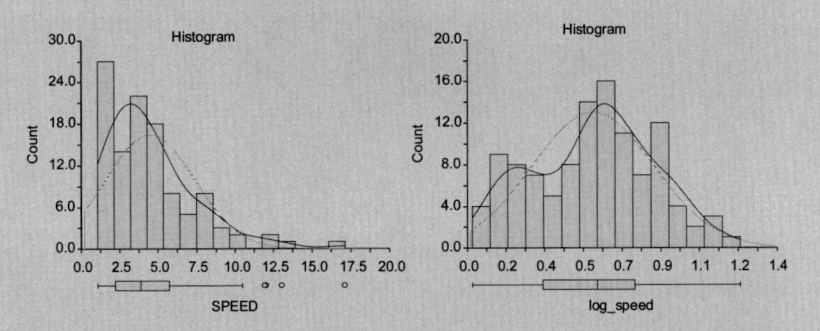

Another visual method for detecting non-normality is to graph what is called a *normal Q-Q plot* (the Q-Q stands for Quantile-Quantile). This plots the percentiles for the observed data against the percentiles for the standard normal distribution (see Cleveland 1995 for more detailed discussion; also see Lane 2007, http://onlinestatbook.com/2/advanccd_graphs/ q-q_plots.html). If the pattern for the observed data follows a normal distribution, then all the points on the graph will fall approximately along a diagonal line.

Figure 5.30 shows the normal Q-Q plots for the original **speed** variable and the transformed **log-speed** variable, produced using the SPSS *Explore...* procedure. The diagnostic diagonal line is shown on each graph. In the left-hand plot, for **speed**, the plot points clearly deviate from the diagonal in a way that signals positive skewness. The right-hand plot, for **log_speed,** shows the plot points generally falling along the diagonal line thereby conforming much more closely to what is expected in a normal distribution.

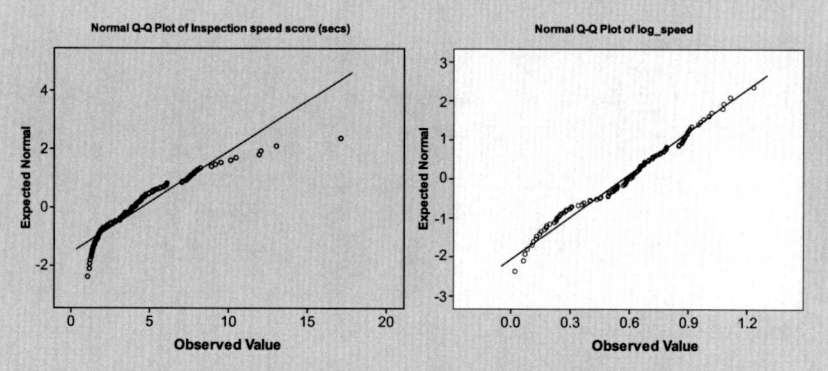

Fig. 5.30 Normal Q-Q plots for the original **speed** variable and the new **log_speed** variable

---

In addition to visual ways of detecting non-normality, there are also numerical ways. As highlighted in Chap. 1, there are two additional characteristics of any distribution, namely *skewness* (asymmetric distribution tails) and *kurtosis* (peakedness of the distribution). Both have an associated statistic that provides a measure of that characteristic, similar to the mean and standard deviation statistics. In a normal distribution, the values for the skewness and kurtosis statistics are both zero (skewness $= 0$ means a symmetric distribution; kurtosis $= 0$ means a mesokurtic distribution). The further away each statistic is from zero, the more the distribution deviates from a normal shape. Both the skewness statistic and the kurtosis statistic have *standard errors* (see *Fundamental Concept VIII*) associated with them (which work very much like the standard deviation, only for a statistic rather than for observations); these can be routinely computed by almost any statistical package when you request a descriptive analysis. Without going into the logic right now (this will come in *Fundamental Concept V*), a rough rule of thumb you can use to check for normality using the skewness and kurtosis statistics is to do the following:

Fundamental Concept II: The Normal Distribution

- Prepare: Take the standard error for the statistic and multiply it by 2 (or 3 if you want to be more conservative).
- Interval: Add the result from the Prepare step to the value of the statistic and subtract the result from the value of the statistic. You will end up with two numbers, one low - one high, that define the ends of an interval (what you have just created approximates what is called a 'confidence interval', see *Procedure 8. 3*).
- Check: If zero falls inside of this interval (i.e. between the low and high endpoints from the Interval step), then there is likely to be no significant issue with that characteristic of the distribution. If zero falls outside of the interval (i.e. lower than the low value endpoint or higher than the high value endpoint), then you likely have an issue with non-normality with respect to that characteristic.

Visually, we saw in the left graph in Fig. 5.29 that the **speed** variable was highly positively skewed. What if Maree wanted to check some numbers to support this judgment? She could ask SPSS to produce the skewness and kurtosis statistics for both the original **speed** variable and the new **log_speed** variable using the *Frequencies...* or the *Explore...* procedure. Table 5.6 shows what SPSS would produce if the *Frequencies...* procedure were used.

Table 5.6 Skewness and kurtosis statistics and their standard errors for both the original **speed** variable and the new **log_speed** variable

Statistics		speed	log_speed
N	Valid	111	111
	Missing	1	1
Mean		4.4801	.5676
Std. Deviation		2.88751	.27491
Skewness		1.487	-.050
Std. Error of Skewness		.229	.229
Kurtosis		3.071	-.672
Std. Error of Kurtosis		.455	.455

Using the 3-step check rule described above, Maree could roughly evaluate the normality of the two variables as follows:

For **speed**:

- *skewness*: [Prepare] $2 \times .229 = .458 \rightarrow$ [Interval] $1.487 - .458 = 1.029$ and $1.487 + .458 = 1.945 \rightarrow$ [Check] zero does not fall inside the interval bounded by 1.029 and 1.945, so there appears to be a significant problem with skewness. Since the value for the skewness statistic (1.487) is positive, this means the problem is positive skewness, confirming what the left graph in Fig. 5.29 showed.

(continued)

- **kurtosis**: [Prepare] 2 × .455 = .91 → [Interval] 3.071 − .91 = 2.161 and 3.071 + .91 = 3.981 → [Check] zero does not fall in interval bounded by 2.161 and 3.981, so there appears to be a significant problem with kurtosis. Since the value for the kurtosis statistic (1.487) is positive, this means the problem is leptokurtosis—the peakedness of the distribution is too tall relative to what is expected in a normal distribution.

For **log_speed**:

- *skewness*: [Prepare] 2 × .229 = .458 → [Interval] −.050 − .458 = −.508 and −.050 + .458 = .408 → [Check] zero falls within interval bounded by −.508 and .408, so there appears to be no problem with skewness. The log transform appears to have corrected the problem, confirming what the right graph in Fig. 5.29 showed.
- *kurtosis*: [Prepare] 2 × .455 = .91 → [Interval] −.672 − .91 = −1.582 and −.672 + .91 = .238 → [Check] zero falls within interval bounded by −1.582 and .238, so there appears to be no problem with kurtosis. The log transform appears to have corrected this problem as well, rendering the distribution more approximately mesokurtic (i.e. normal) in shape.

There are also more formal tests of significance (see *Fundamental Concept V*) that one can use to numerically evaluate normality, such as the *Kolmogorov-Smirnov test* and the *Shapiro-Wilk's test*. Each of these tests, for example, can be produced by SPSS on request, via the *Explore...* procedure.

References

References for Procedure 5.1

Allen, P., Bennett, K., & Heritage, B. (2019). *SPSS statistics: A practical guide* (4th ed.). South Melbourne, VIC: Cengage Learning Australia Pty. ch. 3.

George, D., & Mallery, P. (2019). *IBM SPSS statistics 25 step by step: A simple guide and reference* (15th ed.). New York: Routledge. ch. 6 and 8.

Useful Additional Readings for Procedure 5.1

Agresti, A. (2018). *Statistical methods for the social sciences* (5th ed.). Boston: Pearson. ch. 3.

Argyrous, G. (2011). *Statistics for research: With a guide to SPSS* (3rd ed.). London: Sage. ch. 4–5.

De Vaus, D. (2002). *Analyzing social science data: 50 key problems in data analysis*. London: Sage. ch. 28, 32.

Glass, G. V., & Hopkins, K. D. (1996). *Statistical methods in education and psychology* (3rd ed.). Upper Saddle River, NJ: Pearson. ch. 3.
Gravetter, F. J., & Wallnau, L. B. (2017). *Statistics for the behavioural sciences* (10th ed.). Belmont, CA: Wadsworth Cengage. ch. 2.
Steinberg, W. J. (2011). *Statistics alive* (2nd ed.). Los Angeles: Sage. ch. 3.

References for Procedure 5.2

Chang, W. (2019). *R graphics cookbook: Practical recipes for visualizing data* (2nd ed.). Sebastopol, CA: O'Reilly Media. ch. 2–5.
Jacoby, W. G. (1997). *Statistical graphics for univariate and bivariate data*. Thousand Oaks, CA: Sage.
McCandless, D. (2014). *Knowledge is beautiful*. London: William Collins.
Smithson, M. J. (2000). *Statistics with confidence*. London: Sage. ch. 3.
Toseland, M., & Toseland, S. (2012). *Infographica: The world as you have never seen it before*. London: Quercus Books.
Wilkinson, L. (2009). Cognitive science and graphic design. In SYSTAT Software Inc (Ed.), *SYSTAT 13: Graphics* (pp. 1–21). Chicago, IL: SYSTAT Software Inc.

Useful Additional Readings for Procedure 5.2

Argyrous, G. (2011). *Statistics for research: With a guide to SPSS* (3rd ed.). London: Sage. ch. 4–5.
De Vaus, D. (2002). *Analyzing social science data: 50 key problems in data analysis*. London: Sage. ch. 29, 33.
Field, A. (2018). *Discovering statistics using SPSS for windows* (5th ed.). Los Angeles: Sage. ch. 5.
George, D., & Mallery, P. (2019). *IBM SPSS statistics 25 step by step: A simple guide and reference* (15th ed.). Boston, MA: Pearson Education. ch. 5.
Glass, G. V., & Hopkins, K. D. (1996). *Statistical methods in education and psychology* (3rd ed.). Upper Saddle River, NJ: Pearson. ch. 3.
Hintze, J. L. (2012). *NCSS 8 help system: Graphics*. Kaysville, UT: Number Cruncher Statistical Systems. ch. 141–143, 155, 161.
StatPoint Technologies, Inc. (2010). *STATGRAPHICS Centurion XVI user manual*. Warrenton, VA: StatPoint Technologies Inc.. ch. 4.
Steinberg, W. J. (2011). *Statistics alive* (2nd ed.). Los Angeles: Sage. ch. 4.
SYSTAT Software Inc. (2009). *SYSTAT 13: Graphics*. Chicago, IL: SYSTAT Software Inc. ch. 2, 5.

References for Procedure 5.3

Cleveland, W. R. (1995). *Visualizing data*. Summit, NJ: Hobart Press.
Jacoby, W. J. (1998). *Statistical graphics for visualizing multivariate data*. Thousand Oaks, CA: Sage.
SYSTAT Software Inc. (2009). *SYSTAT 13: Graphics*. Chicago, IL: SYSTAT Software Inc. ch. 6.

Useful Additional Readings for Procedure 5.3

Hintze, J. L. (2012). *NCSS 8 help system: Graphics*. Kaysville, UT: Number Cruncher Statistical Systems. ch. 162.
Kirk, A. (2016). *Data visualisation: A handbook for data driven design*. Los Angeles: Sage.
Knaflic, C. N. (2015). *Storytelling with data: A data visualization guide for business professionals*. Hoboken, NJ: Wiley.
Tufte, E. (2001). *The visual display of quantitative information* (2nd ed.). Cheshire, CN: Graphics Press.

Reference for Procedure 5.4

Field, A. (2018). *Discovering statistics using SPSS for windows* (5th ed.). Los Angeles: Sage. ch. 6.

Useful Additional Readings for Procedure 5.4

Agresti, A. (2018). *Statistical methods for the social sciences* (5th ed.). Boston: Pearson. ch. 3.
Argyrous, G. (2011). *Statistics for research: With a guide to SPSS* (3rd ed.). London: Sage. ch. 9.
De Vaus, D. (2002). *Analyzing social science data: 50 key problems in data analysis*. London: Sage. ch. 30.
George, D., & Mallery, P. (2019). *IBM SPSS statistics 25 step by step: A simple guide and reference* (15th ed.). New York: Routledge. ch. 7.
Glass, G. V., & Hopkins, K. D. (1996). *Statistical methods in education and psychology* (3rd ed.). Upper Saddle River, NJ: Pearson. ch. 4.
Gravetter, F. J., & Wallnau, L. B. (2017). *Statistics for the behavioural sciences* (10th ed.). Belmont, CA: Wadsworth Cengage. ch. 3.
Rosenthal, R., & Rosnow, R. L. (1991). *Essentials of behavioral research: Methods and data analysis* (2nd ed.). New York: McGraw-Hill Inc. ch. 13.
Steinberg, W. J. (2011). *Statistics alive* (2nd ed.). Los Angeles: Sage. ch. 5.

References for Procedure 5.5

Field, A. (2018). *Discovering statistics using SPSS for windows* (5th ed.). Los Angeles: Sage. ch. 6.

Useful Additional Readings for Procedure 5.5

Agresti, A. (2018). *Statistical methods for the social sciences* (5th ed.). Boston: Pearson. ch. 3.
Argyrous, G. (2011). *Statistics for research: With a guide to SPSS* (3rd ed.). London: Sage. ch. 11.
De Vaus, D. (2002). *Analyzing social science data: 50 key problems in data analysis*. London: Sage. ch 15.

Glass, G. V., & Hopkins, K. D. (1996). *Statistical methods in education and psychology* (3rd ed.). Upper Saddle River, NJ: Pearson. ch. 6.
Gravetter, F. J., & Wallnau, L. B. (2012). *Statistics for the behavioural sciences* (9th ed.). Belmont, CA: Wadsworth Cengage. ch. 5.
Rosenthal, R., & Rosnow, R. L. (1991). *Essentials of behavioral research: Methods and data analysis* (2nd ed.). New York: McGraw-Hill Inc. ch. 13.
Steinberg, W. J. (2011). *Statistics alive* (2nd ed.). Los Angeles: Sage. ch. 8.

References for Fundamental Concept I

Smithson, M. J. (2000). *Statistics with confidence*. London: Sage. ch. 4.

Useful Additional Readings for Fundamental Concept I

Agresti, A. (2018). *Statistical methods for the social sciences* (5th ed.). Boston: Pearson. ch. 4.
Glass, G. V., & Hopkins, K. D. (1996). *Statistical methods in education and psychology* (3rd ed.). Upper Saddle River, NJ: Pearson. ch. 9.
Gravetter, F. J., & Wallnau, L. B. (2017). *Statistics for the behavioural sciences* (10th ed.). Belmont, CA: Wadsworth Cengage. ch. 6.
Howell, D. C. (2013). *Statistical methods for psychology* (8th ed.). Belmont, CA: Cengage Wadsworth. ch. 5.
Steinberg, W. J. (2011). *Statistics alive* (2nd ed.). Los Angeles: Sage. ch. 10.

References for Procedure 5.6

Norušis, M. J. (2012). *IBM SPSS statistics 19 guide to data analysis*. Upper Saddle River, NJ: Prentice Hall. ch. 7.
Field, A. (2018). *Discovering statistics using SPSS for Windows* (5th ed.). Los Angeles: Sage. ch. 5, section 5.5.
Hintze, J. L. (2012). *NCSS 8 help system: Introduction*. Kaysville, UT: Number Cruncher Statistical System. ch. 152, 200.
StatPoint Technologies, Inc. (2010). *STATGRAPHICS Centurion XVI user manual*. Warrenton, VA: StatPoint Technologies Inc..
SYSTAT Software Inc. (2009a). *SYSTAT 13: Graphics*. Chicago, IL: SYSTAT Software Inc. ch. 3.
SYSTAT Software Inc. (2009b). *SYSTAT 13: Statistics - I*. Chicago, IL: SYSTAT Software Inc. ch. 1 and 9.

Useful Additional Readings for Procedure 5.6

Glass, G. V., & Hopkins, K. D. (1996). *Statistical methods in education and psychology* (3rd ed.). Upper Saddle River, NJ: Pearson. ch. 3.

Hartwig, F., & Dearing, B. E. (1979). *Exploratory data analysis*. Beverly Hills, CA: Sage.
Howell, D. C. (2013). *Statistical methods for psychology* (8th ed.). Belmont, CA: Cengage Wadsworth. ch. 2.
Leinhardt, G., & Leinhardt, L. (1997). Exploratory data analysis. In J. P. Keeves (Ed.), *Educational research, methodology, and measurement: An international handbook* (2nd ed., pp. 519–528). Oxford: Pergamon Press.
Rosenthal, R., & Rosnow, R. L. (1991). *Essentials of behavioral research: Methods and data analysis* (2nd ed.). New York: McGraw-Hill, Inc.. ch. 13.
Smithson, M. J. (2000). *Statistics with confidence*. London: Sage. ch. 3.
Tukey, J. W. (1977). *Exploratory data analysis*. Reading, MA: Addison-Wesley Publishing. [This is the classic text in the area, written by the statistician who developed most of the standard EDA techniques in use today].
Velleman, P. F., & Hoaglin, D. C. (1981). *ABC's of EDA*. Boston: Duxbury Press.

Useful Additional Readings for Procedure 5.7

Agresti, A. (2018). *Statistical methods for the social sciences* (5th ed.). Boston: Pearson. ch. 3.
Argyrous, G. (2011). *Statistics for research: With a guide to SPSS* (3rd ed.). London: Sage. ch. 10.
De Vaus, D. (2002). *Analyzing social science data: 50 key problems in data analysis*. London: Sage. ch. 30.
George, D., & Mallery, P. (2019). *IBM SPSS statistics 25 step by step: A simple guide and reference* (15th ed.). New York: Routledge. ch. 7.
Glass, G. V., & Hopkins, K. D. (1996). *Statistical methods in education and psychology* (3rd ed.). Upper Saddle River, NJ: Pearson. ch. 5.
Gravetter, F. J., & Wallnau, L. B. (2017). *Statistics for the behavioural sciences* (10th ed.). Belmont, CA: Wadsworth Cengage. ch. 4.
Rosenthal, R., & Rosnow, R. L. (1991). *Essentials of behavioral research: Methods and data analysis* (2nd ed.). New York: McGraw-Hill Inc. ch. 13.
Steinberg, W. J. (2011). *Statistics alive* (2nd ed.). Los Angeles: Sage. ch. 6.

References for Fundemental Concept II

Cleveland, W. R. (1995). *Visualizing data*. Summit, NJ: Hobart Press.
Diebold, F. X., Schuermann, T., & Stroughair, D. (2000). Pitfalls and opportunities in the use of extreme value theory in risk management. *The Journal of Risk Finance, 1*(2), 30–35.
Lane, D. (2007). *Online statistics education: A multimedia course of study*. Houston, TX: Rice University. http://onlinestatbook.com/.

Useful Additional Readings for Fundemental Concept II

Argyrous, G. (2011). *Statistics for research: With a guide to SPSS* (3rd ed.). London: Sage. ch. 11.
De Vaus, D. (2002). *Analyzing social science data: 50 key problems in data analysis*. London: Sage. ch. 11.

References

Glass, G. V., & Hopkins, K. D. (1996). *Statistical methods in education and psychology* (3rd ed.). Upper Saddle River, NJ: Pearson. ch. 6.

Howell, D. C. (2013). *Statistical methods for psychology* (8th ed.). Belmont, CA: Cengage Wadsworth. ch. 3.

Keller, D. K. (2006). *The tao of statistics: A path to understanding (with no math)*. Thousand Oaks, CA: Sage. ch. 10.

Steinberg, W. J. (2011). *Statistics alive* (2nd ed.). Los Angeles: Sage. ch. 7, 8, 9.

Chapter 6
Correlational Statistics for Characterising Relationships

The second broad category of statistics we illustrate and discuss concerns *correlational statistics* for characterising relationships. The purposes of the procedures and fundamental concepts in this category are quite varied ranging from providing a simple summary of the relationship between two variables to facilitating an understanding of complex relationships among many variables. By 'characterising relationship', we are referring to measuring the tendency for scores on different variables to either increase (and decrease) together, move in opposition to each other (one increases while another decreases), or to increase and/or decrease in no systematically connected fashion as we scan across the entities being studied. A statistical relationship is thus a pattern or an association which exists between two or more variables.

Earlier, in *Procedure 5.7*, we alluded to the patterns that might be visually detected by scanning down the set of z-scores for two different variables, noting the tendency for values of like signs to occur together or in opposition as we move from person to person in the sample. However, visual detection of associational patterns in complex numerical data is not easy if the association is rather weak or largely haphazard. Thus, we employ the statistical concept of correlation to summarise, in a single number, the nature of this patterned relationship or association between two variables. No matter how many variables are involved or how sophisticated the analysis is, all correlational procedures depend upon measuring and then analysing the relationships between pairs of variables.

> Think back to the QCI database. Consider the following relational questions which Maree could ask regarding relationships between/among certain variables in the database.

(continued)

- What is the relationship between the speed with which an inspector makes a quality control decision and the accuracy of that decision? [see *Procedure 6.1* – Assessing correlation.]
- Is there a relationship between gender of inspector and the industry in which they are employed? [see *Procedure 6.2* – Assessing association in contingency tables.]
- Can an inspector's mental ability, job satisfaction and/or perceptions of working conditions be used to predict his or her decision accuracy? [see *Procedure 6.3* – Simple linear regression & prediction and *Procedure 6.4* – Multiple regression & prediction.]
- Can the nine attitude items be represented more succinctly by a smaller number of constructs or composites which capture essentially the same general information? [see *Procedure 6.5* – Exploratory factor analysis.]
- Can a general relationship between measures of inspection performance and measures of ability, satisfaction, and working conditions be identified? [see *Procedure 6.8* – Canonical & set correlation.]
- Can groups of inspectors who show similar patterns of responses to the attitudinal items be discovered? [see *Procedure 6.6* – Cluster analysis.]
- Can patterns be detected in the relationships between the type of company an inspector works for and the level of stress they report in their workplace? [see *Procedure 6.7* – Multidimensional scaling and correspondence analysis.]

If you compare these questions to those posed at the beginning of *Chap. 5*, you will see that the complexity of the questions has dramatically increased which means that more sophisticated analytical techniques will be needed to satisfactorily address them. In this chapter, you will explore various procedures that can be employed to answer simple or complex relational or associational questions about data like those posed above. In addition, you will find a more detailed discussion of the fundamental concepts of correlation and partial and semi-partial correlation which will provide necessary foundation material for understanding the discussions to come later in the chapter.

Fundamental Concept III: Correlation and Association

Let's focus more closely on the relationship between inspection **accuracy** scores and perceptions of working conditions, **workcond**, in the QCI database. A scatterplot (*Procedure 5.2*) is useful for representing the scores of the

(continued)

Fundamental Concept III: Correlation and Association

112 inspectors on the two variables simultaneously (see Fig. 6.1, produced using SPSS). Note that the plot looks like a cloud of points which has a roughly elliptical shape (see superimposed dotted ellipse) and has an upward tilt generally suggesting that as **accuracy** scores increase **workcond** scores also tend to be higher (i.e. less stressful) and vice-versa. This increasing or positive trend in the relationship is by no means perfect; there are some low **accuracy** scores which occur in high quality work environments and some high **accuracy** scores which occur in low quality work environments. That is, there is some looseness in the relationship which is characterised by the fact that the ellipse we have drawn around the points has some width to it.

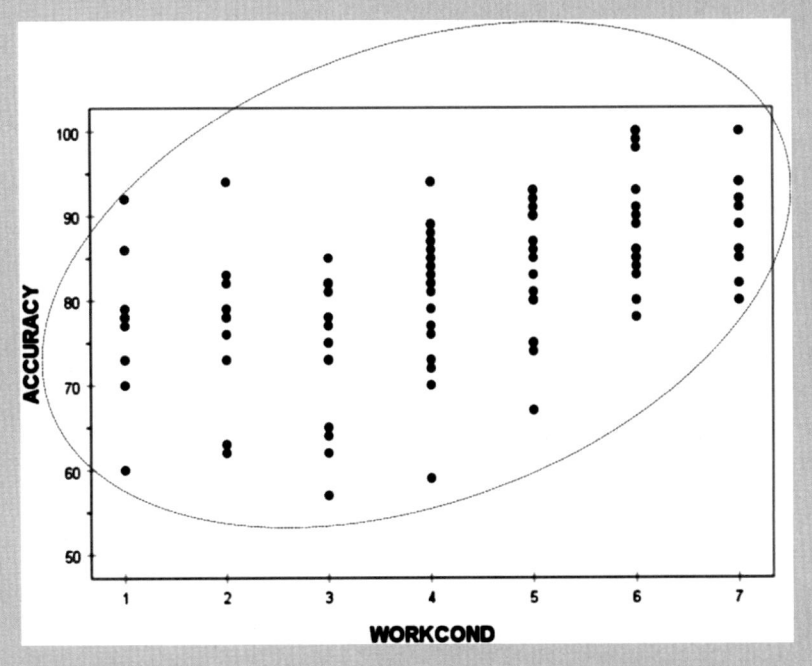

Fig. 6.1 Scatterplot of the relationship between accuracy and workcond QCI scores

A perfectly legitimate question to ask is: is there a single index by which we can express both the 'strength' and the 'direction' of the relationship between these two variables? The idea of strength concerns the magnitude of association (strong, moderate, weak) between the two variables. The idea of direction is concerned with the tendency of the two variables to move in the same direction (i.e. positive association: as one variable increases, the other tends also to increase) or opposite direction (i.e. negative association: as one variable increases, the other tends to decrease).

6 Correlational Statistics for Characterising Relationships

A *correlation coefficient* is a single number that summarises both the strength and direction aspects of variable relationship. It is an index that ranges from −1.0 (perfect negative correlation) through 0.0 (no correlation) to +1.0 (perfect positive correlation). The sign of the correlation coefficient indicates direction; its magnitude relative to either −1.0 or + 1.0 (or to 0.0) indicates how strong (or weak) the relationship is. Thus, a correlation that approaches ±1.0 is very strong; one that approaches 0.0 is very weak to nonexistent.

Below is a series of scatterplots sketching different relationships between generic measures X and Y, i.e., showing different degrees of correlation. The series gives a feeling for what different values of the correlation coefficient reflect about the relationship between two measures. Each scatterplot shows a scatter of data points surrounded by an ellipse. The tilt of each plot immediately indicates the direction, positive or negative, of the relationship. An upward tilt to the right indicates a positive (large score associated with large score; small score associated with small score) relationship. A downward tilt to the right indicates a positive (large score associated with small score; small score associated with large score) relationship. The width of the ellipse gives an immediate impression of the strength of the relationship. The relationship is weaker as the ellipse approaches a circular shape and is stronger as the ellipse approaches a straight line.

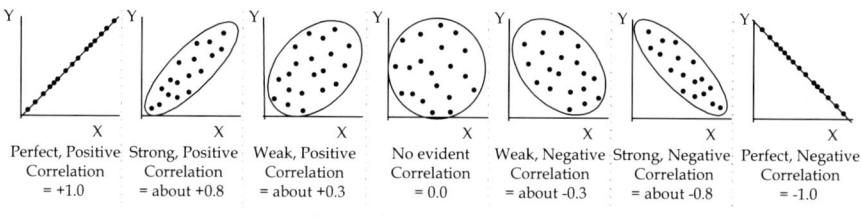

| Perfect, Positive Correlation = +1.0 | Strong, Positive Correlation = about +0.8 | Weak, Positive Correlation = about +0.3 | No evident Correlation = 0.0 | Weak, Negative Correlation = about −0.3 | Strong, Negative Correlation = about -0.8 | Perfect, Negative Correlation = −1.0 |

Degrees of correlation

The fact that the strength of correlation has a link to straight line geometry should not be ignored here. Correlation measures the extent of linear or straight-line relationship between two variables. This means that a perfect correlation is one where all the data points fall along a straight line. In such a case, the relationship between values of each variable is exactly one-to-one and, given knowledge of a person's score on one variable, we would be able to perfectly predict that person's score on the other variable—we just use the line. In the behavioural and social sciences, one almost never encounters a perfect positive (correlation = +1.0) or perfect negative (correlation = −1.0) relationship between any two measures. Human behaviour is just too variable, our measures of those behaviours are less than perfectly valid and reliable and behaviours, in general, are subject to dynamic influences from numerous factors, many of which cannot be controlled. This means that, realistically, relationships may approach the straight line ideal to some degree, but there will always be some inconsistency or error in the relationship that leads to the summary ellipse exhibiting some width.

Note that the absence of any relationship between two measures yields a circular plot (which is the fattest possible ellipse). Here, the correlation is exactly 0.0 indicating that there is no pattern to the linear relationship between the scores on the two measures. One measure has nothing in common with the other. This also means that having knowledge of a person's score on one of the measures will tell us nothing about their likely score on the other measure—one measure cannot predict the other.

Symbolically, and when publishing results from its computation, a correlation coefficient, formally known as the *Pearson correlation coefficient* (or *Pearson correlation*), is represented simply as *r*. However, as we will see in *Procedure 6.1*, there are different types of correlations for use with different scales of measurement, so *r* will often have a subscript to indicate the type of correlation it is meant to represent.

Interpretive Guidelines for a Correlation Coefficient

Many people who read published behavioural research are puzzled as to how to make sense of a reported correlation coefficient. Most reported correlations are accompanied by some indication of their statistical significance (whether the value of the correlation can be considered to be significantly different from a value of 0.0 in the population; see *Fundamental Concept VI* for an overview of the process for establishing statistical significance). However, it is equally important to get some feel for the strength of the relationship being reported. No hard and fast interpretive rules exist but it is possible to provide some useful conceptual guidelines. Figure 6.2 provides a 'correlometer' which can help in the interpretation of the value of a measure of correlation. It should be noted that this correlometer is simply a heuristic device for interpreting correlation coefficients and should not be considered as a substitute for more substantive interpretive work and possible significance testing.

For example, the correlation between **accuracy** and **workcond** in Fig. 6.1 is +.51—a moderate positive correlation.

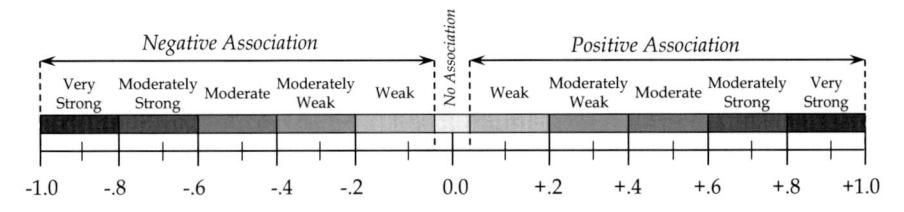

Fig. 6.2 The correlometer for conceptually interpreting a correlation coefficient

Data Patterns that Can Influence or Distort Correlations

There are several important caveats that accompany the use and interpretation of correlation coefficients, signalled by specific types of data patterns:

1. Correlation coefficients are sensitive to distributional differences between the two variables being related. If an undistorted assessment of their relationship is to be achieved, you must assume that the normal distribution describes the behaviour of each variable in the population.

> If either or both variables are strongly skewed, platykurtic or leptokurtic or one variable has widely differing variability in data points at differing values of the second variable, the relationship may produce a fan-shaped plot (denoted by the dotted lines) like that shown in Fig. 6.3—a condition called *heteroscedasticity* by statisticians. Note that this type of relationship emerges from two very positively skewed variables (note the boxplots for X and Y in the margins of the figure). In such cases, the value for the correlation coefficient may not give an accurate summary of the relationship. In Fig. 6.3, the correlation (*r*) between the two highly positively skewed X and Y variables is +.801 and when both X and Y have been log-transformed (see *normalising transformations* in Fundamental Concept II) to reduce their level of positive skewness, the correlation is +.759.

Fig. 6.3 non-normal data reflecting relationship heteroscedasticity

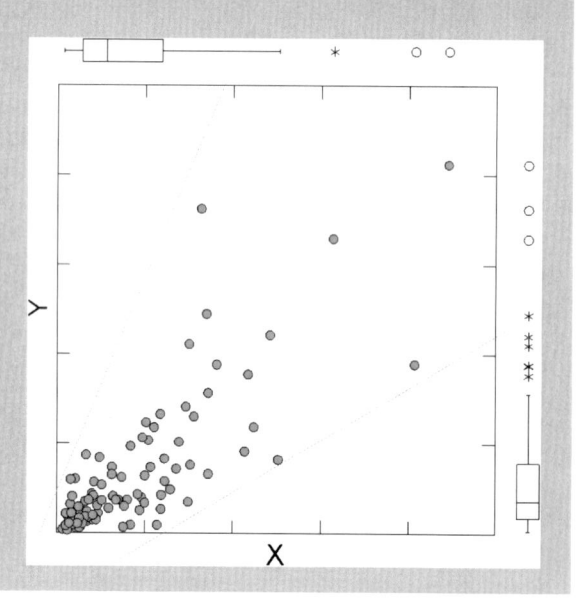

2. Correlation measures only linear (i.e. straight line) relationship. If an undistorted assessment of a correlational relationship is to be achieved, you must assume that a straight line provides the most appropriate description of the relationship.

If the scatterplot looks nonlinear, as illustrated in Fig. 6.4, then the correlation coefficient severely under-estimates the extent of relationship. The correlation for the data in this figure is, in fact, very nearly 0.0 (the linear relationship line (dashed) is nearly flat, reflecting a very weak negative correlation of −.180) which, if we hadn't looked at the plot, might lead us to conclude that the two variables were not related. However, there is obviously a systematic relationship between the variables denoted by the dark solid curve —it is just not linear in form; the (multiple) correlation between X and Y when the nonlinear relationship has been accounted for is +.691 (the method for finding this correlation is a version of multiple regression called 'power polynomials' – see *Procedure 6.4*). The type of nonlinear relationship (inverted U-shape) displayed in Fig. 6.4 is not atypical; it can be observed in the relationship between measures of stress and performance or between the cost of a product and consumer preference for that product.

Fig. 6.4 Curvilinear relationship between X and Y

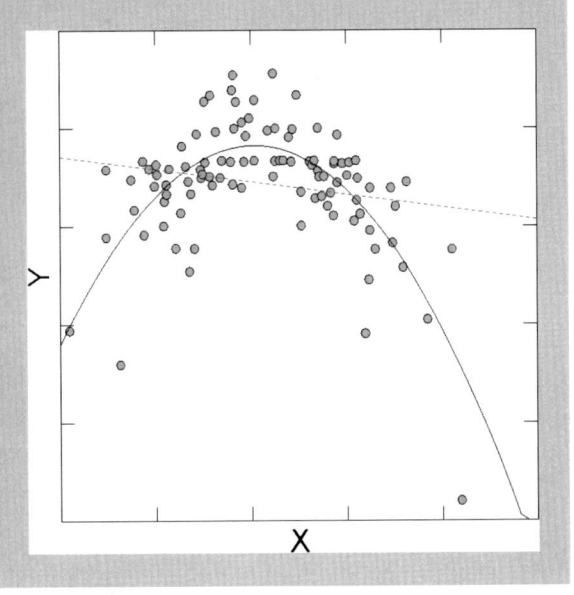

3. Correlation coefficients are sensitive to the presence of extreme data points or outliers. In a small sample, it may take only one or two extreme values in a variable to distort the correlation. For an undistorted depiction of correlation, any outliers would have to be dealt with first, either by arguing that they reflect data errors and should be deleted or by concluding that they are real and transforming

the variable to make the distribution more normal in form (see *normalising transformations* in *Fundamental Concept II*).

> In Fig. 6.5, a correlation computed using all of the data points is strong and positive ($r = +.832$ as suggested by the grey dotted ellipse), whereas if the three extreme data points were ignored, the correlation is very close to 0.0 ($r = -.098$ = no relationship as suggested by the black dashed circle). Clearly, the relationship with the three outliers included is not descriptive of the overall relationship; a clearer view would emerge if the outliers are ignored or the variables transformed.
>
> **Fig. 6.5** influence of outliers

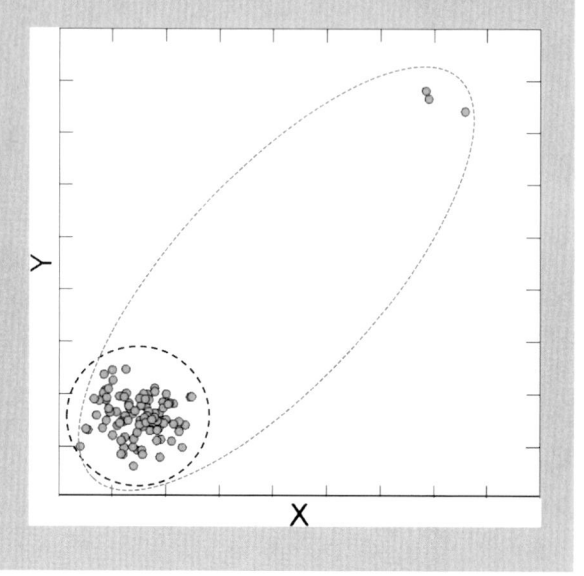

4. Correlation coefficients are also sensitive to the presence of subgroups within which the relationship between the two variables differs. If the relationship between the two variables differs for different subgroups in a sample, then computing a single correlation for the entire sample, ignoring the subgroups, will likely give a very misleading impression of how the two variables correlate.

> In Fig. 6.6, a correlation computed using all of the data points is very weak ($r = +.195$). However, for the subgroup represented by circles, the correlation is strongly positive ($r = +.812$) and for the subgroup represented by xs, the correlation is moderately strongly negative ($r = -.601$). Thus, you would tell very different stories about the relationship between the X and Y variables depending on whether the correlation was reported for the whole sample or reported separately for the two subgroups.

(continued)

Fig. 6.6 influence of the presence of subgroups

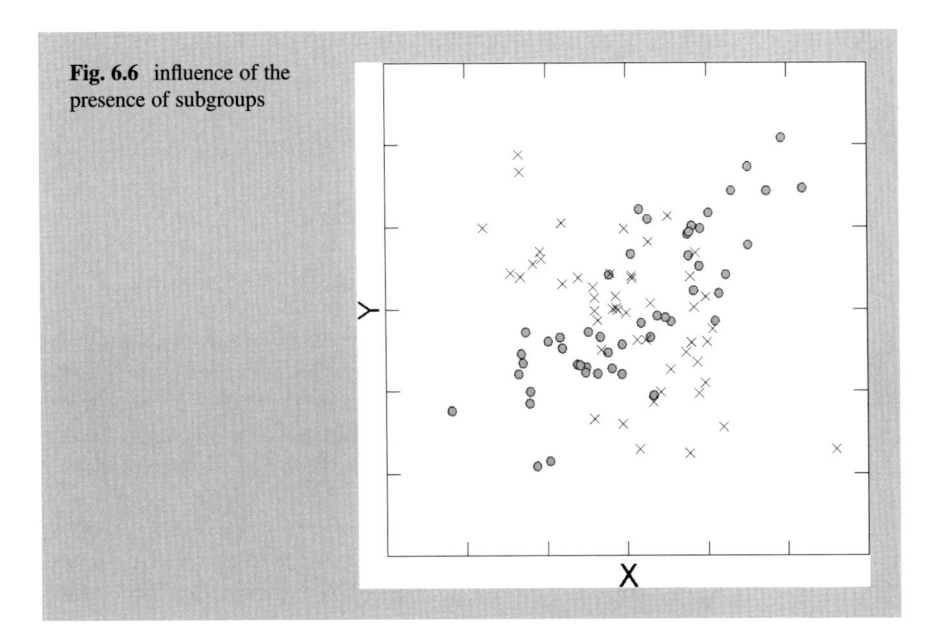

[scatter plot with X and Y axes showing two overlapping subgroups marked with × and ○ symbols]

Two other important issues should be mentioned here. Firstly, correlation is only meaningful if it is computed between variables measured on the same individuals in a sample. Secondly, a strong correlation coefficient in no way implies that one variable causes the other variable. All we know from computing a correlation is how two variables are related to each other. To conclude causality requires special research designs and statistical methods such as inferential group comparison methods like analysis of variance (see *Procedures 7.6* and *7.10*) or structural equation models (see *Procedure 8.7*).

Correlation is a quantity, having no units of measurement, which implicitly transforms each variable to z-scores (see *Procedure 5.7*) so that the two sets of scores are on a comparable footing. In fact, if you are interested and undertake the effort, a correlation coefficient can be easily calculated just from a listing of the z-scores for the two variables (similar to that shown in Table 5.5 in *Procedure 5.7*, but for the entire sample): merely multiply the respective z-scores for the two variables together for each person in the sample, add up all the resulting products, and divide by the number of people in the sample. The result would be the correlation coefficient. It is defined as the average cross-product of the z-scores between two variables.

Procedure 6.1: Assessing Correlation

Classification   Bivariate; correlational.
                         Note: Pearson's correlation is parametric when tested for significance; Spearman's rho is nonparametric.
Purpose   To provide an index which summarises the strength and direction of the relationship between scores or rankings for two variables measured on the same sample of individuals.
Measurement level   *Pearson correlation*: correlates two variables, each measured at the interval or ratio-level although, technically, all of the other forms of correlation to be discussed are special cases of Pearson Correlation for other scales of measurement.
                         Spearman's rho correlation: designed specifically to correlate two ordinal-scale variables (i.e. variables whose values are ranked in order from lowest to highest or vice-versa).
                         Point-biserial correlation: correlates a dichotomous nominal scale variable with an interval or ratio-scale variable.
                         Biserial correlation: correlates one artificially dichotomised ordinal-scale variable (with an assumed underlying normal distribution) with an interval- or ratio-scale variable.
                         Tetrachoric correlation: correlates two artificially dichotomised ordinal-scale variables (each with an assumed underlying normal distribution).

Pearson Correlation

This measure was introduced in *Fundamental Concept III* and is only briefly reviewed here. The *Pearson correlation coefficient* (also known by statisticians as the *Pearson product-moment correlation coefficient*) is a single statistical index which relates two variables measured at the interval or ratio-level of measurement. The Pearson correlation coefficient (*r*) ranges in magnitude from -1.0 to $+1.0$. Pearson correlations are often computed amongst pairs of a number of variables of interest. This results in the creation of a *correlation matrix* which provides a compact display of a multitude of relationships.

> Consider the QCI study context. It would be quite reasonable for Maree Lakota, in her quest to understand the workplace factors associated with the accuracy and speed of quality control inspector's inspection decisions, to

(continued)

explore the relationships between **accuracy, speed, workcond** and **jobsat**. SPSS computes these all of these pairs of correlations quite handily, yielding the correlation matrix shown in Table 6.1 (edited to remove unnecessary details).

Table 6.1 Pearson correlations amongst accuracy, **speed**, **jobsat** and **workcond**

Correlations

	accuracy	speed	jobsat	workcond
accuracy	1	.185	.235	.508
speed	**.185**	1	.002	.332
jobsat	**.235**	**.002**	1	.296
workcond	**.508**	**.332**	**.296**	1

A correlation matrix has a clear structure: variables anchor the rows and columns; the intersection of each row and column (called a 'cell') displays the correlation between the two variables and each variable correlates perfectly with itself giving 1's in the diagonal cells. Note that the correlations in bold below the diagonal cells are the same as the unbolded correlations above the diagonal cells. This pattern will always occur in a correlation matrix: it will be square with 1 s in the diagonal cells and mirror image correlations below and above the diagonal. This mirror image property is called *symmetry* and reflects the fact that correlating X with Y is the same as correlating Y with X. Normal practice in many journals would be to report and interpret only the bold correlations below the diagonal (what is known as the 'lower triangle' of the matrix).

The Pearson correlation between inspection **accuracy** and inspection decision **speed** is **+.185** which can be interpreted as a weak positive relationship indicating perhaps a slight tendency for more accurate inspections to take longer to make.

We should note here that since we know, from our earlier descriptive graphical displays, that the distribution of the **speed** variable is highly positively skewed, this correlation may in fact be providing a distorted indication of the degree of relationship present. The Pearson correlation gives the best and most unbiased depiction of the relationship between two variables only when both variables are at least approximately normally distributed. Because we make this assumption in order to have a clear interpretation, we say that the Pearson correlation is a 'parametric' measure.

The Pearson correlation between inspection **accuracy** and the quality of working conditions, **workcond**, is **+.508** which can be interpreted as a moderate positive relationship indicating a reasonable tendency for more accurate

(continued)

> inspections to be associated with working conditions which inspectors perceive to be more favourable (i.e. less stressful). The Pearson correlation between inspection **accuracy** and **jobsat** is **+.235** which can be interpreted as a moderately weak positive relationship indicating perhaps a slight tendency for more accurate inspections to be associated with inspectors who have a higher level of job satisfaction.
>
> **Speed** and **jobsat** are virtually uncorrelated (**+.002**) whereas **speed** and **workcond** are moderately weakly but positively correlated (**+.332**). **Jobsat** and **workcond** are moderately weakly but positively correlated (**+.296**).
>
> Note that each of our interpretations addresses only tendencies and association, not causality. Knowing, for example, that **accuracy** scores and **workcond** scores are correlated **+.508** cannot be taken to mean that it is the good working conditions that cause more accurate inspection decisions to be made, no matter how much we might wish to conclude this.

Statistical packages make it quite easy to compute Pearson correlations between large numbers of pairs of variables. This is advantageous from a data exploration point of view as shown above, but is disadvantageous in that, if one really gets carried away, computing large numbers of correlations produces too many relationships to clearly understand and can often lead to the identification of spurious relationships.

> To illustrate, the matrix in Table 6.1 correlated just 4 variables and produced 6 unique correlations in the lower triangle. If we correlated 10 variables, 45 unique correlations would be produced; correlating 25 variables, not unreasonable for many questionnaires, would produce 600 unique correlations.

It is for this reason that many of the multivariate correlational procedures such as factor analysis (see *Procedure 6.5*), canonical correlation (see *Procedure 6.8*), and multiple regression (see *Procedure 6.3*) have evolved as systematic and defensible ways to process the information arising from many correlations – virtually all of these procedures use a symmetric correlation matrix as the starting point for the analysis.

Spearman's Rho Correlation

Spearman's rho correlation coefficient is a version of the Pearson correlation coefficient designed specifically for use on rank ordered data. The data may be originally collected as ranks or may comprise interval or ratio-level data that have been converted to ranks for some reason. The most common reason for converting

Procedure 6.1: Assessing Correlation

interval or ratio data into ranks (i.e. ordinal data) is that, in their original form, either one or both variables are non-normal in distribution or have problems with extreme outliers.

As we have seen, such problems will distort the Pearson correlation (which is a parametric technique requiring the assumption of normality). Spearman's rho is nonparametric because we don't have to make such an assumption with ranks. If you compute both the Pearson and Spearman's rho correlations between a pair of variables and there is a substantial difference, this suggests that the Pearson correlation has been influenced by anomalies in the non-ranked data.

> A hospital administrator might obtain a supervising physician's rankings of 15 internists on diagnostic ability and effectiveness of treatments in order to see how the supervisor's ranking judgments in the two areas are correlated. Here the data are obtained in the form of ranks to start with.
>
> Alternatively, the hospital administrator may have access to actual scores for the 15 internists on tests of medical diagnostic ability and ability to design effective patient treatment regimes, but because the scores are highly negatively skewed (two internists scored appallingly poorly on both tests), she converts the scores to ranks before correlating them. The end result is the same, a correlation between two ranking systems, but the starting point and underlying logic for computing the rank-order correlation differs.

In both examples, all that Spearman's rho does is correlate the rankings. To the extent that the rankings tend to agree, Spearman's rho will approach +1.0; to the extent that the rankings tend to move in opposite directions, Spearman's rho will approach −1.0. If the rankings are unrelated, Spearman's rho will equal zero. Thus, the interpretation of Spearman's rho is exactly the same as for Pearson correlation—all that has changed is the measurement level of the data in the computation. Symbolically, and when publishing results from its computation, Spearman's rho correlation is often represented as r_s.

> Table 6.2 (edited to remove unnecessary and redundant detail) presents the relationships between **accuracy**, **speed**, **jobsat** and **workcond**, but recomputed by SPSS as Spearman rho correlations. Notice that most of the correlations do not change substantially in value, thus, data anomalies were generally not a problem.
>
> However, we noted in Table 6.1 that the Pearson correlation between inspector's **accuracy** and **speed** scores was **+.185** but with the caveat that **speed** was known to be a highly positively skewed variable. By converting the original **accuracy** and **speed** scores into ranks (so each variable would now have scores running from 1 (attached to the lowest value) to 112 (attached to

(continued)

the highest value)), we find a value for the Spearman's rho correlation in Table 6.2 of **+.243** which suggests a somewhat stronger, though still moderately weak, positive relationship where more highly ranked **accuracy** scores (= more accurate quality control decisions) tend to be associated with more highly ranked **speed** scores (= slower decision times). Note how, when interpreting, we must keep track of the original scaling direction of the scores for our interpretation to make sense.

Table 6.2 Spearman's rho correlations amongst **accuracy, speed, jobsat** and **workcond**

Correlations

		accuracy	speed	jobsat	workcond
Spearman's rho	accuracy	1.000			
	speed	.243	1.000		
	jobsat	.249	.007	1.000	
	workcond	.539	.398	.330	1.000

Point-Biserial Correlation

The *point-biserial correlation coefficient* is a version of the Pearson correlation coefficient that is useful for relating a dichotomous variable (having two categories such as gender or success/failure) to an interval or ratio-level variable (such as performance test scores or attitude ratings). Point-biserial correlation ranges between −1.0 and close to +1.0 (this version of the correlation can never actually attain a value of +1.0) and has the same general interpretation as the Pearson correlation. The only restriction is that the dichotomous variable should have one category of the variable scored as '0' (e.g. males) and the other category scored as '1' (e.g. females). Thus, there are two series of people observed on the interval or ratio variable: a series of '0' people and a series of '1' people. Symbolically, and when publishing results from its computation, the point-biserial correlation can be represented as r_{pb}.

> In the QCI database, Maree could compute the correlations between the **educlev** of quality control inspectors and the **accuracy, speed, jobsat** and **workcond** scores for those same inspectors. To do so, she simply recodes the High School Only inspectors (originally scored '1') to be scored '0' and the Tertiary Qualified inspectors (originally scored '2') to be scored '1' in **educlev**.

(continued)

Procedure 6.1: Assessing Correlation

Table 6.3, computed by SPSS (edited to remove unnecessary detail and redundant information), shows these correlations. The correlation between **educlev** and **accuracy**, for example, is **−.252** indicating a moderately weak tendency for higher **accuracy** scores to be associated with being in the High School Only **educlev** category (scored as '0'). The sign of the correlation points to which category tends to have higher scores—correlation will be positive if the category scored '1' has the generally higher scores and negative if the category scored '0' tends to have higher scores.

Educlev is also moderately weakly and negatively correlated with **speed** (**−.226**) and with **workcond** (**−.353**) where High School Only-educated inspectors slightly tend to make slower decisions and rate their working conditions as more favourable and is weakly positively correlated with **jobsat** (**+.105**), where Tertiary Qualified inspectors very slightly tend to have higher job satisfaction.

Table 6.3 Point-biserial correlations between **educlev** and **accuracy**, **speed**, **jobsat** and **workcond**

Correlations

	educlev
educlev	1
accuracy	-.252
speed	-.226
jobsat	.105
workcond	-.353

Biserial Correlation

The *biserial correlation coefficient* is less frequently used by behavioural researchers. However, it is useful when one wants to relate an interval or ratio-level variable to a variable that has been artificially dichotomised (reduced to a binary two-category variable) but which, theoretically, would produce a range of scores following the normal distribution. The goal is to estimate what the Pearson correlation between these two variables would be if the full range of scores on the dichotomised variable were available.

Suppose a teacher wants to relate amount of time spent studying chemistry to students' ability to balance chemical equations. While 'equation-balancing'

(continued)

> ability scores would probably follow a normal distribution if an extensive achievement test were used, the teacher may only have time to administer 1 test item—a single chemical equation. The students would either get the item right (scored '1') or wrong (scored '0'), but theoretically achievement would be distributed normally.
>
> Biserial correlation would be appropriate here to relate study time (in hours per week) and whether or not they passed the chemical equation item. Another instance where biserial correlation would be appropriate would be where an employer wants to relate the number of days of sick leave taken by assembly line workers to their overall work performance (which is theoretically and typically normally distributed among workers), but the supervisors of these workers only have enough time and resources to make a broad performance judgment of either 'satisfactory' or 'unsatisfactory'.

Biserial correlation ranges from -1.0 to $+1.0$ and is interpreted similarly to the point-biserial correlation. However, it can stray outside those limits if the normality assumption is unwarranted (as it would likely be with a measure like decision time or reading speed) or if the sample size is too small (e.g. less than 100). Symbolically, and when publishing results from its computation, the biserial correlation can be represented as r_{bis}.

Tetrachoric Correlation

The *tetrachoric correlation coefficient* is similar to biserial correlation except that both variables being correlated are artificially dichotomised but where each, theoretically, would produce a range of scores following the normal distribution. Again, the goal is to estimate what the Pearson correlation would be if the full range of scores on both variables were available.

> A human resource manager might want to estimate the relationship between employee performance and job satisfaction but may only have the time and/or the resources to utilise a very crude dichotomous measure of each variable: for work performance—adequate (scored '1') and inadequate (scored '0'); for job satisfaction—satisfied (scored '1') and dissatisfied (scored '0'). Theoretically, both work performance and job satisfaction are likely to be normally distributed in the population if one takes the time and effort to use measures appropriate to reflecting this distribution. However, if, due to constraints, only limited dichotomous measurement is possible, then tetrachoric correlation offers a way to estimate the relationship between the measures.

Procedure 6.1: Assessing Correlation

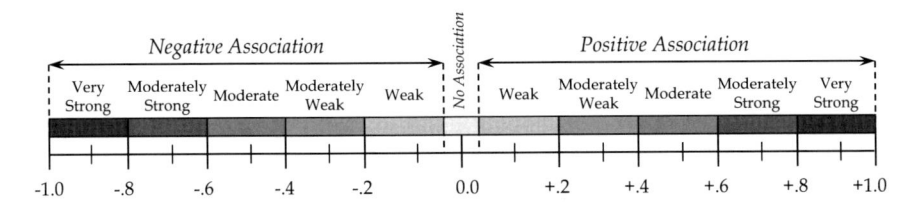

Fig. 6.7 The correlometer for conceptually interpreting a correlation coefficient

The tetrachoric correlation ranges from −1.0 to +1.0 with an interpretation similar to that for the biserial correlation. However, unless the sample size is quite large (in excess of 200), the tetrachoric correlation is a rather poor estimate of the Pearson correlation that would be achieved if both variables had been more precisely measured. Symbolically, and when publishing results from its computation, the tetrachoric correlation can be represented as r_{tet}.

Advantages

Pearson correlation provides a convenient statistical index that summarises the strength and direction of the relationship between two variables (as a reminder, Fig. 6.7 re-displays the 'correlometer' to facilitate interpretation of a correlation coefficient). It is used in many statistical methods as the basic index of association between variables. Correlation is a standardised statistic, without units of measurement, so that the variables being related can vary widely in measurement units. Efficient displays of a number of pairwise variable relationships can be achieved through the construction of a correlation matrix, something that SPSS, STATGRAPHICS, SYSTAT, R and NCSS are all quite effective in producing.

Spearman's rho is much simpler to compute than Pearson correlation (although computers make the difference essentially irrelevant). It can be used when data are unsuitable for parametric correlation analysis using Pearson correlation as well as in cases where the data have been obtained in the form of rankings.

The mixed-scale correlation indices, point-biserial correlation, biserial correlation, and tetrachoric correlation, are often useful as time-saving computations of relationships or as ways of handling data from highly constrained research contexts where full and precise measurement of one or more variables is difficult or impossible. Each index is specialised to a particular type of information. The point-biserial correlation gives an exact index of Pearson correlation between a dichotomous variable and an interval/ratio variable. Both biserial correlation and tetrachoric correlation have been used as alternative methods for approximating the Pearson Correlation in situations where complete variable information is unavailable.

Disadvantages

Correlation indices of any variety measure only linear (straight line) relationships between variables. If the relationship is nonlinear in some fashion (recall *Fundamental Concept III*), then the Pearson correlation coefficient may produce a severely distorted depiction of a relationship. Pearson correlation works best under conditions where both variables are approximately normally distributed. If either or both variables severely violate this condition, the result will be a distorted indication of relationship. For this reason, behavioural researchers are often admonished to examine scatterplots of their data in order to obtain visual evidence for any problems with the data satisfying the linearity or normality assumptions. Furthermore, any index of correlation provides only a descriptive index of association; it cannot tell you which variable causes which if such a conclusion is desired.

While Spearman's rho is a useful nonparametric correlation coefficient, its simplicity tends to lead to it being overused. If data are originally measured at the interval or ratio-level, some information is necessarily lost in converting the data to ranks (i.e. the meaningfulness of the intervals between scores is lost). If this conversion is unnecessary, then Spearman's rho will give a less accurate measure of correlation than would a Pearson correlation based on the original raw scores.

For the point-biserial correlation, the disadvantages are the same as for the Pearson correlation. However, for both the biserial and tetrachoric correlations, there are additional disadvantages. By artificially dichotomising variables, information is necessarily sacrificed, which permits only approximations to the actual Pearson correlation relationship. These approximations become extremely crude and inaccurate in smaller samples. The biserial correlation has the additional problem of possibly producing values outside the expected range of -1.0 to $+1.0$ for correlations. Such out-of-range values are uninterpretable except to say that either the sample was too small or the assumption of an underlying normal distribution for the dichotomised variable was unwarranted.

Where Is This Procedure Useful?

Many types of experimental, quasi-experimental and survey designs employ correlation measures to summarise relationships among variables (e.g. questions on a survey, performance measures in an experiment). While scatterplots are useful visualising tools, correlations provide more economical summaries of interrelationships when many variables are studied. Observational studies may also make use of correlational measures if the observed variables are at the appropriate level of measurement.

Any design where individuals have ranked a set of objects, items, events, and so on, can make good use of Spearman's rho correlation. Certain types of questionnaires may yield ranking data which are amenable to rank order correlation.

Spearman's rho might also be applied in designs where the resulting variables follow very non-normal distributional patterns.

Point-biserial correlation is generally useful in any design where naturally dichotomised variables must be related to interval or ratio-scale variables. Biserial or tetrachoric correlation should only be used in special cases. They should not be used as short-cut methods for estimating Pearson correlation if one has access to full information on a variable.

An unfortunate trend occasionally crops up in the behavioural science literature where a researcher artificially dichotomises a continuously measured variable using the median as the splitting point for two categories, low and high. This process is termed a 'median split' and is undertaken in the hopes of both simplifying and clarifying relationships. However, such a technique sacrifices valuable information and usually provides a more distorted rather than more accurate picture of a relationship. Generally speaking, median splits should be avoided; the full information on variables should be employed where available and where the primary assumptions are reasonably satisfied.

One issue that must be considered when generating a correlation matrix is how to handle missing observations. For most statistical packages, there are two typical choices: listwise deletion of cases and pairwise deletion of cases. *Listwise deletion* means that a case missing an observation for any variable in the analysis is dropped from the entire analysis. This effectively reduces the sample size for all correlations to the number of cases having no missing observations on any variable. Depending upon the extent of missing data in the sample, this reduction could be quite severe and could affect the quality of the correlation coefficient estimates. *Pairwise deletion* means that a case is dropped from consideration only if it is missing an observation on at least one of the two variables being correlated at any one time. This option means that each correlation is based on the largest possible sample size. The drawback is that pairwise deletion can create instabilities in multivariate analyses that use the correlation matrix as the starting point for the analysis (e.g. multiple regression *(Procedure 6.4)*, factor analysis *(Procedure 6.5)* and canonical correlation *(Procedure 6.8)*).

Software Procedures

Almost any software package worth its salt will be capable of computing either Pearson correlations or Spearman's rho correlations. Point-biserial correlations can be computed by any computer procedure designed to compute Pearson correlations as long as the dichotomous variable has been scored on a 0, 1 basis. However, most commercially available software packages lack the capacity for computing biserial or tetrachoric correlations although formulas for such computations are readily available (see, for example, Glass and Hopkins 1996, chapter 7). One exception is that SYSTAT can compute tetrachoric correlations.

Application	Procedures
SPSS	*Analyze → Correlate → Bivariate* ... and select the variables you want to analyse, then choose either 'Pearson' or 'Spearman'. Pressing the *'Options'* button allows you to choose is how to handle missing data: 'listwise deletion' or 'pairwise deletion'.
NCSS	*Analysis → Correlation → Correlation Matrix* and select the variables you want to analyse, and then select the correlation type as either 'Pearson Product Moment' or 'Spearman-Rank' correlation type (or both). You can also select a 'Missing Data Removal' option: 'Row Wise' (equivalent to SPSS listwise deletion) or 'Pair Wise'.
SYSTAT	*Analyze → Correlations → Simple* ... and select the variables you want to analyse, and then select either 'Continuous data' and 'Pearson' or 'Rank order data' and 'Spearman' or 'Binary data' and 'Tetra'. You also select 'Listwise' deletion or 'Pairwise' deletion of missing data.
STATGRAPHICS	*Relate → Multiple Factors → Multiple Variable Analysis (Correlations)...* and select the variables you want to analyse, then hit *'OK'* and in the next dialog box that pops up, choose either 'Complete cases' (= listwise deletion of missing data) or 'All data (pairwise deletion); then under TABLES, select 'Correlations' and/or 'Rank Correlations'. By default, a scatterplot matrix (*Procedure 5.3*) will also be produced.
R Commander	*Statistics → Summaries → Correlation matrix...* and select the variables you want to analyse *and choose 'Pearson product moment' or 'Spearman rank-order'*. By default, **R** Commander only uses listwise deletion of missing data.

Procedure 6.2: Assessing Association in Contingency Tables

Classification   Bivariate; correlational; nonparametric.
Purpose   To provide correlational measures for relating frequency data from categorical variables summarised in a crosstabulated contingency table.
Measurement level   The phi correlation, Cramer's V statistic, and proportional reduction in error (PRE) measures are generally appropriate for nominal level variables (categorical measurements), but are often used for ordinal-level variables having a small number of discrete categories. Some measures such as the Gamma statistic, Kendall's Tau statistic, and Somer's D statistic are specifically designed for ordinal-level data organised into a contingency table.

Procedure 6.2: Assessing Association in Contingency Tables

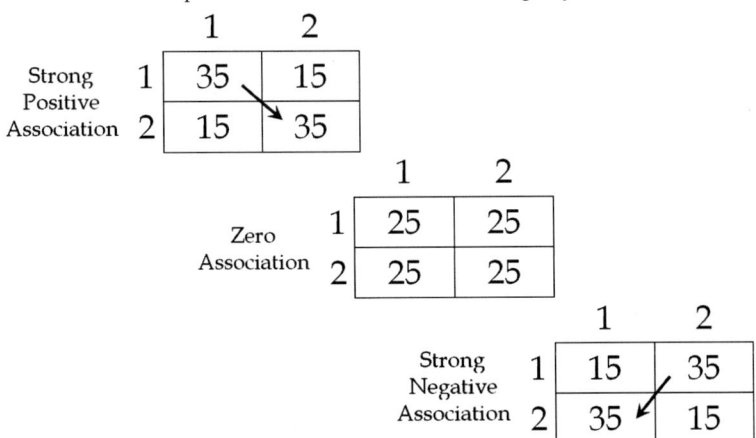

Table 6.4 Possible patterns of association in 2 × 2 contingency tables

		1	2
Strong Positive Association	1	35	15
	2	15	35

		1	2
Zero Association	1	25	25
	2	25	25

		1	2
Strong Negative Association	1	15	35
	2	35	15

Phi Coefficient and Cramer's V

The *phi coefficient* is a version of Pearson correlation that is appropriate for relating dichotomous variables summarised in a 2 rows × 2 columns crosstabulation table (see *Procedure 5.1*). When relationships between categorical variables are being explored, such a crosstabulation is referred to as a 'contingency table'. The phi correlation for a 2 × 2 contingency table ranges between -1.0 (perfect negative association) and $+1.0$ (perfect positive association) with 0.0 reflecting no association (i.e. the variables are independent).

Table 6.4 shows, using three hypothetical mini-tables, what a positive, zero, and a negative association might look like in terms of frequency counts in a 2 × 2 table. What creates the correlation in such tables is a higher frequency of co-occurrence of certain pairs of categories contrasted with a lower frequency of co-occurrence in the others. Table 6.4 makes it clear that what goes on in the 1–1 and 2–2 or the 1–2 and 2–1 diagonal cells of the table (those cells linked by an arrow) shapes both the strength and direction of the relationship.

> Suppose Maree wants to see if the **gender** of inspectors in the QCI database is related to the educational level (**educlev**) attained by inspectors. The resulting 2 × 2 contingency table of frequencies, produced by SPSS, is shown in Table 6.5. The pattern in the **gender** by **educlev** table looks most like the zero-association pattern in Table 6.4 and this visual impression is supported by the value of the phi coefficient which is **+.045** indicating no evident

(continued)

association between the gender of quality control inspectors and their educational level.

The additional rows in the contingency table help to display any trends in the relationship by reporting percentages for categories within a row or column variable. For example, of all Tertiary Qualified inspectors in the sample, exactly 50% were male and 50% were female; nearly the same trend was observed for High School Only educated inspectors (54.5% versus 45.5%). This supports a clear indication of no relationship—genders are not differently distributed across levels of education. Maree would arrive at the same conclusion if she focused on the distribution of education levels within the gender categories.

Table 6.5 2 × 2 contingency table relating **educlev** with **gender**

			gender		
			1 Male	2 Female	Total
educlev	1 High School Only	Count	30	25	55
		% within educlev	54.5%	45.5%	100.0%
		% within gender	53.6%	49.0%	51.4%
	2 Tertiary Qualified	Count	26	26	52
		% within educlev	50.0%	50.0%	100.0%
		% within gender	46.4%	51.0%	48.6%
Total		Count	56	51	107
		% within educlev	52.3%	47.7%	100.0%
		% within gender	100.0%	100.0%	100.0%

Cramer's V statistic is a generalisation of the phi coefficient to contingency tables larger than 2 × 2. Cramer's V cannot be negative; it ranges from 0 to +1.0 with 0 indicating no association and +1.0 indicating perfect association. The magnitude of Cramer's V is somewhat more difficult to interpret than the Pearson correlation because it is not really a correlation coefficient at all. At best, the interpretation must be based on its closeness to 0.0 or 1.0 (so the right side of the correlometer in Fig. 6.7 might be of some assistance here). However, Cramer's V gives no indication of directionality of the relationship, which should not be surprising given that nominal scale variables are being related.

Maree could examine the Cramer's V statistic for a contingency table that relates the type of company in which quality control inspectors work with their education level. Table 6.6, produced using SPSS, presents the contingency

(continued)

table relating **company** and **educlev**, which yields a value for Cramer's *V* of .415 suggesting a moderate level of association. Here the row and column percentages can help Maree tease out where this relationship comes from. If she focuses within companies, the percentages tell her that the majority of inspectors are Tertiary Qualified in Large Electrical Appliance and Automobile manufacturers (66.7% and 81%, respectively), whereas the reverse is true for the other three types of companies.

Table 6.6 5 × 2 contingency table relating **company** with **educlev**

			company * educlev Crosstabulation			
				educlev		
				1 High School Only	2 Tertiary Qualified	Total
company	1 PC Manufacturer	Count		16	7	23
		% within company		69.6%	30.4%	100.0%
		% within educlev		28.6%	13.2%	21.1%
	2 Large Electrical Appliance Manufacturer	Count		7	14	21
		% within company		33.3%	66.7%	100.0%
		% within educlev		12.5%	26.4%	19.3%
	3 Small Electrical Appliance Manufacturer	Count		15	6	21
		% within company		71.4%	28.6%	100.0%
		% within educlev		26.8%	11.3%	19.3%
	4 Large Business Computer Manufacturer	Count		14	9	23
		% within company		60.9%	39.1%	100.0%
		% within educlev		25.0%	17.0%	21.1%
	5 Automobile Manufacturer	Count		4	17	21
		% within company		19.0%	81.0%	100.0%
		% within educlev		7.1%	32.1%	19.3%
Total		Count		56	53	109
		% within company		51.4%	48.6%	100.0%
		% within educlev		100.0%	100.0%	100.0%

Proportional Reduction in Error (PRE) Statistics

The difficulties associated with interpreting a statistic like Cramer's *V* have led many researchers to prefer using ***Proportional Reduction in Error (PRE) statistics*** to summarise the relationship between nominal scale variables having more than two categories. There are a variety of PRE measures (the best known of which are the Goodman & Kruskal lambda statistics) which look at association from a different perspective than Pearson correlation.

PRE statistics attempt to assess the association between two nominal scale variables by reflecting the percentage improvement in our ability to predict the category or value of one variable given that we know the category or value of the second variable involved in the two-way contingency table. The value for a PRE statistic depends on which variable is chosen as the known (or 'predictor') variable. For this reason, there are both symmetric and asymmetric forms of a PRE statistic. An asymmetric PRE statistic requires you to name one of the two contingency table variables as the dependent variable to be predicted. A symmetric PRE statistic docs not require this and will produce an intermediate value between the two asymmetric versions (each one predicting one of the two contingency table variables). Most PRE statistics range in magnitude from 0 (no improvement in prediction – maximal error) to 1.0 (complete improvement in prediction – nil error).

> In the 5 × 2 contingency table shown in Table 6.6, suppose Maree wanted to know how well she could predict the type of company an inspector worked for (the dependent variable of interest) given that she knew the inspector's educational level. The PRE value (using the Goodman & Kruskal lambda statistic, which SPSS can calculate on request) would be **.116** or **11.6%** which means that knowing the attained education level of a quality control inspector allows would allow her to predict the type of company they worked for with 11.6% less error than she would make if the inspector's educational level was unknown.
>
> She could easily ask the question in reverse, whereupon she would note that knowing the type of company an inspector worked for allows her to predict the inspector's educational level (now the dependent variable of interest) with **37.7%** less error than she would make if the inspector's company type was unknown. The percentages within company in Table 6.6 make it clear why the prediction error is reduced so much: if we knew an inspector came from a Large Electrical Appliance manufacturer or especially from an Automobile manufacturer, her chances of correctly predicting an inspector's educational level would be greatly in favour of being Tertiary Qualified and vice-versa for the remaining three types of companies.
>
> If she did not wish to explicitly specify a dependent variable, the symmetric form of the lambda statistic could be reported. For Table 6.6, symmetric lambda is **.216** meaning that knowing where an inspector is with respect to either variable allows Maree to reduce her error in predicting where that inspector is with respect to the other variable by **21.6%**.

Gamma Statistic, Kendall's Tau Statistic, and Somer's D Statistic

The *gamma* statistic, *Kendall's Tau* statistic, and *Somer's D* statistic are measures of association that have been explicitly designed for use in contingency tables involving variables whose categories can be considered to be ordinal, rather than just nominal in measurement.

> The variable **educlev** in the QCI database is technically an ordinal-level categorical variable because being Tertiary Qualified means that a person has more education (but we can't say precisely how much more) than a person with a High School Only education.
>
> Other examples of ordinal-level categorical variables might be any variable which categorises people in terms of being low, medium, or high on some characteristic. For instance, we could solve the problem of the skewness of the **speed** variable, not by doing a log-transform or by rank-ordering times (as Spearman's rho would do,) but by transforming it to an ordinal-scale categorical variable where, say, the fastest 20% of the sample are categorised as 1, the next fastest 20% as 2, and so on down to the slowest 20% categorised as 5. In this transformed variable, the order of the category numbers still reflect speed of decision making; only in the form of a simpler ordinal scale (higher numbers signal longer decision times, but not how much longer - which is the information sacrificed by converting **speed** to a lower level of measurement).

The gamma statistic ranges between -1.0 and $+1.0$ although it is, strictly speaking, not a correlation measure. Gamma measures the extent to which there is a tendency for people falling in the higher order categories of one variable to also be the people who fall in the higher order categories on the second variable. Gamma will approach $+1.0$ if this tendency is very strong. It will approach -1.0 if this tendency is actually reversed where people in higher order categories on one variable tend to be those in the lower order categories on the other variable. Gamma will approach 0.0 if there is no associational tendency at all.

Both the Kendall tau statistic (there are two versions of this statistic referred to as *tau-b* for square tables and *tau-c* for more general tables) and the Somer's *D* statistic are more restrictive forms of the gamma statistic. Somer's D behaves a bit like the Goodman-Kruskal lambda PRE measure in that there are asymmetric forms, requiring the researcher to specify one of the variables as the predictor and the other as the dependent variable, and a symmetric form that does not have this requirement. Of these three measures, gamma appears to be the most popular to report for ordinal-level contingency tables.

> If Maree were to relate the **educlev** variable to an ordinally-transformed **speed** variable (called **cat_speed**) described earlier, the contingency table would appear as shown in Table 6.7 and the gamma statistic would equal **−.427**. This indicates that increasing education level is moderately associated with faster decision speeds—the pattern of frequency counts makes the nature of the relationship fairly clear. The relationship is negative because our category

(continued)

systems for both variables represent ordinal increases in amount of the variable being displayed (either education level or time taken to make a decision).

The value for Kendall's *tau-c* statistic (the table is not square) is $-.353$, showing the same sort of trend as gamma. Somer's *D* for predicting **educlev** is $-.220$ and for predicting **cat_speed** is $-.353$. The value for symmetric Somer's *D* is $-.271$. Somer's *D* reflects the overall negative relationship as well, but additionally shows that the relationship is stronger for predicting **cat_speed** from knowledge of **educlev** than vice-versa.

Table 6.7 5 × 2 contingency table relating **cat_speed** (ordinal category version of speed) with **educlev**

cat_speed * educlev Crosstabulation

Count

		educlev		Total
		1 High School Only	2 Tertiary Qualified	
cat_speed	1 Very fast	5	15	20
	2 Moderately fast	8	14	22
	3 Average	13	10	23
	4 Moderately slow	17	4	21
	5 Very slow	13	9	22
Total		56	52	108

Advantages

The statistics described above provide useful summaries of the degree of association between two nominal or ordinal scale variables in a contingency table. The phi coefficient can be interpreted in exactly the same way as the Pearson correlation for 2 × 2 contingency tables. Cramer's *V* is a more general measure applicable to contingency tables of any dimensionality but cannot be interpreted in quite the same way as a correlation coefficient. PRE statistics allow the assessment of the predictive utility of each of the variables and have a logical interpretation in terms of percentage of error reduction associated with that prediction. Finally, the ordinal association statistics, gamma, Kendall's tau, and Somer's *D*, permit the assessment of association in contingency tables involving ordinal-level variables. Most reputable statistical packages can produce each of these measures upon request; it is left up to the user to decide which measures are most appropriate to request given the nature of the data. There are actually many more measures of association available for assessing relationships between categorical variables (especially dichotomous or 'binary' variables) than have been described here (see Everitt 1977; Liebetrau 1983 or Reynolds 1984, for example) – we have just focused on some of the main ones in use.

Disadvantages

Unfortunately, the phi coefficient and Cramer's V are sensitive to oddities in the marginal (row and column totals) distributions of the contingency table. This means that it may be impossible for either measure to reflect a perfect relationship of +1.0. The size of the table can also influence the magnitude of Cramer's V. Cramer's V has no easy directional interpretation in the sense that the phi coefficient does. PRE measures are generally less sensitive to these problems. The ordinal association measures also have some problems reflecting a perfect relationship if there are great disparities in the marginal totals of the contingency table. They are also more difficult statistics to understand and interpret. The sheer multitude of possible association measures for contingency tables can be a bit bewildering when it comes to deciding which measures are most appropriate. Sometimes the choice, particularly with respect to the ordinal association measures, needs to be based upon the type of relationship one expects to find in the data.

Where Is This Procedure Useful?

Contingency table measures of association are most often used in survey, interview, and observational studies where categorical variable relationships are to be assessed. They can prove particularly useful in understanding relationships among various demographic characteristics of a sample because these variables are often obtained in categorical form. Many in fact are ordinal (e.g. income, education, age). This is especially true where the behavioural researcher is concerned about the sensitivity of or participant reactivity to obtaining precise indications of certain types of information (such as age and income). In such cases, the researcher may seek only categorical indications rather than exact numbers from participants, which puts them more at ease with divulging such information. Relationships would then be explored through contingency tables and relationship strength evaluated using one or more of the measures described here. In particularly long questionnaires, one may resort to the use of many categorical variables simply because they demand less mental effort, on the part of the participant, in making responses.

We also showed that it may be possible to address problems with non-normality in an interval or ratio scale variable by transforming it into an ordinal scale variable and assessing associations at that level. Finally, there is a school of thought which holds that attitudinal measures such as Likert-type (e.g. simple 1 to 5) scales should really be treated as ordinal categorical measures rather than as interval-level measures because one can never be sure that the differences between successive scale points are equal all along the scale (which is an assumption that interval scales require one to make). If you are aligned with this school of thought, the measures discussed here are useful for the analysis of associations amongst such variables.

Software Procedures

Application	Procedures
SPSS	*Analyze → Descriptive Statistics → Crosstabs* ... and choose the variables to define the rows and columns of the table, then click on the '*Statistics*' button and choose your desired measures (all measures described here are available).
NCSS	*Analysis → Descriptive Statistics → Contingency Tables (Cross Tabulation)* and select the desired row and column variables, click the '*Reports*' tab and in the dialog box option labelled 'Chi-Sqr Stats', choose 'All' to obtain all of the measures described here (Somer's *D* is not available).
SYSTAT	*Analyze → Tables → Two-Way* ... and select the desired row and column variables, then select the '*Measures*' tab and choose the measures of association you want to see (all of the measures described here are available; just note that SYSTAT refers to 'Kendall's tau-c' as 'Stuart's tau-c').
STATGRAPHICS	*Describe → Categorical Data → Crosstabulation* ... and select the desired row and column variables, then when the 'Tables and Graphs' dialog box opens up, make sure 'Summary Statistics' is ticked (all the measures described here are produced).
R Commander	*R Commander* cannot produce measures of association for a contingency table. However, in the *R* program itself, if you have the *vcd* package (*Visualizing Categorical Data*), you can compute some of these measures (including the phi coefficient and Cramer's V) for a contingency table.

Procedure 6.3: Simple Linear Regression & Prediction

Classification   Bivariate; correlational; parametric.
Purpose   To summarise the linear relationship between a single independent variable (predictor) and a single dependent variable (criterion) by fitting a predictive line through the data points.
Measurement level   Theoretically, any level of measurement may be used for the predictor variable although an interval or ratio-level measure is often used. The criterion should be measured at the interval or ratio-level.

Procedure 6.3: Simple Linear Regression & Prediction

Simple linear regression is designed to provide a summary of the linear relationship between an independent variable or 'predictor' and a dependent variable or 'criterion'. Furthermore, linear regression produces a mathematical equation which indicates precisely how information from the predictor can be used to compute predictions about the criterion. In this regard, linear regression is related to but goes one step beyond Pearson correlation.

Recall that Pearson correlation (*Procedure 6.1*) summarised the degree to which two variables tended to vary in the same direction, the opposite direction, or independently of each other. Linear regression is based on the Pearson correlation but goes further in that it statistically fits a single straight line through the scatter of data points. This *regression line* is positioned in such a way that it provides the best summary of the linear relationship between the two variables. We use the equation mentioned above to help us draw the line through the scatter of points.

> Maree wants to examine the linear predictive relationship between quality control inspectors' perceptions of working conditions, **workcond** (the predictor), and their **accuracy** in making inspection decisions (the criterion). Figure 6.8 shows a scatterplot of this situation for the 112 inspectors in the QCI database (convention is the criterion anchors the Y-axis). The 'best' fitting regression line is also shown.
>
> **Fig. 6.8** Scatterplot of the relationship between **accuracy** and **workcond** showing the fitted regression line (produced using SYSTAT)
>
> Since the task of regression analysis is to fit the 'best' line through the data points, it is useful to understand what 'best' means. If we find the vertical distance to each data point from the regression line, square each one, and add all these squared distances up, the resulting number will be smallest for the regression line than for any other line we could draw through the points. For this reason, linear regression is often referred to as a 'least squares' method.

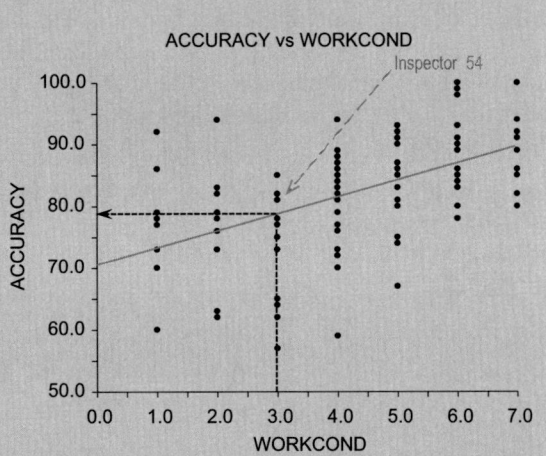

(continued)

> The Pearson correlation coefficient is used to provide a measure of how strongly related **workcond** and **accuracy** are. The value of *r* for the data in Fig. 6.8 is **+.508**, which indicates a moderate positive relationship. The interpretation is relatively straightforward: more highly rated working conditions (i.e. more stress-free) are associated with, and therefore predictive of, higher quality control inspection **accuracy**. However, this relationship is certainly not perfect since there is a wide scattering of data points around the plotted line.

Frequently in regression analysis, the value of the Pearson correlation is squared to produce an index of 'goodness-of-fit' called, not surprisingly, *r-squared*. The measure 'r-squared' has a special interpretation in regression analysis and turns out to be a very important index for many statistical procedures.

Figure 6.9 shows, using Venn diagrams, how *r*-squared (or r^2) is interpreted in simple linear regression analysis. Each circle in the diagram symbolises the total variance of the scores on one variable; think of each circle as having an area of 1.0 (or 100%)—a so-called unit circle. Each circle thus represents 100% of the variance of a variable, encompassing all the possible reasons why scores would differ from one case or participant to another (see *Procedure 5.5*). If the variance circles for two variables overlap, we say that the two variables share some of their variance in common, i.e., they are correlated (see the crosshatched overlap area of Fig. 6.9). That is, knowing something about the behaviour of one variable gives us some insight into at least one of the possible reasons why the second variable behaves in the way it does.

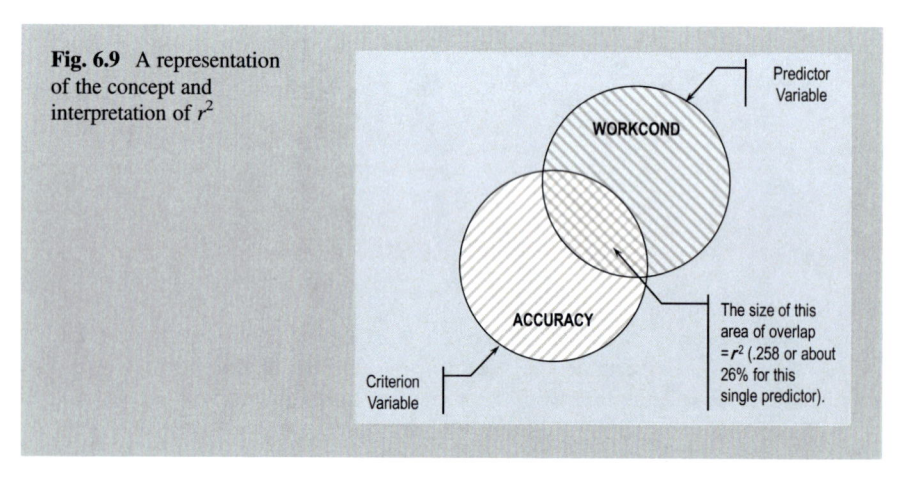

Fig. 6.9 A representation of the concept and interpretation of r^2

Pearson correlation merely tells us that two variables are related. In simple linear regression, we go further and name one of the variables as a predictor for the values of the other variable. The degree of overlap in the variance of two variables can therefore be indexed by the proportion or percentage of the area in the dependent (or criterion) variable's circle that is overlapped by the predictor's circle.

Procedure 6.3: Simple Linear Regression & Prediction

It turns out that the size of this overlapping area is exactly equal to the square of the Pearson correlation, hence r^2. As a measure of 'goodness-of-fit' for a regression equation, r^2 ranges from 0.0 (or 0% variance explained in the criterion) to 1.0 (or 100% variance explained in the criterion). The closer to 1.0, the stronger and more accurate the prediction system will be (and the more condensed the cloud of data points will be around the line). The best of all possible prediction systems will occur if all the data points fall exactly along the regression line.

> For the data plotted in Fig. 6.8, we observed that the value for r was **+.508** which means the value for r^2 was **.258**. We can interpret this value for r^2 by saying "that 25.8% of the variability in inspectors' decision **accuracy** scores can be explained by knowing how they perceived their working conditions" (as reflected in their score on the **workcond** variable).

The equation produced using linear regression analysis tells us how the best fitting line should be drawn. The equation includes a weight (called a 'regression coefficient', 'regression weight' or 'slope') which multiplies the predictor variable and a constant (which is added to that product), which tells us where the line should 'intercept' or cross the Y-axis (i.e. criterion axis) on the scatterplot. The regression weight tells us how much change in the criterion we predict occurs for a one-unit increment in the predictor.

> For the data in Fig. 6.8, the regression weight is **+2.714** which can be interpreted as saying that if we increase **workcond** by one unit value anywhere along its scale (say from 5 to 6), we would predict an increase of 2.714% in an inspector's **accuracy** score (recalling that **accuracy** is measured in percentage score units). The value of the intercept point is **70.31**, which can be seen in Fig. 6.8 as the point at which the regression line crosses the vertical or Y-axis. This tells us that if an inspector's **workcond** score is 0, our best prediction for their inspection **accuracy** would be 70.31%.

The regression equation, and the regression line it describes, provides us with an additional capacity that Pearson correlation could not. For any value of the predictor, we can use the regression line (or equivalently, the regression equation) to predict what the value of the criterion should be. We could thus generate a predicted **accuracy** score for every inspector in the sample.

> One of the data points in Fig. 6.8 (for inspector 54 in the QCI database – indicated by the dashed grey arrow) had a **workcond** score of 3. Using the linear regression line, we could predict that a quality control inspector having this specific score for **workcond** should have an inspection **accuracy** score of

(continued)

> 78.45%. Thus, if we find a score of 3 on the horizontal axis in Fig. 6.8 and proceed vertically up from that point, we will intersect the regression line. Then move horizontally left back to the criterion axis and where you intersect that axis, you can read off the predicted **accuracy** score of 78.45% (this prediction process is depicted by the dotted arrow; obviously using the prediction equation will be more accurate than using visual inspection of the line on a graph).
>
> Note, however, that inspector 54 had an observed **accuracy** score of 81%, so our prediction was close but too low. This prediction error can be quantified by subtracting the predicted value of 78.45% from the observed criterion value of 81% to produce a 'residual' or prediction error of +2.55%. In the QCI database, 11 different inspectors actually reported a **workcond** score of 3 but only one of these inspectors had an **accuracy** score anywhere near the value of 78.45% predicted by the regression line (inspector 92 - the data point nearest the line at the value of 3 for **workcond** - with an **accuracy** score of 78%).

For most observations, the predicted criterion score for any point along the predictor variable's scale will be somewhat different than the actual observed score. To the extent that the Pearson correlation between the two variables is less than perfect (not equal to 1.0 or -1.0), our predictions will always have some residual amount of error associated with them. The regression line is the best one we can use for the data at hand, so the residual errors made are as small as they can be, but nonetheless some error still exists. If a straight line is truly the best way to summarise the data and the data can be assumed to be normally distributed, then errors of prediction for different predictor values will overshoot, undershoot, or fall right on target in a random or unpatterned fashion.

This leads to an interesting way to think about a linear regression model. We observe actual scores on a criterion through collecting data on a sample, giving us 'observed' values. We predict values for the criterion using linear regression analysis information from a predictor of interest, giving us 'predicted' values. And our predictions often miss the mark leaving 'residual' error values unaccounted for. Statistically, what linear regression accomplishes is to construct a model of the world such that each observed value is understood as being composed of a predicted value (using predictor information) plus a residual. Therefore, what we observe is broken into two parts: one part (predicted value) we can explain by knowing something about a relevant predictor and the other part (residual) we can't explain.

> In terms of the previous example, we could summarise the predictive regression equation as follows:
>
> (continued)

> *Predicted **accuracy** = 70.31% + (2.714% × **workcond**)*
>
> The full model for the observations could be conceptually depicted as:
>
> Observed **accuracy** = Predicted **accuracy** + residual.
>
> In Fig. 6.8, we learned that we could only explain a little over 25% of the variability in **accuracy** scores using information on **workcond**. This leaves nearly 75% of the variability unexplained (meaning it is locked up in the residuals), which is why our regression line will not make very good predictions on an inspector by inspector basis.

This shows that a linear regression model can only ever be as good as the predictor you include in the model allows. If you leave out or don't have access to an important predictor, then your explanatory and predictive power will be reduced. For this reason, linear regression models often need to incorporate multiple predictors in order to do good predictive and explanatory work. *Procedure 6.4* explores further this concept in the form of multiple regression analysis.

Advantages

Linear regression provides a useful summary of the linear relationship that exists between two variables; particularly if there is interest in predicting one variable from the other (note that causality still cannot be concluded here). In addition to correlational information, linear regression locates the best-fitting line through the data points, which can then be used for predictive purposes. The line is drawn using a regression equation or model which describes (1) the weight that the predictor should be given in predicting the value of the criterion and (2) what to predict if the predictor value is zero. Variance in the criterion variable explained by the predictor is indexed by r^2. It provides a way of evaluating the 'goodness-of-fit' of the regression line and consequently indicates how good (in terms of predictive accuracy) the overall prediction model is.

Linear regression is a general technique in that it is related to the t-test and ANOVA (see *Procedures 7.2* and *7.6*) if an ordinal or nominal scale predictor is employed. In fact, the general logic of regression analysis forms the statistical foundation for many of the statistical procedures reviewed in this book. It does this through a representation called the *general linear model* (see Cohen et al. 2003 and Miles and Shevlin 2001 for two of the better introductions to the more general application of correlation/regression analysis; we will also explore this concept in *Fundamental Concept VI*).

Disadvantages

If predictor-criterion relationship is not totally linear (e.g. the scatterplot shows perhaps a U-, inverted U- (recall Fig. 6.4) or S- shaped relationship), then linear regression will not provide an adequate predictive model for the relationship between the two variables. To analyse such data, multiple regression analysis (see *Procedure 6.4*) may be needed or the data may need to be transformed to make the relationship more linear in form. [Note that close inspection of the scatterplot in Fig. 6.8 might suggest that a very slight U-shaped relationship might be present— **accuracy** scores tend to dip downward for the very low **workcond** scores of 1 and 2 then increase steadily from **workcond** scores of 3 or higher.]

In order to avoid distortion or predictive bias in the regression model, the criterion must, at least approximately, follow a normal distribution in the population. Another factor that may bias the regression process will be differential variability of criterion scores for different values of the predictor variable leading to a fan-shaped plot like that observed earlier in Fig. 6.3. The assumption of roughly equal variability in criterion scores across the range of predictor scores goes by the technical term of 'homoscedasticity' and its violation, as we saw earlier, is termed 'heteroscedasticity'.

Where Is This Procedure Useful?

Certain types of experimental and quasi-experimental designs (perhaps having only a single group as in our example above), where the intent is to obtain a predictive understanding of the data, may find linear regression a useful method. Linear regression is also useful in various kinds of survey or observational research where relationships between one predictor and one criterion are to be assessed. Research domains that frequently rely heavily upon linear regression include economics, personnel psychology and educational psychology.

For example ... *Industrial psychologists may employ regression equations to make predictions about future job success of job applicants when making selection decisions. Educational researchers also make heavy use of linear regression concepts in such areas as predicting potential students' likely success at university from some type of tertiary entrance score or standardised test score. In economics, price of a product may be predicted from the quantity of product held. In each of these examples, single predictor regression models would be possible, but in reality, rather rare since, in most cases, several predictors would be available (leading directly to the construction of multiple regression models, see* Procedure 6.4 *and* 7.13).

Software Procedures

Application	Procedures
SPSS	*Analyze → Regression → Linear* ... and choose the dependent variable (= criterion to be predicted) and the independent variable (= predictor); by default, SPSS will provide the statistics and indicators needed to assess the predictive relationship and diagnostic plots of residuals can be requested. To graph the scatterplot with the regression line plotted (similar to Figure 6.8) using SPSS, choose *Graph → Chart Builder...* and from the Gallery, select the 'Simple Scatter' template from the Gallery 'Choose from' list, choose the X and Y variables and produce the graph; then open the graph in the Chart Editor, select *Elements → Fit Line at Total* and choose the 'Linear' option.
NCSS	*Analysis → Regression → Linear Regression and Correlation* and select the dependent variable (Y) and the independent variable (X); by default, NCSS will produce a range of statistics and indicators, graph the scatterplot with the regression line plotted and plot the residuals; NCSS will also produce a summary statement telling the story of what the analysis shows.
SYSTAT	*Analyze → Regression → Linear → Least Squares* ... and select the dependent variable and the independent variable; by default, SYSTAT will produce a range of statistics and indicators, graph the scatterplot with the regression line plotted (along with two types of confidence limits – see *Procedure 8.3*) and plot the residuals.
STATGRAPHICS	*Relate → One Factor → Simple Regression...* and select the Y variable and the X variable; by default, STATGRAPHICS will produce a range of statistics and indicators, graph the scatterplot with the regression line plotted (along with two types of confidence limits – see *Procedure 8.3*) and plot the residuals; The StatAdvisor will also produce a summary statement telling the story of what the analysis shows.
R Commander	*Statistics → Fit Models → Linear Regression...* and choose the 'response' variable (the criterion) and the 'explanatory' variable (the predictor); R Commander provides only the minimal statistics needed to interpret the predictive regression relationship. To get the scatterplot with the regression line plotted, choose *Graphs → Scatterplot...* and select the x-variable and the y-variable and make sure that the 'Least-squares line' option box is ticked.

Procedure 6.4: Multiple Regression & Prediction

Classification Multivariate; correlational; parametric.
Purpose To summarise the linear relationship between two or more independent variables (predictors) and a single dependent measure (criterion) by developing a predictive equation or model that relates the predictors to the criterion.

Measurement level Theoretically, any level of measurement may be used for the predictor variables although interval or ratio-level measures are often used. The criterion should be measured at the interval or ratio-level.

Multiple regression is simple linear regression (see *Procedure 6.3*) extended to handle multiple predictor variables measured on the same sample of individuals. The basic objective remains the same: to generate a prediction equation that summarises the relationship between the predictors and the criterion. Like linear regression, this equation represents a model of the world where observed criterion scores are analysed into two components: one part (predicted value) we can explain by knowing something about relevant predictors and the other part (residual) we can't explain. However, since multiple regression deals with two or more predictors, the regression equation describes, not the best fitting line, but the best fitting multidimensional surface with 'best' being defined in a 'least squares' fashion similar to that for simple linear regression (i.e. the surface is placed in such a way that the sum of the squared discrepancies (along the criterion or Y-axis) between the surface and the actual data points is made as small as it can be).

The components of a multiple regression model include a *regression coefficient* or *unstandardised regression weight*, B *weight* (or *slope* which describes the extent of tilt along each predictor's axis) for each predictor and the regression surface 'intercept' where it intersects the criterion axis when all predictors are set to a value of zero. The regression model allows us to make predictions about the value of the criterion for given values of the predictors. Each regression weight tells us how much change in the criterion to expect if we change the value of that predictor by an amount of one unit *while leaving all other predictors unchanged*.

We may be interested in extending the simple linear regression analysis performed in *Procedure 6.3* by incorporating **jobsat** as an additional predictor to **workcond** in the prediction of inspection **accuracy** scores. Since we now have a total of three separate variables (two predictors, one criterion) involved, a simple 2-dimensional scatterplot will not suffice to represent the data. We must represent the three variables simultaneously—hence the 'three-dimensional' scatterplot shown in Fig. 6.10 for the 112 quality control inspectors.

The fact that there are two predictors also means that a straight line will no longer suffice as the predictive device for relating the variables. Instead, we must fit a regression surface to the data points which, in the case of the data represented in the scatterplot shown in Fig. 6.10, amounts to a plane depicted by the flat grid pattern passing through the cloud of data points as shown in Fig. 6.11. The regression plane has a tilt in two dimensions, although the tilt is far steeper along the **workcond** axis than it is along the **jobsat** axis. This suggests that perhaps **workcond** is the more useful of the two predictors (the plane runs very nearly parallel to the **jobsat** axis which indicates that **jobsat**

(continued)

information is much less useful in helping us predict different **accuracy** scores). If neither predictor were useful, the regression plane would be level along both the **workcond** and **jobsat** axes.

Note that if we had three or more predictors, we could not graph the data in a single plot since we are limited, visually and practically, to three-dimensional representations on paper. Theoretically, though, any number of predictors can be accommodated in multiple regression analysis providing the sample is of adequate size.

For the surface fitted to the data in Fig. 6.11, the regression weight for **workcond** is +2.568 and for **jobsat** is +.515 [the intercept point is an **accuracy** score of 68.775%]. Thus, for a unit increase in **workcond** (say from 3 to 4), we would expect a 2.568% increase in **accuracy** scores if we kept the **jobsat** score at a fixed value. Similarly, for a unit increase in **jobsat** (say from 6 to 7), we would expect a .515% increase in **accuracy** scores if we kept the **workcond** score at a fixed value.

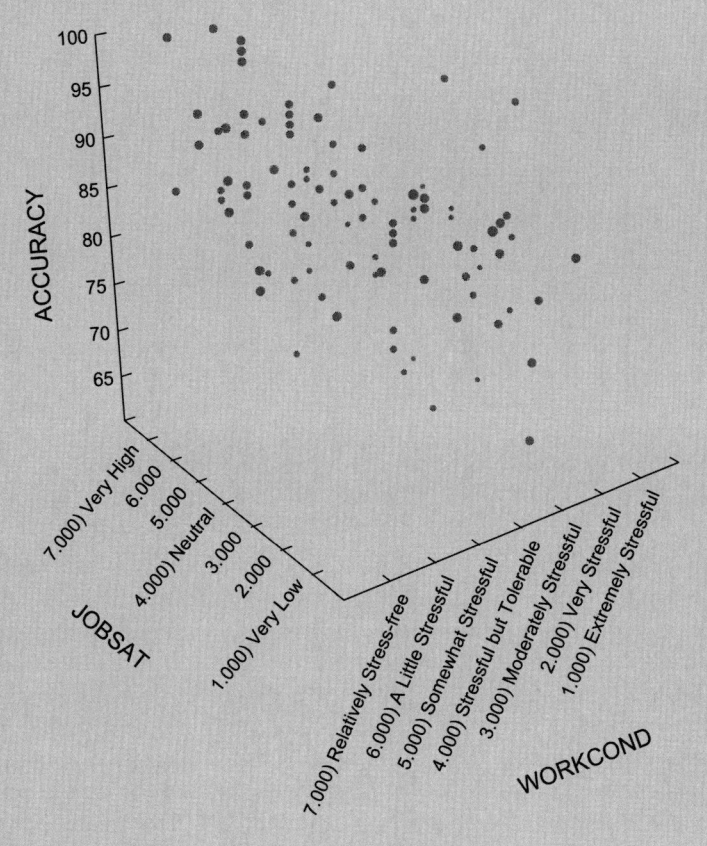

Fig. 6.10 3-D Scatterplot of the relationship between **accuracy**, **jobsat** and **workcond** before fitting the regression surface (produced using SYSTAT)

(continued)

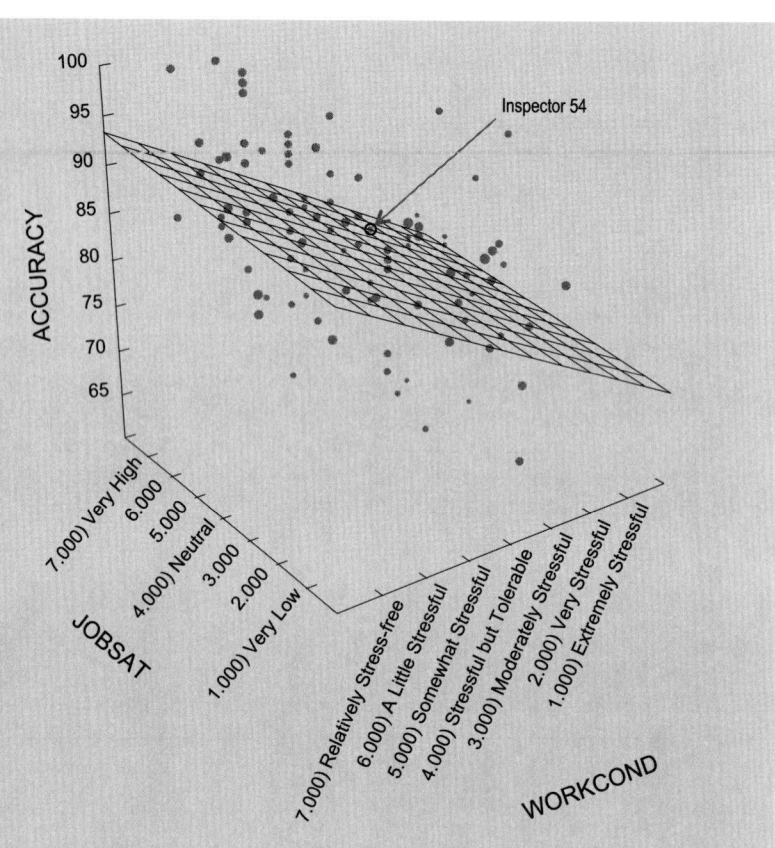

Fig. 6.11 3-D Scatterplot of the relationship between **accuracy**, **jobsat** and **workcond** jobsat with the fitted regression surface (produced using SYSTAT)

We could summarise the predictive regression model as follows:

$$\text{Predicted } \textbf{accuracy} = 68.775\% + (2.568\% \times \textbf{workcond}) + (.515\% \times \textbf{jobsat})$$

We can plug any two reasonable numbers in for **workcond** and **jobsat** and obtain a predicted score for **accuracy**. For example, one of the data points in Fig. 6.11 (for inspector 54 in the QCI database) had a **workcond** score of 3 and **jobsat** score of 6. Using the regression model, we could predict that a quality control inspector having these two specific scores should have an inspection **accuracy** score of 79.569%. However, inspector 54 had an observed **accuracy** score of 81%, so our prediction was close but too low. This prediction error is quantified by subtracting the predicted value of 79.569% from the observed

(continued)

criterion value of 81% to produce a 'residual' error of 1.431%. This reinforces the idea, discussed in *Procedure 6.3* with respect to simple linear regression, that prediction errors are inevitable (few, if any, residual values will be exactly 0.0) when imperfect predictor-criterion relationships are analysed. We could therefore depict the full model for **accuracy** observations as:

Observed **accuracy** = *Predicted* **accuracy** + *residual*

Notice that the addition of the **jobsat** predictor to the regression prediction system produced a predicted value closer to the observed value for inspector 54, resulting in a smaller residual value (+2.1% shown in *Procedure 6.3* versus +1.431% shown here). This illustrates the primary purpose for using multiple predictors—they help reduce residual prediction errors by improving prediction accuracy, providing they each have some relationship with the criterion being predicted.

In most multiple regression analyses, it is not possible to use the regression weights as described above to indicate how important each predictor is to the prediction of the criterion. This is because each regression weight reflects not only differences in predictor importance but also differences in the units of measurement and the scaling (in terms of means and standard deviations) of the predictors and the criterion.

If we conducted a multiple regression analysis to predict **accuracy** scores from **speed** and **mentabil** scores, we would note that the scores for **speed** are expressed in units of seconds, whereas the scores for **mentabil** are expressed in units of IQ points. Furthermore, the **speed** and **mentabil** variables differ dramatically in their means and standard deviations (see Table 5.3 in *Procedure 5.4*) and these differences strongly influence the sizes of the respective regression weights (the regression weight for **speed** in such an analysis is +.593 whereas the weight for **mentabil** is −.012). We cannot compare these two weights and automatically conclude that **speed** was the more important predictor—the weights are contaminated by the scaling differences.

In *Procedure 5.7* we described a process whereby measurement artefacts, such as differing means and standard deviations, could be removed from variables, namely by transformation into *z*-scores. We could apply the same logic to multiple regression analysis and transform all the variables in the analysis (including the criterion) to *z*-scores before conducting the analysis. The result would be that *standardised regression weights* (or *beta weights*) are produced. These standardised regression weights are interpreted in a similar fashion to the unstandardised regression weights (also called 'B weights') described earlier except that all of the units of measurement are now expressed in *z*-score (i.e. standard deviation) units. However, because the weights are standardised, it is now permissible to compare them to form an impression of predictor importance in the regression equation—larger standardised regression weights tend to signal predictors that contribute more strongly to prediction of the criterion variable (however, note that Judd et al. 2017, pp. 130–131, provide contrary arguments suggesting that even this interpretation of standardised weights can be problematic).

> If we focus on the above-mentioned multiple regression analysis to predict **accuracy** scores from **speed** and **mentabil** scores, the standardised regression weight for **speed** would be +.187 and for **mentabil** would be −.011. The weight for **speed** has been greatly reduced when measurement scales have been standardised. However, now it is acceptable to conclude that, of the two predictors, **speed** appears to be the stronger contributor to prediction of **accuracy** scores.
>
> We can interpret the standardised regression weight for **speed** as follows: if we increase the **speed** score by one unit of standard deviation (keeping the **mentabil** *z*-score constant), we would predict an increase of .187 standard deviation units in the **accuracy** *z*-score.

One of the statistics you obtain from a multiple regression analysis is the correlation between the actual criterion values and the predicted values for the criterion (obtained using the regression model). This statistic is called a *multiple correlation coefficient* (symbolised as ***R***) because it is based on more than one predictor. However, it is really nothing more than a simple Pearson correlation; it is just that one of the variables being correlated was formed by combining information from several predictors (the regression equation tells us how to combine this information to produce 'predicted values').

The multiple correlation ranges from 0 to 1.0 and can be tested for significance (see *Procedure 7.13*) just like an ordinary Pearson correlation. Multiple correlation is also interpreted in the same way as Pearson correlation. To the extent that the multiple correlation is less than perfect (i.e. less than 1.0), you will have errors in prediction (i.e. non-zero residuals) made by the regression equation. The closer to zero this correlation is, the larger the residual errors of prediction will be. A multiple correlation of zero tells you that the predictors you have used do not relate at all, in a linear fashion, to the criterion. [Note: nonlinear relationships can be explored in a multiple regression analysis using specialised variations in approach, but this is beyond the scope of the current discussion.]

As with simple linear regression, the multiple correlation can be squared to give an index known as R^2 that summarises how much of the variability in the criterion scores can be explained by knowing about the variability in the values of all of the predictors taken together. Figure 6.12 employs the Venn diagram representation,

Procedure 6.4: Multiple Regression & Prediction

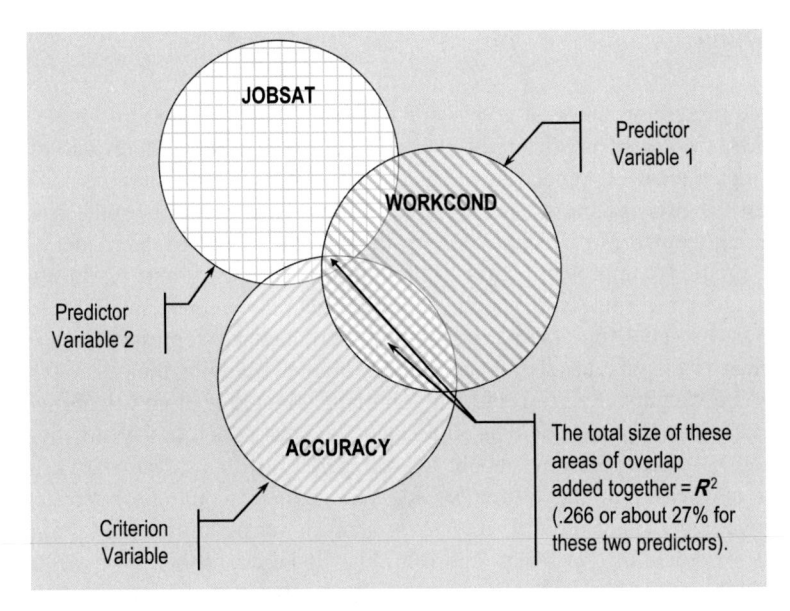

Fig. 6.12 A representation of the concept and interpretation of R^2

[Figure 6.12 shows three overlapping circles labelled JOBSAT (Predictor Variable 1), WORKCOND, and ACCURACY (Criterion Variable), with Predictor Variable 2. Annotation: "The total size of these areas of overlap added together = R^2 (.266 or about 27% for these two predictors)."]

introduced in *Procedure 6.3*, to illustrate the concept and interpretation of R^2. R^2 summarises the totality of the overlap that all predictors share with the criterion. Notice that the unit variance circles for the two predictors also overlap each other indicating that predictors are correlated as well. This will most often be the case in behavioural data: the predictors we employ will be intercorrelated with each other as well as with the criterion.

In multiple regression analysis, correlations among the predictors are not problematic unless they attain very high levels (i.e. in excess of ±.9) which creates a problem called 'multicollinearity' that can severely bias the magnitudes and variability of the regression weights. Tabachnick and Fidell (2019, ch. 4 and 5) discuss this problem of multicollinearity which can cause standardised regression weights to be unreliable indicators of predictor importance.

> For the data shown in Figs. 6.10 and 6.11, the value for **R** is .516 which means the value for R^2 is .266. We can interpret this as indicating that nearly 27% of the variability in **accuracy** scores can be accounted for by knowing an inspector's perception of his/her working conditions and his/her level of job satisfaction. [The correlation between **jobsat** and **workcond** is +.296 which means that the two predictors overlap each other by about 8.8% (equals $.296^2 \times 100\%$).]

Advantages

Multiple regression analysis allows for the examination of multiple independent variables as potential predictors of a criterion of interest. It facilitates two different yet complementary approaches: prediction and explanation. The chief difference between the two approaches lies in how the regression model is built up through controlling the order in which predictors are included as part of the model.

The predictive approach focuses on looking for the best predictors to use when predicting the criterion. In this approach, all predictors of interest are included in the model at the same time (sometimes called 'simultaneous regression analysis' or 'standard regression analysis') to see which variables are doing the predictive work. Regression weights of zero would tell you that a variable is of no use in predicting the criterion, and, hence, could be eliminated from the model. In the end, the goal is to establish the best predictive model you can, perhaps with a view toward using it to predict new cases in the future. Banks, for example, could be interested in a predictive approach to multiple regression analysis for building a model to predict the risk of defaulting on loans. The resulting model could then be used to make predictions about new loan applications, with a known rate of prediction error.

The explanatory approach to using multiple regression analysis focuses on understanding predictor contributions. Here, researchers may be interested in using multiple regression analysis to establish which predictors contribute most to explaining the behaviour of a criterion. The goal here is not the prediction model itself but its underlying structure and composition, which is related to variable importance. The explanatory approach can be a useful strategy for testing theoretical propositions regarding variables and how they are interrelated. This is generally accomplished by building up a regression model one predictor (or one set of conceptually-related predictors) at a time in a specific order so that the sequential contribution of those variables to prediction can be examined. When predictors are entered hierarchically in this way, we say that the explanatory approach is implemented through 'hierarchical regression analysis' (see *Fundamental Concept IV*).

Multiple regression is a very general technique in that it is related to ANOVA (see, for example, *Procedures 7.6* and *7.10*) in cases where ordinal or nominal scale predictors are used. Cohen et al. (2003) and Judd et al. (2017) talk at length about the unified and powerful nature of the multiple regression approach to data analysis, using the general linear model framework. What this means is that multiple regression analysis is a very flexible data analytic system which can be used to pursue many different types of analyses and research questions.

[Note: If nominal or ordinal categorical variables are to be used as predictors, they first must be appropriately transformed into new variables indicating the specific category each observation falls into—a process called **dummy coding** (see Cohen et al. 2003, chapter 8 discusses how to represent categorical predictors in a manner appropriate for regression analysis; *Fundamental Concept VI* also illustrates dummy coding).]

Disadvantages

If the predictor-criterion relationships are not totally linear, then multiple regression will not yield an adequate predictive model. [Note: There is a technique for evaluating nonlinear relationships using multiple regression analysis called *power polynomials*—see Chapter 6 in Cohen et al. (2003); see also *Procedure 9.8*.]

If there are many predictors (e.g. 10 or more), the number of cases/participants needed in the sample in order to compute meaningful and stable regression weights for the equation gets quite large. A general rule of thumb is to have an absolute minimum of 5 but preferably 15 to 20 or more cases/participants per predictor (Hair et al. 2010, p. 175). Thus, if one has 15 predictors, the minimal number of cases/participants needed is 5×15 or 75 but the preferred number is 225 (15×15) to 300 ($= 20 \times 15$).

One important limitation of the predictive approach to multiple regression analysis is that the prediction model is optimised only for the sample on which it was constructed. The predictive power of the model will automatically decline when it is applied to other samples and cases. For this reason, some authors advocate the use of *cross-validation* as a strategy for building robust prediction models. Cross-validation methods generally involve randomly splitting the sample in half and building the model using data from one half but applying the model to make predictions in the other half. This allows for direct checking of the loss in predictive ability when moving from sample to sample. Of course, this means that you need double the starting sample size to ensure that meaningful and stable regression weights can be estimated.

An important limitation of the explanatory approach to multiple regression is that, if hierarchical predictor entry (or inclusion) is used (see *Fundamental Concept IV*), the outcomes are entirely dependent upon the order of predictor entry. Changing predictor entry order will change the story, thus you must ensure that the order of predictor entry is strongly justified before you start, based on theoretical propositions, clear logic or some other determiner such as time-based ordering. Once you have argued for and established an entry order for the predictors, it is inappropriate to change that order after the fact.

The other disadvantages mentioned for simple linear regression (see *Procedure 6.3*), such as assumptions of normality and homoscedasticity, also apply to multiple regression. Multiple regression analysis is also sensitive to the presence of outliers or extreme observations which can produce either distortions in the magnitudes of the regression weights or cause the regression equation to severely over- or under-predict the criterion value for certain participants. Recall that a residual is the difference between a participant's actual observed criterion value and the value the regression model predicts they should have obtained; a large residual value reflects a large error in prediction. Fortunately, there are numerous graphical and numerical approaches available for detecting these problems, most of which do so by systematically examining patterns and magnitudes of the residuals (see Cohen et al. 2003, ch. 4 and 10, Hair et al. 2010, ch. 4, Judd et al. 2017, ch. 13, and Tabachnick and Fidell 2019, ch. 5 for reviews of these diagnostic techniques; *Procedure 7.13* provides some concrete illustrations).

Where Is This Procedure Useful?

Multiple regression would be useful in certain types of experimental and quasi-experimental designs (perhaps having only one group) where the intent is to obtain a predictive understanding of the data. Multiple regression is also useful in various kinds of survey or observational research where the relationships between several predictors and a criterion are of interest. If a stable and useful regression equation can be derived for prediction (the predictive approach), the model might then be used as an actual prediction device for new cases.

For example ... *Personnel psychologists often employ multiple regression models to assist organisations in their selection of appropriate applicants for specific jobs.*

Multiple regression analysis has great potential for use in the testing of specific theoretical propositions, usually but not always in quasi-experimental or general survey-type design contexts. In this theory-testing type of explanatory approach, the quality of the final prediction model is of lesser importance than understanding how the various predictor variables contribute to the explanation of variability in the criterion. In *Procedure 8.7*, we will see that there is a direct connection between multiple regression models and the testing of theory-based structural equation models or so-called path or causal models.

For example ... *A health researcher might be interested in understanding how a patient's demographic characteristics, past medical history, current symptoms, and treatment characteristics explain the variability in the extent of patient recovery from illness. Theoretically, the researcher might argue that demographic characteristics (gender, ethnic origin) exist prior to any medical problems the patient might have exhibited, so these variables are considered first to predict extent of recovery. Past medical history variables might be considered next, then current symptom measures, and finally measures of treatment characteristics. In this example, variables are ordered for consideration in the multiple regression analysis according to their theorised temporal sequencing in the lives of patients.*

Software Procedures

Application	Procedures
SPSS	*Analyze → Regression → Linear* ... and select the dependent variable and the desired independent variables; SPSS allows for very precise control over the order of entry of predictors into the regression equation, so that either the predictive approach or explanatory approach may be pursued. The default is simultaneous entry of all predictors of interest (predictive approach) and a range of statistics and diagnostic plots of residuals can be requested. SPSS can produce a 3-D graph similar to Fig. 6.10 (via *Graphs → Chart Builder* and selecting the 'Simple 3-D Scatter' template from the Gallery 'Choose from' list) but cannot show the fitted regression surface.

(continued)

Application	Procedures
NCSS	*Analysis → Regression → Multiple Regression* and select the Y or dependent variable, any numeric independent variables (Xs) and any categorical independent variables (Xs); NCSS is generally oriented toward the predictive approach, allowing tests of simultaneous predictor entry models or variations which try to choose statistically optimal subsets of predictor variables (what are called 'stepwise' regression models). A range of statistics and diagnostic plots of residuals can be requested.
SYSTAT	*Analyze → Regression → Linear → Least Squares . . .* and select the desired dependent and independent variables; SYSTAT is generally oriented toward the predictive approach, allowing tests of simultaneous predictor entry models or variations which try to choose statistically optimal subsets of predictor variables (what are called 'stepwise' regression models). However, SYSTAT does permit testing of some types of very general hierarchical regression models through a procedure called *Set Correlation* (see *Procedure 6.8*). A range of statistics and diagnostic plots of residuals can be requested. To produce a figure like Fig. 6.10, use *Graph → Scatterplot* and choose an X, Y and Z variable (the Z-variable is the dependent variable for this type of plot). To show the fitted regression surface similar to Fig. 6.11, click on the '*Smoother*' tab and choose 'Linear'. The properties of the plotted surface can be changed by editing the graph.
STATGRAPHICS	*Relate → Multiple Factors → Multiple Regression...* and select the desired dependent and independent variables; STATGRAPHICS is generally oriented toward the predictive approach, allowing tests of simultaneous predictor entry models or variations which try to choose statistically optimal subsets of predictor variables (what are called 'stepwise' regression models). A range of statistics and diagnostic plots of residuals can be requested. STATGRAPHICS can also produce a 3-D graph like Fig. 6.10 using *Plot - → Scatterplots → X-Y-Z Plots* and choose an X, Y and Z variable (the Z-variable is the dependent variable for this type of plot); however, it cannot draw the fitted regression surface.
R Commander	*Statistics → Fit Models → Linear Regression...* and choose the 'response' variable (the criterion) and the 'explanatory' variables (the predictors); **R** Commander provides only the minimal statistics needed to interpret the predictive multiple regression relationship.

Fundamental Concept IV: Partial & Semi-Partial Correlation (Hierarchical Regression)

The methodology of multiple regression analysis provides a general framework within which to understand the fundamental concept of *partialling*; a process that facilitates precise statistical control for variation (Pedhazur 1997). Statistical control of variation is accomplished through the processes of semi-partial and partial correlation, using an approach called *hierarchical regression*. Statistical partialling is a fundamental process underpinning the explanatory approach to multiple regression analysis described in *Procedure 6.4*.

While we will focus our discussion on semi-partial and partial correlation in the context of hierarchical multiple regression analysis, the general concept and process of partialling is used either explicitly or implicitly in a wide range of statistical techniques such as simultaneous multiple regression (*Procedure 6.4*), univariate and multivariate analysis of variance (*Procedures 7.10* and *7.14*), factor analysis (*Procedures 6.5* and *8.4*), canonical and set correlation and discriminant analysis (*Procedures 6.8* and *7.17*), analysis of covariance (*Procedure 7.15*), and structural equation models (*Procedure 8.7*).

We will rely upon the Venn diagram representations introduced in *Procedures 6.3* and *6.4* to anchor our discussion of the partialling concept and to clarify the distinction between *semi-partialling* and *full partialling*. The general process of partialling, in its simplest form, involves the statistical removal of the influence of one or more variables (the *partialling set*) from the correlational relationship between two variables. By statistical removal, we essentially mean that the variability of the variables in the partialling set is held constant in order to examine the relationship between two focal variables of interest.

> With respect to the QCI database, Maree might be interested in seeing how well inspection **accuracy** is predicted by inspector **jobsat** once she has controlled for differences in perceived working conditions, **workcond**.
>
> Such a question may be of importance if she wishes to make sure that any conclusion she might draw about the predictive or explanatory relationship between the accuracy and job satisfaction does not depend upon the quality of working conditions perceived by the inspector. For instance, it could be the case that inspectors in poorer working conditions also provide lower ratings of job satisfaction whereas inspectors in higher quality working conditions might also provide higher ratings of job satisfaction. If this occurs, then **workcond** and **jobsat** would be correlated and it would be unclear as to which variable might be influencing decision accuracy. Thus, to obtain a clearer interpretation, she wants to statistically control for the influence of **workcond** (the partialling set) by partialling it out of the regression relationship between **accuracy** (the criterion) and **jobsat** (the predictor).

In this concrete context, we will now examine two types of partialling processes that yield specific types of correlation: semi-partial correlation and partial correlation.

Semi-Partial Correlation

Semi-partial correlation (also called *part correlation*) involves the statistical removal of the influence of the partialling set variable(s) only from the variability of the predictor variable(s) of interest. The variability of the criterion is left unaltered by the semi-partialling process. It is easier to visualise what the semi-partialling process entails by considering two sequential multiple regression analyses (computationally, we don't actually have to go through this process since statistical software automates the computations, but it is easier to grasp the concept of partialling if we imagine this is the process).

For example ... *In the first regression analysis, we would employ* **accuracy** *as the criterion but only the* **workcond** *variable as a predictor.*

We could call the outcome from this regression analysis the *reduced model* because only the variable from the partialling set has been explicitly included.

In the second regression analysis, we would employ would again employ **accuracy** *as the criterion but would now include both the* **jobsat** *and* **workcond** *variables as predictors.*

We could call the outcome from this regression analysis the *full model* because it has all the variables under consideration included in it. It is important to note that the R^2 value from the full model regression analysis will always be as large or larger than the R^2 from the reduced model analysis. What we have done here is essentially a hierarchical regression analysis where the partialling set variables to be controlled for are included in the model first, then the actual predictor variable(s) of interest are added to the model in a second stage; the regression model is therefore constructed in a deliberately ordered way.

Once we have the R^2 values from these two regression analyses, we would then subtract the reduced model R^2 from the full model R^2. This subtraction is what statistically removes the shared variance between **workcond** and **jobsat**. The resulting value is the square of the semi-partial correlation between **accuracy** and **jobsat** when **workcond** variability has been partialled out of the **jobsat** variable's variability. Note that taking the square root of the squared semi-partial correlation and attaching the sign of the regression weight for **jobsat** from the full model regression analysis gives the semi-partial correlation, which can be interpreted in the same way as an ordinary Pearson correlation. In terms of interpretation, what a squared semi-partial correlation shows is the amount of variance in the criterion that can be *uniquely* explained by the partialled variable(s) of interest over and above what the partialling variable(s) can explain.

> If we conduct the initial analysis predicting **accuracy** from **workcond** scores, we would observe that the reduced model R^2 value was **.258**. If we conduct the second analysis predicting **accuracy** from **workcond** and **jobsat** scores, we would observe that the full model R^2 value was **.266**. If we subtract .258 from .266, we would find a squared semi-partial correlation value of **.008**. This represents a minuscule unique explanation of just .8% of the variability in **accuracy**.

(continued)

We also could take the square root of this value and attach the positive sign of the regression weight for **jobsat** from the full model analysis (which was +.515) to obtain a semi-partial correlation value of +.089, which is a weak correlation.

In short, once we have controlled for the relationship between **jobsat** and **workcond**, there is very little left in the predictive relationship between **jobsat** and **accuracy**. Equivalently (and putting a more positive spin on the finding), we could say that inspectors' job satisfaction uniquely explains .8% of the variance in inspection decision accuracy over and above what inspectors' working conditions can explain.

As a comparison, we could have observed that the original Pearson correlation between **accuracy** and **jobsat** scores without any partialling was +.235: once we partial the influence of working conditions out of the **jobsat** scores, we are left with a paltry +.089. Where we might have originally concluded that the correlation between **accuracy** and **jobsat** was a moderately weak positive one, we now see that this relationship was at least partly attributable to the relationship between working conditions and **jobsat** and once that relationship had been controlled for, the original relationship nearly evaporated.

Figure 6.13 provides a visual illustration of the semi-partialling process using Venn diagrams. The left diagram in Fig. 6.13 shows what the reduced model regression analysis conceptually accomplishes, and the right diagram shows what the full regression model conceptually accomplishes and identifies that portion of the overlap with **accuracy** which describes the squared semi-partial correlation of **jobsat** (partialled by **workcond**) with **accuracy**. The semi-partialling process is employed by many statistical procedures as a way of unambiguously allocating explanatory variance to different predictors while leaving the full variance of the criterion available for explanation.

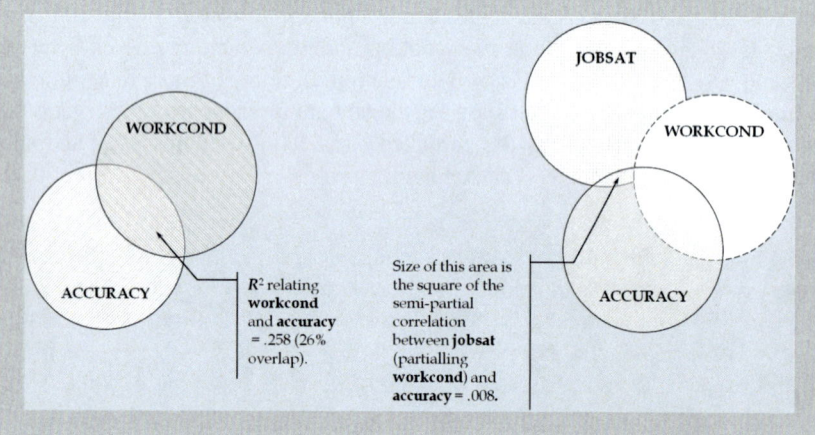

Fig. 6.13 A representation of the process for finding the semi-partial correlation

Partial Correlation

Partial correlation involves the statistical removal of the influence of the partialling set variable(s) from both the variability of the predictor variable(s) of interest and the criterion of interest. It is easier to visualise what this process entails by considering the following three-step multiple regression process (again, computationally, we don't actually have to go through this process since statistical software automates the computations, but it is easier to grasp the concept of partialling if we imagine this is the process).

1). The first step involves computing a regression analysis that employs **accuracy** scores as the criterion variable and the partialling variable, **workcond,** as the sole predictor. Once this analysis has been completed, we compute the residuals (the differences between observed criterion values and the predicted criterion values) which arise from applying the resulting prediction equation to every inspector in the QCI database and we save these residuals in a new variable called **accu_part**. Note that the subtractive process involved in finding the residuals effectively removes whatever relationship **workcond** has with **accuracy** scores leaving only what is left over (this process is called 'residualising').
2). The second step involves computing a regression analysis that employs **jobsat** scores as the criterion variable and the partialling variable, **workcond,** as the sole predictor. Once this second analysis has been completed, we compute the residuals from applying the resulting prediction equation to every inspector in the QCI database and we save these residuals in a new variable called **js_part**.
3). The third step involves computing the simple Pearson correlation between these two new 'residualised' variables, **accu_part** and **js_part**. The number produced is the *partial correlation* between **accuracy** and **jobsat** controlling for the influence of **workcond** on both variables.

If we execute this three-step partialling process for correlating **accuracy** and **jobsat**, controlling for **workcond**, the resulting partial correlation value is **+.102** (recall the original correlation between **accuracy** and **jobsat** was +.235). The square of the partial correlation, .010, represents the proportion (or percentage = 1%) of predictive overlap between **accuracy** and **jobsat** that is completely uncontaminated by the influence of quality of working conditions: a very small amount indeed. It appears that the relationship between **accuracy** and **jobsat** can primarily be explained through reference to a third mediating variable, namely **workcond**.

> Figures 6.14 and 6.15 provide a visual illustration of the full partialling process using Venn diagrams. Figure 6.14 shows what the first and second steps conceptually accomplish. Figure 6.15 shows how the outcomes from the first two steps are combined to identify that portion of the predictive overlap between the residualised **accu_part** and **js_part** variables which describes

(continued)

the squared partial correlation of **jobsat** (partialled by **workcond**) with **accuracy** (partialled by **workcond**). The partialling process is employed by a few specialised statistical procedures, e.g. the analysis of covariance (see *Procedure 7.15*), as a way of statistically controlling the influence of extraneous variables on both sides of the relationship between predictors and a criterion.

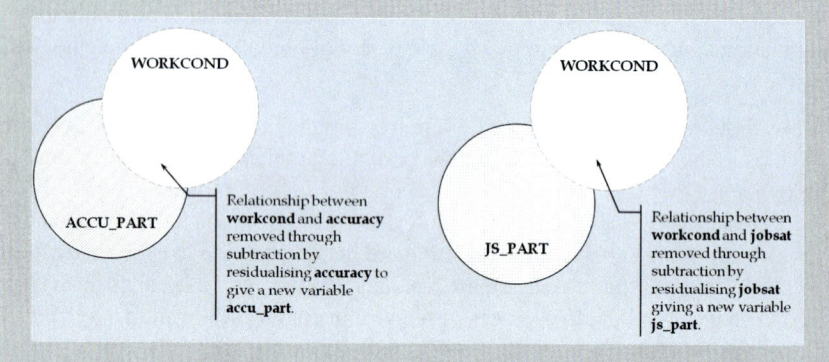

Fig. 6.14 A representation of the first two steps for finding the partial correlation

Fig. 6.15 A representation of the final step for finding the partial correlation

Size of this area is the square of the partial correlation (= .010). Note that this area is *proportionally* larger compared to the squared semi-partial relationship because the **accu_part** circle is of less than unit size.

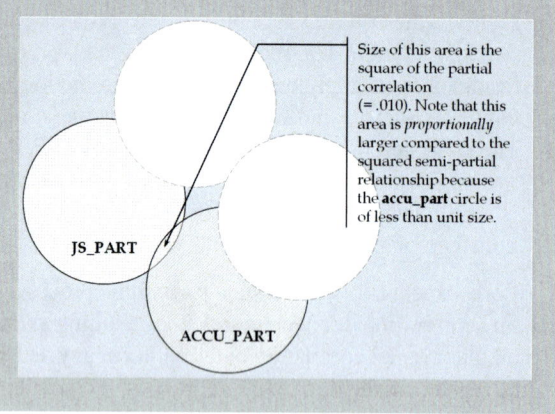

What Can Be Partialled?

Thus far, our discussion has focused on the use of a partialling or semi-partialling process to statistically remove the influence of a third variable from the relationship between two focal variables. Cohen et al. (2003) have shown that the logic of partialling and semi-partialling can be easily and readily extended to the statistical control of the influence of several variables on a specific bivariate relationship. In this way, we can consider partialling out the influence of entire sets of variables such as demographic measures (most questionanaires yield several such measures) or measures of ability (which are typically multidimensional involving verbal, spatial, and numerical domains, for example).

Furthermore, it is possible that the bivariate relationship of direct interest may itself be based on sets of variables. Multiple regression analysis, for example, results in a bivariate relationship between a single dependent variable and a composite variable formed by the weighted combination of several variables in the predictor set (the predicted values).

> In the QCI database, we could therefore address the quite reasonable but more sophisticated research question "how well do the quality of working conditions, **workcond**, and job satisfaction, **jobsat**, predict quality control inspection **accuracy** once the influences of both demographic (**educlev** and **gender**, appropriately coded to use as predictors; see *Fundamental concept VI*) and mental ability (**mentabil**) measures have been semi-partialled out?" The answer to this question is a squared semi-partial correlation of .185 and a semi-partial correlation of +.430. Thus, once **educlev**, **gender** and **mentabil** have been controlled for, **workcond** and **jobsat** uniquely explain about 19% of the variance in **accuracy** scores.

Set correlation (see *Procedure 6.8*) provides the most flexible way to deal with sets of variables on both sides of a relationship, dependent variables, independent or predictor variables as well as control or partialling variables (which may be different) for each side. Set correlation is a specific innovation of the famous statistical psychologist, Jacob Cohen (described in Cohen et al. 2003).

Procedure 6.5: Exploratory Factor Analysis

Classification	Multivariate; correlational; parametric.
Purpose	To summarise and represent the interrelationships amongst variables in a set using a smaller number of composite variates.
	Principal Components Analysis attempts to accomplish this task by explicitly combining variables in a weighted fashion to form 'components' which account for the maximum amount of variability in the variables' scores.
	Common Factor Analysis attempts to accomplish this task by identifying a smaller number of latent variables or constructs (common factors) that best reproduce the original observed correlations among the variables.

Measurement level Generally, the variables employed in exploratory factor analysis are measured at the interval or ratio-level, but this need not always be the case. Variables measured at the ordinal or nominal-level can be accommodated as long as an appropriate measure of corrclation is cmployed.

There are many varieties of *factor analysis* (e.g. principal components analysis, image analysis, maximum likelihood factor analysis, common factor analysis) but they all share the same general aim. Factor analysis seeks to condense the number of observed variables into a smaller number of composite variates which highlight or represent the underlying structure of the domain from which the variables were derived (or were designed to measure). This 'condensation' may be mechanistic and practical in orientation in that a small number of composite variables is sought (done simply to reduce the number of variables you have to deal with) or it may be theoretical in orientation, seeking to demonstrate construct validity through discovering the underlying dimensionality of items on an instrument.

Thus, we might use factor analysis to:

- condense a set of 25 attitude items into a small number of composite factors;
- establish the dimensionality of a self-report questionnaire measure of job satisfaction or personality inventory; or
- test whether there are five specific factors underlying a specific personality inventory or attitude measurement scale.

There are two broad classes of factor analysis techniques: *exploratory*, which answers questions about the number and nature of the underlying dimensions or composites but does not restrict the structure of the solution in any way—a discovery-oriented approach consistent with the first two dot points above; and *confirmatory*, which restricts the structure of the solution sought, thereby permitting tests for specific patterns of relationships and construct alignments among the variables—useful for addressing the third dot point above (to be addressed in *Procedure 8.4*). All factoring approaches work by identifying and characterising patterns among the correlations of all relevant pairs of variables. In this procedure, we focus on exploratory factor analysis.

> Consider Table 6.8 which reports the matrix of correlations among the nine work attitude measures collected from the 112 quality control inspectors in the QCI database (Table 6.8 and the subsequent analyses of it were produced using SPSS). A simple visual inspection of this matrix gives a general, perhaps vague, feeling that some variables seem to be more highly correlated with each other than with other variables, but the nature of the pattern is difficult to detect.

(continued)

Procedure 6.5: Exploratory Factor Analysis

Table 6.8 Correlations among the nine work attitude items for quality control inspectors

Correlation Matrix

		cultqual	acctrain	trainqua	qualmktg	mgmtcomm	qualsat	trainapp	satsuprv	polsatis
Correlation	cultqual	1.000	.258	.014	.528	.397	.509	.111	.269	.254
	acctrain	.258	1.000	.252	.223	-.043	.245	.372	.206	.024
	trainqua	.014	.252	1.000	.144	.063	.034	.404	.097	-.091
	qualmktg	.528	.223	.144	1.000	.387	.473	.141	.315	.204
	mgmtcomm	.397	-.043	.063	.387	1.000	.106	.169	.336	.337
	qualsat	.509	.245	.034	.473	.106	1.000	.161	.221	.173
	trainapp	.111	.372	.404	.141	.169	.161	1.000	.098	.195
	satsuprv	.269	.206	.097	.315	.336	.221	.098	1.000	.492
	polsatis	.254	.024	-.091	.204	.337	.173	.195	.492	1.000

Note: You may have noticed that the correlations above the diagonal line of 1.000 values are mirror images of those below the diagonal line. This is true of all correlation matrices and reflects the property called 'symmetry', as initially highlighted in *Procedure 6.1*.

Exploratory factor analysis offers a range of possible methods (called *extraction methods*) for discovering the underlying structure in a set of many variables. One method is termed *principal components analysis* which attempts to accomplish this task by explicitly combining variables in a weighted fashion to form 'components' that account for the maximum amount of variability in the variables' scores. In most statistical software packages, this is the default method for factor analysis, although strictly speaking, the technique should be referred to as *component analysis* rather than factor analysis. If you are looking to simply reduce or condense a larger number of variables into a smaller number of composite variables, then principal components analysis is the most appropriate tool for the job. Principal components analysis is probably the most commonly reported type of factor analysis.

Other varieties of exploratory factor analysis come under the general heading of *common factor analysis* and these techniques attempt to discover the underlying structure by identifying a smaller number of latent (i.e. not directly observed) dimensions or constructs (called 'common factors') that best reproduce the original observed correlations among the variables (a process akin to the generation of predicted values in multiple regression analysis). Virtually all computer packages offer a widely used form of common factor analysis called *principal axis factoring* which works well and is relatively stable under a variety of data anomalies, but technically does not provide statistically optimal estimates. Principal axis factor analysis is second in popularity with researchers compared to principal components analysis.

The most appropriate form of common factor analysis, from a statistical estimation perspective, is *maximum likelihood factor analysis*. Maximum likelihood factor analysis provides statistically optimal estimates as well as a chi-square hypothesis test for the appropriate number of factors to examine. However, it suffers from the drawback that it is extremely sensitive to anomalies in the data, including missing data patterns, which may mean it won't be able converge on an interpretable solution. Furthermore, maximum likelihood factor analysis is not available in all statistical packages. NCSS and STATGRAPHICS, for instance, do not offer maximum likelihood factor analysis, but do offer principal axis factor analysis. These inherent instabilities and constraints on availability have ensured that maximum likelihood factor analysis lags behind both principal components and principal axis as factor analytic methods of choice. Irrespective of the specific 'extraction' method employed, the goal remains the same: analyse many variables to hopefully identify a few interpretable components or factors.

There are many decisions that a social/behavioural researcher must make in order to get the best results from exploratory factor analysis. The most important decisions are:

- *Is it worth factoring this correlation matrix?* Some computer packages (e.g. SPSS) report indices that signal whether the correlation matrix is likely to have discoverable factors or components in it. Two of these indices are the Kaiser-Meyer-Olkin Measure of Sampling Adequacy (KMO index—look for a value greater than .60) and Bartlett's Test of Sphericity (look for a statistically significant chi-square test, but note that it is highly sensitive to large sample sizes)—see Field (2018, p. 799 and Tabachnick and Fidell 2019, section 13.3.2.6, for more detail.
- *Which factor extraction method to use?* Many researchers uncritically use the default principal components method in their computer package without thinking about its appropriateness to the problem at hand. If it is important to keep a theoretical basis in mind as a reference point for interpreting solutions, a common factor exploratory approach may be more appropriate. If simple variable condensation into a smaller number of composites is the main goal, principal components analysis may be more appropriate.
- *How many components or factors to accept as the final number to interpret?* There are many tools available, both graphical, such as scree plots of eigenvalues, and numerical, such as significance tests (only for the maximum likelihood method) and inspection of residual covariances, but none is definitive. It is best to consider these tools as heuristics rather than hard-and-fast rules.
- *How to obtain the most interpretable solution?* This usually entails a decision as to how to 'rotate' the factor solution so as to produce either uncorrelated factors (called 'orthogonal rotation', of which the 'varimax' procedure is most commonly used method) or correlated factors (called 'oblique rotation'—different software packages employ different algorithms, but the 'promax' method is generally recommended). Rotation is done to make solutions more interpretable by trying to achieve what is called 'simple structure' where only a few variables define each factor while simultaneously contributing minimally to all other factors. [Note that some statistical packages like NCSS and STATGRAPHICS do not offer 'oblique rotation' options.]

Procedure 6.5: Exploratory Factor Analysis

Virtually all exploratory factoring methods produce similar types of output statistics, when correlated factors are being examined (the recommended strategy):

- *eigenvalues*
  These indices summarise how much variance each factor explains in the variables out of total available. Factors are 'extracted' one at a time in decreasing order of eigenvalue (i.e. the first factor explains the most variance in the variables; the second factor the next most and so on). If eigenvalues are plotted against the number of factors using a 'scree' plot, they can give a rough indication as to the number of factors to interpret – the general rule is choose the number of factors that occurs at or immediately prior to the 'elbow' in the plot beyond which the curve is generally much flatter (see Gorsuch 1983 for details). Note that there are always as many eigenvalues as there are variables in the analysis, which is hardly helpful in terms of reducing the number of things to interpret. The trick is to decide where the point of diminishing return occurs indicating we should stop interpreting any more factors or components—the fewer factors or components generally the better, unless we are looking for a specific number.

  [*Technical note*: To understand what an eigenvalue means in a principal component analysis, consider the implications of standardising all variables in the analysis. Standardising each variable transforms its variance to 1.0. If there are 10 variables being analysed, then there are 10 units of variance to potentially explain. An eigenvalue is simply how many of these 10 units of variance a single factor can explain. Eigenvalues are often converted into percentages of the total variance which are then interpreted in a fashion similar to R^2.]

  A general rule of thumb, implemented in many computer packages, is the 'eigenvalue greater than 1.0' rule which says we should accept only those factors that have an eigenvalue larger than 1.0 (meaning that the factor accounts for more than a single variable's worth of variance—variables are implicitly transformed into *z*-scores prior to commencing a factor analysis). After a certain number of factors have been 'extracted', the eigenvalues will drop below 1.0, indicating a diminishing return for extracting a larger number of factors. The point at which this drop below 1.0 occurs is taken to signal the number of factors to interpret. The 'eigenvalue greater than 1.0' rule should supersede the scree value plot indication if the 'elbow' in the plot occurs before the eigenvalues have dropped below 1.0.

- *communalities*
  These indices summarise how much variance in each variable is shared in common with the variance of the set of factors in the final solution. If principal components analysis is used, the communalities are set to 1.0, meaning that all the variance of each variable is available for analysis.

  If principal axis factoring or other extraction method is used, communalities are estimated as the proportion of variance that each variable shares with all other variables. This means they will always be less than 1.0 and the difference between 1.0 and the communality estimate signals the amount of variance that is unique to each variable (i.e. not shared; sometimes called a 'uniqueness' value).

- a rotated *factor pattern* matrix
  This matrix provides a set of standardised regression weights for predicting variable scores on the basis of the factors identified in the final solution. Pattern coefficients typically range between -1.0 and $+1.0$, with larger magnitude coefficients indicating greater alignment between a variable and a factor. This is the best matrix to interpret when you have correlated factors.
- a rotated *factor structure* matrix
  This matrix provides a set of Pearson correlations between each variable and each factor in the final solution.
- a matrix of *factor correlations*
  This matrix shows the estimated degree of correlation which exists between the factors in the population.

If orthogonal rotation (e.g. varimax) is opted for, then uncorrelated factors are obtained, the factor pattern and factor structure matrices will be identical (most software packages will report this 'combined' pattern/structure matrix as a matrix of 'factor loadings'), and the matrix of factor correlations will be non-existent. When correlated factors are reported, many researchers focus upon interpreting just the factor pattern matrix and the matrix of factor correlations. Generally speaking, it is good practice to always look for correlated factors first; if they are not substantially correlated, the correlation matrix among the factors will show this and a simpler solution extracting uncorrelated factors can then be interpreted.

> Table 6.9 shows the eigenvalues, the Scree Plot, the component Pattern Matrix and the Component Correlation Matrix from a principal component analysis (conducted using SPSS) of Maree's correlation matrix in Table 6.8. Bartlett's Test of Sphericity showed a significant chi-square value of 195.504 and the KMO index was an acceptable .649, indicating that there were substantive components that could be discovered in the correlation matrix. The results strongly suggested a 3-component solution and oblique 'promax' rotation of the Pattern Matrix was requested in order to produce correlated components. [Note that this component analysis provides virtually the same interpretive outcomes as a principal axis common factor analysis of the same correlation matrix – something that often occurs.]
>
> Three components were accepted because the first three factors had eigenvalues of 2.908, 1.530 and 1.182, respectively. All remaining eigenvalues were less than 1.0 – a conclusion partly supported by the Scree Plot (where the eigenvalue of 1.0 cut-off is shown as a horizontal dashed line; note however, that the elbow occurs at two factors, so Maree allows the eigenvalue greater than 1.0 heuristic to override the visual elbow heuristic). The component Pattern Matrix has been sorted so that the variables that contributed most strongly to a specific component appear together and only those values larger than .40 or −.40 are shown (these are controllable options within SPSS).

(continued)

Procedure 6.5: Exploratory Factor Analysis

Table 6.9 Principal components analysis outcomes for the three correlated components solution for the nine QCI work attitude variables

Total Variance Explained

Component	Initial Eigenvalues			Rotation Sums of Squared Loadings[a]
	Total	% of Variance	Cumulative %	Total
1	2.908	32.310	32.310	2.426
2	1.530	17.002	49.312	2.141
3	1.182	13.137	62.450	1.811
4	.899	9.989	72.439	
5	.705	7.833	80.271	
6	.617	6.852	87.124	
7	.446	4.953	92.077	
8	.405	4.500	96.576	
9	.308	3.424	100.000	

Extraction Method: Principal Component Analysis.

a. When components are correlated, sums of squared loadings cannot be added to obtain a total variance.

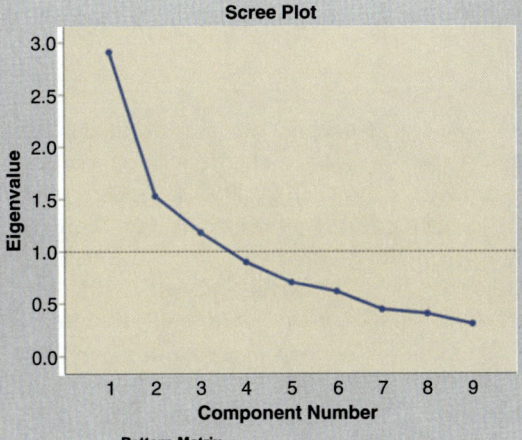

Scree Plot

Pattern Matrix

	Component		
	1	2	3
qualsat	.859		
cultqual	.799		
qualmktg	.722		
polsatis		.842	
satsuprv		.708	
mgmtcomm		.693	
trainqua			.806
trainapp			.801
acctrain			.603

Component (or Factor) 1: Quality-related attitudes → **QUALITY**

Component (or Factor) 2: Organisational satisfaction attitudes → **ORGSATIS**

Component (or Factor) 3: Training-related attitudes → **TRAINING**

Component Correlation Matrix

Component	1	2	3
1	1.000	.340	.235
2	.340	1.000	.106
3	.235	.106	1.000

Researchers differ as to whether they prefer to interpret pattern coefficients or structure coefficients. Generally speaking, pattern coefficients provide a somewhat simpler and easier interpretation, especially when components or factors are correlated, whereas structure correlations provide more stable indicators of relationship across samples. We will focus here on pattern coefficients.

A 'pattern coefficient' summarises the relationship each variable shares with a component or factor. The larger the pattern coefficient, the more the variable helps to define that factor. The pattern matrix is interpreted by:

- identifying those variables that relate most strongly, above a certain threshold value (say, above a pattern coefficient value of ±.40), with each factor. There should be at least three such variables if a stable component or factor is to be identified and each variable should be highly related with only one component or factor if the simplest possible interpretation is to be achieved; and
- examining the content of the variables that define each component or factor to identify the common theme or idea which has drawn those variables together (this, in effect, identifies the construct that the component or factor may be tapping into).

> For Component 1 (the first column of numbers in the Pattern Matrix in Table 6.9), the first three variables, which focus on various aspects of the inspectors' perspective on product quality, have high pattern coefficients associated with this factor (with corresponding low coefficients, not shown because they were less than .40 on the other two factors). Maree might therefore label this component as 'quality-related' or simply **quality**.
>
> Component 2 is defined by a second set of three variables, each focusing on some aspect of inspectors' satisfaction with various aspects of their situation at work. Maree might therefore label this component as tapping into 'organisational satisfaction' or simply **orgsatis**.
>
> Component 3 appears to be defined by a third set of three variables, each focusing on some aspect of training within the company. Maree might therefore label this component as 'training-related' or simply **training**.
>
> The **quality** and **training** components appear to be moderately weakly positively correlated (.235). The **quality** and **orgsatis** factors, as one might have expected, are moderately positively correlated (.340) and the **training** and **orgsatis** factors are only weakly correlated (.106). Altogether, this indicates that the three components tend to define distinctive but somewhat correlated ways in which inspectors can express their attitudes toward their companies.

We have noted that there are many different heuristic tools for identifying the number of components or factors to interpret. However, there is one overriding principle that should be used, namely interpretability. Irrespective of what any heuristic tool says should be the number of components or factors to interpret, if you cannot make interpret or make sense of a specific factor or component, then it should be rejected.

An important follow-up step for any component or factor analysis is the computation of 'internal consistency reliability' for each of the components or factors identified (see the discussion in *Procedure 8.1*). Assuming the identified components or factors are reliable, we could then calculate a score (see discussion below) for each inspector in the QCI database for the **quality**, **orgsatis**, and **training** components and these three new 'composite' variables could then be used, in place of the original nine items, when exploring work attitude relationships with other variables. Thus, principal components analysis or factor analysis provides an empirical and structured way to simplify multivariate data so that the exploration of relationships can be facilitated without undue complexity.

Advantages

Exploratory factor analysis provides a class of techniques useful for condensing many variables into a smaller, more manageable and more reliable subset of dimensions or factors. Factor analysis explicitly utilises all of the available correlational information amongst a set of variables and the various rotational procedures can be used to provide the simplest structure possible for interpreting a solution. Employing factor analysis to reduce the dimensionality of multivariate data is a much more defensible approach than combining variables simply because the researcher thinks they ought to go together.

Principal component analysis is most useful when all that is desired is a straightforward condensation of variables into components that explain the most variance possible in the variables. Component analysis is also most useful when the data being analysed exhibit certain types of instability (including missing observations), extremely high levels of intercorrelation or low numbers of sample participants relative to the number of variables (e.g. less than a 10:1 ratio), because a solution can always be computed.

Common factor analysis is most useful when population estimates of correlations, based on a specific number of factors, are desired. Common factor analysis has the additional advantage of directly producing estimates of error (the 'uniqueness') associated with each variable, meaning that there is explicit acknowledgment that our individual variables (e.g. survey items) are most likely somewhat unreliable and reflect, at least in part, some measurement error.

What is interesting to note is that, at a practical level for 'well-behaved' data, there is little real difference in the solutions provided by principal components analysis and common factor analysis. Of all the possible techniques for reducing the dimensionality of data (e.g. factor analysis, cluster analysis, multidimensional scaling, and correspondence analysis), exploratory factor analyses are probably the most widely reported in the behavioural and social science literature.

Disadvantages

Acquiring a fairly deep knowledge of factoring methods requires a reasonable degree of statistical sophistication. Factor analysis is a complex set of statistical procedures that work best if a relatively high ratio of cases or participants to variables (at least 10 to 1 or more, see Hair et al. 2010, p.102) is available for the analysis. There are many varieties, each of which makes certain assumptions. The most commonly used default procedure, principal component analysis, implicitly makes the assumption that the variables being analysed are measured without error—a questionable assumption in behavioural science data—but is highly stable under a wide variety of problematic data conditions. Common factor analysis, which makes no such assumption about lack of measurement errors, circumvents this problem but at the cost of potential instabilities in solutions or even an inability to produce a solution.

How missing data are handled has implications for the choice of factoring method. Listwise deletion of missing cases (i.e. a case is dropped if missing a value on any variable in the analysis) could greatly reduce your sample size but will yield a very stable correlation matrix, which common factor analysis will be able to cope better with. Pairwise deletion (i.e. a case is dropped only if missing a value on one of two variables being correlated at any one time) creates a more unstable correlation matrix, which principal components analysis can cope better with.

Common factor analyses invariably involve iterative estimation techniques which can break down if the data are 'ill-conditioned' (see Tabachnick and Fidell 2019, ch. 13 for a discussion) or if poor starting points for the estimation procedure are chosen (usually done automatically by the software package being used).

Additionally, the most statistically appropriate common factoring technique, maximum likelihood estimation, typically requires:

- large number of cases/participants relative to the number of variables (a ratio much larger than 10 to 1 is preferable) which, oddly enough if the total sample size is very large (over 500, say), has the effect of reducing the utility of the chi-square test of significance which the technique provides;
- at least three defining variables per factor; and
- the researcher to nominate a specific number of factors to estimate.

Many software packages provide the facility to compute so-called factor scores, but the temptation to use these computed scores, particularly in common factor analysis, should be avoided wherever possible because:

- a unique set of scores cannot be computed (except in the case of principal components analysis);
- the weights used in computing the scores are not stable across changes in samples (i.e. the weights do not cross-validate well); and
- factor scores are produced by weighting every variable in the calculation, no matter how small their contribution to the factor, which tends to dilute the contribution of the major defining variables.

Procedure 6.5: Exploratory Factor Analysis

The simple solution for computing a 'factor score' is to average the scores for all of the variables the researcher has identified as defining the component or factor—the resulting scores are termed *unit-weighted scores* because the defining variables get a weight of 1 whereas all non-defining variables get a weight of 0 (Hair et al. 2010 call these 'summated' scales, see pp. 126–128).

> To produce a unit-weighted score for each inspector in the QCI database on the 'organisational satisfaction' component, Maree would simply average each inspector's scores on the **mgmtcomm**, **satsupvr**, and **polsatis** variables. Similarly, to create a component score for the 'quality-related attitudes', Maree would average each inspector's scores on the **cultqual**, **qualmktg**, and **qualsat** variables. Finally, to create a component score for the 'training-related attitudes', Maree would average each inspector's scores on the **trainapp**, **trainqua**, and **acctrain** variables. In this way, three new component-focused variables, **quality**, **orgsatis**, and **training**, can be created for use in subsequent analyses. Such scores have two advantages: they are in the original units of measurement of the constituent items and missing observations can easily be handled (only non-missing item scores are averaged for any participant).

Where Is This Procedure Useful?

Exploratory factor analysis is useful in any experimental, quasi-experimental or survey design where a number of variables have been measured on the same sample of individuals. These techniques are particularly useful in analysing test or rating items when a smaller more dependable set of variables is desired prior to pursuing further analyses and/or when one wants to closely examine the pattern of variable relationships which seem to best explain the data (a question specifically addressed by common factor, not principal component, analysis). Typically, factor analyses have been used to assess the dimensionality of tests, inventories, and rating scales. It is a key analytical strategy for construct validation.

In any reporting of a factor analysis, you should expect to see:

- a description of the specific methods of factor extraction and rotation used;
- a discussion of the information used to help decide on the proper number of factors to interpret;
- a table of pattern coefficients (often authors will only show pattern coefficients above a certain magnitude in their table, as shown in Table 6.9, in order to simplify interpretation and presentation); and
- if correlated factors are reported, the factor correlation matrix.

If a common factor analysis and/or an oblique factor rotation is performed, one should not see a report of the variance explained by each factor, after extraction, as these are misleading indices of factor importance, (i.e. ignore the '% of Variance' and 'Cumulative %' columns of numbers under the 'Extraction Sums of Squared Loadings' portion of the Total Variance Explained table in Table 6.9).

Software Procedures

Application	Procedures
SPSS	*Analyze → Dimension Reduction → Factor* ... choose the variables you want to analyse; hit the '*Statistics*' button and select 'Correlation' and 'KMO and Bartlett's test of sphericity'; then hit the '*Extraction*' button and select either 'Principal components' (the default) or Principal axis factoring' ('Maximum Likelihood' is also available as are other methods), select to see the 'Scree plot' and choose how you want to select the number of factors to extract (eigenvalue greater than a specific value (1.0 is default) or name a number to extract); then hit the '*Rotation*' button and choose the rotation method you want to use: SPSS offers both 'varimax' (orthogonal) and 'promax' (oblique) rotation strategies, amongst others; finally hit the '*Options*' button to choose the missing data handling method ('listwise' or 'pairwise') and to choose your preferred 'Coefficient Display Format' (choices include sorting coefficients by size and listing only those coefficients larger than a specific absolute value).
NCSS	*Analysis → Multivariate Analysis → Principal Components Analysis* or *Factor Analysis* (the only common factor analysis method NCSS offers is principal axis factoring) ... choose the variables you want to analyse. NCSS only offers 'varimax' and 'quartimax' orthogonal rotation strategies; it cannot extract correlated components or factors. Also, NCSS only offers listwise deletion of missing data (or it can estimate the missing data values – see *Procedure 8.2* for discussion of this type of process).
SYSTAT	*Analyze → Factor Analysis* ... choose the variables you want to analyse; then choose from 'Principal components (PCA)', 'Iterated principal axis (PA)' or 'Maximum likelihood (MLA)'. SYSTAT offers different types of orthogonal rotation strategies, including 'varimax', and offers the 'oblimin' oblique rotation strategy instead of 'promax'. Sorted loadings can be selected as an option.
STATGRAPHICS	*Describe → Multivariate Methods → Principal Components...* or *Factor Analysis...* choose the variables you want to analyse and hit the '*OK*' button; in the '*Options*' dialog box that opens, choose the method for handling missing data and how you want the number of components or factors to be decided and hit the '*OK*' button; then choose your desired 'Tables and Graphs' (the Scree Plot is produced by default).
R Commander	*Statistics → Dimensional analysis → Principal-components analysis...* and choose the variables to be analysed, tick if you want the correlation matrix analysed and if you want to see the scree plot (presented as a bar graph). R Commander will not permit component rotation nor can you specify a specific number of components to be retained (i.e. does not use the eigenvalue greater that 1.0 rule); it extracts as many components as there are

(continued)

Application	Procedures
	variables and you just have to which to interpret on the basis of the scree plot. Generally, for component analysis, **R** Commander would not be the preferred software choice. *Statistics* → *Dimensional analysis* → *Factor analysis*... (**R** Commander will only perform maximum likelihood factor analysis) and choose the variables you want to analyse and choose your desired rotation method (either 'Varimax' or 'Promax'). The package will then ask you how many factors you want to be extracted and will provide a hypothesis test for that number of factors. However, Field et al. (2012) indicated that this approach does not always work and that figuring out why it doesn't work can be difficult. Field et al. (2012) show how to do other types of factor analysis, using available **R** packages (such as *psych*), rather than **R** Commander.

Procedure 6.6: Cluster Analysis

Classification   Multivariate; correlational; nonparametric.
Purpose   To form relatively homogeneous groups of either observed participants/cases in a sample or, less commonly, variables in a sample (similar to factor analysis), based on the extent of similarity or dissimilarity between them.
Measurement level   The variables employed in a cluster analysis can be measured at any level as long as the measure of association, dissimilarity or correlation used to relate each participant (variable) to other participants (variables) is appropriate for that level of measurement.

Cluster analysis refers to a collection of statistical procedures designed to cluster participants/cases or variables into homogeneous subsets of an entire sample. Very simply, a 'cluster' is a grouping of participants/cases (or variables) formed in such a way that members of a specific cluster are more similar to (or correlate more highly with) each other than with any other participants (variables) in other clusters. There are two broad classes of clustering approaches: *nonhierarchical clustering* and *hierarchical clustering*. Relationships between entities may be assessed using a correlation-type index (which measures similarity such that a larger number reflects greater similarity between two participants/cases or two variables (two identical cases, for example, will have a correlation of 1.0) or a distance-type index (which measures dissimilarity such that a larger number reflects greater dissimilarity or difference between two participants/cases or two variables (two identical cases, for example, will have a distance index of 0.0; see the discussions in Aldenderfer and Blashfield 1984; Blashfield and Aldenderfer 1988; and Lorr 1983).

In *nonhierarchical cluster analysis*, clustering occurs in a single iterative stage where a specific number of cluster starting points (called 'cluster seeds') is identified, all entities are then classified into a cluster on the basis of the cluster seed they are most similar (or least dissimilar) to, and the cluster memberships are rechecked for optimal entity allocations in an iterative fashion (entities may be reallocated to another cluster they become closer to during such iterations) until no further changes in cluster membership occur. Nonhierarchical cluster analysis offers an additional benefit in that once the clusters have been identified, new observations can be easily allocated (i.e. classified) to the cluster to which they are most similar.

In *hierarchical cluster analysis*, the clustering process generally involves two distinct phases. The first phase involves assessing the similarity or dissimilarity between all pairs of entities. The second phase occurs using a sequence of merges where each merge joins the two most similar (or least dissimilar) entities into a cluster using a specific mathematical algorithm. The merging process continues until all entities have been successively grouped into one large cluster. The distinctive outcome from the hierarchical clustering process is a *tree diagram* or *dendrogram* which traces the history of all of the merges.

In nonhierarchical clustering, the researcher must specify the number of clusters to construct whereas, in hierarchical clustering, the researcher must decide where, in the hierarchy of merges depicted in the dendrogram, the appropriate number of clusters is to be found. While there are numerous varieties of both nonhierarchical and hierarchical clustering algorithms, the most widely used (and most commonly available in statistical packages) nonhierarchical cluster algorithm appears to be the *K-Means* algorithm and the most widely employed hierarchical clustering algorithm (also commonly available in software packages) appears to be *Ward's Method* (also known as the *Minimum Variance Method*).

When cluster analysis is applied to variables, it performs a function similar to that of exploratory factor analysis (see *Procedure 6.5*). In variable clustering, the appropriate measure of similarity to use is Pearson correlation (or some variant of it appropriate for the level of measurement of the variables—recall *Procedures 6.1* and *6.2*).

> We could have performed a cluster analysis of the nine work attitude items from the QCI database to identify a smaller number of clusters containing those items that had similar patterns of responses across the sample of inspectors (i.e. highly correlated with each other). Each resulting cluster would represent a different facet of work attitudes.

Cluster analysis of variables does not use the same mathematical processes as factor analysis and the two approaches may produce differing outcomes for the same set of data. Cluster analysis is less mathematically rigorous, requires less stringent assumptions about the distributions of the variables, does not require a large sample size to achieve stable results and offers an alternative (not necessarily a substitute for) to the factor analysis of variables.

Procedure 6.6: Cluster Analysis

[Note: If you are looking at clustering variables, the most defensible procedure is the *ICLUST* procedure designed by Revelle (1978, 1979, 2019; http://cran.r-project.org/web/packages/psych/psych.pdf, accessed 9th Sept 2019; see also https://www.personality-project.org/r/html/ICLUST.html, accessed 9th Sept 2019) and more recently discussed in Cooksey and Soutar (2006). *ICLUST* is a hierarchical clustering process designed explicitly for the purpose of clustering variables using psychometric criteria (i.e. reliability information). It is available as part of a psychometric package for **R** called *psych*.]

Cluster analysis makes its greatest contribution to multivariate statistical analysis when applied to the clustering of participants or cases in a sample. The resulting clusters are often referred to as a 'typology' and each cluster represents a specific 'type' of participant or case. We do not know what the types are to begin with, so we employ cluster analysis to help us discover them--which makes cluster analysis one of the many types of statistical procedures collectively referred to as *data mining* techniques (see also *Procedure 9.9*). When clustering participants, it has generally been found that a measure of dissimilarity, called squared Euclidean distance, is most appropriate for relating participants' patterns of response to each other. Frequently, the variables to be used in the clustering process are transformed to z-scores (see *Procedure 5.7*) before clustering to ensure that all variables have the same mean and standard deviation. This has the effect of maximising the opportunity for each variable to contribute to the creation of clusters; in the absence of such standardisation, the variables with the largest standard deviations exert the strongest influence on cluster formation.

> Maree could perform a cluster analysis of the 112 quality control inspectors using information from the two inspection performance measures, **accuracy** and **speed** in order to discover if there are certain 'types' of inspectors who hold distinctive patterns of decision performance (e.g. slow but accurate; slow but inaccurate; fast and accurate).
>
> Figure 6.16 shows the dendrogram, produced using SYSTAT, summarising the hierarchical clustering merges produced by Ward's method. The entire sample of QCI inspectors was clustered using standardised (to z-scores) versions of inspection **accuracy** and **log_speed** (the **log_speed** measure was used because we know that the original **speed** measure was highly positively skewed and this could bias any measures of relationship—recall the discussion of a normalising transformation in *Procedure 5.7*).
>
> [Note: This dendrogram includes 111 of the 112 QCI inspectors; one inspector (#13) was missing data for the **log_speed** variable and was therefore deleted from the analysis. Squared Euclidean distance was the dissimilarity measure used for the analysis, plotted along the horizontal axis. The individual inspector IDs appear vertically at the end of merging branches and identify where each inspector was located in the cluster solution. The 4-cluster solution, where dashed lines cut the tree diagram, is superimposed; it is not part of the original SYSTAT tree diagram.]

(continued)

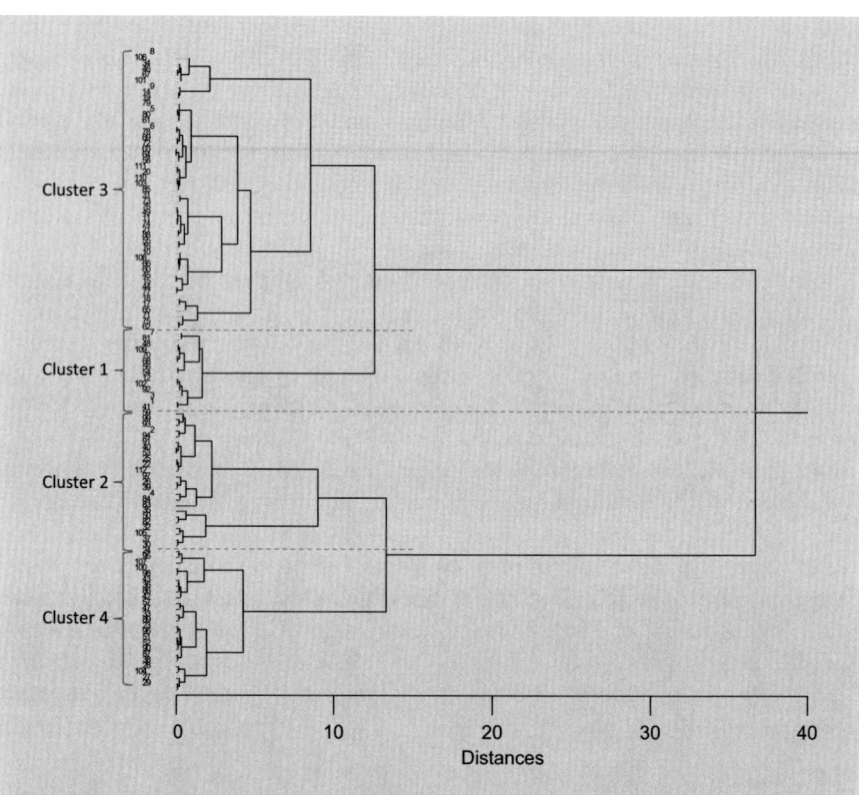

Fig. 6.16 SYSTAT cluster tree or dendrogram summarising Ward's method clustering of inspectors using standardised accuracy and log_speed scores

As you read the dendrogram in Fig. 6.16 from left to right, the inspectors being merged into clusters are becoming progressively more dissimilar to each other in terms of their standardised **accuracy** and **log_speed** scores.

The task of identifying the appropriate number of clusters to interpret generally comes down to visual inspection of the dendrogram. [Note that SYSTAT does offer some 'validity' criteria to help with this decision, such as the root-mean-square standard deviation (RMSSTD), pseudo F-test and Dunn's measure of cluster separation (see SYSTAT Software Inc 2009), however, they are not often conclusive and can provide conflicting information.] Looking at the right-hand side of the plot, you work backward to the left through the tree, splitting successive joins back to the point where it looks like very similar entities are being combined. Every split adds an extra cluster to interpret. The plot in Fig. 6.16 clearly suggests the presence of four distinct clusters of inspectors (separated by dotted grey lines which cut the tree into four main branches); once we cut the plot into four clusters, each inspector becomes a member of one and only one cluster.

(continued)

Procedure 6.6: Cluster Analysis

Fig. 6.17 SYSTAT scatterplot of standardised **accuracy** and **log_speed** scores for the 4 clusters of inspectors (cluster outlines and centroids drawn by researcher)

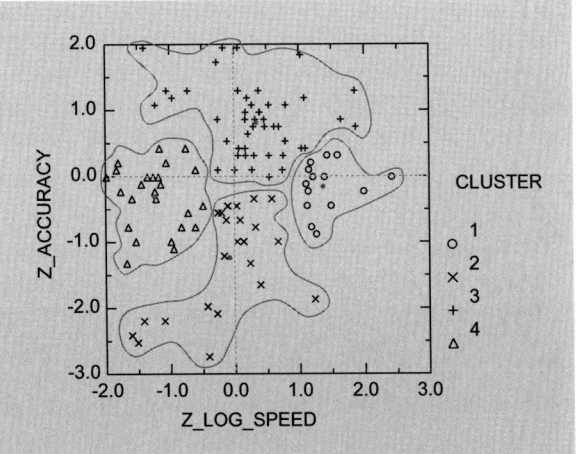

[Scatterplot showing Z_ACCURACY (y-axis, -3.0 to 2.0) vs Z_LOG_SPEED (x-axis, -2.0 to 3.0) with 4 clusters labeled 1 (○), 2 (×), 3 (+), 4 (△)]

Figure 6.17 shows a scatterplot of the standardised **accuracy** and **log_speed** scores but with the members of each cluster being clearly denoted by a different plotting symbol and with the total membership of the cluster being surrounded by a black outline. The grey filled circle inside each cluster region indicates where the mean for each of the two variables falls within the cluster (each grey circle is called a 'centroid'). To facilitate interpretation, clusters can be given descriptive 'names' to summarise their positioning on the two variables (remember that the variables used to construct these clusters were standardised to z-scores, so 0.0 becomes the reference point for interpreting each variable—denoted by the dashed lines).

Inspectors in cluster 1 (symbolised by the open circles) tended to take well above an average amount of time to make their inspection decisions with such decisions being just slightly below average in accuracy, compared to the other clusters. Maree could label this cluster 'slow decision makers'. Inspectors in cluster 2 (represented by the x symbols) tend to make their inspection decisions with about average speed but with well below average accuracy, compared to the other clusters. Maree could label this cluster 'low accuracy inspectors'. Inspectors in cluster 3 (represented by the + symbols) tended to make their decisions with just above average speed but with well above average accuracy, compared to the other clusters. Maree could label this cluster 'high accuracy inspectors'. Inspectors in cluster 4 (symbolised by the open triangles) tended to make very fast inspection decisions (i.e. very fast) with just below average accuracy. Maree could label this cluster 'fast decision makers'.

We can see that what results from the clustering of participants is a classification typology. While the typology itself is useful in helping us understand the nature of the participants in a sample, it could also be important for us to know whether or not the typology is useful for helping us understand how the clusters relate to other variables and measures not employed in the clustering process. This is a process known as **external validation** (see Aldenderfer and Blashfield 1984) and it focuses on the external validity and utility of the cluster structure, namely, is the cluster typology useful for predicting differences between inspectors on other measures?

> Once the 111 QCI inspectors have been clustered, Maree could see if the resulting clusters differed in terms of their gender, educational level, mental ability, job satisfaction, perceptions of their working environment and general attitudes, using other procedures described in this book, e.g. *Procedures 7.1, 7.6*, and *7.16*. However, she can also display the story graphically.
>
> Figure 6.18 shows a bar chart summarising the crosstabulation (recall *Procedure 5.1*) of inspectors' membership in a specific cluster (**cluster** is a new variable created by the SYSTAT clustering procedure) with their educational level (**educlev**). The strong trend in this figure reveals that the 'High Accuracy' cluster comprises a much larger percentage of High School Only-educated inspectors whereas the percentage of Tertiary Qualified inspectors is relatively higher in the 'Fast Decisions' cluster. There was a smaller tendency for Tertiary Qualified inspectors to out-number High School Only-educated inspectors in the 'Slow Decisions' and 'Low Accuracy' clusters.
>
> Figure 6.19 shows a bar chart summarising the crosstabulation of inspectors' membership in a specific cluster with their gender (**gender**). The trend here is much less pronounced compared to the **educlev** pattern, with females slightly out-numbering males in the 'Low Accuracy' and 'Fast Decisions' clusters and vice-versa for the other two clusters.
>
> **Fig. 6.18** SYSTAT comparison of clusters with respect to **educlev**

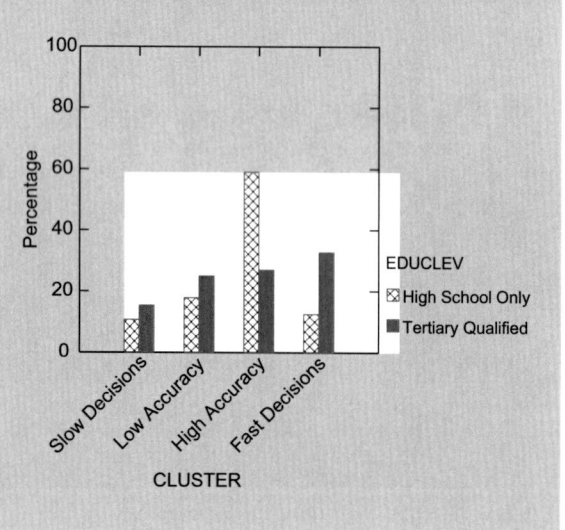

(continued)

Fig. 6.19 SYSTAT comparison of clusters with respect to **gender**

[Bar chart: Percentage (y-axis, 0–100) by CLUSTER (Slow Decisions, Low Accuracy, High Accuracy, Fast Decisions) with Male and Female bars for GENDER]

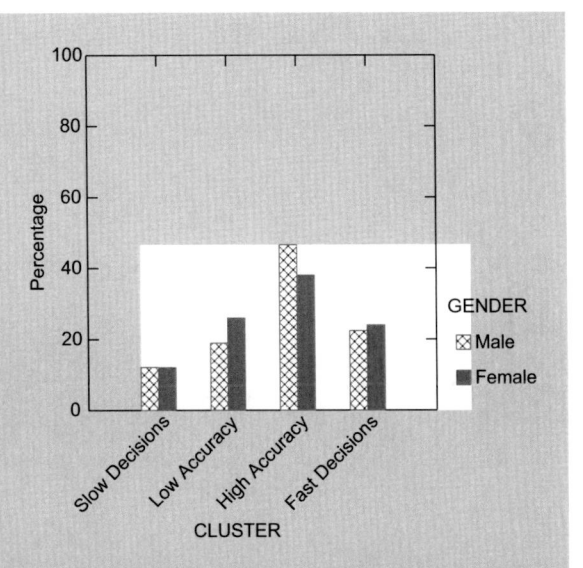

Figure 6.20 presents a profile line graph, produced using SPSS, comparing clusters in terms of their mean standardised scores on the **mentabil**, **jobsat** and **workcond** as well as the **quality**, **training** and **orgsatis** components. The graph shows how the clusters differ in terms of abilities and perceptions of conditions in their workplace. The largest differences between the four clusters appear to occur on the **jobsat**, **workcond** and **orgsatis** variables; differences on the **workcond** variable were especially pronounced. 'High Accuracy' inspectors (who also showed slightly slower than average decision speeds, recall Fig. 6.17) reported the most positive perceptions of their working conditions, followed by the 'Slow Decision' inspectors. The 'Low Accuracy' cluster showed the least favourable perceptions of their working conditions, followed by the 'Fast Decision' cluster (which also had lower than average accuracy scores, again recall Fig. 6.17). This may not mean that **workcond** levels caused lower **accuracy** scores and slower decision speeds in the clusters, but there certainly seems to be some connection. In terms of job satisfaction, 'High Accuracy' inspectors reported an above average level of job satisfaction, whereas the other three clusters all displayed a below average level of **jobsat**.

Figure 6.20 also shows strong cluster differences in terms of organisational satisfaction levels (recall this was one of the component scores created as a part of the component analysis reported in *Procedure 6.5*). Here, the 'Slow Decision' inspectors reported a well above average sentiment regarding their satisfaction with the processes and policies in their organisation; the 'High Accuracy' cluster was just above average and the other two clusters were below average (the 'Fast Decision' inspectors particularly so). Interestingly, the 'Slow Decision' cluster displayed the most positive attitudes on all three component dimensions: **quality**, **training** as well as **orgsatis**.

(continued)

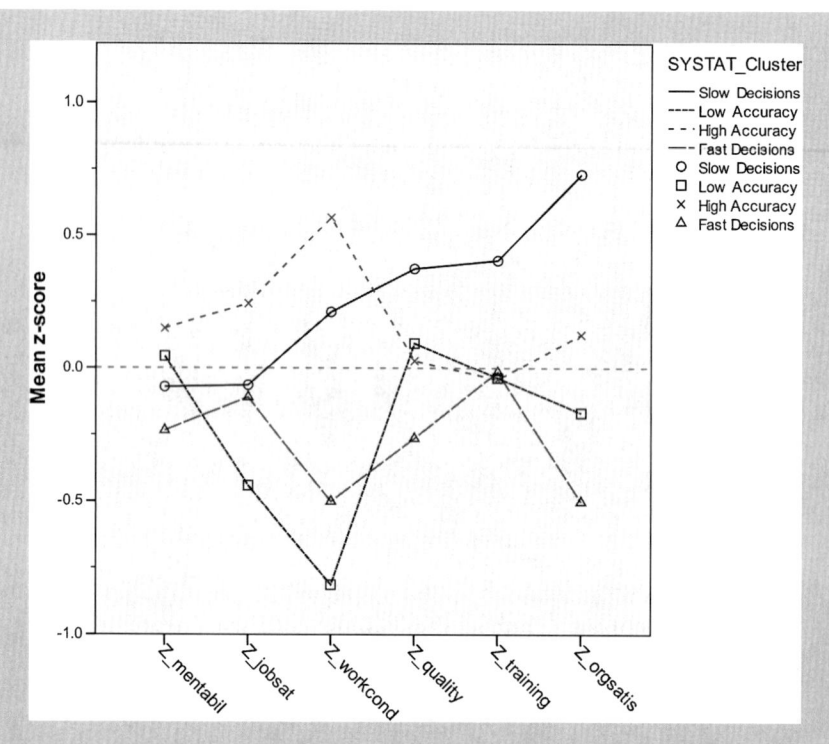

Fig. 6.20 SPSS comparison of clusters using mean z-scores on six 'external' validation variables

Overall, the story appears to be that the best performing inspectors (those in the 'High Accuracy' cluster) tended to be High School Only-educated and expressed quite favourable sentiments about their job and working conditions. The 'Slow Decisions' cluster also tended to report favourable working conditions as well as very high satisfaction with their organisation (but only average satisfaction with their jobs). In contrast, the 'Low Accuracy' cluster reported very poor working conditions and low job satisfaction. Interestingly, both the 'Low Accuracy' and 'Fast Decisions' clusters were populated by a greater proportion of Tertiary Qualified inspectors and somewhat more females than males. It does appear that the four QCI clusters are useful in helping us to understand some of the patterns on other variables.

Advantages

When applied to participants or cases, cluster analysis serves to produce a typology of homogeneous groups of individuals that are well-differentiated across the clustering variables used. Cluster analysis provides an empirical way to discover sub-groups or data patterns within a sample which may then be used in further analytical and interpretive work including the subsequent classification of new observations.

Clusters can be externally validated by relating participant or case membership back to other relevant variables not explicitly used by the clustering procedure. The major advantage of nonhierarchical clustering procedures is their capacity to handle large sample sizes and the fact that the cluster solution is iterated until stable memberships are achieved leading to an optimal classification with respect to the algorithm employed. The major advantage of hierarchical procedures is that an entire typology is built up, comprising a complete history of all clusters formed, which can be succinctly represented in a tree diagram or dendrogram. Clusters and related sub-clusters become readily apparent in this method (in a manner similar to the biological taxonometric system of phylum-class-order-family-genus-species).

Cluster analysis can also provide an alternative method to exploratory factor analysis for the reduction of multivariate data when applied to variables. The resulting 'clusters' could be interpreted in a fashion similar to, but not identical with, the interpretation of 'factors'. However, to do a proper job of clustering variables, specialised software should be used, such as *ICLUST*.

Disadvantages

There are a number of choices that must be made before implementing a cluster analysis of either participants/cases or variables:

- which clustering algorithm (out of the many possible) to use;
- whether or not to standardise variables before clustering;
- how best to interrelate the entities being clustered (there are numerous possible indices of association, similarity, or dissimilarity); and
- how many clusters to interpret.

Often these choices have to be made somewhat subjectively as cluster analysis is a branch of multivariate statistics that is still evolving, and the number of algorithms keeps on growing. For example, NCSS offers 'fuzzy clustering' (where cluster membership is graded according to probability of membership) and 'regression clustering' (which assesses predictive relationships through a clustering process); SPSS offers 'two-step clustering' (which can handle both categorical and continuous variables) and 'nearest neighbour clustering' (a type of nonhierarchical method) and SYSTAT offers 'density clustering' methods (which form a class of hierarchical clustering algorithms).

With respect to whether to standardise, the decision should be based on whether you want the variables with the largest standard deviations to dominate the cluster building process. If you do, then you should not standardise the variables before analysis. However, if you want the variables to be given equal emphasis in the analysis, then you should standardise the variables (usually a z-score transformation, although other standardising methods are available) before clustering.

In terms of measuring similarity or dissimilarity between participants or cases, there is debate about whether a correlation (which reflects similarity) or distance (which reflects dissimilarity) index should be used. Correlation only reflects pattern differences (ups and downs) across the variables whereas distance reflects not only pattern differences but also differences in mean level and variability of the variables being used to assess the relationship between entities. Knowing this, distance (most often in the form of squared Euclidean distance) is preferred as it uses more of the information in the data (see the discussion in Chapter 3 of Lorr 1983).

One disadvantage of nonhierarchical procedures is that a specific number of clusters must be sought, leading to the problem of how best to choose a final solution. The final choice is often subjective. Another problem with nonhierarchical clustering procedures is how to choose the starting points for the number of clusters to be constructed. Different software packages and different algorithms implement different rules for determining starting points, which means that solutions can be very software- or algorithm-dependent. You can improve things in this regard by using a hierarchical cluster analysis to give a preliminary indication of the number of clusters to examine, and then use the means of the variables in those clusters as the starting 'seed' points for nonhierarchical clustering.

One disadvantage of hierarchical clustering procedures is that once an entity is clustered, it is never relocated, even if moving it to another cluster would produce a more optimal solution. The clustering solution is entirely driven by the ordered sequence of merging steps. This specific disadvantage can be counteracted by the two-stage hierarchical → nonhierarchical approach described in the previous paragraph. This hybrid approach allows for mis-classified cases in the dendrogram to be re-allocated to other clusters to which they are more similar.

Another problem is that the interpretation of the dendrogram gets rather subjective when it comes to visually deciding on the proper number of clusters to interpret. Even though there are a number of possible 'validity' indicators one could use to help make this decision, none is definitive. If you have a large number of entities to cluster, then hierarchical procedures become very unwieldy to implement and interpret. In such cases, nonhierarchical cluster analysis would be the preferred approach.

If you have no prior knowledge of the underlying grouping typology in a sample, cluster analysis is one of the few approaches available to discover those groups (other approaches are discussed in *Procedure 9.9*). But if such prior knowledge does exist, discriminant analysis (see *Procedure 7.17*) should be considered as a more mathematically rigorous alternative.

Where Is This Procedure Useful?

Multivariate research that obtains numerous measurements from a sample of participants, via mail questionnaire, interview, or observational methods, can make ready and practical use of cluster analysis if the formation of a participant typology or taxonomy is of interest. Typologies are generally more useful as research outcomes simply because it is far easier to discuss and make use of a small number of highly differentiated clusters that reflect common patterns of response than it is to keep focusing on individual participant patterns. Any research design where a behavioural researcher would consider using factor analysis could conceivably consider the use of cluster analysis as an alternate method, particularly where assumptions are violated, or the sample size is much smaller than recommended for factor analysis.

Cluster analysis has proven to be an invaluable technique for psychologists interested in devising empirically justifiable mental disorder classification systems based on measurements of patient symptoms. Once a taxonomy of symptom patterns has been identified, the task of diagnosing disorders and devising treatments for them can be made much simpler and more effective. Cluster analysis also enjoys wide use by marketing researchers interested in identifying sub-groups of consumer populations who share similar preferences for products and/or product attributes. Thus, it is one important approach to the methodology known as *market segmentation*. Once a market has been segmented into clusters or 'segments', different advertising strategies, product packaging options, and so on, can be designed to more precisely target specific segments. Cluster analysis is also a useful data mining techique for exploring patterns of entities in large databases (see *Procedure 9.9*).

Software Procedures

Application	Procedures
SPSS	*Analyze* → *Classify* → *Hierarchical Cluster* ... then select the variables to cluster on, select a labelling variable to 'Label Cases By' (for the dendrogram), make the choice of clustering algorithm (including 'Ward's Method'), similarity or dissimilarity measure and variable standardisation option ('*z*-scores') by selecting the '*Method...*' button and chose 'Dendrogram' after hitting the '*Plots*' button. By using the '*Save...*' button, you can opt to save the identifier numbers, for a specified number of clusters, into the data file to use for further analyses. *Analyze* → *Classify* → *K-Means Cluster* ... then select the variables to cluster on, select a labelling variable to 'Label Cases By' and indicate the 'Number of Clusters' desired; hitting the '*Options*' button opens a dialog box for selecting the statistics you want reported. This procedure will not standardise clustering variables for you, so you will need to do this before you run the procedure. By using the '*Save...*' button, you can opt to save the

(continued)

Application	Procedures
	identifier numbers, for a specified number of clusters, into the data file for further analyses.
NCSS	*Analysis → Clustering → Hierarchical Clustering/Dendrograms* and select the variables to cluster on (you can choose variables with different scales of measurement) and choose the desired 'Linkage Type' ('Ward's Minimum Variance' is the default). This procedure will not standardise clustering variables for you, so you will need to do this before you run the procedure. The 'Storage' tab allows you to name a variable for storing cluster ID numbers for further analyses. *Analysis → Clustering → K-Means Clustering* and select the variables to cluster on and indicate the minimum and maximum number of clusters you want to look at and you need to name the final number of clusters to report on (usually this means you will have to run this procedure twice to get a final analysis). This procedure will not standardise clustering variables for you, so you will need to do this before you run the procedure. The 'Storage' tab allows you to name a variable for storing cluster ID numbers for further analyses. NCSS offers some interesting types of cluster analysis, not offered by SPSS, SYSTAT, STATGRAPHICS or **R** Commander, including 'fuzzy clustering' and 'regression clustering'. NCSS also gives you finer control over the shape and size of the dendrogram, including options for zooming in on a portion of the tree.
SYSTAT	*Analyze → Cluster Analysis → Clustering → Hierarchical...* then select the variables to cluster on, select the 'Linkage' method, whether or not you want to 'Save' cluster IDs to a variable for further analyses and, under the '*Options*' tab, you can select one or more 'Validity' indices to see. This procedure will not standardise clustering variables for you, so you will need to do this before you run the procedure. SYSTAT differs from SPSS, NCSS, STATGRAPHICS or **R** Commander in offering the option to cluster both variables and participants simultaneously and display the results in a hybrid double dendrogram-shaded matrix graphic. *Analyze → Cluster Analysis → K-Clustering...* then select the variables to cluster on, choose how many 'Groups' to report on, whether or not you want to 'Save' cluster IDs to a variable for further analyses and, under the '*Initial Seeds*' tab, you can select how SYSTAT chooses the cluster seed points to start the clustering process (options include using a hierarchical clustering process to identify the cluster seed points). This procedure will not standardise clustering variables for you, so you will need to do this before you run the procedure.
STATGRAPHICS	*Describe → Multivariate Methods → Cluster Analysis...* then select the variables to cluster on, select a labelling variable to be the 'Point Label' (for the dendrogram, for example); when you hit '*OK*', the '*Cluster Analysis Options*' window opens and you can select the clustering method to use (a hierarchical method or *k*-Means), indicate if you want to cluster observations or variables, indicate how many clusters to report on and whether or not you want to standardise the variables before analysis; then hit '*OK*' again to bring up the '*Tables and Graphs*' window and choose the types of output you want to see.
R Commander	*Statistics → Dimensional analysis → Cluster analysis → Hierarchical cluster analysis...* then select the variables to cluster on, choose the number

(continued)

Application	Procedures
	of clusters to report on, indicate which clustering method you want to use and which distance measure you want to use. To get statistics on a specific number of clusters and to add the cluster identifier variable to the database, you have to run two additional procedures: *Statistics → Dimensional analysis → Cluster analysis → Summarize hierarchical clustering...* and *Statistics → Dimensional analysis → Cluster analysis → Add hierarchical clustering to data set...* This procedure will not standardise clustering variables for you, so you will need to do this before you run the procedure. *Statistics → Dimensional analysis → Cluster analysis → K-means cluster analysis...* then select the variables to cluster on, choose the number of clusters to report on, indicate which reports you want to see and if you want the program to create a new variable with the cluster identifier for each case in the sample. This procedure will not standardise clustering variables for you, so you will need to do this before you run the procedure. Everitt and Hothorn (2006, ch. 15) illustrate how to conduct cluster analyses using functionalities and packages within the **R** program.
ICLUST	This is a dedicated software procedure for clustering variables using psychometric decision criteria, available in the psych package for **R**. For information on the ICLUST procedure, read Cooksey and Soutar (2006) and visit the R Project website to download the psych package: https://cran.r-project.org/web/packages/psych/index.html, accessed 9th Sept 2019. Extensive and up-to-date documentation is available in http://cran.r-project.org/web/packages/psych/psych.pdf, accessed 9th Sept 2019.

Procedure 6.7: Multidimensional Scaling and Correspondence Analysis

Classification	Multivariate; correlational; nonparametric.
Purpose	The purpose of multidimensional scaling is to explore the dimensionality of a set of entities whose nature can vary from simple attitude ratings to implicit or explicit assessments of similarity or preference.
	The purpose of correspondence analysis is to accomplish a similar task to multidimensional scaling in the context of contingency tables by showing how two category systems relate to each other.
Measurement level	For multidimensional scaling, the entities in the analysis may be measured at virtually any level, but the technique is often applied where only ordinal-level data are present.
	For correspondence analysis, the entities are typically measured at the nominal- or ordinal-level at best.

Multidimensional Scaling

Multidimensional scaling is a multivariate technique that can be used to explore the dimensionality of a set of entities. However, the nature of the entities can be much more exotic than those which are appropriate for factoring or clustering. It is possible to use multidimensional scaling to establish the dimensionality of items measured in a Likert-type scaling format (such as the work attitude items in the QCI database). It is also possible that the entities represent actual products, organisations, or any other objects which humans might perceive and compare.

For example ... Market researchers often use multidimensional scaling to understand the dimensionality of human perceptions of products like soft drinks or cars. They accomplish this by first having people rate the degree of perceived similarity or preferences between all possible pairs of a set of products Then the resulting sets of similarity or preference ratings are analysed to identify the underlying dimensions. When used in this type of research context, multidimensional scaling may be referred to as 'perceptual mapping'. If such ratings of similarities or preferences are analysed on a person-by-person basis, the approach is called 'individual differences scaling' and is useful for looking at individual variations in preference structures.

When multidimensional scaling is applied to entities assumed to be measured at the interval or ratio-level, it is called '*metric multidimensional scaling*'. When it is applied to entities assumed to be measured at the ordinal-level, it is called '*nonmetric multidimensional scaling*'. One interesting contribution that multidimensional scaling makes to behavioural research is the potential for discovering the subjective dimensions that people use for making comparisons among holistic objects without the researcher needing to prescribe or even know what the relevant attributes of those objects are.

Another interesting methodology is made possible using multidimensional scaling techniques, namely the dimensional analysis of how people sort various stimuli (e.g. photographs, products, employees) into distinct categories or groups. Rather than have the sorters try to articulate the criteria they used to produce the groups (which they frequently have difficulty doing), multidimensional scaling can be used to try and recover the dimensions used in making judgments regarding which stimuli to place together in a group and which to keep in separate groups. This methodology allows us to quantitatively explore how people subjectively categorise and dimensionalise stimuli (where they are in effect doing an intuitive multidimensional scaling in their heads when they sort things).

When a multidimensional scaling analysis is performed, it is typical for the researcher to examine solutions of varying dimensionality to see which solution fits best and is most interpretable. To assist in the detection of the proper number of dimensions for interpretation, a measure called 'Stress' is employed. Stress ranges from 0.0 to 1.0 with numbers approaching 0.0 indicating better fit of the solution to the data. Stress is related to R^2 (proportion of variability amongst the objects being scaled explained by the specific dimensional solution); most software packages will

Procedure 6.7: Multidimensional Scaling and Correspondence Analysis

report both values. It is important to realise that solutions having 3, 4 or 5 (the maximum allowed in most computer packages) dimensions differentiating the entities are rather tricky to interpret because they are more difficult to visualise. Interpreting a multidimensional scaling solution is largely a visual task.

Every multidimensional scaling solution results in an array of stimulus coordinates, one set for each dimension, which shows where each entity is located along that dimension (these are somewhat similar in conceptualisation to loadings in factor analysis, see *Procedure 6.5*). Examining the content or nature of the entities which are at the extremes of each dimension or which tend to group together along one or more dimensions will help the researcher to interpret the meaning of the dimension. These coordinates bear no actual relationship to factor pattern coefficients, and in some ways, may be considered arbitrary numbers. For this reason, many researchers prefer to examine their solutions from all different angles, not just following the two axes produced by the statistical software program, in an attempt to discover what is differentiating the different clouds of points.

Another use for the stimulus coordinates is as predictors in multiple regression analyses that try to predict ratings or other measures (external to the scaling analysis) on the objects being analysed. This exercise, which is technically called 'property-fitting' (see Schiffman et al. 1981) amounts to a type of external validation for the scaling solution and it may facilitate interpretation of the dimensions as well.

For example ... *If we have a manager sort his/her subordinates into groups which represent poor, average, and good levels of work performance, we might use the resulting multidimensional scaling coordinates to predict those subordinates' absenteeism levels, overtime levels, and/or accident or error rates. In this way, the scaling dimensions can be directly related to certain measurable external properties of each scaled object.*

Multidimensional scaling, when applied to the scaling of attitude rating variables, can allow you to recover the dimensional structure of the variables without having to assume that the items have been measured on an interval scale. If you recall, from ch. 2, we flagged that not all researchers were convinced that attitude measurement items such as Likert-type scales, formed an interval scale. This was because we really can't be sure if the differences between successive numbers on the scale (e.g. between 1 = strongly disagree and 2 = moderately disagree and 3 = slightly disagree) represent equal increments in the amount of attitude. If making the equal intervals assumption is a concern, then multidimensional scaling provides a way forward.

> Maree can use multidimensional scaling to analyse the similarities among the nine work attitude items in the QCI database to accomplish an analysis somewhat similar in nature to what factor analysis (see *Procedure 6.5*) or cluster analysis (see *Procedure 6.6*) would accomplish. However, for this analysis, we will assume that the attitude measures are ordinal in nature so that the use of nonmetric multidimensional scaling can be illustrated.

(continued)

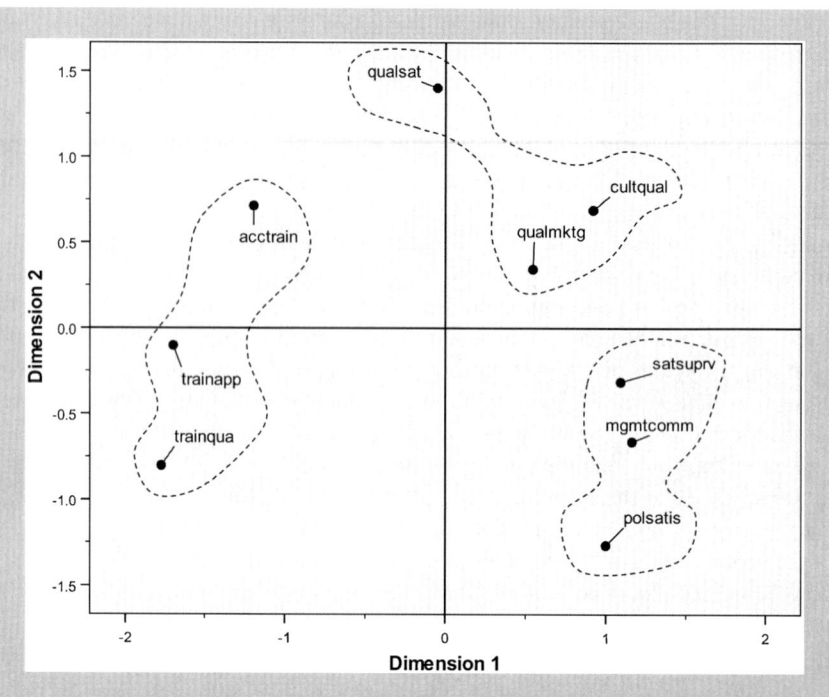

Fig. 6.21 Plot summarising the multidimensional scaling analysis of the nine QCI attitude items in two-dimensional space

Figure 6.21 shows the results of a nonmetric multidimensional scaling analysis of the nine QCI attitude items in two dimensions. Two dimensions were chosen because the value for the 'Stress' measure was .088 (which translates to an approximate R^2 of .953) which was less than the typically recommended level of .10 for a 'good' solution coupled with the fact that there was minimal improvement in stress with three dimensions—see Kruskal and Wish (1978). The plot locates points according to the stimulus coordinates which were produced by the *Multidimensional Scaling (ALSCAL)...* procedure in SPSS.

It seems clear that the two dimensions distinguish well the disparities between three clouds of variables (surrounded by dashed lines, drawn by the researcher): the quality-related variables, the training-related variables, and the organisational satisfaction variables (an interpretive outcome similar to principal component analysis but requiring only two dimensions rather than three factors to depict). The two axes (shown as the black lines at the zero point of each axis in Fig. 6.21) appear to be interpretable where they are placed. Dimension 1 seems to be differentiating work attitude items on the basis of being training-related versus non-training related whereas Dimension 2 seems to be differentiating work attitude items on the basis of being inward-focused (how the work facet reflected in the item impacts on inspectors at an individual

(continued)

or personal level) versus outward-focused (how the work facet reflected in the item impacts on inspectors at an organisational level).

Note that unlike factor or cluster analysis, it would theoretically be possible to produce a multidimensional scaling of these nine attitude items for each individual inspector (thus producing 112 individual scalings of the attitude items). We could then examine the results for perceptual similarities or differences (this type of multidimensional scaling, alluded to earlier, is called INDSCAL for Individual Differences Scaling—see Schiffman et al. 1981—and is available via SPSS).

Correspondence Analysis

Correspondence analysis is essentially a multidimensional scaling procedure designed primarily to be used in the analysis of crosstabulation or contingency tables. Thus, correspondence analysis is very useful for understanding the dimensionality of the relationships between the category systems for two nominal-scale variables. However, ordinal-level measures can also be analysed using correspondence analysis. The technique explores the 'correspondence' between the specific categories of the two variables being related in multidimensional space. The actual data analysed are counts from a contingency table which are then statistically converted into a measurement system that facilitates the recovery of dimensional structure.

Like multidimensional scaling, correspondence analysis produces dimensions of category coordinates which help to locate the specific categories of each variable in relation to each other. The value for learning that correspondence analysis adds over a standard analysis of contingency table association (recall *Procedure 6.2*) is that the categories for each variable can be plotted in a common dimensional space to more clearly show their relationships. The degree-of-fit for a correspondence analysis solution of a specific number of dimensions is called 'Inertia' (somewhat analogous to the 'Stress' measure in multidimensional scaling). If more than two categorical variables are to be related in the analysis, the approach is called *multiple correspondence analysis*.

Maree conducts a correspondence analysis, produced using SPSS, between the **company** variable and the quality of working conditions (**workcond**) variable from the QCI database. The **company** variable is nominal, and for purposes of this analysis, we consider **workcond** to be an ordinal variable. Figure 6.22 shows the joint correspondence plot relating the 2-dimensional coordinates for the **company** variable to those of the **workcond** variable.

(continued)

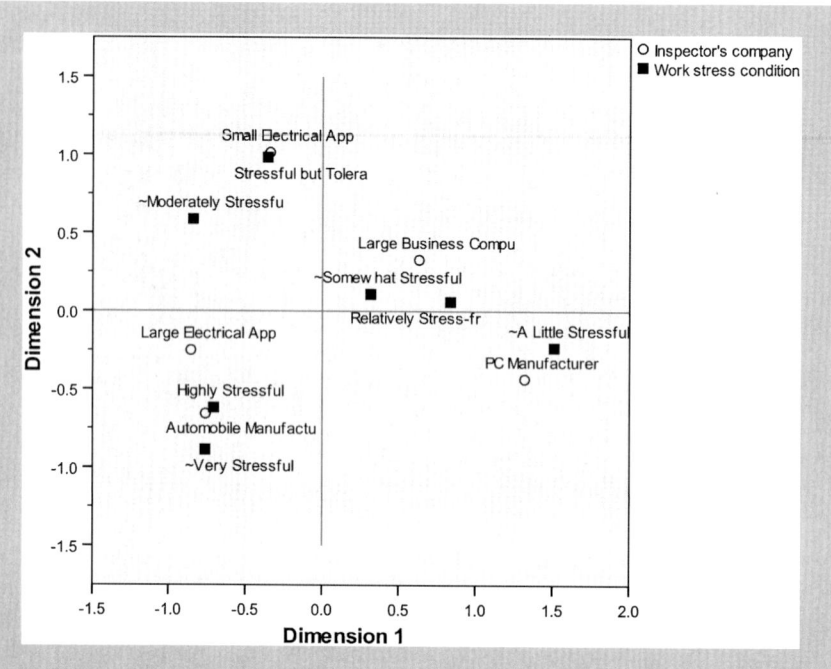

Fig. 6.22 Plot of the 2-dimensional space resulting from a correspondence analysis of the **company** and **workcond** variables

Two dimensions were chosen because the value for inertia was around .82. This equated to explaining about 83% of the variance leaving little to be gained by considering the 3-dimensional solution. Labels have been attached to each plotted point to more clearly identify what specific category each point represented. Additionally, the points for categories of the **workcond** variable have been coded as black squares and company categories have been coded as open circles: this helps to highlight the associational linkages between **workcond** categories and the different types of **company**. [The labels attached to each category are shown for both variables. For **workcond**, the '~' symbol is used as the first character in category labels that only approximate the meaning of the rating number since only ratings of 1, 4 and 7 had specific verbal anchor labels in Maree's survey.]

We can see from this plot that working in companies that manufacture small electrical appliances was most closely associated to working conditions which are around the moderate level (equivalent to the rating categories of 4 ('stressful but tolerable') and 3 ('~moderately stressful') on the **workcond** scale). Working for either type of computer company (large business or PC) was most closely associated with nearly or totally stress-free working conditions (equivalent to ratings of 5 ('~somewhat stressful', 6 ('~a little stressful') or 7 ('relatively stress-free') on the **workcond** scale). Finally, working for

(continued)

automobile or large electrical appliance manufacturers was most closely associated with 'highly stressful' or '~very stressful' working conditions (equivalent to a rating of 1 or 2 on the **workcond** scale).

It seems apparent that this is a much more detailed and informative analysis than merely reporting that the overall association between **company** and **workcond** was a Cramer's V value of .466 (moderate association). For the **company** variable, Dimension 1 appears to differentiate companies on the basis of technological complexity (computers are highly complex; electrical appliances and automobiles are much less technically complex) whereas Dimension 2 appears to differentiate companies on the basis of product size (larger versus smaller products). For the **workcond** variable, Dimension 1 appears to differentiate stress levels on the basis of severity (relatively lower versus relatively higher) whereas Dimension 2 appears to differentiate stress levels on the basis of adaptability (too little or too much stress versus just the right amount to facilitate performance).

Advantages

Multidimensional scaling has the advantage of being able to recover dimensional structure from a wide variety of types of data that may not be suitable for factor or cluster analysis. Additionally, it opens the door for new and innovative approaches to be used to gain insights into human perceptions and preferences. Data can be at virtually any level of measurement and the attributes of the objects being compared need not be known in order to recover the dimensional structure associated with judgments or perceptions about those objects. Multidimensional scaling can be conducted at the individual or at the group level which paves the way for very fine comparisons between the perceptions of different people.

Correspondence analysis also has the advantage of being able to recover dimensional structure from categorical types of data that cannot be handled by factor analysis. Correspondence analysis provides greater insights into the specific associations between categories of contingency table variables as well as reflecting the dimensionality of those associations.

Disadvantages

Neither multidimensional scaling nor correspondence analysis can test specific hypotheses about the possible dimensionality of data. In this sense, they are both purely exploratory techniques. The interpretation of the outcomes from both types of analysis is largely subjective, heavily visually reliant and can be challenging to accomplish. This is particularly so because the placement of the dimensional axes (which are positioned by the magnitudes and directions of the stimulus coordinates) is often arbitrary and individual researchers may differ in their capacity to conceptualise the potential differentiating criteria for each dimension.

The interpretive process can be facilitated in multidimensional scaling using what is called property-fitting analysis, but no such equivalent technique exists for correspondence analysis. As both types of analysis are iterative in their computations, there is always the risk that a solution for a specific number of dimensions cannot be found (what is called a 'degenerate solution'). There is little one can do if faced with such a 'non'-solution.

Where Is This Procedure Useful?

Multidimensional scaling enjoys important utility in market research for experimental investigations into consumer product preferences. It has proven very useful for experimental investigations into human perception and cognition assessing how people cognitively group together a diverse set of stimuli, e.g. how do children categorise animals, how do farmers categorise photographs of landscapes, how do employers categorise employees, how do men and women categorise potential partners. Finally, multidimensional scaling has been employed as an alternative to factor analysis, particularly where the data are not seen to be suitable for the more rigorous factor analytic techniques.

Correspondence analysis has increasingly been used to examine the relationships between products and their attributes where such are known. It is also useful in survey research where several categorical variables are measured, and we wish to see how the specific categories of one variable relate to the specific categories of another variable. This approach can be especially useful in circumstances where the assumption of equal intervals for attitude scales is untenable. Correspondence analysis simply treats each scale point as a separate category.

Software Procedures

Application	Procedures
SPSS	*Analyze → Scale → Multidimensional Scaling (ALSCAL)...* and choose the variables or objects you want to scale; tick that you want to 'Create distances from data'; press the '*Measure*' button and select Squared Euclidean distance as the interval measure (for example) and indicate if you want to standardise the variables first, then hit the '*Continue*' button; go to the '*Options*' button and choose 'Group plots' as a 'Display' option. SPSS offers the best and most flexible procedure for multidimensional scaling; a

(continued)

Procedure 6.7: Multidimensional Scaling and Correspondence Analysis 223

Application	Procedures
	variety of different scaling approaches can be implemented, including the INDSCAL (individual differences scaling) model. *Analyze → Scale → Multidimensional Scaling (PROXSCAL)...* and *Analyze → Scale → Multidimensional Unfolding (PREFSCAL)...* compute two other types of multidimensional scaling (see Schiffman et al. 1981). For simple correspondence analysis, use *Analyze → Dimension Reduction → Correspondence Analysis ....* and choose one categorical variable to be the 'Row variable' ('Define range' for this variable, indicating the lowest and highest category number to include) and another categorical variable to be the 'Column variable'('Define range' for this variable, indicating the lowest and highest category number to include); under the '*Model*' button, choose the 'Euclidean' model. For multiple correspondence analysis, use *Analyze → Dimension Reduction → Optimal Scaling....*
NCSS	NCSS can perform both metric and nonmetric multidimensional scaling via *Analysis → Multivariate Analysis → Multidimensional Scaling* and choosing the variables/objects to be scaled. All other relevant output is set by default, except for the number of dimensions to evaluate and specific number to interpret, which you can select. However, one major limitation in NCSS here is that it cannot compute the necessary similarities (except for correlations) or distances from the raw data –you need to input the upper triangle of an actual dissimilarity or similarity matrix computed some other way. NCSS can perform simple correspondence analysis via *Analysis → Multivariate Analysis →Correspondence Analysis* and select the two categorical variables to be related. However, one major limitation in NCSS here is that it cannot run the analysis from the raw data; you must first produce the two-way crosstabulation table and provide the resulting counts as the input to the procedure.
SYSTAT	*Advanced → Multidimensional Scaling...* and select the variables/objects you want to scale and choose relevant options to control the scaling process. SYSTAT is not quite as flexible in some scaling models as SPSS, but it does offer some distinct advantages including several approaches to preference mapping, which can be used to carry out 'external property fitting' via *Advanced → Perceptual Mapping...* SYSTAT can perform both simple and multiple correspondence analyses via *Analysis → Tables → Correspondence Analysis...* select your desired categorical dependent and independent variable(s) (select only multiple dependent variables for a multiple correspondence analysis).
STATGRAPHICS	STATGRAPHICS can conduct a multidimensional scaling, but not as a stand-alone analysis; it requires **R** to be installed. However, it can perform a stand-alone correspondence analysis: *Describe → Multivariate Methods → Correspondence Analysis...* (or *Multiple Correspondence Analysis...*) and choose the desired 'Row variable' and 'Column variable' (or choose multiple variables as the 'Columns' for the multiple correspondence analysis), select your desired analysis 'Options' and your desired 'Tables and Graphs'.
R Commander	**R** Commander does not offer the capability to conduct multidimensional scaling or correspondence analyses; however, these analyses can be conducted using functions and packages in **R** (see http://www.statmethods.net/advstats/mds.html and http://www.statmethods.net/advstats/ca.html, both accessed 9th Sept 2019; see also the discussion and illustrations in Everitt and Hothorn 2006, ch. 14).

Procedure 6.8: Canonical & Set Correlation

Classification	Multivariate; correlational; parametric.
Purpose	The purpose of canonical correlation is to correlate one set of variables with a second set of variables where each set contains at least two variables.
	The purpose of set correlation is also to correlate one set of variables with a second set of variables where each set contains at least two variables, but where the influence of other sets of one or more variables have been controlled for (by partialling).
Measurement level	Typically, both canonical correlation and set correlation relate variables measured using interval or ratio-scales, but special methods for transforming or coding data will permit the use of variables measured using ordinal or nominal scales.
	In addition, set correlation can control for variables measured on any scale, if they are appropriately coded or transformed. In this sense, set correlation is one of the most flexible multivariate techniques available.

Canonical Correlation

Canonical correlation is a natural extension of multiple regression analysis (see *Procedure 6.4*) to the case where there are several dependent variables (or criteria) instead of just one.

For example ... In a survey study, we might want to relate a number of demographic characteristics of a sample to political attitudes and preferences or to educational abilities. A behavioural researcher in a marketing organisation may be interested in how the personality characteristics of a sample of participants (assessed perhaps by measures of sociability, emotional intelligence, decision style, and locus of control) relate to their preferences for certain attributes in living arrangements (assessed perhaps by measures of cost, amount of maintenance and upkeep required, convenience of access to shopping and work, prestige of the geographical location, spaciousness and suitability of the living area design and closeness to neighbours). This researcher would be interested in how the personality characteristics, as a set, relate to the living arrangement preferences, as a set. The key question might be "what pattern of personality profile is associated with what pattern of preferred living arrangements"?

Canonical correlation is specifically designed to address this general type of research question: given two sets of variables, what is the correlation between the sets and which specific variables within each set contribute most to that correlational relationship. In a rather crude sense, canonical correlation produces a kind of double-

Procedure 6.8: Canonical & Set Correlation

sided factor analysis with the restriction that a 'factor' derived for one set of variables must be correlated, to the highest extent possible, with a 'factor' derived for the other set of variables. Technically, from a canonical correlation perspective, it makes no difference which set of variables contains the 'dependent' variables and which contains the 'independent' variables. Since all that canonical correlation assesses is correlational relationship, it is convenient to refer to the sets of variables simply as the *Dependent Set* and the *Independent Set* (or sometimes, depending upon the software package used, Set 1 and Set 2, or Set X and Set Y).

A canonical correlation analysis produces two main types of correlational information. Firstly, the largest possible canonical correlation between the two sets of variables is found. This is the maximum Pearson correlation that can be achieved using a specific way of combining the variables from the Dependent Set in relation to a specific way of combining the variables from Independent Set. Each 'way of combining variables' defines a composite which is called a *canonical variate*. Secondly, the relationships between each individual variable and the canonical variate, formed while searching for the largest correlation, are reported.

These relationships (which amount to a weighting system) can take several forms, but the most interpretable form is the 'canonical loading' or 'canonical structure coefficient' which measures the correlation between each variable and the canonical variate it contributes to (these coefficients are very similar to factor structure correlations discussed in *Procedure 6.5*). In a fashion similar to factor analysis, canonical loadings can be 'rotated' to produce a simpler and more interpretable solution (however, this is not often done and only SYSTAT offers the option for such rotation). Canonical loadings or structure coefficients reflect how important each variable was in forming the canonical variate which gave rise to the canonical correlation. For every canonical correlation, there is an associated pair of canonical variates (= two systems of loadings or structure coefficients) produced, one for the Dependent Set and one for the Independent Set.

Some behavioural researchers have argued for the use of a third type of information to be included in any report of a canonical correlation analysis: *redundancy indices*. A *redundancy index* is simply the average squared structure correlation between each variable in the Dependent Set and the canonical variate for the Independent Set and vice-versa. In other words, a redundancy index summarises the extent of overlap (in terms of proportion of variance shared) or redundancy which exists between the two sets of variables; larger values for redundancy indicate large degrees of overlap which will be reflected in a much stronger canonical correlation value. Redundancy indices are asymmetric in that the variables in the Dependent Set may overlap more (or less) with the canonical variate of the Independent Set than the variables of the Independent Set overlap with the canonical variate of the Dependent Set.

There are several possible canonical correlation relationships (and corresponding pairs of canonical variates) in any set of multivariate data. The actual number of possible relationships is equal to the number of variables in the smaller of the two sets being related. Each canonical correlation and its corresponding pair of canonical variates is uncorrelated with all other canonical relationships and can be interpreted separately from all others. Canonical correlation analysis always identifies the strongest multivariate relationship first, then partials this relationship out (recall the discussion in *Fundamental Concept IV*) and identifies the next strongest multivariate relationship that remains. Each multivariate relationship can be tested for statistical significance to determine which relational systems should be interpreted.

Consider the following problem within the context of the QCI database.

> Suppose Maree was interested in exploring the relationship(s) between the two performance measures, **accuracy** and normalised **log_speed** (the *Dependent Set* variables) and the work perception variables, **jobsat**, **workcond**, and the inspectors' scores on the three factors, **quality**, **training**, and **orgsatis**, produced and discussed in *Procedure 6.5* (the *Independent Set* variables). This canonical correlation analysis is summarised in Table 6.10 where, for simplicity of interpretation, only the canonical correlations, canonical loadings for each canonical variate for each variable set and redundancies for each canonical variate are reported (this analysis was computed using SYSTAT, then edited for inclusion here).
>
> Note that there are two canonical correlation relationships reported. It turns out both correlations are statistically different from zero (see *Fundamental concept V*); hence each defines a substantive canonical relationship. To interpret each canonical variate, Maree generally looks for structure correlations in excess of about $\pm.40$ in magnitude as indicating variables that contribute substantively to the relationship (these specific substantive relationships are shown in boldface italics in Table 6.10).
>
> The first canonical correlation relationship is moderate in strength with a correlation value of **.570**. To find out which variables in each set are contributing to this relationship, Maree could look at the canonical loadings. For this first (and largest) canonical correlation, she interprets the first variate for each set and inspects the correlations between the variables in that set and the canonical variate itself (these correlations listed under the '1' headings). The first canonical variate for the Dependent Set shows that **accuracy** and **log_speed** contributed extremely strongly and positively and the first canonical variate for the Independent Set shows that both **workcond** and **orgsatis** contributed substantively and positively, with **workcond** being well and truly the stronger contributor to the relationship.
>
> Maree could interpret this pattern of results as suggesting that as inspectors' accuracy scores increased and decision times got longer, there was a moderate tendency for both the quality of working conditions and satisfaction with aspects of the organisation to be higher. Note: The 'redundancy index' for canonical variate 1 for the Dependent Set was **.202** ($= 20.2\%$ of variance shared between the variables in the Dependent Set and the 1st canonical variate for the Independent Set). Conversely, the 'redundancy index' for canonical variate 1 for the Independent Set was **.094** ($= 9.4\%$ of variance shared between the variables in the Independent Set and the 1st canonical variate for the Dependent Set).

(continued)

Table 6.10 Canonical correlation analysis relating the performance set of variables to the work perceptions set of variables

Canonical Correlations	
1	2
0.570	0.347

Canonical Loadings (y variable by factor correlations)

	1	2
ACCURACY	0.812	-0.584
LOG_SPEED	0.764	0.645

Canonical Redundancies for Dependent Set	
1	2
0.202	0.046

Canonical Loadings (x variable by factor correlations)

	1	2
JOBSAT	0.238	-0.542
WORKCOND	0.980	-0.179
QUALITY	0.226	0.467
TRAINING	0.173	0.282
ORGSATIS	0.583	0.585

Canonical Redundancies for Independent Set	
1	2
0.094	0.023

Maree would repeat this process to interpret the second canonical relationship which has a moderately weak correlation value of **.347**. For this second relationship, and having due regard for the signs attached to each loading (the correlations listed under the '2' headings), she would conclude that as inspection decisions take longer to make but become less accurate, there is a moderate tendency for inspectors' job satisfaction to be lower but also a moderate tendency for their attitudes toward product quality as something important to pursue and their satisfaction with aspects of the organisation to be more favourable (in other words, lower job satisfaction but better work attitudes are somewhat associated with making slower less accurate inspection decisions). Note: The 'redundancy index' for canonical variate 2 for the Dependent Set was **.046** (= 4.6% of variance shared between the variables in the Dependent Set and the 2nd canonical variate for the Independent Set). Conversely, the 'redundancy index' for canonical variate 2 for the Independent Set was **.023** (= 2.3% of variance shared between the variables in the Independent Set and the 2nd canonical variate for the Dependent Set).

Set Correlation

Cohen et al. (2003) showed that canonical correlation could be logically extended to create a more flexible analytical approach called *set correlation*. Using set correlation principles, it becomes possible to consider relating three or more sets of variables by using some sets of variables as control variables to be partialled out of the relationship between two focal sets of interest. In this broader context, canonical correlation could be seen simply as a special case of set correlation where there were no partialling variables being controlled.

For example ... *We might be interested in how consumer personality characteristics relate to product attribute preferences but first we may wish to partial out of that relationship (again recall* Fundamental concept IV*) any influence that may be due simply to the demographic characteristics of the sample (including factors such as gender, income, education level, marital status, and number of children).*

Set correlation is a extremely flexible data analytical system in that it can handle variables at any scale of measurement, as long as each variable has been appropriately represented, transformed and/or coded (in the case of categorical variables) for the analysis (Cohen et al. 2003 provide an exhaustive review these transformational and coding methods for variables; also see *Fundamental concept VI*). The flexibility of set correlation is further enhanced by allowing the same (Set A) or different sets of variables (Set A and Set B) to be partialled from the Independent and Dependent Sets, respectively. Given this flexibility, there are five different relational designs for controlling partialling variables:

1. Correlate the Independent Set (not partialled) with the Dependent Set (not partialled)—yields a simple canonical correlation analysis;
2. Correlate the Independent Set, partialling out control variable set A, with the Dependent Set (not partialled)—yields a semi-partialled canonical correlation;
3. Correlate the Independent Set (not partialled) with the Dependent Set, partialling out control variable set A—this is analogous to a multivariate analysis of covariance (see *Procedure 7.16*);
4. Correlate the Independent Set, partialling out control variable set A, with the Dependent Set, partialling out control variable set A—yields a fully partialled canonical correlation; and
5. Correlate the Independent Set, partialling out control variable set A, with the Dependent Set, partialling out control variable set B—yields a 'bi-partialled' canonical correlation.

The outcomes from a set correlation analysis are interpreted in exactly the same way as the outcomes for canonical correlation analysis, with the same types of correlational statistics being reported. The only difference is that, if one or more sets of variables have had their influence partialled out, the correlational results may change (sometimes dramatically) compared to the situation where the partialling variables had not been controlled for. If the difference was substantial, this would indicate that it was indeed very important to control for the partialling variables in order to obtain a clearer and less ambiguous interpretation of the canonical relationship.

Consider this somewhat more complicated scenario within the QCI database context.

> Suppose Maree was interested in exploring the relationship(s) between the two performance measures, **accuracy** and normalised **log_speed** (the *Dependent Set* variables) and the inspector's overall perceptions of their workplace as measured by **jobsat** and **workcond** (the *Independent Set* variables). However, Maree wants to be sure that the performance variables are not contaminated by general inspector characteristics such as **gender, educlev** or **mentabil** (Control Variable Set B). Furthermore, Maree wants to ensure that the workplace perceptions are not contaminated by any more general halo (or 'rusty halo') effects due to the general work attitudes, **quality, training,** and **orgsatis** (Control Variable Set A). This means that the relational design being implemented in this set correlation analysis is the 'bi-partialled' design (5) described above.
>
> The bi-partialled set correlation analysis, computed using SYSTAT, is summarised in Table 6.11 and shows the same types of statistics as the earlier canonical correlation analysis (some outcomes were edited to remove unncecessary detail). [Note: This analysis emerged **after** the influences of gender, educlev and mentabil had been partialled out of the performance variables (Set Y) and the work attitude variables had been partialled out of the jobsat and workcond variables (Set X).]
>
> Only one substantive canonical correlation emerged: canonical correlation 1 which was **.395** (very close to moderate). The other canonical correlation was .158, which basically signalled a weak association at best. Inspection of the loadings on the first canonical variate for the partialled Dependent Set and the partialled Independent Set suggested the following interpretation to Maree. The sole and very strong contributor to the first canonical variate for the partialled Dependent Set was **accuracy**. Both workplace perception variables were substantive contributors to the first canonical variate for the partialled Independent Set, with **workcond** being more important compared to **jobsat**; all loadings were positive. Thus, more accurate quality control decisions were associated with a combination of more favourable ratings with respect to working conditions and job satisfaction. As a final bit of interpretation: the 'redundancy index' for canonical variate 1 for the partialled Dependent Set was **.081** ($= 8.1\%$ of variance shared between the variables in the partialled Dependent Set and the 1st canonical variate for the partialled Independent Set). Conversely, the 'redundancy index' for canonical variate 1 for the Independent Set was **.094** ($= 9.4\%$ of variance shared between the variables in the partialled Independent Set and the 1st canonical variate for the partialled Dependent Set).

(continued)

Table 6.11 Bi-partial set correlation analysis

Bipartial Set Correlation Analysis (Y|YPARTIAL vs. X|XPARTIAL)

Number of Cases on which Analysis is based: 98

Dependent Set y Partialled by these Variables
 EDUCLEV
 GENDER
 MENTABIL

Independent Set x Partialled by these Variables
 QUALITY
 TRAINING
 ORGSATIS

Canonical Correlations	
1	2
0.395	0.158

Canonical Loadings (y variable by factor correlations)		
	1	2
ACCURACY	1.000	-0.020
LOG_SPEED	0.195	0.981

Canonical Redundancies for Dependent Set	
1	2
0.081	0.012

Canonical Loadings (x variable by factor correlations)		
	1	2
JOBSAT	0.616	0.787
WORKCOND	0.911	-0.412

Canonical Redundancies for Independent Set	
1	2
0.094	0.010

Advantages

Canonical correlation facilitates the analysis and interpretation of the relationships between two sets of variables. It is a more effective analytic approach than performing separate multiple regression analyses for each separate dependent variable. Relying upon several multiple regression analyses would effectively ignore any correlations between the variables comprising the dependent variable set; vital information would thus be thrown away. Canonical correlation explicitly utilises all the available information regarding relationships amongst the variables. Variables at all levels of measurement can be accommodated in a canonical correlation analysis if they are correctly coded.

Set correlation extends canonical correlation by adding very flexible partialling capabilities. This allows for the statistical control of nuisance and extraneous influences by partialling them out of the variables they are thought to potentially contaminate. Interestingly, other multivariate procedures such as multiple regression, univariate and multivariate analyses of variance and covariance (see *Procedures 7.6, 7.10, 7.15* and *7.16*), and discriminant analysis (see *Procedure 7.17*) can all be considered to be special cases of the more general set correlation technique. All that really differs among these techniques is the way variables in the analysis are scaled (even contingency table associations can be tested using set correlation). In this sense, set correlation can be seen as a sort of 'unified theory' for multivariate statistical analysis.

Disadvantages

Canonical correlation is a complex statistical technique and can produce results that can be difficult to interpret depending upon the structure correlation patterns which emerge. The presence of a mixture of positive and negative structure correlations in the description of a canonical variate can be particularly problematic from an interpretive standpoint (sometimes rotating a solution will help clear things up and sometimes it won't). From the standpoint of testing research hypotheses, canonical correlation can really only address very general relational questions; precise predictive relationships or theoretical propositions are very difficult to evaluate in canonical correlation and are best left to the more tractable derivative techniques such as multivariate ANOVA, multiple regression analysis, discriminant analysis, and structural equation models. If there are many variables being considered in one or both of the sets being related, the size of the sample should be large relative to this number (Hair et al. 2010 and Tabachnick and Fidell 2019 generally suggest a minimum ratio of 10 observations per variable analysed but a ratio of 15:1 would be considered more defensible). If such a sample size is precluded, the accuracy and stability of the canonical structure coefficients becomes somewhat questionable.

Set correlation is a similarly complex technique that has all the same drawbacks as canonical correlation except that of not being able to test predictive hypotheses. The capacity to partial and control for influences does allow set correlation to make more precise tests of relationships. However, there needs to be sound theoretical or logical justification for including variables in a control set for either the Dependent Set or the Independent Set or both. Without sound theoretical justification, the analysis becomes a shotgun approach, putting things together hit-or-miss and hoping for the best. Set correlation has not yet enjoyed wide application in behaviour and social research, partly because it is generally only available via SYSTAT and therefore remains largely unknown and partly because it is a much more complicated procedure to understand and interpret—flexibility does not always mean simplicity.

Where Is This Procedure Useful?

Canonical correlation analysis may be most useful in quasi-experimental and survey-type research where one might have sets of variables from different domains to be related to each other. If you wish to control for other variables you have measured, then set correlation is one procedure of choice. As multivariate procedures, canonical correlation and set correlation permit the testing of numerous associations simultaneously. Thus, they provide far more efficient approaches to the analysis of many variables at one time.

Software Procedures

Application	Procedures
SPSS	A full and proper canonical correlation analysis can only be performed using a special macro that can be included in the installation of SPSS. The macro is called *Canonical correlation.sps* and is somewhat fiddly to run. It is a command-based macro, not a windows-based procedure, and it is less stable (more error-prone) than the standard SPSS analyses (see the discussion and examples in Tabachnick and Fidell 2019, section 12.7.2). Alternatively, a canonical correlation analysis (without all of the possible desirable statistics) can be run using the older MANOVA procedure, but this has to be done via SPSS syntax, rather than via a procedural dialog box (this procedure can be asked to rotate a canonical correlation solution). However, SPSS cannot perform set correlation analyses.
NCSS	*Analysis → Multivariate Analysis → Canonical Correlation* and select the desired Y and X variables. NCSS can run a standard canonical correlation analysis but it can also run a very limited version of set correlation, namely only the fully partialled design (described in design (4) above), if a set of 'Partial variables' is included.
SYSTAT	*Analyze → Correlations → Set and Canonical ...* and select your desired independent and dependent variable sets; for set correlation, you can choose the partialling variable sets for the independent and dependent variables separately; if any variables are categorical, then you can indicate these variables to SYSTAT via the 'Category' tab and SYSTAT will automatically code those variables for inclusion; the '*Options*' tab allows you to tell SYSTAT that you want to rotate a specific number of canonical variates. SYSTAT can run all of the possible designs for set correlation, including a standard canonical correlation analysis (if there are no partially variables for either the dependent or independent variable sets).
STATGRAPHICS	*Describe → Multivariate Methods → Canonical Correlations...* and select the variables for the 'First Set of Variables', which has to be the larger of the two sets, if the sets differ in size, and the variables for the 'Second Set of Variables'. STATGRAPHICS only gives a basic canonical correlation analysis; it will not produce canonical loadings (structure correlations) or redundancy indices and cannot rotate the solution. STATGRAPHICS cannot perform set correlation analysis.

(continued)

Application	Procedures
R Commander	**R** Commander cannot perform a canonical correlation analysis. However, the CCA package, which can be installed within **R** (https://cran.r-project.org/web/packages/CCA/index.html, accessed 9th Sept 2019) can be used, in conjunction with the cancorr **R** function, to conduct a canonical correlation analysis (but not a set correlation analysis).

References

Useful Additional Readings for Fundamental Concept III

Argyrous, G. (2011). *Statistics for research: With a guide to SPSS* (3rd ed.). London: Sage. ch. 12.
De Vaus, D. (2002). *Analyzing social science data: 50 key problems in data analysis*. London: Sage. ch. 35.
Glass, G. V., & Hopkins, K. D. (1996). *Statistical methods in education and psychology* (3rd ed.). Upper Saddle River, NJ: Pearson. ch. 7.
Gravetter, F. J., & Wallnau, L. B. (2017). *Statistics for the behavioural sciences* (10th ed.). Belmont, CA: Wadsworth Cengage. ch. 15.
Howell, D. C. (2013). *Statistical methods for psychology* (8th ed.). Belmont, CA: Cengage Wadsworth. ch. 9.
Meyers, L. S., Gamst, G. C., & Guarino, A. (2017). *Applied multivariate research: Design and interpretation* (3rd ed.). Thousand Oaks, CA: Sage. ch. 4A.
Steinberg, W. J. (2011). *Statistics alive* (2nd ed.). Los Angeles: Sage. ch. 34–36.
Thompson, B. (2006). *Foundations of behavioral statistics: An insight-based approach*. New York: The Guilford Press. ch. 5.

Reference for Procedure 6.1

Glass, G. V., & Hopkins, K. D. (1996). *Statistical methods in education and psychology* (3rd ed.). Upper Saddle River, NJ: Pearson. ch. 7.

Useful Additional Readings for Procedure 6.1

Allen, P., Bennett, K., & Heritage, B. (2019). *SPSS statistics: A practical guide* (4th ed.). South Melbourne, VIC: Cengage Learning Australia Pty. ch. 12.
Argyrous, G. (2011). *Statistics for research: With a guide to SPSS* (3rd ed.). London: Sage. ch. 12.
Chen, P. Y., & Popovich, P. M. (2002). *Correlation: Parametric and nonparametric approaches*. Thousand Oaks, CA: Sage.
De Vaus, D. (2002). *Analyzing social science data: 50 key problems in data analysis*. London: Sage. ch. 36.
Field, A. (2018). *Discovering statistics using SPSS for Windows* (5th ed.). Los Angeles: Sage. ch. 8.
Field, A., Miles, J., & Field, Z. (2012). *Discovering statistics using R*. Sage. ch. 6.
Gravetter, F. J., & Wallnau, L. B. (2017). *Statistics for the behavioural sciences* (10th ed.). Belmont, CA: Wadsworth Cengage. ch. 15.

Howell, D. C. (2013). *Statistical methods for psychology* (8th ed.). Belmont, CA: Cengage Wadsworth. ch. 9.
Meyers, L. S., Gamst, G. C., & Guarino, A. (2017). *Applied multivariate research: Design and interpretation* (3rd ed.). Thousand Oaks, CA: Sage. ch. 4B.
Steinberg, W. J. (2011). *Statistics alive* (2nd ed.). Los Angeles: Sage. ch. 34–36.
Thompson, B. (2006). *Foundations of behavioral statistics: An insight-based approach.* New York: The Guilford Press. ch. 5.
Thorndike, R. M. (1997). Correlation methods. In J. P. Keeves (Ed.), *Educational research, methodology, and measurement: An international handbook* (2nd ed., pp. 484–493). Oxford: Pergamon Press.

References for Procedure 6.2

Everitt, B. S. (1977). *The analysis of contingency tables.* London: Chapman & Hall. ch. 3.
Liebetrau, A. M. (1983). *Measures of association.* Beverly Hills, CA: Sage.
Reynolds, H. T. (1984). *Analysis of nominal data* (2nd ed.). Beverly Hills, CA: Sage.

Useful Additional Readings for Procedure 6.2

Agresti, A. (2018). *Statistical methods for the social sciences* (5th ed.). Boston: Pearson. Ch. 8.
Argyrous, G. (2011). *Statistics for research: With a guide to SPSS* (3rd ed.). London: Sage. ch. 6 and 7.
Chen, P. Y., & Popovich, P. M. (2002). *Correlation: Parametric and nonparametric approaches.* Thousand Oaks, CA: Sage.
De Vaus, D. (2002). *Analyzing social science data: 50 key problems in data analysis.* London: Sage. ch. 36.
Gibbons, J. D. (1993). *Nonparametric measures of association.* Newbury Park, CA: Sage.
Hardy, M. (2004). Summarizing distributions. In M. Hardy & A. Bryman (Eds.), *Handbook of data analysis* (pp. 35–64). London: Sage. (particularly the section on bivariate distributions).
Howell, D. C. (2013). *Statistical methods for psychology* (7th ed.). Belmont, CA: Cengage Wadsworth. ch. 6.

References for Procedure 6.3

Cohen, J., Cohen, P., West, S. G., & Aiken, L. S. (2003). *Applied multiple regression/correlation analysis for the behavioral sciences* (3rd ed.). Mahwah, NJ: Lawrence Erlbaum Associates. ch. 1 and 2.
Miles, J., & Shevlin, M. (2001). *Applying regression & correlation: A guide for students and researchers.* London: Sage. ch. 1.

Useful Additional Readings for Procedure 6.3

Agresti, A. (2018). *Statistical methods for the social sciences* (5th ed.). Boston: Pearson. Ch. 9.

Argyrous, G. (2011). *Statistics for research: With a guide to SPSS* (3rd ed.). London: Sage. ch. 12.
De Vaus, D. (2002). *Analyzing social science data: 50 key problems in data analysis*. London: Sage. ch. 37.
Field, A. (2018). *Discovering statistics using SPSS for Windows* (5th ed.). Los Angeles: Sage. ch. 9.
Field, A., Miles, J., & Field, Z. (2012). *Discovering statistics using R*. Sage. ch. 7.
George, D., & Mallery, P. (2019). *IBM SPSS statistics 25 step by step: A simple guide and reference* (15th ed.). New York: Routledge. ch. 15.
Glass, G. V., & Hopkins, K. D. (1996). *Statistical methods in education and psychology* (3rd ed.). Upper Saddle River, NJ: Pearson. ch. 8.
Gravetter, F. J., & Wallnau, L. B. (2017). *Statistics for the behavioural sciences* (10th ed.). Belmont, CA: Wadsworth Cengage. ch. 16.
Hardy, M. A., & Reynolds, J. (2004). Incorporating categorical information into regression models: The utility of dummy variables. In M. Hardy & A. Bryman (Eds.), *Handbook of data analysis* (pp. 209–236). London: Sage.
Howell, D. C. (2013). *Statistical methods for psychology* (8th ed.). Belmont, CA: Cengage Wadsworth. ch. 9.
Judd, C. M., McClelland, G. H., & Ryan, C. S. (2017). *Data analysis: A model-comparison approach* (3rd ed.). New York: Routledge. ch. 5.
Lewis-Beck, M. S. (1995). *Data analysis: An introduction*. Thousand Oaks, CA: Sage. ch. 6.
Meyers, L. S., Gamst, G. C., & Guarino, A. (2017). *Applied multivariate research: Design and interpretation* (3rd ed.). Thousand Oaks, CA: Sage. ch. 4A, 4B.
Pedhazur, E. J. (1997). *Multiple regression in behavioral research: Explanation and prediction* (3rd ed.). South Melbourne, VIC: Wadsworth Thomson Learning. ch. 2.
Schroeder, L. D., Sjoquist, D. L., & Stephan, P. E. (1986). *Understanding regression analysis: An introductory guide*. Beverly Hills, CA: Sage. ch. 1.
Steinberg, W. J. (2011). *Statistics alive* (2nd ed.). Los Angeles: Sage. ch. 37–38.
Stolzenberg, R. M. (2004). Multiple regression analysis. In M. Hardy & A. Bryman (Eds.), *Handbook of data analysis* (pp. 165–207). London: Sage.

References for Procedure 6.4

Cohen, J., Cohen, P., West, S. G., & Aiken, L. S. (2003). *Applied multiple regression/correlation analysis for the behavioral sciences* (3rd ed.). Mahwah, NJ: Lawrence Erlbaum Associates. ch. 3, 4, 5 and 8.
Hair, J. F., Black, B., Babin, B., & Anderson, R. E. (2010). *Multivariate data analysis: A global perspective* (7th ed.). Upper Saddle River, NJ: Pearson Education. ch. 4.
Judd, C. M., McClelland, G. H., & Ryan, C. S. (2017). *Data analysis: A model-comparison approach* (3rd ed.). New York: Routledge. ch. 6, 7, 8 and 13.
Tabachnick, B. G., & Fidell, L. S. (2019). *Using multivariate statistics* (7th ed.). New York: Pearson Education. ch. 5.

Useful Additional Readings for Procedure 6.4

Agresti, A. (2018). *Statistical methods for the social sciences* (5th ed.). Boston: Pearson. Ch. 11, 12.
Allen, P., Bennett, K., & Heritage, B. (2019). *SPSS statistics: A practical guide* (4th ed.). South Melbourne, VIC: Cengage Learning Australia Pty. ch. 13.
Argyrous, G. (2011). *Statistics for research: With a guide to SPSS* (3rd ed.). London: Sage. ch. 13.
De Vaus, D. (2002). *Analyzing social science data: 50 key problems in data analysis*. London: Sage. ch. 46–49.

Field, A. (2018). *Discovering statistics using SPSS for Windows* (5th ed.). Los Angeles: Sage. ch. 9, sections 9.9 onward.

Field, A., Miles, J., & Field, Z. (2012). *Discovering statistics using R.* Sage. ch. 7, sections 7.6 onward.

George, D., & Mallery, P. (2019). *IBM SPSS statistics 25 step by step: A simple guide and reference* (15th ed.). New York: Routledge. ch. 16.

Glass, G. V., & Hopkins, K. D. (1996). *Statistical methods in education and psychology* (3rd ed.). Upper Saddle River, NJ: Pearson. ch. 8.

Grimm, L. G., & Yarnold, P. R. (Eds.). (1995). *Reading and understanding multivariate statistics.* Washington, DC: American Psychological Association. ch. 2.

Hardy, M. A., & Reynolds, J. (2004). Incorporating categorical information into regression models: The utility of dummy variables. In M. Hardy & A. Bryman (Eds.), *Handbook of data analysis* (pp. 209–236). London: Sage.

Howell, D. C. (2013). *Statistical methods for psychology* (8th ed.). Belmont, CA: Cengage Wadsworth. ch. 15.

Lewis-Beck, M. S. (1995). *Data analysis: An introduction.* Thousand Oaks, CA: Sage. ch. 6.

Meyers, L. S., Gamst, G. C., & Guarino, A. (2017). *Applied multivariate research: Design and interpretation* (3rd ed.). Thousand Oaks, CA: Sage. ch. 5A, 5B, 6A, 6B.

Miles, J., & Shevlin, M. (2001). *Applying regression & correlation: A guide for students and researchers.* London: Sage. ch. 2–5.

Pedhazur, E. J. (1997). *Multiple regression in behavioral research: Explanation and prediction* (3rd ed.). South Melbourne, VIC: Wadsworth Thomson Learning. ch. 5.

Schroeder, L. D., Sjoquist, D. L., & Stephan, P. E. (1986). *Understanding regression analysis: An introductory guide.* Beverly Hills, CA: Sage.

Spicer, J. (2005). *Making sense of multivariate data analysis.* Thousand Oaks, CA: Sage. ch. 4.

Steinberg, W. J. (2011). *Statistics alive* (2nd ed.). Los Angeles: Sage. ch. 37–38.

Stolzenberg, R. M. (2004). Multiple regression analysis. In M. Hardy & A. Bryman (Eds.), *Handbook of data analysis* (pp. 165–207). London: Sage.

Tatsuoka, M. M. (1997). Regression analysis of quantified data. In J. P. Keeves (Ed.), *Educational research, methodology, and measurement: An international handbook* (2nd ed., pp. 648–657). Oxford: Pergamon Press.

References for Fundamental Concepts IV

Cohen, J., Cohen, P., West, S. G., & Aiken, L. S. (2003). *Applied multiple regression/correlation analysis for the behavioral sciences* (3rd ed.). Mahwah, NJ: Lawrence Erlbaum Associates. ch. 3, 5.

Pedhazur, E. J. (1997). *Multiple regression in behavioral research: Explanation and prediction* (3rd ed.). South Melbourne, VIC: Wadsworth Thomson Learning. ch. 7 and 9.

Useful Additional Readings for Fundamental Concepts IV

Glass, G. V., & Hopkins, K. D. (1996). *Statistical methods in education and psychology* (3rd ed.). Upper Saddle River, NJ: Pearson. ch. 8.

Grimm, L. G., & Yarnold, P. R. (Eds.). (1995). *Reading and understanding multivariate statistics.* Washington, DC: American Psychological Association. ch. 2.

Howell, D. C. (2013). *Statistical methods for psychology* (8th ed.). Belmont, CA: Cengage Wadsworth. ch. 15.

Judd, C. M., McClelland, G. H., & Ryan, C. S. (2017). *Data analysis: A model-comparison approach* (3rd ed.). New York: Routledge. ch. 6.

Meyers, L. S., Gamst, G. C., & Guarino, A. (2017). *Applied multivariate research: Design and interpretation* (3rd ed.). Thousand Oaks, CA: Sage. ch. 5A.

Tabachnick, B. G., & Fidell, L. S. (2019). *Using multivariate statistics* (7th ed.). New York: Pearson Education. ch. 5.

Thorndike, R. M. (1997). Correlation methods. In J. P. Keeves (Ed.), *Educational research, methodology, and measurement: An international handbook* (2nd ed., pp. 484–493). Oxford: Pergamon Press.

References for Procedure 6.5

Field, A. (2018). *Discovering statistics using SPSS for Windows* (5th ed.). Los Angeles: Sage. ch. 18.

Field, A., Miles, J., & Field, Z. (2012). *Discovering statistics using R.* Sage. ch. 17.

Gorsuch, R. L. (1983). *Factor analysis* (2nd ed.). Hillsdale, NJ: Lawrence Erlbaum Associates.

Hair, J. F., Black, B., Babin, B., & Anderson, R. E. (2010). *Multivariate data analysis: A global perspective* (7th ed.). Upper Saddle River, NJ: Pearson Education. ch. 3.

Tabachnick, B. G., & Fidell, L. S. (2019). *Using multivariate statistics* (7th ed.). New York: Pearson Education. ch. 13.

Useful Additional Readings for Procedure 6.5

Allen, P., Bennett, K., & Heritage, B. (2019). *SPSS statistics: A practical guide* (4th ed.). South Melbourne, VIC: Cengage Learning Australia Pty. ch. 15.

Dunteman, G. H. (1989). *Principal components analysis.* Newbury Park, CA: Sage.

George, D., & Mallery, P. (2019). *IBM SPSS statistics 25 step by step: A simple guide and reference* (15th ed.). New York: Routledge. ch. 20.

Grimm, L. G., & Yarnold, P. R. (Eds.). (1995). *Reading and understanding multivariate statistics.* Washington, DC: American Psychological Association. ch. 4.

Kim, J., & Mueller, C. W. (1978a). *Introduction to factor analysis.* Beverly Hills, CA: Sage.

Kim, J., & Mueller, C. W. (1978b). *Factor analysis.* Beverly Hills, CA: Sage.

Meyers, L. S., Gamst, G. C., & Guarino, A. (2017). *Applied multivariate research: Design and interpretation* (3rd ed.). Thousand Oaks, CA: Sage. ch. 10A, 10B.

Pett, M. A., Lackey, N. R., & Sullivan, J. J. (2003). *Making sense of factor analysis: The use of factor analysis for instrument development in health care research.* Thousand Oaks, CA: Sage.

Spearitt, D. (1997). Factor analysis. In J. P. Keeves (Ed.), *Educational research, methodology, and measurement: An international handbook* (2nd ed., pp. 528–539). Oxford: Pergamon Press.

References for Procedure 6.6

Aldenderfer, M. S., & Blashfield, R. K. (1984). *Cluster analysis.* Beverly Hills, CA: Sage.

Blashfield, R., & Aldenderfer, M. S. (1988). The methods and problems of cluster analysis. In J. R. Nesselroade & R. B. Cattell (Eds.), *Handbook of multivariate experimental psychology* (2nd ed., pp. 447–474). New York: Plenum Press.

Cooksey, R. W., & Soutar, G. N. (2006). Coefficient beta and hierarchical item clustering: An analytical procedure for establishing and displaying the dimensionality and homogeneity of summated scales. *Organizational Research Methods, 9*(1), 78–98.
Everitt, B. S., & Hothorn, T. (2006). *A handbook of statistical analyses using R*. Boca Raton, FL: Chapman & Hall/CRC.
Lorr, M. (1983). *Cluster analysis for social scientists*. San Francisco, CA: Jossey-Bass Publishers.
Revelle, W. R. (1978). ICLUST: A cluster analytic approach to exploratory and confirmatory scale construction. *Behavior Research Methods & Instrumentation, 10*, 739–742.
Revelle, W. R. (1979). Hierarchical cluster analysis and the internal structure of tests. *Multivariate Bahavioral Research, 14*, 57–74.
Revelle, W. R. (2019). *Package: 'psych': Procedures for psychological, psychometric and personality research*. Department of Psychology, Northwestern University. http://cran.r-project.org/web/packages/psych/psych.pdf. Accessed 9th Sept 2019.
SYSTAT Software Inc. (2009). *SYSTAT 13: Statistics I*. Chicago, IL: SYSTAT Software Inc.

Useful Additional Readings for Procedure 6.6

Anderberg, M. R. (1973). *Cluster analysis for applications*. New York: Academic.
Bailey, K. D. (1994). *Typologies and taxonomies: An introduction to classification techniques*. Newbury Park, CA: Sage.
Everitt, B. S. (1980). *Cluster analysis* (2nd ed.). London: Heinemann Educational Books.
Everitt, B. S. (1997). Cluster analysis. In J. P. Keeves (Ed.), *Educational research, methodology, and measurement: An international handbook* (2nd ed., pp. 466–472). Oxford: Pergamon Press.
George, D., & Mallery, P. (2019). *IBM SPSS statistics 25 step by step: A simple guide and reference* (15th ed.). New York: Routledge. ch. 21.
Grimm, L. G., & Yarnold, P. R. (Eds.). (2000). *Reading and understanding more multivariate statistics*. Washington, DC: American Psychological Association. ch. 5.
Hair, J. F., Black, B., Babin, B., & Anderson, R. E. (2010). *Multivariate data analysis: A global perspective* (7th ed.). Upper Saddle River, NJ: Pearson Education. ch. 9.
Meyers, L. S., Gamst, G. C., & Guarino, A. (2017). *Applied multivariate research: Design and interpretation* (3rd ed.). Thousand Oaks, CA: Sage. ch. 17A, 17B.

References for Procedure 6.7

Everitt, B. S., & Hothorn, T. (2006). *A handbook of statistical analyses using R*. Boca Raton, FL: Chapman & Hall/CRC.
Kruskal, J. B., & Wish, M. (1978). *Multidimensional scaling*. Beverly Hills, CA: Sage.
Schiffman, S. S., Reynolds, M. L., & Young, F. W. (1981). *Introduction to multidimensional scaling: Theory, methods, and applications*. New York: Academic.

Useful Additional Readings for Procedure 6.7

Clausen, S. E. (1998). *Applied correspondence analysis: An introduction*. Thousand Oaks, CA: Sage.

Davison, M. L. (1983). *Multidimensional scaling*. New York: Wiley.
Dunn-Rankin, P., & Zhang, S. (1997). Scaling methods. In J. P. Keeves (Ed.), *Educational research, methodology, and measurement: An international handbook* (2nd ed., pp. 790–798). Oxford: Pergamon Press.
George, D., & Mallery, P. (2019). *IBM SPSS statistics 25 step by step: A simple guide and reference* (15th ed.). New York: Routledge. ch. 19.
Grimm, L. G., & Yarnold, P. R. (Eds.). (1995). *Reading and understanding multivariate statistics*. Washington, DC: American Psychological Association. ch. 5.
Hair, J. F., Black, B., Babin, B., & Anderson, R. E. (2010). *Multivariate data analysis: A global perspective* (7th ed.). Upper Saddle River, NJ: Prentice Hall. ch. 10.
Henry, G. (1997). Correspondence analysis. In J. P. Keeves (Ed.), *Educational research, methodology, and measurement: An international handbook* (2nd ed., pp. 493–497). Oxford: Pergamon Press.
Hwang, H., Tomiuk, M. A., & Takane, Y. (2009). Correspondence analysis, multiple correspondence analysis, and recent developments. In R. E. Milsap & A. Maydeu-Olivares (Eds.), *The SAGE handbook of quantitative methods in psychology* (pp. 219–242). London: Sage.
Meyers, L. S., Gamst, G. C., & Guarino, A. (2017). *Applied multivariate research: Design and interpretation* (3rd ed.). Thousand Oaks, CA: Sage. ch. 16A, 16B.
Takane, Y., Jung, S., & Oshima-Takane, Y. (2009). Multidimensional scaling. In R. E. Milsap & A. Maydeu-Olivares (Eds.), *The SAGE handbook of quantitative methods in psychology* (pp. 243–264). London: Sage.
Weller, S. (1990). *Metric scaling correspondence analysis*. Thousand Oaks, CA: Sage.

References for Procedure 6.8

Cohen, J., Cohen, P., West, S. G., & Aiken, L. S. (2003). *Applied multiple regression/correlation analysis for the behavioral sciences* (3rd ed.). Mahwah, NJ: Lawrence Erlbaum Associates. ch. 16, which discusses set correlation.
Hair, J. F., Black, B., Babin, B., & Anderson, R. E. (2010). *Multivariate data analysis: A global perspective* (7th ed.). Upper Saddle River, NJ: Prentice Hall. ch. 5.
Tabachnick, B. G., & Fidell, L. S. (2019). *Using multivariate statistics* (7th ed.). New York: Pearson Education. ch. 12.

Useful Additional Readings for Procedure 6.8

Grimm, L. G., & Yarnold, P. R. (Eds.). (2000). *Reading and understanding more multivariate statistics*. Washington, DC: American Psychological Association. ch. 9.
Meyers, L. S., Gamst, G. C., & Guarino, A. (2017). *Applied multivariate research: Design and interpretation* (3rd ed.). Thousand Oaks, CA: Sage. ch. 7A, 7B.
Thompson, B. (1984). *Canonical correlation analysis: Uses and interpretation*. Beverly Hills, CA: Sage.
Thomson, J. D., & Keeves, J. P. (1997). Canonical analysis. In J. P. Keeves (Ed.), *Educational research, methodology, and measurement: An international handbook* (2nd ed., pp. 461–466). Oxford: Pergamon Press.

Chapter 7
Inferential Statistics for Hypothesis Testing

The third broad category of statistics we discuss concerns *inferential statistics* for hypothesis testing. The procedures and fundamental concepts in this category can help to accomplish either or both of the following goals:

- evaluate the statistical and practical significance of the difference between a specific statistic (e.g. a proportion, a mean, a regression weight, or a correlation coefficient) and its hypothesised value in the population; or
- evaluate the statistical and practical significance of the difference between some combination of statistics (e.g. group means) and some combination of their corresponding population parameters.

The first goal involves the testing of a hypothesis regarding a population value estimated by a specific statistic (usually a descriptive statistic or a correlational statistic). The second goal involves the testing of hypotheses about group differences. When we test a statistical hypothesis, we employ a coherent system of rules for statistical inference where we use empirical quantitative evidence gathered in a research investigation to make inferences about the population(s) from which the sample(s) in the study were drawn. Hypothesis testing is therefore a decision-making process where the researcher must judge whether the available evidence is sufficient for drawing certain conclusions about the population. As we will see, such judgments cannot be made without the risk of being incorrect; therefore, much about the process of statistical inference is linked to notions of probability (recall *Fundamental concept I*).

In experimental, quasi-experimental and some types of survey or observational research, specific groups of research participants may have either been treated differently or identified *a priori* as being different on some independent variable. In such contexts, the typical inference of interest is in whether what was done to the participants or what was known about the participants had any effect on some dependent variable(s) we have measured. Thus, we focus on assessing whether group membership significantly influences or relates to observed changes in the dependent variable. In other types of survey and observational research, we may be more interested in inferences about the existence of relationships between variables.

Here, we focus on assessing whether a significant correlational or predictive relationship exists.

Think back to the QCI database.

> Consider the following inferential or hypothesis testing questions which Maree might ask about the relationships between certain variables in the database.
>
> - Is there a significant relationship between type of company worked for and the educational level attained by inspectors who work in those companies? [see *Procedure 7.1* – Contingency tables: tests of significance.]
> - Is the quality of inspectors' working conditions and the extent of their job satisfaction significantly predictive of the accuracy of their inspection decisions? [see *Procedure 7.13* – Multple regression: tests of significance.]
> - Do inspectors who possess tertiary qualifications differ significantly in inspection decision speed and accuracy from inspectors who have a high school-level education? [see *Procedure 7.2* – *t*-test for two independent groups and *Procedure 7.3* – Mann-Whitney *U*-test for two independent groups.]
> - Do inspectors from different types of computer manufacturing companies significantly differ with respect to their mental ability and how strong is this effect? [see *Procedure 7.6* – One-way analysis of variance (ANOVA) and *Procedure 7.7* – Assessing effect sizes, *Procedure 7.8* – Planned comparisons and posthoc multiple comparisons, and *Procedure 7.9* – Kruskal-Wallis rank test for several independent groups.]
> - Do inspectors differ in their workplace attitudes according to their gender and education level? Does controlling for mental ability alter the story to be conveyed here? [see *Procedure 7.10* – Factorial between-groups ANOVA, *Procedure 7.15* – Analysis of covariance (ANCOVA), and *Procedure 7.16* – Multivariate analysis of variance & covariance.]
> - Can a set of predictors from the QCI database be identified which will allow Maree to significantly predict which type of company an inspector works for? [see *Procedure 7.17* – Discriminant analysis.]

If you compare these questions to those posed at the beginning of Chap. 6 – *Correlational statistics for characterising relationships*, you will see that the complexity of the questions has remained fairly constant and the issues still turn on the concept of relationship, but now we have injected the word 'significantly' into each question. This is meant to directly invoke the need for a set of rules for making such an evaluative judgment. In this chapter, you will explore the basic rules for statistical inference as well as various procedures that can be employed in different hypothesis testing situations and research designs to inform judgments of significance. You will also learn that statistical significance is not the only way to address hypotheses—practical significance is almost always relevant as well; in some cases, even more relevant.

Fundamental Concept V: The Logic and Rules of Statistical Inference for Hypothesis Testing

Hypothesis testing, using statistical tests of significance, has several important concepts underpinning it. The process of hypothesis testing, irrespective of the specific type of test being conducted, has a consistent and logical structure (Field 2018 has a particularly lucid account of this logic in section 2.9 of his chapter 2). For purposes of clarification, the discussion below will focus on tests of group differences, but you should be aware that the logic of the process remains the same if a hypothesis about a correlational or regression-based relationship is being tested. In actuality, the two types of tests are really one and the same (see, for example, Judd et al. 2017, or Cohen et al. 2003 as well as the discussion of the General Linear Model in *Fundamental concept VI*), but it is conceptually easier to come to grips with the overall logic of the process if we concentrate on testing group differences. The most important thing to realise about statistical inference is that, in the end, a human being, not a computer or a formula, makes the decision to claim significance or non-significance with respect to a hypothesis – it is always a judgment call by the researcher and is always open to scrutiny by any critical consumer of the research. The intention of statistical inference is always to use sample-based information (i.e. statistics, like sample means) to draw generalising conclusions about characteristics of relevant populations (i.e. parameters, like population means). Thus, if a researcher concludes that two groups are 'significantly different', what he or she is inferring is that, because the sample evidence shows that the two groups differ, it is justifiable to conclude that the two groups will be different in the population(s) from which the samples were drawn.

In order to make the judgment about statistical significance defensible, an initial reference point must be established. In a two-group comparison context, that reference point is to assume that there is no difference between the two groups in the population (which means that the independent variable defining the two groups has no impact on the dependent variable of interest). In a real sense, testing a statistical hypothesis about group differences is analogous to being a member of the jury in a court case deciding the guilt or innocence of a defendant: that is, 'the defendant is considered to be innocent until proven guilty, based on evidence, beyond a reasonable doubt' which translates, in the current context, to 'the two groups are considered **not** to be different on a specific dependent variable until shown to be different, based on evidence, beyond a reasonable doubt'. This initial presumption is termed the 'null hypothesis' (typically identified as H_0) and the goal of a test of significance is to provide evidence that will allow a researcher to reject the null hypothesis in favour of the 'alternative hypothesis' (which states the researcher's actual expectation; typically identified as H_a) that the two groups do, in fact, differ on the dependent variable of interest.

> Suppose Maree has a specific hypothesis of interest in her QCI research project, namely, that male and female quality control inspectors will differ significantly in the accuracy with which they make their quality control decisions. Maree believes, on the basis of past research and her own hunches regarding work in various industries, that there will be a gender difference in decision accuracy. However, the burden is upon her to amass sufficient evidence to publicly conclude this. Until evidence to the contrary is obtained, she must initially assume that the two gender groups do not differ. Maree then draws her samples (hopefully, the samples are randomly drawn – see *Fundamental concept VII*), executes her investigation, and collects her dependent variable measurements on the two groups for subsequent analysis.

However, there is a constraint imposed on this judgmental logic. The researcher must acknowledge that he or she will not be able to reject the null hypothesis in favour of their alternative research hypothesis with 100% certainty of being correct. There will always be some chance of error associated with their decision to reject the null hypothesis in that they could erroneously reject it when it was actually the true state of affairs in the world. This is called the *alpha (α) error* or *Type I error*. With this additional concept, the full courtroom analogy for hypothesis testing can be represented as:

Courtroom analogy (from the perspective of a prosecutor):

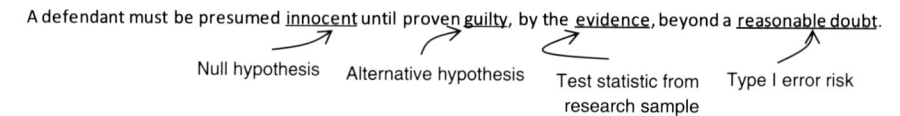

A defendant must be presumed <u>innocent</u> until proven <u>guilty</u>, by the <u>evidence</u>, beyond a <u>reasonable doubt</u>.

                 Null hypothesis   Alternative hypothesis   Test statistic from   Type I error risk
                                                                                         research sample

Since certainty is not possible, a researcher should at least want his or her risk of making the Type I error to be as small as is feasible. Typically, a 5% chance of making the alpha error is considered acceptable—this is often referred to as the *level of significance*. It is critical to note here that the researcher has complete responsibility for choosing the level of error he or she is willing to live with; no magic formula exists to produce this probability. Actually, the 5% level (equals a probability of .05 for making a Type I error) mentioned above is merely an academic convention that has been promulgated by editorial policies and behavioural research expectations over many years. However, individual researchers can argue for an increase or decrease in this level of error depending upon the purposes of their research.

For example, a behavioural researcher whose research has the potential for influencing changes in social or public policy may wish to have a greater level of certainty associated with any claims of significance and may therefore argue that a stricter margin of error (say a 1% or .1% chance of Type I error) should be used.

Alternatively, a behavioural researcher who is investigating phenomena in a new area which has enjoyed very little previous research may argue that a more relaxed position on Type I error is appropriate (say a 10% chance of Type I error) because he or she is merely searching for any hints of potential relationships to investigate in further research. Thus, the research context has a strong bearing on a researcher's choice of an acceptable error level for deciding about statistical significance.

Researchers must recognise that any specific statistical test used to evaluate a null hypothesis will be either more or less likely to detect a difference if one truly exists. This likelihood is called the *power of the test*, reflecting the chances of correctly concluding that the null hypothesis should be rejected in favour of the alternate hypothesis. This likelihood or probability is influenced by interrelated considerations of sample size, alpha error and effect size (see Cohen's 1988 classic text, for details on these issues).

First, the researcher must obtain an appropriate sample size in order to have a reasonably good chance of detecting true group differences. Other things being equal, larger samples give more precise and smaller estimates of the error associated with any specific test statistic, thereby making it easier to reject a null hypothesis.

Second, the alpha error level must be small but not too small. This is the reason 5% has become a conventional value—it represents an acceptable compromise between too much and too little margin for error. Other things being equal, it is easier to reject the null hypothesis using a larger rather than a smaller level of significance.

Third, the researcher should be aware of what *effect size* (e.g. size of group difference) they are looking for. With a given sample size, power level and level of significance, a small effect size (e.g. a small difference between groups) is much harder to find statistical evidence for compared to a medium or large effect size. In general, larger samples provide more powerful hypothesis tests just like a more powerful microscope resolves smaller objects: larger samples make it easier to detect smaller differences.

[Note that statistical packages like SYSTAT and STATGRAPHICS do have procedures to help researchers calculate the appropriate samples sizes needed to achieve a given level of power in their research. NCSS is associated with a separate software package called PASS, which is designed to help researchers with power- and sample size-related questions, see https://www.ncss.com/software/pass/. There is also a well-known freeware software package called *G*Power* (Paul et al. 2007) which facilitates such calculations in a relatively straightforward manner; see http://www.psychologie.hhu.de/arbeitsgruppen/allgemeine-psychologie-und-arbeitspsychologie/ gpower.html and https://www.softpedia.com/get/Science-CAD/G-Power.shtml). This package is often used for calculating the power and samples sizes needed for clinical trials in medical research, for example.]

The actual decision problem confronting all behavioural researchers when they test statistical hypotheses is displayed as shown in Fig. 7.1. It should be noted that this table shows the existence of a second type of decision error, the *Type II error* or *beta (β) error*, which is defined as the probability of failing to reject the null hypothesis when it is in fact false. The beta error is directly related to the power of

		Possible States of Reality (only ONE can be correct!)	
		Null Hypothesis is **TRUE**	Null Hypothesis is **FALSE**
Researcher's decision about reality	"**FAIL TO REJECT** the Null Hypothesis"	Correct Decision (defined as $1 - \alpha$)	Type II Decision Error (β)
	"**REJECT** Null Hypothesis"	Type I Decision Error (α) **Level of Significance**	Correct Decision (defined as $1 - \beta$) **Power of the Test**
Probabilities for any ONE state of reality add to 1.0		$\alpha + (1 - \alpha) = 1.0$	$\beta + (1 - \beta) = 1.0$

Fig. 7.1 Possible outcomes from a behavioural researcher's decision about the likely state of reality based on his/her empirical observations on a sample

a statistical test since the power of a test (its ability to detect real group differences) is defined as 1 minus the beta error (remember, statistical errors are expressed as probabilities).

Figure 7.1 makes it clear that the researcher's fundamental problem in testing a statistical hypothesis is deciding which one of two possible states of reality is most likely to be true. Any such decision can have only one of two possible outcomes: it can be correct with some probability or it can be in error with some probability and both probabilities must add up to 1.0 since they exhaust the possible decision outcomes for any specific state of reality.

The Role of Sampling Distributions in Hypothesis Testing

A key part of the logic of statistical inference is the concept of a *sampling distribution*. Using this concept, it becomes possible to construct a reference distribution for the null hypothesis. The easiest way to understand a sampling distribution in a two group comparison context, for example, is to imagine (i.e. mentally simulate) a researcher repeating his or her study, say 500 times, under conditions where the groups do not actually differ; each time using the same design, same random sampling strategy and same group sizes. For each study replication, the difference between the two group means would be computed; it would likely be a different value each time because of random errors associated with individual differences associated with the people in the samples. The resulting 500 mean differences would form a distribution (recall *Procedure 5.1*) describing how the two groups would naturally differ in a world where the null hypothesis was true – this distribution would then constitute a sampling distribution for the difference between the two group means in a population where there was no difference. The standard deviation of a sampling distribution is called the *standard error* (*se*) and represents the amount

Fundamental Concept V: The Logic and Rules of Statistical Inference for Hypothesis Testing 247

of error associated with estimating a specific statistic (like the difference between two group means).

Figure 7.2a visually displays one version of such a scenario. In this scenario, two groups, each containing 250 randomly sampled participants, are compared and the summary statistics for each group are shown in the upper right-hand portion of the text block beside the figure (i.e. for Group 1 → the observed mean is 102; the observed standard deviation (sd) is 10; for Group 2 → the observed mean is 103; the observed sd is 10). The solid curve depicts the sampling distribution of the differences between the two group means arising from 500 replications of this study scenario where the population mean difference is 0.0; each time taking a random sample of 250 participants in each group and measuring the dependent variable. The solid curve has a mean of 0.0, which is the group difference identified in the null hypothesis (H_0).

If the null hypothesis was true, then the observed difference of 1.0 between the two group means would simply be an observation falling under the solid curve. However, if the true state of the world is the alternative hypothesis (H_a) value of 1.0, then the entire sampling distribution would be relocated to be centred on this value as the mean. This possibility is depicted by the dashed curve and vertical line for mean at 1.0 in Fig. 7.2a. The choice the researcher now faces is which version of reality (solid curve or dashed curve) is most likely to be the case? The decision must start by assuming that H_0 is true. What the researcher then looks for is how rare or unlikely

Scenario showing a non-significant difference in a large sample 2-group comparison study [where the standard deviations for each group are larger]

Group 1: Mean = 102
sd = 10
n = 250
Group 2: Mean = 103
sd = 10
n = 250

H_0: pop mean diff = 0.0
H_a: pop mean diff = 1.0
se_{diff} = .894

Decision criterion: α = .05

Minimum difference needed to claim a significant difference in the population is = H_0 value + (2 x se_{diff}) = 0.0 + 1.788 = 1.788

Observation: H_a difference of 1.0 does *not* exceed the required minimum difference of 1.788; this means that probability of the value of H_a in the distribution of H_0 is *greater* than p = .05.

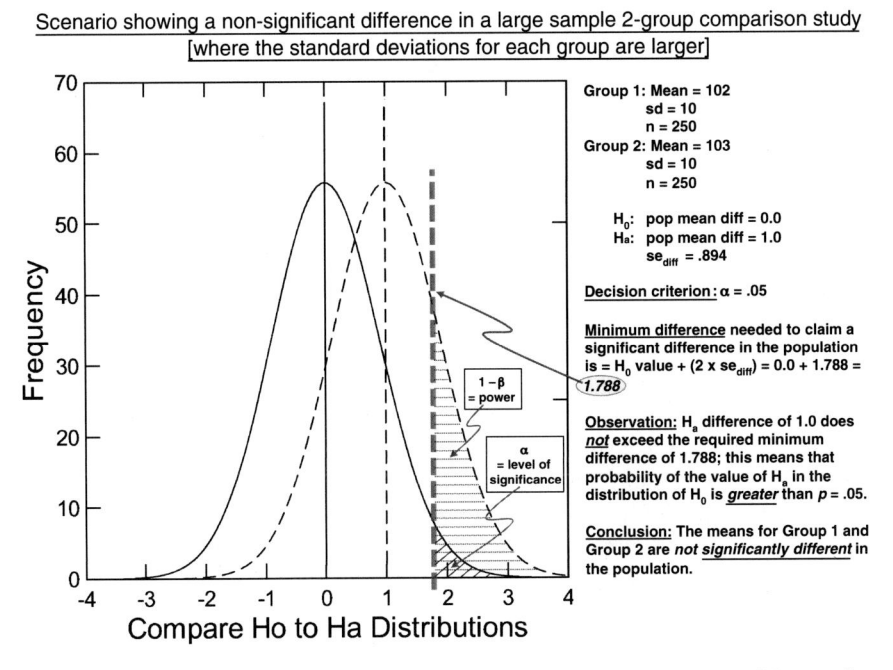

Conclusion: The means for Group 1 and Group 2 are *not significantly different* in the population.

Compare Ho to Ha Distributions

Fig. 7.2a Two-group comparison hypothesis testing scenario, showing two possible sampling distributions for group differences

the observed research outcome would be in this version of reality. The level of significance or alpha error (shown by the cross-hatched area in the right tail of the solid curve) provides a way to quantify what is to count as rare enough in order to conclude that the observed group difference is more likely to have come from the H_a version of reality than from the H_0 version. The alpha level cuts the H_0 curve at the vertical dashed thick grey line in the figure. [In a normal curve like that shown in the figure, this cut-off can be located by adding twice the standard error to the mean value of the distribution – which, for the null hypothesis, is 0.0. This logic is the same as that for relating z-scores to a normal distribution; recall Fig. 5.28 in *Fundamental concept II*]. Note that once this cut-off point has been identified, it not only cuts the H_0 curve (chopping off the small tail), but also the H_a curve, giving the horizontally shaded area under the dashed curve which has a size equal to 1 minus beta or *power* (which reflects the probability of an observation falling under the H_a curve when the researcher correctly concludes that H_a actually describes reality). If the observed group difference falls at or beyond this cut-off point, then the observed group difference is more likely to come from the H_a version than the H_0 version of reality and the researcher could then reject H_0. Figure 7.2a shows that this is **not** the case, so the researcher should fail to reject the null hypothesis and conclude there is insufficient evidence that the two groups differ in the population.

Figure 7.2b shows the same scenario, but with a different hypothesised value for H_a. This time, a group mean difference of 3.0 is observed and this difference falls well beyond the cut-off level established in the H_0 curve. The researcher would be justified in claiming a significant difference between the two groups. Note also the very large amount of power associated with this size of difference (the effect size).

Figure 7.2c shows how sample size can strongly influence the outcomes of a statistical inference test. Here the scenario involves groups of 50 participants (one-fifth the sample size in each group used in Figs. 7.2a and 7.2b). If we assume the standard deviations stay the same, this means the standard error increases from .894 to 2.0, giving a much more variable sampling distribution. In Fig. 7.2c, the observed group mean difference of 1.0 is clearly non-significant. In Fig. 7.2d, even when the observed difference is 3.0, instead of 1.0, the difference is not significant (whereas it was significant in the scenario version shown in Fig. 7.2b, with the larger group sample sizes). In both cases, the small sample sizes provide insufficient power to detect a reasonable effect size.

Finally, Fig. 7.2e shows a revised scenario, nearly identical to that in Fig. 7.2c, but where the group standard deviations are 2.5 instead of 10 (i.e. one-quarter of the variability in observations inside each group). This shows the influence that reducing the amount of variability in one's observations (perhaps through using better measures or more highly controlled data gathering circumstances) can dramatically improve the power of a hypothesis test (note that the value of se_{diff} shrank from 2.0 to 0.5). In Fig. 7.2e, the difference of 1.0 between the two group means just meets

Scenario showing a significant difference in a large sample 2-group comparison study
[where the standard deviations for each group are larger]

Group 1: Mean = 102
sd = 10
n = 250

Group 2: Mean = 105
sd = 10
n = 250

H_0: pop mean diff = 0.0
H_a: pop mean diff = 3.0
se_{diff} = .894

1 − β = power

α = level of significance

Decision criterion: α = .05

Minimum difference needed to claim a significant difference in the population is = H_0 value + (2 x se_{diff}) = 0.0 + 1.788 = **1.788**

Observation: H_a difference of 3.0 *exceeds* the required minimum difference of 1.788; this means that probability of the value of H_a in the distribution of H_0 is *less* than p = .05.

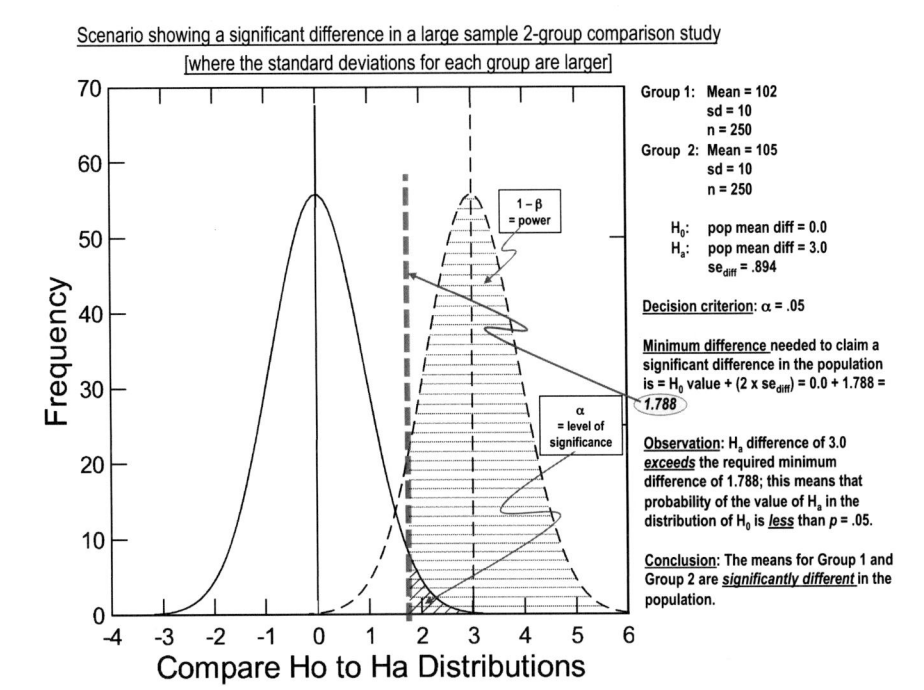

Conclusion: The means for Group 1 and Group 2 are *significantly different* in the population.

Compare Ho to Ha Distributions

Fig. 7.2b Two-group comparison hypothesis testing scenario, showing a significant group difference

Scenario showing a non-significant difference in a small sample 2-group comparison study
[where the standard deviations for each group are larger]

Group 1: Mean = 102
sd = 10
n = 50

Group 2: Mean = 103
sd = 10
n = 50

H_0: pop mean diff = 0.0
H_a: pop mean diff = 1.0
se_{diff} = 2.0

Decision criterion: α = .05

Minimum difference needed to claim a significant difference in the population is = H_0 value + (2 x se_{diff}) = 0.0 + 4.0 = **4.0**

Observation: H_a difference of 1.0 does *not* exceed the required minimum difference of 4.0; this means that probability of the value of H_a in the distribution of H_0 is *greater* than p = .05.

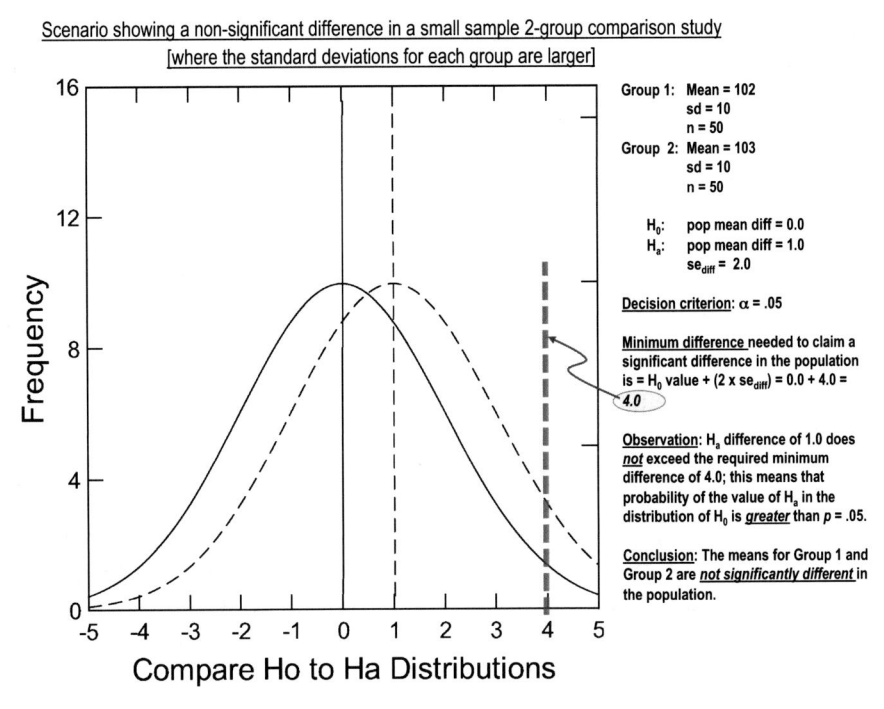

Conclusion: The means for Group 1 and Group 2 are *not significantly different* in the population.

Compare Ho to Ha Distributions

Fig. 7.2c Two-group comparison hypothesis testing scenario, showing a non-significant group difference in small sample groups differing to a smaller extent

Scenario showing a non-significant larger difference in a small sample 2-group comparison study
[where the standard deviations for each group are larger]

Group 1: Mean = 102
sd = 10
n = 50
Group 2: Mean = 105
sd = 10
n = 50

H_0: pop mean diff = 0.0
H_a: pop mean diff = 3.0
se_{diff} = 2.0

Decision criterion: α = .05

Minimum difference needed to claim a significant difference in the population is = H_0 value + (2 x se_{diff}) = 0.0 + 4.0 = 4.0

Observation: H_a difference of 3.0 does *not* exceed the required minimum difference of 4.0; this means that probability of the value of H_a in the distribution of H_0 is *greater* than p = .05.

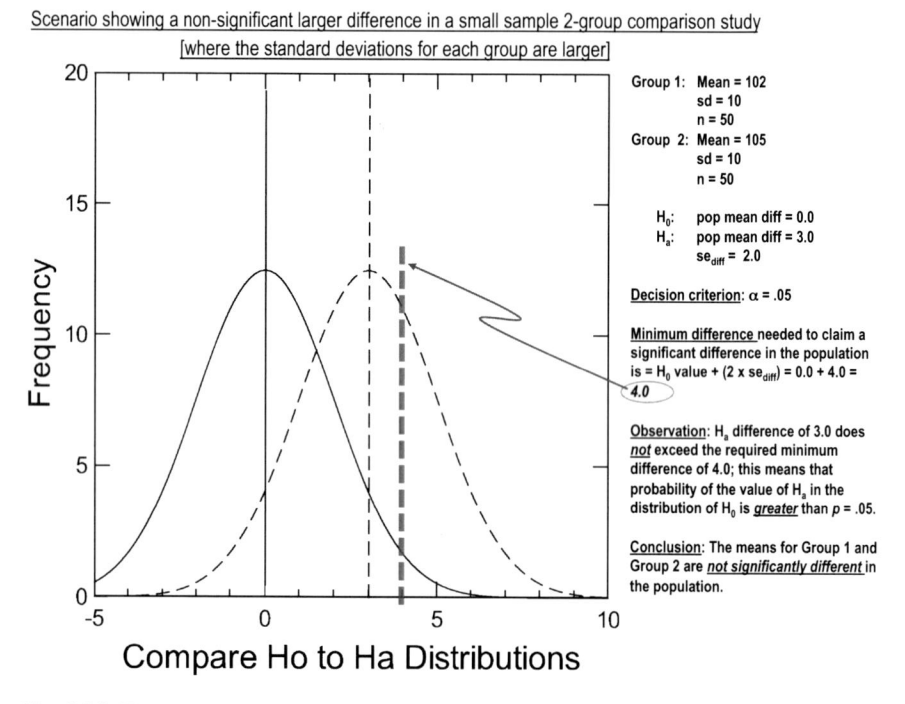

Conclusion: The means for Group 1 and Group 2 are *not significantly different* in the population.

Compare Ho to Ha Distributions

Fig. 7.2d Two-group comparison hypothesis testing scenario, showing a non-significant group difference in small sample groups differing to a larger extent

Scenario showing a just significant difference in a small sample 2-group comparison study
[where the standard deviations for each group are much smaller]

Group 1: Mean = 102
sd = 2.5
n = 50
Group 2: Mean = 103
sd = 2.5
n = 50

H_0: pop mean diff = 0.0
H_a: pop mean diff = 1.0
se_{diff} = 0.50

Decision criterion: α = .05

Minimum difference needed to claim a significant difference in the population is = H_0 value + (2 x se_{diff}) = 0.0 + 1.0 = 1.0

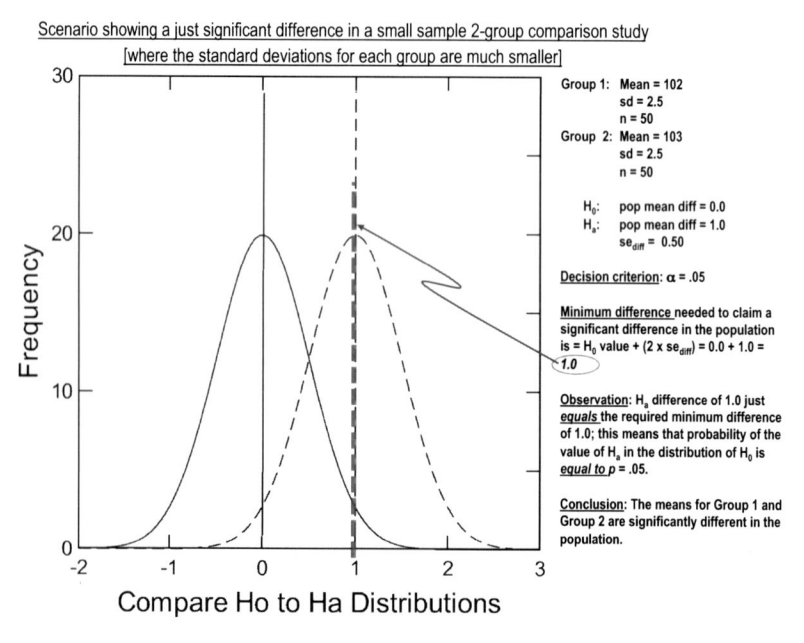

Observation: H_a difference of 1.0 just *equals* the required minimum difference of 1.0; this means that probability of the value of H_a in the distribution of H_0 is *equal to* p = .05.

Conclusion: The means for Group 1 and Group 2 are significantly different in the population.

Compare Ho to Ha Distributions

Fig. 7.2e Two-group comparison hypothesis testing scenario, showing a just significant difference in small sample groups differing to a lesser extent but with less variable observations

Fundamental Concept V: The Logic and Rules of Statistical Inference for Hypothesis Testing 251

the significance criterion value. The researcher could justifiably conclude a significant difference between the two groups in this case, whereas in Fig. 7.2c, the decision wasn't close.

It is important to note that while the conceptual logic of the above illustrations applies, irrespective of whether the researcher is comparing group means, evaluating a correlational relationship (e.g. testing a correlation coefficient for significance) or evaluating a predictive relationship (e.g. testing a regression coefficient), there are specific computational and statistical details associated with each type of hypothesis test, which means that a specific standardised test statistic would be compared to a tailored sampling distribution, like the *t*-distribution, the *F*-distribution or the χ^2- ('chi-squared') distribution. For example, in the scenarios represented in Figs. 7.2a, 7.2b, 7.2c, 7.2d and 7.2e, the precise test statistic that would be used to evaluate the null hypothesis would be the *independent groups t-test* (see *Procedure 7.2*). Two final caveats: (1) In practice, alternate hypotheses seldom specify an exact value, rather they simply state the alternative version of reality in terms of being 'not equal', 'less than' or 'greater than' the null hypothesised value. We only set a specific value for H_a in the above examples for conceptual illustration purposes. (2) All the illustrations above only considered one tail of the distribution for comparison purposes, yielding a so-called *one-tailed test*. If you don't know which tail to use (often the case), you can conduct a *two-tailed test* and this effectively splits the alpha level evenly between the two tests (note that one-tailed tests are more powerful, because all of alpha is in one tail, pushing the cutoff point closer to the mean of the distribution; but they are much harder to defend and disastrous if you pick the wrong tail!).

The Statistical Inference Process in Practice

The statistical inference process which Maree would employ to carry out a test of significance, for comparing male and female inspectors in terms of their decision accuracy, would unfold in the following way.

1. Maree would compute specific relevant summary statistics (e.g. sample means and standard deviations) for the groups being compared.
2. Maree would then compute a *test statistic* (a special standardised index—in this particular scenario, it would be an *independent-groups t-test*, see *Procedure 7.2*) using these group summary statistics. Any such test statistic has associated with it a certain number of *degrees of freedom* which are a function of the sample or group sizes; larger sample sizes mean more degrees of freedom (df) which in turn means a more powerful hypothesis test.

3. Maree would then compare her computed test statistic from Step 2 to a specific standard reference distribution of possible values for that test statistic under the assumption that no group differences exist (i.e. assuming the null hypothesis is true). Within that standard reference distribution, there is a *critical value* (analogous to the vertical thick dashed grey line in Figs. 7.2a, 7.2b, 7.2c, 7.2d and 7.2e) for the test statistic that is established by the degrees of freedom and the *alpha error*, e.g. 5%.
4. Maree's decision processes would then go something like this: If the test statistic she computed from her sample data in Step 2 was sufficiently large relative to the standard reference distribution for that test statistic, she would make the decision to reject her null hypothesis. What counts as 'sufficiently large' would be determined by the *critical value* identified in Step 3.

The decision rule in statistical inference is therefore straightforward: if Maree's computed test statistic exceeds the *critical value* for the test statistic in the standard reference distribution, then she would be justified in concluding that:

- a test statistic as large as the one she found would be found only 5% of the time if the groups really did not differ, i.e. if the null hypothesis was true, therefore
- the more likely conclusion is that the groups, in fact, do differ.

Thus, Maree could reject the null hypothesis in favour of her alternative (research) hypothesis if the above conditions obtained. That is, she would have enough evidence for concluding that the difference between the two groups is 'statistically significant'. She would then be justified in inferring that the sample differences she has detected generalise to the population from which the sample was drawn—provided, of course, that her methodology and analysis were correctly executed. If the above conditions do not obtain, Maree can conclude only one thing: that she has failed to reject her null hypothesis. This is not the same as concluding that the null hypothesis must be true; it merely acknowledges that evidence was insufficient for concluding it was not true.

It is important to understand that virtually all statistical software packages produce an exact probability (called a *p-value* or just *p*) associated with every statistical test of significance. This probability provides a precise picture of just how likely it would be to find a value for the test statistic as large as or larger than the one computed using the sample data, under the assumption that the null hypothesis was true. A very small probability would indicate that the test statistic was highly unlikely in the state of reality where the null hypothesis was true, therefore the researcher would be justified in rejecting the null hypothesis. The decision regarding significance can thus be easily made by directly comparing the *p*-value, computed by the software package, for the test statistic to the probability of the alpha error, e.g. .05

established by the researcher. If the *p*-value is equal to or smaller than the researcher's established value for the alpha error, then the researcher can justifiably reject the null hypothesis as being the more likely state of reality in the population.

The Risk of Inflating the Alpha or Type I Error

There is one more important thing to understand about statistical inference testing and this has to do with making multiple hypothesis tests within the same data set. Statisticians have shown that when a researcher makes many hypothesis tests in the same data set, each one evaluated using say an alpha level of significance of .05, the effective risk of making a false claim of significance (i.e. making the Type I error) across the entire set of tests is much higher than .05. For example, if a researcher was to conduct 10 separate two-group comparisons tests using the same sample of data, the effective Type I error rate would 'inflate' to an alpha value of .40, rather than .05! The more tests done, the worse the problem becomes (e.g. for 25 such tests in the same data set, alpha would effectively be .72 or a 72% chance of making a false claim of significance!). There are several ways to handle this 'inflation of alpha' problem.

One method is fairly simple and involves simply dividing the desired overall alpha level (say, .05) by the number of hypothesis tests planned, and using the new number as the alpha level for evaluating each separate test (this is called the *Bonferroni correction* of alpha; see, for example, Field's 2018, section 2.9.7 discussion of inflated error rates). So, if we plan to conduct 10 independent groups *t*-tests in the same data set, we could divide alpha = .05 by 10 and evaluate each individual *t*-test at an alpha level of .005. This would obviously greatly reduce power making it much harder to find a significant difference, but over the 10 tests, it would keep alpha capped at .05.

A second and generally preferred method is to use multivariate statistical procedures (e.g. multiple regression analysis (see *Procedure 7.13*), multivariate analysis of variance (see *Procedure 7.16*)) which allow researchers to evaluate many differences/relationships simultaneously. Multivariate hypothesis testing employs a hurdle system of logic involving at least two stages. For example, if we wish to compare groups on several dependent variables (a multivariate analysis of variance (MANOVA) scenario), the hurdle logic would unfold as:

- *Stage 1*: group differences are evaluated on all dependent variables at the same time with a single multivariate test (sometimes called an 'omnibus test'). The researcher progresses to the second stage **if and only if** the multivariate test is significant. This is because a significant multivariate test signals that somewhere, amongst the various dependent variables, there are specific group differences worth looking for. If the multivariate test is not significant, interpretation stops at

stage 1 with the conclusion that there are no significant multivariate differences between groups.
- *Stage 2*: group differences are evaluated for each individual dependent variable. Only those identified as showing significant differences are interpreted.

In hierarchical multiple regression (see *Procedure 7.13*), a similar hurdle logic is used to evaluate the contributions of predictor sets (the Stage 1 omnibus tests) and individual predictor contributions within those sets (the Stage 2 tests). There are even some univariate testing procedures such as comparing three or more groups using the analysis of variance with multiple comparisons (e.g. see *Procedure 7.6* and *7.8*) that employ this hurdle logic: Stage 1 involving the omnibus test for differences amongst all groups and Stage 2 involving comparisons between specific pairs of groups.

The hurdle logic works because the Stage 1 test acts as a screening test (which also has the handy benefit of controlling for any interrelationships between dependent variables; something the Bonferroni correction does not do). The researcher only looks deeper into the data (i.e. drilling down to look at specific differences) if there is enough evidence suggesting that deeper exploration is justified. Thus, the Stage 2 tests are not even examined unless and until the Stage 1 screening hurdle has been successfully negotiated. In a nutshell, you can't make a Type I error if you never look at a test that could lead you to make one! Finally, it is possible to combine the multivariate testing hurdle logic with the Bonferroni correction logic, which can provide even greater control over the inflation of the alpha error rate. Some software packages, such as SPSS and SYSTAT offer this capability for certain procedures.

Assumptions About the Data Required by Hypothesis Testing

Every test of a statistical hypothesis that uses sample statistics to estimate population parameters requires the researcher to make certain assumptions about the behaviour of his or her data in the population. It is for this reason that we call such tests *parametric* and the assumptions are necessary in order to ensure that the statistical inference decision rules work fairly, and the resulting test is therefore unbiased.

Most parametric statistical tests that focus on group comparisons require the following minimal set of assumptions about the data:

- The dependent variable, on which the groups are to be compared, follows a normal distribution (recall *Fundamental concept II*) in the population from which the sample has been drawn (*assumption of normality*) [if there are several different dependent variables in a parametric multivariate analysis, then *multivariate normality* is assumed, see Tabachnick and Fidell (2019)];

- The variances of the dependent variable on which the groups are to be compared are approximately the same size in each of the groups (*assumption of homogeneity of variance*); and
- The errors associated with scores from different individuals in each group on the dependent variable are not correlated with each other (*assumption of independently distributed errors*).

Hypothesis tests that evaluate the significance of correlations or regression coefficients require an additional assumption that the criterion variable, in the population, has a linear relationship with the predictor(s) that is being tested (*assumption of linearity*).

If a researcher's data are not in a form that can reasonably meet the requirements of one of these assumptions (easily checked using a range of descriptive statistics, assumption tests and/or graphs; see also the discussion of data screening in *Procedure 8.2*), then one of four strategies are possible:

- mathematically transform the dependent variable to better meet the required assumptions (as was demonstrated in *Fundamental concept II* using the log transformation on the **speed** variable in the QCI database) before conducting the statistical hypothesis test (a strategy that doesn't always work and can make interpretating outcomes more difficult); or
- include additional terms in one's model to permit assumption violations to be explicitly handled (e.g. using squared values for a predictor to handle a non-linear relationship in multiple regression (*Procedure 7.13*) or using predictors that reflect the correlations between successive errors in time series analysis (*Procedure 8.5*); or
- use a nonparametric test, appropriate to the research design used in the study, which requires fewer assumptions but typically at the cost of reducing the level of measurement for the dependent variable from interval- or ratio-level to ordinal-level, i.e. ranks; or
- use another type of nonparametric process called *bootstrapping* (see *Procedure 8.4*), which treats the data set as if it was a population in its own right and repeatedly samples within this 'population' to build up an empirical sampling distribution that is used to evaluate hypotheses (basically harnessing the mental simulation process shown in Figs. 7.2a, 7.2b, 7.2c, 7.2d and 7.2e, but as an actual empirical process rather than a mental exercise).

Most parametric statistical tests are robust against mild violations of one or more assumptions meaning that they will still produce relatively unbiased outcomes. However, if violations are severe, then biased outcomes are likely and an alternative strategy for testing the hypothesis should be pursued.

Fundamental Concepts VI: The General Linear Model

Building Models

We will unfold the logic of the general linear model and hypothesis testing as a kind of story, which builds on conceptual ideas originated by Judd et al. (2017, pp. 1–4). This story provides an advance organiser for many of the procedures you will encounter in this chapter, pulling them together into a single integrative framework. The story commences when we hypothesise and construct a *MODEL* of the world, based on specific research questions or hypotheses; a process captured diagrammatically in Fig. 7.3. Such a *MODEL* comprises one or more independent variables or predictors, representing putative cause(s), which may be continuous (interval or ratio scale), categorical (nominal or ordinal scale) or a mixture. The *MODEL* is used to predict or explain a dependent variable (the **DATA**) representing the putative effect or variable to be explained.

Using the information in a *MODEL*, we are trying to understand why the values of the **DATA** change as they do when we look from case to case or participant to participant. However, a *MODEL* is imperfect and will always make prediction errors. Thus, ERROR is what is left over after a *MODEL* does all of the predictive work it can. ERROR reflects unpredictable random influences as well as systematic influences such as the lack of predictive fit and other *MODEL* deficits, such as the

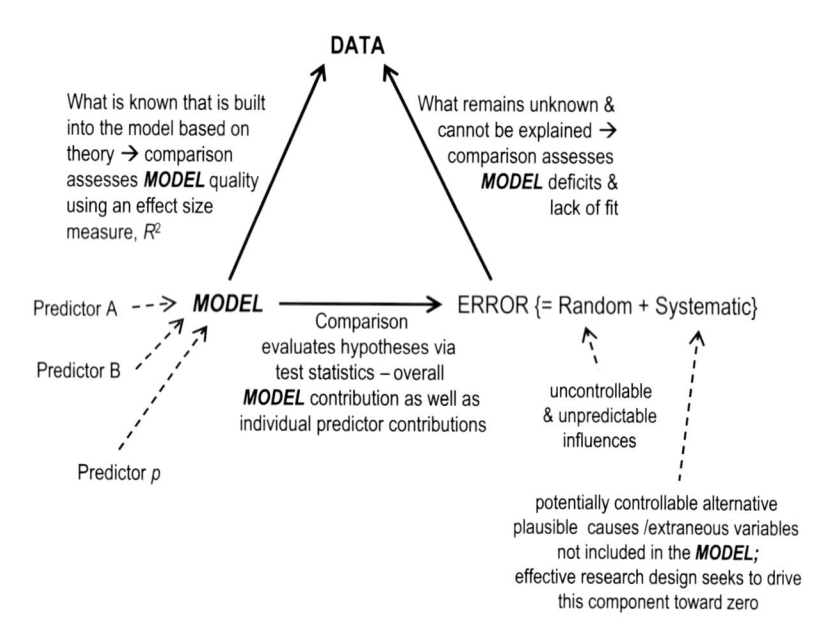

Fig. 7.3 Conceptual representation of the *MODEL* building and testing logic

influence of omitted variables, nonlinear or moderating relationship patterns. This means that the **DATA** can only be fully understood by combining the predictions made by a *MODEL* with the ERRORs associated with those predictions. We can use the following conceptual representations to show this:

$$\textbf{DATA} = \textit{MODEL} + \text{ERROR}$$

and

$$\text{ERROR} = \textbf{DATA} - \textit{MODEL}.$$

Conceptually, if we looked inside of a *MODEL* it would generally look like:

constant + (weight 1 × predictor 1) + (weight 2 × predictor 2) + ...
+ (weight p × predictor p),

where p is the number of predictors included in the *MODEL*, and the constant is the predicted value generated if all predictors are set to zero. [Technically, this *MODEL* is a linear model because the predictor-related components in parentheses are weighted and added together to generate *MODEL* predictions.] A *MODEL* is statistically estimated in a way that ensures the prediction ERRORs are minimised. Typically, this is done using the method of *least squares* where the statistical weights for combining predictor information in the *MODEL* are estimated such that when ERRORs are squared and added across all of the cases or participants in a sample, the resulting sum is as small as it can be.

We can compare the behaviour of the *MODEL* predictions to the behaviour of the **DATA** to assess how well the *MODEL* works to explain the **DATA**. This gives a measure of *effect size* called R^2 which can be conceptually represented as:

$$R^2 = \textit{Effect size} \text{ (reflects model quality)}$$
$$= \textit{MODEL} \text{ behaviour}/\textbf{DATA} \text{ behaviour},$$

where 'behaviour' is measured by the variance statistic (recall *Procedure 5.5*). This measure of effect size reflects the proportion of variability in the **DATA** that can be explained using the *MODEL*, or in other words, what our *MODEL* contributes to understanding of the **DATA**.

We can also design a test for the viability of a *MODEL* by comparing its behaviour to the behaviour of **ERROR**. This conceptually compares what we do know to what we don't know, and we look for a large ratio. This ratio defines a test statistic as:

$$\text{Test Statistic} = \textit{MODEL} \text{ behaviour}/\text{ERROR behaviour}.$$

How large is large enough for this ratio is the central issue for statistical inference. It turns out that this conceptualisation for a test statistic works for a wide variety of statistical tests including *z*-tests, *t*-tests and *F*-tests. Even the chi-square test works in roughly the same conceptual way.

It is important to observe that positivist guiding assumptions, research design and statistical analysis are all brought together in this conceptual model-building system. We want our **MODEL** to do a large amount of work for us, relative to what it fails to do for us. We can achieve a better test if we reduce any systematic components of ERROR through better control and measurement (discussed in *Fundamental concept VII*). There are at least three potential strategies available to us for accomplishing this: (1) incorporate relevant extraneous variables into the **MODEL** (meaning they are no longer be extraneous; they become part of the theory being tested), (2) hold relevant extraneous variables constant for all cases and conditions (so they cannot change when the **DATA** values change; a type of procedural control) and/or (3) measure relevant extraneous variables (now labelled as *covariates*) and statistically remove (i.e., partial) their influence on the **DATA** prior to testing research hypotheses (a process of statistical control).

In addition, we can build our **MODEL** in three different ways to enhance our control: (1) include all predictors in the **MODEL** at the same time (*simultaneous*); (2) add individual predictors or groups of predictors into the **MODEL** in a specific theoretical order (*hierarchical*); or add individual predictors into the **MODEL** in an order determined by the statististical relationships between those predictors and the **DATA** (*stepwise*; this approach is the least preferred because the researcher essentially gives up control, over how their **MODEL** is built, to a statistical rule). For any **MODEL** involving multiple predictors, the contributions of the individual predictors can be individually assessed using a variation of the above test statistic logic, but only after the overall **MODEL** has been shown to work (reflecting the hurdle logic of hypothesis testing reviewed in *Fundamental concept V*).

Configuring Predictors

In order to build a statistical **MODEL** (depicted at the top of Fig. 7.4) where the estimated statistical weights make conceptual sense, it is important that predictor information be represented in the correct way. For continuous predictors (interval or ratio scale measurement), the actual numerical values can be used and a straightforward interpretation of the resulting statistical weight, b, emerges (see the left-hand column in Fig. 7.4). However, for categorical predictors (nominal or ordinal scale measurement, where the categories are often 'scored' as 1, 2, 3 etc), there is an intermediate step to be taken prior to including the predictor information in the **MODEL**. That step is called *coding*. Without this intermediate step, simply using the

Fundamental Concepts VI: The General Linear Model

$$\underbrace{Y}_{DATA} = \underbrace{b_0 + b_1 X_1 + b_2 X_2 + \ldots + b_k X_k}_{MODEL} + \underbrace{e}_{ERROR}$$

Continuous Predictor (interval or ratio scale)

X_1 entered in **MODEL** as is

X_1 interpreted in **MODEL** as

Increase X_1 by 1 unit; predicts a b_1 unit increase in Y

Categorical Predictor (nominal or ordinal scale)

First, need to recode to represent information appropriately

Dummy Coding or **Effect Coding**

Dummy Coding:
Choose reference category to always be scored 0

X	→	New X_1
1	→	1
2	→	0

X	→	New X_1	New X_2
1	→	1	0
2	→	0	1
3	→	0	0

Effect Coding:
Choose reference category to always be scored -1

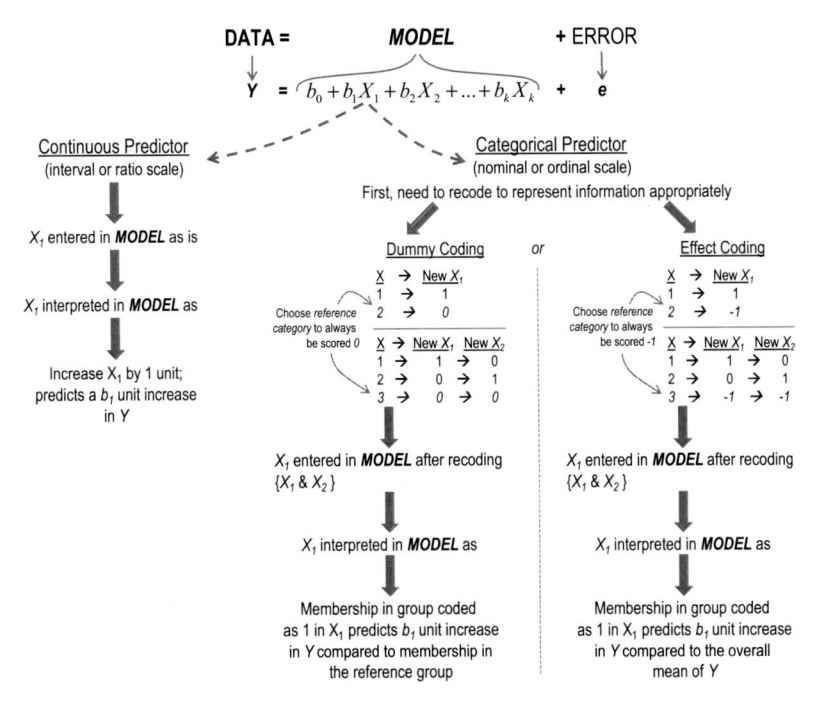

X	→	New X_1
1	→	1
2	→	-1

X	→	New X_1	New X_2
1	→	1	0
2	→	0	1
3	→	-1	-1

X_1 entered in **MODEL** after recoding $\{X_1 \& X_2\}$

X_1 interpreted in **MODEL** as

Membership in group coded as 1 in X_1 predicts b_1 unit increase in Y compared to membership in the reference group

X_1 entered in **MODEL** after recoding $\{X_1 \& X_2\}$

X_1 interpreted in **MODEL** as

Membership in group coded as 1 in X_1 predicts b_1 unit increase in Y compared to the overall mean of Y

Fig. 7.4 Illustration of predictor information coding and the resulting interpretations in a **MODEL**

categorical predictor as is would lead the statistical estimation process to incorrectly use the numbers as if they measured a continuous scale. Instead, we must code the predictor categories using a specific system.

Two such systems are *dummy coding* and *effect coding*. [Note there are actually more than two coding systems that could be used, see Cohen et al. 2003, but these two will serve for our discussion here.] The two right-most columns in Fig. 7.4 illustrate the different processes for coding and interpreting categorical predictor information in a **MODEL**. Coding means that we create new predictors where the number of predictors needed is one less than the number of categories. Figure 7.4 shows how dummy coding and effect coding would work for both a two-category (or dichotomous) predictor and a three-category predictor. Dummy coding uses only two numbers, 0 and 1, and one category is selected as the reference category to always be scored 0. Then for every new predictor created, one category is chosen to be scored 1 (i.e. all members of that category receive a score of 1) and all others, including the reference category are scored 0. Effect coding employs three numbers,

0, 1 and − 1, and one category is selected as the reference category to always be scored −1. Then for every new predictor created, one category is chosen to be scored 1 and all others, *except the reference category*, are scored 0.

In either coding scheme, our goal is to use the new predictors to identify one target category that a case or participant potentially belongs to. You might ask, why not have the same number of new predictors as categories? The answer is, it would be redundant. For a two-category dummy-coded or effect-coded predictor, if a case or participant is not a member of the category coded 1, then, by definition, they must be a reference category member. For a dummy-coded or effect-coded predictor with three or more categories, if a case or participant is not a member of any of the categories coded 1, then, by definition, they must be a reference category member. In the end, the only essential difference between the two coding systems is in how we interpret the statistical weight that gets attached to each new predictor when the **MODEL** is estimated. That interpretation is shown in Fig. 7.4, for new predictor X_1.

How General Is the General Linear Model?

A linear model becomes a general linear **MODEL** when we understand that predictors at any level of measurement may be used, if the information they contain is correctly represented (as illustrated in Fig. 7.4). We can extend the generality even further by understanding that there may be more than one dependent variable on the **DATA** side of the equation, which means that multivariate **MODEL**s can be constructed and tested as well. Throughout this chapter as well as the next two chapters, we will explore specific procedures that go by many different names. However, most of those procedures will just be different forms of the same underlying statistical model. Table 7.1 shows how this logic works and connects the form of the General Linear Model to a specifically named procedure covered in this book.

Limits to a General Linear MODEL

In the above discussion of the General Linear **MODEL**, we did not look very critically at some of the limitations associated with its use. As we shall see throughout this book, procedures for drawing statistical inferences and testing relationships critically rely upon the use of linear **MODEL**s. They are therefore a popular, statistically efficient and often parsimonious approach to handling research inferences. It is, however, important that you realise that relying on such models is not without some important drawbacks. Some of these drawbacks provide one important reason behind why mixed methods research where both quantitative and qualitative

Table 7.1 Different forms of the General Linear Model that go by different names (dummy coding used for categorical variables)

Univariate test	Procedure(s)	DATA = *MODEL*[a] + ERROR
t-test	7.2	**DATA** continuous; one predictor in the *MODEL* coded as 0 or 1 depending upon group membership
Analysis of variance (ANOVA)	7.4, 7.6, 7.10, 7.11, 9.4	**DATA** continuous; several predictors in the *MODEL* coded to represent group membership (additional predictors in the form of interactions are included in designs involving more than one independent design variable – see *Fundamental concept IX*)
Analysis of Covariance (ANCOVA)	7.15	**DATA** continuous; some predictors in the *MODEL* are continuous or categorical but controlled for (these are covariates); other predictors in the *MODEL* code group membership; the *MODEL* is built hierarchically
Bivariate & Partial Correlation & Multiple & Multi-Level Regression	6.3, 6.4, 7.13, 9.5	**DATA** continuous; predictors in the *MODEL* are continuous and/or code group membership and *MODEL* can be built hierarchically
Logistic regression	7.14	Categorical **DATA** coded as 0 or 1 depending upon response category; predictors in the *MODEL* are continuous and/or code group membership
Log-linear models	7.18	**DATA** comprises the log of cell frequency counts; predictors in the *MODEL* code category membership
Time series analysis	8.5	**DATA** continuous; predictors in the *MODEL* are observation times (e.g. days, years) as well as predictors representing the relationships between successive observation periods and, optionally, a predictor coding the presence/absence of an intervention
Multivariate test	Procedure(s)	Multiple DATA[b] = *MODEL* + ERROR
Multivariate ANOVA	7.16	**Multiple DATA** interval or ratio scale; several predictors in the *MODEL* coded to represent group membership as well as interaction information
Multivariate ANCOVA	7.16	**Multiple DATA** interval or ratio scale; some predictors in the *MODEL* interval or ratio scale or categorical (nominal or ordinal) but controlled for (covariates); categorical predictors in the *MODEL* code group membership
Discriminant analysis	7.17	**Multiple DATA** code group membership; predictors in the *MODEL* are interval or ratio or coded group membership
Canonical correlation Set correlation	6.8	**Multiple DATA** interval or ratio scale and/or coded group membership; predictors in the *MODEL* are interval or ratio scale and/or coded group membership; set correlation permits control for up to two additional sets of variables
Principal components analysis	6.5	**Multiple DATA** are unobserved scores on "components" which maximise the variance explained by the variables; predictors in the *MODEL* are the actual individual variables in the analysis; ERRORs (e) assumed = 0.
Common factor analysis	6.5	**Multiple DATA** are the actual individual variables in the analysis; predictors in the *MODEL* are unobserved scores on "factors" which are used to reproduce correlations between variables; ERRORs (e) (or uniquenesses) are estimated.

(continued)

Table 7.1 (continued)

Multivariate test	Procedure(s)	Multiple DATA[b] = ***MODEL*** + ERROR
Confirmatory factor analysis	8.6	**Multiple DATA** are the actual individual variables in the analysis; predictors in the ***MODEL*** are unobserved scores on hypothesised "factors" which are used to reproduce correlations between variables; ERRORs (e) are estimated.
Structural equation modelling	8.7	**Multiple DATA** are the actual individual variables in the analysis (which may be combined to represent hypothesised "factors"); predictors in the ***MODEL*** are scores on unobserved hypothesised "factors" and/or observed variables which are used to reproduce correlations between variables; specific patterns of causal relationship between predictors and **Multiple DATA** are hypothesised; ERRORs (e) are estimated.

[a]*The ***MODEL*** is represented as constant + (weight 1 × predictor 1) + (weight 2 × predictor 2) + ... + (weight p × predictor p) for p predictors*
[b]*Multiple **DATA** is represented as (weight 1 × **DATA** 1) + (weight 2 × **DATA** 2) + ... + (weight d × **DATA** d) for d dependent variables. The weights indicated here are estimated separately, and therefore differ, from the weights for predictors in the ***MODEL****

data are gathered have become increasingly important in social and behavioural research. Figure 7.5 presents a mindmap outlining the limits to General Linear ***MODEL***s and highlighting some of the complexities they may create for researchers.

General linear ***MODEL***s are sensitive to how they are specified, meaning which predictors are included and how they are connected to each other (in a linear model, the connection is by simple addition). A ***MODEL*** can easily be so simplified that it essentially becomes meaningless. If key variables have been omitted in an analysis, then the ***MODEL*** will not be appropriately estimated. Often, there is pressure to create simpler rather than more complex ***MODEL***s (the law of parsimony in action), even when the more appropriate ***MODEL*** would be the more complex one. A linear ***MODEL*** is always more parsimonious than a non-linear ***MODEL*** (and is much easier to statistically estimate and interpret), yet the world does not always work in a linear fashion.

It is important to understand that the ERROR term (*e*) contains everything that the researcher does not explicitly account for in the ***MODEL***; these may be systematic or random influences and they may be operating in complex, perhaps even chaotic ways. The statistics that are computed when you estimate a linear ***MODEL*** provide no clear way to tear apart and diagnose the contents of the ERROR term, so it generally remains uncritically examined by researchers. This is where mixed methods research can be of great benefit, using qualitative data to help understand what might be hidden in the ***MODEL*** ERRORs. Fitting a general linear ***MODEL*** requires absolute consistency in how measures are defined and quantified. However, if the understanding of constructs via the measurement systems is ambiguous or interpreted differently by different participants, cases or organisations, this will affect the adequacy of the resulting ***MODEL***.

Linear ***MODEL***s are completely insensitive to dynamic feedback loops and contextual nuances and only ever focus on minimising errors, rather than on

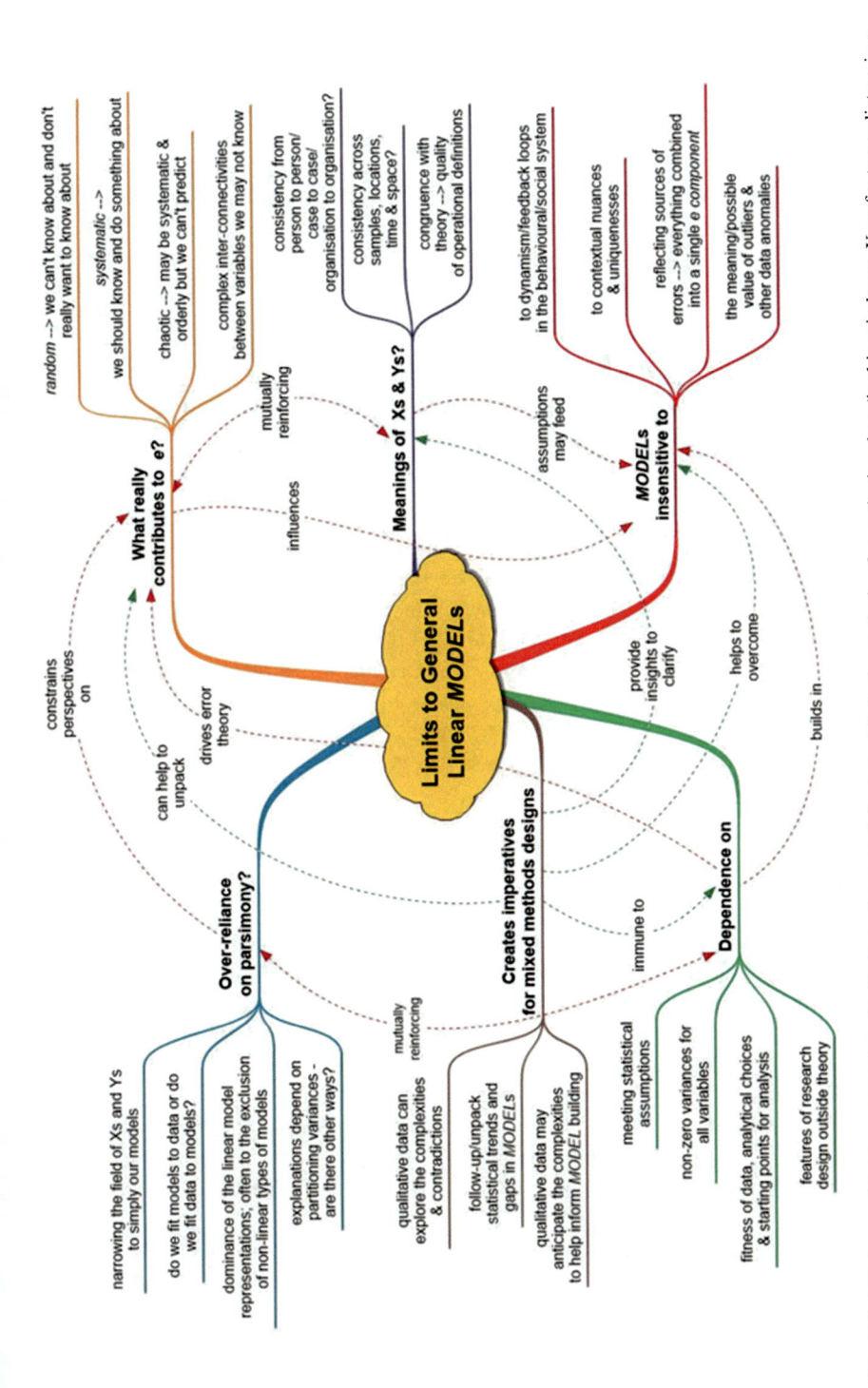

Fundamental Concepts VI: The General Linear Model

Fig. 7.5 Mindmap of issues surrounding the limits to linear statistical models in the analysis of quantitative data (in this mindmap, Xs refer to predictors in a *MODEL*; Ys refer to **DATA** measures and *e* refers to ERROR)

understanding ERRORs. In order for a linear ***MODEL*** to work, all variables involved must have non-zero variances (actually the further from zero, the better), or else statistical estimates cannot be computed (that is, you can't partition variance for ***MODEL*** testing if there is no variance to partition!). Inferential tests of a linear ***MODEL*** require certain assumptions to be met, including, for example, normality in dependent variable distributions and independence of errors from case to case. Not meeting an assumption may mean a ***MODEL*** cannot be properly estimated without changing something (including, perhaps, the entire analytical approach), thereby contributing indirectly to ERRORs.

Fundamental Concept VII: Principles of Experimental & Quasi-Experimental Design

Experimental and quasi-experimental methods generally involve explicit manipulation, by the researcher, of events (such as treatment programs, training programs, specific problems or scenarios) which the research participants are permitted to experience in order to observe the impact of the manipulation on some measure of interest. For these reasons, experimental and quasi-experimental methods are considered to involve interventionist techniques. The goal is to arrange the interventions in such a way that only the independent variables (causal factors) of interest are permitted to change and impact on a dependent variable (the effect) while simultaneously keeping the influences of extraneous variables (alternative plausible causes) from contaminating the story.

To build a clear case for a cause-effect linkage, the researcher must arrange circumstances so that causal factors change before the effect changes (*temporal priority*) and must show that when the causal factor(s) change, the effect also changes (*covariation*). However, the most difficult task for the researcher is to think logically, anticipate the potential role that contaminating extraneous influences could play in the research and take steps to minimise the most critical of those potential contaminating roles (*ruling out alternative plausible causes*). If the goal of the research is to maximise internal validity (i.e. capacity to unambiguously conclude that the intended causal or associational linkages have been found), then sound experimental design practices and choices provide the pathway. No amount of sophistication in statistical analysis can offset poor experimental design. Statistics help build the case for covariation and, in some cases, can help rule out alternative plausible causes, but it is sound experimental design that builds the foundation required for concluding cause and effect.

The practical side of experimental design generally involves making trade-offs between achieving a high level of internal validity in the research and achieving a high level of external validity (i.e. capacity to generalise the relationships and linkages found in the research context to some population of interest and to other places, times and contexts). Making design decisions intended to maximise internal validity will almost invariably mean sacrificing a strong capacity to achieve high external validity and vice-versa. These two types of validity, both of which are desirable features of 'good' positivist quantitative research, generally work against each other, which is why experimental design is such a challenging endeavour. There is no experimental design which can maximise both types of validity simultaneously. True experimental research designs afford the greatest possible control over extraneous factors while sacrificing contextual realism whereas quasi-experimental research designs sacrifice a high degree of control over context in order to obtain greater contextual realism, feasibility and generalisability in the research.

A frequently asked question in experimental design is what is the primary difference between a true experiment and a quasi-experiment? The answer is quite straightforward. If the researcher can randomly determine which participants experience which specific aspects of an independent variable of interest (primarily through random assignment to experimental groups), then the research design constitutes a true experiment. If the researcher does not have this capacity (either because he or she is prohibited from doing so by contextual circumstances or because the independent variable of interest is a measured characteristic of participants or some aspect of their lives, not a manipulated experience; e.g. gender, company they work for, or living through some natural event), then the research design constitutes a quasi-experiment. A true experiment provides the strongest context for achieving a high level of internal validity at the expense of high external validity (i.e. greater opportunity to claim cause and effect but in a context that is highly controlled, often contrived and therefore artificial and less representative of real life). A quasi-experimental design, because random assignment to conditions is not possible or viable, provides a weaker context for achieving high internal validity but a stronger context for external validity (i.e. less opportunity to defend causal conclusions but with a greater opportunity for finding generalisable results under more realistic conditions). Either type of design, true experiment or quasi-experiment, can potentially be used under laboratory or field conditions, but it is more likely that true experiments are done in laboratory contexts and quasi-experiments are done in naturalistic contexts.

The Concept of Control Over Extraneous Variables

The concept of control is central to effective experimental design as it is through the exercise of control that the researcher can minimise the most critical of the potential contaminating influences. There are many different types of extraneous variables that may be relevant in experimental design. Consequently, there are a number of different ways to control for their potential influence. Good experimental design is all about making sensible and defensible design choices, in full recognition of what you gain or lose by making each choice. From a statistical model building and testing perspective, contextual control creates opportunities to remove systematic influences from estimates of model errors, ideally reducing model errors to only unpredictable and uncontrollable random influences. Some control procedures will work to improve internal validity and others will work to improve external validity. Sometimes, the choice of one control procedure or design alternative creates circumstances where another extraneous variable becomes a potential liability, and this means that another control procedure must be put in place. Design choices can thus create cascading or ripple effects in some cases.

Control in experimental/quasi-experimental design can be achieved in several ways (Cooksey and McDonald 2019, pp. 653–654):

- *Control through measurement*: Control over data quality can be achieved through effective quantification/measurement strategies. Here the goal is to minimise measurement errors by utilising effective operational definitions to maximise relevant types of validity (e.g. construct, content, predictive) and reliability. Control through measurement attempts to ensure that independent and dependent variables will permit effective specification of statistical models and tests of hypotheses.
- *Control by design*: A researcher can obtain control over certain extraneous variables through explicitly incorporating them in his or her experimental or quasi-experimental design. This is usually accomplished by adding additional independent variables or groups into the design so that their influence on the dependent variable can be explicitly evaluated.

  Control by design is accomplished by elaborating experimental designs using control groups or making an extraneous variable (such as prior experience, ethnic background, gender, ability, attitude) an explicit independent variable in the design. In some cases, control by design is achieved by testing for the influence of a moderating (a variable that conditionally influences the direct relationship between an independent variable and a dependent variable) or a mediating variable (an intervening variable through which the independent variable exerts an indirect influence on the dependent variable). Control by design works by explicitly tracking the influence of one or more extraneous variables on the dependent variable *while* the dependent variable is measured.
- *Procedural control*: A researcher can obtain control over certain extraneous variables through the implementation of specific procedures, such as random sampling of research participants, random assignment of participants to experimental conditions, counterbalancing of experimental conditions and holding environmental and task conditions constant. Procedural control works by removing the opportunity for one or more extraneous variables to influence the dependent variable *before* the dependent variable is measured.
- *Control through statistical model specification*: This type of control focuses on the model specification process by ensuring that all necessary variables and, where relevant, interrelationships among those variables and functional forms of variables are incorporated into the model the researcher is using to test theoretical propositions and hypotheses. This control process also extends to the specification and incorporation of appropriate error structures in the model being tested.
- *Control through statistical analysis*: A researcher can obtain control over certain extraneous variables through explicitly measuring them along with the dependent variable and then using statistical procedures to adjust the dependent variable for their influence before testing for the effects of the independent variables of interest. Measuring an extraneous variable in this way gives rise to a specific

type of variable used in statistical analysis—the *covariate*. Thus, control through statistical analysis works by statistically removing (a partialling process) the influence of extraneous variables (covariates) on the dependent variable *after* the dependent variable has been measured but before the influence of the independent variables is assessed.

Table 7.2 summarises a range of possible extraneous variables that can operate in an experimental or quasi-experimental context and describes a range of control procedures for handling them. The table focuses on those extraneous variables which can have a *systematic influence* on a dependent variable at the same time as an independent variable may be exerting its influence, thereby producing a *confounding*. There is another class of extraneous variable influences – *unsystematic influences* – which introduce random errors into measurements and behaviours. Such unsystematic or random influences may include: a participant having a 'bad day' or being tired, equipment malfunctions, participant forgets to answer a survey question, poor participant attitudes toward the research and data recording errors. These influences generally cannot be anticipated or predicted, are almost impossible to control for and, consequently, researchers simply must live with them as contributors to the random error that statistical hypotheses are evaluated against.

Table 7.2 Extraneous variables as sources of potential confounding in behavioural and social research and associated control procedures

Extraneous influences	Control procedures
Environmental: Uncontrolled changes in environmental conditions such as distractions, light level, temperature, dust, humidity, noise level, time of day, season, and/or day of week co-occur with changes in the independent variable.	**Holding constant**: Hold experimental conditions as constant as possible for all groups; more feasible in a laboratory setting.
Task: Changes in task complexity or consistency co-occur with changes in the independent variable.	**Holding constant**: Hold measurement contexts as constant as possible for all groups; more feasible in a laboratory setting.
Instructions: Changes (perhaps subtle) in instructions to participants co-occur with changes in the independent variable.	**Standardisation:** Use recorded, computer-delivered or scripted instructions to participants.
Instrumentation: Changes in the measurement characteristics of testing instruments co-occur with changes in the independent variable.	**Standardisation:** Use highly structured, highly reliable and construct valid instruments and tests, with clear and unambiguous response formats, under consistent administration conditions; use back-up systems for technological instrumentation.
Participant-related*:* Individual differences in participant characteristics such as gender, ethnic background, intelligence, manual dexterity, reactivity to stress, motivation, attitudes, etc. covary with changes in the independent variable. Non-random selection of participants, including use of volunteers (whether paid or unpaid), is often involved here (e.g. having	**Random sampling**: Ensuring that participants are selected from the population using some type of probability-based random sampling scheme (see *Fundamental concept VIII*).
	Random assignment: Ensuring the participants are randomly assigned to experimental conditions or groups.

(continued)

Table 7.2 (continued)

Extraneous influences	Control procedures
more males than females in the sample when gender is related to scores on the dependent variable of interest).	**Use participant as his or her own control**: Use a repeated measures design where participants are measured on multiple occasions (as in pretest-posttest designs) or in all experimental conditions. Use of this control procedure may create new problems related to arrangement of conditions (see below) that may then need to be controlled for.
	Matching participants: Matching participants on one or more characteristics known to influence the dependent variable (a less preferred control procedure, unless genetically-identical and similarly raised and schooled individuals are available (e.g. twins), as it requires a large pool of participants to identify matches from and it is virtually impossible to match people on all potentially relevant features).
	Covariate measurement: Explicitly measure characteristics of participants that are anticipated or theorised to be influential on the dependent variable, but which are considered extraneous to the hypotheses being tested (i.e. covariates). Then statistically partial the influence of these covariates from the dependent variable before assessing the impact of any independent variable(s). See *Fundamental concept IV* and *Procedure 7.15*.
Arrangement of conditions: A problem that emerges only when the researcher employs repeated measures designs using participants (as their own control) in different experimental conditions (not relevant to time-based repeated measures designs like pretest-posttest designs). Two types of arrangement effects can emerge:	
Sequence effects where participants all participate in experimental conditions in the same order (e.g. all participants taste 5 different cola drinks in the same order in a taste test experiment); and	**Counterbalancing**: Systematically varying the ordering of conditions that different participants experience through the study. Counterbalancing may be designed according some scheme (like a **Latin square design**) where all or some rational subset of possible condition orders are used (preferred because order effects can be explicitly evaluated in such designs) or counterbalancing may be random where every participant gets a random ordering of conditions.
Order (or carry-over) effects where one condition may have effects that carry over into the condition that follows it (e.g. in a taste test experiment on cola drinks, tasting a very sweet	**Carry-over effect cancellation** or **negation procedure**: introducing an intervention between conditions designed to not influence the dependent variable but to counteract

(continued)

Table 7.2 (continued)

Extraneous influences	Control procedures
cola followed by a not-so-sweet cola may make the not-so-sweet cola taste sweeter than if the order of tasting was reversed).	possible carry-over effects, e.g. having participants rinse their mouths out with soda water between each cola tasted.
Multiple response effects: A potential problem in studies where multiple measurements are obtained on the same participants. There are three general effects:	
Practice effects: Where improvements on the dependent variable over time and/or measurements can be attributed to participants simply getting more familiar with and better at the experimental task(s) they are performing.	**Threshold training** or **training to a criterion level**: Give all participants practice trials or experience on the experimental task before conducting the actual experiment in order to get them all up to an equivalent starting performance level.
Fatigue effects: Where decrements in performance over time and/or measurements can be attributed to participants being over-worked or increasingly tired. A risk in any research that asks many questions (including long questionnaires), involves long participation times or uses very complex testing tasks.	**Fatigue breakers**: Design in rest breaks or vary experimental tasks to reduce participant workloads and keep interest in performing high. Use of incentives for continued focused performance can work in some circumstances. Design shorter and easier-to-read instruments if questionnaires are used to gather the data.
Habituation effects: where decreased effectiveness of independent variable manipulation in influencing the dependent variable can be attributable to the fact that participants become adapted to or used to the manipulation conditions, e.g. drug dosages, light levels, noise levels, stressful conditions, etc. Often a problem in repeated measures designs but may also be a problem in any design where environmental conditions are manipulated.	**Habituation disruption**: Separate experimental conditions by some constant interval of time to allow the adaptation to dissipate. The introduction of novel stimuli may also help to disrupt adaptation, but you need to be careful it does not interfere with the manipulation itself or influence the dependent variable. Also, designing shorter exposures to the experimental conditions may help avoid longer-term adaptation.
Demand characteristics: A problem where cues in the experimental situation provide participants with information about the experiment that may subtly alter their performance (the danger is greater when these 'cues' are actually irrelevant or imagined yet the participants act on them anyway). There are four general types of demand characteristics:	
Experimenter bias: where the behaviour of the researcher toward the participant inadvertently or subtly influences the participant's responses. A risk when the researcher strongly believes his or her hypothesis is correct and communicates those expectations, usually non-verbally via body language, word intonations or different instructional emphases, to participants. Such effects may also be reflected in inadvertent mis-recording of observations.	**Double blind control procedure**: Design the study so that neither the researcher nor the participant knows which condition they are participating in. Can be facilitated by using a dispassionate 'experimenter' who has no vested interest in how the study comes out. Only works for those experimental manipulations whose nature is not immediately obvious (e.g. drug dosages, certain environmental changes, task complexity).

(continued)

Table 7.2 (continued)

Extraneous influences	Control procedures
Guinea pig effect: A problem that arises because participants are people who think and analyse and will try to figure out what the researcher is up to (e.g. they know they are being studied). Such participants may be motivated to try to give the researcher the information they think is desired (**acquiescence bias**) or to give contrary indications or responses (**sabotage**).	**Single blind control procedure**: Keeping the participant in the dark regarding their participation in a research study (covert research) can avoid this problem but has severe negative ethical implications. Keeping the participant blind as to which condition they are participating in can work, but only for those experimental manipulations whose nature is not immediately obvious or detectable, e.g. drug dosages, certain environmental changes, task complexity.
Hawthorne effect: named for the Hawthorne industrial plant (specifically the Hawthorne plant at the Western Electric Company in Chicago in the 1920s) where the effect was first noted. Refers to the problem that participants will often show improvements on productivity-related dependent variables simply because they know they are being specially or differently treated (e.g. their environment is changing, they are participating in a new program while getting time away from their normal job). Almost anything novel that happens to people can produce short-term gains on many performance- and motivation-related variables.	**Control group design**: Elaborate the research design by including one or more control groups design to show whether this effect is occurring. Such a group will usually receive an identical level of special treatment or attention as do participants in an experimental group, but in activities that will not have any impact on the dependent variables of interest.
Placebo effect: This is a variation of the Hawthorne effect where participants respond to a control intervention (a placebo; a sugar pill in drug studies or an irrelevant training program, for example) as if they had received a treatment or intervention. It is linked to participants' belief that they are receiving something beneficial and respond accordingly and, perhaps, even genuinely.	**Single blind control procedure**: keep the participant ignorant of the nature of the condition they are serving in or treatment they are receiving. One control group, which receives the placebo condition (a null or irrelevant treatment), should be included in the design to counteract this alternative plausible cause.

Common Types of Experimental and Quasi-Experimental Designs

There are many design configurations possible in experimental and quasi-experimental research. Reference texts such as Campbell and Stanley (1966), Cook and Campbell (1979), Keppel and Wickens (2004), Kirk (2013) and Shadish et al. (2001) provide very comprehensive coverage of different types of experimental and quasi-experimental designs and of principles associated with their use. Table 7.3 shows a few of the more commonly used designs and provides a brief synopsis of

Fundamental Concept VII: Principles of Experimental & Quasi-Experimental Design

Table 7.3 Some commonly used experimental and quasi-experimental designs for social and behavioural research

Advantages	Disadvantages
True experimental designs[a]	

Pretest-posttest control group design

		Pre		Post
Experimental Group:	R	O_1	X	O_2
Control Group:	R	O_1		O_2

Advantages	Disadvantages
Controls for history and maturation effects – natural changes over time and due to growth and learning; random assignment spreads idiosyncratic individual differences of participants across groups, making groups theoretically equivalent (however, the design permits a test of this equivalence); control group used for comparison purposes. See *Procedure 7.11* and *Fundamental concept VIII*.	At risk of test sensitisation effect where the pretest alerts participants to what aspects of behaviour are of interest; also at risk of testing/instrumentation reactivity if participants react poorly to questions asked or measures obtained; also at risk of participant 'mortality' or leaving the study between pretest and posttest.

Posttest only control group design

Experimental Group:	R	X	O_2
Control Group:	R		O_2

Advantages	Disadvantages
A very commonly used design; takes less time and fewer resources than above pretest-posttest design while still providing control group comparison and benefits of random assignment; low 'mortality' risk. See *Procedure 7.2 and 7.6*.	At risk of history & maturation effects as well as testing/instrumentation reactivity effects.

Solomon four group design

Experimental Group 1:	R	O_1	X	O_2
Control Group 1:	R	O_1		O_2
Experimental Group 2:	R		X	O_2
Control Group 2:	R			O_2

Advantages	Disadvantages
Combines both previous designs to give best control over most extraneous variables, enjoys benefits of random assignment, two control groups used for comparison purposes and allows specific tests for history, maturation, and testing effects—most powerful experimental design for showing change due to X. See *Procedure 7.10* and *Fundamental concept VIII*.	Very resource intensive design; requires large sample size to run effectively; some 'mortality' risk in repeated measures part of the design (between O_1 and O_2).

(continued)

Table 7.3 (continued)

Advantages	Disadvantages
True experimental designs[a]	

Factorial posttest only control group design

	Indep. Var. 1 →	Control Group	Experimental Group
Indep. Var 2 →	Control Group	R	R
	Experimental Group	R	R

Advantages	Disadvantages
Very common design choice when control by design (adding extra groups and independent variables) is used; enjoys benefits of random assignment and can test for moderator (i.e. interaction) effects; control group for each independent variable used for comparison purposes. See *Procedure 7.10* and *Fundamental concept VIII*.	Resource intensive design; requires large sample size to run effectively; analysis is more complicated; not a good design for inferring change over time.

Quasi-experimental designs

Single-shot posttest only design (many surveys)

	Pre	Post
Experimental Group:	X	O_2

Advantages	Disadvantages
Simple design; resource efficient; often used in survey studies; low mortality risk. Analysable primarily using descriptive statistics. See *Procedure 5.1* to *5.7*.	Limited control over any extraneous variables including no random assignment of participants to groups; at risk of history, maturation, testing reactivity and other plausible alternative causes such as biased response rates from different types of participants; cannot show change over time; no comparative control group—weakest of the quasi-experimental alternatives.

Single group pretest-posttest design (many program evaluations)

Experimental Group: O_1 X O_2

Advantages	Disadvantages
Relatively simple design; resource efficient; allows some inference about change over time due to X; often used in program evaluation research. See *Procedure 7.4*.	Relatively limited control over any extraneous variables including no random assignment of participants to groups; at risk of history, maturation, testing, mortality and reactivity and other plausible alternative causes; can show change over time but not strongly attributable to X; no comparative control group—a rather weak but commonly used quasi-experimental alternative.

(continued)

Fundamental Concept VII: Principles of Experimental & Quasi-Experimental Design

Table 7.3 (continued)

Quasi-experimental designs

Non-equivalent static group design

Experimental Group: X O_2
Control Group: O_2

Resource efficient and popular design; control group used for comparison; low mortality risk. See *Procedure 7.2* and *7.6*.	Cannot support inferences about change over time; at some risk of history, maturation, testing and reactivity and other plausible alternative causes.

Non-equivalent control group pretest-posttest design

Experimental Group: O_1 X O_2
Control Group: O_1 O_2

Controls for history and maturation effects – natural changes over time and due to growth and learning; design permits a test of group non-equivalence; control group used for comparison purposes. See *Procedure 7.11* and *Fundamental concept VIII*.	No random assignment means that inferences about changes over time attributable to X are on weaker footing; some risk of testing effects, mortality effects and reactivity effects.

Interrupted time series (single participant) design

$O_1 O_2 O_3 O_4 O_5 \; X \; O_6 O_7 O_8 O_9 O_{10}$

Powerful design for program intervention evaluation on single or small numbers of participants or for social/economic indicator research; power for causal inference comes from long chain of pre and post observations; causal inference can be strengthened further by elaborating the design to remove X and add X back in (what is sometimes called an ABAB design), each time making a series of observations/measurements; there is a control group variant as well where some participants do not receive the X treatment or condition. See *Procedure 8.5*.	High risk of participant mortality and testing effects due to multiple measurement occasions; risk of instrumentation/measurement changes over time and other historical events, not related to the research, occurring through the series of observations can confound causal inferences.

[a]R indicates random assignment of participants to groups or conditions; O indicates an observational occasion where **DATA** measurements are obtained, and X indicates an explicit treatment or intervention, serving as an independent variable for the ***MODEL***

their major advantages and disadvantages (see also Cooksey and McDonald 2019, section 14.3.2). [As there is a lot of jargon that accompanies the use of experimental and quesi-experimental designs, you may also find it useful to look at Cooksey and McDonald 2019, pp. 676–677, entitled *Appendix: Clarifying Experimental/Quasi-experimental Design Jargon*.]

Fundamental Concept VIII: Principles of Sampling

Sampling is a crucial process for ensuring that defensible generalisations can be made from the participants chosen for your sample(s) to the population(s) from which those participants were selected. Thus, sampling processes have a strong influence on the external validity of a study. As well, appropriate sampling can improve the power of a study to detect significant effects and improve the precision of statistical estimates of population parameters. In the logic of statistical inference (recall *Fundamental concept V*), the goal of the statistical inference process is to generalise from the relationship or difference found in a sample to relationships or differences in the population.

Two key aspects of sampling concern the *size of the sample* (often gauged relative to the size of the intended population), and the *representativeness of the sample*.

Sample size has a simple and direct relationship to statistical power and precision of estimates: larger samples yield greater power and more precise statistical estimates. Representativeness of a sample refers to the degree to which the composition of the sample reflects the composition of the population to which generalisations are intended. It is representativeness, more so than sample size, that has major implications for external validity/generalisability: biases in sample selection or final composition will negatively impact on one's ability to make appropriate generalisations, even if the sample size is appropriate.

Types of Sampling Schemes

The most fundamental concept associated with sampling as it relates to statistical analysis is *random sampling*. By *random* is meant that (1) every member of the relevant population has an equal chance of being selected in your sample and (2) every sample of a given size has an equal chance of being selected. A random sample thus depends on the rules of probability (recall *Fundamental concept I*) and the 'law of large numbers' which governs the likelihood of occurrences in the long term to ensure sample representativeness. In principle, random sampling works in the same way as a lottery draw.

There are two broad classes of sampling schemes: probability sampling schemes and non-probability schemes. The difference between the two types of scheme is simple: in probability sampling schemes, the probability of any member of the population being selected for a sample is known; in non-probability sampling schemes, the probability is either unknown or irrelevant to the purposes of the research. For probability schemes to work correctly, you need to have a very clear idea of the intended population of interest and an available listing of all population members from which you will draw your sample.

Table 7.4 lists and briefly describes several probability and non-probability sampling schemes (a number of these schemes as well as more depth of discussion about focal criteria for sampling appear in Cooksey and McDonald 2019, chapter 19 and in Fink 2002). Probability sampling schemes offer the best statistical basis on which to defend generalisations to a population of interest; non-probability sampling schemes are always weaker in their capacity to support generalisations beyond the boundaries of the research. However, a good probability sample is hard and generally costly to achieve; non-probability samples tend to be much more feasible and resource-efficient for researchers, but at the cost of reduced capacity to generalise.

Table 7.4 Different types of probability and non-probability sampling schemes

Schemes	Description[a]
Probability sampling schemes	
Simple random sample	A sampling scheme that randomly selects participants from a single listing of the members of the intended population such that each person has an equal chance of being selected and each sample of a given size has the same chance of being selected. Each member of the population is assigned an identifying number and a computerised random number generator can be used to create a random list of identification numbers for the sample. For example, randomly sampling 1000 certified practicing accountants from 9000 accountants registered with a CPA professional association within NSW would constitute a simple random sample. For such a scheme to work, the researcher would have to gain access to the listing of all registered accountants with the NSW branch of that CPA professional association.
Stratified random sample	A sampling scheme that begins by identifying key segments or layers (*strata*) within a population of interest, layers that would likely have an impact on dependent variables of interest, and randomly sampling a certain number of participants from within each stratum. Stratified random sampling is often used where researchers want to ensure that their sample composition is representative in ways that a simple random sample cannot guarantee. Simple random sampling can fail to provide a representative sample simply because of how probabilities operate: by chance alone. For example, a random sample could contain too many non-Indigenous Australians and too few or maybe no Indigenous Australians. If indigeneity could be important to understand as it impacts on dependent variables of interest or if we

(continued)

Table 7.4 (continued)

Schemes	Description[a]
	wanted to have a sample that was Indigenous representative, we would then have a biased sample. For example, suppose a researcher knew that the population of CPAs in NSW comprised 6000 accountants and 3000 auditors (a 2 to 1 ratio). If she wished to ensure a representative sample of 1200 CPAs proportionally comprised of CPAs who worked as accountants and CPAs who worked as auditors in NSW, she could use professional orientation to define the strata and randomly sample 800 accountants and 400 auditors.
Cluster sample	A sampling scheme that begins by segmenting the population according to some specific criterion of interest. This segmentation would identify clusters in the population. We could then randomly sample a specific number of clusters and survey all members of the population contained in each selected cluster. Cluster sampling is often used where, for reasons of feasibility in conducting the research, the researcher wants to segment the population by geographic region or by industry and then randomly sample a small number of clusters to conduct their research in. For example, a researcher may only have the time and resources to conduct his research in 5 geographic locations within Queensland. He could use electoral boundaries to define clusters and randomly sample 5 out of the total number of clusters to work within.
Multistage sample	Multistage sampling combines two or more of the different types of probability sampling schemes. For example, a researcher may have limited resources so that only 6 geographical regions could be visited for the research. However, she may split the clusters into metropolitan, provincial or rural strata and randomly sample two of each type of cluster to conduct her research in. Furthermore, once she has sampled her 6 clusters, she could do a simple random sample inside each cluster to select her final list of participants. In doing this, she will have used simple random, stratified random and cluster sampling in a multistage process.
Systematic sample	This type of sampling produces a probability sample, but the sample is not random. The scheme runs very simply: randomly sample a participant to begin with and then select every k^{th} participant after that on a population list until the desired number of participants has been sampled. Such a scheme could work in a mall intercept marketing study where a shopper is randomly identified to interview, and then every 10th shopper after that is sampled. Another popular systematic sampling scheme is to pick a random page in the phone book, pick a random name on the page, then sample say every 5th name after that. Such schemes are probabilistic in that once the starting point has been identified, the probability of each person being in the sample is known (e.g. every 10th person has a probability of 1.0 of being chosen; those people between every 10th person have a 0 probability of being selected). Systematic samples are often convenient to assemble, but they are almost guaranteed *not* to be representative.
Non-probability sampling schemes	
Quota sample	This sampling scheme is quite popular and generally straightforward. It is the non-probability version of stratified random sampling. Strata are identified within the population of interest and a quota of a specific number of

(continued)

Table 7.4 (continued)

Schemes	Description[a]
	people to sample within each stratum is set. Then, people continue to be sampled (usually using another non-probability scheme such as convenience sampling) until each quota has been filled. For example, a researcher may be interested in a sample that covers male and female bank tellers as well as tellers from different ethnic backgrounds (indigenous, white Australian, non-white Australian . . .). Quotas would be set for every combination of gender and ethnic background and sampling would continue until each quota was achieved. This ensures representativeness on the two strata variables but is not a random sample—it is therefore less generalisable than a stratified random sampling scheme.
Convenience sample	This sampling scheme simply targets those potential participants that the researcher can get easy access to (such as shoppers, managers, nurses or teachers in the town where the researcher lives). Sampling may continue until some quota has been achieved. Resource constraints will often restrict a researcher to this type of non-random sampling scheme. Generalisations to a larger population are very hard to defend with such samples.
Purposive sample	This sampling scheme works from the perspective that the researcher knows specific people or specific types of people he wishes to include in his sample and those are the people approached for participation. This type of (non-random) sampling is relevant if the point of the research is to explore the thinking and/or behaviours of people in specific roles or who possess specific knowledge, skills, expertise or perspectives. However, generalisability potential is very weak.
Volunteer sample	This type of sampling is really a variation of convenience sampling and consists of advertising for and enlisting the participation of people in the research. People decide whether to participate based on what they have been told or have heard about the study. Technically speaking, given the modern-day ethical requirement for informed consent, every sampling process (probability or otherwise) depends on volunteers. Such samples are non-random and will probably lack representativeness relative to the general population because research has shown that people who volunteer to participate in research tend to have different motivations, attitudes and personality characteristics compared to people who don't volunteer. If these motivations, attitudes and personality characteristics are potentially influential on dependent variables of interest, then a sample of volunteers will necessarily be biased and hard to generalise from.
Snowball sample	This type of sampling is more often used in qualitative research but can be used in quantitative research as well. It commences with the researcher approaching a few people he or she knows have useful information for the research (a purposive sampling logic here) and asking those people to recommend who else would be good to approach for their participation in the study. While this type of sampling scheme can build up a reasonably large (but non-random) sample fairly rapidly, the researcher will never be sure on what basis each participant will be recommending others to approach. Thus, sampling biases will almost certainly be present and may, in fact, increase as more and more people are included. The ethics committees at some universities may not approve snowball sampling plans because of perceived risks to the right to privacy.

[a]Note that the table focuses on sampling people, but the principles extend to sampling cases, organisations, documents, farms or any other type of sampling unit.

There is one important caveat about random sampling that needs to be reinforced. In social and behavioural research involving human participants, modern ethical principles governing the conduct of research on humans reinforce that participants have:

- the right to give their informed consent to participate in the research (i.e. they must know what they are getting into before they agree to participate); and
- the right to decline to participate or withdraw their participation at any time.

These two human rights supersede any researcher's requirement for a random sample. This means that while a researcher may fully intend to draw a random sample from the population so that he or she can make defensible generalisations, every person to be potentially included in the sample has the final say over their own participation (every person is, therefore, their own gatekeeper). Thus, the sample may start out being random, but by the time the study is completed, the sample may not be random due to some intended participants declining to participate or withdrawing from the research after initially agreeing to participate (i.e. mortality). If there is something systematically different about the people who decline or withdraw from the study, especially if their decision is based on something about the research process itself, then selection biases in the final sample are likely and this will severely limit generalising capabilities.

The Concept of Standard Error

Sampling plays a fundamental role in the statistical inference process through its influence on the *standard error* for statistical estimates of population parameters – a concept briefly introduced in *Fundamental concept V* with respect to the logic of statistical inference. When we estimate a population parameter like a proportion, mean, standard deviation, correlation or regression coefficient using sample information, we need an indicator of how precise our estimate is. In other words, we need a way of measuring how much error is associated with our estimate.

> *For example* ... *In a political poll, we may be interested in estimating the proportion of the Australian population who would vote Labour in the next federal election. Using a random sample, we may find that 35% of people in the sample indicate they would vote Labour. This 35% is a statistic that estimates the value of that proportion (a parameter) in the population.*

(continued)

Fundamental Concept VIII: Principles of Sampling

> *However, we know that there is some amount of error associated with this estimate, since we did not measure the whole population. This amount of error is estimated by the* standard error *and allows us to make statements that place an error 'band' around our estimate; e.g. '35% of Australians polled said they would vote Labour in the next federal election, plus or minus 2.5%'.*

How can we calculate a standard error for a statistical estimate? Imagine conducting your study 1000 times using the same random sampling scheme and the same size of sample each time. In each study, you would calculate your statistical estimate (e.g. a proportion or a mean). Over the 1000 repetitions of the study, a range of values for the statistical estimate would be calculated forming a sampling distribution of values. The standard deviation of this sampling distribution of statistical estimates becomes the standard error for that estimate. In practice, we don't have to repeat our study 1000 times to calculate a standard error, we can estimate it directly, for example, using the sample standard deviation and dividing it by the square root of the sample size. Knowing this, it becomes rather easy to see why larger samples give more precise statistical estimates: dividing a standard deviation by a larger sample size (even if it is the square root of that sample size) will give you a smaller value for the standard error. Every test of a statistical hypothesis involves the use of some form of standard error.

In the QCI database, we found the mean decision accuracy for the sample of 111 inspectors (one was missing an **accuracy** score) to be 82.14 with a corresponding standard deviation of 9.17 (refer to *Procedure 5.5*). We could easily find the standard error for this mean by dividing 9.17 by the square root of 111, which gives a value of 0.87. Thus, we could say that our estimate of mean decision accuracy in the population of quality control inspectors is 82.14 plus or minus 0.87. If Maree's data follow a normal distribution, we could go further and say that the chances are approximately 68% that the population mean would fall between 81.27 (mean minus one amount of the standard error) and 83.01 (mean plus one amount of the standard error).

If Maree had obtained a sample of 250 inspectors instead of 111, the standard error would have been 0.60—a smaller estimate of error, which would give a narrower error band around the mean. If the sample size had been 500, the standard error would have been an even smaller value of 0.41.

Procedure 7.1: Contingency Tables: Test of Significance

Classification Bivariate; inferential; nonparametric.
Purpose To evaluate the statistical significance of the association between the categories of two variables represented in a contingency table.
Measurement level This test of significance is appropriate for variables measured at the nominal- or ordinal-level.

Recall (from *Procedure 6.2*) that contingency tables are summaries of the joint membership of observations in one mutually exclusive category from each of two nominal or ordinal scale variables. Contingency tables can vary in dimension from 2 categories × 2 categories (the smallest possible) to *j* categories by *k* categories (the different letters, *j* and *k*, signal that the number of categories need not be the same for both variables and that there is no theoretical upper limit on either). Furthermore, the simplest contingency table is 2-dimensional (relating only two categorical variables – often referred to as a *2-way table*), but it is possible to have 3-dimensional (relating three variables simultaneously – a *3-way table*) or higher tables (an analytical procedure, called *log-linear models*, for testing complex multi-dimensional contingency tables is illustrated in *Procedure 7.18*).

We already discussed, in *Procedure 6.2*, several measures of association that can be derived from 2-way contingency tables; such measures quantify the effect size of the association. Now the discussion turns to the question 'is the association between the two variables statistically significant?'. In the case of measures of association, 'statistically significant' generally means that the magnitude of association between the two variables is non-zero in the population (the null hypothesis is that the association is zero in the population).

The test of significance for 2-way contingency tables is called *Pearson's chi-square* (χ^2) *test*. Technically, the chi-square evaluates whether two variables are independent of each other (i.e. not associated) such that knowing the category of an observation on one variable tells us nothing about the category for that observation on the other variable. If the computed chi-square statistic is large enough in value, then the notion of independence can be rejected, and we can conclude the two variables are significantly associated. The chi-square test works by assessing the size of the differences between the observed frequencies (counts) of joint category membership and the frequencies we would expect to see if the two variables were, in fact, independent. Large differences will lead to larger chi-square values and to greater likelihood of significance.

Suppose Maree wants to evaluate the significance of the relationship between the company that the inspectors worked for (**company**) and their educational level (**educlev**) in the QCI database. She adopts the conventional .05

(continued)

Procedure 7.1: Contingency Tables: Test of Significance

probability level for the alpha error associated with conducting this test. Table 7.5, generated using the SPSS *Crosstabs* procedure, displays the contingency table at the top, which results from associating these two variables. Table 7.5 shows the observed 'Count' in each table block (or 'cell') as well as the percentage of inspectors observed in each cell relative to each of the two variables ('% within educlev' and '% within company').

Table 7.5 Contingency table analysis relating **company** and **educlev**[a]

company * educlev Crosstabulation

			educlev		Total
			1 High School Only	2 Tertiary Qualified	
company	1 PC Manufacturer	Count	16	7	23
		Expected Count	11.8	11.2	23.0
		% within company	69.6%	30.4%	100.0%
		% within educlev	28.6%	13.2%	21.1%
		Adjusted Residual	2.0	-2.0	
	2 Large Electrical Appliance Manufacturer	Count	7	14	21
		Expected Count	10.8	10.2	21.0
		% within company	33.3%	66.7%	100.0%
		% within educlev	12.5%	26.4%	19.3%
		Adjusted Residual	-1.8	1.8	
	3 Small Electrical Appliance Manufacturer	Count	15	6	21
		Expected Count	10.8	10.2	21.0
		% within company	71.4%	28.6%	100.0%
		% within educlev	26.8%	11.3%	19.3%
		Adjusted Residual	2.0	-2.0	
	4 Large Business Computer Manufacturer	Count	14	9	23
		Expected Count	11.8	11.2	23.0
		% within company	60.9%	39.1%	100.0%
		% within educlev	25.0%	17.0%	21.1%
		Adjusted Residual	1.0	-1.0	
	5 Automobile Manufacturer	Count	4	17	21
		Expected Count	10.8	10.2	21.0
		% within company	19.0%	81.0%	100.0%
		% within educlev	7.1%	32.1%	19.3%
		Adjusted Residual	-3.3	3.3	
Total		Count	56	53	109
		Expected Count	56.0	53.0	109.0
		% within company	51.4%	48.6%	100.0%
		% within educlev	100.0%	100.0%	100.0%

Chi-Square Tests

	Value	df	Asymp. Sig. (2-sided)
Pearson Chi-Square	18.778[a]	4	.001
N of Valid Cases	109		

a. 0 cells (0.0%) have expected count less than 5. The minimum expected count is 10.21.

Symmetric Measures

		Value	Approx. Sig.
Nominal by Nominal	Cramer's V	.415	.001
N of Valid Cases		109	

a. Not assuming the null hypothesis.
b. Using the asymptotic standard error assuming the null hypothesis.

[a]Note: Output from SPSS has been edited slightly for simplicity of presentation

(continued)

Thus, 16 inspectors worked for the PC manufacturer and had only a high school level education; this was 28.6% of all inspectors with only a high school education and 69.6% of inspectors working for PC manufacturers. If these two variables were not associated, that is, independent of each other, Maree should expect to see the boldface frequencies in each cell. Thus, these 'Expected Count' values reflect what Maree should have observed if the null hypothesis provided the correct description of the real relationship between these two variables in the population (namely, no relationship).

Note that there are discrepancies between the boldface expected counts and the observed counts. So, for example, 16 actual inspectors worked for the PC manufacturer and had only a high school level education but under the null hypothesis, Maree should have only seen 11.8 inspectors – a discrepancy of 4.2. It is the size of the squares of these discrepancies that Pearson's chi-square (χ^2) test evaluates to determine the significance of the association. The value for Pearson's chi-square (χ^2) test statistic for this contingency table (bottom left subtable in Table 7.5) is 18.778 (the test statistic has 4 degrees of freedom or 'df') which is statistically significant (meaning that Maree has less than a 5% chance of being incorrect—see *Fundamental concept V*). Maree can make this decision because the printed *p*-value of .001 (under the column labelled 'Asymp. Sig.'—SPSS will often label *p*-values as 'Sig' standing for 'Significance level') is less than Maree's adopted alpha error probability level of .05. Therefore, she can conclude that the educational level of a quality control inspector and the type of company he or she works for are significantly associated (the null hypothesis of 'no association' in the population is rejected). The strength (i.e. effect size = practical significance) of the relationship is measured by Cramer's V which reports a moderate relationship of .415 (bottom right subtable in Table 7.5). Note that Cramer's V has the same *p*-value as Pearson's chi-square (χ^2) test which is to be expected since Cramer's V is based on the χ^2 measure.

The final value shown in each cell of the contingency table (top table of Table 7.5) is called an 'Adjusted Residual' (such residuals can usually be requested as optional output in many statistical packages). These are measures that reflect how strongly each cell contributed to the overall relationship in the table. These residuals are actually *z-scores* (recall *Procedure 5.7*). If the magnitude of an adjusted residual exceeds about ±2.0, we can conclude that the cell reported a significantly large discrepancy between the number expected and the number observed (thus, a large residual means a large discrepancy). This will help Maree to clarify more precisely which categories

(continued)

for each variable appear to contribute most strongly to the significance of the relationship. Note that, consistent with the hurdle logic outlined in *Fundamental concept V* for handling inflation of the alpha error in statistical inference, Maree should only look at these individual cell-level residual statistics if she finds a significant overall relationship between the two variables, which she has indeed shown with the significant χ^2 statistic.

In the top table of Table 7.5, it is clear that the residuals (shown in bold-face italics) for inspectors who work in PC, Small Electrical Appliance and Automobile manufacturing companies reflect the greatest discrepancies between observed and expected counts. Thus, significantly fewer High School-only educated inspectors were observed working in Automobile manufacturing companies than the null hypothesis said should have been expected and significantly more Tertiary-qualified inspectors than expected were observed working in those companies. The residuals for the PC and Small Electrical Appliance manufacturers suggest that significantly more High School-only educated inspectors were observed than expected from these companies and significantly fewer than expected Tertiary-qualified inspectors were observed—the reverse trend from Automobile manufacturers.

Advantages

The Pearson chi-square test permits a test of the statistical significance of the association between two categorical variables crosstabulated in a 2-way contingency table. It is relatively simple and easy to interpret. The test can be applied to virtually any size of contingency table and it has a direct linkage to the adjusted residuals which can be used to help clarify the interpretation of a significant association. The interpretation of this test does not change if ordinal-level variables rather than nominal-level variables are being tested. The Pearson chi-square test can also be used as a direct test for the significance of either the Phi Coefficient (only for 2×2 tables) or Cramer's V measure of association. However, Pearson's chi-square test should not be used to evaluate the significance of PRE measures of association such as Goodman-Kruskal's lambda (recall the discussion of these measures in *Procedure 6.2*) as each PRE measure has its own associated test of significance (which can be produced on request by most statistical software packages, including SPSS).

Disadvantages

The Pearson chi-square test can be biased if some of the cells in the contingency table have very small expected counts (typically, less than 5, which is the reason for footnote *a.* under the Chi-Square Tests subtable in Table 7.5), very low observed counts, and/or zero counts (empty cells where the cell categories are never observed together). Maree's analysis did not have this problem; the crosstabulation table in Table 7.5 reported that the smallest expected frequency was 10.2.

Some combining of cell categories may have to be done in order to reduce or eliminate low-frequency cells. For 2×2 contingency tables, Pearson's chi-square value must be adjusted before evaluating it for significance (many software packages report the adjustment automatically, including SPSS). There is an alternative form of chi-square test called the 'likelihood ratio' test which is sometimes considered to be more accurate than the Pearson chi-square test under certain conditions (particularly for multi-way contingency tables). The Pearson chi-square test can only be employed in contingency tables where each individual in a sample can be counted as falling in one and only one cell. Thus, data involving repeated measurements of the same individuals under different conditions (e.g. analysing political party voting patterns for the same sample of voters over three different elections) cannot be analysed using the traditional Pearson chi-square test described here. Strategies for analysing such repeated measurement contingency table scenarios are discussed in Everitt (1992).

Where Is This Procedure Useful?

The Pearson chi-square test of a 2-way contingency table is most useful in survey or observational studies where tests of associations between nominal- or ordinal-level variables (which characterise most measures of demographic characteristics) are desired. This means that contingency table analysis and tests of association provide very useful input into the demographic story of a sample that a researcher would need to write in their research report or article. For example, Pearson's chi-square may be useful for testing the significance of the relationship between occupation and gender or between religion and social class obtained by survey methods. Some behavioural researchers prefer to analyse the relationship between Likert-type attitude survey items (such as the work attitude items in the QCI database) using chi-square contingency table analysis rather than traditional Pearson correlations because they are uncomfortable treating such data as being measured at the interval-level. Data from experiments or quasi-experiments tend not to be amenable to

Procedure 7.1: Contingency Tables: Test of Significance

chi-square analysis simply because the samples for such studies are typically configured so that associations between categorical variables are minimised.

Software Procedures

Application	Procedures
SPSS	*Analyze → Descriptive Statistics → Crosstabs* ... select the Row and Column variables to be related; then press the '*Statistics*...' button to select the 'Chi square' test and 'Phi and Cramer's V' and hit 'Continue'; then press the '*Cells*...' button to select 'Expected' counts, 'Row' and 'Column' percentages and 'Adjusted standardized' residuals and hit 'Continue'.
NCSS	*Analysis → Descriptive Statistics → Contingency Tables (Cross Tabulation)* and select the desired row and column variables; then select the '*Reports*' tab and choose the analysis outputs you wish to see reported, plotted or both. Note that while NCSS can be asked to report standardised residuals, they will not be the same as the adjusted standardised residuals reported by SPSS.
SYSTAT	*Analyze → Tables → Two-Way* ... and select the desired row and column variables; then select the desired 'Tables' outputs to be reported (which can include 'Standardized deviates') and on the '*Measures*' tab, select the 'Pearson chi-square' statistic and 'Cramer's V' or other desired measures of association. Note that while SYSTAT can be asked to report standardised residuals, they will not be the same as the adjusted standardised residuals reported by SPSS.
STATGRAPHICS	*Describe → Categorical Data → Crosstabulation...* and select the desired 'Row variable' and 'Column variable'; hit '*OK*' and in the '*Tables and Graphs*' dialog box, tick 'Analysis Summary', 'Frequency Table' and 'Tests of Independence'. STATGRAPHICS can also produce various graphical displays, which may be interesting to look at as well. STATGRAPHICS automatically includes adjusted standardised residuals (same as SPSS) in the Frequency Tables.
R Commander	*Statistics → Contingency tables → Two-way tables...* and select the desired 'Row variable' and 'Column variable'; under 'Compute Percentages' choose the desired output (note you cannot choose to see both 'Row percentages' and 'Column percentages' at the same time); under 'Hypothesis Tests', choose 'Chi-square test of independence', 'Components of chi-square statistic' and 'Print expected frequencies'. **R** Commander will not produce adjusted standardised residuals; it will, if requested, produce components of the chi-square statistic, which can provide a descriptive feel for which cells contributed most to the chi-square statistic.

Procedure 7.2: *t*-Test for Two Independent Groups

Classification   Univariate; inferential; parametric.
Purpose   To compare two independent (unrelated) groups of participants on a single dependent measure.
Measurement level   Any level of measurement may be used for the variable that defines the two groups. The dependent measure must be measured at the interval- or ratio-level.

The *independent groups t-test* is useful for evaluating the difference between two groups of sample participants on a single dependent variable (see, for example, Field 2018, chapter 10). The information used by the *t*-test is the dependent variable mean and standard deviation for each of the two groups as well as the sample sizes of the two groups (which need not be equal). For the *t*-test, there is a single dimension (e.g. participant gender; treatment group vs. control group) which serves to define the nature of the two groups—this dimension therefore defines the independent variable for the analysis.

'Independent groups' refers to the idea that participant membership in either group is mutually exclusive and that there is no logical or statistical connection between the membership of the two groups. Thus, for example, each participant would be either a male or a female or would be in either the treatment group or the control group. The *t*-test compares the observed difference between the sample means of the two groups to an estimate of the variability of such mean differences in the population, under the assumption that the group means do not really differ (i.e. assuming the null hypothesis of 'no difference between the groups' to be true—recall *Fundamental concept V*). In this sense, the *t*-test operates much like a *z*-score (recall *Procedure 5.7*) and can take on positive or negative values depending upon which group has the larger mean.

> With respect to the QCI database, Maree might be interested in comparing Tertiary qualified inspectors with High School educated inspectors in terms of their quality control decision accuracy. The dependent measure is therefore **accuracy** scores and the two groups are defined by the two levels of the **educlev** variable (i.e. the independent variable). Her research hypothesis would be that the two groups do, in fact, differ in their decision accuracy, but she must start the inference process by assuming that the two groups do not differ.
>
> Figure 7.6 provides an error bar plot of the mean **accuracy** scores for the two groups (the error bars show the span of scores between two standard errors above each mean and two standard errors below each mean). Visually, it looks like the two groups do differ, as their error bars do not overlap very much indicating that the two sample means are likely to be estimating two different

(continued)

population means. However, Maree does need the assurance of a statistical test to help her decide if the evidence is sufficient for her to reject the null hypothesis of 'no difference between the groups' in the population.

The relevant analysis information, produced using SPSS, is summarised in Table 7.6. The 'Group Statistics' subtable of Table 7.6 provides relevant descriptive statistics for each of the two education-level groups. The *t*-test evaluates whether the difference between the mean **accuracy** scores for Tertiary qualified inspectors and High School educated inspectors is sufficiently large to be considered 'statistically significant', i.e. non-zero.

Maree already knows, from her earlier graphical inspection of the **accuracy** distribution, that **accuracy** scores are approximately normally distributed – one of the assumptions of the independent groups *t*-test. However, another important assumption is *homogeneity of variance* (i.e. the variance of scores in both groups is approximately the same). SPSS produces a statistic called Levene's *F*-Test that Maree can check to evaluate this assumption. This test, shown in Table 7.6 on the left-side of the 'Independent Samples Test' subtable, is not significant because the *p*-value ('Sig.') of .727 is much greater than her alpha error decision criterion of .05. This means that the assumption is satisfied, and Maree can safely interpret the 'equal variances assumed' version of *t*-test.

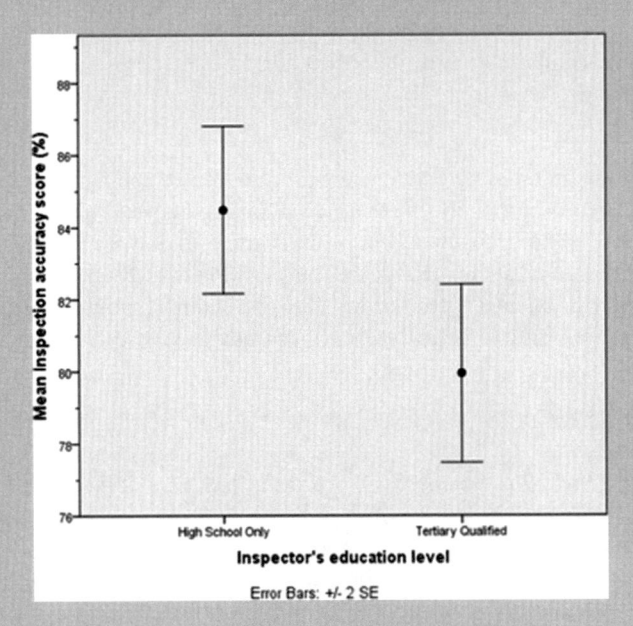

Fig. 7.6 Error bar plot of mean decision **accuracy** scores for Tertiary qualified and High School educated inspectors
Note: Error bars extend to a two standard error distance either side of each mean.

(continued)

Table 7.6 Relevant summary statistics and *t*-test outcomes comparing groups defined by **educlev** categories in terms of their decision **accuracy**

Group Statistics

	educlev	N	Mean	Std. Deviation	Std. Error Mean
accuracy	1 High School Only	56	84.48	8.676	1.159
	2 Tertiary Qualified	52	79.94	8.901	1.234

Independent Samples Test

		Levene's Test for Equality of Variances		t-test for Equality of Means				
		F	Sig.	t	df	Sig. (2-tailed)	Mean Difference	Std. Error Difference
accuracy	Equal variances assumed	.123	.727	2.683	106	.008	4.540	1.692
	Equal variances not assumed			2.681	104.942	.009	4.540	1.693

Note: This table has been edited for simplicity and clarity.

The observed difference between the two group means is +4.54 and value of the *t*-test for evaluating the size of this difference is +2.683 with an associated *p*-value of .008. Since the reported *p*-value (.008) is less than her alpha error decision criterion of .05, Maree is justified in rejecting the null hypothesis and concluding that inspectors who have different levels of educational attainment significantly differ in their inspection decision accuracy. More precisely, inspectors who have only a High School education appear to make significantly more accurate inspection decisions than do inspectors with a Tertiary qualification.

It is very important to note that whether or not Maree can attribute this observed significant difference (the 'effect') to a causal linkage between educational level (the putative 'cause') and decision accuracy has nothing to do with the *t*-test itself—the *t*-test can only show that the mean scores between the two groups significantly differ within a certain tolerance for error. The test cannot show that the grouping (or independent) variable actually caused the changes observed in the dependent variable.

In the case of the QCI database, what determines Maree's ability to infer a causal relationship is the appropriateness and tightness of her research design and methodological procedures, not her statistical analysis (recall *Fundamental concept VII*).

Since the QCI data were obtained using, at best, quasi-experimental field research techniques (where very little control over extraneous variables was possible), Maree's capacity to infer cause would be quite limited. This is further compounded by the fact that the educational level of the inspectors in her sample was **not** under her control. She could not randomly assign

(continued)

Procedure 7.2: *t*-Test for Two Independent Groups

> educational levels to inspectors; it was a variable whose values were predetermined by the individual life histories of each inspector well before she ever undertook the study. It could well be that what is being reflected in the accuracy differences in the QCI data is not the influence of educational level per se, but the potential influence of other unmeasured demographic characteristics, cognitive ability, social and economic factors which help to determine whether or not an inspector would pursue higher levels of education.

The independent groups *t*-test for two groups yields exactly the same conclusions as would a one-way analysis of variance (see *Procedure 7.6*) on these two groups because the two procedures are related. In fact, the *t*-statistic outcome from an independent groups *t*-test is simply the square root of the *F*-statistic outcome from an analysis of variance performed on the two groups. The major difference is that the independent groups *t*-test can only be used to compare two groups whereas the analysis of variance is a more general technique which can be used to compare mean scores from any number of groups.

Advantages

The independent groups *t*-test is a relatively straightforward and easy to interpret procedure—the groups either differ in terms of their mean dependent variable scores or they do not. The *t*-test does not require the groups to have equal numbers of participants (although having equal sample sizes is a preferred state of affairs because this makes it easier to meet the homogeneity of variance assumption).

Additionally, it is possible to test directional predictions using the *t*-test where, for example, the researcher states, before any analysis, that he or she expects an experimental group to perform better (have a higher mean on the dependent variable) than a control group. This is termed a *one-tailed test* where the direction of expected difference is predicted (recall *Fundamental concept V*). If a one-tailed prediction can be made, a more powerful hypothesis test results as long as you predict the correct tail. The *p*-values for *t*-tests produced by statistical software packages are always *two-tailed* values (refer back to Table 7.6), so to evaluate a one-tailed hypothesis, you simply divide the printed *p*-value by 2 and compare the new value to your desired alpha level (while making sure the difference points in the direction you predicted).

> If Maree had predicted, on the basis of some theory, that High School educated inspectors would be more accurate than Tertiary qualified inspectors, then she could have used a *p*-value of .004 (= .008/2) to evaluate her test, as long as the mean difference correctly favoured the High School educated group.

Disadvantages

We have indicated that the *t*-test makes two important assumptions about the distribution of the data obtained from two groups which may or may not be reasonable for a specific situation.

- Firstly, it assumes that the distribution of dependent variable scores (at least in the population of interest) is approximately normal.
- Secondly, it assumes that the variances (or standard deviations) of dependent variable scores within each group are approximately equal (the 'homogeneity of variance' assumption).

Of the two assumptions, the more serious one to violate is the homogeneity of variance assumption. If the data distributions for each group are skewed or otherwise non-normal, there may not be a substantial problem with violating the normality assumption as long as the two group distributions are non-normal is the same way—for example, if the scores in both groups are somewhat negatively skewed.

Many software computing packages produce tests for these assumptions which can provide insights into whether either has been severely violated (as SPSS did in Table 7.6). SPSS also produces two versions of the *t*-test, the 'equal variances assumed' version and the 'Equal variances not assumed' version (also known as the 'Welch' two-sample *t*-test). If the Levene's test turns out to be significant, you would then interpret the 'Equal variances not assumed' version. However, by comparing the outcomes from both versions, you can see how much impact violating the assumption makes on the outcome. If the two tests would lead you to different decisions, then the assumption has been violated enough to make a difference and you should interpret the 'Equal variances not assumed' version.

When either or both of these two assumptions are severely violated, the *t*-test may become biased and less powerful for detecting real group differences. In such cases, it is usually recommended that a nonparametric test such as the *Mann-Whitney U-test* (see *Procedure 7.3*) be used. An additional disadvantage is that the independent groups *t*-test is restricted to two group 'single-shot' research designs whereas the analysis of variance is a more generally applicable procedure.

Where Is This Procedure Useful?

The *t-test* can be used in almost any experimental or quasi-experimental design involving two independent groups being compared. It has the strongest chance of supporting a causal claim in true experiments where the researcher has control over which group the participant is a member of. The *t*-test can sometimes be useful in survey or observational research applications when comparing responses on some measure of interest from two demographic categories (such as males and females) but with a corresponding decrease in one's capacity to infer causality.

Software Procedures

Application	Procedures
SPSS	*Analyze → Compare Means → Independent Samples T Test* ... and choose the 'Grouping Variable' and 'Test Variable' (= dependent variable) accordingly; for the 'Grouping Variable', hit the *'Define Groups'* button and indicate which category numbers identify each of the two groups to be compared.
NCSS	*Analysis → T-Tests → T-Test – Two-Sample* and choose the 'Group Variable' and 'Response Variable' (= dependent variable) accordingly; NCSS also allows you to choose the null hypothesised value to evaluate (0 is default) and the alpha level to use to evaluate the test for significance. You also have the option, under the *'Reports'* tab, to have NCSS report a nonparametric version of the *t*-test as well – the Mann-Whitney test (see Procedure 7.3).
SYSTAT	*Analyze → Hypothesis Testing → Mean → Two Sample t-Test* ... and choose the 'Grouping variable' and 'Selected Variables' (= dependent variable) accordingly; SYSTAT offers the option of using the Bonferroni correction if you run *t*-tests on a number of dependent variables.
STATGRAPHICS	*Compare → Two-Sample → Independent Samples...* and choose the 'Input' as 'Data and Code Columns', then choose the 'Sample Code' variable (= independent variable) and the 'Data' (= dependent variable); hit *'OK'* and in the *'Tables and Graphs'* dialog box, tick what analyses and graphs you want to see ('Comparison of Means' should at least be ticked to obtain *t*-tests; ticking 'Comparison of Standard Deviations' will test the homogeneity of variance assumption; ticking 'Comparison of Medians' will give a non-parametric Mann-Whitney test comparison of the two groups); The StatAdvisor will provide detailed interpretive comments about each analysis.
R Commander	*Statistics → Means → Independent samples t-test...* and pick your 'Groups' variable and your 'Response Variable' (= dependent variable); R Commander will let you choose whether you evaluate a two-tailed hypothesis test or a one-tailed test (you have to pick the direction of difference anticipated) and you will need to indicate if you want to 'Assume equal variances?' – 'Yes' or 'No' (choosing 'No' will produce the Welch version of the *t*-test).

Procerdure 7.3: Mann-Whitney U-Test for Two Independent Groups

Classification   Univariate; inferential; nonparametric.
Purpose   To evaluate the difference between two independent groups of participants on the basis of ranked scores on a dependent variable rather than scores at the interval- or ratio-level of measurement.
Measurement level   The grouping variable can be defined by any level of measurement. The dependent measure may be initially obtained in terms of ordinal ranks or, more commonly, may be measured at the interval- or ratio-level but converted to ordinal ranks prior to analysis.

The *Mann-Whitney U-test* for two independent groups is a nonparametric analog of the independent groups t-test (see *Procedure 7.2*). It evaluates the significance of the difference between two independent groups on the basis of mean ranks rather than mean scores in some other units of measurement. One use of this test is apparent when behavioural data are actually collected in the form of ranks.

> *For example* ... We might look at the online database reporting the 2019 Times Higher Education overall reputation rankings (subjective judgments by experts) of the top 100 world universities to compare the average ranking of universities versus institutes of technology and polytechnics. The ranks from the database could be sorted into the two groups and mean ranks then compared.

Another frequently occurring scenario for the Mann-Whitney U-test emerges when the original measurements of a dependent variable are distributed in some highly distorted fashion so as to violate the basic assumptions of the t-test (normality and/or homogeneity of variance).

> Suppose Maree wants to compare quality control inspectors who differ in their educational level (**educlev**) within the QCI database in terms of their inspection decision **speed**. We have seen in earlier discussions (recall *Procedure 5.7*) that **speed** is a highly positively skewed variable (as are most time-based measures of human behaviour) which may be problematic enough to violate the normality assumption.

(continued)

Procerdure 7.3: Mann-Whitney U-Test for Two Independent Groups

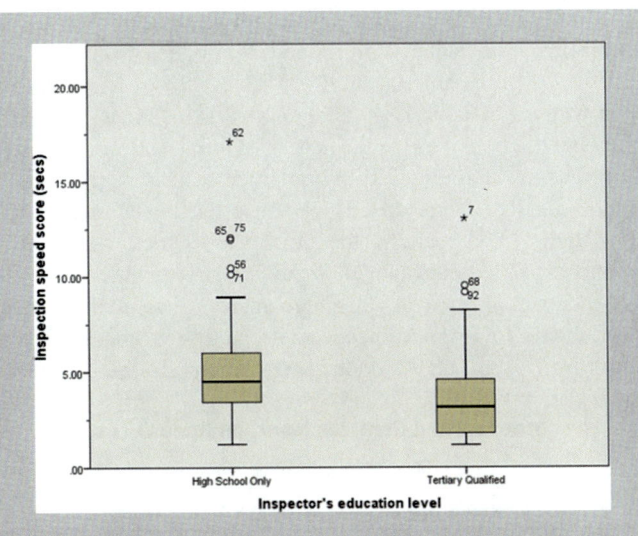

Fig. 7.7 Boxplots comparing the distributions of **speed** scores for the two education levels of quality control inspectors

Figure 7.7 shows the box plots for the **speed** variable within the two education groups. The plot shows not only that the two groups may significantly differ but also that their distributions are highly positively skewed, each group with one extreme outlier (inspectors 62 and 7).

Instead of the parametric t-test, Maree can opt to conduct the Mann-Whitney U-test. The U-test is conducted in three simple steps:

1. The scores in the original units of measurement (e.g. **speed** measured in seconds) are ranked from 1 to N (the total sample size) ignoring each participant's membership in a group defined by the grouping variable. Thus, **speed** scores would be ranked in order from 1 (fastest) to 108 (slowest); four inspectors were missing their **speed** scores.
2. The new set of ranks is then sorted back into the two groups whose membership is defined by the grouping variable (resulting in 52 ranks in the Tertiary-qualified group and 56 ranks in the High School-educated group).
3. An independent groups t-test is essentially conducted using the ranks as data and the two groups are compared in terms of their average rank scores. The mean rank for inspectors in the Tertiary qualified group is 45.27 and the mean rank for inspectors in the High School educated group is 63.07. The Mann-Whitney U-test then simply evaluates the significance of the difference (equals -17.80) between these two mean ranks.

The outcome from the Mann-Whitney U-test conducted on these data using SPSS (see Fig. 7.8) is a value for the U-statistic of 976 (converted to a z-score of -2.951) with an associated p-value of .003. This means Maree can conclude that High School Only-educated inspectors had significantly higher ranked **speed** scores than did Tertiary qualified inspectors. Since the ranking system was such that faster speeds garnered the lower ranks, this finding means that High School Only-educated inspectors took significantly longer to make their inspection decisions than did Tertiary-qualified inspectors. This finding is visually confirmed, as is the high degree of positive skewness, by the dual histogram plot (of original **speed** scores, similar to what was shown in the boxplots in Fig. 7.7) in Fig. 7.8 comparing the two groups.

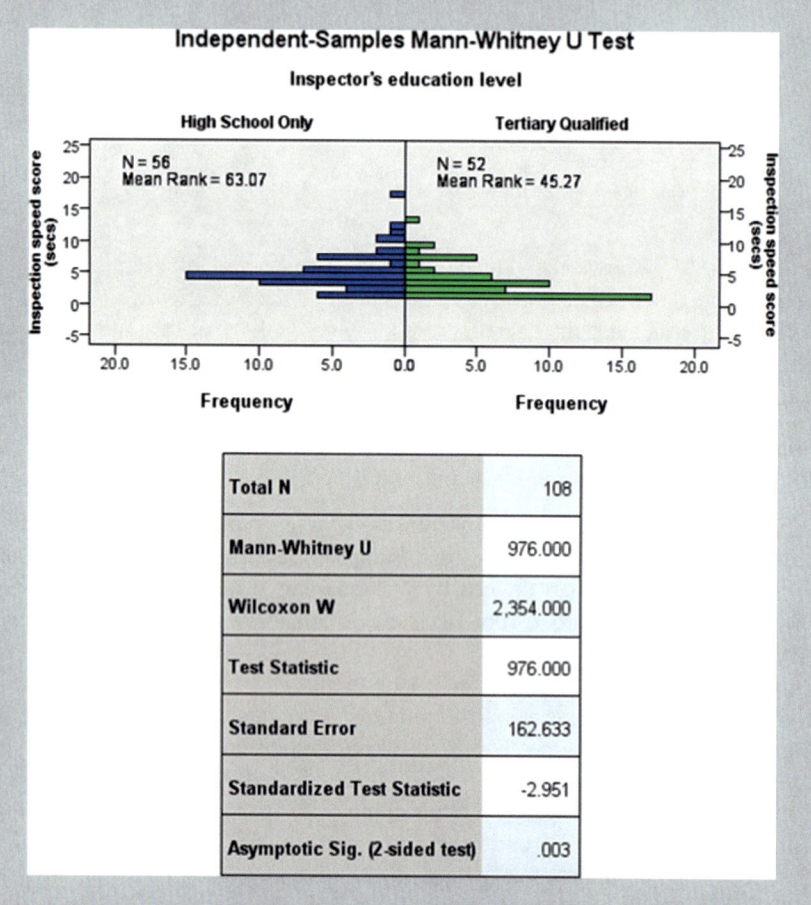

Total N	108
Mann-Whitney U	976.000
Wilcoxon W	2,354.000
Test Statistic	976.000
Standard Error	162.633
Standardized Test Statistic	-2.951
Asymptotic Sig. (2-sided test)	.003

Fig. 7.8 Mann-Whitney U-test comparing inspector's education level groups in terms of inspection speed

(continued)

Procerdure 7.3: Mann-Whitney U-Test for Two Independent Groups

> Maree could compare this result to what would have been obtained had she run a parametric *t*-test to see if her conclusions differed between the two tests. If her inference decision was different for the two tests or if the *p*-values differed dramatically between the two tests, then assumption violations have probably distorted the parametric test and she should interpret the Mann-Whitney *U*-test outcome. This is the case here with the **educlev** comparison on **speed** scores where the Mann-Whitney *U*-test *p*-value is .003 and the *t*-test *p*-value is only .019. If the inference decision remains unchanged or the *p-values* are similar, then the assumption violation was not severe enough to distort the outcome and we should report the parametric test because it utilises more of the original information in the sample.

The Mann-Whitney *U*-test is also known as the *Wilcoxon Rank Sum test* or *Wilcoxon W test* (shown in the SPSS analysis reported in Fig. 7.8) and is closely related to the Kruskal-Wallis *H*-test discussed in *Procedure 7.9*. There are several other alternative forms of nonparametric test for differences between two independent groups of which the most frequently used appears to be the *Kolmogorov-Smirnov test* (see Siegel and Castellan Jr 1988 pp. 144–151).

Advantages

The Mann-Whitney *U*-test is particularly useful when the assumptions of the parametric *t*-test cannot be or are not reasonably fulfilled by the data. It permits analysis of such data by reducing the data in measurement complexity to the ordinal level and then conducting the test on the resulting ranks. The test is easy to compute and interpret and requires no underlying assumptions about the distribution of the data. In comparisons involving large samples, the *U*-statistic is converted to a *z*-score and interpreted accordingly. The Mann-Whitney *U*-test also provides a simultaneous test for significant difference between the medians of the two independent groups. NCSS reports the Mann-Whitney *U*-test this way.

Disadvantages

Unless the data rather severely violate *t*-test assumptions, it is usually better to use the parametric test rather than the Mann-Whitney *U*-test. The reason is that when assumptions are even moderately violated, the *t*-test is a more powerful test for a given sample of data than the Mann-Whitney *U*-test. That is, it is more likely to detect group differences if they really exist. However, if the violations are serious, then the Mann-Whitney *U*-test becomes the more powerful test. Also, since the original data, in many applications, are converted to ranks, some information is necessarily lost. Such a loss can hamper meaningful interpretation of the observed group differences.

Where Is This Procedure Useful?

The Mann-Whitney *U*-test can be applied in any experimental, quasi-experimental, or survey research investigation where comparison of two independent groups of participants is of interest and parametric assumptions are strongly violated. Many software packages offer specific tests for assumption violation which renders the choice of analytical technique easier to make. If the data are obtained in the form of ranks from the start, then the Mann-Whitney *U*-test is automatically the test of choice. As with the parametric *t*-test, the statistical conclusion offered by the Mann-Whitney *U*-test relates solely to observed group differences. Whether a causal conclusion is warranted (i.e. that the independent variable difference between the two groups caused the significant difference in the group mean ranks) is a separate decision made considering the quality of one's research design and methodological procedures.

Software Procedures

Application	Procedures
SPSS	*Analyze → Nonparametric Tests → Independent Samples ...* and on the '*Objective*' tab, choose 'Compare medians across groups'; on the '*Fields*' tab, choose the variable to define the 'Groups' and the dependent variable ('Test Fields'); on the '*Settings*' tab, choose 'Mann-Whitney U (2 Samples)'. This procedure in SPSS is unusual in that you must double click on the output table (which only reports the *p*-value) to get access to the visualisation similar to Fig. 7.8.
NCSS	*Analysis → Nonparametric → Mann-Whitney U/Wilcoxon Rank-Sum (2 Sample)...* and choose the 'Grouping variable' and 'Response variable(s)' (= dependent variable) accordingly; the default options produce independent groups *t*-test information, nonparametric test information and tests of the normality and homogeneity of variance assumptions as well as a boxplot comparison similar to Fig. 7.7.
SYSTAT	*Analyze → Nonparametric Tests → Kruskal-Wallis ...* and choose the 'Grouping variable' and the dependent variable as a 'Selected variable(s)'; if the grouping variable only has two levels, SYSTAT will automatically calculate the Mann-Whitney *U*-test. SYSTAT converts the Mann-Whitney *U* statistic to a chi-square value, rather than a *z*-score.

(continued)

Procedure 7.4: *t*-Test for Two Related Groups

Application	Procedures
STATGRAPHICS	*Compare → Two-Sample → Independent Samples...* and choose the 'Input' as 'Data and Code Columns', then choose the 'Sample Code' variable (= independent variable) and the 'Data' (= dependent variable); hit '*OK*' and in the '*Tables and Graphs*' dialog box, tick what analyses and graphs you want to see (ticking 'Comparison of Medians' will give a nonparametric Mann-Whitney test comparison of the two groups; ticking 'Box-and-Whisker Plot' will produce a boxplot comparison similar to Fig. 7.7). The StatAdvisor will provide detailed interpretive comments about the analysis but note that STATGRAPHICS refers to the test as the Mann-Whitney (Wilcoxon) *W* test.
R commander	*Statistics → Nonparametric tests → Two-sample Wilcoxon test...* and choose the variable to define the 'Groups' and the 'Response variable' (= dependent variable); **R** Commander will let you choose whether you evaluate a two-tailed hypothesis test or a one-tailed test (you have to pick the direction of difference anticipated) and, under 'Type of Test', choose 'Normal approximation', if you have a reasonable (greater than 30) sample size, otherwise, tick 'Default'.

Procedure 7.4: *t*-Test for Two Related Groups

Classification	Univariate; inferential; parametric.
Purpose	To compare scores on a single dependent measure obtained on the same sample (or a matched sample) of individuals under two conditions, at two different times or in two different contexts. In this sense, the groups being compared are considered related.
Measurement level	Any level of measurement may be used for the variable that defines the two conditions or contexts. The dependent measure must be measured at the interval- or ratio-level.

The *related groups t-test* (also known as the *dependent samples t-test* or the *paired t-test*) is useful for evaluating the difference between two experimental conditions or contexts for a sample of participants on a single dependent variable (Field 2018, section 10.9). For the related groups *t*-test, there is a single dimension (e.g. Time 1 versus Time 2; pretest versus posttest; treatment versus control) which serves to define the nature of the two conditions—this dimension therefore defines the independent variable for the analysis.

'Related groups' refers to the idea that participant membership in the two conditions is determined in one of two ways.

- Firstly, participants may be repeatedly measured in that each individual is measured under both conditions (with the exception of time-based definitions

of the two conditions, measures are obtained in some sort of counterbalanced or random order so as to control for the extraneous variables of carry-over effects and order-of-participation-in-condition effects – recall *Fundamental concept VII*).
- Secondly, participants may be matched in pairs (twins are ideal for this) according to one or more extraneous variables which the researcher suspects could potentially influence the dependent variable at the same time as the conditions change. The members of each pair are then randomly assigned to participate in one of the two conditions.

> ***For example*** ... *We might match children in terms of their mental ability before measuring their reading comprehension achievement under two different experimental reading programs. Here, matched participants are treated as if they were the same person being measured under both conditions.*

The related groups t-test compares the observed difference between the sample means of the two conditions to an estimate of the variability of such mean differences in the population under the assumption that the condition means do not really differ, i.e. assuming the null hypothesis of 'no difference between the conditions' to be true. In this sense, the t-test operates much like a z-score (recall *Procedure 5.7*) and can take on positive or negative values depending upon which condition has the larger mean.

Since the data in the QCI database were not obtained using a design involving matching or repeated measurement of quality control inspectors under different conditions, we will illustrate the use of the related groups t-test in a hypothetical scenario extension of the QCI research context.

> Suppose Maree obtains a supplemental contract to explore the effects of a newly designed quality control training program on inspection decision **accuracy** and **speed**. She obtains permission to run a small experiment where she identifies quality control inspectors, from her original sample of 112, who are in the bottom 25% of inspectors with respect to decision accuracy (most inaccurate inspectors) or in the top 25% with respect to decision speed (slowest inspectors). This selection process yields a sample of the 60 most 'poorly performing' quality control inspectors on at least one aspect of quality control inspection performance.
>
> Maree then randomly assigns the 60 inspectors into either a control group (N = 30 who don't undertake the new training program) or an experimental group (N = 30 who do undertake the new training program). She already has pretest scores for the inspection decision **accuracy** and **speed** of these 60 inspectors from her original sample, obtained of course, prior to their

(continued)

undertaking the new training program. Continuing with the same methodology and measures as in the large QCI research project, Maree conducts her experiment where experimental group inspectors work through the new program and have their decision **accuracy** and **speed** measured again once they have completed the program (the control group inspectors are given equivalent experiences in terms of time away from work, but doing activities unrelated to quality control inspection).

This research design illustrates the classical pretest-posttest control groups design involving repeated measurements on the same sample of participants. This type of research design, or some variant of it (a common variant being that no control group is available or used), is often used in program evaluation research. The repeated measurements nature of the design ensures that whatever is idiosyncratic about each inspector is present (which means they are, in effect, held constant for each inspector) for the measurements obtained on both two occasions. For purposes of illustrating the related groups t-test here, we will focus on the decision **accuracy** results from just the experimental group.

Table 7.7 shows the resulting data for the experimental group inspectors along with relevant summary statistics for each occasion. The dependent variable is **accuracy** scores and the two measurement occasions are defined by two variables called **pre_acc and post_acc**. Maree's research hypothesis is that inspectors should show higher decision **accuracy** scores under the Posttest condition than under the Pretest condition (i.e. the training program works), but she must start the inference process by assuming that the two conditions do not differ (assume Maree has set her alpha decision criterion at .05). [Note: In this scenario, Maree has generated a one-tailed or directional hypothesis which runs the risk of being dramatically in error if the training program decreases decision accuracy. This highlights the idea that directional hypotheses are not without their dangers; the researcher must be reasonably certain he/she has predicted the correct direction for the difference, or all bets are off.]

The related groups t-test will indicate whether the difference between the mean **accuracy** scores achieved in the Pretest and Posttest occasions is sufficiently large to be considered statistically significant, i.e. non-zero.

Figure 7.9 provides an error bar plot of the mean **accuracy** scores for the two occasions (the error bars show the span of scores between 2 standard errors above each mean and two standard errors below each mean). Visually, it looks like the means for the two occasions do differ in the predicted direction, but their error bars somewhat overlap indicating that a portion of the estimates for the population mean in each occasion would be of the same magnitude. Thus, Maree needs a statistical test to help her decide if the evidence is sufficient for her to reject the null hypothesis of 'no difference between the occasions' in the population.

(continued)

Table 7.7 Listing of the decision **accuracy** scores and summary descriptive statistics on the pretest and posttest occasions for 30 inspectors undertaking the new training program

Case summaries

		inspecto	pre_acc	post_acc
1		1	74.00	85.00
2		3	75.00	89.00
3		10	83.00	83.00
4		12	80.00	88.00
5		15	86.00	87.00
6		17	99.00	100.00
7		22	76.00	73.00
8		30	75.00	71.00
9		32	62.00	64.00
10		43	75.00	82.00
11		45	87.00	94.00
12		50	64.00	71.00
13		53	77.00	84.00
14		54	81.00	83.00
15		59	75.00	82.00
16		61	89.00	99.00
17		62	82.00	89.00
18		68	85.00	93.00
19		70	84.00	87.00
20		77	92.00	98.00
21		81	82.00	89.00
22		82	62.00	77.00
23		83	57.00	67.00
24		92	78.00	86.00
25		95	70.00	77.00
26		97	73.00	79.00
27		102	78.00	85.00
28		108	85.00	80.00
29		109	83.00	88.00
30		112	71.00	73.00
Total	N	30	30	30
	Mean	58.17	78.0000	83.4333
	Std. Deviation	34.159	9.24009	9.09218

(continued)

Procedure 7.4: *t*-Test for Two Related Groups

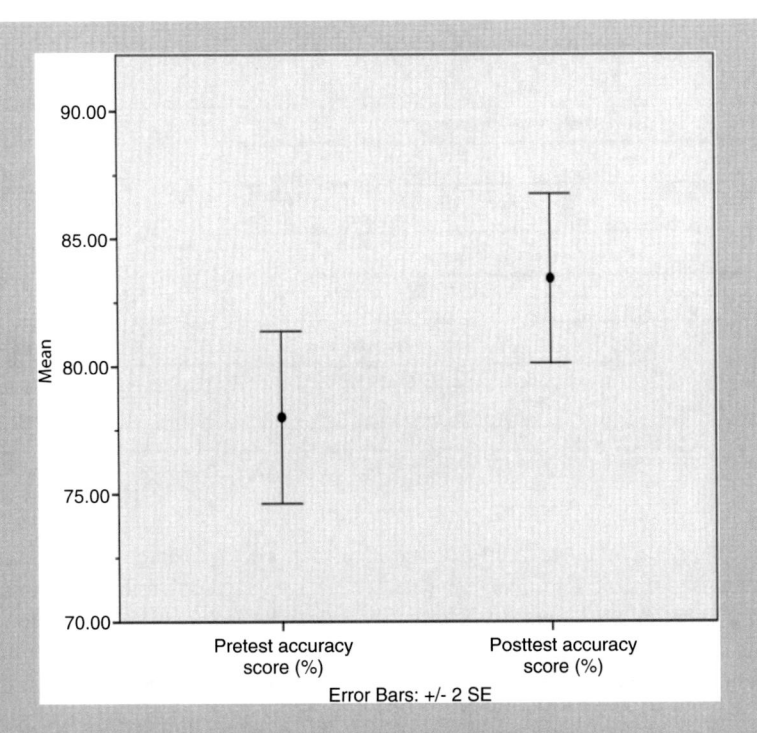

Fig. 7.9 Error bar plot of mean decision **accuracy** scores for the pretest and posttest occasions for 30 inspectors undertaking the new training program
Note: Error bars extend to a two standard error distance either side of each mean.

Table 7.8 shows the outcomes from the paired *t*-test analysis of this small experimental design. Note that the scores on the two occasions were correlated .865, a very strong positive correlation, usually expected in a repeated measures-type of experimental design involving pretest-posttest conditions. The observed difference between the two occasion means is +5.43 and value of the *t*-test for evaluating the size of this difference is +6.24 with an associated one-tailed *p*-value of less than .001 (remember a 1-tailed *p*-value is found by dividing the 2-tailed *p*-value by 2). Since the reported *p*-value is less than her alpha error decision criterion of .05, Maree would be justified in rejecting the null hypothesis and concluding that the inspectors who undertook the new training program did significantly improve in their inspection decision accuracy from the Pretest (mean = 78.00) to the Posttest occasion (mean = 83.43). Her predicted direction for the differences over the occasions was thus confirmed by the observed pattern of mean differences.

(continued)

Table 7.8 Related (paired) *t*-test analysis (using SPSS) of the experimental design data in Table 7.7

Paired Samples Statistics

		Mean	N	Std. Deviation	Std. Error Mean
Pair 1	pre_acc	78.0000	30	9.24009	1.68700
	post_acc	83.4333	30	9.09218	1.66000

Paired Samples Correlations

		N	Correlation	Sig.
Pair 1	pre_acc & post_acc	30	.865	.000

Paired Samples Test

		Paired Differences			t	df	Sig. (2-tailed)
		Mean	Std. Deviation	Std. Error Mean			
Pair 1	pre_acc - post_acc	-5.43333	4.76831	.87057	-6.241	29	.000

It is very important to reinforce the idea that whether Maree can attribute the observed significant improvement (the 'effect') to a causal linkage between the training program (the putative 'cause') and decision accuracy has nothing to do with the *t*-test itself—the *t*-test can only show that the mean scores between the two occasions significantly differed within a certain tolerance for error; the test cannot show that the training program (or independent) variable actually caused the changes observed in the dependent variable. In the case of the above scenario, what determines Maree's ability to infer a causal relationship is the appropriateness and tightness of her research design and methodological procedures, not her statistical analysis (see *Fundamental concept VII*).

The scenario data were obtained using a true experiment in a field setting (where Maree had some control (e.g. random assignment of inspectors to the experimental or control group) over extraneous variables). However, by ignoring the control group in this specific *t*-test analysis, Maree's capacity to infer cause is somewhat limited. This is because Maree has no way of knowing how inspectors' decision accuracy would have changed over the time period between the pretest and the posttest without the interposition of the training program. That is, she has not used a comparative standard against which to unambiguously gauge why her means differed in the way they did. This shows the inherent research design limitations associated with the related groups *t*-test; it is restricted to comparing one group of participants (repeatedly measured or matched) on two occasions or under two conditions.

The only way Maree could overcome this problem would be to re-analyse her experiment and include her control group of participants who were measured on both occasions but did not undergo the training program—such an analytical approach is explored further in *Procedure 7.11*. It could well be that what was being reflected in the accuracy differences in the scenario as analysed here was not the influence of the training program per se, but the potential influence of other unmeasured factors such as extent of prior quality control experience, learning ability, and working conditions which may influence whether or not an inspector would show improvements in decision accuracy over time anyway.

The t-test for two related groups yields exactly the same conclusions as would a one-way repeated measures analysis of variance (see *Procedure 7.11*) on these two occasions; the two procedures are related. In fact, the t-statistic outcome from a related groups t-test is simply the square root of the F-statistic outcome from the one-way repeated measures analysis of variance performed on the two occasions. The major difference is that the related groups t-test can only be used to compare two groups whereas repeated measures analysis of variance is a more general technique that can be used to compare mean scores across any number of conditions or occasions.

Advantages

The related groups t-test is a relatively straightforward procedure and it is relatively easy to interpret its results—the conditions either differ or they do not. The real advantage of using the related groups t-test over the independent groups t-test is research design-based: the related groups t-test is associated with a research design that controls for individual differences between participants by having them participate under each condition. In a repeated measures design, each individual thus serves as their own control – effectively holding constant across conditions any background characteristics, idiosyncrasies and individual differences they bring with them into the research context. Using matched pairs effectively accomplishes the same goal as long as the matching process is properly executed and uses one or more variables that are actually related to the dependent variable.

This control over individual differences effectively reduces the amount of error associated with the test of significance making it easier to identify differences between the conditions (i.e. a more powerful test results, usually with the use of fewer participants). Additionally, it is possible to test one-tailed or directional predictions using the related groups t-test where, for example, the researcher states, before any analysis, that they expect performance under an experimental condition to be better (i.e. a higher dependent variable mean) than performance under a control condition.

Disadvantages

The related groups t-test makes two important assumptions about the nature of the data obtained under the two conditions which may or may not be reasonable for a specific situation. Firstly, it assumes that the distribution of the differences between scores in the two conditions (at least in the population of interest) is normal. Secondly, it assumes that controlling for individual differences either through repeated measurements or matching is appropriate for the research context.

In other words, the related groups t-test only works well when there is a strong correlation between the scores obtained under the different conditions (this was in fact observed in Maree's scenario data in Table 7.8 where the correlation between the two conditions was very strong at .865). If there are no large individual differences (which translates into no correlation between participants' scores in the two conditions) on the dependent variable of interest, then utilising a repeated measures research design can render the hypothesis test weaker than it would have been without such control. If matching is employed, the use of one or more inappropriate matching variables or the lack of sufficient numbers in the original sample from which to form adequately matched pairs will damage one's chances of achieving a useful hypothesis test. No nonparametric alternative can rescue a researcher from the effects of such poor research design. However, when the normality assumption is severely violated, the related groups t-test may become biased and less powerful for detecting real group differences. In such cases, it is usually recommended that a nonparametric test such as the Wilcoxon Signed Ranks test (see *Procedure 7.5*) be used. An additional disadvantage is that the related groups t-test is restricted to a single matched or repeated measures two-group design whereas repeated measures analysis of variance is more general.

Where Is This Procedure Useful?

The related groups *t-test* can be used in almost any simple experimental or quasi-experimental design involving repeated measures or matched comparisons, but only under two conditions (Field 2018, section 10.9). It has the strongest chance of supporting a causal claim in true experiments where the researcher has control over which condition the participant participates in first and second. Pretest-posttest repeated measures designs are vulnerable to history effects between the two measurement occasions, unless a control group is used through the same time period.

The related groups t-test can also be useful in survey or observational research applications when comparing responses on some measure of interest on two different occasions or before and after some treatment program (a pretest-posttest design) but with some corresponding decrease in one's capacity to infer causality linked to the time dimension. To handle this problem in pretest-posttest designs, the design should be expanded to incorporate a second group of participants who are measured on two occasions but who do not receive the treatment program received by the first group (this effectively becomes a control group which can facilitate a more unambiguous interpretation of observed changes from pretest condition to posttest condition due to possible treatment effects).

Software Procedures

Application	Procedures
SPSS	*Analyze → Compare Means → Paired Samples T Test* ... and choose the variables to define the two conditions or occasions accordingly – 'Variable 1' and 'Variable 2' (multiple pairs can be specified, if desired).
NCSS	*Analysis → T-Tests → T-Test – Paired* and choose the variables to be 'paired' to define the two conditions or occasions accordingly – 'Pair 1 Variable(s)' and 'Pair 2 Variable(s)'; NCSS also allows you to choose the null hypothesised value to evaluate (0 is default) and the alpha level to use to evaluate the test for significance. You also have the option, under the '*Reports*' tabs, to have NCSS report nonparametric versions of the *t*-test as well – the Wilcoxon Signed-Ranks and Sign tests (see *Procedure 7.5*).
SYSTAT	*Analyze → Hypothesis Testing → Mean → Paired t-Test* ... and choose the variables to define the two conditions or occasions under 'Selected variables'; SYSTAT allows you to choose whether a two-tailed or one-tailed test (including the direction) is conducted and, if you are conducting a number of paired *t*-tests, allows you to choose the Bonferroni correction to be applied.
STATGRAPHICS	*Compare → Two Samples → Paired Samples...* and choose the 'Sample 1' variable (measure for condition 1 and the 'Sample 2' variable (measure for condition 2); hit '*OK*', then choose the '*Tables and Graphs*' you want to see (tick the 'Hypothesis Tests' option under 'Tables' at a minimum). STATGRAPHICS will automatically produce nonparametric alternative tests (see *Procedure 7.5*) as well, when you choose the 'Hypothesis Tests' option; the StatAdviser will help you interpret each test.
R Commander	*Statistics → Means → Paired t-test...* and choose the 'First variable' (measure for condition 1) and the 'Second variable' (measure for condition 2); **R** Commander offers the opportunity to choose a two-tailed test or a one tailed test and to pick the direction tested.

Procedure 7.5: Wilcoxon Signed Ranks Text for Two Related Groups

Classification   Univariate; inferential; nonparametric.

Purpose   To evaluate the ranked differences in scores on a single dependent measure obtained on the same sample (or a matched sample) of individuals under two conditions, at two different times or in two different contexts. In this sense, the groups being compared are considered related.

Measurement level   The grouping variable can be defined by any level of measurement. The dependent measure consists of differences originally measured at the interval- or ratio-level but converted to ordinal ranks prior to analysis.

The *Wilcoxon Signed Ranks test* is a nonparametric analog of the related groups *t*-test (see *Procedure 7.4*). It evaluates the significance of the difference between two related groups on the basis of mean ranks rather than mean scores in some other units of measurement (Siegel and Castellan Jr 1988). As with the related groups *t*-test, the conditions may be rendered related by virtue of measuring each participant under two conditions (repeated measures design) or by matching participants on one or more extraneous variables which are thought to be influential on the dependent variable (matched pairs design). The use of the Wilcoxon Signed Ranks test is frequently entertained when the original measurements of a dependent variable are distributed in some highly distorted fashion so as to violate the basic normality assumption of the related groups *t*-test.

> Suppose we reconsider the experimental scenario set out in *Procedure 7.4* where Maree is evaluating the effectiveness of a new quality control training program. Maree had also collected **speed** data in addition to the **accuracy** data in her pretest-posttest design. For the purposes of illustrating the Wilcoxon Signed Ranks test, we will focus only on the experimental group's results.
>
> The **speed** data for the experimental group on the two occasions are shown in Table 7.9. Maree is still be interested in evaluating the effectiveness of the training program but now focuses on differences in their inspection decision **speed** before and after undergoing the training program (Maree wishes to make no directional prediction about the possible effect of the training program on **speed**).
>
> We have seen in earlier discussions (e.g. recall *Procedure 5.7*) that **speed** is a highly positively skewed variable (as are most time-based measures of human behaviour) which may be problematic enough to violate the normality assumption. Figure 7.10 shows the box plots for the **speed** variable for the pretest and posttest occasions; the plot shows not only that the two occasions may significantly differ (the posttest median is lower than the pretest median) but also that their data tend to be rather strongly skewed.
>
> Instead of employing the parametric related groups *t*-test with its associated assumption of normality, Maree opts to conduct the Wilcoxon Signed Ranks test. This test is conducted in three simple steps:
>
> 1. The pretest scores in the original units of measurement (i.e. **speed** measured in seconds) are subtracted from the posttest scores for each participant forming a new variable which contains the values of the resulting differences.
> 2. These differences are then ranked from 1 to N (the total number of inspectors or matched pairs) in order from 1 (smallest difference) to 30 (largest), ignoring the sign of the difference; and
> 3. The signs of the differences are re-attached to the resulting ranks and converted into a test statistic called Wilcoxon's W (W is simply the smaller

(continued)

of the two totals of ranks of like sign). Wilcoxon's *W* is used to evaluate the null hypothesis that the negatively signed ranks balance out the positively signed ranks. If the balance is dramatically shifted in favour of say a preponderance of positively signed ranks, then there is evidence for a difference between the two conditions or occasions.

Table 7.9 Listing of the decision **speed** scores and summary descriptive statistics on the pretest and posttest occasions for 30 inspectors undertaking the new training program

Case summaries			inspecto	pre_spe	post_spe
1			1	8.24	8.08
2			3	7.85	8.21
3			10	5.91	4.74
4			12	7.61	6.94
5			15	7.12	6.62
6			17	7.07	5.47
7			22	3.40	1.91
8			30	1.30	1.06
9			32	1.51	1.27
10			43	2.19	1.54
11			45	6.09	8.06
12			50	2.82	2.70
13			53	3.14	2.67
14			54	7.45	6.56
15			59	4.52	3.82
16			61	5.71	5.37
17			62	17.10	14.19
18			68	9.17	14.06
19			70	7.82	7.20
20			77	6.11	3.79
21			81	8.93	8.01
22			82	1.85	1.58
23			83	2.85	1.86
24			92	9.52	8.35
25			95	1.28	1.81
26			97	3.78	3.14
27			102	7.53	5.68
28			108	5.96	5.12
29			109	7.66	4.97
30			112	3.33	1.67
Total		N	30	30	30
		Mean	58.17	5.8273	5.2155
		Std. Deviation	34.159	3.33676	3.42392

(continued)

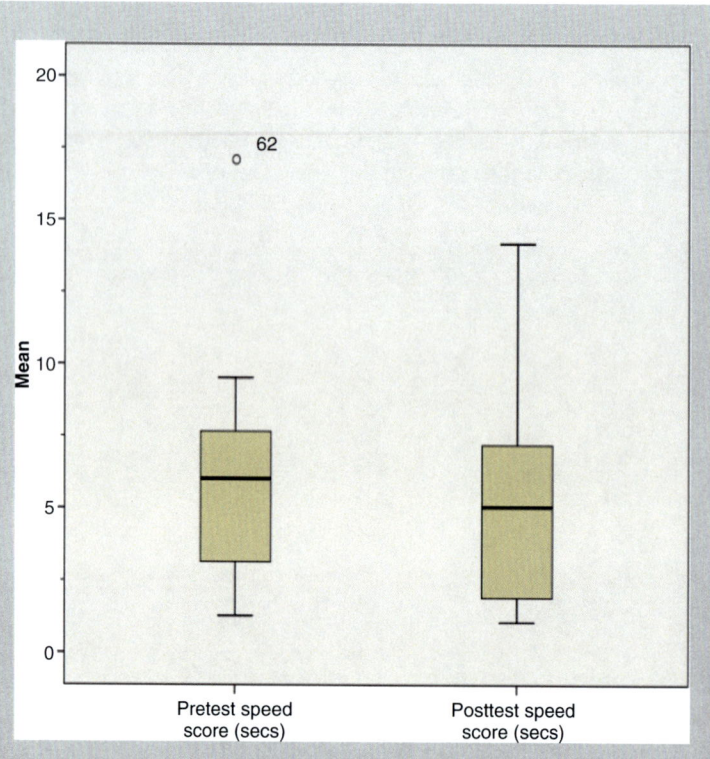

Fig. 7.10 Boxplots of mean **speed** scores for the pretest and posttest occasions for 30 inspectors undertaking the new training program

Figure 7.11 shows the outcomes from the Wilcoxon Signed Ranks test conducted (using SPSS) on the data in Table 7.9. The histogram in Fig. 7.11 clearly shows the distribution of differences to be non-normal. The test statistic (the W statistic) shows a total of 73 for positively ranked differences. Maree knows W is associated with positively ranked differences because only four cases showed positive differences. This value for W is then converted to a z-score of −3.28 which has an associated p-value of .001. This means Maree can conclude that the discrepancies between the posttest and pretest decision speed of inspectors were significantly oriented toward negative ranks. Since the difference ranking system is such that far more negative ranks than positive ranks were produced, this implies that the posttest decision **speed** scores of inspectors were significantly faster compared to the pretest decision **speed** scores.

(continued)

Procedure 7.5: Wilcoxon Signed Ranks Text for Two Related Groups

Related-Samples Wilcoxon Signed Rank Test

Positive Differences (N=4)
Negative Differences (N=26)
(Number of Ties = 0)

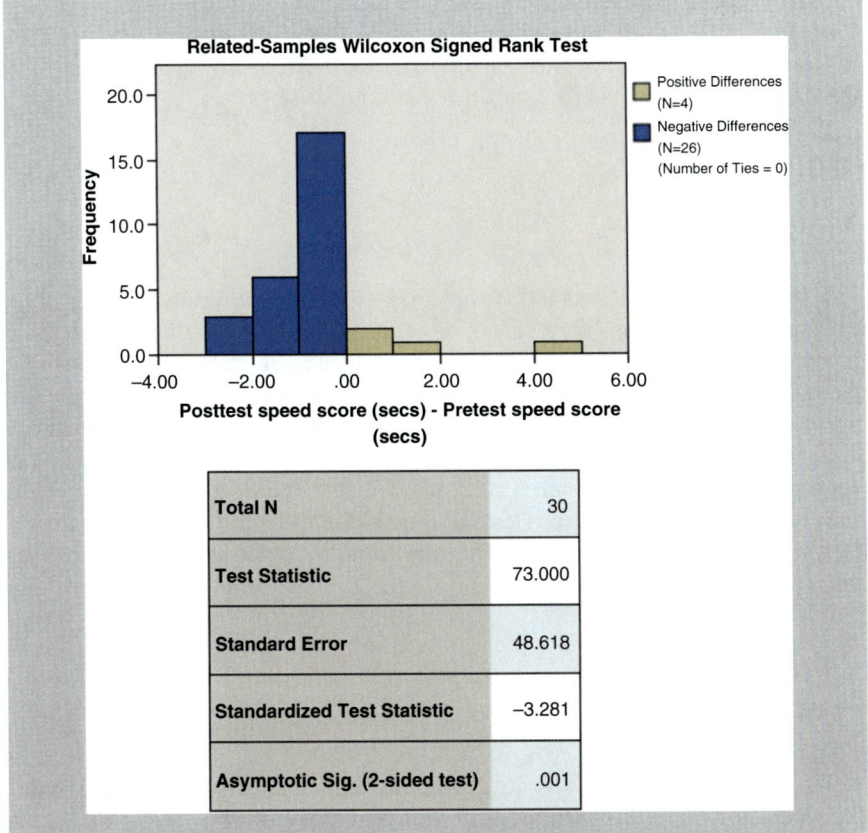

Posttest speed score (secs) - Pretest speed score (secs)

Total N	30
Test Statistic	73.000
Standard Error	48.618
Standardized Test Statistic	−3.281
Asymptotic Sig. (2-sided test)	.001

Fig. 7.11 Related (paired) samples Wilcoxon Signed Ranks analysis (using SPSS) of the experimental design data in Table 7.9

Maree could compare this result to what would have been obtained if she had run the parametric related groups *t*-test to see if her conclusions would differ between the two tests. If the inference decision was different for the two tests or if the *p*-values differed dramatically between the two tests, then assumption violations have likely distorted the parametric test and she should interpret the Wilcoxon Signed Ranks test outcome. This is the case here with the pretest-posttest comparison on **speed** scores where the Wilcoxon Signed Ranks test *p*-value is .001 and the related groups *t*-test *p*-value is only .025. If the inference decision remains unchanged or the *p*-values are similar, then assumption violation is not severe enough to distort the outcome and Maree should report the parametric test because it utilises more of the original information in the sample.

The Wilcoxon Signed Ranks test is related to the *Friedman Rank test* discussed in *Procedure 7.11*. There are several alternative forms of nonparametric test for differences between two related groups. The most frequently used appears to be the *Sign test* (see Siegel and Castellan Jr 1988, pp. 80–87).

Advantages

The Wilcoxon Signed Ranks test is particularly useful when the assumptions of the parametric related groups t-test cannot be or are not reasonably fulfilled by the data. It permits analysis of such data by reducing the them in measurement complexity to the ordinal level and then conducting the test. The test is easy to compute and interpret and has no underlying assumptions about the distribution of the data which need to be satisfied. In comparisons involving large samples, the W-statistic is converted to a z-score and interpreted accordingly. The Wilcoxon Signed Ranks test also provides a simultaneous test for a significant difference between medians of the two conditions. NCSS reports outcomes from the Wilcoxon Signed Ranks test this way.

Disadvantages

Unless the data rather severely violate t-test assumptions, it is usually better to use the parametric test rather than the Wilcoxon Signed Ranks test. The reason is that when assumptions are even moderately violated, the related groups t-test is a more powerful test for a given sample of data than the Wilcoxon Signed Ranks test; that is, it is more likely to detect condition differences if they really exist. However, if the violations are serious, then the Wilcoxon Signed Ranks test becomes the more powerful test. Also, since the original data are converted to ranks, some information is necessarily lost. Such a loss can hamper meaningful interpretation of the observed condition differences.

Where Is This Procedure Useful?

The Wilcoxon Signed Ranks test can be applied in any experimental, quasi-experimental, or survey research investigation where comparison of participants measured under two conditions is of interest and parametric assumptions are strongly violated. Many software packages offer specific tests for assumption violation which renders the choice of analytical technique easier to make.

As with the parametric related groups *t*-test, the statistical conclusion offered by the Wilcoxon Signed Ranks test relates solely to observed differences between two conditions. Whether or not a causal conclusion is warranted (i.e. that the variation in the definition of the two conditions or what happened between two occasions is what caused the significant difference in the signed ranks) is a separate decision which must be considering the quality of one's research design and methodological procedures.

Software Procedures

Application	Procedures
SPSS	*Analyze → Nonparametric Tests → Related Samples* ... and, on the '*Fields*' tab, choose the two variables to be used to define the 'pairs' or groups/occasions ('Test Fields').
NCSS	*Analysis → Nonparametric → Wilcoxon and Sign Test (Paired Sample)* and choose the variables to be 'paired' to define the two conditions or occasions ('Pair 1 Variable(s)' and 'Pair 2 Variable(s)') and on the '*Reports*' tab, make sure that 'Nonparametric Reports' is ticked.
SYSTAT	*Analyze → Nonparametric Tests → Wilcoxon* ... and choose the two variables to be used to define the 'pairs' or groups/occasions ('Selected variable (s)').
STATGRAPHICS	*Compare → Two Samples → Paired Samples*... and choose the 'Sample 1' variable (measure for condition 1 and the 'Sample 2' variable (measure for condition 2); hit '*OK*', then choose the Tables and Graphs you want to see (tick the 'Hypothesis Tests' option under 'Tables' at a minimum). STATGRAPHICS will automatically produce nonparametric alternative tests (see *Procedure 7.5*), when you choose the 'Hypothesis tests' option; the StatAdviser will help you interpret each test.
R Commander	*Statistics → Nonparametric tests → Paired samples Wilcoxon test*... and choose the 'First variable' (measure for condition 1) and the 'Second variable' (measure for condition 2); **R** Commander offers the opportunity to choose a two-tailed test or a one tailed test (you will need to pick the direction tested); under 'Type of Test', choose 'Normal approximation' if you have a reasonable (greater than 30) sample size, otherwise, tick 'Default'.

Procedure 7.6: One-Way Analysis of Variance (ANOVA)

Classification	Univariate; inferential; parametric.
Purpose	To compare two or more independent groups of participants on a single dependent variable.
Measurement level	The grouping variable can be defined at any level of measurement. The dependent variable must be measured at the interval- or ratio-level.

The *one-way analysis of variance (ANOVA)* provides a test for significant differences among two or more independent groups of participants on a single dependent measure (Field 2018; Iversen and Norpoth 1987). In one-way ANOVA, there is a single dimension or variable which serves to define the groups being compared and participants appear in one and only one of the groups defined by a specific value of that variable. This grouping variable thus defines the independent variable for the analysis.

> *For example* . . .*In an experimental study of human performance under environmental heat stress (e.g. varying levels of temperature), four experimental conditions (very cold, cold, hot and very hot conditions) and one control condition (room temperature condition) may define five groups to be compared, each group containing its own sample of participants. Ideally, the participants should be randomly assigned to one of the groups if the researcher is to have the best chance of drawing causal conclusions, but in quasi-experimental designs or survey designs such control may be impossible, as when the groups being compared are defined by some demographic characteristic of the participants (e.g. religious or ethnic background).*

One-way ANOVA simultaneously evaluates the size of the mean differences among the various groups by comparing the observed variability of these group means to the random variability that would be observed if the null hypothesis (i.e. no group differences in the population) were true. This comparison produces an F-test which indicates whether the variability of the group means is sufficiently large to warrant concluding that there are significant differences among the means. This focus on the variability across the entire set of group means is what gives rise to the name 'analysis of variance'. Two key assumptions associated with the one-way ANOVA are: (1) the dependent variable is approximately normally distributed and (2) homogeneity of variances amongst the groups.

Procedure 7.6: One-Way Analysis of Variance (ANOVA)

Consider an analysis of the differences in **mentabil** scores which might exist between the five different types of manufacturing companies (**company**) within the QCI database. Maree's research hypothesis would be that there would be significant differences among the five types of company whereas she must initially assume the null hypothesis of no significant differences among group means. Table 7.10, produced using SPSS, shows the relevant summary statistics for **mentabil** scores within each of the five types of **company**. The top subtable displays relevant descriptive statistics for each company. The middle subtable provides Levene's test of the homogeneity of variance assumption, which with a *Sig.* value of .240, is non-significant, indicating that the assumption is satisfied. [Note that this test must be specifically requested for this type of analysis in SPSS.]

Table 7.10 Analysis of variance comparing five types of manufacturing companies on **mentabil** scores

Descriptives

mentabil

	N	Mean	Std. Deviation	Std. Error	Minimum	Maximum
1 PC Manufacturer	22	111.50	8.667	1.848	93	131
2 Large Electrical Appliance Manufacturer	23	110.30	10.623	2.215	90	135
3 Small Electrical Appliance Manufacturer	21	105.05	8.634	1.884	85	117
4 Large Business Computer Manufacturer	23	112.70	7.813	1.629	92	126
5 Automobile Manufacturer	21	109.71	6.010	1.311	95	119
Total	110	109.93	8.753	.835	85	135

Test of Homogeneity of Variances

mentabil

Levene Statistic	df1	df2	Sig.
1.396	4	105	.240

ANOVA

mentabil

	Sum of Squares	df	Mean Square	F	Sig.
Between Groups	734.941	4	183.735	2.533	.045
Within Groups	7616.477	105	72.538		
Total	8351.418	109			

(continued)

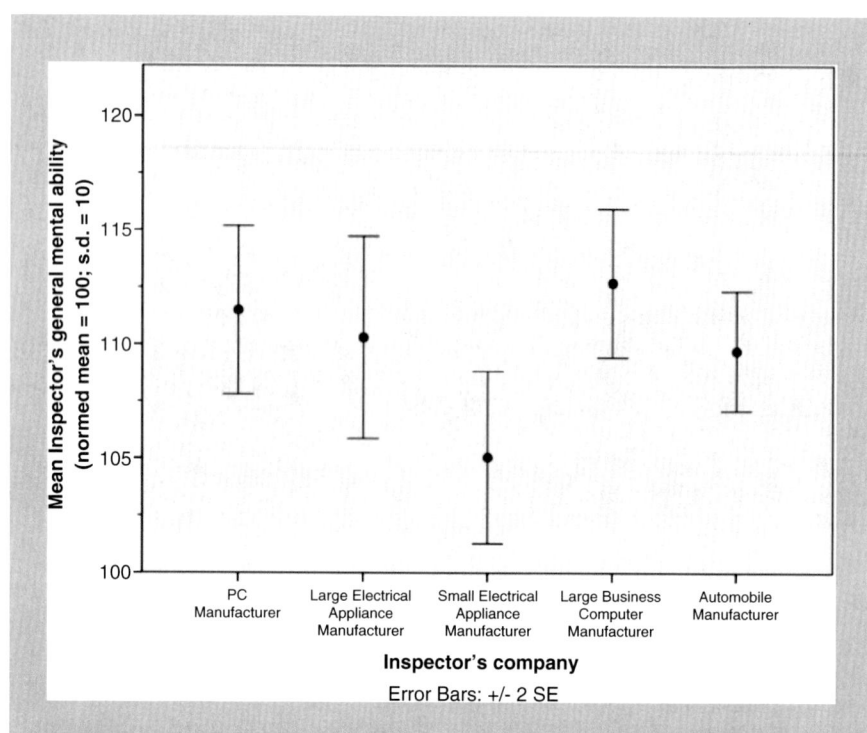

Fig. 7.12 Error bar plot of mean **mentabil** scores for inspectors in the five types of company
Note: Error bars extend to a two standard error distance either side of each mean.

The hypothesis testing outcome from the one-way ANOVA on the data shown in the bottom subtable of Table 7.10 is an F-statistic value of 2.53 (having 4 and 105 degrees of freedom) with associated p-value of .045. As this p-value is less than the desirable alpha error probability of .05, Maree is justified in rejecting the null hypothesis that, in the population, the means for these five types of companies do not differ. Maree can therefore conclude that **mentabil** does significantly differ between the different types of manufacturing companies.

Figure 7.12 shows an error bar plot of the five group means where each error bar is defined by a 2 × standard error band associated with each mean. Visual inspection of Fig. 7.12 suggests that quality control inspectors' mental ability in small electrical appliance manufacturing companies may be substantively lower than in the remaining types of companies.

It must now be emphasised that the fact that Maree obtained a significant univariate ANOVA F-test tells her nothing about which specific group means actually differ from each other. In other words, a significant F-test only indicates that somewhere in the set of group means there is at least one significant difference but not where precise differences are located. We can get some hints regarding specific group differences by visually inspecting the group means as shown in Fig. 7.12, but such visual inspection is subjective and not very defensible statistically.

The statistically appropriate way to proceed after obtaining a significant F-test is to employ multiple comparison (posthoc) procedures such as *Tukey's HSD test* or *Scheffé test* (see *Procedure 7.8*) to isolate exactly which group means differ from which other group means. Thus, obtaining a significant F-test is just the first hurdle we must pass on the way to a detailed search for group differences; if we fail to pass this first hurdle, we are not justified in proceeding with a more detailed search for differences (recall the hurdle logic for significance testing discussed in *Fundamental concept V*). It should be noted that specific comparisons between certain groups of interest can be specified *a priori* (prior to data collection and analysis) if such comparisons can be theoretically derived or justified on the basis of past research findings. These comparisons are called *planned contrasts* (see *Procedure 7.8*) and allow one to bypass the overall ANOVA F-test to obtain a more powerful test of a specific hypothesis.

It is also very important to note that whether or not Maree can attribute a significant ANOVA difference (the 'effect') to a causal linkage between working for a particular type of company (the putative 'cause') and mental ability (the putative 'effect') has nothing to do with the F-test itself—the F-test can only show that the mean scores between the various groups significantly differ within a certain tolerance for error. The test cannot show that the grouping (or independent) variable caused the changes observed in the dependent variable. As before, in the case of the QCI database, what determines Maree's ability to infer a causal relationship is the appropriateness and tightness of her research design and methodological procedures, not her statistical analysis (recall *Fundamental concept VII*).

Since the QCI data were obtained using, at best, quasi-experimental field research techniques (where very little control over extraneous variables was possible), Maree's capacity to infer cause is quite limited. This is further compounded by the fact that the type of company the inspectors in her sample worked for was **not** under her control. She could not randomly assign inspectors to companies; it was a variable whose values were predetermined by the individual backgrounds and work choices of each inspector well before she ever undertook the study. It could well be that what is being reflected in the **mentabil** differences in the QCI data is not the influence of type of company per se, but the potential influence of other demographic, personal and social characteristics, such as vocational interests, education level and past performance in similar jobs, which may help to determine whether or not an inspector with a specific level of mental ability would pursue work within a particular type of company. They may also reflect the influence of procedures for recruiting and selecting quality control inspectors in each industry.

One-way ANOVA is closely related to both the *t*-test (see *Procedure 7.2*) when only two independent groups are involved and to multiple regression methods (see *Procedure 6.4* and *Fundamental concept VI*). One-way ANOVA also provides the fundamental statistical principles required for handling more complex experimental research designs involving multiple grouping variables, namely Factorial ANOVA (see *Procedure 7.10*) and Repeated Measures ANOVA (see *Procedure 7.11*).

Advantages

One-way ANOVA permits several group means to be compared simultaneously on the same dependent variable. This is a much more satisfactory approach to multiple group testing than the testing of all possible pairs of groups using independent groups *t*-tests. The reason is that conducting multiple hypothesis tests on the same sample of data has the net effect of driving the effective alpha error probability for the study upward toward 1.0, even if the researcher thinks he or she is operating at the conventional .05 level.

> **For example** ... To completely analyse the differences between five groups on a pairwise basis would require 10 separate independent groups t-tests instead of one ANOVA F-test and the true alpha error rate would be near .40 which represents a 40% chance of incorrectly concluding that any two groups significantly differ.

The more groups there are in the research design, the worse the problem becomes. ANOVA controls for this problem by testing all groups at once and sacrificing testing precision. If desired, one-way ANOVA will allow a researcher to test for trends across groups (such as mean scores consistently increasing from group to group) using a specific type of planned contrast called an 'orthogonal polynomial' (see *Procedure 7.8*).

Disadvantages

One-way ANOVA will not work as well as it should in detecting group differences if the data being analysed do not reasonably satisfy the normality and homogeneity of variance assumptions. While ANOVA is relatively robust to mild violations of these two assumptions, more severe violations, particularly of the homogeneity of variance assumption, would suggest that a nonparametric alternative such as the *Kruskal-Wallis test* be considered (see *Procedure 7.9*) or that the data need to be transformed or rescaled in some way so as to bring them more into line with the assumptions. Depending upon the software package used, there are also variations of the F-test, such as the Welch test, that can be obtained to use in situations where the homogeneity of variance assumption is significantly violated.

Violation of the normality assumption is not too problematic as long as each group violates it in roughly the same way. Unlike the t-test, ANOVA often provides only a first stage general analysis; one that requires additional follow-up tests, conditional on achieving a significant F-test, in order to tell a complete story.

Where Is This Procedure Useful?

One-way ANOVA is most useful in experimental and quasi-experimental research designs involving a single grouping or independent variable. Certain comparisons among demographically or socio-economically-defined groups (e.g. groups defined by gender, ethnic background, income level, or occupation) in surveys can also be compared using one-way ANOVA but with a corresponding loss of capacity to conclude that a causal relationship exists between different values of the grouping (i.e. the independent) variable and any observed differences in values of the dependent variable.

More complex experimental, quasi-experimental, and survey designs involving multiple grouping factors can be handled using extended versions of the ANOVA technique. ANOVA (both the one-way version and its extensions) is a popular analytical procedure that has long served many psychologists, sociologists, educationists, marketing researchers and management researchers in the pursuit of experimental and social comparisons.

Software Procedures

Application	Procedures
SPSS	*Analyze* → *Compare Means* → *One-Way ANOVA* ... and choose the appropriate dependent variable and an independent variable (labelled as 'Factor:' in the ANOVA dialog box); hit the '*Options*' button to select analysis options such as the 'Homogeneity of variance test', obtaining versions of the F-test to use if that assumption is violated ('Welch' test or 'Brown-Forsythe' test) and a 'Means plot' (this option does not produce an error bar plot like Fig. 7.12, however).
NCSS	*Analysis* → *Analysis of Variance (ANOVA)* → *One-Way Analysis of Variance* and choose your dependent variable (called 'Response Variable') and an independent variable (called 'Factor Variable'); under the '*Reports*' tab, select the output options desired (including the nonparametric 'Kruskal-Wallis Report' and 'Assumptions Report'); under the '*Plots*' tab, choose the plots you want to see ('Means Plot' line chart or 'Box Plot').

(continued)

Application	Procedures
SYSTAT	*Analyze → Analysis of Variance (ANOVA) → Estimate Model ...* and choose your dependent variable ('Dependent(s)') and a single 'Factor(s)' variable; you have the option of selecting either 'Dummy' coding or 'Effect' coding to represent the factor variables; under the '*Options*' tab, choose the assumption tests you wish to see. SYSTAT will automatically produce relevant graphs.
STATGRAPHICS	*Compare → Multiple Samples → One-Way ANOVA ...* and choose the desired 'Dependent Variable' and the desired 'Factor' variable; hit '*OK*' and in the '*Tables and Graphs*' dialog box, select the 'Analysis Summary', 'Summary Tables', 'ANOVA Table' and 'Tables of Means' options and select any graphs you wish to see. The StatAdvisor provides interpretive comments on all requested aspects of the analysis.
R Commander	*Statistics → Means → One-Way ANOVA...* and select the independent variable ('Groups') and dependent variable ('Response Variable') desired. **R** Commander provides a minimalist one-way ANOVA report, with no options to check assumptions or produce group-comparison plots – these would have to be obtained using other functions in the main **R** program (see Field et al. 2012).

Procedure 7.7: Assessing Effect Sizes

Classification   Bivariate; correlational.
Purpose   To provide a measure of the strength or magnitude of the differences established by parametric tests for group differences such as the t-test and ANOVA F-test which can be used to gauge the importance or meaningfulness of a significant difference (i.e. practical significance).
Measurement level   The grouping variable can be defined at any level of measurement. The dependent variable must be measured at the interval- or ratio-level.

Any parametric test of hypothesis that focuses upon differences between group means, whether it is based on the t-test or ANOVA F-test, has an associated correlational measure that can be used to index the strength of the difference the test has evaluated. These correlational measures, which are called *measures of effect size* or *measures of strength of association*, are interpreted in the same way as a squared correlation or squared multiple correlation (recall the discussions in *Fundamental concepts III, IV* and *VI*). They indicate what proportion (or percentage) of the variability in the dependent variable can be explained by knowing which group a participant came from.

Over the past two decades or so, social and behavioural scientists have come to realise that merely reporting the significance level, i.e. p-value, for a hypothesis test

Procedure 7.7: Assessing Effect Sizes

is no longer sufficient to establish the importance of a research finding. The reasons for this are primarily twofold:

- with a large enough sample, almost any difference between groups, no matter how small, can be found to be significant (because of the extremely high power afforded by large sample sizes); and
- as a consequence of the first reason, it is improper to infer the importance of a statistical test result on the basis of a very small p-value (e.g. .0005) as many behavioural scientists tend to do.

Behavioural scientists needed a way to index the practical importance of a finding, summarising how much we have learned from the result, that was not directly linked to how improbable the result was, assuming the null hypothesis was true. This opened a debate with respect to which research outcome was more important: statistical significance or practical significance. It turns out both are needed for a complete story to be told and more and more academic journals now expect the results of hypothesis tests to include both level of significance and effect size information.

Consider the following thought experiment.

> Medical researcher A experimentally tests a new drug for treating melanomas using two groups (one treated with the drug; one receiving a placebo treatment) of 1000 patients and finds that his t-test has a p-value of .0000035. Researcher A rejoices and claims that his new drug is highly effective in reducing the symptoms of melanoma.
>
> However, medical researcher B comes along and computes a measure of effect size for A's data and finds that the new drug accounts for .01% of the variability in reduction of skin cancer symptoms in patients. Researcher B says the drug might be effective statistically but for all practical purposes, it accounts for such a minuscule proportion of the variability in patient symptom reduction that it is not worth investing in the marketing of the new drug. Researcher B maintains that it would be far more productive to search for the factors that can help explain the other 99.99% of the variability in patient symptom reduction.
>
> What has happened? Researcher A's study had so much power that it was able to reveal a very small improvement to be highly significant. Researcher B showed that although the improvement was statistically non-zero it might as well have been zero for all the practical difference it made. Researcher A could have mounted a much stronger case if, in his original study, the new drug had been able to account for perhaps 15–20% of the variability in patient symptom reduction.
>
> Thus, there is a trade-off relationship between statistical significance, measured by p-values and practical significance, assessed by correlational measures of effect size.

Two measures of effect size for group-based hypothesis tests, which have enjoyed increased use in the behavioural science research literature, are *eta-squared* (η^2) and *omega-squared* (ω^2). Eta-squared is generally considered to be a descriptive measure of effect size whereas omega-squared is designed to estimate the population value of variance explained. Eta-squared is typically the most frequently reported measure and is relatively easy to compute. Conceptually, eta-squared is the ratio of the variability of the group means to the variability of all the data (ignoring group membership) in the sample. It is analogous to R^2 when group differences are considered. If the proportion is high, then the differences between group means explain much about why participants' scores differ as they do between groups. If eta-squared is close to zero, then other perhaps extraneous factors, possibly not measured in the study (thereby remaining unidentifiable as part of error), explain the variability in dependent variable scores. Omega-squared (Hays 1988, p. 375) has a similar interpretation, but will always provide a more conservative estimate of variability explained compared to eta-squared.

> Recall the ANOVA that was reported in *Procedure 7.6* which compared the five types of **company** in the QCI database in terms of inspectors' mean decision **mentabil** scores. The *F*-test in this analysis was significant at a *p*-value equal to .045. To check the practical importance of this finding, Maree should report that the eta-squared value associated with this particular ANOVA result was .088 (or 8.8%) which says that variability in the type of **company** an inspector works for accounts for almost 9% of the total variability in inspectors' **mentabil** scores (Tabachnick and Fidell 2019, section 3.4 shows the formula for computing this value from the ANOVA results; see also Appendix A). In writing up this finding, Maree could say 'A one-way ANOVA test revealed significant differences between types of manufacturing company in terms of the mental ability of quality control inspectors (F (4, 105) = 2.53, p = .045, η^2 = .09)'. The value for omega-squared for this analysis is .053 or about 5% of variance in **mentabil** scores in the population can be explained by knowing the type of **company** a participant worked for (Tabachnick and Fidell 2019, section 3.4 also shows the formula for computing this value from the ANOVA results).
>
> For social and behavioural science data, explaining a large proportion (e.g. an eta-squared value around .25 – what Cohen 1988, terms a 'large' effect) of the variability in a dependent variable is a desirable goal as it can point to an important and practically useful result. In many cases, researchers would be pleased to be able to explain 5–10% of the variability in a dependent variable using some grouping variable (what Cohen 1988, refers to as a 'medium' effect) since we know that the behaviour of human beings is notoriously hard to predict with any measure. However, what must always
>
> (continued)

Procedure 7.7: Assessing Effect Sizes

> be remembered is that these measures of effect size are simply descriptive measures. Any causal inferences would need to be justified by considering the character and quality of the research design itself. In the case here, the causal argument could **not** plausibly be made that working for different types of companies causes differences in mental ability.

We will see later on in *Procedure 8.8* that the concept of *effect size* can be defined in a somewhat different way to provide a standardised measure of the physical size of the differences or relationships identified in virtually any type of research, be it correlational or experimental, i.e. group-comparison focused, in nature. Field (2018, sections 7.4.5, 7.5.5 and 7.6.7) discusses how you can compute measures of effect size for various nonparametric tests such as the Mann-Whitney test (*Procedure 7.3*), Wilcoxon Signed Ranks Test (*Procedure 7.5)* and Kruskal-Wallis test (*Procedure 7.9).*

Advantages

A measure of effect size, such as eta-squared or omega-squared, provides a useful and practical correlational index of the strength of group differences identified as significant using a statistical inference test. These indices are interpreted in exactly the same fashion as a squared correlation coefficient in terms of summarising the proportion of total variability in a dependent measure that can be explained by variability in the independent or grouping variable. Eta-squared is a simple measure to compute, even by hand, and can be used as a descriptive index for virtually any analysis of variance design, no matter how complex (an alternative for of eta-squared, called *partial eta-squared*, exists as well for more complex ANOVA designs, see Tabachnick and Fidell 2019, section 3.4). Another interesting feature of eta-squared, as a correlational measure of effect size, is that it is not restricted to summarising only linear relationships between the variable that defines the groups and the dependent variable.

Disadvantages

There does tend to be something of an inverse relationship between measures of effect size and statistical significance p-values when sample sizes are large. This means that, in very large studies, while it is very easy to achieve very small p-values, it is very difficult to achieve very high proportions of variance explained. In studies where sample sizes are small, a large eta-squared value may be associated with a non-significant test result.

A second disadvantage is that many statistical computing software programs do not report measures of effect size and the researcher is left to do the calculation by hand (this is especially the case with *t*-test results, where formulas given in Hays (1988, pp. 311–313) would have to be used; see also Appendix A). Omega-squared has an additional problem in that the formulas for computing it are complex and, in fact, do not exist for more complicated types of ANOVA designs, including repeated measures designs. One drawback to eta-squared is its descriptive nature; it tends to provide a very biased (i.e. overly liberal) estimate of what the proportion of variance explained would be in the population.

Where Is This Procedure Useful?

Any experimental, quasi-experimental, or survey research design where group comparisons are being made should always report the results of the test of significance in tandem with a measure of effect size. This gives any potential reader or consumer of the research enough information in the results to decide on the implications of the trade-off between statistical significance and practical importance.

Software Procedures

Application	Procedures
SPSS	*Analyze → Compare Means → Means* ... and select your dependent variable(s) ('Dependent List') and grouping ('Independent List') variables; then click the '*Options* ...' button and tick the option 'ANOVA table and eta'. This will produce a one-way ANOVA table for every dependent variable and independent you choose and report the eta and eta-squared for each. *Analyze → General Linear Model → Univariate* ... and select your dependent variable ('Dependent Variable') and independent variables ('Fixed Factors(s)'); click on the '*Options* ...' button and tick the 'Estimates of effect size' option. The version of eta-squared produced by this SPSS procedure is called 'partial eta-squared' and is calculated slightly differently from traditional eta-squared (see Tabachnick and Fidell 2019, section 3.4; see also Appendix A).
NCSS	NCSS does not compute eta squared measures in its ANOVA routines; you must compute them by hand, using formulas given in Tabachnick and Fidell (2019, section 3.4; see also Appendix A) and values given in the analysis output.
SYSTAT	*Analyze → Analysis of Variance → Estimate Model* ... and select your dependent variable(s) ('Dependent(s)') and independent variables ('Factor (s)'); you have the option of selecting either 'Dummy' coding or 'Effect' coding to represent the factor variables. *Analyze → General Linear Model → Estimate Model...* and select your

(continued)

Application	Procedures
	dependent variable(s) ('Dependent(s)') and independent variables ('Independents(s)'); under the '*Category*' tab, choose which variable(s) are categorical and choose whether to use 'Dummy' or 'Effect' coding. For either procedure, SYSTAT does not report eta-squared but it does report 'Squared Multiple R' which can be interpreted exactly the same way.
STATGRAPHICS	STATGRAPHICS does not compute eta squared measures in its ANOVA routines; you must compute them by hand, using formulas given in Tabachnick and Fidell (2019, section 3.4; see also Appendix A) and values given in the analysis output.
R Commander	**R** Commander does not compute eta squared measures in its ANOVA routines; you must compute them by hand, using formulas given in Tabachnick and Fidell (2019, section 3.4; see also Appendix A) and values given in the analysis output.

Procedure 7.8: Planned Comparisons and Posthoc Multiple Comparisons

Classification   Univariate; inferential; parametric.

Purpose   Planned comparisons involve the testing of specific hypotheses regarding differences amongst a set of dependent variable means for three or more groups without the need for a significant overall F-test.

Posthoc multiple comparisons are employed to identify specific differences amongst three or more group means conditional upon achieving a significant overall F-test result in an analysis of variance.

Measurement level   The grouping variable can be defined at any level of measurement except for planned orthogonal polynomial tests for trend where the grouping variable should be defined at the interval-level (group definition can be ordinal in a pinch, but this is not preferred). The dependent variable must be measured at the interval- or ratio-level.

Planned comparisons and *posthoc multiple comparison tests* are procedures designed to evaluate the significance of differences between specific combinations of group means in an ANOVA-type analysis. Planned comparisons are generally conducted in lieu of an overall ANOVA F-test whereas posthoc multiple comparisons are conducted only after a significant ANOVA F-test has been achieved. Both types of comparisons can be performed in the context of virtually any ANOVA design, univariate or multivariate.

Planned Comparison Tests

A priori Contrasts

A priori contrasts permit the testing of virtually any combination of three or more group means. However, there are certain rules attached to the use and testing of *a priori* contrasts.

1. They must be specified before the data are gathered which means that their formation must be guided by theory or by past research findings. Testing more *a priori* contrasts than permitted or changing the nature of the hypotheses (and therefore the structure of the contrasts) after the data have been gathered constitute ethical abuses of statistical technology and are temptations to be avoided.
2. The number of such contrasts that can be tested is limited to the number of groups less one (which equals the degrees of freedom for the independent variable defining the groups).

> *For example* ... *in an ANOVA design involving the comparison of four groups of participants, only three* a priori *contrasts can be tested. This is only an upper limit; the researcher may have fewer, perhaps just one or two, specific contrasts of interest.*

3. The contrasts must be structured in such a way that they are uncorrelated with each other (for this reason, some authors refer to these as orthogonal contrasts; see Keppel and Wickens 2004 and Kirk 2013). *Uncorrelated* or *orthogonal contrasts* are separate contrasts that avoid combining and comparing specific group means in exactly the same way twice.

A priori contrasts are formed by combining the means for all groups using specific weighting systems. These weighting systems can be designed by the researcher to test specific hypotheses about patterns of group mean differences (some weights can be set to zero; a group receiving a 0 weight is effectively ignored in the contrast). *A priori* contrasts are tested using a special version of a *t*-test (which means that one-tailed hypotheses can be tested). One further constraint attached to building contrasts is that the weights assigned by the researcher for combining groups must add to 0—the contrast weights must balance out.

For example, in an investigation involving four groups, one possible *a priori* contrast might be to compare groups 1 and 2 combined against groups 3 and 4 combined—a useful *a priori* comparison if, for example, groups 1 and 2 were experimental groups and groups 3 and 4 were control groups and theory dictated that the experimental treatments should alter behaviour relative to control treatments.

Procedure 7.8: Planned Comparisons and Posthoc Multiple Comparisons

One weighting system for group means which could represent this contrast would be to give groups 1 and 2 each a weight of +1 and groups 3 and 4 each a weight of −1. This weighting system would effectively add the means for groups 1 and 2 and subtract, from that total, the means for groups 3 and 4. [Note that dummy coding and effect coding, discussed in *Fundamental concept VI*, actually represent two specialised types of contrasts; however, they are non-orthogonal or correlated contrasts. What we are talking about here are additional ways one could code groups in a General Linear Model in order to test specific hypotheses about mean differences (see Cohen et al. 2003, ch. 6 and 8, for more details.)]

If we were also interested in comparing the two experimental groups (groups 1 and 2) against each other, we could build a second contrast using a weighting system of +1 for group 1, −1 for group 2, and 0 for each of groups 3 and 4. This contrast would be uncorrelated with ('orthogonal to') the first contrast since the means for groups 1 and 2 were not being compared in the same way.

A simple check to ensure that two contrasts are orthogonal is to cross-multiply respective weights for each group and add up the resulting products. If the total is 0, then the two contrasts are orthogonal. In our case, we would have $(+1 \times +1) + (+1 \times -1) + (-1 \times 0) + (-1 \times 0)$; these cross-products do add up to 0 ($= +1 + -1 + 0 + 0$), so our two contrasts are indeed orthogonal. [Note that to be technically correct, contrasts will be empirically orthogonal only if all the groups being compared have equal sample sizes.]

Suppose Maree was interested in comparing **company** groups in terms of mean decision **accuracy** scores. On the basis of past research findings, she might be interested in comparing the two computer-oriented manufacturing companies (PC and Large Business Computer) against the two electrical appliance manufacturing companies (Large and Small). This would be a legitimate contrast that deliberately ignored the automotive companies.

Maree could quantify this contrast by giving the PC and Large Business Computer manufacturing groups each a weight of +1, the Large and Small Electrical Appliance manufacturing groups each a weight of −1, and the Automobile manufacturing group a weight of 0. This is shown in Table 7.11 in the row for Contrast 1 in 'Contrast Coefficients' subtable of the SPSS output from the One-Way ANOVA procedure. Furthermore, Maree might also be interested in directly comparing the two types of computer company to each other and the two types of electrical appliance manufacturing company to each other. The contrast weights that she could use to define these latter two contrasts are shown in row '2' and row '3' of the Contrast Coefficients table, respectively. Try the cross-products rule for any two rows and you will see that all three of these contrasts are orthogonal to each other.

(continued)

Table 7.11 also shows the results of the statistical test for each contrast in the lower subtable. Two versions of each test are offered; one that assumes the groups have equal variances (i.e. the homogeneity of variance assumption is satisfied) and the other can be used if that assumption is not satisfied.

Table 7.11 Orthogonal contrast tests looking for specific hypothesised differences between the five types of **company**

Contrast Coefficients

		company				
Contrast	1 PC Manufacturer	2 Large Electrical Appliance Manufacturer	3 Small Electrical Appliance Manufacturer	4 Large Business Computer Manufacturer	5 Automobile Manufacturer
1	1	-1	-1	1	0
2	1	0	0	-1	0
3	0	1	-1	0	0

Contrast Tests

		Contrast	Value of Contrast	Std. Error	t	df	Sig. (2-tailed)
accuracy	Assume equal variances	1	17.13	3.276	5.230	105	.000
		2	-1.83	2.303	-.794	105	.429
		3	-4.25	2.330	-1.826	105	.071
	Does not assume equal variances	1	17.13	3.291	5.206	66.101	.000
		2	-1.83	2.168	-.843	26.699	.407
		3	-4.25	2.476	-1.718	39.681	.094

We'll focus here on the 'Assume equal variances' version. For her first *a priori* contrast, the value of the contrast itself [when the weights are used to combine the respective group means in the following computational way: $(+1 \times 86.95) + (+1 \times 88.78) + (-1 \times 77.17) + (-1 \times 81.43) + (0 \times 75.71)$] is 17.13; the contrast *t*-test value is +5.23 and its associated *p*-value is less than .001. This indicates a significant contrast result: inspectors in the two computing manufacturing companies considered together show significantly higher inspection **accuracy** scores than do inspectors in the two electrical appliance manufacturing companies considered together. Contrast 2, comparing the two types of computer company, is not significant at $p = .429$ and contrast 3, comparing the types of electrical appliance manufacturing company, is suggestive of a possible difference at $p = .071$, but is not statistically significant.

Orthogonal Polynomial Tests for Trend

Orthogonal polynomial tests for trend are the statistician's mouthful of a way to label a procedure that can test for the existence of specific trends across the group means. Orthogonal again means uncorrelated and polynomial refers to the way the contrasts are constructed. The groups should be defined by some quantitative interval-level measure (e.g. drug dosage level, years of job experience, number of weeks of training, salary level or age) and the spacing between the levels defining the various groups should be equal. The simplest way to explain such tests is by example.

> Consider a medical researcher who is interested in establishing the most effective dosage level for a new depression treatment drug about to be marketed. This researcher sets up an experiment where patients exhibiting equivalent symptoms of depression are randomly assigned to one of five treatment groups defined by drug dosage level (in equal increments from group to group).
>
> - Group 1 receives 0 ccs of the drug, receiving instead, placebo injections of saline each day of a weekly period;
> - Group 2 receives 2 ccs each day;
> - Group 3 receives 4 ccs;
> - Group 4 receives 6 ccs; and
> - Group 5 receives 8 ccs.
>
> At the end of the treatment week, all patients are measured using the Beck depression inventory. What orthogonal polynomials allow the researcher to do is to test, up front, for the existence of specific mathematical trends in the depression scores as the dosage levels increase across the groups.
>
> Figure 7.13 illustrates some of the trends that orthogonal polynomials can be used to evaluate. In practice, orthogonal polynomial testing is conducted exactly the same way as the testing of contrasts; all that changes is the weighting system used to combine the group means. Certain reference texts (e.g. Cohen et al. 2003, Kirk 2013 and Keppel and Wickens 2004) have tables that provide the contrast weights to be used for testing different trends for various numbers of groups. Figure 7.13 shows the contrast weights, vertically corresponding to the dosage group each would apply to, one would use to form the linear, quadratice and cubic trend tests. [Note that it is possible to do orthogonal polynomial tests using an ordinal grouping variable, such as low, medium and highly stressed employees, for example, but the results will only be approximate and inferences correspondingly weaker.]

(continued)

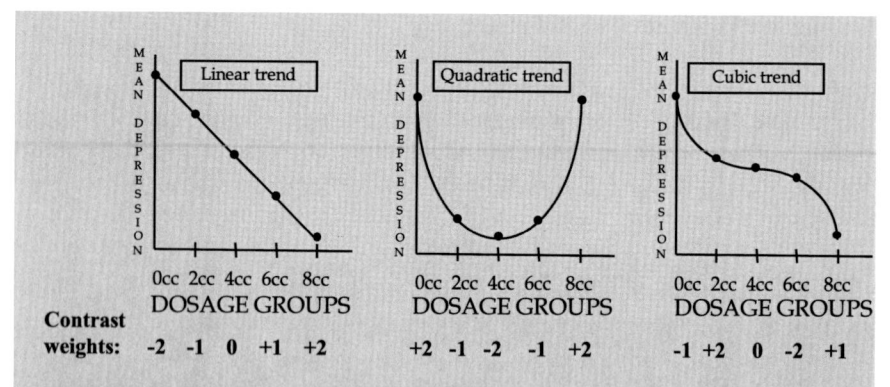

Fig. 7.13 Examples of the possible trends across dosage group means that orthogonal polynomials can test

Posthoc Multiple Comparison Tests

There are numerous *posthoc multiple comparison tests* available to social and behavioural researchers and experts differ as to which test is the best one to use. We will discuss only two of the more commonly employed tests here. Both tests lie toward the conservative end of the possible types of multiple comparisons in that they will tend not to identify too many significant differences—these tests trade away some degree of power (ability to detect true group differences) in order to achieve greater control over problems associated with the inflation of the alpha error level due to conducting multiple hypothesis tests on the same sample of data. Posthoc multiple comparison tests are permissible only after the overall ANOVA *F*-test comparing the groups of interest has been shown to be significant. If the ANOVA *F*-test is not significant, then multiple comparisons tests cannot be conducted. In this way, posthoc multiple comparisons become a critical final step in the hurdle logic for significance testing where more than two groups are being compared (recall *Fundamental concept V*).

Tukey's HSD Test

Tukey's HSD multiple comparison test (HSD means 'Honestly Significant Difference') evaluates all possible pairs of group means in order to locate those means that differ by more than a critical amount from each other. This critical amount is called the HSD criterion and it establishes the minimum difference that needs to exist between any two group means before those means can be called significantly different at the alpha error level desired by the researcher. The alpha level used for the Tukey HSD test should be the same alpha level used to evaluate the overall F-test for significance in the ANOVA.

When testing all possible pairs of means, one thing should be noted. As the number of groups being compared increases, the number of pairs of group means that can be compared increases even faster. With three groups, there are three possible pairs of means (1 with 2; 1 with 3; and 2 with 3). With four groups, there are 6 possible pairs (1 with 2; 1 with 3; 1 with 4; 2 with 3; 2 with 4; and 3 with 4). With 5 groups, there are 10 possible pairs and so on. The reason t-tests aren't typically used to test all these pairs of means is that as the number of t-tests (or any statistical tests one performs) increases within a single set of data, the more likely one is to claim a significant difference that is false (i.e. the true alpha error rate increases). Tukey's HSD test is specifically designed to control this problem of alpha error inflation irrespective of the number of pairs of means being evaluated.

To illustrate the use of Tukey's HSD multiple comparison test, suppose that Maree did not have *a priori* hypotheses when comparing **company** groups in terms of mean decision **accuracy** scores. Table 7.12 shows an excerpt from the SPSS One-Way ANOVA output containing the results of overall ANOVA F-test as well as a subtable reporting the Tukey's HSD multiple comparison tests. A significant F-test of 12.45 with an associated p-value of less than .001 was found. This would allow Maree to conclude that there were significant differences among the five company means with respect to mean decision **accuracy** scores, but she would be unable to pinpoint precisely where those differences occurred. By employing Tukey's HSD test, she could proceed to a posthoc examination of all possible pairs of **company** means for significant differences.

All possible non-redundant paired comparisons of group mean differences are represented in the Multiple Comparisons subtable in Table 7.12 (redundant aspects of this table have been edited out to simplify the table). Significant mean differences are indicated, by SPSS, with an asterisk (∗). Maree could conclude the following from Table 7.12: inspectors from PC computer manufacturers make significantly more accurate inspection decisions than the inspectors from Large Electrical Appliance and Automobile manufacturers; inspectors from Large Electrical Appliance manufacturers make significantly less accurate inspection decisions than inspectors from Large Business Computer manufacturers; Small Electrical Appliance manufacturers make significantly less accurate inspection decisions than inspectors from Large Business Computer manufacturers; and Large Business Computer manufacturers make significantly more accurate inspection decisions than the inspectors from Automobile manufacturers.

(continued)

Table 7.12 Tukey's HSD multiple comparison test following a significant one-way ANOVA *F*-test

ANOVA

accuracy

	Sum of Squares	df	Mean Square	F	Sig.
Between Groups	2969.490	4	742.373	12.453	.000
Within Groups	6259.601	105	59.615		
Total	9229.091	109			

Multiple Comparisons

Dependent Variable: accuracy
Tukey HSD

(I) company	(J) company	Mean Difference (I-J)	Std. Error	Sig.	95% Confidence Interval	
					Lower Bound	Upper Bound
1 PC Manufacturer	2 Large Electrical Appliance Manufacturer	9.781*	2.303	.000	3.39	16.17
	3 Small Electrical Appliance Manufacturer	5.526	2.356	.139	-1.01	12.06
	4 Large Business Computer Manufacturer	-1.828	2.303	.932	-8.22	4.56
	5 Automobile Manufacturer	11.240*	2.356	.000	4.70	17.78
2 Large Electrical Appliance Manufacturer	3 Small Electrical Appliance Manufacturer	-4.255	2.330	.364	-10.72	2.21
	4 Large Business Computer Manufacturer	-11.609*	2.277	.000	-17.93	-5.29
	5 Automobile Manufacturer	1.460	2.330	.971	-5.01	7.93
3 Small Electrical Appliance Manufacturer	4 Large Business Computer Manufacturer	-7.354	2.330	.017	-13.82	-.89
	5 Automobile Manufacturer	5.714	2.383	.124	-.90	12.33
4 Large Business Computer Manufacturer	5 Automobile Manufacturer	13.068*	2.330	.000	6.60	19.54

*. The mean difference is significant at the 0.05 level.

Scheffé Contrasts

Scheffé contrasts offer the most general type of multiple comparison test compared to any other available test, including the Tukey's HSD; however, this generality is achieved at the cost of the Scheffé test being the least powerful of all the available tests. Scheffé contrasts permit the testing of virtually any number and combination of group means, not just pairs of group means as for Tukey's HSD and other multiple comparison tests. These contrasts take the same structure and form as *a priori* contrasts, but without the restrictions of the contrasts having to be either orthogonal or limited in number to the number of groups less one.

Procedure 7.8: Planned Comparisons and Posthoc Multiple Comparisons

> *For example* ... *With four groups, one possible Scheffé contrast could be to compare groups 1 and 2 combined against groups 3 and 4 combined—a useful comparison if, for example, groups 1 and 2 were experimental groups and groups 3 and 4 were control groups.*

> In the context of the Maree's ANOVA comparing **company** types in terms of mean **accuracy** scores, she might be interested in comparing the two computer-oriented manufacturing companies (PC and Large Business Computer) against the two electrical appliance manufacturing companies (Large and Small). This is a legitimate contrast that deliberately ignores the automotive companies. Scheffé contrasts are formulated and tested in exactly the same way as *a priori* contrasts with one important exception: the critical value used to establish the significance of the Scheffé contrast is much higher than for the corresponding *a priori* contrast. A stricter comparative standard is employed so as to protect the researcher from the alpha error inflation problem.
>
> Most software packages will not compute Scheffé contrasts for anything other than pairwise comparisons of means, so posthoc contrasts such as the **company**-related posthoc contrast described above one must be computed by hand. For the above contrast, the value of the Scheffé contrast test statistic is +17.13, the Scheffé contrast *t*-test value is +5.23 and its associated *p*-value is less than .001 which indicates a significant contrast result: inspectors in the two computing companies considered together show significantly higher inspection **accuracy** scores than do inspectors in the two electrical appliance manufacturing companies considered together. [Note, however, that in order to be considered significant, the value of this Scheffé contrast needed to exceed a critical value of 4.47 as opposed to a critical value of value of 2.24 for the *a priori* version of the contrast.]

If the researcher is only interested in evaluating pairwise comparisons among group means, then Tukey's HSD test would be preferred over Scheffé contrasts because the HSD test is more powerful. However, if the researcher wishes to assess, posthoc, a more complex relationships among the group means, then Scheffé contrasts must be used.

Advantages

The advantages of conducting planned comparisons are twofold: they permit powerful tests of precise hypotheses about group differences to be made and they force the researcher to be focused in what he or she is looking for in the data. Tests for trend are particularly useful in characterising the patterns of changes in group means as one proceeds from group to group in order of increasing amount of the independent variable (e.g. in order of increasing drug dosages, stress levels or ages).

Posthoc multiple comparison tests are a necessary follow-up to any significant ANOVA F-test performed to compare three or more independent groups. These tests facilitate obtaining a more detailed understanding of the precise nature of group differences. Tukey's HSD test is more powerful when pairs of group means are being compared; Scheffé contrast tests are the only available posthoc test for evaluating more complex non-pairwise comparisons among group means. Both methods adequately control for the inflation of the alpha error rate which occurs when multiple tests are conducted on the same set of data, but at the expense of some power to detect true differences. There are many other types of posthoc multiple comparison tests, each having their own advantages and disadvantages (see Klockars 1986 or Toothaker 1993); some tests such as the Games-Howell test have been designed to work where the homogeneity of variance assumption is not viable.

Disadvantages

Planned comparisons do suffer from two major disadvantages. First, they are highly constrained tests (recall the rules which govern the number and orthogonal structuring of contrasts) which may limit open-minded exploration of the data. Second, they must be justified before the research is commenced (to some, this is actually an advantage) which means that strong reasons for their testing must be found. It is very tempting to succumb to the ethical abuse of statistical technology by ignoring these constraints.

There are several other multiple comparison tests available besides Tukey's HSD and Scheffé contrasts. These include tests such as Duncan's Multiple Range Test, the Student-Newman-Keuls test, the Games-Howell test and Fisher's LSD test, and deciding which one to employ can sometimes be difficult. Depending upon your choice, you may be more or less likely to detect true significant differences. In more complex ANOVA designs, such as those described in *Procedures 7.10* and *7.11*, multiple comparison tests become somewhat trickier to conduct and very few software packages offer such tests for anything beyond one-way ANOVAs (meaning that such tests often must be computed by hand for more complex ANOVAs). Also, all the available posthoc multiple comparison tests are generally less powerful in detecting group differences than if the researcher can specify, justify and test only a small number of *a priori* contrasts.

Where Is This Procedure Useful?

A priori contrasts or orthogonal polynomial tests for trend can be employed in any experimental, quasi-experimental, or survey research design which evaluates differences among three or more groups where such tests can be justified and set up before the collection of data commences. These procedures provide the most precise basis for testing hypotheses about multiple group differences. However, once the data are in, it is not 'cricket' for a researcher to visually inspect the data and then derive an *a priori* contrast to test; such behaviour amounts to an ethical abuse of statistics. Orthogonal polynomial tests for trend are most useful in studies involving a quantitatively defined independent variable where developmental trends or differential effectiveness outcome patterns are hypothesised.

Some type of multiple comparison test should be employed in any experimental, quasi-experimental, or survey research design where an ANOVA *F*-test indicates that significant differences among three or more groups exist—particularly if *a priori* contrasts cannot be justified or are of no interest. For Scheffé contrasts, it is entirely appropriate to base the construction of the contrasts on visual inspection of patterns among the group means since the test is structured to prevent the over-inferencing that accompanies the use of such an approach. It is truly a technique for data snooping (but is *only justified if a significant overall ANOVA F-test is achieved*). Merely reporting that a significant *F*-test is achieved in these designs is not sufficient, except in the special case where only two groups are involved in the design. Exploration of specific group differences is required.

Software Procedures

Application	Procedures
SPSS	*Analyze* → *Compare Means* → *One-Way ANOVA* ... and choose the appropriate dependent variable and an independent variable (labelled as 'Factor:' in the ANOVA dialog box); click on the '*Post Hoc* ...' button and choose the desired test (you can also set the desired alpha level). If you want Tukey's HSD test, choose the 'Tukey' option rather than 'Tukey's – b'. The 'Scheffé' option only produces pairwise tests of mean differences. If you click on the '*Help*' button, you can see brief descriptions of each of the posthoc multiple comparison tests that SPSS offers. To set up *a priori* contrasts, click on the '*Contrasts* ...' button and select either the 'Polynomial' option or enter the weights you want for each of your contrasts (weights are entered in the same order as the ordering of the groups being compared). *Analyze* → *General Linear Model* → *Univariate* ... and select your dependent variable ('Dependent Variable') and independent variables ('Fixed Factors(s)'); click on the '*Post Hoc* ...' button and choose the desired test. If you want Tukey's HSD test, choose the 'Tukey' option rather than 'Tukey's – b'. The 'Scheffé' option only produces pairwise tests of

(continued)

Application	Procedures
	mean differences. If you click on the '*Help*' button, you can see brief descriptions of each of the posthoc multiple comparison tests that SPSS offers. If you want to test orthogonal polynomial contrasts, click on the '*Contrasts* ...' button and choose the 'Polynomial' option under 'Change Contrast'. The GLM procedure will not let you define your own orthogonal contrasts; it simply offers a range of pre-defined possibilities – all of which are pairwise in nature. Use the *One-Way ANOVA* approach described above if you want to define your own contrasts.
NCSS	*Analysis → Analysis of Variance (ANOVA) → One-Way Analysis of Variance* and choose your dependent variable (called 'Response Variable") and an independent variable (called 'Factor Variable'); under the *Comparisons* option in the 'Planned Comparisons' section, choose 'Polynomials' if desired or choose 'Custom' and set up the custom *a priori* contrasts you want to test; under the '*Reports'* tab, choose the multiple comparison test you want (if you want the Tukey HSD test, choose the 'Tukey-Kramar Test'). Under the '*Reports'* tab, NCSS also lets you choose the alpha level you want to use for evaluating the multiple comparisons tests.
SYSTAT	Step 1: *Analysis → Analysis of Variance (ANOVA) → Estimate Model ...* and choose your dependent variable and a single 'Factor' variable; this analysis must be run before the next step can be undertaken. Step 2: *Analysis → Analysis of Variance (ANOVA) → Pairwise Comparisons* ... and choose the desired 'Available effects' variable and click the '*Add*' button to move it into the 'Groups' window; then choose 'Tukey' or 'Scheffe' as appropriate. Clicking on the '*?*' will open a Help windows that describes each of the available tests. This procedure can only be run after the *Estimate Model* ... procedure has been run. or Step 2: *Analysis → Analysis of Variance (ANOVA) → Hypothesis Test ...* and for the 'Hypothesis' option, choose 'Effects' and select the independent variable you want to set up contrasts for; then go to the '*Contrast'* tab, tick the 'Use contrast' box and select either 'Polynomial' or 'Custom' and enter your desired weights. This procedure can only be run after the *Estimate Model* ... procedure has been run.
STATGRAPHICS	*Compare → Multiple Samples → One-Way ANOVA* ... and choose the desired 'Dependent Variable' and the desired 'Factor' variable; hit '*OK*' and in the '*Tables and Graphs*' dialog box, select the 'Analysis Summary', 'Summary Tables', 'ANOVA Table' and 'Tables of Means' options and select any graphs you wish to see. To produce multiple comparison tests, also choose 'Multiple Range Tests'. By default, STATGRAPHICS produces the LSD ('Least Significant Difference') multiple comparisons test. To change the test it conducts, place the mouse inside the Multiple Range Test output window and right-click; choose 'Pane Options' and a list of other multiple comparison tests will pop up – choices include 'Tukey' and 'Scheffe' and you can set the desired alpha level. The StatAdvisor provides interpretive comments on all requested aspects of the analysis. STATGRAPHICS does not offer the capability to test defined *a priori* contrasts or orthogonal polynomials.
R Commander	*Statistics → Means → One-Way ANOVA...* and select the independent variable ('Groups') and dependent variable ('Response Variable') desired. Tick the 'Pairwise comparison of means' box to produce multiple

(continued)

Application	Procedures
	comparison tests as well as a graphical comparison of all pairwise group differences. **R** Commander only offers the Tukey HSD test for multiple comparisons. **R** Commander offers no options for conducting orthogonal contrast or orthogonal polynomial tests – these would have to be obtained using other functions in the main **R** program (see Field et al. 2012, pp. 443–454).

Procedure 7.9: Kruskal-Wallis Rank Test for Several Independent Groups

Classification Univariate; inferential; nonparametric.
Purpose To evaluate the difference between three or more independent groups of participants on the basis of ranked scores on a dependent variable rather than scores in some specific unit of measurement.
Measurement level The grouping variable can be defined by any level of measurement. The dependent measure may be initially obtained in terms of ordinal ranks or, more commonly, may be measured at the interval- or ratio-level but converted to ordinal ranks prior to analysis.

The *Kruskal-Wallis Rank test* is the nonparametric analog of one-way ANOVA (see *Procedure 7.6*). It evaluates the significance of the difference between three or more independent groups on the basis of mean ranks rather than mean scores in some other units of measurement. One use of this test is apparent when behavioural data are actually collected in the form of ranks.

> *For example* ... *We might ask the Managing Director of a company to rank all 35 of his/her managers in terms of performance using the numbers from 1 (the best manager) to 35 (the worst manager). We might then be interested in addressing the question of whether the managers from the Personnel Department, Finance Department, and Marketing Department differ in the average rank their Managing Director assigns them.*

Another frequently employed scenario for the Kruskal-Wallis Rank test emerges when the original measurements of a dependent variable are distributed in some highly distorted fashion so as to violate the basic assumptions of one-way ANOVA (normality and/or homogeneity of variance).

Suppose Maree wishes to compare quality control inspectors from the five different types of manufacturing companies (**company**) within the QCI database in terms of their inspection decision **speed**.

We have seen in earlier discussions (e.g. recall *Procedure 5.7*) that **speed** is a highly positively skewed variable (as are most time-based measures of human behaviour) which may be problematic enough to violate the normality assumption. Figure 7.14 shows the boxplots for the **speed** variable within the five types of **company**; the plot shows not only that the five groups may significantly differ but also that their data are generally positively skewed to different degrees in the different groups. Instead of conducting a potentially biased parametric one-way ANOVA F-test, Maree opts to conduct the Kruskal-Wallis Rank test.

The Kruskal-Wallis Rank test is conducted in three straightforward steps:

1. The scores in the original units of measurement (i.e. **speed**) are ranked from 1 to N (the total sample size) ignoring each participant's membership in the groups defined by the grouping variable **company.** Thus, **speed** scores would be ranked in order from 1 (fastest) to 111 (slowest; one inspector was missing a **speed** score).
2. The new set of ranks is then allocated back into the five groups whose membership is defined by the grouping variable.
3. A one-way ANOVA is essentially conducted using the ranks as data and the five groups are compared in terms of their average rank scores. The mean rank for the 22 inspectors in PC manufacturing companies is 73.59; the mean rank for 23 inspectors in Large Electrical Appliance manufacturing companies is 24.83; the mean rank for 21 inspectors in Small Electrical Appliance manufacturing companies is 55.10; the mean rank for 23 inspectors in Large Business Computer manufacturing companies is 79.65; and the mean rank for 21 inspectors in the Automotive manufacturing companies is 44.10. The Kruskal-Wallis Rank test then simply evaluates the significance of the differences between these five mean ranks.

The Kruskal-Wallis Rank test is based on the chi-squared statistic and the outcome for the test conducted on these data is a value for the chi-squared statistic of 44.22 with an associated p-value of less than .0001. This means we can conclude that quality control inspectors within the five types of manufacturing companies had significantly different means for ranked **speed** scores.

Since the ranking system was such that faster speeds garnered the lower ranks, visual inspection of Fig. 7.14 suggests that inspectors from both types of computer manufacturing company took substantially longer to make their inspection decisions than did inspectors from the remaining three types of company.

(continued)

Procedure 7.9: Kruskal-Wallis Rank Test for Several Independent Groups

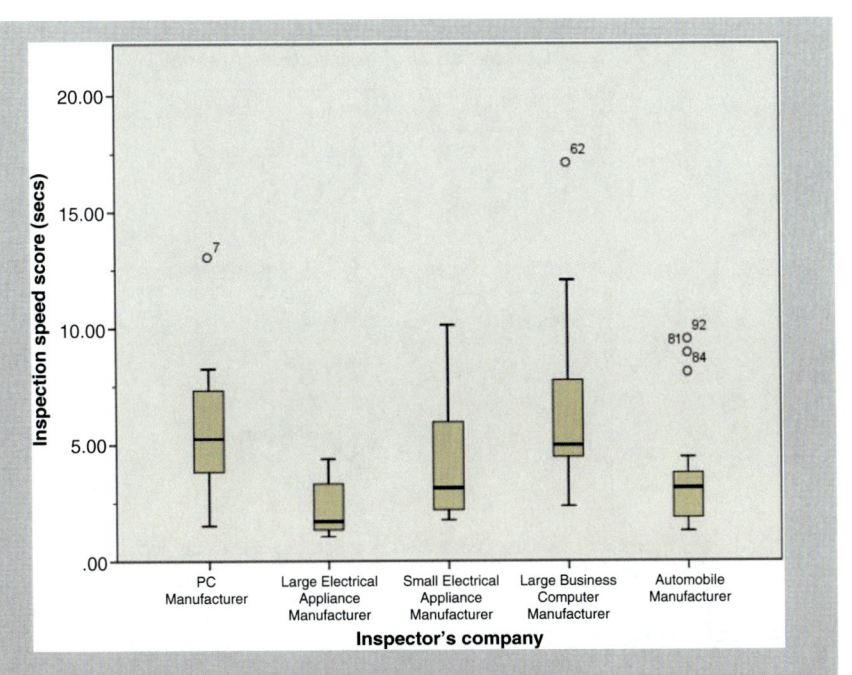

Fig. 7.14 Boxplots comparing the distributions of **speed** scores for quality control inspectors in the five types of manufacturing **company**

The Kruskal-Wallis Rank test, if statistically significant, simply tells us that the groups do differ in terms of average ranks and, hence, in terms of **speed** scores. It does not tell us exactly which groups actually differ; it only gives an indication that among the set of groups being compared there is at least one significant difference. Further data exploration with nonparametric rank multiple comparison tests, used in a posthoc fashion analogous to the multiple comparison tests described in *Procedure 7.8* (see Siegel and Castellan Jr 1988, pp. 213–215), must be done to isolate the precise group differences. Figure 7.15 shows how SPSS presents the results of such a pairwise multiple comparison test (in this case, Dunn's test) following the significant Kruskal-Wallis test comparing companies on their mean ranked decision **speed** scores.

The table at bottom of Fig. 7.15 summarises all the pairwise comparisons and highlights those comparisons that are significant with gray shading (focus on the **Adj. Sig.** column, which adjusts the *p*-values for the number of tests being conducted). The graph at the top of Fig. 7.15 displays the mean ranked decision **speed** scores for each of the five company types and summarises the outcomes that are numerically summarised in the table. Those companies linked by a lighter gray line are significantly different from each other; those connected by a black line are not significantly different from each other.

(continued)

Pairwise Comparisons of Inspector's company

Automobile Manufacturer
44.10

Small Electrical Appliance Manufacturer
55.10

PC Manufacturer
73.59

Large Electrical Appliance Manufacturer
24.83

Large Business Computer Manufacturer
79.65

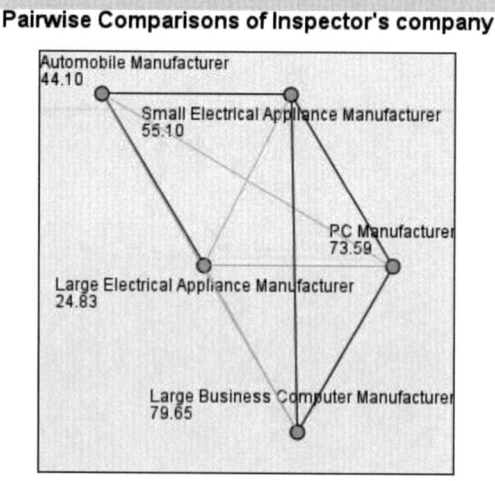

Each node shows the sample average rank of Inspector's company.

Sample1-Sample2	Test Statistic	Std. Error	Std. Test Statistic	Sig.	Adj.Sig.
Large Electrical Appliance Manufacturer-Automobile Manufacturer	-19.269	9.627	-2.001	.045	.453
Large Electrical Appliance Manufacturer-Small Electrical Appliance Manufacturer	-30.269	9.627	-3.144	.002	.017
Large Electrical Appliance Manufacturer-PC Manufacturer	48.765	9.512	5.126	.000	.000
Large Electrical Appliance Manufacturer-Large Business Computer Manufacturer	-54.826	9.406	-5.829	.000	.000
Automobile Manufacturer-Small Electrical Appliance Manufacturer	11.000	9.844	1.117	.264	1.000
Automobile Manufacturer-PC Manufacturer	29.496	9.731	3.031	.002	.024
Automobile Manufacturer-Large Business Computer Manufacturer	35.557	9.627	3.693	.000	.002
Small Electrical Appliance Manufacturer-PC Manufacturer	18.496	9.731	1.901	.057	.573
Small Electrical Appliance Manufacturer-Large Business Computer Manufacturer	-24.557	9.627	-2.551	.011	.107
PC Manufacturer-Large Business Computer Manufacturer	-6.061	9.512	-.637	.524	1.000

Each row tests the null hypothesis that the Sample 1 and Sample 2 distributions are the same.
Asymptotic significances (2-sided tests) are displayed. The significance level is .05.

Fig. 7.15 Pairwise multiple comparisons of ranked **speed** scores between pairs of companies

An additional useful strategy for a researcher to pursue is to compare the Kruskal-Wallis analysis result to the result that would have been obtained if a parametric ANOVA F-test were run on the same data in order to see if the conclusions would differ between the two tests. If the inference decision is different for the two tests or if the p-values differed dramatically between the two tests, then assumption violations have likely distorted the parametric test and the researcher should interpret the Kruskal-Wallis Rank test outcome. If the inference decision remains unchanged or the p-values are similar, then assumption violation was not severe enough to distort the test and the researcher should report the parametric test because it utilises more of the original information in the sample and it has access to more powerful multiple comparison strategies (see *Procedure 7.8*).

With respect to Maree's **company** comparison on **speed** scores, her Kruskal-Wallis Rank test p-value was less than .001. If she had conducted a parametric one-way ANOVA comparing companies on their untransformed **speed** scores, she would have found the resulting p-value for the F-test to also be less than .001. In this case, the positive skewness of **speed** scores apparently does not substantively distort the parametric outcome (likely because all groups were positively skewed).

Advantages

The Kruskal-Wallis Rank test is particularly useful when the assumptions of the parametric one-way ANOVA F-test cannot be or are not reasonably fulfilled by the data. It permits analysis of such data by reducing the data in measurement complexity to the ordinal level and then conducting the test. The test is easy to compute and interpret and requires no underlying assumptions about the distribution of the data which need to be satisfied. The Kruskal-Wallis Rank test also provides a simultaneous test for a significant difference among the medians of three or more independent group conditions. The Kruskal-Wallis Rank test as well as its nonparametric rank multiple comparison test can easily compare groups of unequal sizes.

Disadvantages

Unless the behavioural data rather severely violate ANOVA F-test assumptions, it is usually better to use the parametric test rather than the Kruskal-Wallis Rank Test. The reason is that when assumptions are even moderately violated, the F-test is a more powerful test for a given sample of data than the Kruskal-Wallis Rank test; that is, it is more likely to detect group differences if they really exist. However, if the violations are serious, then the Kruskal-Wallis Rank test becomes the more powerful test. Also, since the original data, in many applications, are converted to ranks, some information is necessarily lost. Such a loss can hamper meaningful interpretation of the observed group differences.

Where Is This Procedure Useful?

The Kruskal-Wallis Rank test can be applied in any experimental, quasi-experimental, or survey research investigation where comparison of three or more independent groups of participants is of interest and parametric assumptions are strongly violated. Many software packages offer specific tests for assumption violation which renders the choice of analytical technique easier to make. If the data are obtained in the form of ranks from the start, then the Kruskal-Wallis Rank test is automatically the test of choice. As with the parametric F-test, the statistical conclusion offered by the Kruskal-Wallis Rank test relates solely to observed group differences. Whether or not a causal conclusion is warranted (i.e. that the variation in the definition of the various groups is what caused the significant difference in the group mean ranks) is a separate decision which must be made in light of the quality of one's research design and methodological procedures.

Software Procedures

Application	Procedures
SPSS	*Analyze* → *Nonparametric Tests* → *Independent Samples* . . . and under the '*Objective*' tab, choose 'Customize analysis'; under the '*Fields*' tab, choose your desired independent variable ('Groups') and dependent variable ('Test Fields'); under the ;'*Settings*' tab, choose 'Customize tests' and tick the 'Kruskal-Wallis 1-Way ANOVA (k samples)' box and in the 'Multiple Comparisons' drop-down menu, choose 'All pairwise'. When you run the test, only a brief overall summary table for the analysis appears in the output window. You must double-click on this summary table to open up the '*Model Viewer*' window to see more detailed results (right-side of the window that opens). At the bottom of the right-hand '*Model Viewer* 'window, there is a drop-down menu labelled 'View' where you can select which analysis view you want to look at: 'Independent Samples Test View' or 'Pairwise Comparisons' view. The version of multiple comparison test reported by SPSS is called Dunn's test.
NCSS	*Analysis* → *Nonparametric* → *Kruskal-Wallis Test (ANOVA)* and choose the independent grouping variable ('Factor Variable') and dependent variable ('Response Variable(s)') accordingly and on the '*Reports*' tab, make sure that 'Kruskal-Wallis Report' is ticked and under 'Multiple Comparisons', make sure that 'Kruskal-Wallis Z-test (Dunn's Test)' is ticked.

(continued)

Application	Procedures
SYSTAT	*Analyze → Nonparametric Tests → Kruskal-Wallis* ... and choose the 'Grouping variable' and the dependent variable as a 'Selected variable(s)'; you can also select which type of pairwise multiple comparison test you want reported (the default is the 'Dwass-Steel-Chritchlow-Fligner Test' which is similar to what SPSS reports as Dunn's test). Strangely, SYSTAT does not report the mean rank for each group; instead it reports the group sample size ('Count') and the 'Rank Sum'. To get the mean rank for each group, simply divide the 'Rank Sum' for that group by the 'Count' for that group.
STATGRAPHICS	*Compare → Multiple Sample → Multiple-Sample Comparisons...* and choose the 'Input' as 'Data and Code Columns', then choose the 'Level Code' variable (= independent variable) and the 'Data' (= dependent variable); hit '*OK*' and in the '*Tables and Graphs*' dialog box, tick what analyses and graphs you want to see ('Kruskal-Wallis test'; ticking 'Box-and-Whisker Plot' will produce a boxplot comparison similar to Fig. 7.14). The StatAdvisor will provide detailed interpretive comments about the analysis. STATGRAPHICS only provides the overall Kruskal-Wallis test result; multiple comparisons are not an option offered. However, if you ensure the Box-and-Whisker plot is produced, you can make visual multiple comparisons. Right-click on the resulting boxplot graph, choose 'Pane Options' and tick the option labelled 'Median Notch' in the 'Features' section, the boxplot will be redrawn with notches (based on 95% confidence intervals (see *Procedure 8.3*), which corresponds to an alpha level of .05). If the notches for the boxplots for any two groups do not overlap each other on the graph, you can infer that the two groups are significantly different in their medians or mean ranks.
R Commander	*Statistics → Nonparametric tests → Kruskal-Wallis test...* and choose the variable to define the 'Groups' and the 'Response variable' (= dependent variable); **R** Commander does not offer an option to compute posthoc multiple comparisons for a Kruskal-Wallis analysis. However, Field et al. (2012, pp. 681–684) describe how a multiple comparison test for a Kruskal-Wallis analysis can be accomplished using commands in the main **R** program.

Procedure 7.10: Factorial Between-Groups ANOVA

Classification	Univariate; inferential; parametric.
Purpose	To compare several unrelated groups of participants, systematically classified with respect to two or more distinct grouping (independent) variables, on a single dependent measure.
Measurement level	Any level of measurement can be used to define the grouping variables. The dependent measure must be measured at the interval- or ratio-level.

Factorial between-groups ANOVA provides tests for significant differences on a single dependent measure among several unrelated groups, each group being jointly defined by two or more distinct independent variables. [Note that if you have several dependent measures to be analysed in the context of the same ANOVA design, then multivariate analysis of variance is the way to proceed (see *Procedure 7.16*).] In factorial between-groups ANOVA there is more than one classification dimension (e.g. sex of participant, occupation, experimental treatment versus control, experience level) which serves to define the nature of the groups to be compared (see Field 2018, ch. 14). Factorial between-groups ANOVA is also known as *n-way ANOVA* where the '*n*' in *n*-way refers to the number of classification dimensions (i.e. independent variables) used. A 2-way factorial ANOVA design, where each independent variable had only two levels, would thus constitute the simplest type of *n*-way factorial ANOVA design.

In a factorial between-groups ANOVA, each category (or level) of each grouping dimension is paired with every category (or level) of all other grouping dimensions. This way of forming all possible combinations of categories for the grouping dimensions yields what is termed a *completely-crossed factorial between-groups ANOVA design*. Each combination of categories from the *n* grouping dimensions is called a **cell** (recall the usage of the term 'cell' in earlier discussions of contingency tables in *Procedure 6.2* and *Procedure 7.1*). The general rule of thumb for calculating the number of cells in any completely crossed factorial between-groups ANOVA design is simply to multiply the number of levels of each independent or grouping variable. Thus, a 2 by 2 completely crossed factorial design has 2×2 or 4 cells. A 6 by 3 by 2 completely crossed factorial design would have $6 \times 3 \times 2$ or 36 cells.

The term *between-groups* is used to signal that there are distinct samples of participants in the cells of the design; all comparisons are thus between different groups of participants. If the same participants or matched sets of participants are used in all of the levels of at least one of the independent variables (e.g. as would be the case with a pretest-posttest control groups design—recall *Fundamental concept VII*), the design changes to a *within-groups factorial design* or *repeated measures design* (see *Procedure 7.11*).

> A 2-way between-groups factorial ANOVA in the context of the QCI database might involve Maree considering two classification dimensions for comparing the inspection **accuracy** of quality control inspectors: the type of company they work for (**company** having five different categories) and the level of education they have achieved (**educlev** having two different categories). This design would be characterised as a 5 (**company**) by 2 (**educlev**) completely-crossed factorial between-groups ANOVA design. There are 10 separate cells in this ANOVA design and each cell has a different sample of inspectors.

(continued)

Procedure 7.10: Factorial Between-Groups ANOVA

Table 7.13 The design layout for the 5 (**company**) by 2 (**educlev**) completely crossed between-groups factorial ANOVA

				EducLev		Company Statistics
				1 High School Only	2 Tertiary Qualified	
Company	1 PC Manufacturer	accuracy	Mean	87.75	84.83	86.95
			Std Dev	9.72	9.54	9.53
			Valid N	16	6	22
	2 Large Electrical Appliance Manufacturer	accuracy	Mean	75.57	78.86	77.76
			Std Dev	8.89	9.40	9.15
			Valid N	7	14	21
	3 Small Electrical Appliance Manufacturer	accuracy	Mean	82.07	79.83	81.43
			Std Dev	5.66	9.43	6.76
			Valid N	15	6	21
	4 Large Business Computer Manufacturer	accuracy	Mean	89.86	87.11	88.78
			Std Dev	3.55	3.22	3.62
			Valid N	14	9	23
	5 Automobile Manufacturer	accuracy	Mean	77.25	75.35	75.71
			Std Dev	8.34	7.62	7.58
			Valid N	4	17	21
EducLev Statistics		accuracy	Mean	84.48	79.94	82.30
			Std Dev	8.68	8.90	9.04
			Valid N	56	52	108

Note: This table was created using SPSS Custom Tables and formatted using Excel

The layout of the cells defined by this design is depicted in Table 7.13. Table 7.13 shows basic descriptive statistics (valid (i.e. non-missing) *N*, mean **accuracy** scores and standard deviations) for inspectors in the design. Note that Table 7.13 shows the different ways that mean **accuracy** scores and standard deviations can be computed, which relates to the different types of hypothesis tests that can be conducted to compare the means. Thus, you can see mean **accuracy** scores and standard deviations inside each cell of the design (white area) as well as mean **accuracy** scores and standard deviations along the bottom row and down the final column (lightly shaded) of the table. Descriptive statistics for the entire sample in the analysis are shown in the darker shaded cell in the bottom right corner of the table.

In *n*-way factorial ANOVA, several *F*-test statistics are produced to help evaluate group differences. Each *F*-test in a factorial ANOVA essentially compares the variability of a specific set of means to the random variability of **accuracy** scores within the cells, assuming a null hypothesis of no effect in the design.

That there is such random variability in ANOVA designs can be seen in Table 7.13 by looking at the standard deviations inside each of the white cells.

None of these standard deviations is zero, which tells us that even inspectors, classified identically according to the two independent variables will have differing scores, largely because of uncontrolled individual differences (e.g. High School only-educated inspectors working in PC Manufacturing companies have a standard deviation of 9.72). Any ANOVA F-test looks for group-level differences to be much greater than would be expected simply because of individual differences between inspectors.

Let's consider more closely the types of F-tests that emerge from a 2-way ANOVA. We'll focus our discussion on the analysis of the 2-way ANOVA design in Table 7.13, but the principles we discuss will generalise to n-way ANOVAs. In a 2-way ANOVA, we can evaluate group differences in three separate ways, each using a separate F-test. All three tests are necessary in helping us fully understand group differences.

Firstly, we can evaluate group differences based on the **company** dimension by itself (this is called the *main effect* of **company**). Here, we ignore **educlev** dimension for classifying inspectors, and concentrate on comparing groups defined only by type of **company**; the mean **accuracy** scores, standard deviations and Valid Ns for these five groups, ignoring educational level appear in the final lightly shaded column of statistics in Table 7.13. This main effect F-test will tell us if there are one or more significant differences among the five **company** means.

Secondly, we can evaluate group differences based on the education level dimension alone (the **educlev** *main effect*, ignoring the **company** dimension); the mean **accuracy** scores, standard deviations and Valid Ns for these two groups, ignoring type of **company**, appear in the lightly shaded bottom row of statistics in Table 7.13. This second main effect F-test would tell us if there is a significant difference between the two **educlev** means.

Finally, we can evaluate group differences based on both the **company** and **educlev** dimensions considered jointly. This third test called a test of the *interaction* effect. This interaction effect F-test would tell us if there are one or more significant differences among the 10 **company** by **educlev** cell means over and above what the main effects can tell us (thus, the interaction tests for any systematic patterning among the means in the white cells of Table 7.13). Any interaction involving two independent variables is called a *two-way interaction*.

To summarise, using our example, analysing a 2-way factorial ANOVA design will produce three separate F-tests: one for the **company** main effect; one for the **educlev** main effect; and one for the interaction of **company** and **educlev** (called the **company** by **educlev** (or, more simply **company**∗**educlev**) interaction). In more complex n-way ANOVAs, there are more F-tests necessary to completely evaluate group differences. For example, in a 3-way ANOVA (let's label the three dimensions generically as A, B, and C, but in the QCI database, they could easily be **company**, **educlev** and **gender**), seven F-tests are needed to completely analyse the design: A main effect; B main effect; C main effect; A by B interaction; A by C interaction; B by C interaction; and the A by B by C interaction (this last is called a *three-way interaction*). This logic extends to any number of 'ways' for a factorial

ANOVA design. In a 4-way ANOVA, there would be 15 possible *F-tests:* four main effects, six 2-way interactions, four 3-way interactions and one 4-way interaction.

Interpretation of any *n*-way ANOVA should start with the most complex interaction (in the case of Table 7.13, this would be the **company** by **educlev** interaction) and work its way toward interpretation of the simpler main effects. If a significant interaction is encountered, then interpretation of that interaction is imperative. The interpretation of a significant interaction is a bit tricky and is an exercise to which we will devote some time in *Fundamental concept IX*.

Table 7.14 shows the results, produced by SPSS, for Maree's analysis of the 2-way ANOVA design in Table 7.13. Note that the table reports not only the three *F*-tests, but also the 'Partial Eta Squared' associated with each effect (recall *Procedure 7.7*). Maree would start her interpretations with the 2-way interaction and then move on to consider the main effects. In terms of how programs like SPSS present the outcomes, this generally means working from the bottom of the summary table upward.

The **educlev** * **company** interaction *F*-test has a value of 0.534 with an associated *p*-value of .711 which is not significant. This essentially says that Maree can safely move on to consider the two main effects and tell a simpler and unambiguous story about their potential influences on inspection **accuracy** scores if such influences can be shown to be statistically significant.

The interpretation of any significant factorial ANOVA *main effect* is straight-forward: if the *F*-test is significant, the groups differ across the mean values of that specific grouping dimension. Remember though, from our discussion in *Procedure 7.6*, that where three or more groups are involved, a significant *F*-test only signals the existence of group differences. It does not tell us where those differences are located among the groups (the next logical step would be to proceed with a posthoc multiple comparison test—see *Procedure 7.8*).

Table 7.14 Results of the SPSS analysis of mean **accuracy** differences in the completely crossed between-groups factorial ANOVA design shown in Table 7.13

Tests of Between-Subjects Effects

Dependent Variable: accuracy

Source	Type III Sum of Squares	df	Mean Square	F	Sig.	Partial Eta Squared
Corrected Model	2964.254[a]	9	329.362	5.592	.000	.339
company	2095.238	4	523.810	8.893	.000	.266
educlev	36.250	1	36.250	.615	.435	.006
company * educlev	125.871	4	31.468	.534	.711	.021
Error	5772.264	98	58.901			
Corrected Total	8736.519	107				

a. R Squared = .339 (Adjusted R Squared = .279)

Note: The table has had non-essential details edited out

(continued)

> For the analysis in Table 7.14, the *F*-test for **educlev** has a value of .615 with an associated *p*-value of .435 which is not significant and which tells Maree that there is no significant difference between mean **accuracy** scores for High School-only-educated inspectors and Tertiary-qualified inspectors.
> Finally, the *F*-test for **company** has a value of 8.893 with an associated *p*-value less than .001 which is significant, and which tells Maree that there is at least one significant difference between mean **accuracy** scores for the different types of **company**. The strength of this effect can be measured by partial eta-squared which has a value for **company** of .266 meaning that 26.6% of the variability in **accuracy** scores can be explained by knowing of what type of **company** an inspector works for. Maree could use the Tukey HSD multiple comparison test (recall *Procedure 7.8*) to isolate more precisely which companies differed from which other companies.
> If Maree were to conduct such a multiple comparison test, which can be requested in SPSS, she would find, for example, that inspection decisions in the PC and Large Business Computer manufacturing companies were significantly more accurate than those made in either the Large Electrical Appliance or Automotive manufacturing companies.

Note that *n*-way ANOVA is closely related to multiple linear regression (see *Procedure 6.3* and *Procedure 7.13*) within the context of the *general linear model* (see, for example, Cohen et al. 2003, or Judd et al. 2017; also recall *Fundamental concept VI*). We can see this logic in play if we look more closely at Table 7.14. The table includes three rows, labelled: 'Corrected Model', 'Error' and 'Corrected Total'. These correspond roughly to reflections of the behaviour of ***MODEL*, ERROR** and **DATA** in the general linear model conceptual representation in *Fundamental concept VI*. The table has a footnote that reports *R*-square is .339 – this means that 33.9% of the variability of the **DATA** (**accuracy** scores in this case) can be explained by the predictors in the ***MODEL*** (which would be all of the predictors needed to represent the two main effects and the interaction). ERROR accounts for the remaining 76.1% of the variability in the **DATA**. The partial eta-squared values and associated *F*-tests allow us to understand what each ANOVA design effect uniquely explains in the **DATA** over and above all the other effects in the design. Thus, technically, **company** explains 26.6% of the variability in **accuracy** scores over and above what the 2-way interaction and the **educlev** main effect can explain.

Advantages

Factorial between-groups ANOVA permits the simultaneous assessment of the influence of several grouping variables (i.e. independent variables) on a single dependent measure. This is both in terms of variable main effects and in terms of interactions between various combinations of those variables. It permits a researcher to have greater experimental control over variables by explicitly testing for their effects in the ANOVA. Very precise tests of both simple and complex grouping variable effects can be teased out using ANOVAs. In many cases, an appropriate factorial between-groups ANOVA design that includes relevant grouping variables (even if some are essentially nuisance or extraneous variables included to control explicitly for their influence—a 'control by design' strategy, refer back to *Fundamental concept VII*) may strengthen the researcher's claims to have isolated specific causal factors.

Although many authors advise that the best ANOVA results are obtained when all cells have an equal number of participants in them, modern-day computing algorithms have rendered this requirement essentially obsolete. Unequal cells sizes (as we have in Table 7.13) are automatically handled by most statistical packages using a semi-partialling process (recall *Fundamental concept IV*) since the effect of unequal cell sizes is to produce correlations between the various ANOVA effects, which the semi-partialling process can control for (see the discussion in Cohen et al. 2003, ch. 3, 5). In Table 7.14, this process is reflected in the label 'Type III Sum of Squares', to signal that every ANOVA F-test has been automatically adjusted, by SPSS, for any correlations with any other design effects.

Disadvantages

The same assumptions apply for n-way factorial between-groups ANOVA as for one-way ANOVA (see *Procedure 7.6*) and if these assumptions are not reasonably met, ANOVA may not be the statistical technique to use. Interpretation becomes much more complicated in factorial ANOVAs where three or more grouping variables are involved, particularly if one or more interaction F-tests emerge as significant (see *Fundamental concept IX*). The more independent variables included in an ANOVA design, the more rapidly the number of F-tests grows, creating an alpha error inflation problem. Additionally, if the number of grouping dimensions is large ($n = 5$ or more grouping variables) or if the number of categories for the different classification dimensions increases, then a very large sample of participants will be required to implement and successfully evaluate a fully-crossed factorial between-groups design.

> If a behavioural researcher has three grouping dimensions or independent variables in his/her factorial between-groups ANOVA design where one variable has three classification categories, one has four classification categories, and the third has five classification categories, then the total number of separate cells to which participants must be assigned, or in which participants must be measured, is $3 \times 4 \times 5$ or 60 cells.

(continued)

> A rule of thumb for achieving stable and reliable statistical results from an n-way ANOVA is to ensure that there are at least 10 (Keppel and Wickens 2004), but preferably 15 or more, participants per cell. Thus, a minimum of 10×60 or 600, but preferably 900 participants would be needed to run the study—a number that the researcher may or may not have access to or have the resources to obtain. If the design is run as a true experiment, then the researcher can take steps to ensure there are enough participants per cell through the process of random assignment. However, in a cross-sectional survey design, where measured demographic characteristics define the independent variables, there may be too few or even no participants in some cells depending upon who returns a questionnaire, making the ANOVA model more difficult to statistically estimate.

There may be types of research where a completely crossed factorial design makes no sense because of a hierarchical relationship between two or more of the independent variables. Such designs are called *nested ANOVA designs* or *hierarchical ANOVA designs* (see Kirk 2013, ch. 11, and Keppel and Wickens 2004, ch. 25, for analytical details for such designs). In such designs, a factorial ANOVA analysis is inappropriate and may lead to misleading conclusions.

> A marketing researcher may be interested in running an experiment to look for differences in perceptions of product quality and brand equity for different brands of products in three product categories. She may be interested in the product categories of Blu-ray/DVD players, jeans and cars. For Blu-ray/DVD players, she tests the Sony and Panasonic brands; for jeans, she tests the Just Jeans and Jeanswest brands and for cars, she tests the Ford and Toyota brands. In such a design, it makes no sense to consider the two levels of brands to be crossed with three products in a factorial design because the brands differ within (or are 'nested within') each of the product categories. This design would require an ANOVA that treated brands as nested within product category.
>
> In educational research, a common problem faced by researchers into classroom behaviour is that classrooms are nested within schools, such that it may be unreasonable to expect the Year 4 and Year 6 classes in School A to be identical to the Year 4 and Year 6 classes in School B (due to teacher and curricular differences and perhaps socio-economic factors associated with the neighbourhood each school is located in). This could mean that main effect comparisons between year levels would be problematic to interpret, unless their nesting within schools was acknowledged in the analysis.

Where Is This Procedure Useful?

Factorial between-groups ANOVAs are most effectively employed in experimental and quasi-experimental designs where several grouping variables might be of interest. Factorial designs are particularly popular in psychological, industrial-organisational and educational research where the researcher has randomly assigned participants to groups defined by explicit experimental manipulations or treatments. However, these designs have also enjoyed some major use in organisational and marketing research as well. In large social surveys, some comparisons between various demographic variable classifications on some scale of interest (perhaps an attitude scale) can be done with n-way ANOVA, but with a corresponding reduction in internal validity (i.e. loss of capacity to infer cause-and-effect relationships). As there is a lot of jargon that accompanies the use of experimental and quesi-experimental ANOVA designs, you may find it useful to look at Cooksey and McDonald 2019, pp. 676–677, entitled *Appendix: Clarifying Experimental/Quasi-experimental Design Jargon*.

Market researchers have frequently employed a special type of n-way ANOVA methodology to gather data on consumer preferences and decision making. This special ANOVA methodology goes by the name *conjoint measurement* or *conjoint analysis* (see *Procedure 9.4*). One important feature of conjoint methods is that the completely crossed factorial requirement is relaxed in order to pursue designs employing large numbers of grouping variables (e.g. 11 or more grouping variables are not uncommon in conjoint analysis). Such designs are called fractional factorial designs because only a certain fraction of the combinations of independent variables are required. However, such designs require more sophisticated ANOVA procedures, using the general linear model. Another typical feature of conjoint analysis is that aspects of the design may involve repeated measurements of the same sample of participants (see *Procedure 7.11*).

Software Procedures

Application	Procedures
SPSS	*Analyze* → *General Linear Model* → *Univariate* ... and select your 'Dependent Variable' and the different grouping or independent variables ('Fixed Factor(s)') of interest; click the '*Post Hoc* ...' button to select a multiple comparison test for any main effect involving three or more levels and/or click '*Contrast* ...' to set up any contrasts of interest for particular main effects; click '*Options* ...' to indicate you want to see 'Descriptive Statistics' and 'Estimates of Effect Size' and indicate that you want to see the 'Estimated Marginal Means' for all of the effects in your ANOVA design. If you press the '*Plots*' button, you can instruct SPSS to graph specific interactions (only up to a 3-way interaction) to aid in their interpretation.

(continued)

Application	Procedures
NCSS	*Analysis → Analysis of Variance (ANOVA) → General Linear Models (GLM)* and choose your dependent variable (called 'Response Variable") and two or more independent variables (called 'Factor Variables'); under the *Comparisons* drop-down menu for each factor, choose 'Polynomial' if desired or set up the *a priori* contrasts you want to test; under the *'Reports'* tab, choose the multiple comparison test you want (if you want the Tukey HSD test, choose the 'Tukey-Kramar Test'); under the *'Plots'* tab, indicate which main effect and 2-way interaction plots you want to see.
SYSTAT	*Analysis → Analysis of Variance → Estimate Model ...* and choose your dependent variable ('Dependent(s)') and two or more 'Factor(s)' variables and select your desired coding scheme, 'Effect' or 'Dummy'. *Analysis → Analysis of Variance → Hypothesis Test ...* and for the *Hypothesis* option, choose 'Effects' and select the independent variable you want to set up contrasts for; then go to the *'Contrast'* tab and select either 'Polynomial' or 'Custom' and enter your desired weights. This procedure can only be run after the *Estimate Model ...* procedure has been run. *Analysis → Analysis of Variance → Pairwise Comparisons...* and choose the 'Effect' you want to analyse and choose the multiple comparison tests you want to see. This procedure can only be run after the *Estimate Model ...* procedure has been run.
STATGRAPHICS	*Compare → Analysis of Variance → Multifactor ANOVA ...* and choose your 'Dependent Variable' and the desired 'Factors' variables; hit *'OK'* and the *'Options'* window will open – indicate the maximum order of interaction you want to test (e.g. for a 2-way ANOVA, this would be '2'); hit *'OK'* and in the *'Tables and Graphs'* window, tick all options in the 'Tables' section and make sure you tick 'Interaction plot' in the 'Graphs' section.
R Commander	*Statistics → Means → Multi-way ANOVA ...* and choose your desired 'Response Variable' (dependent variable) and 'Factors' variables. **R** Commander does not offer options to conduct posthoc multiple comparison tests or planned comparisons – these would have to be managed through normal **R** program commands (see Field et al. 2012, pp. 518–520, 528–530). *Graphs → Plot of means ...* can be used to produce an interaction plot; choose your desired 'Response Variable' (dependent variable) and 'Factors' variables (should be the same as for the ANOVA analysis) and choose what type of error bars you want plotted and what alpha level should be used.

Fundamental Concept IX: The Concept of Interaction

Interaction in ANOVA Designs

Designing and conducting factorial ANOVAs involving more than one grouping variable or way of classifying participants (recall *Procedure 7.10*) leads directly to the possibility of several distinct grouping variables jointly influencing the

Fundamental Concept IX: The Concept of Interaction

variability of group means for a dependent variable. This joint influence is what constitutes an interaction. A significant interaction from a n-way ANOVA tells us that groups defined using one grouping dimension behave differently depending upon which category of the other grouping dimensions involved are being considered.

In a significant 2-way ANOVA interaction, the effect of the first grouping dimension on the dependent variable is different for each category of the second grouping dimension (and vice versa). The most efficient and effective way to interpret any interaction, whether it is significant or not, is by using a graph of the interaction means. The general structure of a 2-way interaction graph is to use the Y-axis to anchor the dependent variable mean scores; select one grouping variable from the ANOVA which is involved in the interaction to anchor the X-axis of the graph (usually, the grouping variable having the most categories); and plot one line of mean scores for each category of the remaining grouping variable involved in the interaction. Generally speaking, if a 2-way interaction is non-significant, then the graph should reveal parallel (or roughly parallel) lines indicating a consistent pattern of responses for one grouping variable across the categories of the other grouping variable.

To focus our discussion, consider the simplest type of 2-way ANOVA design involving two grouping variables, call them factors **A** and **B**, each of which have only two categories (thus we would have a 2 by 2 factorial ANOVA design). Speaking hypothetically, the graphs in Fig. 7.16 provide three different idealised theoretical representations of what a non-significant interaction might look like with respect to these two variables for a particular dependent variable, call it **Y**. Parallel or nearly parallel lines, indicating a non-significant interaction effect, are the order of the day in each of the three graphs in Fig. 7.16. A slope in the lines would signal a likely significant main effect due to the grouping variable **A** anchoring the X-axis and substantial separation between lines would indicate a likely significant main effect due to the grouping variable **B** defining the categories for the different lines. Thus, a non-significant interaction signals that the main effects are consistent in how they play out, irrespective of what the other independent variable is doing.

- Graph 1 in Fig. 7.16 shows a pattern where the interaction is non-significant, the **B** main effect is non-significant, and the **A** main effect is significant. *A2* participants show a significantly higher overall **Y** mean compared to *A1* participants, but *B1* and *B2* participants do not differ significantly.
- Graph 2 in Fig. 7.16 shows a pattern where the interaction is non-significant, the **B** main effect is significant, and the **A** main effect is non-significant. *B2* participants show significantly higher **Y** means than *B1* participants, but *A1* and *A2* participants do not differ significantly.
- Graph 3 in Fig. 7.16 shows a pattern where the interaction is non-significant, the **B** main effect is significant, and the **A** main effect is significant. *B2* participants

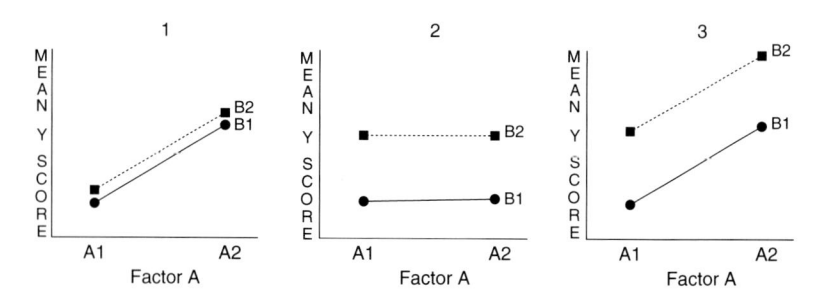

Fig. 7.16 Idealised graphical representations of different types of non-significant interaction patterns

show significantly higher **Y** means than *B1* participants and *A2* participants show significantly higher **Y** means than *A1* participants.

A significant 2-way interaction indicates inconsistency in group responses across the various categories of the grouping dimensions. Such inconsistency tells us that the main effects for each classification dimension involved in the interaction change within the different categories of the other dimension. Thus, if an interaction is significant, one must be cautious in interpreting any significant main effects for the grouping dimensions involved in that interaction. This is the main reason that interpretation of a factorial ANOVA summary table should progress from the bottom of the table (starting with the most complex interaction effect) upward. If an interaction is significant, the lines will, in general, be non-parallel (they may, in fact, dramatically cross each other). Note that posthoc multiple comparison tests, as discussed in *Procedure 7.8*, can be used to further explore the precise nature of group differences indicated in a significant interaction.

The graphs in Fig. 7.17 all illustrate possible significant interaction patterns (although the likelihood of significance would generally be rather less for the pattern in Graph 3).

- Graph 1 in Fig. 7.17 shows a pattern where the interaction is significant, the **B** main effect is non-significant, and the **A** main effect is non-significant. This is what statisticians term a *classic* or *disordinal* interaction pattern where the lines physically cross on the graph. In this specific type of interaction, the fact that the main effects are non-significant obscures the real pattern: the effect of one variable depends upon which category of the other variable you choose to examine.

  Overall, *A1* participants do not significantly differ from *A2* participants (you can tell this by visually 'averaging' the *B1* and *B2* points for each of *A1* and *A2*) and *B1* and *B2* inspectors also do not differ significantly. However, if we look only at

Fundamental Concept IX: The Concept of Interaction

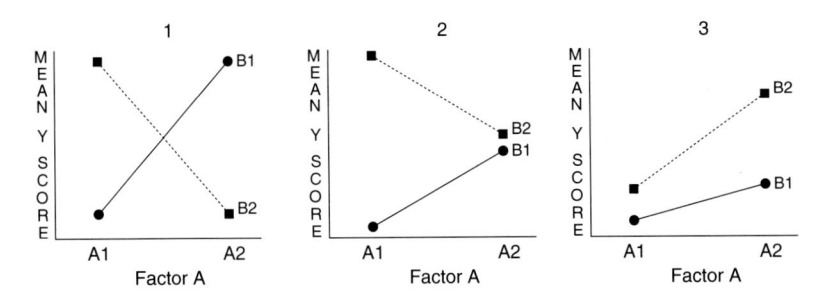

Fig. 7.17 Idealised graphical representations of different types of significant interaction patterns

the *B1* participants, it is clear that there is a strong difference between the **A** categories (*A2* **Y** mean is higher than the *A1* **Y** mean). The same is true if we focus on the *B2* participants, but the direction of the difference is directly opposed to the direction for the *B1* difference. Thus, we would tell a completely reversed story for the effects of **A** and **B** on **Y** depending upon which level of **A** or **B** we chose to look at. It is for this reason that any significant interaction should be interpreted before main effects are interpreted.

- Graph 2 in Fig. 7.17 shows a pattern where the interaction is significant, the **B** main effect is significant, and the **A** main effect is non-significant. The lines don't physically cross on the graph but the trend in results still suggests the presence of a strong interaction. *B2* participants show a significantly higher overall **Y** mean compared to *B1* participants, but *A1* participants and *A2* participants do not differ significantly overall. However, like Graph 1 in Fig. 7.17, we tell a diametrically opposed story when we focus on *B1* participants as opposed to *B2* participants.

Graph 3 in Fig. 7.17 shows a pattern where the interaction might be significant depending upon error variability in the sample, the **B** main effect would likely be significant, and the **A** main effect would also likely be significant. This type of interaction is termed *ordinal* by statisticians because the lines do not cross on the graph and the general trend in the data pattern remains relatively stable across categories. *B2* participants are likely to show a significantly higher **Y** mean compared to *B1* participants and *A2* participants are likely to show a significantly higher **Y** mean compared to *A1* participants. This interaction pattern tells us, for example, that the differences across the levels of the **A** factor are somewhat more dramatic for *B2* participants than for *B1* participants.

Suppose Maree is interested in conducting two separate 2-way factorial between-groups ANOVAs using **company** and **educlev** as independent variables and inspection decision **accuracy** and **log_speed** (correcting for the non-normality of **speed** scores – recall *Fundamental concept II*). After running each ANOVA, she compares the 2-way interaction graphs for the two analyses.

Figure 7.18a shows the graph of the 2-way interaction of **company** and **educlev** on **accuracy** scores (note that this outcome is from the 2-way ANOVA conducted in *Procedure 7.10* – the data for this interaction graph are the 10 cell means from Table 7.13). This graph illustrates a non-significant interaction because, with only one minor exception (Large Electrical Appliance manufacturing companies), the lines for the two **educlev** categories are largely parallel as we look from one type of **company** to another. This visual impression is backed up by the statistical evidence that showed a nonsignificant *F*-test (p-value $= .711$) for the interaction.

It appears that, irrespective of the type of company an inspector works for, having a Tertiary qualification tends to be associated with lower inspection decision accuracy levels relative to what High School-only educated inspectors achieve in the same companies. The one exception with respect to Large Electrical Appliance manufacturing companies does not represent a statistically reliable differential trend in the data. That is, the minor crossing of the lines for Large Electrical Appliance manufacturing companies is more likely to be the result of sampling fluctuations rather than a genuine trend.

Figure 7.18b shows the graph of the 2-way interaction of **company** and **educlev** on **log_speed** scores. This graph shows a significant interaction (e.g. non-parallel lines) largely because the lines either cross-over or diverge from each other for certain groups in the graph. This visual impression is backed up by the statistical evidence that shows a significant *F*-test for the interaction ($F(4, 98) = 3.056, p = .020$, partial $\eta^2 = .111$).

The significant interaction pattern can be interpreted as follows. High School Only-educated inspectors made substantially slower inspection decisions and Tertiary-Qualified inspectors made substantially faster inspection decisions for Automobile Manufacturers. The lines also crossed-over for inspectors from PC Manufacturers where High School Only-educated inspectors were much faster and Tertiary-Qualified inspectors were relatively slower. For the other three types of companies, High School Only-educated inspectors were relatively slower and Tertiary-Qualified inspectors were relatively faster or they were equal in the case of Large Business Computer Manufacturers.

(continued)

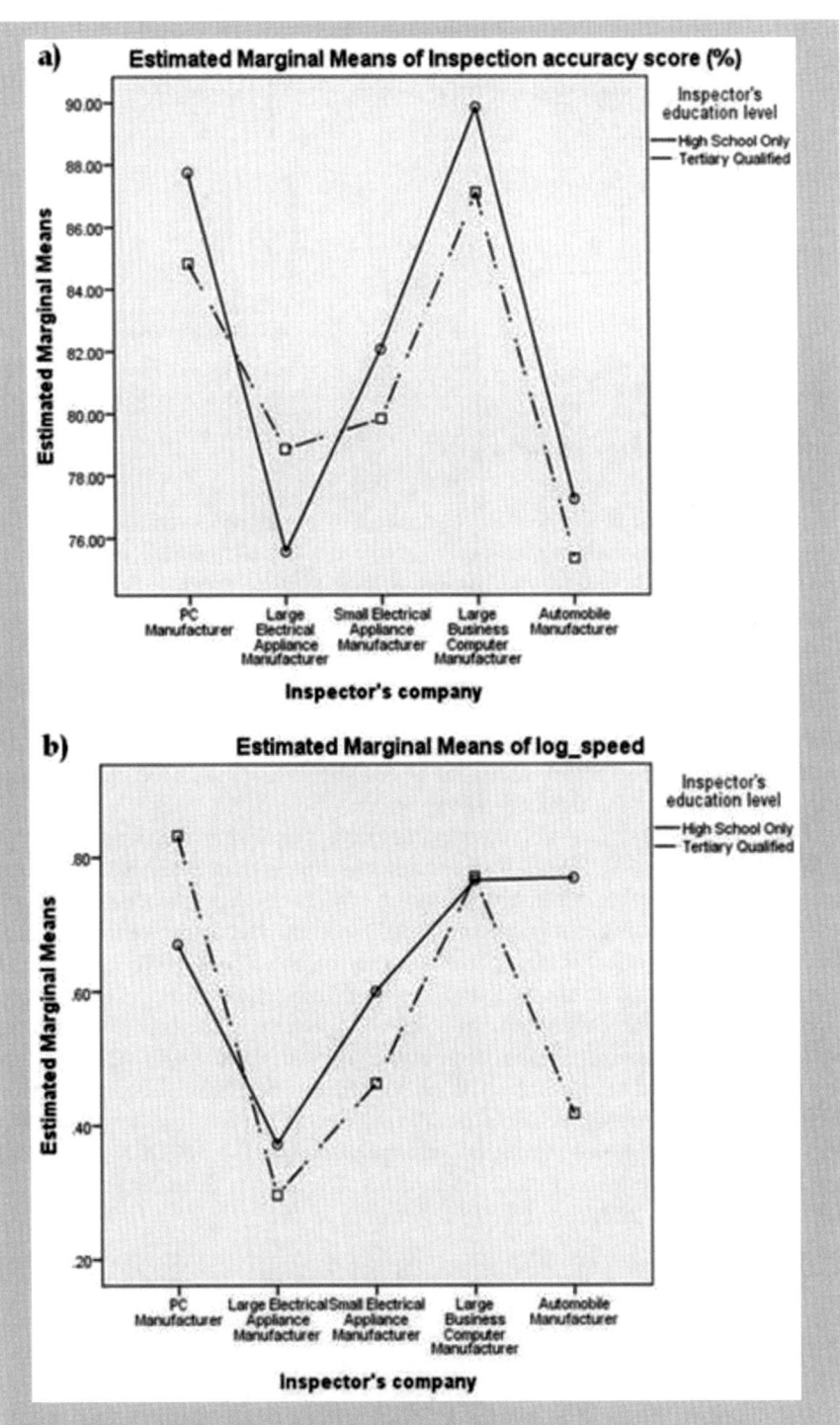

Fig. 7.18 Plots of the **company** by **educlev** 2-way interaction from the 2-way ANOVAs of inspector **accuracy** and **log_speed** scores from the QCI database
Note: Plots produced by the SPSS General Linear Model procedure.

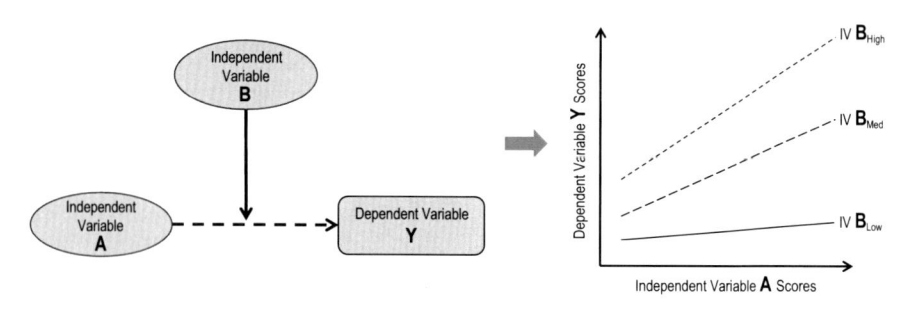

Fig. 7.19 Diagrammatic representation of a moderator effect

Interactions as Moderator Effects

There is a more general way to view the concept of an interaction consistent with the general linear model view of statistical analysis (recall *Fundamental concept VI*). In quantitative research design, the concept of a moderator effect is one where the relationship between an independent variable and a dependent variable is altered (i.e. *moderated*) by the behaviour of another independent variable. Figure 7.19 illustrates this concept diagrammatically using generic variables. Independent variable (IV) **B** moderates (solid line) the relationship (dashed line) between IV **A** and Dependent Variable **Y**. Graphically, the interaction might look like a set of lines with different slopes for different values of IV **B** (shown as Low, Medium and High values) as shown on the right side of Fig. 7.19.

However, this description of a moderator effect is really just the same as how we describe the nature of an ANOVA interaction: the effect of one independent variable on a dependent variable depends upon which level of another independent variable one examines (meaning that graph lines will be non-parallel as shown in Fig. 7.19). Thus, a moderator effect is simply a type of interaction effect (Hayes 2018, ch. 7; Miles and Shevlin 2001, ch. 7). Often, the label 'moderator effect' is reserved for interactions involving continuous variables or a mixture of categorical and continuous variables, whereas the label 'interaction' is often reserved for ANOVA-type interactions between categorical variables. Moderator effects are generally estimated using multiple regression models (see *Procedure 7.13*) because to represent a moderator effect, you need to include new predictors in the ***MODEL*** that are the products of two or more separate predictors involved in the moderating relationship (Miles and Shevlin 2001, pp. 174–187).

> For example, suppose Maree want to build a research ***MODEL*** using job satisfaction (**jobsat**) and working conditions (**workcond**) to predict quality inspection **accuracy**. However, she may also want to test for one possible moderator effect: **workcond** moderating the relationship between **jobsat** and

(continued)

accuracy (where poor working conditions might reduce the relationship between job satisfaction and accuracy whereas good working conditions might increase the relationship between job satisfaction and accuracy). To test this *MODEL*, Maree would need to recode both **workcond** and **jobsat** using effect-coding (recall *Fundamental concept VI*) and then create a new predictor by multiplying recoded **workcond** times recoded **jobsat** (after doing what is called 'centreing' – subtracting each respective predictor's mean from each inspector's scores on the two predictors – see Cohen et al. 2003, pp. 261–267; see also *Procedure 9.8*). She would then include that new predictor in the *MODEL* along with the centred versions of both the **jobsat** and **workcond** predictors.

Procedure 7.11: Repeated Measures ANOVA

Classification	Univariate (but can be treated as multivariate); inferential; parametric.
Purpose	To compare scores on a single dependent variable obtained on the same (or 'matched' sample) individuals under several experimental conditions, at several different times or in several contexts, either alone or in combination with other grouping (independent) variables.
Measurement level	Any level of measurement may be used for the variables defining the repeated measures conditions as well as any other grouping (independent) variables. The dependent measure must be measured at the interval- or ratio-level.

Repeated measures ANOVA is a special form of ANOVA that provides for the analysis of experimental and quasi-experimental designs that obtain multiple measurements on the same individuals, or participants matched on one or more characteristics, under different experimental conditions, at several different times or in several different contexts. Such designs permit us to explicitly control for individual differences among participants in a behavioural study (e.g. differences due to extraneous variables such as interest level, motivation, fatigue, prior experience, skill, ability). This sort of control procedure is particularly useful if individual differences are expected to be relatively large with respect to the dependent variable of interest.

A wide variety of repeated measures designs is available ranging from the simplest one-way repeated measures design to more complex factorial ANOVA designs where one or more of the grouping variables is a repeated measure variable involving the same (or matched) participants being measured under all conditions or in all categories. A *one-way repeated measures ANOVA design* is an extension of the related groups *t*-test design (see *Procedure 7.4*) to three or more conditions. A one-way repeated measures ANOVA provides the same *F*-test for group (experimental condition) differences that one-way between groups ANOVA (see *Procedure 7.6*) does except that, now, each group contains the same (or equivalent) participants. For this reason, a repeated measures design is sometimes called a 'within groups' design. Similarly, a factorial ANOVA that involves one or more repeated measures dimensions produces *F*-tests for the same effects (i.e. main effects and interactions) using the same type of logic as ordinary between groups factorial ANOVA designs such as those described in *Procedure 7.10*.

> Since the data in the QCI database were not obtained using a design involving matching or repeated measurement of quality control inspectors under different conditions, we will illustrate the use of repeated measures ANOVA in a hypothetical scenario extension of the QCI research context. Suppose Maree obtains a supplemental contract to explore the effects of a newly designed quality control training program on inspection decision **accuracy** and **speed**. She obtains permission to run a small experiment where she identifies quality control inspectors, from her original sample of 112, who are in the bottom 25% of inspectors with respect to decision **accuracy** (most inaccurate inspectors) or in the top 25% with respect to decision **speed** (slowest inspectors). This selection process yields a sample of the 60 most 'poorly performing' quality control inspectors on at least one aspect of quality control inspection performance.
>
> Maree then randomly assigns the 60 inspectors into either a control group (N = 30 inspectors who don't undertake the new training program) or an experimental group (N = 30 inspectors who do undertake the new training program). However, before the start of her supplementary experiment, six of the inspectors assigned to the control ('no training') group declined to participate in the training evaluation study, leaving 24 control group inspectors. Maree already had pretest scores for the inspection decision **accuracy** and **speed** of these 54 inspectors from her original sample, obtained of course, prior to their undertaking the new training program. Continuing with the same methodology and measures as in the large QCI research project, Maree conducts her experiment where experimental group inspectors work through the new program and have their decision **accuracy** and **speed** measured again once they have completed the program; the control group inspectors are given equivalent experiences in terms of time away from work, but do activities completely unrelated to quality control inspection. For purposes of illustrating repeated measures ANOVA, we will focus on the decision **accuracy** results from both the experimental ('training') and control ('no training') groups.

(continued)

Procedure 7.11: Repeated Measures ANOVA

Maree's research design for her supplemental contract illustrates the classical *pretest-posttest control groups design* (recall *Fundamental concept VII*) involving repeated measurements on the same sample of participants (the design can also be referred to as a 2-way repeated measures factorial ANOVA with one within-groups factor (pretest-posttest occasion) and one between-groups factor (training group)). This type of research design, or some variant of it (a common variant being that no control group is available or used, in which case it becomes a one-way repeated measures ANOVA design), is often used in program or change intervention evaluation research. The repeated measurement nature of the design ensures that whatever is idiosyncratic about each inspector is present (which means, in effect, held constant) for the measurements obtained on the pretest and posttest occasions. Note that if Maree had been unable to randomly assign inspectors to the two training groups, then her design would be quasi-experimental in nature and would be identified as a *non-equivalent control group pretest-posttest ANOVA design* where the experimental and control groups could not be presumed to be equivalent at the outset (random assignment is the methodological procedure through which group equivalence can theoretically be obtained). The design would still be analysed in the same way using repeated measures ANOVA.

Table 7.15 shows the hypothetical data for this scenario along with relevant means for each condition. The dependent measure is **accuracy** scores, the two experimental conditions are denoted by a variable called **training group**, and the two measurement times are defined by two levels of a variable called **occasion**. In this design, the experimental manipulation of group conditions happens between the pretest and the posttest measurement occasions. Maree's primary research hypothesis is that the 'Training' group of inspectors should show higher decision **accuracy** scores under the Posttest condition than under the Pretest condition whereas the 'No Training' group of inspectors should not show much change in their **accuracy** scores over time (assume Maree has set her alpha significance level at .05).

Maree's research design involves two separate grouping variables, **training_condition**, which compares two different samples of inspectors (often referred to as a 'between-groups factor'), and **occasion**, which compares the same inspectors over time (the repeated measure or a 'within groups factor'). With two distinct independent variables, Maree should expect to obtain separate F-tests for a main effect due to **training_condition** (comparing the two means shown in the final column in Table 7.15 in dark gray), a main effect due to **occasion** (comparing the two means shown in the bottom row in Table 7.15, highlighted in dark gray), and the 2-way interaction between **training_condition** and **occasion** (comparing the four means shown in the summary rows below each cell of inspector scores in Table 7.15, highlighted in light gray).

(continued)

Table 7.15 Decision accuracy scores and summary descriptive statistics on the pretest and posttest occasions for the 24 'no training' group and 30 'training' group inspectors involved in the training program evaluation experiment

		Measurement OCCASION		Statistics for TRAINING GROUP Main Effect
	Inspector #	Pretest Accuracy Score (%)	Posttest Accuracy Score (%)	
No Training Group (n = 24)	4	67	73	Mean = 76.92 Std. Dev. = 9.64
	7	80	80	
	18	93	89	
	24	73	72	
	25	77	74	
	26	73	72	
	35	63	68	
	36	77	72	
	37	59	63	
	39	70	72	
	41	73	72	
	44	86	90	
	48	82	78	
	56	85	80	
	60	87	91	
	65	94	96	
	71	90	93	
	75	89	84	
	84	65	71	
	98	75	75	
	99	76	74	
	100	73	82	
	105	60	63	
	107	72	69	
	Mean	76.63	77.21	
	Std. Dev.	10.06	9.22	
Training Group (n = 30)	1	74	85	Mean = 80.72 Std. Dev. = 9.17
	3	75	89	
	10	83	83	
	12	80	88	
	15	86	87	
	17	99	100	
	22	76	73	
	30	75	71	
	32	62	64	
	43	75	82	
	45	87	94	
	50	64	71	
	53	77	84	
	54	81	83	
	59	75	82	
	61	89	99	
	62	82	89	
	68	85	93	
	70	84	87	
	77	92	98	
	81	82	89	
	82	62	77	
	83	57	67	
	92	78	86	
	95	70	77	
	97	73	79	
	102	78	85	
	108	85	80	
	109	83	88	
	112	71	73	
	Mean	78.00	83.43	
	Std. Dev.	9.24	9.09	
Statistics for OCCASION Main Effect	Mean	77.39	80.67	
	Std. Dev.	9.55	9.58	

(continued)

It is important to note that in this type of pretest-posttest control group design, the critical test is the *interaction* test because Maree fully expects the Training program to work; therefore, she expects the 'Training' and 'No Training' groups to behave differently from Pretest to Posttest. She would thus want to see non-parallel lines in the interaction plot (while the main effects are certainly tested, they are of much less interpretive interest for this specific type of research design). Note how this expectation is reflected in the earlier statement about Maree's research hypothesis.

The ANOVA on these data, performed using the SPSS *General Linear Model → Repeated Measures* ... procedure, reveals that there is a statistically significant **training_condition** by **occasion** interaction with an F-test value of 15.75 (with 1 and 52 degrees of freedom) having an associated p-value of less than 0.001. Formally, when writing up this statistical test, Maree would display the result as '$(F(1, 52) = 15.75, p < .001,$ partial $\eta^2 = .233)$'. The significant interaction explains a respectable 23.3% of the variability in inspectors' **accuracy** scores (recall the discussion of eta-squared (η^2) in *Procedure 7.7*).

Figure 7.20 shows a graph of the **training_condition** by **occasion** interaction with respect to mean decision **accuracy** scores. The figure provides visual verification of the nature of the significant interaction: the 'Training' group showed a substantive gain in mean decision **accuracy** from the Pretest occasion to the Posttest occasion whereas the 'No Training' group showed only very slight improvement (the precise differences between these four means could be further explored using multiple comparison tests—refer to *Procedure 7.8*).

The fact that the main effect for **training_condition** was non-significant ($F(1, 52) = 2.318, p = .134,$ partial $\eta^2 = .043$) and the main effect for **occasion** was significant ($F(1, 52) = 24.245, p < .001,$ partial $\eta^2 = .318$) is of relatively minor interest here since the major hypothesis focused on the expectation of a significant interaction. However, this would not be the case for other types of repeated measures designs where the repeated measure variable does not focus on a time-based pretest-posttest type of dimension.

Note that the inclusion of the 'No Training' group helps Maree counter the claim that inspectors naturally improved their inspection decision **accuracy** over time which was a possible alternative causal explanation for the results obtained in her original design which used the related groups t-test (refer to the discussion in *Procedure 7.4*) but had no control group available for comparison.

Since Maree could randomly determine which inspectors were members of each of her two groups, she is on relatively strong footing regarding a causal claim that attributes the observed significant comparative improvement in **accuracy** of the 'Training' group relative to the 'No Training' group (the

(continued)

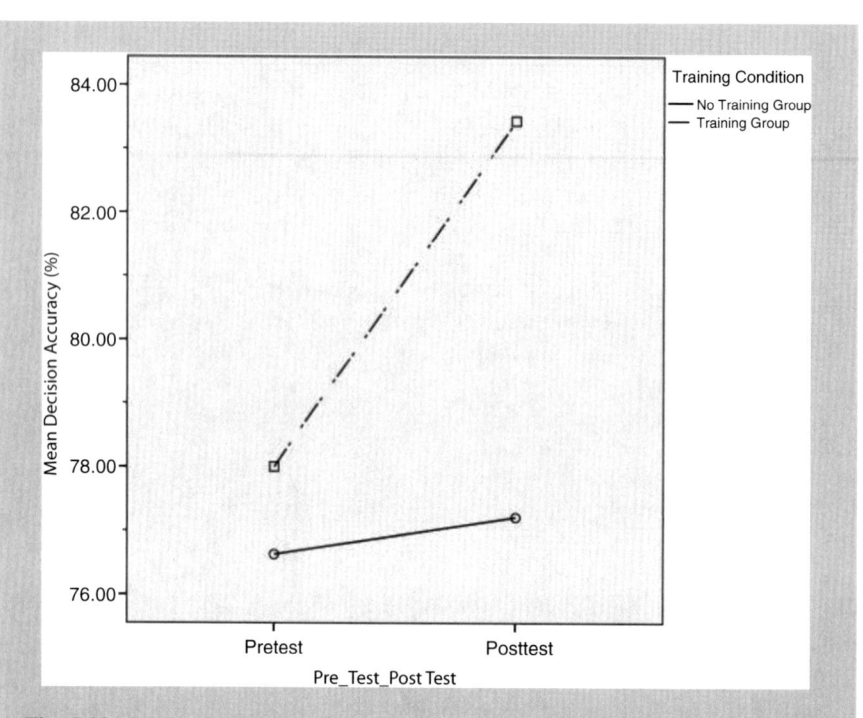

Fig. 7.20 Graph of the mean **accuracy** levels for the significant group by occasion interaction from the repeated measures ANOVA of the data in Table 7.15

'effect') to the actual training program itself (the putative 'cause'). Her grounds for the claim are not as strong as they would have been in a laboratory test of the program since there may still be other extraneous factors operating in the company which could have differentially influenced the performance of each group. For example, by random chance alone, it may be that the inspectors in one group know each other better than those in the other group which could lead to some sort of social facilitation or competitive enhancement in their performance.

The one-way repeated measures ANOVA design is also known as the *single factor within-subjects* or *Randomised blocks ANOVA design* (the latter applies particularly if matched participants are used in the different experimental conditions; see Kirk 2013, ch. 10). The pretest-posttest types of factorial repeated measures ANOVA designs are known in more sophisticated statistical circles as *Split Plot, Mixed* or *Between-Within ANOVA designs* (see, for example, Field 2018, ch. 15; Keppel and Wickens 2004, ch. 19–20 and Kirk 2013, ch. 12). In these latter designs, it is possible to have several repeated measures factors involved, which would mean that each participant is measured under all combinations of conditions in a factorial ANOVA design.

Advantages

Repeated measures ANOVA designs permit behavioural researchers to obtain additional control in their experiment or quasi-experiment by explicitly recognising and eliminating extraneous effects due to individual differences among participants. In many cases, this results in more powerful tests of hypotheses concerning the repeated measures components of these designs. In addition, when compared to traditional one-way or factorial ANOVA designs, repeated measures ANOVA designs require fewer participants to achieve the same level of hypothesis testing power if there truly are large individual differences on the dependent variable that can be controlled (which, in general linear model terms, translates to being removed from ERROR and incorporated into the ***MODEL***). In effect, each participant serves as his/her own control in a repeated measures design.

> Let's compare two different ways of designing an experiment involving two independent variables or factors. Call the factors *A* and *B* for purposes of the illustration and assume each factor has three levels. This will give an ANOVA design that has 9 cells (3 levels of factor *A* times 3 levels of factor *B*).
>
> If we assume that a good rule of thumb is to have at least 15 participants in each cell of a factorial ANOVA design, then a full factorial ANOVA version of this design that is completely between-groups (as described in *Procedure 7.10*) would require 15 × 9 or 135 participants. If, however, we could make factor *B* a repeated measures factor and measure each participant under all three of its levels, then we would require only 15 × 3 (levels of factor *A*) or 45 participants—a third of the number required for the completely between-groups version of the design. Of course, this gain in design efficiency assumes that it makes sense to implement factor *B* as a repeated measures independent variable. Furthermore, this advantage disappears if matched participants are used, rather than repeatedly measuring each participant.

Disadvantages

A major disadvantage of repeated measures ANOVA designs is that if participants do not differ substantially in their patterns of dependent variable responses to conditions or occasions, the power of the method suffers greatly. If matched participants are used, rather than repeatedly measuring participants (recall that matching, in effect, identifies participants that are considered to be equivalent for testing purposes), then the number of repeated measurement conditions determines how many participants have to be matched to randomly assign to conditions. This can create the need for a very large sample pool of participants to build the matched sets from. Additionally, if participants are matched on some criterion that is irrelevant to the dependent measure being used, then all the advantages of using a repeated measures design are lost.

> In a repeated measures design involving four distinct conditions or levels, using matching as the control procedure would require building sets of four participants matched on one or more criteria thought to influence, as extraneous variables, the dependent variable of interest. Each member of a matched set of four would then be randomly allocated to one of the four conditions of the repeated measures factor.
>
> For sound statistical tests, we would want at least 15 such sets of participants, which means we would need a starting sample pool size many times the size of the final 60 participants we want to end up with. The more variables you want to match people on, the worse the problem becomes. Thus, matching, in most behavioural science investigations, becomes an ill-advised control strategy when a repeated measures factor has more than two levels. Furthermore, if you match using extraneous variables that are unrelated to the dependent variable of interest, you actually pay a penalty by sacrificing power in the statistical tests because the ERROR term in the ***MODEL*** would not be effectively reduced.

You should be aware of potential testing and memory carry-over effects which could arise from repeatedly measuring the same individuals under different conditions. It is for this reason that counterbalancing control procedures have become important. For example, in market research, a common research design is to expose participants to a number of different brands of product (e.g. brands of cola or Blu-ray/DVD player) and have them rate their level of preference or some other quality for each one. If all participants experience the products in the same order, the researcher will be unable to counter an argument that observed differences may have been caused by the order in which participants experienced the stimulus objects or treatment conditions. By using a counterbalancing procedure (e.g. ensuring each participant experiences stimulus objects in a different order), this argument is negated, but at the price of a more complex research design. In some cases, a specific experimental design, called a 'latin square' design (see, for example, Kirk 2013, ch. 14) can be used to control the ordering of conditions to be experienced by each participant.

Finally, repeated measures ANOVA designs require a very stringent assumption about the nature of the data being analysed in addition to the typical normality and homogeneity of variance assumptions. This assumption is referred to as 'sphericity' which reflects the expectations that the correlations between scores in all possible various pairs of the repeated measurement conditions are of the same magnitude and that, if separate groups of participants are repeatedly measured, the pattern of correlations remains approximately the same. The sphericity assumption is often violated in practice, but it can be effectively circumvented by analysing the repeated measures ANOVA design using multivariate ANOVA methods (see *Procedure 7.16* and see Tabachnick and Fidell 2019, ch. 8, for a discussion).

Where Is This Procedure Useful?

A repeated measures ANOVA design should be considered in any experimental or quasi-experimental context where participants have been matched or repeatedly measured under different conditions and where large individual differences on the dependent measure are expected to exist. In research where programs are being evaluated, repeated measures designs provide the most efficient and defensible path to a claim of program effectiveness, irrespective of whether the research is conducted under highly controllable laboratory conditions or under less controllable natural conditions in the field. The time-based pretest-posttest (with perhaps a third follow-up testing occasion) dimension is probably the most frequently employed repeated measure variable in behavioural science research. Many *conjoint measurement* designs (see *Procedure 9.4*) employed in marketing research use complex repeated measures factorial or fractional factorial designs.

Software Procedures

Application	Procedures
SPSS	*Analyze → General Linear Model → Repeated Measures* ... and define one or more 'Within-Subjects Factors' (you must provide a name and specify the number of levels for each one); hit the '*Define*' button; then choose the 'Between-Subjects Factors' and the 'Within-Subjects Variables' of interest (each Within-Subjects Variables selected constitutes a dependent variable that represents measurements under one condition of a repeated measure or 'Within-Subjects Factor'); hit the '*Plots*' button to configure the interaction plots you want to see; hit the '*Options*' button and indicate you want to see 'Estimated marginal means' for all design effects and you want 'Descriptive Statistics' and 'Measures of effect size' reported. Note that SPSS can only report posthoc multiple comparison tests for between-groups factors, which is not always much help in a repeated measures design.
NCSS	*Analysis → Analysis of Variance (ANOVA) → Repeated Measures Analysis of Variance* and define the 'Response Variable', up to three 'Between-Subjects Factors' and up to three 'Within-Subjects Factors'; you can also choose relevant 'Reports' and 'Plots' you want to see. You must also have a separate variable, called the 'Subject Variable', which uniquely identifies every participant in the design, for every repeated measurement condition. This means that your database needs to be configured differently from the way it would be in SPSS. Each line of the NCSS database would contain the measurements from one participant under one repeated measurement

(continued)

Application	Procedures
	condition and each participant would have as many rows of data as they have repeated measurement conditions they participated in (e.g. for a pretest-posttest design, each participant would have two rows of data in the NCSS database. This is why you need the additional variable that holds a participant (i.e. 'subject') number.
SYSTAT	*Analyze → Analysis of Variance → Estimate Model* ... select your dependent variables ('Dependents' means those variables that contain measurements for each condition of a repeated measure factor); select one or more between groups 'Factors'; click on the '*Repeated Measures*' tab, tick the box 'Perform Repeated Measures analysis' and set up the repeated measures design amongst the dependent variables by naming each repeated measurement factor and indicating the number of levels it has. If you want to see a graph of an interaction, you have to graph this separately within SYSTAT using *Graph → Line Chart...* and configuring the graph with 'Repeated Trials'. After the analysis has been run, you can do *Analyze → Analysis of Variance → Post Hoc Test for Repeated Measures...* and select the repeated measures factor and type of posthoc test desired. It is only worth going the extra step if your repeated measures factor has more than two levels.
STATGRAPHICS	STATGRAPHICS does not have the direct capability to analyse repeated measures ANOVA designs. However, Polhemus (2006) provides a description of how STATGRAPHICS can be used to analyse a repeated measures design using the *Compare → Analysis of Variance → General Linear Models* ... procedure. This process requires the relevant design predictors to be configured in a specific way (similar to NCSS) within STATGRAPHICS, which Polhemus (2006) demonstrates. In order to make this process work, you need to understand a bit more deeply how the effects in a repeated measures design need to be represented; something that goes beyond what we can cover in this book.
R Commander	**R** Commander does not have the capability to analyse repeated measured ANOVA designs; it is limited to between-groups factorial ANOVA designs. However, Field et al. 2012, ch. 13 and 14) describe and illustrate how repeated measures and mixed ANOVA designs can be analysed using other functionalities and packages within **R**.

Procedure 7.12: Friedman Rank Test for Repeated Measures

Classification Univariate; inferential; nonparametric.
Purpose To evaluate the ranked differences in scores on a single dependent measure obtained on the same sample (or a matched sample) of individuals under three or more conditions or in three or more different contexts. In this sense, the groups being compared are considered related.

Procedure 7.12: Friedman Rank Test for Repeated Measures

Measurement level The grouping variable can be defined by any level of measurement. The dependent measure may originally consist of ordinal ranks across the various conditions or of scores that were originally measured at the interval- or ratio-level but converted to ordinal ranks prior to analysis.

The *Friedman rank test* is the nonparametric analog of the one-way repeated measures ANOVA *F*-test (see *Procedure 7.11*). It evaluates the significance of the differences between three or more related conditions on the basis of mean ranks rather than mean scores in some other unit of measurement (Field 2018, section 7.7; Siegel and Castellan 1988, ch. 7). As with the repeated measures ANOVA design, the conditions may be rendered related by virtue of repeatedly measuring each participant under every condition or by matching participants on some extraneous variable which is thought to be influential on the dependent variable then randomly allocating members of the matched sets to each condition.

The use of the Friedman rank test is frequently entertained when the original measurements of a dependent variable are distributed in some highly distorted fashion so as to violate the basic normality or sphericity assumption of the repeated measures ANOVA *F*-test.

> Suppose Maree further extends her supplemental contract scenario set out in *Procedures 7.4* and *7.11*, where she is evaluating the effectiveness of a quality control training program. Suppose Maree wanted to check the durability of the effects of the training program by conducting a 2-month follow-up test after completion of the training program. If she had collected **speed** data on the 30 inspectors that participated in the training program in this Pretest-Posttest-Follow-Up design, the data might look as shown in Table 7.16.
>
> Maree would be interested in comparing the quality control inspectors but would now focus on differences in their inspection decision **speed** before, immediately after, and 2 months after undergoing the training program. We have seen in earlier discussions (recall *Procedure 5.7*) that **speed** is a highly positively skewed variable (as are most time-based measures of human behaviour) which may be problematic enough to violate the normality or sphericity (equal correlations between conditions) assumption.
>
> Figure 7.21 shows the box plots for the **speed** variable for the pretest, posttest, and follow-up occasions. The plot shows not only that the three occasions may significantly differ but also that their data are highly and unevenly skewed. Instead of the parametric one-way repeated measures ANOVA *F*-test, Maree opts to conduct the Friedman Rank Test.
>
> The Friedman rank test is conducted in two simple steps:
>
> - The scores in the original units of measurement (i.e. **speed**) are ranked from lowest to highest across the three occasions for *each* inspector, These ranks
>
> (continued)

Table 7.16 Decision **speed** scores and summary descriptive statistics on the pretest, posttest and 2-month follow-up occasions for the 30 'training' group inspectors involved in the training program durability study

Participant No.		Inspector	Pretest speed	Posttest speed	2 month follow-up speed
1		1	8.24	8.08	5.17
2		3	7.85	8.21	7.93
3		10	5.91	4.74	4.42
4		12	7.61	6.94	12.78
5		15	7.12	6.62	6.17
6		17	7.07	5.47	3.94
7		22	3.40	1.91	2.03
8		30	1.30	1.06	2.04
9		32	1.51	1.27	1.09
10		43	2.19	1.54	1.57
11		45	6.09	8.06	8.19
12		50	2.82	2.70	3.01
13		53	3.14	2.67	3.46
14		54	7.45	6.56	7.64
15		59	4.52	3.82	3.14
16		61	5.71	5.37	6.43
17		62	17.10	14.19	16.02
18		68	9.17	14.06	8.20
19		70	7.82	7.20	5.07
20		77	6.11	3.79	2.32
21		81	8.93	8.01	7.60
22		82	1.85	1.58	1.61
23		83	2.85	1.86	1.70
24		92	9.52	8.35	5.03
25		95	1.28	1.81	1.48
26		97	3.78	3.14	2.11
27		102	7.53	5.68	4.98
28		108	5.96	5.12	4.94
29		109	7.66	4.97	4.41
30		112	3.33	1.67	1.69
Total	N	30	30	30	30
	Mean		5.83	5.22	4.87
	Std. Deviation		3.34	3.42	3.45

(continued)

Procedure 7.12: Friedman Rank Test for Repeated Measures

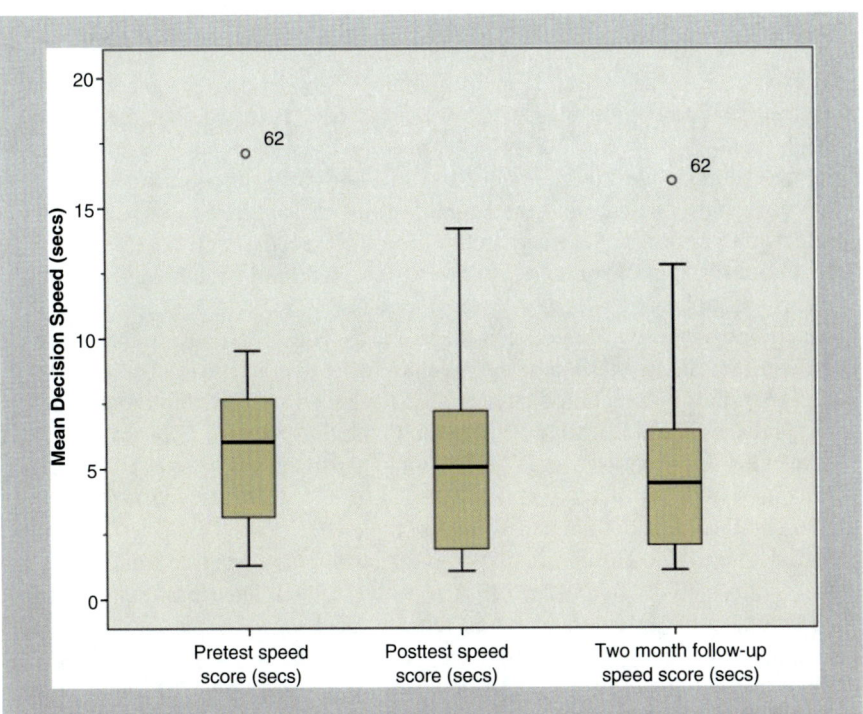

Fig. 7.21 Boxplot of the **speed** scores for the pretest, posttest and 2-month follow-up occasions in Table 7.16

generally range from 1 to K (the total number of occasions or conditions), so in this example the ranks would go from 1 (fastest **speed** score) to 3 (slowest **speed** score).
- A one-way repeated measures ANOVA is essentially conducted on these rank scores, producing a chi-squared statistic which tests for differences between the mean ranks among the three occasions.

The outcome from the Friedman rank test, conducted on the data in Table 7.16, using SPSS is a value for the chi-squared statistic of 14.467 with 2 degrees of freedom and an associated p-value of .001. This means Maree can conclude that there is at least one significant difference between the mean ranks for **speed** scores in the three testing occasions.

There is a nonparametric pairwise multiple comparison test that Maree could request in SPSS as a follow-up to this significant result. This test could help her to identify which occasions differed significantly from which other occasions (see, for example, Siegel and Castellan Jr 1988, pp. 180–183). Since the ranking system was such that slower decision times garnered the

(continued)

higher ranks, the pattern of the multiple comparison tests suggested that the Posttest (mean rank = 1.73) and 2-month Follow-Up (mean rank = 1.70) ranked decision **speed** of inspectors were significantly slower than the Pretest (mean rank = 2.57) ranked decision **speed** and that there was no difference between immediate Posttest and 2-month Follow-Up decision times. Thus, Maree could conclude that the training program appeared to have durable effects on performance as measured by decision **speed**.

Maree could compare this result to what would have been obtained if she had run a parametric one-way repeated measures ANOVA F-test to see if her conclusions would differ between the two tests. If the inference decision was different for the two tests or if the p-values differed dramatically between the two tests, then assumption violations will have likely distorted the parametric test and she should interpret the Friedman rank test outcome. This is the case here with the Pretest-Posttest-Follow-up comparison on **speed** scores: the Friedman rank test p-value was significant at $p = .001$ and the F-test was also significant but at a much greater p-value ($F(2, 58) = 4.614$, $p = .014$, partial $\eta^2 = .137$). The conclusion does not change between the two tests, but the p-values do differ substantially suggesting some influence of assumption violation. If the inference decision remains unchanged or the p-values are similar, then assumption violation was not severe enough to distort the test and Maree should report the parametric test because it utilises more of the original information in the sample.

Advantages

The Friedman Rank Test is particularly useful when the assumptions of the parametric one-way repeated measures ANOVA F-test cannot be or are not reasonably fulfilled by the data. It permits analysis of such data by reducing the data in measurement quality to the ordinal level and then conducting the test. The test is easy to compute and interpret and has no underlying assumptions about the distribution of the data which need to be satisfied. The Friedman Rank Test also provides a simultaneous test for a significant difference between the medians of the various conditions; the approach taken by the NCSS computing package, which reports its results in terms of population medians.

Disadvantages

Unless the behavioural data rather severely violate F-test assumptions, it is usually better to use the parametric test rather than the Friedman Rank Test. The reason is that when assumptions are even moderately violated, the F-test is a more powerful test for a given sample of data than the Friedman Rank Test; that is, it is more likely to detect condition differences if they really exist. However, if the violations are serious, then the Friedman Rank Test becomes the more powerful test. Also, since the original data, in many applications, are converted to ranks, some information is necessarily lost. Such a loss can hamper meaningful interpretation of the observed condition differences. It should be noted that there are no readily available nonparametric test alternatives for any repeated measures design more complex than a one-way repeated measures design or for any factorial ANOVA-type design in general.

Where Is This Procedure Useful?

The Friedman Rank Test can be applied in any experimental, quasi-experimental, or survey research investigation where comparison of participants measured under three or more conditions is of interest and parametric assumptions are strongly violated. Many software packages offer specific tests for assumption violation which makes the choice of analytical technique easier to make. As with the one-way repeated measures ANOVA F-test, the statistical conclusion offered by the Friedman Rank Test relates solely to observed differences between three or more conditions. Whether or not a causal conclusion is warranted (i.e. that the variation in the definition of the various conditions is what caused the significant difference or differences among the mean ranks) is a separate decision which must be made in light of the quality of one's research design and methodological procedures.

Software Procedures

Application	Procedures
SPSS	*Analyze* → *Nonparametric tests* → *Related Samples* ... and under the '*Objective*' tab, choose 'Customize analysis'; under the '*Fields*' tab, choose your desired dependent variables for all repeated measurement conditions ('Test Fields'); under the ;'*Settings*' tab, choose 'Customize tests' and tick the 'Friedman 2-way ANOVA by ranks (k samples)' box and in the 'Multiple Comparisons' drop-down menu, choose 'All pairwise'. When you run the test, only a brief overall summary table for the analysis appears in the output window. You must double-click on this summary table to open up the '*Model Viewer*' window to see more detailed results (right-side of the

(continued)

Application	Procedures
	window that opens). At the bottom of the right-hand '*Model Viewer* 'window, there is a drop-down menu labelled 'View' where you can select which analysis view you want to look at: 'Related Samples Test View' or 'Pairwise Comparisons' view.
NCSS	*Analysis → Analysis of Variance (ANOVA) → Balanced Design Analysis of Variance* and select your 'Response Variable(s)' (which must contain the dependent measurements under all conditions in one long list); you also need two separate variables – one that codes the repeated measurement condition number for each observation and the other codes the participant number for each observation – these variables are entered as 'Factor Variables' (the participant code variable must be set to 'Random' under 'Type'; the condition code variable must be set to 'Fixed'); then on the '*Reports*' tab, make sure that the 'Friedman Report' is ticked.
SYSTAT	*Analyze → Nonparametric tests → Friedman* ... and select the variables that contain the measurements for the various conditions of the repeated measures factor ('Selected Variable(s)' and tick the 'Pairwise Comparisons' box.
STATGRAPHICS	*Compare → Multiple Samples → Multiple-Sample Comparisons...* and choose 'Multiple Data Columns' for the 'Input' window; hit '*OK*' and choose all of the columns containing the dependent variable scores for each repeated measurement condition ('Samples'); hit '*OK*' and under 'Tables', make sure that 'Kruskal-Wallis and Friedman Tests' is ticked at a minimum and choose your desired 'Graphs'; hit '*OK*' and right-click in the 'Kruskal-Wallis Test' pane in the output window and choose 'Friedman Test'. STATGRAPHICS does not offer a multiple comparison procedure for the Friedman test.
R commander	*Statistics → Nonparametric tests → Friedman rank-sum test...* and choose the desired repeated measures variables. The procedure reports the medians for each condition instead of the mean rank. **R** Commander does not offer a multiple comparison procedure for the Friedman test. However, Field et al. 2012, p. 691) illustrates an **R** command that can be used to produce such a multiple comparison test.

Procedure 7.13: Multiple Regression: Tests of Significance

Classification Bivariate/multivariate; inferential; parametric.
Purpose To test specific hypotheses about the values of specific regression weights or sets of regression weights or to test the values of specific partial correlations, semi-partial correlations, or R^2. In multiple regression analysis, these tests may be done simultaneously with all predictors of interest included in the regression model or hierarchically with predictors being sequentially included in the model in a pre-argued order.

Procedure 7.13: Multiple Regression: Tests of Significance 373

Measurement level   The predictors may be assessed at virtually any level of measurement provided they are correctly represented (coded) for proper inclusion in the analysis. The dependent variable should be measured at the interval- or ratio-level.

A special type of regression analysis, called logistic regression, is available for analysing a dichotomous dependent variable – see *Procedure 7.14*. Other more specialised types of regression analysis are discussed in *Procedure 9.8*.

Multiple regression analysis was described in some detail earlier in *Procedure 6.4* and focused on employing independent variables, called *predictors* to make predictions about a dependent variable, called a *criterion*. Multiple regression analysis is also one major manifestation of the general linear model, discussed in *Fundamental concept VI*. The discussion here will focus on the types of statistical hypothesis tests that are available within the multiple regression framework. Generally speaking, there are two broad types of statistical tests available in multiple regression analysis: those that focus on the regression coefficients or weights and those that focus on correlational indices such as partial correlations, semi-partial correlations, or R^2.

If the predictors of interest are all included at the same time in the regression model, the process is called *simultaneous regression analysis* (Cohen et al. 2003; may also be referred to as *predictive regression* or *standard regression*). In the simultaneous mode of regression model building, hypothesis testing focuses on R^2 and on regression coefficients or weights and is often used in predictive exercises where the goal is to build the most useful model for predicting some criterion measure of interest. The resulting model can then be used for predicting criterion values for new cases not included in the original analysis. Hypothesis testing may also focus on partial or semi-partial correlations, which is more typical in explanatory exercises where the goal is to understand how the various predictors contribute to the overall explanation of variability in the dependent variable.

If the predictors of interest are included (either singly or in specific sets) in the regression model in a specific researcher-determined order, the process is called *hierarchical regression analysis* (Cohen et al. 2003 and recall *Fundamental concept IV*). In hierarchical regression model building, hypothesis testing focuses primarily on changes in R^2 to evaluate the evolving model as well as on partial or semi-partial correlations to evaluate individual predictor contributions to the model. In this light, hierarchical multiple regression analysis can provide a powerful theory testing platform and is a more highly focused and more complete type of relational analysis compared to simultaneous regression.

If the predictors of interest are included in the regression model in an order determined by a statistical algorithm (there are usually several to choose from), the process is called *stepwise regression analysis* (Cohen et al. 2003). In stepwise regression, the researcher effectively hands all control over the model building process to the computer program, leaving hypothesis testing outcomes to be influenced by the vagaries of sample-specific predictor-criterion relationships. For this reason, stepwise regression is not generally recommended for social and behavioural research.

Instead, what is recommended is that researchers make the conscious decision to use either a simultaneous regression approach or hierarchical regression approach. That decision would hinge upon whether they felt they had a strong enough theoretical or logical basis on which to specify/justify an order of predictor inclusion into the model to evaluate. The price paid for this choice is that only one ordering can be defended and tested; once specified, the researcher cannot then second-guess the analysis posthoc and change the ordering around. The gain is a more powerful model testing framework in which all relationships can be accounted for. In the absence of such a basis, the researcher's best option is to include all predictors in the model at same time. The price paid for this latter choice is that some information about relationships is discarded.

Another way to think about the process of multiple regression and how the approaches for building regression models can be differentiated is as an exercise in variance partitioning. The goal is to explain the variability of a criterion or dependent variable (to be general, call this variable **Y**) using information available in the predictors (call these variables **X**s). Think back to the discussion of the general linear model in *Fundamental concept VI*: **Y** is the **DATA** part of the system; the **X**s comprise the *MODEL* and whatever the *MODEL* cannot explain in the **DATA** is called ERROR (in multiple regression analysis, the term 'residual' is used as a synonym for ERROR). The variance of **DATA** is thus partitioned into two components – one attributable to what the *MODEL* can predict and one attributable to what the *MODEL* cannot predict, i.e. ERROR. Figure 7.22 illustrates, using two predictors, the difference between simultaneous and hierarchical regression analysis.

The chief and critical difference between the two approaches to regression model building is that the simultaneous approach (top Venn diagram in Fig. 7.22) must ignore (meaning 'throw away') any region of overlapping variance explained in **Y** that is shared between predictors. In the absence of a theory that can help the researcher decide which predictor the overlap area belongs to, the best decision is to ignore it. Thus, in simultaneous regression, what is evaluated is what each predictor uniquely contributes to the regression model over and above all other predictors in the model (i.e. the 'football-shaped areas with bites taken out' for both X_1 and X_2, assessed by the squared semi-partial correlations – recall *Fundamental concept IV*). In hierarchical regression (bottom Venn diagrams in Fig. 7.22), the area of overlap is no longer ambiguous, it is assigned explicitly to the predictor that is included first and is assessed by a squared correlation (i.e. the complete 'football-shaped area' for X_2). Then, the predictor that has been added to the model is evaluated for what it uniquely adds to prediction over and above what the first-entered predictor accounted for and is assessed by a squared semi-partial correlation (i.e. the 'football-shaped area with a bite removed' for X_1).

Procedure 7.13: Multiple Regression: Tests of Significance

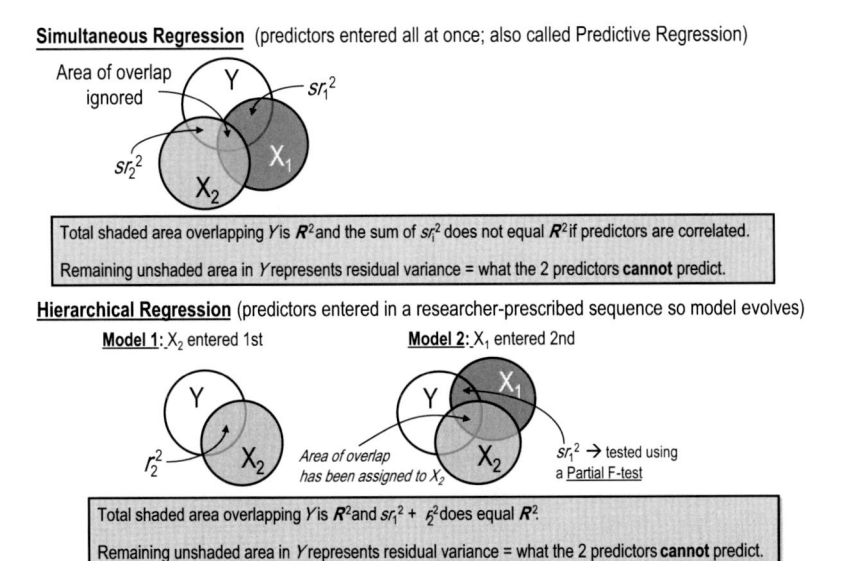

Simultaneous Regression (predictors entered all at once; also called Predictive Regression)

[Venn diagram: two overlapping circles X_1 and X_2 overlapping with Y. Label "Area of overlap ignored" points to the X_1–X_2–Y triple overlap; sr_1^2 labels the X_1-only overlap with Y; sr_2^2 labels the X_2-only overlap with Y.]

Total shaded area overlapping Y is R^2 and the sum of sr_i^2 does not equal R^2 if predictors are correlated.
Remaining unshaded area in Y represents residual variance = what the 2 predictors **cannot** predict.

Hierarchical Regression (predictors entered in a researcher-prescribed sequence so model evolves)

Model 1: X_2 entered 1st   **Model 2:** X_1 entered 2nd

[Two Venn diagrams. Model 1: circles Y and X_2 overlap, with r_2^2 labeling the overlap. Model 2: circles Y, X_2, and X_1; "Area of overlap has been assigned to X_2"; sr_1^2 → tested using a Partial F-test.]

Total shaded area overlapping Y is R^2 and $sr_1^2 + r_2^2$ does equal R^2.
Remaining unshaded area in Y represents residual variance = what the 2 predictors **cannot** predict.

Note: sr^2 refers to a squared semi-partial correlation; r^2 to a squared correlation and R^2 to a squared multiple correlation.

Fig. 7.22 Variance partitioning in simultaneous and hierarchical regression model building

Testing R^2

In conjunction with testing hypotheses regarding specific regression weights in a prediction model in simultaneous regression analysis, it is usual practice to first assess the overall significance of the squared multiple correlation, R^2. This is accomplished using what is called an 'omnibus' *F*-test and it tests the overall quality of the full model. This *F*-test evaluates the significance of R^2 relative to a null hypothesised value of 0.0 and is produced from an analysis of variance table that compares the variance that can be predicted using the 'Regression' model to the variance that is left over as error (called the 'Residual') after prediction. If R^2 is found to be significant, it signals a regression model that can predict a significant proportion of the variance in the criterion variable's scores. Testing R^2 first allows us to take advantage of the hurdle step logic for hypothesis testing (recall *Fundamental concept V*) in order to control for inflation of the alpha error. In simultaneous regression analysis, we would not interpret the contributions of individual predictors to the model unless the overall model R^2 was first shown to be significant.

It should be noted that while R^2 can be tested for significance, it is actually a poor estimate of the population value for R^2. For this reason, behavioural researchers, who are particularly interested in showing how powerful their model is for prediction, often report an 'adjusted' R^2 which counteracts the biased nature of ordinary R^2. The adjusted R^2 is always lower in magnitude than the original R^2 and can be interpreted as the value that the ordinary R^2 would shrink to if the prediction model were to be applied to a new sample of scores for both the predictors and the criterion (which is why adjusted R^2 is sometimes interpreted as an estimate of 'cross-validated' R^2; it is also sometimes called 'shrunken R^2').

Testing Regression Coefficients

Regression coefficients are typically evaluated for significance at a point where all predictors of interest have been included in the equation. In simultaneous regression analysis, this would occur after the overall R^2 had been evaluated and found to be significant. Each regression coefficient or weight has an individual *t*-test of significance associated with it. This *t*-test evaluates the significance of the difference between the observed value of the regression coefficient and the null hypothesised value of 0.0 (other non-zero null hypothesised values can also be tested, but these tests would always have to be computed by hand; virtually all statistical software programs assume a null hypothesis value of 0.0).

If the *t*-test for a specific regression coefficient is found to be significant, then the researcher can conclude that the predictor being evaluated has a statistically non-zero weight in the regression model and is therefore a potentially useful variable contributing toward prediction of the criterion. If there is only one predictor in the equation (i.e. simple linear regression, see *Procedure 6.3*), then the *t*-test for the regression coefficient is identical to the *t*-test for evaluating the significance of the Pearson correlation between the predictor and the criterion. Since *t*-tests are employed to test regression coefficients, it is possible to test one-tailed or direction-predicting hypotheses (e.g. testing the expectation that a specific regression coefficient will be greater than 0.0).

> In the QCI database context, Maree might be interested in developing a model for predicting inspection **accuracy** using, as predictors, **educlev, gender, company, mentabil, log_speed, jobsat,** and **workcond** (the transformed version of the **speed** variable, **log_speed**, is used because **speed** was a highly positively skewed variable, recall *Procedure 5.7*). Suppose she has no access to a defensible theory for deciding on an ordering of predictor entry, so she will use a simultaneous regression approach.
>
> However, in order for her to be able to use the categorical variables, **educlev, gender** and **company**, as predictors in the multiple regression analysis, Maree first has to 'dummy-code' the variables to properly represent their information (recall *Fundamental concept VI*; see also Cohen et al. 2003). For **educlev**, she decides to score the 'High School Only') category as 0 (the reference category) and the 'Tertiary-Qualified' category a 1 in a new predictor she calls **dummy_coded_educlev**. For **gender**, she decides to score 'Males'

(continued)

scored as 0 (the reference category) and 'Females' as 1 in a new predictor called **dummy_coded_gender**. For **company**, Maree might wish to simplify things a bit and simply compare three broad categories of industry: Computer Industries (combining PC and Large Business Computer companies), Electrical Appliance Industries (combining Small and Large Electrical Appliance companies) and Automobile Industry. This makes three categories Maree needs to dummy-code, meaning that she needs to make two new predictors. Maree chooses the Electrical Appliance industries to be the reference category for each new predictor, always coded as 0. For one new predictor, **dummy_coded_computer_ind**, Maree codes Computer Industries as 1 and both Electrical Appliance Industries and Automobile Companies as 0. For the second new predictor, **dummy_coded_automobile_ind**, she codes Automobile Industry as 1 and both Electrical Appliance Industries and Computer Industries as 0.

Table 7.17 presents the slightly edited 'Model Summary', 'ANOVA' and regression 'Coefficients' subtables produced by the SPSS *Linear Regression* procedure for this multiple regression analysis. The 'Model Summary' subtable summarises the statistics related to R^2 (highlighted in boldface) and the 'ANOVA' subtable summarises information relevant to the F-test used to test the significance of R^2 (relevant interpretive information in boldface in this subtable includes the degrees of freedom (df) for the 'Regression' model and the 'Residual', the F-value and the *'Sig.'* (= the p-value)). Here we see that Maree's prediction model, based on the eight QCI predictors, can account for roughly 35% of the variability in inspectors' **accuracy** scores, an amount which is significant at a p-value of less than .001 (Maree could report this result as: '$R^2 = .345$, adjusted $R^2 = .289$, $F(8, 94) = 6.177$, $p < .001$'). R^2, which is automatically produced by any multiple regression analysis, is a measure of 'effect size', analogous to eta-squared for ANOVA (recall *Procedure 7.7*). It reflects how strong the overall regression model is for predicting the criterion.

The value for adjusted R^2 in the 'Model Summary' table is .289. This means that about 29% of the variance in **accuracy** would be explained if this model were to be applied to making predictions on a new sample of inspectors—an estimated reduction of roughly 6% in predictive power.

Now, we will focus on the regression 'Coefficients' subtable in Table 7.17 for Maree's simultaneous regression model, particularly looking at the column of unstandardised regression coefficients, labelled 'B' (the standard errors for each regression coefficient are in the column labelled 'Std. Error'), the column of standardised regression coefficients, labelled 'Beta', the column of t-tests, labelled 't', and the column of p-values, labelled 'Sig.'. Note that the regression constant is significant, but this is almost never of interpretive interest in the social and behavioural sciences and is almost always significant anyway.

(continued)

Table 7.17 Summary tables from a simultaneous regression analysis predicting inspection decision accuracy scores using **mentabil, dummy-recoded gender, dummy-recoded educlev, dummy_coded_computer_ind, dummy_coded_automobile_ind, jobsat, workcond** and **log_speed**

Model Summary[b]

Model	R	R Square	Adjusted R Square	Std. Error of the Estimate	Durbin-Watson
1	.587[a]	.345	.289	7.73476	1.903

a. Predictors: (Constant), log_speed, dummy_coded_gender, jobsat, mentabil, dummy_coded_educlev, dummy_coded_automobile_ind, workcond, dummy_coded_computer_ind
b. Dependent Variable: accuracy

ANOVA

Model		Sum of Squares	df	Mean Square	F	Sig.
1	Regression	2956.333	8	369.542	6.177	.000
	Residual	5623.697	94	59.827		
	Total	8580.029	102			

Coefficients

Model		Unstandardized Coefficients		Standardized Coefficients	t	Sig.	Correlations		
		B	Std. Error	Beta			Zero-order	Partial	Part
1	(Constant)	86.660	10.081		8.596	.000			
	dummy_coded_gender	-.105	1.597	-.006	-.066	.948	-.118	-.007	-.005
	dummy_coded_educlev	-1.271	1.774	-.070	-.717	.475	-.252	-.074	-.060
	dummy_coded_computer_ind	**6.692**	**2.652**	**.361**	**2.523**	**.013**	**.527**	**.252**	**.211**
	dummy_coded_automobile_ind	-2.083	2.361	-.089	-.882	.380	-.338	-.091	-.074
	mentabil	-.095	.092	-.091	-1.030	.306	.011	-.106	-.086
	jobsat	.010	.600	.002	.017	.986	.235	.002	.001
	workcond	**1.291**	**.642**	**.242**	**2.010**	**.047**	**.508**	**.203**	**.168**
	log_speed	-2.264	3.599	-.068	-.629	.531	.239	-.065	-.053

We can see that the only two significant predictors of inspection decision **accuracy** in the context of Maree's prediction model were **dummy_coded_computer_ind** and **workcond** (lines of statistics highlighted in boldface in the table, associated with *p*-values less than .05). To interpret the regression coefficient for the significant **dummy_coded_computer_ind** categorical predictor, Maree would need to remember two things: which coding scheme she used (dummy coding) and which category was coded as 0 (the reference category = Electrical Appliance industries) and 1 (= computer industries) respectively. Knowing this, Maree can now interpret the B weight for **dummy_coded_computer_ind**, which was 6.692, as reflecting a significant mean increase in the prediction of **accuracy** that is observed if a participant is known to work in a Computer industry relative to a participant working in the Electrical Appliance industries. Note that the other dummy-coded predictor associated with the company a participant worked for was **dummy_coded_automobile_ind**, but this predictor was not significant. This effectively means that participants from the Electrical Appliance industries and the Automobile industry did not significantly differ with respect to their **accuracy** scores (note this is exactly the same interpretation as would be made

(continued)

by an independent groups *t*-test comparing these two categories of participants – this is the beauty of the general linear model way of thinking).

The significant regression weight for **workcond** of 1.291 is technically a geometric slope which means that every increase of 1-unit amount in **workcond** predicts a 1.291 unit increase in **accuracy**. The standardised regression coefficient, *beta*, for **workcond** is .242 which means that every increase of 1 *z*-score unit amount in **workcond** predicts a .242 *z*-score unit increase in **accuracy**.

The remaining predictors do not significantly contribute to prediction of **accuracy** (i.e. no other *p*-values are less than .05). However, this interpretation is rendered somewhat ambiguous because the various predictors are correlated with each other. If you recall our earlier discussion, what happens in simultaneous regression analysis (where all predictors of interest are in the equation at the same time) is that any correlations between the predictors which also happen to overlap the criterion measure, even if large, are ignored. Thus, sometimes, an important aspect of the predictive power of one or more predictors is thrown away. This means that simultaneous regression analysis may produce a simpler story about effective predictors than is really warranted if predictors are strongly correlated in their relationships with the criterion.

You may have noticed three additional columns out to the right side of the **Coefficients** subtable in Table 7.17. These three columns contain three types of correlations, produced by SPSS on request (they are not produced by default). 'Zero-order' correlation is the original correlation between the predictor of interest and **accuracy**. Thus, for example, **workcond** correlates .508 with **accuracy**.

'Partial' correlation is the correlation between a predictor of interest and **accuracy**, after all other predictors in the regression model have had their influence statistically removed from *both* the predictor and **accuracy** (recall *Fundamental concept IV*). Thus, for example, **workcond** has a partial correlation of .203 with **accuracy**, after all effects of the other predictors have been controlled for in both variables. The *t*-test for the regression coefficient also tests whether this partial correlation is zero. Since the *t*-test was significant for **workcond**, the partial correlation of .203 estimates a non-zero partial correlation in the population.

Finally, the 'Part' correlation is the semi-partial correlation between a predictor of interest and **accuracy**, after all other predictors in the regression model have had their influence statistically removed from the predictor (recall *Fundamental concept IV*). Thus, for example, **workcond** has a part or semi-partial correlation of .168 with **accuracy**, after all effects of the other predictors have been controlled for with respect to **workcond**. The *t*-test for the regression coefficient also tests whether this part correlation is zero. Since the

(continued)

t-test was significant for **workcond**, the part correlation of .168 estimates a non-zero semi-partial correlation in the population.

Squaring the 'part' correlation will, of course, give you the squared semi-partial correlation that was discussed in *Fundamental concept IV*. For **dummy_coded_computer_ind,** this value is $.211^2 = .045$ and for **workcond**, this value is $.168^2 = .028$. It is the squared semi-partial correlation that gives the most useful measure of individual predictor effect size in multiple regression. Thus, Maree could say that **dummy_coded_computer_ind** uniquely explains 4.5 of the variance in **accuracy** scores over and above all other predictors in the model and **workcond** uniquely explains 2.8% of the variance in **accuracy** scores over and above all other predictors in the model. It has now become common practice for researchers to report these values along with the significant test results. So, for example, Maree could summarise the evidence for the significant **dummy_coded_computer_ind** predictor as 'B = 6.692, beta = .361, $t(94)$ = 2.523, p = .013, sr^2 = .045' and for the significant **workcond** predictor as 'B = 1.291, beta = .242, $t(94)$ = 2.010, p = .047, sr^2 = .003'.

Checking Residual Patterns & Assumptions

An important step in multiple regression analysis is to check that the assumptions it requires are reasonably met. There are graphical as well as numerical ways to do such checking and, in most cases, the diagnostic devices focus on the behaviour of the residuals. The most diagnostic type of graph plots the standardised residuals against the standardised predicted values – a graph that directly compares the behaviour of the regression model with the behaviour of the prediction errors the model makes (see, for example, Fox 1991; Lewis-Beck 1980, pp. 39–41).

Figure 7.23 shows a range of different diagnostic patterns that such a graph might reveal. Figure 7.23a shows a well-behaved prediction model where the residuals are randomly and uniformly scattered around 0 in a horizontal band. Figure 7.23b shows what the residual pattern might look like if the dependent variable being predicted is non-normal; the horizontal band is still there but compressed on one side of 0 and expanded on the other side. Figure 7.23c reveals the existence of an unmodelled non-linear relationship; the band of residuals has a bend in it. Figure 7.23d shows a fan-shaped pattern typical of violation of the homogeneity of variance assumption. Figure 7.23e shows what the impact of a couple of extreme values (influential outliers) might look like; the horizontal band is tilted slightly from horizontal away from the influential points. Figure 7.23f reveals the influence of an important predictor that has not been included in the prediction model; identifying and including that predictor in the model would likely rectify the plot. Finally,

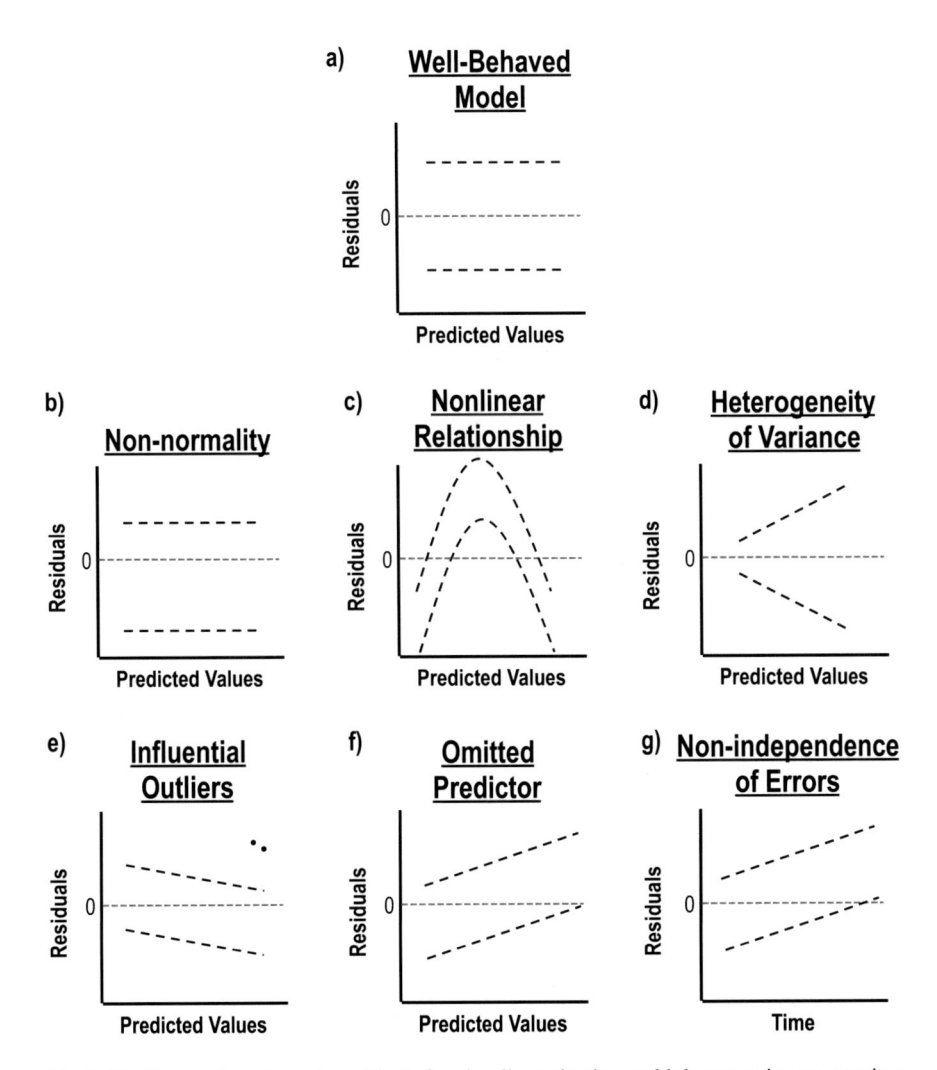

Procedure 7.13: Multiple Regression: Tests of Significance

Fig. 7.23 Diagnostic patterns in residuals for visually evaluating multiple regression assumptions

Fig. 7.23g presents a slightly differently configured plot from the others – the residuals are plotted against time. If the band of residuals is tilted in such a plot, this signals that prediction errors made by the regression model are not independent of each other. This plot would also work where residuals are plotted against case numbers in the database. Other types of diagnostic graphs, especially useful for checking the normality assumption, include a histogram plot of the residuals and a normal Q-Q (or P-P) plot (recall *Fundamental concept II*).

There is also a range of numerical indicators one could examine. For example, the Durbin-Watson test (refer back to the 'Model Summary' subtable in Table 7.17, last entry on the right) evaluates whether or not errors are non-independent (strictly speaking, that successive errors across cases are not correlated). A value near to or exceeding 2 indicates that the assumption of independence of errors is satisfied (as is the case with a value of 1.903 in Table 7.17); a value approaching 1 signals that there may be a problem with non-independent model errors (Field 2018, p. 387). Other indices that could be examined include: Studentised and Studentised deleted residuals, DfBeta, standardised DfBeta, DfFit, standardised DfFit, Cook's D and Leverage values (Fox 1991). Each of these measures assesses a different aspect of influence on regression estimates attached to specific observations, estimated by dropping an observation case from the analysis and re-estimating the regression model each time.

For Maree's simultaneous regression model, analysed in Table 7.17, it would be useful for her to examine several diagnostic plots to see how well certain assumptions have been satisfied. In Fig. 7.24, the top two plots are diagnostic

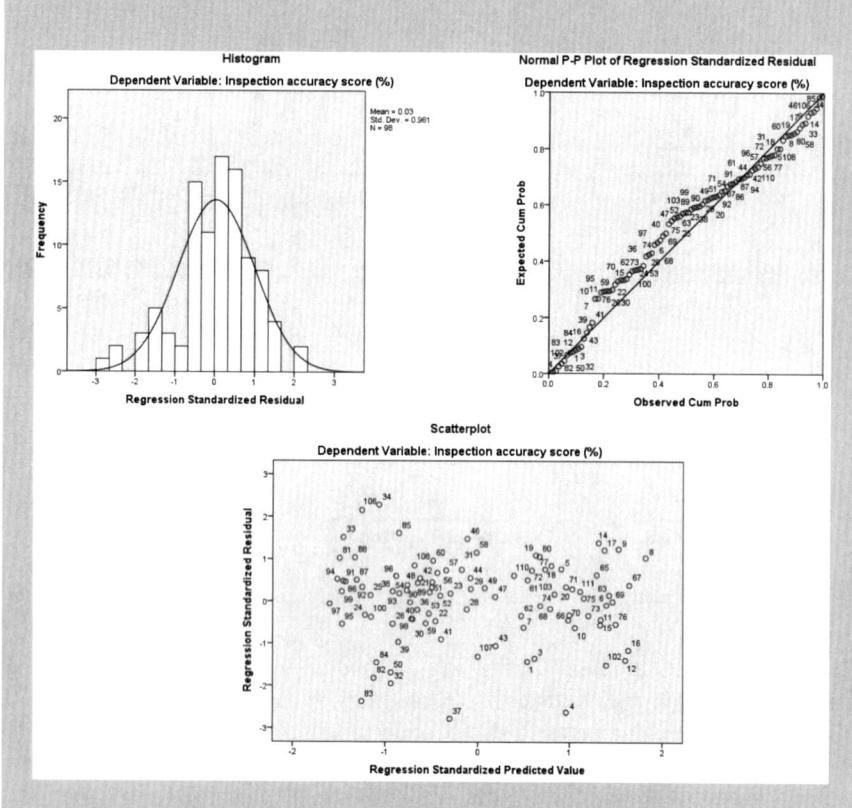

Fig. 7.24 Residual diagnostic plots for Maree's simultaneous regression model

(continued)

of the normality assumption and would be interpreted as described in *Fundamental concept II*. In terms of standardised residuals, Maree should expect to see very few or no residuals larger than 3 standard deviations away from the mean residual value of 0; a conclusion supported by the histogram plot. Thus, it seems Maree could conclude her data broadly satisfy the normality assumption. The bottom plot in Fig. 7.24 plots the standardised residuals against the standardised predicted values (similar to the plots shown in Fig. 7.23).

Broadly speaking, Maree could conclude that her prediction model is relatively well-behaved. The band of residuals resembles Fig. 7.23a: random and a roughly horizontal band around 0 with no obvious fan shape. The homogeneity of variance assumption would appear to be satisfied. However, participants 37 and 4 seem fairly well separated from the rest of the residuals, which might mark them as potential outliers. It could be worth Maree's time to drop these two participants from the analysis and re-estimate the regression model. If the new estimates with the two cases excluded looked highly similar to the estimates with the cases included, the two cases would not be considered to be influential outliers.

Testing Squared Semi-partial Correlations Hierarchically

We have thus far focused on testing hypotheses in the context of simultaneous regression analysis where the goal was to assemble a useful prediction model. However, it is possible to use hierarchical multiple regression analysis as a tool for testing specific theoretical propositions regarding the predictors. This is a more explanatory, as opposed to predictive, emphasis.

Generally speaking, theoretical propositions are tested through a specific ordering of the inclusion of predictors into the regression model and evaluating the gain that accrues as each new predictor enters the system. Predictors may be included singly or in logical sets depending upon the hypotheses being tested. The critical issue in hierarchical regression analysis is how one arrives at the proper ordering. This is done by logical, theoretical, or temporally based arguments. Once an order for considering predictors has been argued, the order cannot be changed in any further considerations of the data sample in question (to do so would constitute another ethical violation of the use of statistical technology similar to that which occurs if *a priori* comparison hypotheses for ANOVAs are specified after you have inspected your data).

> Suppose Maree wishes to build up an explanatory model for job satisfaction. Maree could argue, on the basis of logical and temporal grounds as well as prior job satisfaction theory, that the predictors of job satisfaction should be incorporated in the model in the following order. Gender and mental ability as personal characteristics exist long before any inspector gets hired and so should be considered as a set first. An inspector's education level should be considered next because it is determined after many of their other personal characteristics but generally before any employment. Then the general industry an inspector has chosen to work in should be considered next. Once hired as an inspector within a specific industry, his/her general attitudes toward work within their company should be considered next. Finally, perceptions of his/her general working conditions should be considered as having the most immediate potential influence on job satisfaction over and above all the contributions of the previous predictors.
>
> Thus, Maree's hierarchical order of predictor entry into the regression model for predicting **jobsat** would be: **dummy-recoded_gender** and **mentabil** together as a set first, then **dummy_recoded_educlev**, then **dummy_coded_computer_ind** and **dummy_coded_ automobile_ind** together as a set, then **training**, **quality** and **orgsatis** (the three components from the principal component analysis in *Procedure 6.5*) together as a set and finally **workcond**.

Hierarchical testing of predictor contributions proceeds by focusing on the R^2 values that result from each stage of the regression process. The regression model is built up by successively adding the next predictor(s) in the hierarchical order and observing the R^2 value at each stage. Each stage thus produces a new regression model. The R^2 from the first 'Model' is evaluated using an omnibus F-test. The contributions for predictors added into every 'Model' after the first are assessed by testing the difference between the R^2 from the current 'Model' (where the latest predictor(s) have been included) and the R^2 from the previous 'Model' using a partial F-test (recall Fig. 7.22). For every 'Model' after the first, the difference between the R^2 from the current stage and the R^2 from the previous stage gives the squared semi-partial correlation between the predictor(s) just entered and the criterion. We can interpret such a squared semi-partial correlation as reflecting the proportion (or percentage) of variance in the criterion which is uniquely explained by the current predictor(s) over and above all previously entered predictors but ignoring all predictors yet to come. Thus, a ready measure of the effect size for each predictor set is automatically obtained.

Procedure 7.13: Multiple Regression: Tests of Significance

Table 7.18 shows the two key interpretive subtables for Maree's hierarchical regression model. The analysis was produced by SPSS and unnecessary details have been edited out of the tables to simplify and focus the discussion. The 'Model' rows (Model 1 through Model 5) in the 'Model Summary' table of Table 7.18 shows the logical progression for building up the multiple regression model in the way described above; the footnotes tell you exactly which predictors are included in the model at any particular point (the regression 'constant' is always included at every stage in the hierarchical 'Model' development, but is typically ignored for interpretive purposes). The contributions for predictors for every 'Model' after the first are assessed by testing the difference (called 'R Square Change' in the 'Model Summary' subtable of Table 7.18) between the R^2 from the current 'Model' (where the latest predictor(s) have been included) and the R^2 from the previous 'Model' using a partial F-test (labelled by SPSS as 'F Change').

We can see the outcome from each hierarchical stage of the regression model building process. In 'Model 1' of the hierarchical ordering, the two personal characteristics, **dummy-recoded gender** and **mentabil**, entered together as a set do not contribute significantly, but do come close at

Table 7.18 Model summary of the hierarchical multiple regression analysis of predictor contributions to explanation of variability in job satisfaction scores

Model Summary[f]

Model	R	R Square	Adjusted R Square	Std. Error of the Estimate	R Square Change	F Change	df1	df2	Sig. F Change	Durbin-Watson
1	.233[a]	.054	.035	1.615	.054	2.870	2	100	.061	
2	.254[b]	.065	.036	1.614	.010	1.094	1	99	.298	
3	.576[c]	.332	.297	1.378	.267	19.372	2	97	.000	
4	.582[d]	.339	.283	1.392	.007	.346	3	94	.792	
5	.584[e]	.342	.278	1.397	.003	.384	1	93	.537	1.798

a. Predictors: (Constant), dummy_coded_gender, mentabil
b. Predictors: (Constant), dummy_coded_gender, mentabil, dummy_coded_educlev
c. Predictors: (Constant), dummy_coded_gender, mentabil, dummy_coded_educlev, dummy_coded_automobile_ind, dummy_coded_computer_ind
d. Predictors: (Constant), dummy_coded_gender, mentabil, dummy_coded_educlev, dummy_coded_automobile_ind, dummy_coded_computer_ind, training, quality, orgsatis
e. Predictors: (Constant), dummy_coded_gender, mentabil, dummy_coded_educlev, dummy_coded_automobile_ind, dummy_coded_computer_ind, training, quality, orgsatis, workcond
f. Dependent Variable: jobsat

Coefficients

		Unstandardized Coefficients		Standardized Coefficients			Correlations		
Model		B	Std. Error	Beta	t	Sig.	Zero-order	Partial	Part
1	(Constant)	.239	2.010		.119	.906			
	mentabil	.044	.018	.233	2.383	.019	.228	.232	.232
	dummy_coded_gender	-.156	.321	-.048	-.487	.627	-.024	-.049	-.047
2	dummy_coded_educlev	.333	.319	.102	1.046	.298	.105	.105	.102
3	dummy_coded_computer_ind	1.100	.316	.331	3.479	.001	.423	.333	.289
	dummy_coded_automobile_ind	-1.432	.394	-.343	-3.636	.000	-.381	-.346	-.302
4	quality	-.082	.155	-.052	-.529	.598	.130	-.054	-.044
	training	-.025	.137	-.016	-.179	.858	-.023	-.019	-.015
	orgsatis	-.097	.176	-.058	-.553	.581	.215	-.057	-.046
5	workcond	.073	.118	.077	.620	.537	.296	.064	.052

(continued)

$p = .061$ (which could be considered a suggestive or marginally significant result). At the next stage in 'Model 2', **dummy-recoded educlev** does not add a significant amount to explanation of variability in job satisfaction. At the next stage in 'Model 3' (shown in boldface), the industry dummy-coded predictors, as a set, explain about 26.7% of the variance in **jobsat** scores over and above the personal characteristics and educational level, which is a significant amount ($p < .001$). The remaining variables, added into the hierarchy in 'Model 4' and 'Model 5', contribute non-significant amounts of explanation.

Maree now knows that the only predictors that contribute significantly to explanation of variance in **jobsat** scores are those in the industry set. However, since there were two predictors in this set, Maree has more interpretive work to do, namely understanding which of the two predictors contributed significantly to the explanatory power of the set. Here, the hurdle step logic for hypothesis testing is activated. Maree does not investigate the contributions of predictors inside a set unless the set itself is first shown to contribute significantly to variance explanation.

To interpret the contributions of individual predictors inside a significantly contributing set, Maree must look at the **Coefficients** subtable (which has been heavily edited to remove unnecessary and potentially confusing detail). In that subtable, she must focus on that section of the table relevant to the specific Model being evaluated (Model 3 in this case). There is a line of statistics for each of the dummy-coded industry predictors. Both predictors are shown to contribute significantly to explanation of variance in **jobsat** scores. By squaring the 'Part' correlation for each predictor (yields the squared semi-partial correlation), Maree gets a measure of effect size that tells exactly what each predictor has been able to account for in **jobsat**. For **dummy_coded_computer_ind** predictor, the squared semi-partial correlation is $.289^2 = .084$ and for **dummy_coded_ automobile_ind** predictor, the squared semi-partial correlation is $-.302^2 = .091$. The t-tests and Sig values for each predictor evaluate the significance of their contribution. Maree can therefore conclude that the industry variables, as a set, explain a significant 23.7% of the variability in job satisfaction scores over and above all previously entered predictors. Within that set, knowing that an inspector works in the computer industry explains a significant ($p = .001$) 8.4% of the variability in job satisfaction scores over and above all previously entered predictors and knowing that an inspector works in the automobile industry explains a significant ($p < .001$) 9.1% of the variability in job satisfaction scores over and above all previously entered predictors. [Technical note: if you square the t-test for each predictor you would obtain the partial F-test for that predictor.]

There is one final important issue to discuss with respect to hierarchical multiple regression and this has to do with how best to conduct the *F*-tests for each stage in the analysis. As was shown in *Fundamental concept VI*, the *F*-test in the general linear model is a comparison between the behaviour of the ***MODEL*** to the behaviour of ERROR. In hierarchical regression terms, this essentially means comparing the variance explained by the regression model at a specific step in the model building process to the variance of the residuals. However, there is debate as to which residuals should be used as the comparison reference. Cohen et al. (2003, pp. 171–179) discuss this as the debate between what they call Model I and Model II error terms. Conceptually, Model I error involves using the residuals from the step at which the predictors just added into the regression model are being evaluated and is the default used by programs like SPSS. Model II error involves using the residuals from the very last step of the regression model building process, when all predictors have been included in the model.

If you refer to the Venn diagrams for hierarchical regression in Fig. 7.22, the issue can be visualised as addressing the question of which unshaded area in the ***Y*** variance circle provides the comparative error term for hypothesis testing. Model I error is the unshaded area at each step of the Model, when the predictors of interest have just been entered. Model II error is the unshaded area at the last stage of the model building process when all predictors have been represented in the Venn diagram. Cohen et al. (2003) discuss the advantages and disadvantages of each choice of error term.

Model I error provides a potentially more powerful *F*-test but at the cost of the residuals being potentially contaminated by systematic variance that predictors yet to be included in the regression model could explain. Model II error provides a potentially less powerful *F*-test but with the advantage that the residuals are uncontaminated by any systematic predictor influences (since they are all accounted for in the regression model). Model II error will often give a better and cleaner *F*-test and therefore may be the more appealing choice for the researcher on that basis. However, because statistical packages like SPSS will only use Model I error in computing the hierarchical regression *F*-Change tests (as was done in Table 7.18), using Model II error means recomputing and evaluating all of the *F*-tests by hand.

Appendix A presents the computational formula you would need to recompute a partial *F*-test for hierarchical regression using Model II error; the formula uses quantities produced by programs like SPSS, such as 'R Square', 'R Square Change' (for a Model step in the hierarchy), 'df1' (= degrees of freedom for the number of predictors entered at the Model step being evaluated), 'df2' (degrees of freedom for the residual error term at the last Model step) and the 'Part' correlation (for an individual predictor). Note that in simultaneous multiple regression, the question of Model I vs Model II error does not arise; Model II error is always used.

If Maree wished to evaluate her hierarchical regression model using partial F-tests based on Model II error, she must recompute the F-Change tests in the **Model Summary** table in Table 7.18. Doing this, using the formula given in Appendix A, she would find the following:

Model	R Square change	df1	df2[a]	Model II F change	Sig. F Change[b]
1	.054	2	93	3.816	.026
2	.010	1	93	1.413	.238
3	.267	2	93	18.869	<.001
4	.007	1	93	0.989	.346
5	.003	1	93	0.424	.517

[a]The values for df2 are all the same because Model II error comes from the last model in the hierarchical analysis
[b]These p-values can be found using the FDIST statistical function in Excel.

Notice that, using Model II error, Model 1 in the hierarchical regression analysis, where the **mentabil** and **dummy-coded_gender** predictors were entered, is now significant, in addition to Model 3, where the two industry-related predictors were entered. The **Model Summary** table in Table 7.18 showed that Model 1 was not quite significant at $p = .061$. This shows what impact having a cleaner error term can have on hypothesis tests. Maree's next step, if using Model II error, would be to compute Model II partial F-tests for the individual **mentabil** and **dummy-coded_gender** predictors, using information from the relevant row in the **Coefficients** table of Table 7.18. This would show her that **mentabil** was the only significant contributor to explanation of the variance in **jobsat** at this point in the regression model (Model II partial $F(1, 93) = 7.607$, $p = .007$, $sr^2 = .054$).

Advantages

Multiple regression tests of significance permit the examination of multiple independent variables as potential predictors of a criterion of interest. Predictors may be continuous, group-based or any mixture thereof, which creates a very flexible analytical framework. Regression weights can be tested using a type of t-test, which permits you to look for significant predictors to use when predicting the criterion. This is especially useful in simultaneous regression analysis where a regression weight near to or exactly 0.0 tells you that a variable is of no importance in predicting the criterion and could thus be eliminated from the prediction model. These t-tests also permit the testing of directional hypotheses if desired.

Multiple regression is a very general technique in that it is related to ANOVA as part of the general linear model (see, for example, *Fundamental concept VI*,

Procedures 7.6, 7.10, and *7.11*) if ordinal- or nominal-scale predictors are used. All that is required is that categorical predictors are recoded in specific ways to form proper predictors (e.g. dummy coding, effect coding) for use in the regression analysis.

Cohen et al. (2003), Dunteman (2005), Gill (2000) and Pedhazur (1997) talk at length about variable coding systems and the unified and powerful nature of the multiple regression approach to data analysis, most particularly with respect to hierarchical regression analysis and its capacity for partitioning the variance of a dependent variable or criterion. In hierarchical mode, multiple regression becomes a potent theory-testing technique which permits precise control over the ordering in which predictors get to exert their influence. The influences of early-entering predictors in the model building process are effectively controlled (partialled out, but only from the later-entry predictors, not the criterion, making it a semi-partialling process) when assessing the contribution of later-entering predictors.

Disadvantages

If the relationships between specific predictors and the criterion are not totally linear, then multiple regression will not provide a completely accurate summary of the predictive relationships. There is a more complex hierarchical multiple regression technique which can be used to test nonlinear predictor-criterion relationships called *power polynomials* (see Cohen et al. 2003, ch. 6 for more details; see also *Procedure 9.8*).

If there are many predictors (e.g. 10 or more), the number of participants in the sample needed to compute meaningful and stable regression weights for the model gets large. A general rule of thumb is to have an absolute minimum of 5 but preferably 10 or more participants per predictor (thus, if one has 15 predictors, the minimal number of participants needed would be 5×15 or 75 but the preferred number would be 150).

The other disadvantages mentioned for simple linear regression, such as assumptions of normality, homogeneity of variance (technically called 'homoscedasticity', when talking about multiple regression analysis) and independence of errors or residuals, also apply to multiple regression (recall *Procedure 6.3*; see also Berry 1993). Multiple regression analysis is also sensitive to the presence of outliers or extreme observations that can produce either distortions in the magnitudes of the regression weights or cause the regression model to severely over- or under-predict the criterion value for certain participants.

Fortunately, there are numerous graphical (recall Fig. 7.24) and numerical approaches available for detecting these problems, most of which do so by systematically examining the residuals from regression prediction (see Field et al. 2012, pp. 297–298; Fox 1991; Cohen et al. 2003, chapter 4 and Hair et al. 2010, pp. 220–223 for a review of these techniques). Remember that a **residual** is the difference between a participant's observed criterion value and the value the regression model predicts he/she should have obtained; a large residual value reflects a large error in prediction. There have been some recent nonparametric developments in the area of multiple regression analysis (including *bootstrapping*, see *Procedure 8.4*), which may provide alternative approaches where regression assumptions cannot be reasonably satisfied (see Fox 2000).

Most statistical packages (including SPSS, NCSS, STATGRAPHICS and SYSTAT) offer automatic stepwise predictor-selection options such as *stepwise regression analysis, forward selection regression analysis, backward elimination regression analysis*, and even, in some cases, *all possible subsets* **regression**, which are forms of regression analysis where the statistical software selects the order of entry of predictors into the regression model or makes 'decisions' about which predictors to keep in or reject from the model, based on each predictor's level of significance of contribution at any one stage. In my view, these types of regression analyses, while they may look very attractive, particularly if you have a large number of predictors and don't know which ones to use, are actually very dangerous and facilitate the automated construction of regression models without having to think about what you are doing. In the behavioural and social sciences, under most circumstances, you should make a conscious decision to either analyse all predictors at once or explicitly establish an order for predictor entry that you can control and conceptually/theoretically defend, as opposed to letting the software use simple statistical rules to choose the ordering of predictors and/or which predictors to keep or reject for you (Cohen et al. 2003, pp. 161–162, present some very compelling arguments in this regard).

Where Is This Procedure Useful?

Multiple regression would be useful in certain types of experimental and quasi-experimental designs (perhaps having only one group) where the intent is to obtain a predictive understanding of the data. Multiple regression is also useful in various kinds of surveys or observational studies where the relationships between several predictors and a criterion are of interest. If a stable and useful regression model can be derived for prediction purposes (as indexed by a high value for adjusted R^2), it might then be useful as an actual prediction rule for future cases.

> **For example** ... *Personnel psychologists often employ multiple regression models to assist organisations in their selection of appropriate applicants for specific jobs.*

Hierarchical multiple regression is extremely useful for the testing of specific theoretical propositions, usually but not always in quasi-experimental or general survey-type design contexts. In the theory-testing type of behavioural research endeavour, the quality of the final prediction model is of lesser importance than understanding how the various predictors contribute to the explanation of variability in the criterion variable.

> A health psychologist might be interested in understanding how a patient's demographic characteristics, past medical history, current life circumstances, current symptoms and treatment characteristics explain the variability in the extent of patient recovery from illness. Theoretically, the researcher might argue that demographic characteristics (gender, ethnic origin) exist prior to any medical problems the patient might have exhibited so these variables are used first to predict extent of recovery. Past medical history variables might be considered next, then current life circumstance, then current symptom measures, and finally measures of treatment characteristics. In this example, variables are ordered for consideration in the multiple regression analysis according to their theorised temporal sequencing in the lives of patients.

In *Procedure 8.7*, you will see that there is a direct connection between multiple regression models and the testing of theory-based structural equation models or so-called path or causal models. *Procedures 7.14*, *9.6* and *9.8* discuss and illustrate other types of multiple regression procedures.

Software Procedures

Application	Procedures
SPSS	*Analyze → Regression → Linear* ... and select your 'Dependent' variable. If you want a simultaneous regression analysis, enter all the predictors ('Independent(s)') of interest in one block, click the '*Statistics*' button and choose 'Part and partial correlations' as additional output; under the '*Options*' button, you can choose how SPSS will handle missing data: 'Exclude cases listwise' means a participant missing a score on any variable in the analysis is deleted from the analysis; 'Exclude cases pairwise' means a participant is dropped from a particular correlation calculation if he/she is missing a score on at least one of the variables; 'Pairwise' will maximise the sample size for the analysis but may render the solution unstable. If you want a hierarchical regression analysis, enter your predictors ('Independent(s)') in blocks, clicking 'Next' to enter the next block of predictors; click the '*Statistics...*' button and choose 'R-squared change' and 'Part and partial correlations' as additional output. SPSS is the easiest statistical package to use for hierarchical regression analysis and up to 9 blocks of predictors can be defined. Regression diagnostic graphs and indicators are not produced by default but can be selected as optional output. Click the '*Statistics...*' button and tick 'Durbin-Watson' and 'Casewise Diagnostics' and click the '*Plots*' button, tick 'Histogram' and 'Normal probability plot' and configure a 'Scatter' plot by choosing '∗ZRESID' for 'Y:' and '∗ZPRED' for 'X:'.
NCSS	*Analysis → Regression → Multiple Regression* and choose your dependent variable ('Y: Dependent(s)'), numeric independent variables ('Xs: Numeric Independent Variables') and categorical independent variables ('Xs: Categorical Independent Variables'); on the '*Reports*' tab, you can select from a slew of output options including diagnostic indicators; on the '*Plots*' tab,

(continued)

Application	Procedures
	you can select from a slew of output options including diagnostic plots. The analysis is of the simultaneous variety. One option offered by NCSS is to perform robust regression, which is a form of regression analysis that performs well when assumptions are not satisfied (this is a type of non-parametric regression analysis; see also *Procedure 9.8*). NCSS is not well-configured to conduct hierarchical regression analysis (it does not allow control over order of predictor entry nor does it produce semi-partial correlations).
SYSTAT	*Analyze → Regression → Linear → Least Squares* ... and select your 'Dependent' and 'Independent(s)' variables. By default, SYSTAT conducts simultaneous multiple regression analysis and under the '*Options*' tab, you can select various tests for normality. SYSTAT automatically produces the Durbin-Watson statistic and the residual scatterplot of the form illustrated in Fig. 7.23. SYSTAT is not well-configured to conduct hierarchical regression analysis (it does not allow control over order of predictor entry nor does it produce semi-partial correlations), except in the context of *Set Correlation*; recall *Procedure 6.8*.
STATGRAPHICS	*Relate → Multiple Factors → Multiple Regression...* and choose your desired 'Dependent Variable' and 'Independent Variables'; hit 'OK' and select the 'Fitting Procedure' as 'Ordinary Least Squares' (means a simultaneous regression model is estimated); hit '*OK*' and in the '*Tables and Graphs*' dialog box, select the tables and graphs you want to see – especially useful under 'Tables' is ticking 'Analysis Summary', 'Unusual Residuals' and 'Influential Points' and under 'Graphs', ticking 'Residuals vs Predicted' and 'Residuals vs Row Number'. STATGRAPHICS is not well-configured to conduct hierarchical regression analysis (it does not allow control over order of predictor entry nor does it produce semi-partial correlations).
R Commander	*Statistics → Fit models → Linear regression...* give your regression model a name in 'Enter a name for your model:' choose your dependent variable as 'Response variable'; and choose your predictors as 'Explanatory variables'. A simultaneous regression model analysis is reported, but with minimal statistics, few diagnostic indicators and no residual plots. However, because you have given your model a name, you can use other functionalities within R Commander to explore and diagnose your regression model. For example, the Durbin-Watson test can be produced using *Models → Numerical diagnostics → Durbin-Watson test for autocorrelation...* and ticking 'rho >0' for the 'Alternative Hypothesis'. To obtain basic diagnostic plots of the residuals, use *Models → Graphs → Basic diagnostic plots*.... More extensive multiple regression analyses and diagnostic devices can be produced using other functionalities and packages in the main R program (see Field et al. 2012, ch. 7, pp. 276–298). R Commander can conduct hierarchical regression analysis under the guise of regression model comparisons. To do this, you need to run separate regression analyses (as described above), each one successively including the new predictors to be added at the step in the hierarchical entry order being examined and giving each model a distinct name. Then use *Models → Hypothesis tests → Compare two models...* then choose the 'First model' to be compared (the model at the step previous to the one you are testing) and the 'Second model' to be compared (the model at the step you want to test.

(continued)

Application	Procedures
	Note that **R** Commander will offer all of the models you have built as possible choices for both the first and second models to be compared, so you need to be careful; the second model must be the one with the added predictors to be assessed. **R** Commander will not report squared semi-partial correlations (i.e. R^2 Change, although you could compute them by hand using the R square outcomes from each of the separate regression models you built and named), it will only report the partial F-test (using Model I error).

Procedure 7.14: Logistic Regression

Classification	Bivariate/multivariate; inferential; parametric.
Purpose	To test hypotheses about the values of specific regression weights for predictors of a dichotomous categorical dependent variable. Hierarchical models can also be tested in the context of logistic regression analysis.
Measurement level	The predictors may be assessed at virtually any level of measurement provided they are correctly represented (coded) for proper inclusion in the analysis. The dependent variable is measured at the nominal- or ordinal-level as a binary (two-category or 0, 1) indicator. [Note that there is a variant of logistic regression, called *multinomial logistic regression*, which can deal with predictions of a nominal- or ordinal-scale dependent variable having more than two categories; see Tabachnick and Fidell 2019, section 10.7.3.]

Logistic regression is a variant of ordinary multiple regression that handles predictions of and modelling responses to a dichotomous categorical dependent variable using any type of properly-coded predictors (Cohen et al. 2003; Field 2018, ch. 20). The dependent variable can be represented at either the nominal or ordinal scale of measurement if there are only two categories (e.g. yes/no response, two groups, two choices). The fact that the dependent variable is categorical creates some special problems that ordinary multiple regression, as described in *Procedure 7.13*, cannot handle.

Basically, logistic regression uses information from a set of predictors to predict the probability that any observation will fall into one or the other category of the dependent variable. To do this, logistic regression relies on a nonlinear modelling function, the logistic equation, to predict the probabilities rather than on the linear model as in ordinary regression. The nonlinear logistic model ensures that probability predictions remain realistically scaled between 0.0 and 1.0, whereas ordinary multiple regression would impose no bottom or top limits on such predictions, making it possible for nonsensical probability predictions like −.5 or + 1.5 to emerge. Aside from this fairly fundamental change in the base model, the general logic of the regression process is still the same: the overall quality of the prediction model can be described and tested using a variation of the R^2 measure and regression weights are produced and can be tested for significance.

> Suppose, in the QCI context, a quality control industry expert has convinced Maree that there would be value in trying to predict which inspectors need further training and development to improve their quality inspection performance, using information available in the QCI database. The expert states that industry standards are such that quality control inspection accuracy is four times as important as quality inspection speed.
>
> At the suggestion of the industry expert, Maree converts both **accuracy** scores and **speed** scores to standard z-scores (recall *Procedure 5.7*). Then she creates a new performance measure that combines 80% of the **accuracy** z-score with 20% of the **speed** z-score (after reversing the signs of the **speed** z-scores to account for the fact that slower decision times are indicative of poorer performance).
>
> Maree then classifies inspectors in the QCI database as either (0) meets performance standards (if their new performance measure score was greater than 0.0, equating to 'above average' overall performance) or (1) needs further training and development (if their new performance measure score was less than or equal to 0.0, equating to 'average to below average' overall performance) and represents this classification in a new variable called **inspect_perf**.
>
> Since the goal is to focus on performance with respect to industry, rather than type of company, Maree creates another new variable called **industry** which combines the PC and Large Business Computer manufacturing companies into category 1 (Computer industries); combines Small and Large Electrical Appliance manufacturing companies into category 2 (Electrical Appliance industries) and recodes the Automobile manufacturing companies into category 3 (Automotive industry). Maree is then interested in using logistic regression to predict **inspect_perf** using all of the relevant predictors available in the QCI database (**educlev, gender, industry, mentabil, quality, training, orgsatis, jobsat** and **workcond**).

Logistic regression works by focusing on predicting the probability that each observation in a sample belongs to the dependent variable category coded as 1. Thus, all predictions are in terms of probabilities, rather than actual scores. Predictors may be entered all at one time in the logistic regression model (a simultaneous approach) or may be entered in a series of blocks of predictors (a hierarchical approach, where the entry order is determined by the researcher). Each predictor has an associated regression coefficient and an associated conversion of that regression coefficient into an expression of the 'odds' that higher scores on the predictor predict membership in the category coded as 1 on the dependent variable. Instead of a t-test for each regression coefficient, logistic regression uses a 'Wald test', which has the same function, namely testing if the regression coefficient is significantly different from zero. Predictors with odds ratios near 1.0 will likely be non-significant predictors.

The overall goodness-of-fit of the logistic regression model, indicating how well the resulting logistic regression model predicts membership in the '1' category of the dependent variable, is indexed by a 'pseudo-R^2' measure called *Nagelkerke R Square*. The measure is called a 'pseudo-R^2' because, while it indexes the overall quality of the prediction system, it can't be interpreted as a proportion of variance explained in quite the same way as the usual multiple regression R^2, because categories, not scores, are being predicted. Technically, 'Nagelkerke R Square' reflects the proportion of variance in the predicted probability of category '1' membership that is explained by the predictors.

The 'Nagelkerke R Square' is tested for significance using a chi-square statistic. This test is what is called an *omnibus test of model coefficients* because it tests the significance of all predictor regression weights simultaneously. If the test is significant, it says that at least one predictor has a significant non-zero regression weight in the prediction model. Using the hurdle step logic of hypothesis testing (recall *Fundamental concept V*), a significant omnibus test gives the researcher permission to then explore the contributions of individual predictors to the model.

Table 7.19 shows selected outcomes, produced by SPSS, of Maree's logistic regression analysis predicting **inspect_perf** from the available predictors in the QCI database, including the new **industry** variable. All predictors were entered simultaneously. The logistic regression procedure in SPSS automatically builds any dummy-coded predictors needed to represent categorical variables as shown in the 'Dependent Variable Encoding' and 'Categorical Variables Codings' subtables at the top of Table 7.19. For example, the two **industry** variables **(1)** and **(2)** are dummy-coded predictors created by logistic regression to represent membership in one of the three industry groups (Automotive industry is the reference category). The bottom two subtables in Table 7.19 focus on the omnibus test of the logistic regression model.

The 'Model Summary' subtable in Table 7.19 shows the value for the 'Nagelkerke R Square' is .422, which can be interpreted as the prediction model accounting for about 42% of the variability in probabilities for belonging in the **inspect_perf** category coded as '1' (= 'Performance Below Expectations'). The 'Nagelkerke R Square' is statistically significant at $p < .001$ as shown in the 'Omnibus Test of Model Coefficients' subtable ('Model' row) in Table 7.19 ($\chi^2(10) = 37.282$, $p < .001$, Nagelkerke R-square $= .422$). The significance of this test indicates that at least one predictor has a significant regression weight in Maree's logistic prediction model. Maree can now explore the individual contributions of specific predictors to the model.

(continued)

Table 7.19 Logistic regression outcomes related to categorical variable coding and the omnibus test of the model

Dependent Variable Encoding

Original Value	Internal Value
.00 Performance Meets Expectations	0
1.00 Performance Below Expectations	1

Categorical Variables Codings

		Frequency	Parameter coding (1)	(2)
industry	1.00 Computer industry	39	1.000	.000
	2.00 Electrical appliance industry	39	.000	1.000
	3.00 Automotive industry	20	.000	.000
gender	1 Male	51	1.000	
	2 Female	47	.000	
educlev	1 High School Only	50	1.000	
	2 Tertiary Qualified	48	.000	

Omnibus Tests of Model Coefficients

		Chi-square	df	Sig.
Step 1	Step	37.282	10	.000
	Block	37.282	10	.000
	Model	37.282	10	.000

Model Summary

Step	-2 Log likelihood	Cox & Snell R Square	Nagelkerke R Square
1	98.208[a]	.316	.422

a. Estimation terminated at iteration number 5 because parameter estimates changed by less than .001.

(continued)

The 'Variables in the Equation' subtable at the top of Table 7.20 shows the contribution of every individual predictor in the logistic regression model. The significant predictors have their statistics highlighted in boldface. The column labelled 'B' contains the estimated regression weights for each predictor and the column labelled 'Exp(B)' shows each regression weight transformed into a statement of the odds that an inspector scoring higher on the relevant predictor will be a member of category '1' of **inspect_perf** (namely, the odds that the inspector's performance is below expectations). If 'B' is negative, the 'Exp(B)' odds favour the category coded '0' and will therefore be less than 1.0 (if you take the reciprocal of 'Exp(B)' in such cases, this will give the odds of an inspector belonging in the '0' category, given a higher score on the relevant predictor).

We can see that the predictors **industry(1), industry(2), orgsatis** and **workcond** are all significant. [Note that both **industry** dummy variables are tested simultaneously in the row highlighted in bold italics. This test is significant, so Maree can then go on to interpret each of the industry dummy variables separately.] The **industry(1)** predictor codes inspectors from the computer industries and **industry(2)** predictor codes inspectors from the electrical appliances industries. A third variable is not needed to code for the automotive industry, because if an inspector is not from the computer or electrical appliance industries, they must automatically be from the automotive industry. What the significant negative regression weights show for these two dummy-coded predictors is that membership in the '1' category for **inspect_perf** ('Performance Below Expectations') is less probable if inspectors come from either the computer industries or the electrical appliance industries. The exact odds are, respectively .056 to 1 and .083 to 1. Since both regression weights were negative, we could invert these odds by taking their reciprocal, 1/.056 and 1/.083, and conclude that the odds of being in the '0' category of **inspect_perf** ('Performance Meets Expectations') are 17.86 to 1 (nearly 18 times as likely) and 12.05 to 1 (12 times as likely) for inspectors from the computer and electrical appliance industries, respectively.

The significant regression weight for **orgsatis** is positive and indicates that higher scores on **orgsatis** are associated with a greater likelihood of being in the '1' category for **inspect_perf** ('Performance Below Expectations'). The exact odds of being in the '1' category for **inspect_perf**, for higher **orgsatis** scores, are 3.37 to 1 (over 3 times as likely). Interestingly for Maree, this means that inspectors who are more satisfied with their organisation tend to be the inspectors most in need of further performance development. The significant regression weight for **workcond** is negative indicating that higher scores on **workcond** are associated with a greater likelihood of being in the '0' category for **inspect_perf** ('Performance Meets Expectations')—not a surprising finding. The exact odds of being in the '0' category for **inspect_perf**, for higher **workcond** scores, are, after inverting the 'Exp(B)' odds from the table, 2.36 to 1 (over twice as likely).

(continued)

Table 7.20 Logistic regression outcomes related to individual predictor contribution and overall model performance

Variables in the Equation

		B	S.E.	Wald	df	Sig.	Exp(B)
Step 1[a]	educlev(1)	.626	.615	1.034	1	.309	1.870
	gender(1)	-.353	.513	.472	1	.492	.703
	industry			*7.460*	*2*	*.024*	
	industry(1)	-2.880	1.140	6.382	1	.012	.056
	industry(2)	-2.484	.947	6.876	1	.009	.083
	mentabil	-.026	.029	.820	1	.365	.974
	quality	-.135	.296	.208	1	.648	.874
	training	.352	.253	1.940	1	.164	1.422
	orgsatis	**1.215**	**.375**	**10.488**	**1**	**.001**	**3.370**
	jobsat	.255	.194	1.725	1	.189	1.290
	workcond	**-.860**	**.259**	**11.007**	**1**	**.001**	**.423**
	Constant	1.363	3.478	.154	1	.695	3.907

a. Variable(s) entered on step 1: educlev, gender, industry, mentabil, quality, training, orgsatis, jobsat, workcond.

Classification Table[a]

			Predicted		
			inspect_perf		
			.00 Performance Meets Expectations	1.00 Performance Below Expectations	Percentage Correct
Observed					
Step 1	inspect_perf	.00 Performance Meets Expectations	44	8	84.6
		1.00 Performance Below Expectations	11	35	76.1
	Overall Percentage				80.6

a. The cut value is .500

The 'Classification Table' at the bottom of Table 7.20 shows how well the logistic prediction model works in predicting which category of **inspect_perf** each inspector belongs to by comparing what the model predicts with the actual membership category for each inspector. Thus, 44 inspectors are correctly classified as 'Performance Meets Expectations' and 35 inspectors are correctly classified as 'Performance Below Expectations'—an overall correct classification prediction rate of 80.6%. The 'Classification Table' also shows where the model misclassifies inspectors. Thus, 11 inspectors are predicted to be in category '1' but are actually in category '0' and 8 inspectors are predicted to be in category '0' but are actually in category '1'. So, the prediction model works slightly better at predicting category '1' members. Figure 7.25 shows a

(continued)

Procedure 7.14: Logistic Regression

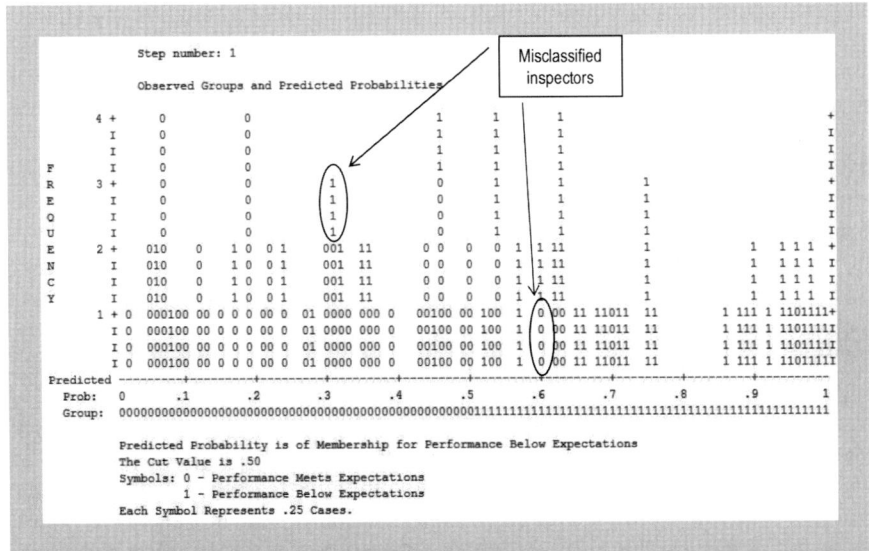

Fig. 7.25 Model probability prediction plot for cases

plot of the predicted probabilities for every case in the analysis against the number of cases for each prediction. Two misclassified inspectors are highlighted: one belonging to performance category 1 but predicted to be in category 0 (by virtue of a predicted probability less than .5) and the other one belonging to performance category 0 but predicted to be in category 1 (by virtue of a predicted probability greater than .5).

Logistic regression analysis is closely related to discriminant analysis (see *Procedure 7.17*) in that both types of analysis try to build models for predicting category or group membership. The chief differences are that logistic regression uses a nonlinear logistic model whereas discriminant analysis relies on a linear model. Logistic regression is somewhat less restrictive in the assumptions that it requires and its similarity to ordinary multiple regression analysis means that interpretation is relatively straightforward.

Advantages

Logistic regression provides a coherent way of building predictive models for categorical dependent variables in a way that maintains logical and interpretive consistency with ordinary multiple regression. Such models can easily be evaluated for their predictive accuracy using a classification table. Predictors at any scale of measurement and in any mixture can be used if the categorical predictors are properly coded for the analysis. Explanatory and hierarchical models can be also evaluated using logistic regression (see, for example, Miles and Shevlin 2001, ch. 6; Tabachnick and Fidell 2019, ch. 10). Logistic regression can provide an analytical solution in cases where there are nonlinear relationships between predictors and a continuous version of the dependent variable. In such cases, dichotomising the dependent variable using some statistical or theoretical rule to define high and low groups, may provide good estimates of predictor contributions and higher predictive accuracy unconfounded by the nonlinear relationships.

Disadvantages

Logistic regression analysis requires larger sample sizes relative to ordinary multiple regression analysis. Pedhazur (1997) has indicated that a 30 to 1 ratio of cases to predictors is preferred as opposed to the 5 or 10 to 1 ratio recommended for ordinary multiple regression analysis. Logistic regression may produce somewhat biased estimates if the number of cases in the category coded '1' is relatively small, especially if the number of predictors being tested is large. Furthermore, if a predictive model is to be used to predict new cases, the researcher should expect some loss of predictive accuracy and predictor efficacy similar to the loss shown in ordinary multiple regression analysis with the adjusted R^2 index.

Logistic regression is restricted to binary or two category dependent variables. More complex statistical models are required for predicting multiple category dependent variables (see Tabachnick and Fidell 2019, section 10.7.3). Such multinomial logistic models are more likely to be required in such areas as choice modelling research in consumer decision making (see *Procedure 9.4*), where, for example, choices amongst a larger set of products are to be evaluated, or in medical and health research where a number of possible diagnoses may need to be predicted using diagnostic indicators. Finally, at a more practical level, some people find the odds ratio interpretations of the regression weights rather confusing, especially when they go below 1.0 (i.e. when the B regression weight is negative) and predictions favour the dependent variable category coded '0'.

Where Is This Procedure Useful?

Logistic regression analysis is becoming an increasingly popular analytical technique in a wide range of research areas such as medical and social research, marketing, business and decision-making research and educational research. Data obtained using experimental, quasi-experimental or survey designs can be handled using logistic regression. The advent of readily available software packages to handle the sophisticated maximum likelihood estimation computations means that researchers now enjoy greater access to this modelling technology.

Procedure 7.14: Logistic Regression

Medical researchers often use logistic regression to predict whether patients have a specific disease (categories would be 'Yes' or 'No') based on the results of various diagnostic tests. Market researchers may use logistic regression in binary choice experiments to predict which of two products people will purchase based on scores for various attributes (e.g. price, size, availability, quality, and brand) of those products.

Social scientists may use logistic regression to predict whether people in a sample will vote Labour in a state/federal election based on various demographic and personality factors known about those people. Educational researchers may use logistic regression to test for factors that predict whether students will pass or fail a course of study. Accounting researchers may use logistic regression to build models using credit and personal history measures to predict loan defaulters.

Software Procedures

Application	Procedures
SPSS	*Analyze → Regression → Binary Logistic* . . . and select your 'Dependent' variable (which must be dichotomous) and the 'Covariates' (synonym for predictors in this case) for the analysis; covariates/predictors can be entered in hierarchical blocks; the default analysis is a simultaneous analysis; if any 'Covariates' are categorical, click the '*Categorical* . . .' button and indicate which variables are categorical and how you want them coded; click the '*Options...*' button to choose the statistics and plots you want to see ('Classification plots' will give you the plot illustrated in Fig. 7.25). SPSS can also perform multinomial logistic regression analyses as well (*Analyze → Regression → Multinomial Logistic* . . . only for this analysis, categorical predictors are called 'Factor(s)' and continuous predictors are called 'Covariate(s)').
NCSS	*Analysis → Regression → Logistic Regression* and choose your 'Y: Group Variable' (the categorical dependent variable), and your desired 'Numeric Independent Variables' and 'Categorical Independent Variables'; under the '*Model*' tab, you can choose the configuration of model you want to test (main effects only or main effects with interactions); under the '*Reports*' tab, choose the statistics you want to see and under the '*Plots*' tab, choose the plots you want to see. If your selected 'Y: Group Variable' has more than two categories, NCSS will automatically conduct a multinomial logistic regression analysis. For any categorical variables, you can choose whether the first or last category is used as the reference group for dummy coding. Note that NCSS only offers simultaneous or stepwise-constructed logistic models; hierarchical logistic regression models where the researcher controls the predictor entry order cannot be analysed in NCSS.

(continued)

Application	Procedures
SYSTAT	*Analyze → Regression → Logistic → Binary...*' and choose your 'Dependent' variable and predictors ('Independent (s)' – the default analysis is simultaneous entry of predictors (SYSTAT calls the 'Complete' estimation); for any categorical predictors, select the '*Category*' tab and identify those variables and choose the desired coding scheme; to see the prediction results of the model, select the '*Results*' tab and choose 'Prediction success table' and 'Classification table' at a minimum). Note that SYSTAT only offers simultaneous or stepwise-constructed logistic models; hierarchical logistic regression models where the researcher controls the predictor entry order cannot be analysed in SYSTAT. SYSTAT can also perform multinomial logistic regression analyses as well (*Analyze → Regression → Logistic →: Multinomial...*).
STATGRAPHICS	*Relate → Attribute Data → Logistic Regression...* and choose your 'Dependent Variable', 'Quantitative Factors' and 'Categorical Factors'; hit '*OK*' and in the '*Logistic Regression Options*' window, accept the default choices, including 'Fit' set at 'All Variables' (= simultaneous analysis); hit '*OK*' and in the '*Tables and Graphs*' window, select your desired tables (make sure 'Goodness of fit' and 'Predictions' are selected in addition to the default choices) and graphs (make sure 'Prediction Histogram' is selected in addition to the default choices). Note that STATGRAPHICS only offers simultaneous or stepwise-constructed logistic models; logistic regression hierarchical models where the researcher controls the predictor entry order cannot be analysed in STATGRAPHICS. STATGRAPHICS does not have the capability to run a multinomial logistic regression analysis.
R Commander	**R** Commander does not have the capability to conduct either logistic or multinomial logistic regression analyses. However, Field et al. 2012, pp.329–356) shows how both types of regression analyses can be run using other functionalities and packages in the main **R** program. See also the discussion and illustrations in Everitt and Hothorn (2006, ch. 6).

Procedure 7.15: Analysis of Covariance (ANCOVA)

Classification Univariate; inferential; parametric.

Purpose To permit control over the influence of specific extraneous variables, called 'covariates', by statistically removing their influence on a single dependent measure when comparing groups of participants in n-way ANOVA designs.

Measurement level The independent (grouping) variable(s) can be at any level of measurement. The dependent measure should be measured at the interval- or ratio-level; and the covariate(s) are typically also measured at the interval- or ratio-level. However, by using proper variable coding procedures, nominal or ordinal variables may also be used as covariates.

Procedure 7.15: Analysis of Covariance (ANCOVA)

Analysis of covariance is a specialised statistical method that integrates n-way ANOVA (see *Procedures 7.6* and *7.10*) and linear/multiple regression (see *Procedure 6.2* and *7.13*). The method is designed to statistically control for potential biasing variables by removing their influence on the dependent measure prior to comparing two or more groups (defined by one or more independent variables) of participants (see, for example, Field 2018, ch. 13; Tabachnick and Fidell 2019; Wildt and Ahtola 1978).

The biasing or extraneous variables are termed *covariates* and, as such, are believed to influence the dependent variable over and above whatever group differences (as defined by the independent variables) exist. Their influence is statistically removed from both the dependent variable and the independent variables through a partialling process (recall *Fundamental concept IV*). If such influence is significant, the resulting F-tests for the normal ANOVA design effects will be more powerful. This is because adjusting for covariates reduces the size of the error term (i.e. statistically controlling for extraneous variables reduces the error term in ANOVA by subtracting out one or more systematic sources of error).

> In a study of gender differences in worker job satisfaction, it could be prudent to measure participants' extent of experience in their current job in order to control for the influence of experience on level of satisfaction. In this case, the measure of experience would be used as a covariate to be controlled for prior to comparing the gender groups in terms of their job satisfaction.
>
> In a study that sets out to compare the academic performance of first-year university students in different degree programs, it could be prudent to obtain each student's Australian Tertiary Admission Rank (ATAR—a system-wide score used to select students into Australian tertiary degree programs at specific institutions), so that ATAR scores could be used as a covariate to be controlled for prior to making any degree program comparisons.
>
> In both examples, covariates are defined as extraneous variables that can theoretically influence the dependent variable at the same time that the independent variable is operating. Removing a covariate's influence on a dependent variable effectively rules out that covariate as an alternative plausible explanation for any group differences observed.

Analysis of covariance is especially useful in removing the influence of one or more extraneous variables from quasi-experimental observations obtained on 'intact' groups, where participants are not randomly assigned to the groups, but have to be treated as a single cohort (such as a hospital ward, classroom or an organisational department). In this case, the analysis adjusts for pre-existing differences among the various groups in the quasi-experimental or observational study. However, care must be taken in such designs to ensure that the covariate(s) cannot influence the category membership participants take with respect to the independent variable(s). For instance, in an educational study it would not be appropriate to use a variable like general mental ability as a covariate in a design where the independent variable for children is defined by low or high reading ability levels because general mental ability certainly has an influence on more specific abilities like reading.

> To illustrate what ANCOVA accomplishes, suppose Maree wishes to compare job satisfaction (**jobsat**) ratings by quality control inspectors within groups defined by **educlev** (High School Only and Tertiary-Qualified) and **industry** (Computer, Electrical Appliance and Automotive). These independent variable classification groups can be considered as intact groups because Maree could not randomly assign inspectors to specific categories.
>
> However, prior to conducting this analysis using a 2-way factorial ANOVA, Maree might consider it important to control for the potential influence of mental ability, gender and perceptions of working conditions. Thus, Maree could use **mentabil, dummy-coded_ gender** (dummy coded as '0' for males and '1' for females) and **workcond**, as covariates for this analysis.
>
> The covariate adjustment portion of ANCOVA is the first step in the analysis and is accomplished using multiple regression procedures to predict the **jobsat** scores using the three covariates. At this stage, the covariates can be tested for significance using standard F-tests (see *Procedure 7.13* where ANOVA F-tests can be used to evaluate the significance of regression predictors) because Maree stands to gain the most if she employs covariates that will significantly adjust scores on the dependent variable.
>
> The residuals (predicted **jobsat** scores subtracted from the observed **jobsat** scores) from this multiple regression analysis are then computed. These residuals are essentially what remain of **jobsat** scores once they have been adjusted for the influence of the covariates (i.e. the influence of the covariates has been partialled out of **jobsat**). It is these adjusted **jobsat** scores which are then analysed for differences with respect to the **educlev** and **industry** independent variables using standard 2-way factorial ANOVA procedures.
>
> Table 7.21 presents an edited version of the relevant SPSS analysis output for the ANCOVA described above. The full general linear model with all effects represented (covariates plus ANOVA effect variables) explains 38% of the variance in **jobsat** scores (shown in footnote **a** of Table 7.21). There are three rows at the top of Table 7.21 (underneath the 'Source' labelled 'Corrected Model'), each reporting a significance test for one of the covariates. Of the three covariates, **mentabil** is the only significant one at $p = .05$ ($F(1, 89) = 3.962$, $p = .05$, partial $\eta^2 = .043$). This means that **mentabil** has a significant influence on **jobsat** that Maree could statistically remove. Even though only one of the three covariates was significant, they still have some adjustment effect on **jobsat** scores which may make it slightly easier to detect significant effects due to the independent variables that define the groups.

(continued)

Procedure 7.15: Analysis of Covariance (ANCOVA)

Table 7.21 Summary ANCOVA table of the **educlev** by **industry** ANOVA design, adjusting for **mentabil**, **dummy-coded_gender** and **workcond** as covariates

Dependent Variable: jobsat

Source	Type III Sum of Squares	df	Mean Square	F	Sig.	Partial Eta Squared
Corrected Model	97.679^a	8	12.210	6.832	.000	.380
dummy_coded_gender	2.400	1	2.400	1.343	.250	.015
mentabil	7.082	1	7.082	3.962	.050	.043
workcond	.167	1	.167	.094	.760	.001
industry	43.279	2	21.640	12.108	.000	.214
educlev	23.851	1	23.851	13.345	.000	.130
industry * educlev	15.323	2	7.662	4.287	.017	.088
Error	159.066	89	1.787			
Corrected Total	256.745	97				

a. R Squared = .380 (Adjusted R Squared = .325)

Below the 'Source' row labelled 'workcond' in Table 7.21, the standard 2-way factorial ANOVA effect tests appear for **educlev**, **industry**, and the **educlev** by **industry** interaction. Note that the interaction is significant (F (1, 89) = 4.287, p = .017, partial η^2 = .088) so Maree should interpret the interaction at this point. Figure 7.26 shows the graph of this significant 2-way interaction. As signalled in the footnote to Fig. 7.26, the mean scores plotted have been adjusted for any influences of the three covariates (the most important of which Maree has already established as **mentabil**). The graph clearly shows that job satisfaction levels are virtually identical for inspectors, at both educational levels, in the Computer industry. However, Tertiary-Qualified inspectors (squares) show substantially higher levels of job satisfaction in both the Electrical Appliance and Automotive industries (especially so for the Automotive industry), compared to their High School-only educated colleagues (circles).

[Note, if Maree had conducted this analysis as a standard ANOVA instead of an ANCOVA with three covariates, the 2-way interaction would have barely been significant at p = .047. This shows that an ANCOVA, where effective covariates are used, will yield more powerful F-tests for the main effects and interactions in the n-way ANOVA design.]

There is a significant main effect difference between the two **educlev** categories (F(1, 89) = *13.345, p < .001*, partial η^2 = .130). There is also a significant main effect difference between the three **industry** types (F (1, 89) = 12.108, p < .001, partial η^2 = .214). A quick glance at the pattern of covariate-adjusted means in Fig. 7.26 gives a feeling for the nature of these significant main effects (although posthoc multiple comparison tests, as described in *Procedure 7.8*, could certainly be performed on the **industry** means to confirm the visual impressions).

(continued)

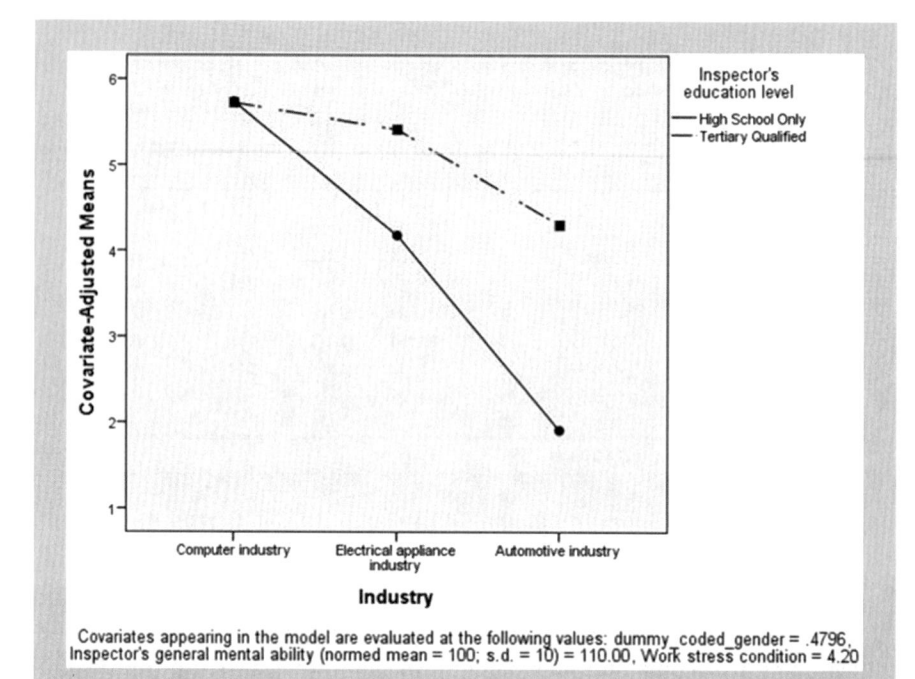

Fig. 7.26 The significant 2-way interaction between educlev and industry after adjusting for mentabil, dummy-coded_gender and workcond covariates

> On average, it looks like Tertiary-Qualified inspectors generally have a higher level of job satisfaction compared to High School-Only educated inspectors (the interaction tells us this isn't strictly true, if one chooses to look at the Computer industry). On average, it looks as if the general level of job satisfaction is lowest in the Automotive industry, next lowest in the Electrical Appliances industry and highest in the Computer industry.

Analysis of covariance can be useful even in formal experiments where participants have been randomly assigned to groups or conditions and the researcher wishes to remove the influences of certain known extraneous variables from the dependent variable in order to obtain more sensitive (powerful) tests for group differences. Many behavioural researchers have also employed ANCOVA as an alternative to the use of repeated measures ANOVA in the analysis of Pretest-Posttest control group designs by using pretest scores as a covariate for adjusting posttest scores. However, this practice is now generally frowned upon because it can lead to certain unresolvable paradoxes in the interpretation of analysis outcomes. The better practice is to use repeated measures ANOVA to analyse such designs.

Advantages

Analysis of covariance permits the statistical control of extraneous variables believed to influence the dependent measure. The partialling process removes any overlap that is shared between the covariates and the dependent variable, which effectively reduces the amount of unexplained variability in the system. Such control permits more sensitive and powerful tests for group differences using ANOVA procedures. Ideally, this advantage will be maximised in the case where significant covariates are employed. The important thing to remember in ANCOVA is that the independent variable mean scores to interpret should be the covariate-adjusted means, not the original descriptive means for the groups.

Disadvantages

Analysis of covariance is a controversial method in that it is often difficult to tell when it can be appropriately applied. In fact, it is a frequently abused procedure. It requires the same assumptions as ANOVA (see *Procedure 7.6*) plus two additional assumptions. One additional assumption concerns the fact that the covariate(s) must relate to the dependent measure to the same degree and in the same way across all comparison groups (termed the 'homogeneity of regression lines' assumption); and the other concerns the fact that the covariate(s) should not influence the conduct or effects of any experimental conditions and, strictly speaking, should be free of measurement error (this is often difficult to satisfy). The typical recommendation is to apply analysis of covariance only if other methods of control and analysis are not possible or are inappropriate and even then to use it with some interpretive caution.

Where Is This Procedure Useful?

Analysis of covariance may be useful in certain experimental or quasi-experimental designs where experimental control of certain influential variables (either control by procedure or control by design; refer back to *Fundamental concept VII*) is not possible yet it is desirable (perhaps necessary) to remove their influence on the dependent measure prior to executing group comparisons. In many cases, the contexts where ANCOVA proves most useful is in quasi-experimental social research where the grouping variables being considered define intact or pre-existing groups whose memberships could not be influenced or determined by the behavioural researcher (most demographically-, educationally-, or occupationally-defined groups, for example, are of this type).

Software Procedures

Application	Procedures
SPSS	*Analyze → General Linear Model → Univariate* ... and select your 'Dependent Variable', the 'Fixed Factor(s)' (i.e. independent variable(s)) in your design, and your 'Covariate(s)' (making sure any categorical covariates have been previously and appropriately coded); configure your desired '*Plots*' for interactions; hit the '*Options*' button and ensure that you 'Display Means For:' all of the relevant 'Factor(s) and Factor Interactions' in the 'Estimated Marginal Means' section (this provides the dependent variable means for each ANOVA design effect, adjusted for any covariates) and indicate, at a minimum, that you want to see 'Estimates of effect size' and 'Descriptive statistics'.
NCSS	*Analysis → Analysis of Variance (ANOVA) → General Linear Model (GLM)* and specify your 'Response Variable' (i.e. dependent variable), your 'Factor Variables' (i.e. independent variables) and your 'Covariate Variable (s)' (making sure any categorical covariates have been previously and appropriately coded); under the '*Reports*' tab, indicate which analysis reports and multiple comparison tests you want to see and the alpha level you want to use; under the '*Plots*' tabs, indicate which plots you want to see.
SYSTAT	*Analyze → Analysis of Variance → Estimate Model...* and select your 'Dependent' variable, your 'Factor(s)' (independent variables and choose the coding system to use) and your 'Covariate(s)' (making sure any categorical covariates have been previously and appropriately coded). *Analysis → Analysis of Variance → Pairwise Comparisons...* and choose the 'Effects' you want to analyse and choose the multiple comparison tests you want to see. This procedure can only be run after the *Estimate Model ...* procedure has been run.
STATGRAPHICS	*Compare → Analysis of Variance → Multifactor ANOVA* ... and choose your 'Dependent Variable', the desired 'Factors' variables and the desired 'Covariate(s)' (making sure any categorical covariates have been previously and appropriately coded); hit '*OK*' and the 'Options' window will open – indicate the maximum order of interaction you want to test (e.g. for a 2-way ANCOVA, this would be '2'); hit '*OK*' and in the '*Tables and Graphs*' window, tick all options in the 'Tables' section and make sure you tick 'Interaction plot' in the 'Graphs' section.
R Commander	**R** Commander does not offer an explicit procedure for computing an ANCOVA. Instead, ANCOVA must be conducted using commands and packages within the main **R** programming environment. Field et al. (2012) demonstrate how ANCOVA can be accomplished using **R** in their chapter 11.

Procedure 7.16: Multivariate Analysis of Variance & Covariance (MANOVA & MANCOVA)

Classification	Multivariate; inferential; parametric.
Purpose	To compare groups of participants classified according to one or more grouping (independent) variables on two or more dependent measures simultaneously. The analysis can easily be extended to include n-way factorial MANOVA designs and multivariate analysis of covariance (MANCOVA) designs.
Measurement level	Any level of measurement can be used to define the independent grouping variables. The dependent measures must be measured at the interval- or ratio-level. Any covariate(s) are typically also measured at the interval- or ratio-level. However, by using proper variable coding procedures, nominal or ordinal variables may also be used as covariates.

Multivariate analysis of variance (*MANOVA*) provides tests for significant differences on a set (two or more) of dependent variables between groups of participants defined by one or more grouping or independent variables (Bray and Maxwell 1985; Hair et al. 2010, ch. 8). If covariates are to be controlled for, the analysis becomes a *multivariate analysis of covariance* (*MANCOVA*). All dependent variables are analysed simultaneously with the fundamental logic following traditional one-way ANOVA, factorial ANOVA, repeated measures ANOVA or factorial ANCOVA lines in terms of the effects (main effects and interactions, if any) being tested (Field 2018, ch. 17; Tabachnick and Fidell 2019, ch. 7).

Figure 7.27 displays the logic of the testing and interpretation pathways available in MANOVA or MANCOVA using a flowchart representation. The gray boxes each represent a step in the hurdle logic, with progression conditional on achieving a significant result at that step. The flowchart also makes clear that there are choices of interpretive pathways available to the researcher, once a significant multivariate result for a specific effect has been achieved – pathways we will review through the discussion below.

One important outcome from a MANOVA is the provision of a multivariate F-test (most commonly based on a statistic called Wilks' lambda—see Tabachnick and Fidell 2019, ch. 7) for each and every effect in the MANOVA design. This multivariate F-test provides an additional step in the significance testing hurdle logic that the researcher must clear before moving to investigate specific differences on specific dependent variables for the effect of interest. A significant multivariate F-test indicates that group differences between the *profiles* of means for all dependent measures are significant for the particular main effect or interaction being tested, once the influences of any covariates have been removed. By introducing this extra hurdle for each MANOVA effect to clear, we control for the possible inflation of the

Testing/interpretation logic for each Main Effect and Interaction effect in the MANOVA/MANCOVA design

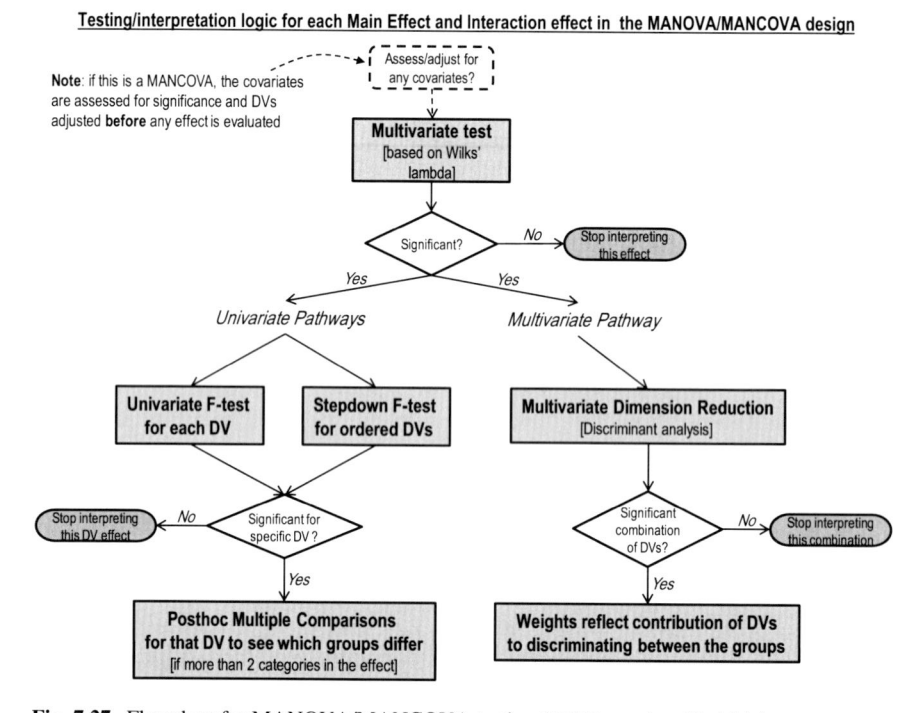

Note: if this is a MANCOVA, the covariates are assessed for significance and DVs adjusted **before** any effect is evaluated

- Assess/adjust for any covariates?
- Multivariate test [based on Wilks' lambda]
- Significant? No → Stop interpreting this effect
- Yes → Univariate Pathways / Multivariate Pathway

Univariate Pathways:
- Univariate F-test for each DV
- Stepdown F-test for ordered DVs
- Significant for specific DV? No → Stop interpreting this DV effect
- Yes → Posthoc Multiple Comparisons for that DV to see which groups differ [if more than 2 categories in the effect]

Multivariate Pathway:
- Multivariate Dimension Reduction [Discriminant analysis]
- Significant combination of DVs? No → Stop interpreting this combination
- Yes → Weights reflect contribution of DVs to discriminating between the groups

Fig. 7.27 Flowchart for MANOVA/MANCOVA testing (*DV* Dependent Variable)

alpha error rate for a study in which multiple ANOVAs (one for each separate dependent variable) might otherwise be contemplated. Additionally, the multivariate test accounts for any correlations which may exist among the various dependent variables (which would be completely ignored if separate univariate ANOVAs on each dependent variable were to be conducted).

> Suppose Maree is interested in examining differences in the general work attitudes of quality control inspectors from the three broad types of manufacturing industries in the QCI database (Computer, Electrical Appliance and Automotive, formed by recoding the five categories of the **company** variable into a new three-category variable called **industry**). Given the exploratory factor analysis of the nine work attitude items in the QCI database conducted earlier in *Procedure 6.5*, Maree identified three distinct components or dimensions of work attitude: attitude toward product quality (**quality**), attitude toward training (**training**), and satisfaction with aspects of the organisation (**orgsatis**). Her MANOVA design would thus involve three dependent variables being compared across the three categories of the grouping variable **industry**. Figure 7.28 plots the means for each type of company on

(continued)

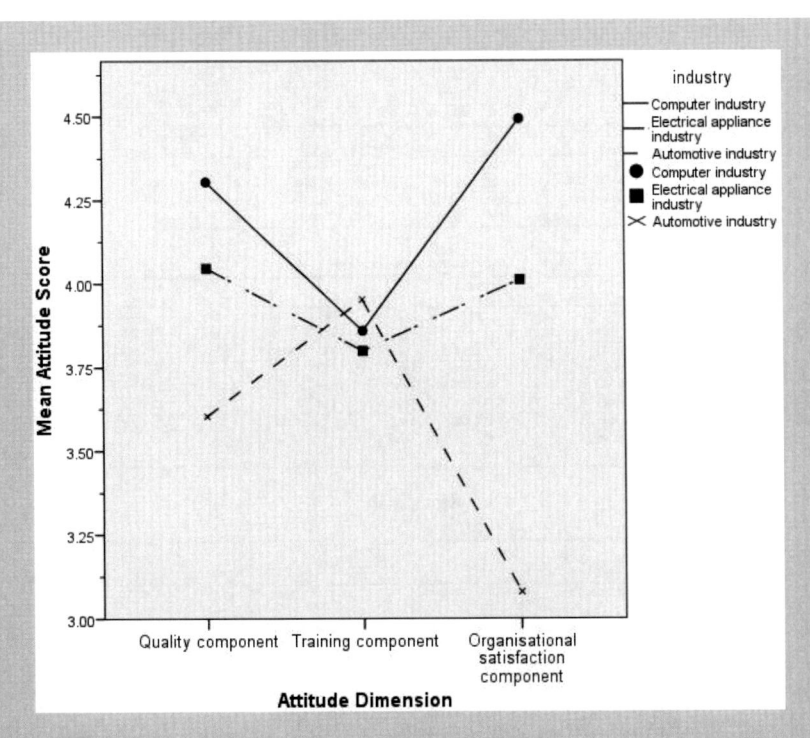

Fig. 7.28 Mean work attitude dimension scores by **industry**

these three dependent measures, yielding three company profiles to compare. This provides a visual idea of the problem MANOVA is designed to solve.

MANOVA examines these three **industry** profiles for evidence of overall significant differences (think of the process as comparing the overall 'average' level of the three profiles in Fig. 7.28). In this way, use of MANOVA provides a more efficient preliminary test for group differences than would conducting three separate one-way ANOVAs (one analysing **quality** differences; one analysing **training** differences; and one analysing **orgsatis** differences). There is also a multivariate version of eta-squared which can be used to summarise the proportion (percentage) of variance in the collection of dependent measures which can be explained by the effect being tested.

Table 7.22 shows some selected and edited subtables from the SPSS MANOVA of this design. The top subtable, labelled 'Multivariate Tests', reports four different multivariate tests (the advantages and disadvantages of each are discussed by Tabachnick and Fidell 2019, section 7.5.2). However, common practice suggests that Maree should focus on the multivariate F-test associated with Wilks' lambda (shown in boldface), which is significant in this

(continued)

Table 7.22 Summary one-way MANOVA tables (SPSS) analysing **industry** differences on the three work attitude dimensions

Multivariate Tests

Effect		Value	F	Hypothesis df	Error df	Sig.	Partial Eta Squared
industry	Pillai's Trace	.285	5.937	6.000	214.000	.000	.143
	Wilks' Lambda	.715	6.450	6.000	212.000	.000	.154
	Hotelling's Trace	.398	6.960	6.000	210.000	.000	.166
	Roy's Largest Root	.396	14.123	3.000	107.000	.000	.284

Tests of Between-Subjects Effects

Source	Dependent Variable	Type III Sum of Squares	df	Mean Square	F	Sig.	Partial Eta Squared
industry	quality	7.129	2	3.565	3.477	.034	.060
	training	.336	2	.168	.147	.864	.003
	orgsatis	28.841	2	14.421	19.798	.000	.268
Error	quality	110.730	108	1.025			
	training	123.816	108	1.146			
	orgsatis	78.665	108	.728			
Corrected Total	quality	117.859	110				
	training	124.152	110				
	orgsatis	107.507	110				

Multiple Comparisons

Tukey HSD

Dependent Variable	(I) industry	(J) industry	Mean Difference (I-J)	Std. Error	Sig.	95% Confidence Interval Lower Bound	Upper Bound
quality	1.00 Computer industry	2.00 Electrical appliance industry	.2589	.21352	.448	-.2485	.7663
		3.00 Automotive industry	.7012	.26667	.026	.0674	1.3349
	2.00 Electrical appliance industry	3.00 Automotive industry	.4423	.26856	.231	-.1959	1.0805
training	1.00 Computer industry	2.00 Electrical appliance industry	.0595	.22578	.963	-.4771	.5960
		3.00 Automotive industry	-.0937	.28198	.941	-.7638	.5764
	2.00 Electrical appliance industry	3.00 Automotive industry	-.1531	.28399	.852	-.8280	.5217
orgsatis	1.00 Computer industry	2.00 Electrical appliance industry	.4814	.17997	.023	.0537	.9091
		3.00 Automotive industry	1.4134	.22476	.000	.8792	1.9475
	2.00 Electrical appliance industry	3.00 Automotive industry	.9320	.22636	.000	.3941	1.4699

Based on observed means.
The error term is Mean Square(Error) = .728.

Note: These tables have been heavily edited to convey a less 'cluttered' story and remove potentially confusing detail

analysis. The value for the multivariate F-test is 6.45 with an associated p-value of less than .001, reflecting an **industry** effect that explains a respectable 15.4% of the variability of scores on all three work attitude dimensions (Wilks' lambda $= .715$, multivariate $F(6, 212) = 6.45, p < .001$, multivariate partial $\eta^2 = .154$). The significant multivariate F-test tells Maree that the three types of **industry** do differ in their average ratings across the set of three work attitude dimensions.

It must be emphasised, however, that this significant multivariate F-test tells Maree nothing about which specific work attitude dimensions contributed to the significant **industry** effect. By looking at Fig. 7.28, she could certainly form a visual impression regarding which work attitude dimensions might have contributed to the significant multivariate F-test (e.g. **orgsatis** seems to show the largest differences across the industries), but she really needs statistical evidence to support this conclusion.

After finding a significant multivariate *F*-test, there are three alternative ways one can progress with follow-up tests. Figure 7.27 displayed these as the *Univariate Pathways* (where there are two sub-pathways) and the *Multivariate* pathway. The simplest follow-up procedure is the 'Univariate *F*-test' pathway (tracing down the left side of Fig. 7.27), which most behavioural and social researchers tend to follow. This procedure involves evaluating, in a posthoc spirit (meaning this step is only taken if a significant multivariate *F*-test for the effect of interest is achieved), group differences on each separate dependent measure using univariate ANOVA *F*-tests (see *Procedures 7.6* and *7.10*). Furthermore, if a significant univariate ANOVA *F*-test is found and three or more groups are being compared, it would then be appropriate to use Tukey HSD or Scheffé multiple comparison tests (see *Procedure 7.8*) to isolate exactly which specific groups differ on the dependent measure of interest. One drawback of this univariate strategy is that it ignores any correlations between dependent variables. If such correlations are high, the resulting tests may be biased.

[It should be noted that specific MANOVA group comparisons can be specified *a priori* if such comparisons can be justified theoretically or on the basis of past research findings. *A priori* specifications of hypotheses permit much more powerful tests for group differences on the dependent measures.]

The univariate *F*-test approach for Maree's MANOVA is summarised in the middle subtable, labelled 'Tests of Between-Subjects Effects', in Table 7.22. This subtable shows that the types of **industry** significantly differ on both the **quality** dimension ($F(2, 108) = 3.565, p = .034$, partial $\eta^2 = .060$) and the **orgsatis** dimension ($F(2, 108) = 14.421, p < .001$, partial $\eta^2 = .268$), but not on the **training** dimension.

The bottom subtable in Table 7.22, labelled 'Multiple Comparisons', reports posthoc Tukey HSD multiple comparison tests (recall *Procedure 7.8*) for pairs of **industry** groups. Here, Maree should only look at the outcomes for the **quality** and **orgsatis** dependent variables (since the univariate *F*-test for **training** was not significant). The HSD tests reveal that: (1) inspectors in the Computer industry have a significantly more positive **quality** attitude than inspectors in the Automotive industry; (2) inspectors in Automotive and Electrical Appliance industries have significantly poorer attitudes compared to inspectors in the Computer industry on the **orgsatis** dimension and (3) the Automotive industry is significantly poorer in **orgsatis** attitude than the Electrical Appliance industry. These are all patterns reflected in Fig. 7.28.

The second strategy for following up a significant multivariate F-test is a variation of the *Univariate Pathway* strategy called 'Stepdown F-tests' (see Tabachnick and Fidell 2019, section 7.5.3.2), which travels down the right-hand fork of the *Univariate Pathway* in Fig. 7.27. This approach is useful if the dependent variables are highly correlated. It involves specifying a hierarchical ordering of importance among the dependent variables. Then a univariate F-test is computed for the most important dependent variable and ANCOVA F-tests are computed for all other dependent variables in order, where each F-test controls for relationships with all previously tested dependent variables (i.e. those tested earlier in the ordering are used as covariates for the F-test at a particular subsequent step). The stepdown strategy is not used very often because specifying an ordering of importance among dependent variables is hard to justify (they are often of equal importance in the researcher's mind). An additional problem confronting the researcher wishing to conduct stepdown F-tests is that as the SPSS statistical package evolves (SPSS is the only package to offer stepdown F-tests), stepdown tests become trickier to perform.

> Although the three work attitude dimensions are not highly correlated, the stepdown F-test strategy could theoretically be used in Maree's MANOVA reported in Table 7.22 if she assumes an ordering of importance among the three work attitude dependent variables.
>
> If she considered the **quality** attitude to be most important, followed by **orgsatis** followed by **training**, then the stepdown strategy would proceed in the following steps:
>
> - a standard univariate ANOVA F-test would be conducted on the **quality** dependent variable;
> - a univariate ANCOVA F-test would be conducted on the **orgsatis** dependent variable, using the **quality** variable as a covariate; and
> - a univariate ANCOVA F-test would be conducted on the **training** dependent variable, using both the **quality** and **orgsatis** variables as covariates.
>
> The stepdown strategy, if applied to the MANOVA in Table 7.22 would show that industries significantly differ on the **quality** dependent variable and significantly differ on the **orgsatis** dependent variable, after controlling for its relationship with the **quality** variable as a covariate. Industries do not significantly differ on the **training** dependent variable, after controlling for both **quality** and **orgsatis** as covariates.

There is a more complicated *Multivariate Pathway* follow-up procedure, travelling down the right-hand side of Fig. 7.27, that involves conducting a 'discriminant analysis' (see *Procedure 7.17* and Tabachnick and Fidell 2019, section 7.5.3.3) to

Procedure 7.16: Multivariate Analysis of Variance & Covariance

isolate which combinations of dependent variables discriminate best between the various groups defined by the effect being tested. The advantage of this multivariate follow-up procedure is that it uses any information that is available regarding correlations between the different dependent variables. The univariate F-test follow-up approach discards this information whereas the stepdown approach requires the researcher to decide how to partition this information among the dependent variables, using a prioritisation system. For this multivariate approach, there will be as many ways of combining dependent variables as there are degrees of freedom for the groups. Each combination, called a *discriminant function*, can be evaluated for significance, each will have its own pattern of weights (or 'standardised coefficients') reflecting the contribution of each dependent variable to that function and each has an associated correlation (called a 'canonical correlation'), which can be squared to give a measure of effect size.

> The 'discriminant analysis' strategy could be used in Maree's MANOVA reported in Table 7.22 to see which of the three work attitude dimensions discriminated best between inspectors from the three different industries. Two different discriminant functions are possible, but only the first one is significant, and it explains 28.4% of the variance in group differences. It turns out that, largely in parallel with the univariate follow-up results, the work attitude dimension that discriminates best between the three industries is **orgsatis**; **quality** and **training** contribute much less to discrimination between the three industries. Thus, with respect to this particular MANOVA in the QCI database, all three follow-up strategies to the significant multivariate F-test for **industry** differences tell the same story: it is on the **orgsatis** variable and, to a lesser extent, the **quality** variable where **industry** differences tend to be focused.

In some instances, the three follow-up approaches will tell different stories, in which case the multivariate approach is to be preferred because it uses more of the information in the data, without requiring or forcing priorities of importance to be set up for the dependent variables. One problem confronting the researcher wishing to conduct the multivariate discriminant analysis approach is that as the SPSS statistical package evolves (SPSS is one of the few packages to offer discriminant analysis as an optional part of the MANOVA process), discriminant analysis as a follow-up test to MANOVA becomes trickier to perform. It may be easier to perform this follow-up strategy using the process described in *Procedure 7.17*.

Advantages

In many studies, particularly experiments, quasi-experiments, and surveys, a number of dependent measures (e.g. questionnaire items, attitude items, performance measures, component or factor scores) are obtained from the same participants. Going straight to conducting univariate analyses of variance on each measure separately would likely lead to at least one false significant result, particularly if the number of dependent measures is large. MANOVA controls for this problem by placing another hurdle in the researcher's path to claiming significance, that hurdle being an analysis of all the dependent measures simultaneously using one multivariate F-test for each MANOVA effect. If a multivariate F-test is significant, then the most common approach to follow-up testing is the univariate ANOVA F-test strategy, which can be used to explore group differences in a posthoc fashion. This approach ignores dependent variable correlations, so if such correlations are high, the stepdown F-test strategy or the multivariate discriminant analysis strategy becomes more viable.

Also, MANOVA takes account of any intercorrelations among the dependent measures (which separate univariate F-tests cannot do) thereby using this additional information to help evaluate group differences. Using the multivariate follow-up approach maintains this logic in the posthoc investigation of group differences. Virtually any ANOVA or ANCOVA design involving any combination of between groups and within groups (repeated measures) variables can have multiple dependent measures analysed using MANOVA or MANCOVA, as appropriate.

Disadvantages

MANOVA is a complex statistical procedure. For valid tests (tests having adequate power to detect group differences) of group differences on many dependent measures, relatively large sample sizes within each group may be required. MANOVA can only be done using a computer, and the available programs are somewhat difficult to learn how to use and tend to produce more output information than you would ever need. This is because most available programs implement a general linear model approach which is general enough to be able to produce many different types of analyses, for a wide variety of MANOVA/MANCOVA designs, but sometimes at the expense of meaningful labelling of output statistics.

If dependent measures having different units of measurement are included, they may bias the significance tests. Finally, MANOVA appears to be more sensitive to violations of the usual ANOVA assumptions of normality (multivariate normality in the case of MANOVA) and homogeneity of group variances) than are the univariate procedures, which makes checking the distributions of each dependent variable an important exercise (see discussions in Tabachnick and Fidell 2019, section 7.3). [Note that some programs, such as SPSS, can be asked to provide multivariate (Box's M test) and univariate (Levene's F-test) tests for the homogeneity of variance assumption.]

Where Is This Procedure Useful?

MANOVA is quite useful in any experimental and quasi-experimental designs where multiple dependent measures have been obtained on the same participants. Certain types of survey comparisons on multiple items may be assessed using MANOVA with intact groups defined by various demographic characteristics (e.g. gender, ethnic background, occupation, educational level obtained). More complex experimental designs can be handled by MANOVA—in fact, any design one can analyse with univariate ANOVA/ANCOVA can, theoretically, be analysed with MANOVA/MANCOVA if multiple dependent measures are used. The MANOVA approach to handling repeated measures designs (a procedure called 'profile analysis', see Tabachnick and Fidell 2019, ch. 8) allows a researcher to circumvent the stringent assumptions associated with the more traditional ANOVA approach to repeated measures designs.

Software Procedures

Application	Procedures
| SPSS | *Analyze* → *General Linear Model* → *Multivariate* ... and select the desired 'Dependent variables', independent variables ('Fixed Factor (s)') and covariates ('Covariate(s)'), if any; click the '*Post Hoc* ...' button to select a posthoc multiple comparison test to apply; click the '*Plots*' button to configure any graphs you want to see (e.g. interaction graphs); and click the '*Options* ...' button to select reporting of 'Descriptive statistics', select 'Estimated Marginal Means' to see (for adjusting effect means for covariate influences), 'Estimates of effect size' and any other desired analysis outputs.<br><br>The *General Linear Model* procedure, described above, cannot provide you with access to stepdown F-tests or the follow-up multivariate discriminant analysis. To implement these strategies, the older style MANOVA command syntax must be used, which means working in the SPSS Syntax window, rather than with a specific analytical procedure on the *Analysis* menu. For example, for Maree's analysis to compute stepdown F-tests and the posthoc discriminant analysis, the syntax of her command looked like:<br><br>```
MANOVA quality orgsatis training BY industry (1,3)
  /OMEANS=TABLES(industry)
  /PMEANS=TABLES(industry)
  /PRINT= SIGNIF(EIGEN, DIMENR, STEPDOWN, EFSIZE) HOMOGENEITY
  /DISCRIM STAN ROTATE ALPHA(.05)
  /DESIGN= industry.
```<br><br>[Note that the ordering of variables after the 'MANOVA' command is important if stepdown tests are desired – the variables must be listed in the hierarchical ordering you want SPSS to use for the tests, most important first, least important last.]<br><br>(continued) |
```

Procedure 7.16: Multivariate Analysis of Variance & Covariance

Application	Procedures
NCSS	*Analysis → Analysis of variance (ANOVA) → Multivariate Analysis of variance (MANOVA)* ... and select the desired 'Response variables' and 'Factor Variable'(s) of interest; under the '*Reports*' tab, select the output statistics you want to see; under the '*Plots*' tab, select the graphs you want to see. By default, the univariate F-test follow-up approach is implemented. NCSS cannot perform stepdown F-tests or follow-up discriminant analysis within the MANOVA process (you would have to run a separate discriminant analysis, as described in *Procedure 7.17*). NCSS cannot perform MANCOVAs.
SYSTAT	*Analyze → MANOVA → Estimate Model* ... and select the desired 'Dependent(s)' variables and 'Independent(s)' variables; on the '*Category*' tab, indicate which independent variables are categorical (usually all of them for a MANOVA, but if some are not categorical, SYSTAT treats them as covariates and thus produces a MANCOVA). By default, the univariate F-test follow-up approach is implemented, and the follow-up discriminant analysis is reported. SYSTAT cannot perform stepdown F-tests. The MANOVA/MANCOVA output from SYSTAT can be difficult to read because of the statistical labelling system it uses for various portions of the output. Also, because of how the SYSTAT MANOVA procedure is set up, it cannot easily handle categorical covariates. The way to manage this is to properly code the categorical variables outside of the MANOVA procedure, then use the recoded variables as numerical independent variables.
STATGRAPHICS	*Compare → Analysis of Variance → General Linear Models...* and select the desired 'Dependent Variables' and 'Categorical Factors' (the independent variables in the MANOVA design); hit '*OK*' and configure the MANOVA design you want analysed (specifying all effects you want to see tested); hit '*OK*' and select the Tables and Graphs you want to see reported. 'Hit '*OK*' to run the analysis. Then place the mouse cursor inside of an output table and right-click; select 'Include MANOVA' and hit '*OK*'. By default, the univariate F-test follow-up approach is implemented. STATGRAPHICS cannot perform stepdown F-tests or follow-up discriminant analysis within the MANOVA process (you would have to run a separate discriminant analysis, as described in *Procedure 7.17*). STATGRAPHICS cannot perform MANCOVAs.
R Commander	**R** Commander does not offer the capability to conduct a MANOVA or MANCOVA. However, Field et al. (2012, pp. 719–746) illustrate how MANOVAs can be run using functionalities and packages within the main **R** program. They also show how to obtain univariate F-tests as well as the follow-up discriminant analysis. However, stepdown F-tests cannot easily be conducted within **R**.

Procedure 7.17: Discriminant Analysis

Classification	Multivariate; inferential; parametric.
Purpose	Provides a method of predicting differences between two or more groups of participants with respect to several variables or predictors simultaneously, especially focusing on deriving specific mathematical rules for predictively classifying participants as members of a specific group based on the multivariate information.
Measurement level	The group indicator variable is usually measured at the nominal level. Although the predictor variables may be measured at any level if they are appropriately coded for the analysis, they are typically measured at the interval- or ratio-level.

Discriminant analysis (Hair et al. 2010; Klecka 1980; Tabachnick and Fidell 2019) is very closely related to MANOVA (see *Procedure 7.16*) and canonical correlation (see *Procedure 6.8*). Huberty (1984) usefully distinguished between two general approaches to using discriminant analysis. One approach, called *descriptive discriminant analysis*, is used to show which variables out of a multivariate set are most important in differentiating between various groups; this is the approach used in the multivariate follow-up test for posthoc investigation of a significant MANOVA *F*-test (recall *Procedure 7.16*). The second approach to using discriminant analysis is called *predictive discriminant analysis* and has the explicit intention of finding the best mathematical equation(s) for predicting group memberships for individual participants.

Conceptually, discriminant analysis is like a MANOVA run in reverse, where the dependent variables become the predictors and group membership becomes the criterion to be predicted. Discriminant analysis also has broadly the same purpose as logistic regression analysis (see *Procedure 7.14*) in the case where membership in either of two groups is being predicted. In contrast to cluster analysis (recall *Procedure 6.6*) where variables are used to discover or create a typology of groups or 'clusters', discriminant analysis takes the groups as a given and tries to find which combinations of variables are best able to tell them apart.

Generally, the aim of discriminant analysis is to construct statistical models that combine predictors in various ways to maximally discriminate between the two or more known groups of individuals in a sample. These models are known as *discriminant functions*, each of which contains a set of standardised weights to be applied to the predictors. These standardised discriminant function weights are analogous to regression weights and reflect how important each predictor variable is to discrimination among groups with that specific model. However, the functions themselves are commonly interpreted based on canonical structure correlations (see *Procedure 6.8*) that reflect how correlated each predictor is with the discriminant function. [Note that structure correlations tend to be more stable indicators of

Procedure 7.17: Discriminant Analysis

predictor importance to group discrimination than the standardised discriminant weights (Hair et al. 2010, pp. 389–390).]

As with canonical correlation analysis, discriminant analysis can yield more than one significant discriminant function (the actual number of functions possible is either the number of groups minus one, or the number of predictors, whichever is smaller). The overall quality of each discriminant function is assessed using a canonical correlation which measures the extent to which the discriminant function relates to specially created (by the analysis procedure) variables that index group membership. Discriminant functions that have a high canonical correlation are more useful (and will be more predictively accurate) than functions with a low canonical correlation. Each canonical correlation can be tested for significance. Descriptive discriminant analysis stops with the substantive interpretation of the significant discriminant functions.

Predictive discriminant analysis pushes things a bit further by using the significant discriminant function(s) to predict the group that each individual in the sample belongs to. The degree to which this model-based classification matches the original known group memberships can be assessed using contingency table analysis (recall *Procedure 6.2* and *Procedure 7.1*). Discriminant analysis is designed so that the optimum models (those that would make fewest classification errors) are found for the sample. Discriminant functions that are highly successful in correctly classifying individuals into their groups may be considered to be substantively useful.

In addition, new individuals, not in the original sample, can be predictively classified using the derived discriminant functions. This last step is frequently undertaken as a way of evaluating how well the models would operate in a non-optimal environment to classify entities that were not used to produce the models. The discriminant functions will always make more classification errors when applied outside the sample context in which they were derived, but if, despite the loss in optimum performance, the models still perform adequately in the new context, we would feel more confident in applying them as general predictive devices. This process of applying mathematical decision rules, optimally computed using information from one sample, to an entirely different sample is called *cross-validation* and is a critical process for behavioural researchers to undertake in predictive discriminant analysis.

> Suppose in the context of the QCI database, Maree is asked to derive a model or models that will predict which general type of industry a quality control inspector works in. To simply the process, Maree combines the five categories of **company** into three types of **industry**: 'Computer' (combining PC and Large Business Computer companies); 'Electrical Appliance' (combining Small and Large Electrical Appliance companies); and 'Automotive'.
>
> Maree wishes to derive models that discriminate employment in those **industry** groups using a set of other measures available in the QCI database:

(continued)

dummy_coded_gender, **dummy_coded_educlev**, **mentabil**, **jobsat**, the scores on the three components identified in *Procedure 6.5* – **quality, training**, and **orgsatis, workcond, accuracy** and **log_speed**. The goal of the exercise is to show that employment in different manufacturing industries can be predicted from knowledge of a quality control inspector's demographic characteristics, work attitudes, work environment and work performance. With three groups and ten predictor variables, Maree can produce and interpret a maximum of two possible discriminant functions. Table 7.23 shows an edited summary of the relevant aspects of this discriminant analysis as produced by SPSS.

Two significant discriminant functions emerge from this analysis, reflected in the statistics shown in the top two subtables (labelled 'Eigenvalues' and 'Wilks' Lambda') in Table 7.23. The tests for significant canonical correlations (which translate to significant discriminant functions) are based on Wilks' lambda and chi-squared and are peculiar in that they are ordered (a stepdown testing strategy).

The first test which was significant $(\chi^2 (10) = 136.300, p < .001)$ evaluated whether there were any significant canonical correlations. The second test which was also significant $(\chi^2 (4) = 26.053, p = .002)$ evaluated whether any significant canonical correlations remained after the first and largest one was partialled out. Because both tests were significant (shown in boldface), Maree needs to interpret two discriminant functions, one of which has a very strong canonical correlation of .839 and the other a moderate canonical correlation of .500.

The bottom subtable in Table 7.23, labelled 'Structure Matrix', reports the canonical structure correlations for each significant discriminant function. The substantive contributors to each discriminant function, shown in boldface, were identified using a researcher-determined cut-off correlation value of 0.3 to gauge variable importance. The QCI variables that contributed most to the first discriminant function were **workcond, accuracy, orgsatis, log_speed** and **jobsat**. The QCI variables that contributed most to the second discriminant function were **orgsatis** (with a substantive negative correlation), **log_speed** and **dummy-coded_educlev**. To understand how these variables influenced predictions of employment in **industry** groups, Fig. 7.29 plots the position of every single inspector in the sample with respect to his or her score predicted using each discriminant function model (discriminant function scores are standardised as z-scores).

Each **industry** is summarised by a 'Group Centroid' point, representing the average discriminant function scores for all members of that specific industry (represented by the black squares in the plot and generally surrounded by employees in that industry). You can see that for discriminant function 1 (horizontal axis), the Computer industry scores highest (right-most black square) and the Automotive industry the lowest (left-most black square). The

(continued)

Procedure 7.17: Discriminant Analysis

Table 7.23 Relevant SPSS summary tables from a discriminant analysis using QCI measures to predict the employment **industry** of inspectors

Eigenvalues

Function	Eigenvalue	% of Variance	Cumulative %	Canonical Correlation
1	2.381[a]	87.7	87.7	.839
2	.334[a]	12.3	100.0	.500

a. First 2 canonical discriminant functions were used in the analysis.

Wilks' Lambda

Test of Function(s)	Wilks' Lambda	Chi-square	df	Sig.
1 through 2	.222	136.300	20	.000
2	.750	26.053	9	.002

Structure Matrix

	Function 1	Function 2
workcond	.598*	.206
accuracy	.408*	.011
jobsat	.343*	-.117
quality	.185*	-.071
dummy_coded_gender	-.133*	-.032
orgsatis	.382	-.437*
log_speed	.363	.417*
dummy_coded_educlev	-.192	.311*
mentabil	.094	.162*
training	.016	.153*

Variables ordered by absolute size of correlation within function.

*. Largest absolute correlation between each variable and any discriminant function

(continued)

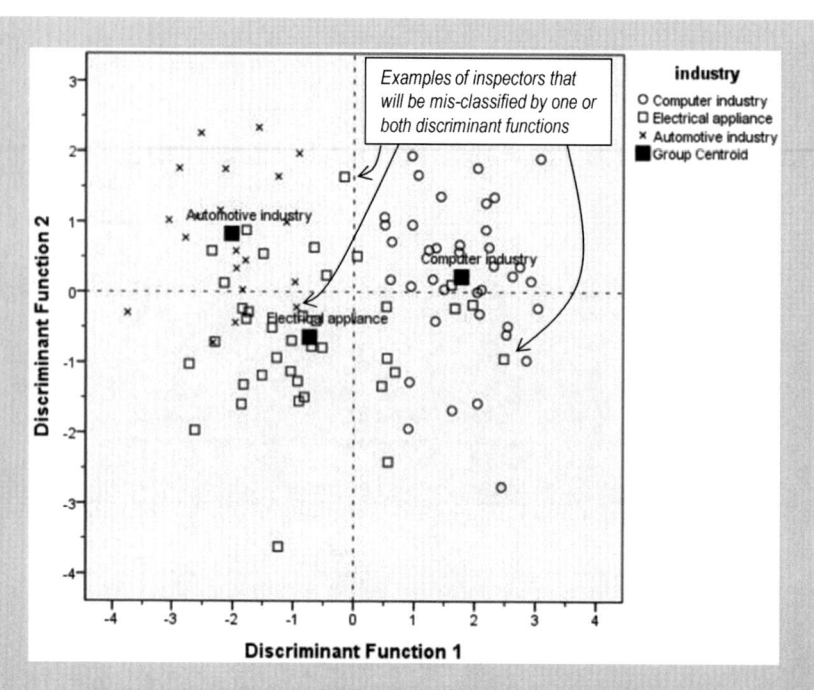

Fig. 7.29 Plot of the two discriminant significant functions for predicting membership in one of the three industry groups

structure correlations suggested that **workcond, accuracy**, **orgsatis, log_speed** and **jobsat** contributed most to this function and all with positive correlations. This means that the Computer industry inspectors can generally be distinguished from Electrical Appliance and Automotive industry inspectors by virtue of having more positive perceptions of their working conditions, higher levels of job and organisational satisfaction and more accurate but slower work performance. Discriminant function 1 does not distinguish very well between the Automotive and Electrical Appliance industries.

You can also see that for discriminant function 2 (vertical axis), the Electrical Appliance industry scores somewhat lower than both the Computer and Automotive industries. The structure correlations suggested that **orgsatis, log_speed** and **dummy-coded_educlev** contributed most to this function and **orgsatis** had a negative correlation whereas **log_speed** and **dummy_coded_educlev** had positive correlations. This means that Electrical Appliance industry inspectors were generally distinguished from Computer and Automotive industry inspectors by virtue of having a more negative attitude toward their own organisation, slower work performance and a tendency to be tertiary-qualified. The fact that the horizontal separation between

(continued)

Procedure 7.17: Discriminant Analysis

the **industry** groups was larger than their vertical separation reflects the greater predictive strength of discriminant function 1 (i.e. it had a much larger canonical correlation).

In order to predict group membership for inspectors, what discriminant analysis does is classify each inspector based on which 'Group Centroid' they are closest to. Figure 7.29 also makes it clear that there were cases where individual inspectors appeared to be in the wrong area of the plot (e.g. some Electrical Appliance squares appeared closer to the Automotive 'Group Centroid'; some Electrical Appliance squares appeared closer to the Computer 'Group Centroid' and some Automotive xs appeared closer to the Electrical Appliance 'Group Centroid'). This indicates that some individual inspectors will be misclassified by the discriminant function models—i.e. the models will not make perfect classification predictions.

Table 7.24 summarises the results of using the two discriminant functions to predict each inspector's **industry** group. The top half of the table, labelled 'Original', shows counts in the upper part and percentages in the lower part with predicted group memberships are identified in the columns and original group memberships are identified in the rows). If the discriminant functions are working well, Maree should see large numbers of observations in the main diagonal cells of the 'Original' table (boldface numbers) and very few observations elsewhere.

Table 7.24 Classification results

Classification Results[a,c]

			Predicted Group Membership			
		industry	1.00 Computer industry	2.00 Electrical appliance industry	3.00 Automotive industry	Total
Original	Count	1.00 Computer industry	**38**	1	0	39
		2.00 Electrical appliance industry	5	**29**	5	39
		3.00 Automotive industry	0	5	**15**	20
	%	1.00 Computer industry	**97.4**	2.6	.0	100.0
		2.00 Electrical appliance industry	12.8	**74.4**	12.8	100.0
		3.00 Automotive industry	.0	25.0	**75.0**	100.0
Cross-validated[b]	Count	1.00 Computer industry	37	2	0	39
		2.00 Electrical appliance industry	8	23	8	39
		3.00 Automotive industry	0	8	12	20
	%	1.00 Computer industry	94.9	5.1	.0	100.0
		2.00 Electrical appliance industry	20.5	59.0	20.5	100.0
		3.00 Automotive industry	.0	40.0	60.0	100.0

a. 83.7% of original grouped cases correctly classified.
b. Cross validation is done only for those cases in the analysis. In cross validation, each case is classified by the functions derived from all cases other than that case.
c. 73.5% of cross-validated grouped cases correctly classified.

(continued)

It is easy to see from Table 7.24 that inspectors from the Computer industry were most accurately classified (97.4%) and those from the Electrical Appliance industries were least accurately classified (74.4%). No one from the Computing industry was incorrectly classified as being from the Automotive industry and no one from the Automotive industry was incorrectly classified as being from the Computer industry—these two industries are thus unambiguously discriminable (Fig. 7.29 makes it clear that discriminant function 1 would be doing most of the work here). However, there was some misclassification of inspectors between the Automotive and Electrical Appliance industries and some misclassification of inspectors between the Electrical Appliance and Computing industries. Overall, Maree's two discriminant functions correctly classified 83.7% of the quality control inspectors (see footnote a in Table 7.24); a very respectable predictive outcome.

The bottom half of Table 7.24, labelled 'Cross-validated', simulates what would happen if Maree's two discriminant functions were used to predict new cases. SPSS simulates this by recalculating both discriminant functions without a specific case and then classifying that case using the resulting functions. SPSS repeats this process for all cases, yielding what is called 'leave-one-out cross-validation' (in some circles, this process is also known as 'jackknifing'; see also *Procedure 8.4*). Note how the performance of the two functions degrades somewhat through this cross-validation process, giving an estimated cross-validated classification rate of 73.5%. So, while the discriminant functions were significant and quite strong statistically, they would not be highly accurate devices for predicting industry employment, especially for those industries which are not Computer-based.

Advantages

Descriptive discriminant analysis has the same general advantages as MANOVA. This approach to discriminant analysis is useful for characterising the importance of variables in differentiating between groups to the maximum extent possible. Predictive discriminant analysis utilises the discriminant function models, consisting of weighted combinations of predictors, to predict the group membership of sample participants. The general accuracy of these predictions can be readily assessed, and more specific information can be obtained about the tendencies for the model(s) to misclassify participants in specific groups.

Procedure 7.17: Discriminant Analysis

Predictive discriminant analysis might seem similar to cluster analysis (see *Procedure 6.6*) but, in cluster analysis, the group memberships are unknown and remain to be discovered whereas, in discriminant analysis, the group memberships are known and remain to be recovered. In *Procedure 7.14*, we noted that logistic regression analysis was often preferred to discriminant analysis for a variety of reasons. However, if a linear function is the most appropriate way to model differences between the two groups being predicted, discriminant analysis will provide a more efficient and interpretable analysis (without requiring the interpretation of 'odds').

Disadvantages

Unfortunately, discriminant analysis also suffers from the same disadvantages as MANOVA. In addition, there can be problems in deriving accurate and stable discriminant weights and performing accurate participant reclassification into groups if the pattern of correlations among the predictors changes from one group to another. All the patterns should be approximately the same (this is called the 'equality of covariance matrices' assumption and is not dissimilar to the sphericity assumption in repeated measures designs). The same rule for sample size as applies to multiple regression (see *Procedures 6.4* and *7.13*) should be applied to discriminant analysis. Some classification problems can arise if the sample sizes of the various groups are vastly different and the researcher ignores this fact by relying on statistical software default settings that typically assume that the groups are of equal size.

As the classification functions that result from discriminant analysis are optimised only for the sample from which the models(s) are derived, cross-validation of the models becomes imperative if actual use of the model(s) as predictive devices is entertained. This usually means that even larger samples are required as one typical approach to cross-validation involves obtaining a large sample of participants, randomly splitting the large sample into two sub-samples, deriving the discriminant functions using one sub-sample, and applying the resulting model(s) to predicting the group memberships of participants in the other sub-sample (see also *Appendix B*). The classification table from this latter analysis can then be closely inspected for patterns that indicate how well the model(s) will perform under non-optimal prediction conditions.

Virtually all statistical packages offer the option of building discriminant functions in a stepwise fashion (the default is usually building 'simultaneous' functions—all variables entered at once). For reasons outlined earlier in *Procedure 7.13*, stepwise selection of predictor variables to enter or leave a model should be avoided as this process gives over control of the model building process to the software and it results in discriminant functions that will lose a great deal of predictive power upon cross-validation.

Where Is This Procedure Useful?

Discriminant analysis may be generally useful in any experimental or quasi-experimental research where prediction of treatment group (or experimental conditions) using multiple measures might be of interest. Descriptive discriminant analysis is primarily used in a multivariate follow-up role for significant MANOVA F-tests rather than as a stand-alone analysis. Survey and observational research might have a use for predictive discriminant analysis if classification rules and participant discrimination are of primary interest.

> **For example** ... *The banking and finance industry has made good use of discriminant analysis to derive classification functions (based on loan application and credit history data including past loan repayment history) for predicting loan applicants who are likely to default on their loans (poor loan risks) as distinct from loan applicants who would be good loan risks.*

Software Procedures

Application	Procedures
SPSS	*Analyze → Classify → Discriminant* ... and select your 'Grouping Variable' and 'Independents' (predictors); clicking on the '*Statistics*...' button will allow you to select statistics of interest and clicking the '*Classify*...' button will let you select classification outcome tables and plots to be shown. You can control whether or not the procedure assumes equal probabilities of group membership or probabilities based on the original group sizes (called 'Prior Probabilities') and you can choose to display the 'Leave one out classification' table for cross-validation. By default, SPSS will enter all predictors into the model together.
NCSS	*Analysis → Multivariate Analysis → Discriminant Analysis* and select your 'Y: Group Variable' and 'X's: Independent Variables'; the '*Reports*' tab lets you select output statistics you want to see and the '*Plots*' tab lets you choose the graphs you want to see. You can control whether or not the procedure assumes equal probabilities of group membership or probabilities based on the original group sizes (called 'Prior Probabilities'). Simultaneous entry of all predictors is the default analysis. NCSS does not offer an option to conduct a cross-validation analysis.
SYSTAT	*Analyze → Discriminant Analysis → Classical* and select your 'Grouping variable' and 'Predictor(s)'; the '*Statistics*' tab lets you select output statistics you want to see. You can control whether or not the procedure assumes equal probabilities of group membership or probabilities based on the original group sizes (called 'Priors'). Simultaneous entry of all predictors is the default analysis. You can also choose to run a 'Quadratic' form of discriminant analysis, which is useful if the equality of covariance matrices assumption is violated (needs a larger sample size for a stable solution to be achieved). SYSTAT automatically produces a 'Jackknifed Classification Matrix' as part of its output. SYSTAT offers another type of discriminant analysis, called *Robust Discriminant Analysis (Analyze → Discriminant Analysis → Robust)*, which can perform better in situations where outliers are present.

(continued)

Application	Procedures
STATGRAPHICS	*Relate → Classification Methods → Discriminant Analysis...* and choose the 'Classification Factor' and the 'Data" (i.e. predictors) to be used to construct the discriminant functions; hit '*OK*' and choose the 'Discriminant Analysis Options' in the dialog box that opens; hit '*OK*' and choose the 'Tables' and 'Graphs' you want to see. If you right-click the mouse in the 'Classification' output pane, this will bring up a 'Classification Options' window where you can choose your 'Prior Probabilities'. Simultaneous entry of all predictors is the default analysis. STATGRAPHICS does not offer the option to conduct a cross-validation analysis.
R Commander	**R Commander** does not offer discriminant analysis as one of its procedures. However, Field et al. (2012, pp. 738–743) illustrate how to conduct a descriptive discriminant analysis using functionality and packages in the main **R** program.

Procedure 7.18: Log-Linear Models for Contingency Tables

Classification   Bivariate/multivariate; inferential; parametric.
Purpose   To provide a systematic approach to analysing frequency count data in multi-way contingency tables involving two or more categorical variables. This approach tests for main effects and interactions in a fashion analogous to ANOVA.
Measurement level   Technically, any of the variables involved in the multi-way contingency table may be measured at any level, but in practice they are typically categorically measured at the nominal- or ordinal-level. The dependent variable in log-linear analysis is the cell frequency count which is technically ratio-scale data.

Log-linear models describe an analytical strategy designed to evaluate precise effects in multi-way contingency tables defined by two or more categorical variables (see, for example, Anderton and Cheney 2004; Knoke and Burke 1980; Tabachnick and Fidell 2019, ch. 16). The models are log-linear because the logarithms of cell counts are analysed, not the actual cell counts themselves. Chi-squared statistics are employed to systematically evaluate the significance of main effects and interactions in a fashion analogous to ANOVA. Thus, in a 2-way contingency table relating generic categorical variables A and B, we would expect to see a test for the main effect of A, the main effect for B, and the A by B interaction. In a 3-way contingency table involving generic categorical variables A, B, and C, we would expect to see the following seven effect tests: main effect for A, main effect for B, main effect for C, A by B interaction, A by C interaction, B by C interaction, and the A by B by C interaction. There are two broad approaches to log-linear analysis (Norušis 2012, ch. 1 and 2): *general* and *logit*. In general log-linear analysis, all the variables in the analysis are considered to be independent variables. In logit log-linear analysis, one of the variables is explicitly designated as a dependent variable.

General Log-Linear Models

In *general log-linear analysis*, there are two possible modes of analysis: *general*, where the researcher considers all desired effects at the same time (a type of simultaneous model) and *hierarchical* (where the researcher enters selected effects in a specified order or uses an algorithm to select the best fitting model). General log-linear analysis can be used much like ANOVA in that all design effects (main effects and interactions) are entered at once. If the model includes all possible effects, it is called 'saturated' and will predict cell frequencies perfectly. Each effect is then evaluated using significance tests of specific model parameters.

Hierarchical log-linear analysis generally begins by estimating a model having all effects represented, yielding a saturated model. Then successive effects are dropped out of the model, starting with the highest order interaction, and the fit of the model is retested at each stage (a process SPSS refers to as 'backward elimination', see Norušis 2012, pp. 20–22). The approach is called hierarchical because each model involving an interaction has all combinations of its defining variables, including main effects, incorporated. For example, testing the A by B interaction model in the hierarchical approach also includes the A and B main effects; testing the A by B by C interaction incorporates all other effects that involve variables A, B, and C.

The goal in the hierarchical approach is to establish the best fitting and simplest model (involving the fewest effects) that can adequately describe the data. The interpretation of a hierarchical log-linear analysis takes place in two stages:

- searching for the best fitting and simplest model; and
- interpreting the resulting best-fitting model.

> To illustrate a hierarchical log-linear model analysis in the QCI database, suppose Maree wishes to analyse a multi-way contingency table defined by the **industry** (created for earlier procedures by combining the five categories of **company** into three **industry** groups; see *Procedure 7.17*, for example), **educlev**, and **gender** variables (this table would have 3 × 2 × 2 or 12 cells total). Maree does not know what the best model for predicting her data would be, so she wants to use hierarchical log-linear analysis to try to find the best fitting model. Using the SPSS *Hierarchical Log-Linear* procedure, the best fitting and simplest model was found to involve the **industry** by **educlev** interaction (and by implication in hierarchical modelling approach, the **industry** and **educlev** main effects).

(continued)

Procedure 7.18: Log-Linear Models for Contingency Tables   431

The 'Partial Associations' subtable at the top of Table 7.25 shows the sequential testing outcomes for all effects except the 3-way interaction (which defines the 'saturated' model for which no test is possible). The 2-way interaction between **industry** and **educlev** (χ^2 (2) = 13.021, $p = .001$, shown in boldface) is clearly significant as is the main effect for **industry** ($\chi^2(2) = 9.961, p = .007$, shown in boldface).

The first column of numbers ('Observed') in the 'Cell Counts and Residuals' subtable in Table 7.25 summarises the observed cell counts for this 3-way log-linear analysis. Using the 'best-fitting' 2-way interaction model, the 'Expected' counts for each cell were estimated and the residuals (how badly the expected counts missed the observed counts) were computed and standardised. Maree can examine the last column of numbers in the 'Cell Counts and Residuals' subtable, which reports the standardised residuals, and look for any residuals larger than ±2.0 or ±3.0. Residuals this large would signal that the fitted model doesn't work well in predicting some portions of the contingency table. None of the residuals in the 'Std Residuals' column come close to being this large indicating that the 2-way interaction model fits the data adequately.

Table 7.25 **Selected** and edited SPSS hierarchical log-linear analysis tables

Partial Associations

Effect	df	Partial Chi-Square	Sig.	Number of Iterations
educlev*gender	1	.036	.849	2
educlev*industry	2	13.021	.001	2
gender*industry	2	4.193	.123	2
educlev	1	.084	.772	2
gender	1	.234	.629	2
industry	2	9.961	.007	2

Cell Counts and Residuals

educlev	gender	industry	Observed Count	%	Expected Count	%	Residuals	Std. Residuals
1 High School Only	1 Male	1.00 Computer industry	19.000	17.8%	14.500	13.6%	4.500	1.182
		2.00 Electrical appliance industry	10.000	9.3%	11.000	10.3%	-1.000	-.302
		3.00 Automotive industry	1.000	0.9%	2.000	1.9%	-1.000	-.707
	2 Female	1.00 Computer industry	10.000	9.3%	14.500	13.6%	-4.500	-1.182
		2.00 Electrical appliance industry	12.000	11.2%	11.000	10.3%	1.000	.302
		3.00 Automotive industry	3.000	2.8%	2.000	1.9%	1.000	.707
2 Tertiary Qualified	1 Male	1.00 Computer industry	9.000	8.4%	7.500	7.0%	1.500	.548
		2.00 Electrical appliance industry	10.000	9.3%	10.000	9.3%	.000	.000
		3.00 Automotive industry	7.000	6.5%	8.500	7.9%	-1.500	-.514
	2 Female	1.00 Computer industry	6.000	5.6%	7.500	7.0%	-1.500	-.548
		2.00 Electrical appliance industry	10.000	9.3%	10.000	9.3%	.000	.000
		3.00 Automotive industry	10.000	9.3%	8.500	7.9%	1.500	.514

Goodness-of-Fit Tests

	Chi-Square	df	Sig.
Likelihood Ratio	5.205	6	.518
Pearson	5.104	6	.531

(continued)

Table 7.26 educlev ∗ industry crosstabulation

			industry			
			1.00 Computer industry	2.00 Electrical appliance industry	3.00 Automotive industry	Total
educlev	1 High School Only	Count	30	22	4	56
		% within educlev	53.6%	39.3%	7.1%	100.0%
		% within industry	65.2%	52.4%	19.0%	51.4%
	2 Tertiary Qualified	Count	16	20	17	53
		% within educlev	30.2%	37.7%	32.1%	100.0%
		% within industry	34.8%	47.6%	81.0%	48.6%
Total		Count	46	42	21	109
		% within educlev	42.2%	38.5%	19.3%	100.0%
		% within industry	100.0%	100.0%	100.0%	100.0%

The 'Goodness-of-Fit Tests' subtable at the bottom of Table 7.25 shows that the **industry** by **educlev** 2-way interaction model does, indeed, provide the best fitting model because the chi-square test of the residuals is not significant (likelihood ratio $\chi^2(6) = 5.205, p = .518$), indicating that there is nothing systematic left to predict amongst the cell frequency counts).

To understand the significant **industry** by **educlev** interaction effect and the **industry** main effect, Maree can compute a 2-way contingency table involving only the **industry** and **educlev** variables and inspect the patterns of percentages in the table. This **educlev** ∗ **industry Crosstabulation** table is shown in Table 7.26. The pattern of percentages down the columns indicates the following trend: High School Only-educated inspectors dominate the Computer industry whereas Tertiary-Qualified inspectors dominate the Automotive industry.

The significant **industry** main effect comes about only because of the very different percentages of inspectors within the three industries—this is just a design artefact of combining the various types of companies together. In all of this, the **gender** variable plays no apparent role in any significant associations.

Logit Log-Linear Models

In *logit log-linear analysis*, one categorical variable is named as the dependent variable (may have two or more categories) and the remaining categorical variables serve as independent variables. The analysis then proceeds to test the fit of the desired effects for the independent variables as they relate to the dependent variable. This model testing process requires all desired effects to be entered simultaneously, but the researcher is free to choose which effects are represented in the model. The fit of the constructed model is assessed using a likelihood ratio chi-squared test. There are also two possible measures of association, somewhat analogous to R^2, which can be obtained (Norušis 2012, pp. 33–35). One measure, 'entropy', reflects the proportional reduction in error in predicting the probabilities in the dependent variable cells using information from the independent variables. The other measure, 'concentration', reflects the extent to which the model explains the distribution of probabilities across the dependent variable cells.

Logit log-linear analysis also reports significance tests for individual model parameters (each linked to a specific main effect or interaction in the design). Each parameter estimate has a log odds interpretation, similar to that for a regression coefficient in logistic regression analysis (recall *Procedure 7.14*). This means that taking the anti-log of the parameter estimate (= raising the mathematical constant e to the power of the estimate) gives the odds of a sampled entity being observed in a specific category of the dependent variable compared to not being observed in that category given that entity has been classified in a specific category (in the case of a main effect) or combination of categories (in the case of an interaction) of the independent variables.

> To illustrate logit log-linear model analysis in the QCI database, suppose Maree wishes to address the following hypothesis: a quality control inspector's overall decision performance can be at least partially explained by the educational background of the inspector as well as by the level of stress they experience in their working conditions. To achieve a simple test of this hypothesis, Maree uses the **educlev** variable and creates two new categorical variables.
>
> First, to characterise performance level, she creates a new variable called **inspect_perf**. As was done in *Procedure 7.14*, Maree converts both **accuracy** scores and **speed** scores to standard z-scores. Then, based on expert advice, she creates a new performance measure by combining 80% of the **accuracy** z-score with 20% of the **speed** z-score (after reversing the signs of the speed z-scores to account for the fact that slower decision times are indicative of poorer performance). Maree then classifies inspectors in the QCI database as either (0) 'performance meets expectations' (if their new performance measure score was greater than 0.0, equating to 'above average' overall performance) or (1) 'performance below expectations' (if their new performance measure score was less than or equal to 0.0, equating to 'average to below average' overall performance).
>
> Second, to characterise stressful and non–stressful working conditions, she creates a new variable called **recoded_workcond** by recoding the **workcond** variable as 1 = 'low stress working conditions' (combining **workcond** rating categories of 1, 2 and 3) and 2 = 'Moderate to high stress working conditions' (combining **workcond** rating categories of 4, 5, 6 and 7).
>
> (continued)

Table 7.27 shows selected and edited tables from the SPSS *Logit* log-linear procedure where the **inspect_perf** was the dependent variable and **educlev** and **recoded_workcond** were the independent variables. To simplify things, Maree decided she was only interested in looking at a model that included the **educlev** and **recoded_workcond** main effects but not their interaction. Note that this is reflected in the analysis summarised in Table 7.27. The model that SPSS constructed considered the dependent variable and the selected independent variable effects together (see footnote b of the 'Goodness-of-Fit Tests' subtable).

The 'Goodness-of-Fit Tests' subtable shows a non-significant likelihood ratio chi-squared test ($\chi^2(1) = 0.859$, $p = .351$). This specific test evaluates whether any significant variability in dependent cell probabilities remains to be explained after the model has been accounted for. A non-significant test signals that there is no more systematic variance to be explained after the model has been fitted.

Table 7.27 Selected and edited SPSS logit log-linear analysis tables with **inspect_perf** as the nominated dependent variable

Goodness-of-Fit Tests[a,b]

	Value	df	Sig
Likelihood Ratio	.869	1	.351
Pearson Chi-Square	.889	1	.346

a. Model: Multinomial Logit
b. Design: Constant + inspect_perf + inspect_perf * recoded_workcond + inspect_perf * educlev

Measure of Association

Entropy	.167
Concentration	.220

Parameter Estimates

Parameter	Estimate	Std. Error	Z	Sig	95% Confidence Interval Lower Bound	Upper Bound
[inspect_perf = .00]	-1.545	.480	-3.220	.001	-2.485	-.604
[inspect_perf = .00] * [recoded_workcond = 1.00]	2.207	.531	4.155	.000	1.166	3.248
[inspect_perf = .00] * [educlev = 1]	.239	.456	.524	.600	-.655	1.133

Cell Counts and Residuals

recoded_workcond	educlev	inspect_perf	Observed Count	%	Expected Count	%	Residual	Standardized Residual	Adjusted Residual	Deviance
1.00 Low stress working conditions	1 High School Only	.00 Performance Meets Expectations	29	69.0%	29.868	71.1%	-.868	-.296	-.943	-1.308
		1.00 Performance Below Expectations	13	31.0%	12.132	28.9%	.868	.296	.943	1.341
	2 Tertiary Qualified	.00 Performance Meets Expectations	20	69.0%	19.132	66.0%	.868	.340	.943	1.332
		1.00 Performance Below Expectations	9	31.0%	9.868	34.0%	-.868	-.340	-.943	-1.288
2.00 Moderate to high stress working conditions	1 High School Only	.00 Performance Meets Expectations	3	30.0%	2.132	21.3%	.868	.670	.943	1.432
		1.00 Performance Below Expectations	7	70.0%	7.868	78.7%	-.868	-.670	-.943	-1.279
	2 Tertiary Qualified	.00 Performance Meets Expectations	3	13.6%	3.868	17.6%	-.868	-.486	-.943	-1.235
		1.00 Performance Below Expectations	19	86.4%	18.132	82.4%	.868	.486	.943	1.333

(continued)

The 'Measure of Association' subtable shows a value for 'entropy' of .167 and a value for 'concentration' of .220. Maree cannot interpret either of these measures like R^2 (i.e. they do not reflect variance explained per se), she simply notes their value as reflecting the goodness-of-fit of the model (values very near zero reflect a poorly fitted model; values near 1.0 signal a good-fitting model). Maree's values are not high, but they aren't zero either and the goodness-of-fit test shows that, once the model has been fitted, there is nothing systematic left to explain.

The 'Parameter Estimates' subtable shows that the **inspect_perf** by **recoded_workcond** interaction was significant (shown in boldface; estimate $= 2.207$, $z = 4.155$, $p < .001$). The labelling of the effect as:

$$[\text{inspect\_perf} = 0] * [\text{recoded\_workcond} = 1.00]$$

signals that the log odds of an inspector being classified as 'Performance Meets Expectations' given that they reported working in 'Low stress working conditions' was 2.207. If Maree calculates e raised to the power of 2.207, she would find that this estimate translates to an inspector, working in low stress conditions, being 9.1 times as likely to display decision performance as meeting expectations than as not meeting expectations.

The main effect for **inspect_perf** was also significant (estimate $= -1.545$, $z = -3.22$, $p = .007$). However, this effect is generally not of interest, since it focuses solely on the dependent variable and signals that, in this case, the probabilities of an inspector being classified into one of the two decision performance categories are not equal.

The 'Cell Counts and Residuals' subtable provides descriptive statistics to help interpret observed and predicted relationship patterns. None of the residuals is substantive, which reinforces the overall goodness-of-fit story that nothing systematic remained to be explained. Maree can also interpret this to mean that her decision not to have her model include the **educlev** by **recoded_workcond** interaction was a good one.

Advantages

Log-linear models are very flexible statistical models that require few assumptions about the nature of the data. They are most appropriate for analysing multi-way contingency tables defined by three or more categorical variables. In this way, log-linear models provide a more efficient and thorough statistical assessment of variable effects than would separate (and numerous) 2-way contingency tables defined by all possible pairs of variables.

Log-linear models can even be employed as a substitute for parametric statistical methods to analyse continuous variables that violate the usual parametric assumptions (this can be done by cutting the continuous variables into category intervals, according to some statistical or theoretical criterion). Since ANOVA-type logic is employed in log-linear models, their interpretation is relatively straightforward. The hierarchical approach to log-linear models helps to ensure that only the most statistically meaningful relationships are examined. The logit approach to log-linear models allows the researcher to specifically test for relationships with a dependent variable.

Disadvantages

Log-linear models based on tables involving four or more variables may be very complicated to interpret, especially if some 3-way or 4-way interactions are significant. Interpreting individual parameter estimates as log odds (as well as when converting them to odds) can be challenging and confusing, particularly if the dependent variable has more than two categories.

Problems can also arise in any contingency table where one or more cells are empty (i.e. have a zero count) or have very small expected frequency counts. Because the data being analysed are the logs of frequency counts, zero counts cannot be tolerated. For this reason, most statistical programs offer a parameter called 'delta' which the researcher can define as a small amount (usually 0.5) which is added to each count before the analysis is conducted. This process introduces slight distortions into the data but does not influence the overall patterns tied to each effect. Small expected frequency counts (i.e. values less than 5.0) are troublesome if there are too many of them (generally, no more than 20% of the cells should have expected frequency counts less than 5.0). This implies that for 4-way or 5-way tables, very large sample sizes may be required in order to ensure that all cells have enough observations in them.

Unless the researcher is specifically interested in examining the saturated model (with all effects represented and cell frequencies perfectly predicted), he/she must make choices about which effects to include in the model. It would be better if the researcher chose which effects to explicitly include, but there may be no theoretical basis for making such choices. In this case, the model selection process available through hierarchical log-linear analysis should be undertaken. However, this approach is not without risks as it turns control over the model building process to a statistical algorithm (much as stepwise regression analysis does), which may produce a disjuncture between the statistically optimal model outcomes and theoretical and practical meaningfulness.

Where Is This Procedure Useful?

Log-linear models are most useful in large-scale social survey, interview, or observational research where large numbers of participants have been measured and associations among different types of categorical variables are to be examined. The generality of the tehnique for examining problematic continuous variables means that sophisticated log-linear analyses can prove useful in studies where oddly distributed measures are obtained. Log-linear models are generally not as useful in experimental or quasi-experimental research simply because group sizes are often pre-determined by sampling designs (and are often equal or near equal in size) which deliberately forces zero or very small associations to exist between grouping variables. However, in those cases where a specific categorical dependent variable is of interest and cell sizes have not been determined by research design or sample-specific quotas, logit log-linear analysis may be of benefit in testing research hypotheses.

Software Procedures

Application	Procedures
SPSS	*Analyze* → *Loglinear* → *Model Selection* ... and select your categorical variables ('Factor(s)') for analysis and define the ranges of levels for each categorical variable; choose whether you want 'Model Building' to occur using 'Backward Elimination' (hierarchical mode) or 'Enter in single step' (simultaneous mode); click the '*Model...*' button if you want to customise the effects to be included in the model; and click the '*Options...*' button to select the statistics and tables you want to see. The default model building process is 'Backward Elimination', which was the process demonstrated in Table 7.25. *Analyze* → *Loglinear* → *General* ... provides a way to assess a general log-linear model where the researcher knows what type of model they wish to evaluate. In this procedure simultaneous entry of all desired effects is performed. The analysis is configured similarly to the *Model Selection...* procedure except here you will typically choose 'Distribution of Cell Counts' to be 'Multinomial' – a choice not available in other procedures (the 'Poisson' choice should only be used if you are modelling the frequencies of rare events). *Analyze* → *Loglinear* → *Logit* ... and select your dependent variable ('Dependent') and your independent variables ('Factor(s)') for analysis; click the '*Model...*' button if you want to customise the effects to be included in the model; and click the '*Options...*' button to select the statistics and tables you want to see, but make sure you tick 'Estimates' in the 'Display' section of the dialog box so that the parameter estimates are produced.
NCSS	*Analysis* → *Proportions* → *Loglinear Models* and select your desired categorical variables (called 'Factor Variables'). By default, NCSS uses backward elimination (NCSS calls this a 'stepdown search') to find the best fitting model, but you can change to simultaneous mode by unticking the

(continued)

Application	Procedures
	'Perform step-down search' box under 'Step-Down Search Options'. By default, NCSS builds the full or 'saturated' model; alternatively, you can choose or customise the model that is constructed by selecting the appropriate option under 'Model Specification'. The '*Report*' tab allows you to select the statistics and tables you wish to see produced. NCSS does not offer a logit log-linear procedure where dependent and independent variables can be differentiated.
SYSTAT	*Analyze* → *Loglinear Model* → *Estimate*... and select the variables you want included in the 'Table' and all of the 'Model terms' you want to test (you need to specify each and every term you want included in the model); the '*Statistics*' tab will allow you to select the statistics you want to see reported, including 'Coefficients' for parameters. SYSTAT does not perform a full hierarchical backward elimination analysis; it expects you to tell it exactly which effects you want tested. However, it will report statistics with and without specific model terms included. SYSTAT's loglinear analysis outcomes are rather more difficult to read and interpret than those produced by either SPSS (the easiest to figure out) or NCSS. [Note that *Analyze* → *Loglinear Model* → *Tabulate*... allows you to produce a simple multiway cross-tabulation table for descriptive purposes.] SYSTAT does not offer a logit log-linear procedure where dependent and independent variables can be differentiated.
STATGRAPHICS	STATGRAPHICS does not offer log-linear analysis capabilities.
R Commander	**R** Commander does not offer log-linear analysis capabilities. However, Field et al. (2012, pp. 829–852) illustrate how to conduct log-linear analyses using functionalities and packages in the main **R** program.

References

References for Fundamental Concept V

Cohen, J. (1988). *Statistical power analysis for the behavioral sciences* (2nd ed.). Hillsdale: Lawrence Erlbaum Associates.

Cohen, J., Cohen, P., West, S. G., & Aiken, L. S. (2003). *Applied multiple regression/correlation analysis for the behavioral sciences* (3rd ed.). Mahwah: Lawrence Erlbaum Associates. Ch. 1, 2.

Field, A. (2018). *Discovering statistics using SPSS for Windows* (5th ed.). Los Angeles: Sage. Section 2.9.

Judd, C. M., McClelland, G. H., & Ryan, C. S. (2017). *Data analysis: A model-comparison approach* (3rd ed.). New York: Routledge. Ch. 4 onward.

Paul, F., Erdfelder, E., Lang, A.-G., & Buchner, A. (2007). G∗Power 3: A flexible statistical power analysis program for the social, behavioral, and biomedical sciences. *Behavior Research Methods, 39*(2), 175–191.

Tabachnick, B. G., & Fidell, L. S. (2019). *Using multivariate statistics* (7th ed.). New York: Pearson Education. Ch. 4.

Useful Additional Reading for Fundamental Concept V

Argyrous, G. (2011). *Statistics for research: with a guide to SPSS* (3rd ed.). London: Sage. Ch. 14, 15, 27.
De Vaus, D. (2002). *Analyzing social science data: 50 key problems in data analysis.* Sage, London: . Ch. 23, 24, 25 and 39.
Glass, G. V., & Hopkins, K. D. (1996). *Statistical methods in education and psychology* (3rd ed.). Upper Saddle River: Pearson. Ch. 10–12.
Gravetter, F. J., & Wallnau, L. B. (2017). *Statistics for the behavioural sciences* (10th ed.). Belmont: Wadsworth Cengage. Ch. 7, 8.
Henkel, R. E. (1976). *Tests of significance.* Beverly Hills: Sage. Ch. 3.
Howell, D. C. (2013). *Statistical methods for psychology* (8th ed.). Belmont: Cengage Wadsworth. Ch. 4, 18.
Lewis-Beck, M. S. (1995). *Data analysis: An introduction.* Thousand Oaks: Sage.
Meyers, L. S., Gamst, G. C., & Guarino, A. (2017). *Applied multivariate research: Design and interpretation* (3rd ed.). Thousand Oaks: Sage. Ch. 2.
Mohr, L. B. (1990). *Understanding significance testing.* Newbury Park: Sage.
Steinberg, W. J. (2011). *Statistics alive* (2nd ed.). Los Angeles: Sage. Ch. 12–15, 19.

References for Fundamental Concept VI

Cohen, J., Cohen, P., West, S. G., & Aiken, L. S. (2003). *Applied multiple regression/correlation analysis for the behavioral sciences* (3rd ed.). Mahwah: Lawrence Erlbaum Associates. Ch. 8.
Judd, C. M., McClelland, G. H., & Ryan, C. S. (2017). *Data analysis: A model-comparison approach* (3rd ed.). New York: Routledge. Ch. 1.

Useful Additional Reading for for Fundamental Concept VI

Haase, R. F. (2011). *Multivariate general linear models.* Los Angeles: Sage.
Hardy, M. A. (1993). *Regression with dummy variables.* Los Angeles: Sage.
Hardy, M. A., & Reynolds, J. (2004). Incorporating categorical information into regression models: The utility of dummy variables. In M. Hardy & A. Bryman (Eds.), *Handbook of data analysis* (pp. 209–236). London: Sage.
Miles, J., & Shevlin, M. (2001). *Applying regression & correlation: A guide for students and researchers.* Los Angeles: Sage. Ch. 1–3.
Pedhazur, E. J. (1997). *Multiple regression in behavioral research: Explanation and prediction* (3rd ed.). South Melbourne: Wadsworth Thomson Learning. Ch. 11.
Tabachnick, B. G., & Fidell, L. S. (2019). *Using multivariate statistics* (7th ed.). New York: Pearson Education. Ch. 18.
Vik, P. (2013). *Regression, ANOVA and the general linear model: A statistics primer.* Los Angeles: Sage.

References for Fundamental Concept VII

Campbell, D. T., & Stanley, J. C. (1966). *Experimental and quasi-experimental designs for research*. Boston: Houghton Mifflin.
Cook, T. D., & Campbell, D. T. (1979). *Quasi-experimentation: Design and analysis issues for field settings*. Chicago: Rand McNally.
Cooksey, R. W., & McDonald, G. (2019). *Surviving and thriving in postgraduate research* (2nd ed., pp. 653–654–676–677). Singapore,. Ch. 14, section 14.3.2: Springer.
Keppel, G., & Wickens, T. D. (2004). *Design and analysis: A researcher's handbook* (4th ed.). Upper Saddle River: Prentice Hall. Ch. 1.
Kirk, R. E. (2013). *Experimental design: Procedures for behavioral sciences* (4th ed.). Thousand Oaks: Sage. Ch. 10 and 12.
Shadish, W. R., Cook, T. D., & Campbell, D. T. (2001). *Experimental and quasi-experimental designs for generalized causal inference* (2nd ed.). Boston: Cengage.

Useful Additional Reading for Fundamental Concept VII

Edmonds, W. E., & Kennedy, T. D. (2013). *An applied reference guide to research designs: Quantitative, qualitative and mixed methods*. Los Angeles: Sage. Ch. 1–8.
Jackson, S. L. (2012). *Research methods and statistics: A critical thinking approach* (4th ed.). Belmont: Wadsworth Cengage Learning. Ch. 9, 11–13.
Levin, I. P. (1999). *Relating statistics and experimental design: An introduction*. Thousand Oaks: Sage Publications.
Rosenthal, R., & Rosnow, R. L. (1991). *Essentials of behavioral research: Methods and data analysis* (2nd ed.). New York: McGraw-Hill. Ch. 4, 5, 6, 16 and 18.
Spector, P. (1981). *Research designs*. Beverly Hills: Sage.

References for Fundamental Concept VIII

Cooksey, R. W., & McDonald, G. (2019). *Surviving and thriving in postgraduate research* (2nd ed.). Singapore: Springer. Ch. 19.
Fink, A. (2002). *How to sample in surveys* (2nd ed.). Thousand Oaks: Sage.

Useful Additional Reading for Fundamental Concept VIII

Argyrous, G. (2011). *Statistics for research: With a guide to SPSS* (3rd ed.). London: Sage. Ch. 14.
De Vaus, D. (2002). *Analyzing social science data: 50 key problems in data analysis*. London: Sage. Ch. 20, 21, 22 and 26.
Fricker, R. D. (2008). Sampling methods for web and e-mail surveys. In N. Fielding, R. M. Lee, & G. Blank (Eds.), *The Sage handbook of online research methods* (pp. 195–217). London: Sage Publications.
Glass, G. V., & Hopkins, K. D. (1996). *Statistical methods in education and psychology* (3rd ed.). Upper Saddle River: Pearson. Ch. 10.
Kalton, G. (1983). *Introduction to survey sampling*. Beverly Hills: Sage.

References

Rosenthal, R., & Rosnow, R. L. (1991). *Essentials of behavioral research: Methods and data analysis* (2nd ed.). New York: McGraw-Hill. Ch. 10.
Scheaffer, R. L., Mendenhall, W., III, Ott, L., & Kerow, K. G. (2012). *Elementary survey sampling* (7th ed.). Boston: Brooks/Cole Cengage Learning.

Reference for Procedure 7.1

Everitt, B. S. (1992). *The analysis of contingency tables* (2nd ed.). London: Chapman & Hall. Ch. 3.

Useful Additional Reading for Procedure 7.1

Agresti, A. (2018). *Statistical methods for the social sciences* (5th ed.). Boston: Pearson. Ch. 8.
Allen, P., Bennett, K., & Heritage, B. (2019). *SPSS statistics: A practical guide* (4th ed.). South Melbourne: Cengage Learning Australia Pty. Ch. 17.
Argyrous, G. (2011). *Statistics for research: With a guide to SPSS* (3rd ed.). London: Sage. Ch. 23.
Field, A. (2018). *Discovering statistics using SPSS for Windows* (5th ed.). Los Angeles: Sage. Ch. 19, (Sections 19.1 to 19.3).
George, D., & Mallery, P. (2019). *IBM SPSS statistics 25 step by step: A simple guide and reference* (15th ed.). New York: Routledge. Ch. 8.
Hildebrand, D. K., Laing, J. D., & Rosenthal, H. (1977). *The analysis of ordinal data*. Beverly Hills: Sage.
Liebetrau, A. M. (1983). *Measures of association*. Beverly Hills: Sage.
Reynolds, H. T. (1984). *Analysis of nominal data* (2nd ed.). Beverly Hills: Sage.
Smithson, M. J. (2000). *Statistics with confidence*. London: Sage. Ch. 9.
Steinberg, W. J. (2011). *Statistics alive* (2nd ed.). Los Angeles: Sage. Ch. 31.

Reference for for Procedure 7.2

Field, A. (2018). *Discovering statistics using SPSS for Windows* (5th ed.). Los Angeles: Sage. Ch. 10 (sections 10.1 to 10.8 and 10.10).

Useful Additional Reading for Procedure 7.2

Allen, P., Bennett, K., & Heritage, B. (2019). *SPSS statistics: A practical guide* (4th ed.). South Melbourne: Cengage Learning Australia Pty. Ch. 5.
Argyrous, G. (2011). *Statistics for research: With a guide to SPSS* (3rd ed.). London: Sage. Ch. 18.
George, D., & Mallery, P. (2019). *IBM SPSS statistics 25 step by step: A simple guide and reference* (15th ed.). New York: Routledge. Ch. 11.
Glass, G. V., & Hopkins, K. D. (1996). *Statistical methods in education and psychology* (3rd ed.). Upper Saddle River: Pearson. Ch. 12.
Gravetter, F. J., & Wallnau, L. B. (2017). *Statistics for the behavioural sciences* (10th ed.). Belmont: Wadsworth Cengage. Ch. 10.
Howell, D. C. (2013). *Statistical methods for psychology* (8th ed.). Belmont: Cengage Wadsworth. Ch. 7.

Rosenthal, R., & Rosnow, R. L. (1991). *Essentials of behavioral research: Methods and data analysis* (2nd ed.). New York: McGraw-Hill. Ch. 15.

Steinberg, W. J. (2011). *Statistics alive* (2nd ed.). Los Angeles: Sage. Ch. 20–21, 23.

Reference for for Procedure 7.3

Siegel, S., & Castellan, N. J., Jr. (1988). *Nonparametric statistics* (2nd ed., pp. 128–137). New York: McGraw-Hill. Ch. 6.

Useful Additional Reading for Procedure 7.3

Allen, P., Bennett, K., & Heritage, B. (2019). *SPSS statistics: A practical guide* (4th ed.). South Melbourne: Cengage Learning Australia Pty. Ch. 17.

Argyrous, G. (2011). *Statistics for research: With a guide to SPSS* (3rd ed.). London: Sage. Ch. 25.

Corder, G. W., & Foreman, D. I. (2009). *Nonparametric statistics for non-statisticians: A step-by-step approach*. Hoboken: Wiley. Ch. 4.

Field, A. (2018). *Discovering statistics using SPSS for Windows* (5th ed.). Los Angeles: Sage. Ch. 7, Sections 7.1 to 7.4.

Gibbons, J. D. (1993). *Nonparametric statistics: An introduction*. Beverly Hills: Sage. Ch. 4.

Glass, G. V., & Hopkins, K. D. (1996). *Statistical methods in education and psychology* (3rd ed.). Upper Saddle River: Pearson. Ch. 12.

Howell, D. C. (2013). *Statistical methods for psychology* (8th ed.). Belmont: Cengage Wadsworth. Ch. 18.

Neave, H. R., & Worthington, P. L. (1988). *Distribution-free statistics*. London: Unwin Hyman. Ch. 5, 6, and 7.

Reference for Procedure 7.4

Field, A. (2018). *Discovering statistics using SPSS for Windows* (5th ed.). Los Angeles: Sage. Ch. 10, Sections 10.9 to 10.11.

Useful Additional Reading for Procedure 7.4

Allen, P., Bennett, K., & Heritage, B. (2019). *SPSS statistics: A practical guide* (4th ed.). South Melbourne: Cengage Learning Australia Pty. Ch. 6.

Argyrous, G. (2011). *Statistics for research: With a guide to SPSS* (3rd ed.). London: Sage. Ch. 20.

George, D., & Mallery, P. (2019). *IBM SPSS statistics 25 step by step: A simple guide and reference* (15th ed.). New York: Routledge. Ch. 11.

Glass, G. V., & Hopkins, K. D. (1996). *Statistical methods in education and psychology* (3rd ed.). Upper Saddle River: Pearson. Ch. 12.

Gravetter, F. J., & Wallnau, L. B. (2017). *Statistics for the behavioural sciences* (10th ed.). Belmont: Wadsworth Cengage. Ch. 11.

Howell, D. C. (2013). *Statistical methods for psychology* (8th ed.). Belmont: Cengage Wadsworth. Ch. 7.

Rosenthal, R., & Rosnow, R. L. (1991). *Essentials of behavioral research: Methods and data analysis* (2nd ed.). New York: McGraw-Hill. Ch. 15.

Steinberg, W. J. (2011). *Statistics alive* (2nd ed.). Los Angeles: Sage. Ch. 22.

Reference for Procedure 7.5

Siegel, S., & Castellan, N. J., Jr. (1988). *Nonparametric statistics* (2nd ed., pp. 87–95). New York,. Ch. 5: McGraw-Hill.

Useful Additional Reading for Procedure 7.5

Allen, P., Bennett, K., & Heritage, B. (2019). *SPSS statistics: A practical guide* (4th ed.). South Melbourne: Cengage Learning Australia Pty. Ch. 17.

Argyrous, G. (2011). *Statistics for research: With a guide to SPSS* (3rd ed.). London: Sage. Ch. 25.

Field, A. (2018). *Discovering statistics using SPSS for Windows* (5th ed.). Los Angeles: Sage. Ch. 6, Section 7.5.

Gibbons, J. D. (1993). *Nonparametric statistics: An introduction*. Beverly Hills: Sage. Ch. 3.

Glass, G. V., & Hopkins, K. D. (1996). *Statistical methods in education and psychology* (3rd ed.). Upper Saddle River: Pearson. Ch. 12.

Howell, D. C. (2013). *Statistical methods for psychology* (8th ed.). Belmont: Cengage Wadsworth. Ch. 18.

Neave, H. R., & Worthington, P. L. (1988). *Distribution-free statistics*. London: Unwin Hyman. Ch. 8.

References for Procedure 7.6

Field, A. (2018). *Discovering statistics using SPSS for Windows* (5th ed.). Los Angeles: Sage. Ch. 12.

Field, A., Miles, J., & Field, Z. (2012). *Discovering statistics using R*. London: Sage. Ch. 10.

Iversen, G. R., & Norpoth, H. (1987). *Analysis of variance* (2nd ed.). Newbury Park: Sage. Ch. 2 and 4.

Useful Additional Reading for Procedure 7.6

Allen, P., Bennett, K., & Heritage, B. (2019). *SPSS statistics: A practical guide* (4th ed.). South Melbourne: Cengage Learning Australia Pty. Ch. 7.

Argyrous, G. (2011). *Statistics for research: With a guide to SPSS* (3rd ed.). London: Sage. Ch. 19.

Everitt, B. S. (1995). *Making sense of statistics in psychology: A second level course*. Oxford: Oxford University Press. Ch. 3.

George, D., & Mallery, P. (2019). *IBM SPSS statistics 25 step by step: A simple guide and reference* (15th ed.). New York: Routledge. Ch. 8.

Glass, G. V., & Hopkins, K. D. (1996). *Statistical methods in education and psychology* (3rd ed.). Upper Saddle River: Pearson. Ch. 15.

Gravetter, F. J., & Wallnau, L. B. (2017). *Statistics for the behavioural sciences* (10th ed.). Belmont: Wadsworth Cengage. Ch. 12.

Howell, D. C. (2013). *Statistical methods for psychology* (8th ed.). Belmont: Cengage Wadsworth. Ch. 11.

Rosenthal, R., & Rosnow, R. L. (1991). *Essentials of behavioral research: Methods and data analysis* (2nd ed.). New York: McGraw-Hill. Ch. 15.

Steinberg, W. J. (2011). *Statistics alive* (2nd ed.). Los Angeles: Sage. Ch. 24 and 25.

References for Procedure 7.7

Cohen, J. (1988). *Statistical power analysis for the behavioral sciences* (2nd ed.). Hillsdale: Lawrence Erlbaum Associates.

Field, A. (2018). *Discovering statistics using SPSS for Windows* (5th ed.). Los Angeles: Sage. Ch. 12, Sections 12.10; see also ch. 7, sections 7.4.5, 7.5.5 and 7.6.7.

Hays, W. L. (1988). *Statistics* (3rd ed.). New York: Holt, Rinehart, & Winston. Ch. 8, pp. 306–313; Ch. 10, p. 369 and pp. 374–376.

Tabachnick, B. G., & Fidell, L. S. (2019). *Using multivariate statistics* (7th ed.). New York: Pearson Education. Ch. 3, Section 3.4.

Useful Additional Reading for Procedure 7.7

Cortina, J., & Nouri, H. (2000). *Effect size for ANOVA designs*. Thousand Oaks: Sage.

Keppel, G., & Wickens, T. D. (2004). *Design and analysis: A researcher's handbook* (4th ed.). Upper Saddle River: Prentice Hall. Ch. 8.

Rosenthal, R., & Rosnow, R. L. (1991). *Essentials of behavioral research: Methods and data analysis* (2nd ed.). New York: McGraw-Hill. Ch. 15, pp. 317–318; Ch. 16, pp. 351–352.

References for Procedure 7.8

Cohen, J., Cohen, P., West, S. G., & Aiken, L. S. (2003). *Applied multiple regression/correlation analysis for the behavioral sciences* (3rd ed.). Mahwah: Lawrence Erlbaum Associates. Ch. 6 and 8.

Field, A., Miles, J., & Field, Z. (2012). *Discovering statistics using R*. London: Sage. Ch. 10.

Keppel, G., & Wickens, T. D. (2004). *Design and analysis: A researcher's handbook* (4th ed.). Upper Saddle River: Prentice Hall. Ch. 4, 5 and 6.

Klockars, A. (1986). *Multiple comparisons*. Beverly Hills: Sage.

Kirk, R. E. (2013). *Experimental design: Procedures for behavioral sciences* (4th ed.). Thousand Oaks: Sage. Ch. 4.

Toothaker, L. E. (1993). *Multiple comparison procedures*. Newbury Park: Sage.

Useful Additional Reading for Procedure 7.8

Field, A. (2018). *Discovering statistics using SPSS for Windows* (5th ed.). Los Angeles: Sage. Section 12.5 and 12.6.

Howell, D. C. (2013). *Statistical methods for psychology* (8th ed.). Belmont: Cengage Wadsworth. Ch. 12.
Rosenthal, R., & Rosnow, R. L. (1991). *Essentials of behavioral research: Methods and data analysis* (2nd ed.). New York: McGraw-Hill. Ch. 21.
Tabachnick, B. G., & Fidell, L. S. (2019). *Using multivariate statistics* (7th ed.). New York: Pearson Education. Ch. 3.

References for Procedure 7.9

Field, A. (2018). *Discovering statistics using SPSS for Windows* (5th ed.). Los Angeles: Sage. Section 7.6.
Field, A., Miles, J., & Field, Z. (2012). *Discovering statistics using R* (pp. 674–686). London,. Ch. 15: Sage.
Siegel, S., & Castellan, N. J., Jr. (1988). *Nonparametric statistics* (2nd ed.). New York: McGraw-Hill. Ch. 8, which also discusses multiple comparison methods.

Useful Additional Reading for Procedure 7.9

Allen, P., Bennett, K., & Heritage, B. (2019). *SPSS statistics: A practical guide* (4th ed.). South Melbourne: Cengage Learning Australia Pty. Ch. 17.
Argyrous, G. (2011). *Statistics for research: With a guide to SPSS* (3rd ed.). London: Sage. Ch. 25.
Gibbons, J. D. (1993). *Nonparametric statistics: An introduction*. Beverly Hills: Sage.
Howell, D. C. (2013). *Statistical methods for psychology* (8th ed.). Belmont: Cengage Wadsworth. Ch. 18.
Neave, H. R., & Worthington, P. L. (1988). *Distribution-free statistics*. London: Unwin Hyman. Ch. 13, which also discusses multiple comparison methods.

References for Procedure 7.10

Cohen, J., Cohen, P., West, S. G., & Aiken, L. S. (2003). *Applied multiple regression/correlation analysis for the behavioral sciences* (3rd ed.). Mahwah: Lawrence Erlbaum Associates. Ch. 5, 6, 8 and 9.
Cooksey, R. W., & McDonald, G. (2019). *Surviving and thriving in postgraduate research* (2nd ed.). Singapore: Springer. Ch. 14, section 14.3.2 and pp. 676–677.
Field, A. (2018). *Discovering statistics using SPSS for Windows* (5th ed.). Los Angeles: Sage. Ch. 14.
Field, A., Miles, J., & Field, Z. (2012). *Discovering statistics using R*. London: Sage. Ch. 12.
Judd, C. M., McClelland, G. H., & Ryan, C. S. (2017). *Data analysis: A model-comparison approach* (3rd ed.). New York: Routledge. Ch. 9.
Keppel, G., & Wickens, T. D. (2004). *Design and analysis: A researcher's handbook* (4th ed.). Upper Saddle River: Prentice Hall. Ch. 10, 11, 12, 13, 14, 21, 22, 25 and 26.
Kirk, R. E. (2013). *Experimental design: Procedures for behavioral sciences* (4th ed.). Thousand Oaks: Sage. Ch. 6, 9, 10 and 11.

Useful Additional Reading for Procedure 7.10

Allen, P., Bennett, K., & Heritage, B. (2019). *SPSS statistics: A practical guide* (4th ed.). South Melbourne: Cengage Learning Australia Pty. Ch. 8.

Brown, S. R., & Melamed, L. E. (1990). *Experimental design and analysis*. Newbury Park: Sage.

George, D., & Mallery, P. (2019). *IBM SPSS statistics 25 step by step: A simple guide and reference* (15th ed.). New York: Routledge. Ch. 8.

Gravetter, F. J., & Wallnau, L. B. (2017). *Statistics for the behavioural sciences* (10th ed.). Belmont: Wadsworth Cengage. Ch. 14.

Howell, D. C. (2013). *Statistical methods for psychology* (8th ed.). Belmont: Cengage Wadsworth. Ch. 13.

Rosenthal, R., & Rosnow, R. L. (1991). *Essentials of behavioral research: Methods and data analysis* (2nd ed.). New York: McGraw-Hill. Ch. 16 and 17.

References for Fundamental Concept IX

Cohen, J., Cohen, P., West, S. G., & Aiken, L. S. (2003). *Applied multiple regression/correlation analysis for the behavioral sciences* (3rd ed.). Mahwah: Lawrence Erlbaum Associates. Ch. 9.

Hayes, A. F. (2018). *Introduction to mediation, moderation and conditional process analysis: A regression-based approach* (3rd ed.). New York: The Guilford Press. Ch. 7.

Keppel, G., & Wickens, T. D. (2004). *Design and analysis: A researcher's handbook* (4th ed.). Upper Saddle River: Prentice Hall. Ch. 12 and 13.

Kirk, R. E. (2013). *Experimental design: Procedures for behavioral sciences* (4th ed.). Thousand Oaks: Sage. Ch. 9.

Miles, J., & Shevlin, M. (2001). *Applying regression & correlation: A guide for students and researchers*. Los Angeles: Sage. Ch. 7.

Useful Additional Reading for Fundamental Concept IX

Brown, S. R., & Melamed, L. E. (1990). *Experimental design and analysis*. Newbury Park: Sage.

Field, A. (2018). *Discovering statistics using SPSS for Windows* (5th ed.). Los Angeles: Sage. Sections 14.6 and 14.7.

Jaccard, J. (1997). *Interaction effects in factorial analysis of variance*. Thousand Oaks: Sage.

Jaccard, J., & Turrisi, R. (2003). *Interaction effects in multiple regression* (2nd ed.). Thousand Oaks: Sage.

Jose, P. E. (2013). *Doing statistical mediation and moderation*. New York: The Guilford Press.

Judd, C. M., McClelland, G. H., & Ryan, C. S. (2017). *Data analysis: A model-comparison approach* (3rd ed.). New York: Routledge. Ch. 7, 9.

Majoribanks, K. M. (1997). Interaction, detection, and its effects. In J. P. Keeves (Ed.), *Educational research, methodology, and measurement: An international handbook* (2nd ed., pp. 561–571). Oxford: Pergamon Press.

Pedhazur, E. J. (1997). *Multiple regression in behavioral research: Explanation and prediction* (3rd ed.). South Melbourne: Wadsworth Thomson Learning. Ch. 12.

Rosenthal, R., & Rosnow, R. L. (1991). *Essentials of behavioral research: Methods and data analysis* (2nd ed.). New York: McGraw-Hill. Ch. 17.

Vik, P. (2013). *Regression, ANOVA and the General Linear Model: A statistics primer*. Los Angeles: Sage. Ch. 10 and 12.

References for Procedure 7.11

Field, A. (2018). *Discovering statistics using SPSS for Windows* (5th ed.). Los Angeles: Sage. Ch. 15 and 16.
Field, A., Miles, J., & Field, Z. (2012). *Discovering statistics using R*. London: Sage. Ch. 12.
Keppel, G., & Wickens, T. D. (2004). *Design and analysis: A researcher's handbook* (4th ed.). Upper Saddle River: Prentice Hall. Ch. 16–20, 23.
Kirk, R. E. (2013). *Experimental design: Procedures for behavioral sciences* (4th ed.). Thousand Oaks: Sage. Ch. 10 and 12.
Tabachnick, B. G., & Fidell, L. S. (2019). *Using multivariate statistics* (7th ed.). New York: Pearson Education. Ch. 8.
Polhemus, N. W. (2006). *How to: Analyze a repeated measures experiment using STATGRAPHICS Centurion*. Document downloaded from http://cdn2.hubspot.net/hubfs/402067/PDFs/How_To_Analyze_a_Repeated_Measures_Experiment.pdf. Accessed 1 Oct 2019.

Useful Additional Reading for Procedure 7.11

Allen, P., Bennett, K., & Heritage, B. (2019). *SPSS statistics: A practical guide* (4th ed.). South Melbourne: Cengage Learning Australia Pty. Ch. 9.
Brown, S. R., & Melamed, L. E. (1990). *Experimental design and analysis*. Newbury Park: Sage.
Cohen, J., Cohen, P., West, S. G., & Aiken, L. S. (2003). *Applied multiple regression/correlation analysis for the behavioral sciences* (3rd ed.). Mahwah: Lawrence Erlbaum Associates. Ch. 15.
George, D., & Mallery, P. (2019). *IBM SPSS statistics 25 step by step: A simple guide and reference* (15th ed.). New York: Routledge. Ch. 8.
Girden, E. R. (1992). *ANOVA repeated measures*. Newbury Park: Sage.
Gravetter, F. J., & Wallnau, L. B. (2017). *Statistics for the behavioural sciences* (10th ed.). Belmont: Wadsworth Cengage. Ch. 14.
Grimm, L. G., & Yarnold, P. R. (Eds.). (2000). *Reading and understanding more multivariate statistics*. Washington, DC: American Psychological Association (APA). Ch. 10.
Howell, D. C. (2013). *Statistical methods for psychology* (8th ed.). Belmont: Cengage Wadsworth. Ch. 14.
Judd, C. M., McClelland, G. H., & Ryan, C. S. (2017). *Data analysis: A model-comparison approach* (3rd ed.). New York: Routledge. Ch. 11.
Rosenthal, R., & Rosnow, R. L. (1991). *Essentials of behavioral research: Methods and data analysis* (2nd ed.). New York: McGraw-Hill. Ch. 18.

References for Procedure 7.12

Field, A. (2018). *Discovering statistics using SPSS for Windows* (5th ed.). Los Angeles: Sage. Section 7.7.
Field, A., Miles, J., & Field, Z. (2012). *Discovering statistics using R* (pp. 686–692). London,. Ch. 15: Sage.
Siegel, S., & Castellan, N. J., Jr. (1988). *Nonparametric statistics* (2nd ed.). New York: McGraw-Hill. Ch. 7, which also discusses multiple comparison methods.

Useful Additional Reading for Procedure 7.12

Allen, P., Bennett, K., & Heritage, B. (2019). *SPSS statistics: A practical guide* (4th ed.). South Melbourne: Cengage Learning Australia Pty. Ch. 17.
Gibbons, J. D. (1993). *Nonparametric statistics: An introduction*. Beverly Hills: Sage.
Howell, D. C. (2013). *Statistical methods for psychology* (8th ed.). Belmont: Cengage Wadsworth. Ch. 18.
Neave, H. R., & Worthington, P. L. (1988). *Distribution-free statistics*. London: Unwin Hyman. Ch. 14, which also discusses multiple comparison methods.

References for Procedure 7.13

Berry, W. (1993). *Understanding regression assumptions*. Beverly Hills: Sage.
Cohen, J., Cohen, P., West, S. G., & Aiken, L. S. (2003). *Applied multiple regression/correlation analysis for the behavioral sciences* (3rd ed.). Mahwah: Lawrence Erlbaum Associates. Ch. 3, 4, 5, 6–9, 10 provide comprehensive coverage of multiple regression concepts at a good conceptual and technical level].
Dunteman, G. (2005). *Introduction to generalized linear models*. Thousand Oaks: Sage.
Field, A. (2018). *Discovering statistics using SPSS for Windows* (5th ed.). Los Angeles: Sage. Ch. 6 and 9.
Field, A., Miles, J., & Field, Z. (2012). *Discovering statistics using R*. London: Sage. Ch. 7.
Fox, J. (1991). *Regression diagnostics: An introduction*. Beverly Hills: Sage.
Fox, J. (2000). *Multiple and generalized nonparametric regression*. Thousand Oaks: Sage.
Gill, J. (2000). *Generalized linear models: A unified approach*. Thousand Oaks: Sage.
Hair, J. F., Black, B., Babin, B., & Anderson, R. E. (2010). *Multivariate data analysis: A global perspective* (7th ed.). Upper Saddle River: Pearson Education. Ch. 4.
Judd, C. M., McClelland, G. H., & Ryan, C. S. (2017). *Data analysis: A model-comparison approach* (3rd ed.). New York: Routledge. Ch. 6.
Lewis-Beck, M. S. (1980). *Applied regression: An introduction*. Newbury Park: Sage.
Miles, J., & Shevlin, M. (2001). *Applying regression & correlation: A guide for students and researchers*. London: Sage. Ch. 2–7 provide comprehensive coverage of multiple regression concepts at a good conceptual level.
Pedhazur, E. J. (1997). *Multiple regression in behavioral research: Explanation and prediction* (3rd ed.). South Melbourne: Wadsworth Thomson Learning. Ch. 3, 5–15 provide comprehensive coverage of multiple regression concepts at a more technical level.

Useful Additional Reading for Procedure 7.13

Agresti, A. (2018). *Statistical methods for the social sciences* (5th ed.). Boston: Pearson. Ch. 12.
Allen, P., Bennett, K., & Heritage, B. (2019). *SPSS statistics: A practical guide* (4th ed.). South Melbourne: Cengage Learning Australia Pty. Ch. 13.
Darlington, R. B., & Hayes, A. F. (2017). *Regression analysis and linear models: Concepts, applications, and implementation*. New York: The Guilford Press.
Grimm, L. G., & Yarnold, P. R. (1995). *Reading and understanding multivariate statistics*. Washington, DC: American Psychological Association. Ch. 2.
Hardy, M. (1993). *Regression with dummy variables*. Thousand Oaks: Sage.
Howell, D. C. (2013). *Statistical methods for psychology* (8th ed.). Belmont: Cengage Wadsworth. Ch. 15.

George, D., & Mallery, P. (2019). *IBM SPSS statistics 25 step by step: A simple guide and reference* (15th ed.). New York: Routledge. Ch. 16 and 28.

Meyers, L. S., Gamst, G. C., & Guarino, A. (2017). *Applied multivariate research: Design and interpretation* (3rd ed.). Thousand Oaks: Sage. Ch. 5A, 5B, 6A, 6B.

Schroeder, L. D., Sjoquist, D. L., & Stephan, P. E. (1986). *Understanding regression analysis: An introductory guide*. Beverly Hills: Sage.

Tabachnick, B. G., & Fidell, L. S. (2019). *Using multivariate statistics* (7th ed.). New York: Pearson Education. Ch. 5.

References for Procedure 7.14

Cohen, J., Cohen, P., West, S. G., & Aiken, L. S. (2003). *Applied multiple regression/correlation analysis for the behavioral sciences* (3rd ed.). Mahwah: Lawrence Erlbaum Associates. Ch. 13.

Everitt, B. S., & Hothorn, T. (2006). *A handbook of statistical analyses using **R***. Boca Raton: Chapman & Hall/CRC.

Field, A. (2018). *Discovering statistics using SPSS for Windows* (5th ed.). Los Angeles: Sage. Ch. 20.

Miles, J., & Shevlin, M. (2001). *Applying regression & correlation: A guide for students and researchers*. London: Sage. Ch. 6.

Pedhazur, E. J. (1997). *Multiple regression in behavioral research: Explanation and prediction* (3rd ed.). South Melbourne: Wadsworth Thomson Learning. Ch. 17.

Tabachnick, B. G., & Fidell, L. S. (2019). *Using multivariate statistics* (7th ed.). New York: Pearson Education. Ch. 10.

Useful Additional Reading for Procedure 7.14

Agresti, A. (2018). *Statistical methods for the social sciences* (5th ed.). Boston: Pearson. Ch. 13.

Allen, P., Bennett, K., & Heritage, B. (2019). *SPSS statistics: A practical guide* (4th ed.). South Melbourne: Cengage Learning Australia Pty. Ch. 14.

Grimm, L. G., & Yarnold, P. R. (Eds.). (1995). *Reading and understanding multivariate statistics*. Washington, DC: American Psychological Association (APA). Ch. 7.

George, D., & Mallery, P. (2019). *IBM SPSS statistics 25 step by step: A simple guide and reference* (15th ed.). New York: Routledge. Ch. 25.

Hair, J. F., Black, B., Babin, B., & Anderson, R. E. (2010). *Multivariate data analysis: A global perspective* (7th ed.). Upper Saddle River: Pearson Education. Ch. 7.

Menard, S. (2002). *Applied logistic regression analysis* (2nd ed.). Thousand Oaks: Sage.

Meyers, L. S., Gamst, G. C., & Guarino, A. (2017). *Applied multivariate research: Design and interpretation* (3rd ed.). Thousand Oaks: Sage. Ch. 9A, 9B.

Pampel, F. (2000). *Logistic regression: A primer*. Thousand Oaks: Sage.

References for Procedure 7.15

Field, A. (2018). *Discovering statistics using SPSS for Windows* (5th ed.). Los Angeles: Sage. Ch. 13.

Field, A., Miles, J., & Field, Z. (2012). *Discovering statistics using R*. London: Sage. Ch. 11.

Tabachnick, B. G., & Fidell, L. S. (2019). *Using multivariate statistics* (7th ed.). New York: Pearson Education. Ch. 6.
Wildt, A. R., & Ahtola, O. T. (1978). *Analysis of covariance*. Beverly Hills: Sage.

Useful Additional Reading for Procedure 7.15

Allen, P., Bennett, K., & Heritage, B. (2019). *SPSS statistics: A practical guide* (4th ed.). South Melbourne: Cengage Learning Australia Pty. Ch. 10.
Judd, C. M., McClelland, G. H., & Ryan, C. S. (2017). *Data analysis: A model-comparison approach* (3rd ed.). New York: Routledge. Ch. 10.
Keppel, G., & Wickens, T. D. (2004). *Design and analysis: A researcher's handbook* (4th ed.). Upper Saddle River: Prentice Hall. Ch. 15.
Kirk, R. E. (2013). *Experimental design: Procedures for behavioral sciences* (4th ed.). Thousand Oaks: Sage. Ch. 13.
Pedhazur, E. J. (1997). *Multiple regression in behavioral research: Explanation and prediction* (3rd ed.). South Melbourne: Wadsworth Thomson Learning. Ch. 15.

References for Procedure 7.16

Bray, J. H., & Maxwell, S. E. (1985). *Multivariate analysis of variance*. Beverly Hills: Sage.
Field, A. (2018). *Discovering statistics using SPSS for Windows* (5th ed.). Los Angeles: Sage. Ch. 17.
Field, A., Miles, J., & Field, Z. (2012). *Discovering statistics using R*. London: Sage. Ch. 16.
Hair, J. F., Black, B., Babin, B., & Anderson, R. E. (2010). *Multivariate data analysis: A global perspective* (7th ed.). Upper Saddle River: Pearson Education. Ch. 8.
Tabachnick, B. G., & Fidell, L. S. (2019). *Using multivariate statistics* (7th ed.). New York: Pearson Education. Ch. 7.

Useful Additional Reading for Procedure 7.16

Allen, P., Bennett, K., & Heritage, B. (2019). *SPSS statistics: A practical guide* (4th ed.). South Melbourne: Cengage Learning Australia Pty. Ch. 11.
George, D., & Mallery, P. (2019). *IBM SPSS statistics 25 step by step: A simple guide and reference* (15th ed.). New York: Routledge. Ch. 23.
Grimm, L. G., & Yarnold, P. R. (1995). *Reading and understanding multivariate statistics*. Washington, DC: American Psychological Association. Ch. 8.
Meyers, L. S., Gamst, G. C., & Guarino, A. (2017). *Applied multivariate research: Design and interpretation* (3rd ed.). Thousand Oaks: Sage. Ch. 18A, 18B.

References for Procedure 7.17

Hair, J. F., Black, B., Babin, B., & Anderson, R. E. (2010). *Multivariate data analysis: A global perspective* (7th ed.). Upper Saddle River: Pearson Education. Ch. 7.

Huberty, C. J. (1984). Issues in the use and interpretation of discriminant analysis. *Psychological Bulletin, 95*(1), 156–171.
Klecka, W. R. (1980). *Discriminant analysis*. Beverly Hills: Sage.
Tabachnick, B. G., & Fidell, L. S. (2019). *Using multivariate statistics* (7th ed.). New York: Pearson Education. Ch. 9.

Useful Additional Reading for Procedure 7.17

Allen, P., Bennett, K., & Heritage, B. (2019). *SPSS statistics: A practical guide* (4th ed.). South Melbourne: Cengage Learning Australia Pty. Ch. 11.
Field, A. (2018). *Discovering statistics using SPSS for Windows* (5th ed.). Los Angeles: Sage. Sections 17.9 to 17.11.
Field, A., Miles, J., & Field, Z. (2012). *Discovering statistics using R*. London: Sage. Ch. 16.
George, D., & Mallery, P. (2019). *IBM SPSS statistics 25 step by step: A simple guide and reference* (15th ed.). New York: Routledge. Ch. 22.
Grimm, L. G., & Yarnold, P. R. (1995). *Reading and understanding multivariate statistics*. Washington, DC: American Psychological Association. Ch. 9.
Lohnes, P. R. (1997). Discriminant analysis. In J. P. Keeves (Ed.), *Educational research, methodology, and measurement: An international handbook* (2nd ed., pp. 503–508). Oxford: Pergamon Press.
Meyers, L. S., Gamst, G. C., & Guarino, A. (2017). *Applied multivariate research: Design and interpretation* (3rd ed.). Thousand Oaks: Sage. Ch. 19A, 19B.

References for Procedure 7.18

Anderton, D. L., & Cheney, E. (2004). Log-linear analysis. In M. Hardy & A. Bryman (Eds.), *Handbook of data analysis* (pp. 285–306). London: Sage.
Field, A., Miles, J., & Field, Z. (2012). *Discovering statistics using R*. London: Sage. Ch. 18.
Knoke, D., & Burke, P. J. (1980). *Log-linear models*. Beverly Hills: Sage.
Norušis, M. J. (2012). *IBM SPSS Statistics 19: Advanced statistical procedures companion*. Upper Saddle River: Prentice Hall. Ch. 1 and 2.
Tabachnick, B. G., & Fidell, L. S. (2019). *Using multivariate statistics* (7th ed.). New York: Pearson Education. Ch. 16.

Useful Additional Reading for Procedure 7.18

Everitt, B. S. (1977). *The analysis of contingency tables*. New York: Wiley. Ch. 5.
Field, A. (2018). *Discovering statistics using SPSS for Windows* (5th ed.). Los Angeles: Sage. Section 19.9 to 19.11.
George, D., & Mallery, P. (2019). *IBM SPSS statistics 25 step by step: A simple guide and reference* (15th ed.). New York: Routledge. Ch. 26 and 27.
Grimm, L. G., & Yarnold, P. R. (1995). *Reading and understanding multivariate statistics*. Washington, DC: American Psychological Association. Ch. 6.
Kennedy, J. J., & Tam, H. K. (1997). Log-linear models. In J. P. Keeves (Ed.), *Educational research, methodology, and measurement: An international handbook* (2nd ed., pp. 571–580). Oxford: Pergamon Press.

Chapter 8
Other Commonly Used Statistical Procedures

In this chapter, we explore some other commonly used but less 'traditional' statistical procedures. While these procedures are now commonly reported in behavioural and social research, they tend not to be well-covered in standard statistical texts. *Reliability analysis & classical item analysis* are procedures useful for obtaining evidence in support of the measurement quality of scales and tests formed by combinations of specific items (e.g. items which measure specific beliefs, attitudes, preferences, opinions, or tests of knowledge, skills and understanding) or for establishing the consistency with which different observers of behaviour agree in their judgments about the nature of those behaviours. *Data screening & missing value analysis* constitute a set of procedures designed to assist the researcher in evaluating statistical regularities and anomalies in data, including assessment of the potential impact of patterns of missing data and estimation of missing data values prior to further analyses. *Confidence intervals* provide an alternative logic for evaluating statistical hypotheses and estimates of parameters, additional to the classical statistical perspective (recall *Fundamental concept V*). *Bootstrapping* and *jackknifing* provide very different types of nonparametric approaches to the analysis of data that can yield statistical tests of hypotheses for situations where parametric assumptions are violated and there exist no other nonparametric analytical alternatives. This approach has wide appeal across disciplines but is especially useful in structural equation modelling.

Time series analysis is useful for statistically analysing data that emerge from longitudinal investigations of an economic, financial, behavioural or social phenomenon over time. Such analysis may focus purely on the essential nature of the series of observations made over time or may focus on assessing the impact of a specific policy, behavioural or social intervention on the nature of a series of observations over time. *Confirmatory factor analysis* and *structural equation (causal path) models* share a common statistical goal, namely the estimation and testing of specific theoretical causal models for behavioural or social phenomena using correlational data. Finally, *meta-analysis* refers to a general systematic statistical approach for analysing and aggregating published quantitative findings in medical, behavioural, educational and social research literature, for example. The procedure is extremely useful for estimating generalisable statistical trends in data amassed over many years by many different researchers using many samples.

> Think back to the QCI database. Consider the following potential research objectives which Maree thinks could be appropriately addressed using one of the techniques to be discussed in this chapter.
>
> - Maree might wish to statistically show that the three general work attitude factors, identified using Exploratory Factor Analysis in *Procedure 6.5*, are sufficiently reliable measures of the work attitudes of quality control inspectors. [See *Procedure 8.1* – Reliability analysis.]
> - Maree might need to show whether or not specific QCI variables: (a) meet the required assumptions for specific parametric statistical procedures, (b) exhibit any outliers that should be dealt with, (c) have problematic patterns of very high correlations or nonlinear relationships or (d) reflect the undue influence of patterns of missing values. [See *Procedure 8.2* – Data screening & Missing value analysis.]
> - Maree might want to place bands of estimate precision around mean decision accuracy and mean decision speed that have a 95% chance of covering their respective population values? [See *Procedure 8.3* – Confidence Intervals].
> - Given the non-normal nature of time-based measures such as speed of decision making, how could Maree adequately estimate and test a multiple regression equation which predicts inspection decision speed on the basis of job satisfaction, working condition perceptions, educational level and general mental ability? [See *Procedure 8.4* – Bootstrapping and Jackknifing.]
> - Maree might wish to evaluate and characterise the influence of a training intervention on the accuracy of an inspector's quality control decisions over time. [See *Procedure 8.5* – Time series analysis.]
> - Maree might wish to test a model specifying three unobserved latent constructs, focusing on quality control inspectors' attitudes towards product quality, training provisions, and aspects of the organisation and its management, which she hypothesises cause the observed response patterns on the nine general work attitude items. [See *Procedure 8.6* – Confirmatory factor analysis.]
>
> (continued)

- Maree might want to test specific hypothesised pathways in a proposed model that causally relates quality control inspectors' general mental ability, gender, industry and educational level to their perceptions of working conditions and job satisfaction as well as to their inspection decision speed and accuracy. [See *Procedure 8.7* – Structural equation (causal path) models (SEM).]
- What would a sample of behavioural and social science journal articles, published within the last 10 years, allow Maree to conclude about the relationship between perceptions of working conditions, job satisfaction, and measures of work performance such as the speed and accuracy of decision making and what are the estimated magnitudes of those relationships? [See *Procedure 8.8* – Meta- analysis.]

Procedure 8.1: Reliability & Classical Item Analysis

Classification   Bivariate/multivariate; correlational.
Purpose   *Internal consistency reliability* measures how consistently a collection of items contributes to a scale which measures a construct they were either designed to measure or have been statistically shown to measure (using a technique such as factor analysis).

Item analysis generally accompanies the assessment of internal consistency reliability and summarises the contribution of each item to the overall scale being evaluated.

Inter-observer reliability (or Inter-rater agreement) measures how well two or more observers of the same behavioural or social phenomena agree in their ratings.

Measurement level   Variables for item analysis and internal consistency reliability assessment are typically measured at the interval- or ratio-level, although dichotomous items can also be analysed (e.g. test items scored as right or wrong or as multiple-choice options).

Inter-observer reliability is typically assessed using categorical nominal- or ordinal-level measures although this need not be the case.

Internal Consistency Reliability

Internal consistency reliability can be assessed for a collection of items designed to measure some construct (e.g. mathematics ability, degree of self-concept, dimension of attitude or belief) in a single testing administration. Internal consistency reliability estimates the overall reliability of a test or scale and can be applied to many situations in experimental, quasi-experimental, and survey research (see, for example, Athanasou 1997; Carmines and Zeller 1979; Thorndike and Thorndike 1997). The most frequently used measure of internal consistency reliability is called *Cronbach's alpha*. It measures the degree to which the items comprising a scale work together in measuring a given construct. It is primarily based on two data characteristics: the average correlation amongst all the items comprising the scale and the number of items comprising the scale.

Typically, attitudinal scales are formed using a number of Likert-type items which are treated as interval-level measures. However, when actual learning or knowledge is being measured using specific test items like multiple choice or true-false items, the items are typically scored dichotomously as right (1) or wrong (0) or, perhaps, true (1) or false (0), and a simpler form of Cronbach's alpha, the KR20 (KR for Kuder-Richardson) formula, is appropriate.

Internal Consistency Reliability assessment is an important follow-up procedure for establishing the measurement quality of dimensions identified using exploratory factor analysis (see *Procedure 6.5*) of items or some other scale construction approach. Cronbach's alpha (or KR20) reliability ranges between 0 and 1.0 with 0 indicating no reliability and 1.0 indicating perfect reliability. Acceptable levels of reliability for researcher-designed scales are considered to be in the .6 to .8 range whereas the expectation for scales and tests designed by professional test development organisations is for reliabilities in the range of .8 to .9. There are two important caveats that apply to the assessment of internal consistency reliability: (1) as the number of items in the scale or test increases, Cronbach's alpha (KR20) also tends to increase as a statistical artefact (adding some items can be a good practice to enhance reliability, but only if they are the same type of items and measure the same construct as the items already comprising the scale), and (2) scale unidimensionality (what is called a 'homogeneous test', see Thorndike 1982) must be assumed as computing Cronbach's alpha (KR20) on a multidimensional scale will produce an artificially inflated value.

> In the QCI database, Maree previously established the existence of three correlated work attitude dimensions: **quality, training**, and **orgsatis** (refer to *Procedure 6.5*). Each of these dimensions or factors can be used to define a specific scale by the same name.
>
> If Maree were to conduct a reliability analysis on the scales constructed by the three work attitude items which contribute to each dimension, she would

(continued)

find that the **quality** scale has a Cronbach's alpha reliability of .76, the **training** scale has a reliability of .64, and the **orgsatis** scale has a reliability of .65 (see the subtables labelled 'Reliability Statistics' in Table 8.1). All three scales consist of three work attitude items (refer to Table 6.9 for details on the items contributing to each factor) and all have an acceptable level of reliability for researcher-designed scales. This would be important information for Maree to report immediately following her report on the exploratory factor analysis results.

Table 8.1 Summary reliability and item analysis statistics for Maree's three work attitude scales

Scale: Quality			Scale: Training			Scale: OrgSatis		
Reliability Statistics			**Reliability Statistics**			**Reliability Statistics**		
Cronbach's Alpha	N of Items		Cronbach's Alpha	N of Items		Cronbach's Alpha	N of Items	
.757	3		.635	3		.651	3	
Item-Total Statistics			**Item-Total Statistics**			**Item-Total Statistics**		
	Corrected Item-Total Correlation	Cronbach's Alpha if Item Deleted		Corrected Item-Total Correlation	Cronbach's Alpha if Item Deleted		Corrected Item-Total Correlation	Cronbach's Alpha if Item Deleted
cultqual	.620	.637	acctrain	.416	.577	mgmtcomm	.388	.654
qualmktg	.578	.684	trainqua	.388	.610	satsuprv	.503	.497
qualsat	.565	.698	trainapp	.537	.396	polsatis	.519	.504

Note: These tables were produced by the SPSS Reliability procedure. They have been edited to remove unnecessary detail.

It should be noted here that internal consistency reliability is only one of several available methods for assessing the reliability of tests or scales (other methods include the split-halves method; the parallel or alternate forms method; and the test-retest method) and it may not be appropriate for all measurement situations (see Cooksey and McDonald 2019, ch. 18).

Classical Item Analysis

Attitude Scales

Most analyses of internal consistency reliability are accompanied by an *item analysis* which provides a detailed analysis of how well each item, used to define a scale, contributes to the measurement quality of that scale. While there are many types of item analysis statistics, for attitude measures and other types of scales where identifying correct responses does not make sense, behavioural researchers frequently emphasise one of two measures:

- the *corrected item-total correlation* (sometimes called *part-whole correlation*) of each item with the scale score (found by adding up or averaging the responses on all items comprising the scale); and
- an assessment of the *Cronbach's alpha if item-deleted* which is the value of Cronbach's alpha that would be observed if each specific item was deleted from the scale and the reliability for the shortened scale recomputed.

The 'corrected item-total correlation' measures the degree to which each item contributes to the entire scale (in this sense, it is similar in meaning to a 'structure correlation' or 'factor loading' from exploratory factor analysis). The 'corrected' aspect of the statistic stems from the fact that correlating an item with the scale it contributes to produces a biased (i.e. inflated) measure of correlation unless that correlation is corrected by computing it **after** first subtracting the score on the item of interest from the scale score. An item may be considered as an important contributor to a scale if it shows a corrected item-total correlation in excess of +.30; if the corrected item-total correlation falls below this value, the researcher should consider dropping the item from the scale.

It is important to note that negative corrected item-total correlations are possible, and if large, signal an important item which must first have its scores reversed before the scale is constructed. Reverse-scoring an item does not change the magnitude of its correlations with other items, it merely reverses their sign. The need for reverse item scoring often arises in the assessment of attitudes using bi-polar Likert-type item rated along an Agree-Disagree or Unsatisfied-Satisfied dimension, that are worded in the negative to prevent response biases from influencing participant ratings. One such bias is for a participant to blindly agree with every item whereas wording some items in the negative forces the participant to pay close attention to item wording and content and use the full range of the response scale. Reverse-scoring a Likert-type item is quite easy to accomplish. Take the highest rating number that a participant could give as a response to the item, add 1 to that number and then subtract their original rating. For example, on a 7-point Likert-type item, a participant who rated the item as '6' would have their score reversed as $(7 + 1) - 6 =$ '2'. Most statistical software packages have computation functions that can do such reverse scoring in a single easy step for all participants in a sample.

The 'Cronbach's alpha if item-deleted' measure shows the extent to which the scale's reliability would change if the item of interest were to be deleted from the scale. A large loss of reliability between the original reliability of the full scale and the scale where the item has been deleted would suggest that the item was important to the overall reliability of the scale. Small losses may signal relatively unimportant items whereas an increase in reliability when an item is deleted may signal a dysfunctional item that ought to be deleted from the scale.

Procedure 8.1: Reliability & Classical Item Analysis 459

> The subtables labelled 'Item-Total Statistics' in Table 8.1 summarise the relevant item analysis statistics for each of the three work attitude scales formed on the basis of the QCI database factor analysis. Note that all items have an acceptable level of corrected item-total correlation and most make reasonably substantive contributions to scale reliability as judged by the Cronbach's alpha if item-deleted values. The one exception is the **mgmtcomm** item, which if deleted from the **OrgSatis** scale, results in a very slight increase (.003) in alpha reliability. However, since the corrected item-total corrected correlation for this item was in the acceptable range, Maree should probably retain it as a contributor to the **OrgSatis** scale. None of the items required reverse scoring.
>
> In summary, Maree appears to have three scale measures of work attitude that exhibit adequate levels of internal consistency reliability and reasonably well-behaved item components.

Ability and Achievement Tests

Item analysis of ability and achievement tests and other instruments, especially multiple-choice tests, where there are definite correct and incorrect responses, requires a focus on some rather different statistics. The classical approach to item analysis of achievement and ability tests, irrespective of the design of the items, relies on two basic measures of item quality (*Procedure 9.1* discusses another more recently-developed type of approach to item analysis called *Rasch model analysis*).

The first important measure used in classical item analysis is *item difficulty* and is simply defined as the proportion (or percentage) of people in a sample who get the item correct. The higher the proportion, the easier the item is for participants to get right. A good test should not be too easy (where the average item difficulty falls above .85) or too difficult (where the average item difficulty falls below .05); no item should be 1.0 (everyone gets it correct) and no item should be 0.0 (no one gets it correct). For a multiple-choice test, the difficulty level for any item should not drop below the random chance level for getting that item correct. The random chance level is easily found by dividing 1.0 by the number of multiple-choice options per item. So, for a 4-option multiple choice item, a performance no better than random chance would be ¼ or .25, which would represent the probability of getting the item correct simply by guessing (for a 3-option item, the chance performance level would be .33; for a 5-option item, it would be .20).

The second measure used in classical item analysis is *item discrimination*. The simplest approach is to divide the group of participants who took the test into three subgroups: a high scoring subgroup (e.g. participants obtaining the top 33.3% of test scores or thereabouts), a low scoring subgroup (e.g. participants obtaining the lowest 33.3% of test scores or thereabouts) and a medium scoring subgroup (the remaining participants). Then find the item difficulty (proportion who get the item correct) for each subgroup and subtract the low subgroup's proportion from the high subgroup's proportion.

This process can be done for each item on the test and will yield an index between −1.0 and 1.0. A test item that discriminates strongly between participants who know the content area well and those who do not will show a large positive value for item discrimination. Athanasou (1997, p. 124) suggests that an item discrimination index that exceeds +.30 indicates a reasonable test item. An item that produces a discrimination index of 0.0 is a poorly performing item that cannot discriminate between participants who do better and those who do worse. A discrimination index that is negative reflects an item that poorly performing participants have a better chance of getting correct. This latter outcome is more problematic and may signal a problem such as an ambiguity regarding which answer is actually correct, confusing item writing or a response scoring error.

If the test items are in the form of multiple-choice questions, there is a third type of diagnostic information that can be obtained from what is termed a *distractor analysis*. Distractors are the available multiple-choice options other than the correct answer and they should be written in such a way as to attract participants who know less about the item content. Distractor analysis uses the high, medium and low scoring subgroups similar to item discrimination analysis but looks at the proportion of participants in each subgroup that chose each option in a multiple-choice question. A good distractor will be chosen by a progressively higher proportion of participants in the medium and low scoring subgroups. A poor distractor will either not be chosen at all (meaning it was too easy to dismiss by everyone as a possible answer) or will be chosen by a higher proportion of the high scoring subgroup (in which case this may signal that the option used as the correct answer may, in fact, not be the correct answer or that the question was worded ambiguously).

We can illustrate an item analysis by extending the QCI research scenario just a bit. Suppose Maree is asked to screen the potential of 20 incoming QCI trainees using a new quality assurance knowledge test comprising 10 multiple-choice questions, each offering three answer options. Maree knows that many of the major statistical programs, including NCSS and STATGRAPHICS, are not well-equipped to conduct classical item analyses of multiple-choice tests, without some effort, and that SYSTAT and the *psych* package for **R** produce only partial classical test analyses. Instead, for the specific purposes of conducting a classical item analysis, Maree uses a specialised SPSS item analysis syntax program, developed as part of Raynard's SPSS Tools (http://www.spsstools.net/en/) (the *SyntaxForItemAnalysisV6.sps* program is available at http://spsstools.net/en/syntax/syntax-index/item-analysis/syntax-for-item-analysis-v6/). Tables 8.2 and 8.3 show the output from the SPSS item-

(continued)

Table 8.2 Item difficulty, discrimination and reliability analyses of the 10-question multiple 3-option choice test, produced by the *SyntaxForItemAnalysisV6.sps* program

Item Statistics

	Mean	Std. Deviation	N
q1	.35	.489	20
q2	.45	.510	20
q3	.55	.510	20
q4	.55	.510	20
q5	.60	.503	20
q6	.85	.366	20
q7	.50	.513	20
q8	.20	.410	20
q9	.50	.513	20
q10	.35	.489	20

Index of Discrimination

	qnum	high	middle	low	index
1	q1	.67	.25	.17	.50
2	q2	.83	.50	.00	.83
3	q3	1.00	.63	.00	1.00
4	q4	1.00	.63	.00	1.00
5	q5	1.00	.75	.00	1.00
6	q6	1.00	1.00	.50	.50
7	q7	.17	.38	1.00	-.83
8	q8	.67	.00	.00	.67
9	q9	1.00	.38	.17	.83
10	q10	.83	.25	.00	.83

Reliability Statistics

Cronbach's Alpha	Cronbach's Alpha Based on Standardized Items	N of Items
.752	.763	10

Item-Total Statistics

	Scale Mean if Item Deleted	Scale Variance if Item Deleted	Corrected Item-Total Correlation	Squared Multiple Correlation	Cronbach's Alpha if Item Deleted
q1	4.55	6.366	.262	.532	.753
q2	4.45	5.524	.612	.805	.701
q3	4.35	5.292	.724	.833	.682
q4	4.35	5.292	.724	.929	.682
q5	4.30	5.379	.695	.758	.688
q6	4.05	6.261	.468	.443	.728
q7	4.40	9.200	-.710	.871	.871
q8	4.70	5.800	.650	.739	.703
q9	4.40	5.832	.467	.628	.724
q10	4.55	5.418	.700	.782	.688

(continued)

Table 8.3 Item distractor analyses, produced by the *SyntaxForItemAnalysisV6.sps* program

Percentage answering distracter * q1 Crosstabulation					
% within Percentage answering distracter					
		1*	q1 2	3	Total
Percentage answering distracter	high 1/3	66.7%	16.7%	16.7%	100.0%
	middle 1/3	25.0%	37.5%	37.5%	100.0%
	low 1/3	16.7%	33.3%	50.0%	100.0%
Total		35.0%	30.0%	35.0%	100.0%

Percentage answering distracter * q2 Crosstabulation					
% within Percentage answering distracter					
		1	q2 2	3*	Total
Percentage answering distracter	high 1/3		16.7%	83.3%	100.0%
	middle 1/3	25.0%	25.0%	50.0%	100.0%
	low 1/3	50.0%	50.0%		100.0%
Total		25.0%	30.0%	45.0%	100.0%

Percentage answering distracter * q3 Crosstabulation				
% within Percentage answering distracter				
		q3 2*	3	Total
Percentage answering distracter	high 1/3	100.0%		100.0%
	middle 1/3	62.5%	37.5%	100.0%
	low 1/3		100.0%	100.0%
Total		55.0%	45.0%	100.0%

Percentage answering distracter * q4 Crosstabulation					
% within Percentage answering distracter					
		1	q4 2*	3	Total
Percentage answering distracter	high 1/3		100.0%		100.0%
	middle 1/3	12.5%	62.5%	25.0%	100.0%
	low 1/3	66.7%		33.3%	100.0%
Total		25.0%	55.0%	20.0%	100.0%

Percentage answering distracter * q5 Crosstabulation					
% within Percentage answering distracter					
		1	q5 2	3*	Total
Percentage answering distracter	high 1/3			100.0%	100.0%
	middle 1/3	12.5%	12.5%	75.0%	100.0%
	low 1/3	66.7%	33.3%		100.0%
Total		25.0%	15.0%	60.0%	100.0%

Percentage answering distracter * q6 Crosstabulation					
% within Percentage answering distracter					
		1*	q6 2	3	Total
Percentage answering distracter	high 1/3	100.0%			100.0%
	middle 1/3	100.0%			100.0%
	low 1/3	50.0%	33.3%	16.7%	100.0%
Total		85.0%	10.0%	5.0%	100.0%

Percentage answering distracter * q7 Crosstabulation					
% within Percentage answering distracter					
		1*	q7 2	3	Total
Percentage answering distracter	high 1/3	16.7%	16.7%	66.7%	100.0%
	middle 1/3	37.5%	37.5%	25.0%	100.0%
	low 1/3	100.0%			100.0%
Total		50.0%	20.0%	30.0%	100.0%

Percentage answering distracter * q8 Crosstabulation					
% within Percentage answering distracter					
		1	q8 2	3*	Total
Percentage answering distracter	high 1/3		33.3%	66.7%	100.0%
	middle 1/3	50.0%	50.0%		100.0%
	low 1/3	50.0%	50.0%		100.0%
Total		35.0%	45.0%	20.0%	100.0%

Percentage answering distracter * q9 Crosstabulation					
% within Percentage answering distracter					
		1	q9 2*	3	Total
Percentage answering distracter	high 1/3		100.0%		100.0%
	middle 1/3		37.5%	62.5%	100.0%
	low 1/3	16.7%	16.7%	66.7%	100.0%
Total		5.0%	50.0%	45.0%	100.0%

Percentage answering distracter * q10 Crosstabulation					
% within Percentage answering distracter					
		1	q10 2	3*	Total
Percentage answering distracter	high 1/3	16.7%		83.3%	100.0%
	middle 1/3	12.5%	62.5%	25.0%	100.0%
	low 1/3	83.3%	16.7%		100.0%
Total		35.0%	30.0%	35.0%	100.0%

analysis syntax program, edited to show only the relevant results for item difficulty and discrimination, test reliability and distractor analyses.

We will focus first on Table 8.2. The top subtable labelled 'Item Statistics' reports the item difficulties in the column labelled 'Mean' (boldfaced numbers, reporting the proportion of the 20 trainees that got each item correct). Item 8 was the most difficult question for the trainees (difficulty = .2 or only 20% of the trainees got it correct) and item 6 was the easiest item (difficulty = .85 or 85% of trainees got the question correct. All item difficulties exceeded the random chance level of .33, except for question 8.

The second subtable, labelled 'Index of Discrimination', reports the item discrimination analysis. To accomplish, the syntax program splits the sample into three subgroups, High, Middle and Low, based on their total test scores then counts the proportion of participants in each subgroup that got the item correct. Item discrimination values are highlighted in bold under the column labelled 'Index'. All values exceed the desired level of +.30 except for

(continued)

question 7, which has a negative discrimination index of −.833. This clearly signals that question 7 is dysfunctional, in that trainees who scored lower on the test tended to get this question correct, much more so than trainees who did well on the test. The item thus works in the opposite direction than desired and signals that the question needs rewriting or less ambiguous distractors.

The subtable labelled 'Reliability Statistics' reports the overall reliability of the test (measured by the KR20 version of Cronbach's alpha) as a respectable .752, well above the accepted minimum of .6. The subtable labelled 'Item-Total Statistics' shows the corrected item-total correlation (boldface figures) for each item, which provides information similar to the discrimination index above. All corrected item-total correlations are positive and substantial except for question 1 (positive but small at .262) and question 7 (large but negative at −.710 – mirroring the negative discrimination index). To help support test revision decisions, the item analysis syntax program drops each item from the total test score and re-calculates the overall test reliability (column labelled 'Cronbach's Alpha if Item Deleted'). When Question 7 is removed, reliability improves dramatically to .871. This is because of its negative corrected item-total correlation and negative discrimination index. Item 1 is weak, but still functional.

Table 8.3 shows the results for the distractor analyses. Each test question has its own summary table, crosstabulating score category (high, middle, low) with multiple-choice distractor choices. In each subtable, the correct answer is starred and the percentage of each scoring subgroup selecting each answer option are shown. You can see in this table that 9 of the 10 items have a correct option that works well – the better test performers tend to choose the correct option (only question 7 reflects an undesirable pattern). For question 10, for example, most participants in the High subgroup tended to choose the correct option 3, participants in the Low subgroup tend to choose option 1 whereas the Medium subgroup tended to choose option 2. For question 9, the Middle subgroup never chose the incorrect option 1, whereas at least one person in the Low subgroup chose that option. This suggests that knowing at least a little bit about the content area allowed the Medium subgroup people to dismiss option 1 as a viable answer, making theirs a two-choice problem. If an incorrect option has no one choosing it, as in option 1 for question 3, this flags an ineffective distractor and indicates a need to refine the option to improve the question. However, Maree knows, from Table 8.2, that question 7 will need to be completely revised; questions 6 and 8 might also require some attention to improve their difficulty levels (i.e., make q6 harder and make q8 easier).

Inter-Observer Reliability (Inter-Rater Agreement)

Inter-observer reliability (Gwet 2012, ch. 2 and 3) is useful to assess in observational studies where observers or raters must make evaluative ratings on or categorical codings of certain behavioural or social phenomena (say, observing children at play or in a classroom, monitoring employees at work, observing members of a group, coding interview protocols). To see how reliable such subjective ratings are, it is useful to have two or more people observing and coding/rating the same situations. The codings/ratings of these multiple observers should agree if we are to conclude that they are, in fact, observing the same events in the same way (using the same judgmental criteria). If codings/ratings don't agree, then the observation recording system needs revision.

Inter-observer reliability can be assessed very simply using standard statistical methods, depending upon the scale of measurement that the observers use in making their ratings. If interval/ratio measurement scales are used by observers (e.g. assessing degree of aggressiveness in the behaviours of specific children during play), then their ratings can be correlated (see *Procedure 6.1*) across a sample of stimuli or individuals which all observers rated. This Pearson correlation between any two observers gives an index of the degree to which their rating patterns agree, thus estimating inter-observer reliability for the dimension of behaviour being observed.

If observers use categorical coding systems (involving the classification of observed behaviour into one of two or more mutually exclusive categories), then inter-observer agreement can be assessed using a specific measure of association, called *Cohen's kappa coefficient of agreement*, implemented using a specialised version of contingency table analysis (recall *Procedure 6.2*). With Cohen's kappa, we are looking for the degree to which two observers agree with respect to their category classifications. However, since some amount of agreement is bound to occur simply by accident or chance, it is important to control for this possibility. Cohen's kappa has been specifically designed to report inter-observer agreement over and above the level of chance agreement one would expect if the two observers were completely unreliable. Kappa ranges from 0 (no better than chance agreement) to 1.0 (perfect agreement). Ideally, we would look for kappa values in excess of .70 before we would consider an observational coding system to be reliable (kappa also has a statistical test associated with it which can be used to test the significance of the agreement relative to chance levels).

> The QCI database does not include data obtained by two observers on the same inspectors. However, we can create a scenario to illustrate the use of Cohen's kappa.
>
> Suppose Maree hired a research associate and both Maree and her associate observed and coded, at the same time but without communicating with each

(continued)

other, a random sample of 50 quality control inspectors with respect to their 'inspection diligence'. Assume that Inspection Diligence is an observational classification system that was developed for use by supervisors in the general appraisal of quality control inspector performance. It involves classifying each inspector's behaviour into one of three mutually exclusive categories: 'Very Diligent' (high level of attention to detail and conformance with inspection policies coupled with a high level of inspection accuracy and reasonable decision speed); 'Moderately Diligent' (adequate attention to detail and reasonable conformance with inspection policies coupled with a adequate level of inspection accuracy and reasonable decision speed); and 'Needs Improvement' (shows problems with attention to detail and/or conformance with inspection policies coupled with a low level of inspection accuracy and very slow decision speeds).

Table 8.4 shows what the agreement data contingency table from such an investigation might look like if Maree and her associate evaluated the classification framework. Note that highly reliable observers (e.g. observers using a well-constructed scheme) should show large numbers of observation counts in the main diagonal cells (boldface numbers) of the table, relative to the expected counts (which reflect chance agreement), and few observations in the off-diagonal cells whereas low reliability observers (e.g. observers using a poorly-constructed scheme) would show the reverse trend.

Table 8.4 Inter-observer reliability, computed using the SPSS Crosstabs procedure

Maree_Codes * Associate_Codes Crosstabulation

			Associate_Codes			Total
			1.00 Very Diligent	2.00 Moderately Diligent	3.00 Needs Improvement	
Maree_Codes	1.00 Very Diligent	Count	**11**	6	0	17
		Expected Count	5.4	4.8	6.8	17.0
	2.00 Moderately Diligent	Count	4	**6**	0	10
		Expected Count	3.2	2.8	4.0	10.0
	3.00 Needs Improvement	Count	1	2	**20**	23
		Expected Count	7.4	6.4	9.2	23.0
Total		Count	16	14	20	50
		Expected Count	16.0	14.0	20.0	50.0

Symmetric Measures

		Value	Asymp. Std. Error	Approx. T	Approx. Sig.
Measure of Agreement	Kappa	.601	.090	5.986	.000
N of Valid Cases		50			

(continued)

> The value for Cohen's kappa in this contingency table is .601 which is significant at p-value less than .001 but is somewhat less than the generally acceptable level of .70. From inspection of off-diagonal cells in Table 8.4 (in italics), it is apparent that the two researchers had difficulty distinguishing clearly between the 'Very Diligent' and 'Moderately Diligent' categories but were highly reliable in classifying those inspectors who 'Need Improvement'. As it stands, the Inspection Diligence framework would not be considered to be a reliable framework for appraising inspector performance unless, perhaps, the two Diligence categories were collapsed into a single 'Doesn't Need Improvement'-type of category.

Advantages

Finding the reliability of a set of test or scale items or observations allows one to empirically evaluate the quality of one's measurement instrument in terms of how consistently it measures what it was designed to measure. The methods discussed here provide specific reliability information for given types of methodologies. Knowledge of reliability helps to make the adoption and use of any measurement instrument more defensible. The advantage of item analysis for knowledge-based tests is that it can provide clear insights into those items that work well and those that work poorly, providing clearer targets for test improvement. The advantage of inter-observer reliability is primarily linked to its capacity to demonstrate that multiple observers can appropriately utilise a common framework for making judgments about observations. Using the crosstabulation table that accompanies the computation of Cohen's kappa, one can 'diagnose' where the classification system seems to be working less well, thereby providing clearly targeted areas for improvement.

Disadvantages

If the test or scale items measure more than one construct, Cronbach's alpha will give a misleading impression of scale reliability, usually in the direction of over-estimating reliability. This is a common problem when pre-existing or pre-designed scales are utilised without the benefit of factor analysis to justify the dimensional structure of the items.

> *For example* ...Behavioural researchers frequently utilise scales designed in the US in an Australian or other cross-cultural research context without checking dimensional structure for cross-cultural compatibility. In such cases, application of factor analytic techniques (see Procedure 6.5) can be of substantial assistance, prior to assessing reliability.

It is also not always desirable to have a reliability measure (Cronbach's alpha) whose magnitude partly depends on the length of the test being used. Thus, you must be careful not to artificially inflate Cronbach's alpha simply by including a large number of items in a scale or test.

One mistake that many researchers make is to assume that reporting scale reliability (such as Cronbach's alpha) is all they need to do to demonstrate scale quality. However, reliability is only one characteristic of a good measure; specific types of validity, such as construct validity (see *Chap. 2 – Measurement issues in quantitative research*), predictive validity or content validity, constitute other important characteristics of good measures and should be considered to be the more crucial qualities to demonstrate.

One limitation to the use of Cohen's kappa is that it will not be able to achieve its maximum value of 1.0 when the distribution of observation classifications is dramatically different for the two observers being compared. This means that even perfect agreement may not be reflected by a kappa value of 1.0, but by a kappa value somewhat less than 1.0. [Note that it is possible to compute the maximum possible value for kappa for any given contingency table and then divide the observed kappa value by this maximum (to express the observed kappa value as a proportion of maximum kappa), but the statistical properties of this refined measure are unknown (Waltz et al. 2010, pp. 155–156.]

Where Is This Procedure Useful?

Internal consistency reliability and item analysis should be considered for any experimental, quasi-experimental, or survey design which uses a measurement instrument composed of items purporting to measure or demonstrably measuring the same construct. Reliability and item analyses are essential follow-up procedures for an exploratory factor analysis as a way of demonstrating a key aspect of the measurement quality of the resulting factors or dimensions. Classical item analysis of ability and achievement tests provides a useful way to demonstrate the overall quality of the tests as well as providing important diagnostic information to help guide revisions of specific items or questions.

In observational studies, results tend to have more credibility if the rating systems can also be shown to be reliable across different observers. Inter-observer reliability is one method for demonstrating this quality in contexts where direct live observations are made as well as in contexts where video or audio recordings of behaviour have been obtained and coded. The lack of inter-observer reliability provides clear indications that an observational recording instrument or coding system needs revision before it is put into general use.

Software Procedures

Application	Procedures
SPSS	For reliability analysis: *Analyze → Scale → Reliability Analysis...* and select the 'Items:' defining a specific scale; provide a 'Scale label:', if desired, and make sure that the 'Model:' selected is 'Alpha'; click the '*Statistics...*' button and in the 'Descriptives for :' section, choose the 'Scale if item deleted' option. This type of analysis is primarily useful for analysing attitude scales, not ability and achievement tests. SPSS does not have a specific procedure for conducting a classical item analysis. However, you can use one of Raynald's SPSS Tools (http://www.spsstools.net/en/), *SyntaxForItemAnalysisV6.sps* program (http://spsstools.net/en/syntax/syntax-index/item-analysis/syntax-for-item-analysis-v6/; accessed 16 Sept 2019) to carry out an item analysis similar to what we illustrated in Tables 8.2 and 8.3 (a short user's guide, showing how to set up a test item file for analysis, is available at http://www.spsstools.net/Syntax/ItemAnalysis/UsingSPSSforItemAnalysis.pdf; accessed 16 Sept 2019; SPSS 1998). This syntax program works in IBM SPSS Statistics up to version 25; just make sure you enter the exact directory path for the data file into the syntax program because the program will place temporary files in that directory; otherwise you will get error messages and an incomplete analysis. For inter-rater reliability: *Analyze → Descriptive Statistics → Crosstabs...* and choose the 'Row(s)' and 'Column(s)' variables to correspond with the ratings or classifications from your two observers; click the '*Statistics...*' button and choose the 'Kappa' statistic; click the '*Cells'* button and, in the 'Counts' section, select 'Expected' in addition to 'Observed'. This will give you an inter-observer reliability analysis, similar to Table 8.4.
NCSS	For reliability analysis: *Analysis → Descriptive Statistics → Item Analysis* and select the 'Item Variables:' comprising the scale you want to evaluate; on the *Reports* tab, make sure that 'Reliability Report and 'Item Detail Report' are selected. This type of analysis is primarily useful for analysing attitude scales, not ability and achievement tests. NCSS does not have a specific procedure for conducting a classical item analysis for ability and achievement tests, however, some sense of item difficulty and discrimination can be gleaned from the output. For inter-rater reliability: *Analysis → Descriptive Statistics → Contingency Table (Cross Tabulation)* and choose the 'Discrete Variables' to define the 'Table Rows' and 'Table Columns' containing the ratings or classifications from your two observers; on the *Reports* tab, select 'All' for 'Chi Sqr Stats:'. This will give you an inter-observer reliability analysis.

(continued)

Application	Procedures
SYSTAT	For reliability analysis: *Analysis → Correlations → Cronbach's Alpha* and select the items comprising the scale you want to analyse and move them to the 'Selected variable(s)' box. This type of analysis is primarily useful for analysing attitude scales, not ability and achievement tests. Note that in this procedure SYSTAT only reports Cronbach's alpha, not corrected item-total correlations or alpha if item deleted statistics. For item analysis: *Advanced → Test Item Analysis → Classical...* and choose each of the 'Test Item(s)' required, entering the correct response code in the 'Key' as you select each item. SYSTAT computes Cronbach's alpha for the whole test as well as for odd-even splits of the test (by default). It also reports the distribution of test scores in raw and standardised form and shows item difficulties ('Mean'), corrected item-total correlations ('Excl Item R') and alpha if item deleted values ('Excl Item Alpha'). Finally, it reports a sort of graphical form of discrimination analysis, using a sideways histogram format summarising the percentage of people getting the item correct at various levels of overall test score. SYSTAT does not report a distractor analysis or discrimination indices. For inter-rater reliability: *Analysis → Tables → Two Way...* and choose the row and column variables to correspond with the ratings or classifications from your two observers; on the *Statistics* tab, select 'Cohen's kappa' under 'r x r tables'. This will produce an inter-observer reliability analysis.
STATGRAPHICS	STATGRAPHICS, prior to version 17, does not include any procedures for conducting reliability analysis or classical item analysis. However, with version 17, there is a new procedure *Reliability Analysis* which can be used to compute Cronbach's alpha, corrected item-total correlations and Cronbach's alpha if item deleted statistics.
R Commander	R Commander does not include any procedures for conducting reliability analysis or classical item analysis. However, there is a package available for use with **R**, called p*sych* that does include several procedures for reliability analysis and item analysis. The *psych* package, developed by Prof William Revelle at Northwestern University, can be downloaded from http://cran.r-project.org/web/packages/psych/index.html and an overview/user's guide is available at http://cran.r-project.org/web/packages/psych/vignettes/over view.pdf.

Procedure 8.2: Data Screening & Missing Value Analysis

Classification   Univariate/multivariate; descriptive.
Purpose   To examine data for irregularities and anomalies, particularly with respect to assumptions required by parametric statistical procedures and to examine data for anomalous, biased or influential missing response patterns.
Measurement level   Most data screening methods examine the distributions and relationships between dependent variables, thus requiring interval- or ratio-level measures. Missing value analysis can handle data at any level of measurement.

Data screening and *missing value analysis* comprise a range of techniques for inspecting one's data for irregularities and anomalies. Many of these techniques are visual (although statistical tests also exist) and some of them you have already seen in previous procedures. However, our focus here is on the use of such techniques for evaluating whether data meet the assumptions required by parametric statistical techniques.

Key assumptions that can be visually examined include:

- normality;
- linearity;
- homogeneity of variance (homoscedasticity);
- multicollinearity (abnormally high variable correlations); and
- the presence of outliers or extreme values.

One type of problem not readily amenable to visual inspection is the impact of patterns of missing data on the quality of the data to be analysed. Patterns of missing data have the potential to adversely impact on the representativeness of one's sample as well as on the capacity of specific statistical methods (such as multiple regression and factor analysis) to produce appropriate and interpretable results. A secondary problem that arises when one considers missing data patterns concerns what to do about them.

Data Screening

Data screening (e.g. Hair et al. 2010, ch. 2; Tabachnick and Fidell 2019, ch. 4) can involve a range of approaches designed to ensure that the quality of data is appropriate for the intended analyses. Such approaches include:

- the use of appropriate preparation procedures to ensure data integrity at the point of data entry;
- the use of graphical techniques to visually detect anomalous patterns and extreme values in the data; and
- the use of statistical tests to explicitly evaluate specific statistical assumptions.

The first and most basic approach to data screening is taking steps to ensure that the data have been correctly entered into whatever statistical package database or spreadsheet is being used. This requires clear specification of the steps to be taken to consistently and numerically translate the data from their original recorded format (completed questionnaire, taped interview, observation rating sheet) into the necessary row-column spreadsheet format. Such steps would include the following.

1. Set out clear rules for how responses to any categorical variables should be numerically represented (this includes considerations for how responses to open-ended questions should be coded and represented).

2. Set out clear rules for how to code missing data and any anomalous responses (such as when a participant ticks two responses for an item where only one was expected). Here we would also include decisions about how to represent responses to item ratings such as 'Not Applicable' or 'Unable to Rate' (where such a rating option is offered to the participant). The researcher needs to decide if he or she wishes such responses to be differentiated from omitted responses when the data are coded. Differentiating omitted responses from 'Not Applicable'/ 'Unable to Rate' responses would allow the researcher to treat both types of response as missing data for some analyses while preserving the option to analyse the 'Not Applicable'/'Unable to Rate' responses separately.
3. Set out clear rules for how to deal with non-serious participants (such as participants who respond randomly or in a discernible pattern down a questionnaire page).
4. Set out clear rules for how to deal with participants who arbitrarily change response formats in ways not intended by the researcher (e.g. by writing in their own numbers) or respond in unexpected ways (e.g. writing a verbal response rather than giving the requested rating).
5. Set out clear expectations for how the data are to be entered (a good practice is entering all of a participant's data before moving to the next participant).

These steps are important when the researcher enters the data him- or herself but are even more important if another person is enlisted to enter the data. In this latter case, training in the data entry process to meet the researcher's requirements is essential. [Note that for web-based data gathering instruments (e.g. online surveys), many of the above data entry decisions can be set up to occur automatically when the data file is produced by the online survey software.]

Once the data have been entered, it is almost certain that some typographic errors will have been made, resulting in incorrect data values (e.g. inadvertently typing a '4' for a male participant whose gender was supposed to be quantified as a '1'). If left unchecked, such typographical errors can introduce unnecessary anomalies into any data analysis and may even produce outliers. In large data sets, these types of errors may go unnoticed, leaving them to exert their subtle and perhaps not so subtle influences on the analyses. To circumvent these problems, it pays to invest time and energy double checking each and every record in the database against the original recorded form of the data.

If the database is large, a random or systematic sampling data verification strategy could be employed (e.g. checking every 5th data record or a 25% random sample of data records). Also, for large databases, certain data screening analyses can be run (see below), which may reveal erroneous data entry values. For categorical variables, the simplest tactic may be to compute frequency tables for each variable (recall *Procedure 5.1*) and look for unintended values. If some are revealed, it is then a simple matter to locate them in the database and correct their value.

Once the database has been verified as correctly entered, one of the most effective processes for data screening against assumptions is to use graphical display procedures to look at the data. Two important graphical methods one can employ are:

- frequency distributions with the normal curve overlaid (recall *Procedure 5.2*); and
- scatterplot matrices (recall *Procedure 5.3*).

The goal in using such visual displays is to look for patterns that signal non-normality. Frequency distributions can provide insights for single variables whereas a scatterplot matrix can provide insights for multiple variables and their relationships. Exploratory data analysis plots may also prove useful to examine (see *Procedure 5.6*).

Part of data screening can also involve more formal statistical testing. Some of these statistical tests are specific to certain analytical procedures (e.g. tests of the homogeneity of variance assumption). However, the most common data screening tests, which are not specific to a particular type of model building exercise, assess the degree to which the data for a variable or variables fit or reflect a normal distribution. There are both univariate (one variable at a time) and multivariate (many variables simultaneously) forms of such tests and different statistical packages will offer different types of tests. For example, SPSS can produce the Shapiro-Wilk and Kolmogorov-Smirnov univariate tests on request in its *Explore* procedure (recall *Procedure 5.6*); SYSTAT can produce the Shapiro-Wilk and the Anderson-Darling univariate tests as well as Mardia's multivariate tests for skewness and kurtosis on request in its *Basic Statistics* procedure and NCSS can produce D'Agostino univariate normality tests (for skewness, for kurtosis and omnibus) and the T^2 Mahalanobis distance multivariate test for outliers (not a normality test as such, but a test for extreme values, which, if there are many, would indirectly signal a non-normal distribution), on request in its *Data Screening* procedure.

For any of these data screening tests, the null hypothesis being evaluated is that the data follow a normal (or multivariate normal) distribution, so that, if the test is significant, it signals that the data do not fit a normal (or multivariate normal) distribution. The one drawback for many of these tests is their tendency to signal false alarms, i.e. giving a significant result when there is really no issue. This can occur if you have a large sample (e.g. greater than 100, say) and/or if you have discrete quasi-interval scale variables with only a few possible values (such as a Likert-type scale). This means that these tests should always be interpreted in conjunction with visual inspection of the data distribution.

> Suppose Maree wishes to be relatively certain that the **mentabil, accuracy, speed, quality, training** and **orgsatis** variables meet the generally required assumptions of normality, homogeneity of variance and linearity and don't have any serious outliers. Figure 8.1, produced using SPSS, shows a frequency distribution plot for each of the six variables, with the normal curve overlaid. These graphs provide a rough visual indication of how well each variable meets the normality assumption.

(continued)

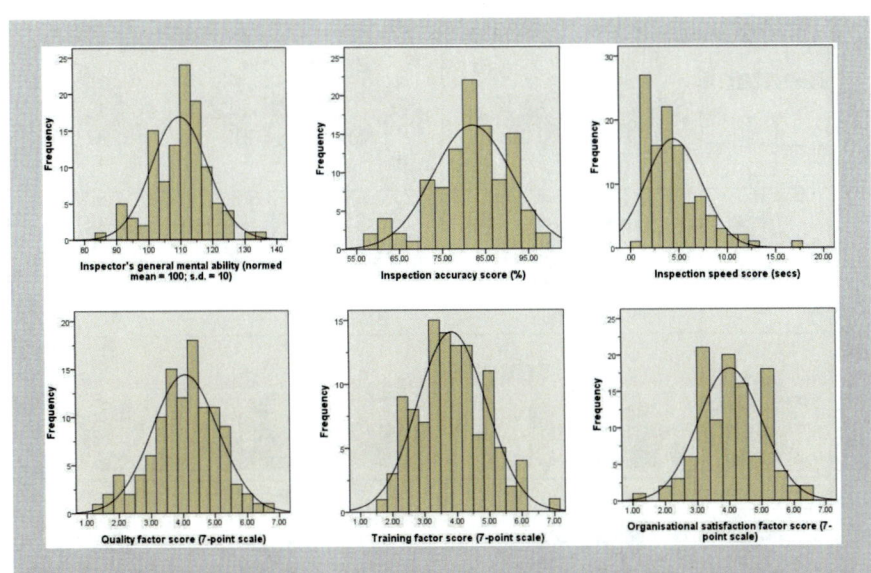

Fig. 8.1 Frequency distributions for visually checking the normality assumption relative to the mentabil, accuracy, speed, quality, training and orgsatis QCI variables

It is clear that the only really problematic variable is **speed** (as we already knew), which is rather strongly positively skewed. Note that **jobsat** looks somewhat negatively skewed. Often an inspection of frequency distributions can provide information that can help select a transformation that would normalise the data. In the case of the **speed** variable, *Procedure 5.7* showed how a logarithmic transformation could normalise a positively skewed variable. Other transformations are also possible, such as taking the square root or the inverse of a variable; see Tabachnick and Fidell 2019 section 4.1.6, and Field 2018, pp. 268–276, for further details).

Figure 8.2, produced using STATGRAPHICS, shows a scatterplot matrix comparing all six variables, in a pairwise fashion. Notice how STATGRAPHICS also places boxplots (recall *Procedure 5.6*) in the diagonal cells of the matrix, providing a picture of the univariate distribution of each variable at the same time. In a scatterplot matrix, nonlinearity will show as a distinct bend (e.g. a U- or inverted U-shape) away from a straight-line relationship. None of the scatterplots in Fig. 8.2 show any strong evidence of nonlinearity. Outliers will stick out as 'lonesome' data points on this plot and signal potential observations that might be worth double checking for data accuracy or some other reason for the extremity. All of the relationships with **speed** appear to show at least one, sometimes two, such outlier points.

(continued)

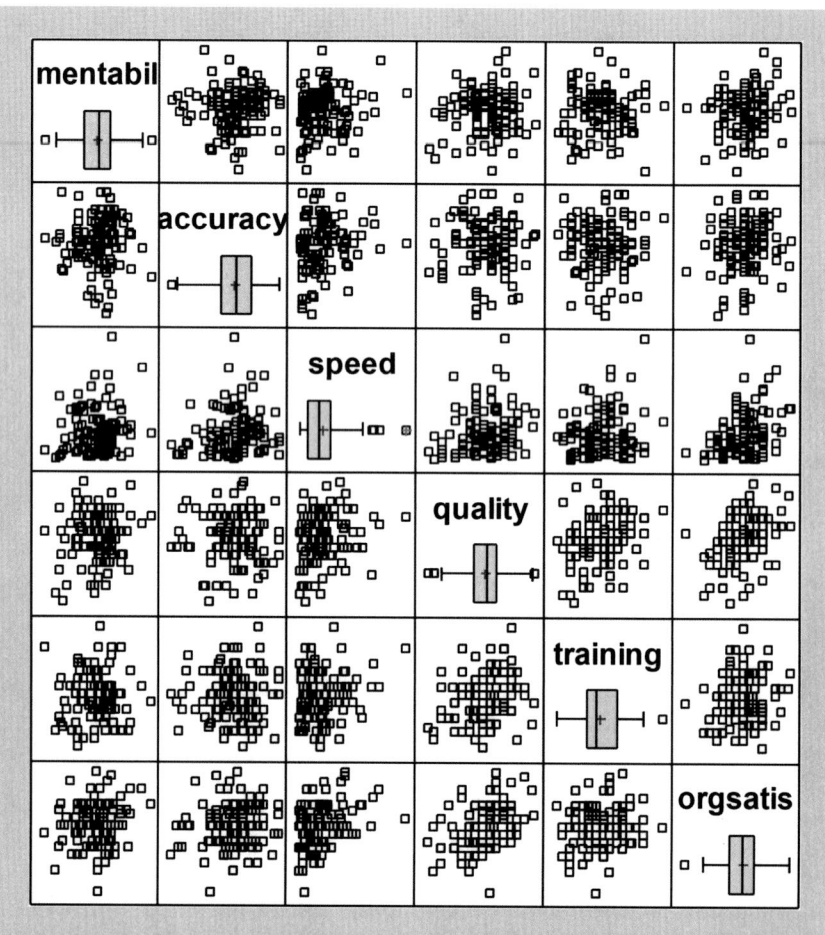

Fig. 8.2 Scatterplot matrix for visually checking the multicollinearity, homogeneity of variance and linearity assumptions as well as outliers relative to the mentabil, accuracy, speed, jobsat and workcond QCI variables

Heteroscedasticity (failure to meet the homogeneity of variance assumption; recall the discussion in *Procedure 7.13*) reveals itself in a fan-shaped scatterplot in Fig. 8.2, shown most clearly in the relationship between **speed** and **accuracy**. The fan shape occurs because **accuracy** scores are much more variable for faster levels of **speed** than for slower levels of **speed**.

Roughly the same sort of pattern emerges in the relationship between **speed** and **mentabil**. What is interesting is that this heterogeneity of variance and the 'lonesome' outliers are most likely caused by the high positive skew of the **speed** variable. Transforming **speed** scores using logarithms rectifies these problems as shown in Fig. 8.3. This shows that one type of anomaly in the data (e.g. non-normality) may create other problems such as non-linearity or heteroscedasticity, so that fixing one problem may fix the other problems as well—the trick is to correctly diagnose the root problem.

(continued)

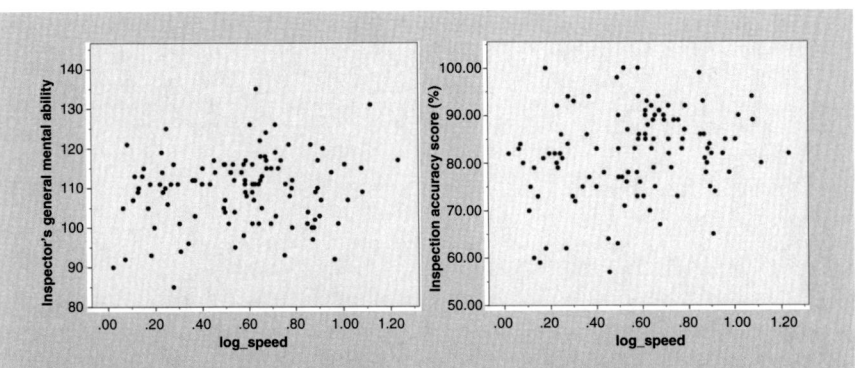

Procedure 8.2: Data Screening & Missing Value Analysis

Fig. 8.3 Scatterplot of mentabil and accuracy against log_speed showing how a transformation can correct data relationship anomalies

Multicollinearity between variables refers to the existence of an abnormally high correlations between them (usually of a magnitude of .90 or higher). Normally, highly correlated variables are desirable, but not in statistical procedures like multiple regression, MANOVA or factor analysis where they can produce dramatic instabilities in the statistical results. Variables that are close to perfectly correlated are nearly identical in what they measure, so that one variable is essentially redundant. In a scatterplot matrix like that shown in Fig. 8.2, two variables would be multicollinear if their scatterplot points looked very close to falling along a straight line. As you can see, none of the graphs in Fig. 8.2 even come close to such a near-perfect relationship.

Maree could also examine data screening tests for these six variables. Table 8.5 shows what SYSTAT would produce for such an examination. The top subtable in Table 8.5 shows basic descriptive statistics for each variable as well as the Shapiro-Wilk and Anderson-Darling tests of normality (and their associated *p*-values). The Shapiro-Wilk tests suggest that there may be non-normality associated with both the **accuracy** and **speed** variables, but much more so with **speed**. The Anderson-Darling test is a more powerful test but may send signals about non-normality that may be false alarms (especially in a sample size over 100, as is the case for the QCI database). In Table 8.5, the Anderson-Darling tests show **mentabil**, **speed**, and **quality** may all have problems with non-normality, but especially so for **speed**. If Maree looked at these tests in conjunction with the histograms in Fig. 8.1, she might reasonably conclude that **speed** was the only variable where non-normality needed to be addressed.

(continued)

Table 8.5 Data screening tests produced by SYSTAT for the six variables of interest

	Mentabil	Accuracy	Speed	Quality	Training	Orgsatis
N of Cases	111	111	111	112	112	112
Median	111.000	83.000	3.890	4.167	3.667	4.000
Arithmetic Mean	109.838	82.135	4.480	4.077	3.857	4.031
Standard Deviation	8.764	9.172	2.888	1.034	1.059	0.985
Shapiro-Wilk Statistic	0.985	0.975	0.880	0.982	0.980	0.986
Shapiro-Wilk p-Value	0.249	0.032	0.000	0.125	0.099	0.296
Anderson-Darling Statistic	0.758	0.686	3.128	0.767	0.682	0.674
Adjusted AndersonDarling Statistic	0.763	0.691	3.150	0.772	0.687	0.679
p-Value	0.047	0.071	<0.01*	0.045	0.073	0.076

*The p-value cannot be precisely computed

	Coefficient	Statistic	p-Value
Mardia's Skewness	5.108	97.795	0.000
Mardia's Kurtosis	48.689	0.370	0.711

The bottom subtable in Table 8.5 shows Mardia's multivariate tests for skewness and kurtosis. The test for skewness is significant; the test for kurtosis is not. This suggests that if all six variables were used together in a single parametric analysis, the data should not be assumed to follow a multivariate normal distribution. In this case, given the patterns associated with the Shapiro-Wilk and Anderson-Darling tests, in conjunction with the visual displays produced in Figs. 8.1 and 8.2, Maree might reasonably conclude that **speed** is the main culprit in this significant multivariate test.

Missing Value Analysis

Missing Value Analysis (*MVA*) is a relatively recent addition to the toolkit of social and behavioural researchers (e.g. Allison 2001; Hair et al. 2010, pp. 42–63; Tabachnick and Fidell 2019, section 4.1.3). It is designed to help researchers examine their data for patterns of missing data and for potential influence of missing data patterns on different group comparisons. Almost all data sets have some degree of missing data in them. Data may be missing for a variety of reasons including: participants don't understand a question or forget or refuse to answer one or more questions, participants withdraw partway through an investigation (a problem with repeated measures designs), equipment breaks down leading to a loss of data or an observer misses something that should have been recorded.

One very important consideration with missing data is whether they seem to be missing randomly or due to some fault in the research design or instruments. Data that are missing at random (called *MCAR*—Missing Completely at Random), as long as they don't constitute a large proportion of the data to hand, are less problematic than data that are not missing at random (called *MNAR*—Missing Not at Random. Some researchers further distinguish between 'missing completely at random' and 'missing at random, called ignorable nonresponse', MAR—see Tabachnick and Fidell 2019, section 4.1.3).

If the data are MNAR, then some feature of the study (such as requiring participants to work too long or hard) or how the data are gathered (such as a question that offends female participants or participants from particular cultures) is likely creating systematic patterns of response omission and this can severely hamper any attempts at generalisation because the representativeness of the sample has been altered. Very often, researchers will assume their data are MCAR and don't think about the possibility that they could be MNAR. Here is where *missing value analysis* can be of assistance.

> Suppose Maree wishes to use missing value analysis to examine her data set for potential problematic patterns. Tables 8.6, 8.7, 8.8, 8.9 and 8.10, produced by SPSS, shows some selected outcomes from a missing value analysis of the original 17 variables in the QCI database.
>
> Table 8.6 provides simple univariate descriptive statistics for each variable and shows how many and what percentage of scores were missing for each variable. SPSS MVA also has another nice feature in that it tries to show where there are extreme values (outliers) that the researcher might wish to investigate further. This table shows that the **trainqua** variable had the greatest percentage of missing data of all variables (9.8% of inspectors did not give a score on this item) followed by **satsuprv** (6.3% of inspectors did not give a score on this item). This outcome could arise, perhaps, from inspectors who felt that the question may have asked them to reveal a potentially negative perception of their company; however, this is only supposition. Overall, there is a relatively low amount of missing data in the QCI database.

(continued)

Table 8.6 Univariate statistics – missing value analysis

Univariate Statistics

	N	Mean	Std. Deviation	Missing Count	Missing Percent	No. of Extremes[a] Low	No. of Extremes[a] High
mentabil	111	109.84	8.764	1	.9	1	1
accuracy	111	82.1351	9.17158	1	.9	2	0
speed	111	4.4801	2.88751	1	.9	0	4
jobsat	109	4.96	1.644	3	2.7	4	0
workcond	106	4.21	1.717	6	5.4	0	0
cultqual	106	3.91	1.342	6	5.4	0	0
acctrain	109	3.60	1.395	3	2.7	0	0
trainqua	101	3.66	1.235	11	9.8	0	0
qualmktg	108	4.46	1.195	4	3.6	6	1
mgmtcomm	108	4.10	1.289	4	3.6	0	0
qualsat	107	3.87	1.190	5	4.5	0	0
trainapp	110	4.21	1.441	2	1.8	0	0
satsuprv	105	3.82	1.392	7	6.3	0	0
polsatis	106	4.15	1.111	6	5.4	0	0
company	111			1	.9		
educlev	109			3	2.7		
gender	109			3	2.7		

a. Number of cases outside the range (Q1 - 1.5*IQR, Q3 + 1.5*IQR).

Table 8.7 Separate variance t-tests comparing groups missing a score on a variable with groups not missing a score

Separate Variance t Tests[a]

		mentabil	accuracy	speed	jobsat	workcond	cultqual	acctrain	trainqua	qualmktg	mgmtcomm	qualsat	trainapp	satsuprv	polsatis
trainqua	t	-.3	.4	-.4	-1.9	.0	.3	-2.0	.	.6	-.5	.1	.1	.7	-.1
	df	12.6	11.7	15.5	14.7	9.6	11.1	11.0	.	10.2	10.6	12.2	11.2	10.5	11.0
	P(2-tail)	.745	.707	.671	.082	.979	.795	.074	.	.560	.649	.888	.960	.512	.946
	# Present	100	100	100	98	97	96	99	101	98	98	96	99	95	95
	# Missing	11	11	11	11	9	10	10	0	10	10	11	11	10	11
	Mean(Present)	109.75	82.26	4.45	4.89	4.21	3.92	3.52	3.66	4.49	4.08	3.87	4.21	3.85	4.15
	Mean(Missing)	110.64	81.00	4.74	5.64	4.22	3.80	4.40	.	4.20	4.30	3.82	4.18	3.50	4.18
satsuprv	t	-.9	-1.2	-1.5	-1.0	-2.6	-.3	.4	.7	.1	-.9	.4	-.2	.	-2.5
	df	6.8	6.9	7.0	6.9	9.6	7.7	7.3	5.7	8.2	7.6	7.0	5.4	.	5.5
	P(2-tail)	.402	.285	.183	.339	.027	.808	.702	.519	.912	.399	.716	.863	.	.052
	# Present	104	104	104	102	99	99	102	95	101	101	100	104	105	100
	# Missing	7	7	7	7	7	7	7	6	7	7	7	6	0	6
	Mean(Present)	109.64	81.88	4.38	4.92	4.14	3.90	3.61	3.68	4.47	4.08	3.88	4.20	3.82	4.08
	Mean(Missing)	112.71	85.86	5.94	5.57	5.14	4.00	3.43	3.33	4.43	4.43	3.71	4.33	.	5.33

For each quantitative variable, pairs of groups are formed by indicator variables (present, missing).
a. Indicator variables with less than 6% missing are not displayed.

Table 8.7 reports the results for separate variance t-tests focusing on quantitative (i.e. continuous) variables where 6% or more of the sample were missing scores on a variable. This test works by constructing two groups: one defined by the cases missing a score on a specific variable (like **trainqua**) and the other defined by the cases not missing a score on that variable. The t-test then compares the means for the two groups on each of the other continuous variables in the analysis. This test operates on the basis that the homogeneity of variance assumption is violated (because the two groups are generally of such disparate sizes), therefore requiring separate variance estimates to be used. Table 8.7 reports separate variance t-tests only for the

(continued)

trainqua and **satsuprv** variables, because only those variables each had in excess of 6% of the sample missing scores. For each test, all relevant descriptive data are presented for interpretation as are the p values ('P(2-tail)') for each test. For the **trainqua** variable, there were no significant differences on any other continuous variable in the database between the group missing a score on that variable and the group not missing a score – signalling no apparent biasing effect due to the missing data pattern. For the **satsuprv** variable, there was only one significant difference between the group missing a score on that variable and the group not missing a score and that focused on the **workcond** variable. Here, the pattern was for the group missing a score on the **satsuprv** variable to report a significantly less stressful work environment than the group not missing a score on the **satsuprv** variable. There is no logical reason for such a pattern to occur; Maree would be justified in concluding that this is more likely to be a false alarm, given the number of t-tests being interpreted. Generally, speaking, Maree would only need to be concerned about potential biasing influences if many of the t-tests were significant.

Table 8.8 shows three crosstabulation tables comparing continuous variables against a categorical variable. These tables show how many and what percentage of participants did ('Present') and did not ('Missing') give a legitimate score on specific variables. Only variables having more than 6% of scores missing are shown in each table. Here, Maree would look for any strong biases in the pattern of missing data across the five companies, the two levels of education and/or the two gender categories. For example, in the top subtable, focusing on **company**, for the **trainqua** variable, the missing data percentages range from 4.8% (Small Appliance Manufacturer) to 17.4% (PC Manufacturer) and, for the **satsuprv** variable the range is 0% to 13%. Maree could well undertake further inquiries to investigate why these variations occurred, e.g. whether the variations could be attributed to differing company cultures. If differing company cultures were the culprit, this would suggest that the data for these variables may be MNAR. The subtables for **educlev** and **gender** do not suggest any differential bias.

Table 8.9 shows all inspectors in the database that were missing a score on at least one variable in the database. Each inspector is listed by ID number ('Case') and an 'S' appears wherever a score was missing. The table has been ordered so that similar missing patterns appear together. Thus, the top five rows of the table show the five inspectors who were missing a score on the **cultqual** variable. This table also shows where specific inspectors have extreme scores (+ = high; – = low) on particular variables. Thus, inspector 62 shows an extremely high score on the **speed** variable. There appears to be no obvious systematic pattern of missing responses shown in this table. Inspector 13 declined to give a response on four of the survey items, the most missing data for any individual inspector. Interestingly, two of the

(continued)

Table 8.8 Missing data patterns across different companies, education levels and genders

					company					Missing
			Total	1 PC Manufacturer	2 Large Electrical Appliance Manufacturer	3 Small Electrical Appliance Manufacturer	4 Large Business Computer Manufacturer	5 Automobile Manufacturer		SysMis
trainqua	Present	Count	101	19	21	20	20	20		1
		Percent	90.2	82.6	91.3	95.2	87.0	95.2		100.0
	Missing	% SysMis	9.8	17.4	8.7	4.8	13.0	4.8		.0
satsuprv	Present	Count	105	21	21	21	20	21		1
		Percent	93.8	91.3	91.3	100.0	87.0	100.0		100.0
	Missing	% SysMis	6.3	8.7	8.7	.0	13.0	.0		.0

Indicator variables with less than 6% missing are not displayed.

				educlev		Missing
			Total	1 High School Only	2 Tertiary Qualified	SysMis
trainqua	Present	Count	101	51	47	3
		Percent	90.2	91.1	88.7	100.0
	Missing	% SysMis	9.8	8.9	11.3	.0
satsuprv	Present	Count	105	52	50	3
		Percent	93.8	92.9	94.3	100.0
	Missing	% SysMis	6.3	7.1	5.7	.0

Indicator variables with less than 6% missing are not displayed.

				gender		Missing
			Total	1 Male	2 Female	SysMis
trainqua	Present	Count	101	52	46	3
		Percent	90.2	89.7	90.2	100.0
	Missing	% SysMis	9.8	10.3	9.8	.0
satsuprv	Present	Count	105	54	48	3
		Percent	93.8	93.1	94.1	100.0
	Missing	% SysMis	6.3	6.9	5.9	.0

Indicator variables with less than 6% missing are not displayed.

missing scores were the inspection performance measures, which could reflect inspector absences on the date the measures were collected (something Maree could perhaps verify by checking her data collection records).

Table 8.10 shows the EM correlation matrix among the continuously measured variables that is produced by first estimating values for the missing data using what is called 'expectation maximisation'. The correlations aren't

(continued)

Table 8.9 Missing patterns (cases with missing values)

Case	# Missing	%	ment	accur	spee	comp	educl	obsat	traina	acctra	polsat	quals	qual	gend	mgmt	cultqu	workc	satsu	traing
12	1	5.9														S			
67	1	5.9														S			
85	1	5.9														S			
95	1	5.9											-			S			
36	2	12														S		S	
18	1	5.9																	S
25	1	5.9																	S
83	1	5.9	-																S
107	1	5.9											-						S
2	2	12															S		S
78	2	12															S		S
27	1	5.9														S			
55	1	5.9														S			
79	1	5.9														S			
112	1	5.9											-			S			
34	1	5.9										S							
62	1	5.9			+							S							
15	2	12									S	S							
30	1	5.9									S								
1	2	12								S	S								
52	3	18								S	S	S							
77	2	12								S									S
70	2	12															S	S	
71	1	5.9															S		
31	1	5.9															S		
14	1	5.9															S		
110	1	5.9															S		
26	2	12						S											S
98	1	5.9						S											
64	1	5.9					S												
35	1	5.9					S												
104	2	12				S	S						-						
105	1	5.9				S													
50	1	5.9														S			
33	1	5.9														S			
10	2	12														S			S
103	2	12										S							S
23	1	5.9										S							
65	1	5.9			+							S							
106	2	12										S	S						
109	1	5.9												S					
101	1	5.9												S					
45	3	18			S	S	-							S					
102	2	12									S							S	
96	2	12									S				S				
13	4	24	S	S	S										S				

- indicates an extreme low value, while + indicates an extreme high value. The range used is (Q1 - 1.5*IQR, Q3 + 1.5*IQR).
a. Cases and variables are sorted on missing patterns.

(continued)

Table 8.10 EM correlations

	ment.	accur.	spee.	jobsat	workc.	cultqu.	acctra.	trainq.	qual.	mgmt.	quals.	traina.	satsu.	polsat.
mentabil	1.0													
accuracy	.01	1.0												
speed	.13	.17	1.0											
jobsat	.23	.27	.02	1.0										
workcond	.15	.48	.32	.31	1.0									
cultqual	.16	.00	.18	.25	.18	1.0								
acctrain	.05	-.1	.05	-.1	.00	.24	1.0							
trainqua	-.1	-.2	.12	.0	.00	.0	.22	1.0						
qualmktg	.06	.08	.18	.12	.19	.52	.22	.12	1.0					
mgmtcomm	.08	.38	.33	.24	.57	.43	.0	.08	.38	1.0				
qualsat	.13	-.1	.13	.08	.07	.50	.26	.05	.47	.15	1.0			
trainapp	-.1	.09	.17	.0	.03	.13	.40	.40	.16	.13	.17	1.0		
satsuprv	.15	.0	.30	.15	.24	.27	.19	.12	.30	.34	.23	.11	1.0	
polsatis	.06	.0	.17	.10	.17	.28	.03	.0	.21	.35	.18	.17	.49	1.0

a. Little's MCAR test: Chi-Square = 319.942, DF = 295, Sig. = .152

really of interest to Maree, but the footnote to the matrix is (the only way Maree can get the footnote is to request the matrix from SPSS). This footnote reports a significance test called 'Little's MCAR test'. This test evaluates the likelihood that the data in the matrix are MCAR. What Maree hopes to see here is a **nonsignificant** test, i.e. $p > .05$. In this case, $p = .152$, which is not significant. She can therefore conclude that there is insufficient evidence in the data to warrant rejecting the assumption that the data were MCAR— a good outcome that means Maree could safely proceed with most analyses without worrying that patterns of missing data would substantively influence the outcomes.

There are several ways of dealing with missing data in the conduct of analyses. For most statistical programs, missing data are handled by default, in one of two passive ways. These methods are passive because they handle the problem by dropping participants who are missing observations from the analysis. One passive method is *listwise deletion*: if participants are missing a score on any variable being analysed by the same procedure at the same time, they are removed from the entire analysis. This method is safest from a statistical estimation perspective but has the drawback that the sample size for the analysis may be reduced to a problematic level if a lot of participants are missing a score on one or more variables. Furthermore, many people think that listwise deletion is too severe and results in throwing away too much legitimate data as well as missing data.

A second passive method for handling missing data is *variablewise* or *pairwise deletion*: if participants are missing a score on any one variable or any pair of variables being analysed by the same procedure at the same time, they are removed from calculations involving just those variables. So, when computing a correlation matrix using pairwise deletion for example, each correlation is based on only those participants who had a legitimate score on both variables. Over the entire matrix, this means that all of the nonmissing information available has been used. This method is safest from an overall sample size and information use perspective. However, it has the drawback, from a statistical estimation perspective, that there is a greater chance of problems occurring during the statistical computation process, because every variable and every correlation is based on a different number of participants. This is especially troublesome for procedures like multiple regression, factor analysis and structural equation modelling.

There are three active methods available for handling missing data. An active method attempts to replace the missing data with some estimate of what the value would have been had it been available. The simplest method is called *prior knowledge*: if a participant is missing a score on a variable, the researcher makes an educated guess as to what the score might have been. While this procedure may seem easy, it is subjective and would be hard to defend in any scientific sense, unless you had a good understanding of the population from which the sample was obtained.

A second active method is called *mean substitution* and only works for interval/ratio level measures: if a participant is missing a score on a variable, that missing score is replaced with the mean for the variable. This leaves the overall variable mean unchanged but changes the standard deviation. However, while this procedure allows you to use the compete sample, mean substitution can have a biasing effect on both standard deviations and correlations. The technique can be refined and improved a bit: if you know which subgroup a participant comes from (e.g. male, tertiary-qualified, PC manufacturing company), replace a missing score by that subgroup's mean on the variable.

The third and more recent innovation for actively handling missing data is called *imputation*: if a participant is missing a score on one or more variables, use the information from other variables to predict or estimate what the score(s) should have been (see, for example, Tabachnick and Fidell 2019, section 4.1.3.2). There are several statistical ways of performing the estimation process: the regression method, the expectation maximisation (EM) method and multiple imputation. Whatever estimation process is used (EM seems to be the most widely accepted method), the resulting estimates are plugged back into the original database and analyses can proceed without any missing observations. These methods depend on the capacity of other variables in the database to predict the variable whose missing data are to be imputed. This is a condition that many databases may not meet primarily because the researcher lacks sufficient and relevant knowledge. These methods are mathematically complex (but with a sound underpinning theory), require the data to be MCAR (or MAR at least) and only work for replacing scores on interval/ratio variables (although categorical variables can be used as predictors).

One practice advocated by Tabachnick and Fidell (2019, section 4.1.3.5) is to repeat analyses with missing data imputed via some method and without missing data being imputed (i.e. using listwise or pairwise deletion) and compare the outcomes. If the stories are largely the same, you could conclude that the missing values imputation process has not substantively distorted the outcomes while, at the same time, giving you a set of 'complete' data to work with. If, however, the results are dramatically different, then the imputation process is introducing some distortions. In that case, the best decision is to use a passive deletion method for dealing with missing data, rather than an active imputation method, if the MCAR assumption is reasonably met.

Advantages

Data screening and missing value analysis are important preliminary processes to undertake before commencing the major analyses in an investigation. Effective data screening can provide increased confidence that the assumptions of parametric statistical techniques are being satisfied, which means that proper inferences and generalisations have a better chance of emerging. Effective missing value analysis can provide increased confidence that whatever patterns of missing data exist in the database, they are not substantively influencing the outcomes of statistical analyses. If there are problems identified by data screening or MVA, the results can provide some clear signals as to what might need to be done to rectify the problem. There is a very compelling argument for every social and behavioural researcher to undertake both data screening and missing value analysis, indicate clearly that such analyses have been conducted, summarise their outcomes and describe the subsequent actions taken in any write-up of their research.

Disadvantages

The main disadvantage of graphical data screening methods is that some types of anomalies, especially those of a multivariate nature, can be very difficult to detect visually. In such cases, statistical tests can sometimes be of assistance. However, even these tests (e.g. tests for normality, homogeneity of variance and for univariate and multivariate outliers—see, for example, Tabachnick and Fidell 2019, 4.2.2.3 and Hair et al. 2010, pp. 64–87) have their difficulties, often tending to show problems as significant when they really aren't. That is, many such tests are overly sensitive and can give a false alarm, especially in large samples.

MVA as a process for analysing patterns really doesn't have any drawbacks. However, the methods that one uses to handle missing data do have important disadvantages. The major disadvantage of passive methods for handling missing data is that disturbances in sample sizes are introduced and, potentially, the stability of certain statistical procedures is placed at risk. Pairwise deletion of missing data, for example, can create some severe estimation problems for any method of factor analysis other than principal components analysis (recall *Procedure 6.5*). Active approaches for handling missing data can work well for interval/ratio dependent variables but are not very useful if data are missing for categorical variables. In cases where categorical data (such as a participant's gender or educational level) are missing, the prior knowledge approach may be the only feasible way to deal with the problem. The mean substitution and imputation methods have their own additional disadvantages as well. Mean substitution can introduce undesirable biases into certain statistical procedures and is often overly simplistic. Imputation methods depend on having other variables in the database that are substantively useful for predicting missing scores—choosing which variables are the best ones to use for the imputation process is sometimes problematic.

Where Is This Procedure Useful?

Data screening and MVA are essential techniques to employ in any investigation that employs quantitative measures. They provide evidence that can help to rule out alternative plausible explanations for why statistical results do or don't show particular outcomes. As such, their use can help a researcher to tell a more convincing story about their investigation and what conclusions the data support. Data screening can help the researcher decide on appropriate transformations to normalise their data and can help identify outliers and other anomalous data patterns that require further attention. MVA also helps the researcher to understand anomalous data patterns, associated with missing responses, and provides approaches for overcoming various types of missing data problems. Here, the goal of MVA is to help the researcher maintain the representativeness and size of their sample so that intended generalisations can be justified.

Software Procedures

Application	Procedures
SPSS	Refer to *Procedures 5.2* and *5.3* for information on obtaining the graphical data screening displays employed here. Specific SPSS procedures, such as multiple regression, ANOVA/MANOVA, and descriptive statistics, have their own options for statistically testing assumptions. The multiple regression procedure also has options to produce some useful diagnostic graphs associated with patterns of residuals (recall *Procedure 7.13*). Analyze → Missing Value Analysis... and select the 'Quantitative Variables:' and 'Categorical Variables:' to be analysed; under 'Estimation', tick the 'EM' box; click the *Patterns...* button and select the 'Tabulated cases ...' and 'Cases with missing values...' option (you can also specify 'Omit patterns with less than X% of cases', by entering the desired percent X in the box; you can also request 'Additional information for:' selected variables – particularly useful is asking for additional information on any categorical variables); click the *Descriptives...* button, select the 'Univariate statistics' and tick **all** boxes under 'Indicator Variable Statistics' (you can also specify 'Omit variables missing less than X% of cases', by entering the desired percent X in the box). An alternative procedure offering a quick exploration of missing value

(continued)

Application	Procedures
	patterns prior to using multiple imputation methods to estimate them can be carried out by *Analyze → Multiple Imputation → Analyze Patterns...* and choosing the variables for which missing value patterns are to be analysed. You can also choose the percentage of the sample that must be missing before a variable will be focused on in the pattern analysis. This procedure is usually followed by *Analyze → Multiple Imputation → Impute Missing Data Values...* which provides methods for imputing missing data values such that a new version of the dataset can be created for analysis.
NCSS	*Analysis → Descriptive Statistics → Data Screening* and select the desired 'Variables to Screen'; under the *Reports* tab, make sure all output options are ticked. This procedure gives you the most comprehensive statistical data screening process of the various packages and includes some features of MVA. NCSS offers two active methods for dealing with missing data, 'Average' (essentially the mean substitution method) and 'Multivariate Normal' (essentially the regression imputation method). Of course, NCSS can also produce a wide range of diagnostic graphs as well (refer to *Procedures 5.2, 5.3* and *7.13*).
SYSTAT	Refer to *Procedures 5.2* and *5.3* for information on obtaining the graphical data screening displays employed here. Specific SYSTAT procedures, such as multiple regression, ANOVA/MANOVA, and descriptive statistics, have their own options for statistically testing assumptions. The multiple regression procedure also has options to produce some useful diagnostic graphs associated with patterns of residuals (recall *Procedure 7.13*). *Advanced → Missing Value Analysis* and choose your 'Selected variable(s):' (all variables must be continuous variables) for analysis (the EM method is the default method and will produce Little's MCAR test). SYSTAT offers only EM or regression imputation methods.
STATGRAPHICS	Refer to *Procedures 5.2* and *5.3* for information on obtaining the graphical data screening displays employed here. Specific STATGRAPHICS procedures, such as multiple regression, ANOVA, and descriptive statistics, have their own options for statistically testing assumptions. The multiple regression procedure also has options to produce some useful diagnostic graphs associated with patterns of residuals (recall *Procedure 7.13*). STATGRAPHICS does not offer missing value analysis capabilities.
R Commander	Refer to *Procedures 5.2* and *5.3* for information on obtaining the graphical data screening displays employed here. The *Models* menu also has options to produce some useful numerical and graphical diagnostics associated with patterns of residuals once a regression model has been constructed (recall *Procedure 7.13*). *Statistics → Summaries → Shapiro-Wilk test of normality...* and choose one variable to conduct the test on (**R** Commander can only deal with one variable at a time for this test). *Statistics → Summaries → Count missing observations* will provide a very simplistic summary of missing values for every variable in the database; no patterns are analysed. Field et al. (2012, ch. 5) describe some procedures and packages within the main **R** software system for data screening activities.

Procedure 8.3: Confidence Intervals

Classification	Univariate/bivariate/multivariate; inferential; parametric.
Purpose	To establish lower and upper limits or bounds around a sample statistic such that the population parameter estimated by that statistic has a specified probability of being captured within such intervals in the long run.
Measurement level	Confidence intervals are normally constructed using variables measured on an interval- or ratio-scale. However, it is possible to construct confidence intervals around proportions as well, so that nominal- or ordinal-scale variables can be accommodated.

Confidence intervals provide an alternative methodology for displaying the accuracy and amount of error associated with a statistical estimate (see, for example, Smithson 2003). Since one goal of calculating a statistic, based on sample data, is to estimate a specific population parameter, a confidence interval establishes a precise probability that the population value will fall between specific lower and upper limits if we were to repeat the study using different samples a large number of times.

> Suppose a political poll is conducted where 1000 randomly sampled voters are asked to indicate if they would vote for a specific candidate, Bill. In a media report, a spokesperson for the poll indicates that they are 95% certain that 49% of polled voters indicated they would vote for Bill, plus or minus 3%. In this scenario, what the spokesperson has done is report a 95% confidence interval around the sample statistic (observed proportion who would vote for Bill). However, what this statement also indicates is that we would expect such intervals to contain the population value 95% of the time if we were to replicate the study over and over, using random samples of 1000 voters under identical conditions.

A confidence interval has two key features:

- a named probability value indicating a level of certainty (or uncertainty, depending upon how you look at it) associated with the interval; and
- the precise upper and lower limits or bounds associated with that probability value.

The probability level is selected by the researcher (in the same way as the level of significance is selected for a hypothesis test). The upper and lower limits are established as a function of:

- the standard error associated with the statistic being reported (computed from the data sample); and

- a standardised critical value from a reference statistical distribution (such as the normal distribution, the t distribution or the F distribution).

The width of an interval (i.e. the distance between the upper and lower limits) is influenced by sample size (larger samples yield narrower confidence intervals) and random error (better research designs that control for more systematic sources of error and/or that use better quality measures will yield narrower confidence intervals).

Confidence intervals have several uses. They can be used to indicate the precision of a statistical estimate in the context of random error, as in the scenario above. They can also serve as alternative method for testing statistical hypotheses. It is possible to place a confidence interval around a proportion, a correlation coefficient, a regression coefficient, a mean or a difference between two means.

Taking the last as a focus for discussion, if we have two groups we wish to compare on some dependent variable of interest, the first step would be to compute the difference between the two group means. Suppose we assume the standard null hypothesis that the two groups don't differ on the dependent variable in the population (i.e. the mean difference is 0 in the population). However, instead of computing a statistical hypothesis test such as the independent groups t-test, we could instead construct a 95% confidence interval around the observed mean difference, using the t-distribution as our reference distribution (any good statistics text, such as Smithson 2000, or Thompson 2006, describes this process). We could then evaluate the hypothesis by simply inspecting our 95% confidence interval upper and lower limits. If our null hypothesised value does **not** fall between these two limits, then we can infer that 95% of the intervals we might compute by replicating the study over and over would not contain a population mean difference of zero (or conversely, that only 5% of such intervals would contain a population mean difference of 0). The null hypothesis could therefore be rejected.

There is, however, more work that can be done with a confidence interval. Not only can we evaluate the standard statistical null hypothesis using a confidence interval, we can evaluate the likelihood of other potential hypothesised values as well. With a 95% confidence interval around a mean difference between two groups, we can be 95% certain that the population mean difference falls somewhere between the lower and upper limits while simultaneously ruling out, with the same degree of confidence, any population values that fall outside of the two limits.

Thus, an effective range of possible differences in the population is identified by a confidence interval. So, if we have a 95% confidence interval around an observed mean difference of 4.1, which has a lower limit of 2.6 and an upper limit of 5.6, then (1) we know that the precision of our estimate of the difference is represented by a band 3 units wide (5.6 – 2.6) and (2) we can conclude that the true population mean difference would fall somewhere between the upper and lower limits in 95% of the confidence intervals we would compute if we replicated our study a large number of times. The same logic would apply for confidence intervals around other statistics such as proportions, correlations and regression coefficients. The typical probability values used to construct confidence intervals are 90%, 95% or 99%. Most software packages that construct confidence intervals will construct 95% confidence intervals by default.

In the QCI database, suppose Maree is interested in constructing two 95% confidence intervals for the mean difference in **accuracy** scores: one for the difference between tertiary qualified inspectors and high school educated inspectors and one between male and female inspectors. SPSS can easily calculate such confidence intervals using the *Independent Samples t-Test* procedure.

The resulting confidence interval for **educlev** has a lower limit of 1.186 and an upper limit of 7.894. The observed mean difference between high school educated inspectors and tertiary qualified inspectors is 4.54. This observed mean difference is Maree's best estimate of the population mean difference and its precision covers a span of 6.71 units of **accuracy**. Additionally, she can infer that the true population mean difference would be captured in 95% of the intervals she would obtain by replicating her study. The standard null hypothesised value of 0 lies outside the lower limit and Maree can therefore infer that the mean difference is significantly different from 0. If, for some reason, Maree had been interested in evaluating whether or not high school educated inspectors and tertiary qualified inspectors differed in their decision **accuracy** by 10% or more, she could also reject this hypothesis since a mean difference of 10 falls outside the upper limit of the confidence interval. Thus, Maree is not restricted to testing simple zero-value 'null' hypotheses.

The resulting confidence interval for **gender** has a lower limit of −1.345 and an upper limit of 5.647. The observed mean difference between male and female inspectors is 2.151. This observed mean difference is Maree's best estimate of the population mean difference and its precision covers a span of 6.745 units of **accuracy**. Additionally, she can infer that the true population mean difference would be captured in 95% of the intervals she would obtain by replicating her study. The standard null hypothesised value of 0 lies between the lower limit and the upper limit and Maree can therefore infer that the mean difference in the population is not significantly different from 0. If, for some reason, Maree had been interested in evaluating whether or not male inspectors and female inspectors differed in their decision **accuracy** by up to 5%, she would also fail to reject this hypothesis since a mean difference of 5 still falls inside the upper limit of the confidence interval.

Maree might also be interested in constructing a 95% confidence around the regression coefficient that results from using **workcond** to predict **accuracy** scores. SPSS can easily calculate such confidence intervals using the *Linear Regression* procedure (computing confidence intervals is an option that must be selected). The resulting confidence interval has a lower limit of 1.839 and an upper limit of 3.661. The observed value for the regression coefficient is 2.75 and its precision covers a span of 1.822 units of **accuracy**. If the null hypothesis is that the regression coefficient is 0, then Maree could reject that hypothesis since 0 falls outside the confidence interval limits. Overall, 95% of

(continued)

> such confidence intervals in replicated samples would not contain a regression coefficient value of 0. Additionally, Maree could theoretically reject any other hypothesised value for the regression coefficient that was less than 1.839 or greater than 3.661 (if she were interested in testing such a value).

Advantages

Confidence intervals provide a distinct advantage in terms of displaying a range of key features about a statistical estimate of a population parameter with a specific known probability. Thus, confidence intervals quantify the precision as well as likely location of population parameter values. Thus, we simultaneously gain a clear idea of the impact of random error on our statistical estimates as well as a clear window within which the population value is likely to be captured. We do not have to test standard null hypotheses that assume a point value of 0. We can be more flexible and evaluate ranges of possible values. Confidence intervals have also proven extremely useful in meta-analysis (discussed later in *Procedure 8.8*), where confidence intervals, calculated from multiple studies in an area, can be 'stacked' in a graph to gain a visual/statistical impression of the general direction and magnitude of a specific effect or relationship across studies.

Furthermore, as shown by Smithson (2003, pp. 12–16), if we establish several confidence intervals for the same measures in different samples (i.e. replicating a study), we can gain a clearer idea of where the population value is likely to fall, compared to conducting separate hypothesis tests. This is especially the case when some tests are significant, and some are not. Cumming and Finch (2001), Smithson (2003) and Thompson (2002) have all argued, as have others, that the computation and reporting of confidence intervals should replace standard null hypothesis testing as the normal approach to statistical inference. Their reasoning is based on the greater level of information afforded by a confidence interval compared to a standard hypothesis test. However, while their arguments are persuasive and the pressure is increasing, the fact is that confidence intervals are still not routinely reported in most business, social and behavioural science research.

Disadvantages

Probably the biggest disadvantage of confidence intervals is their convoluted interpretation. Technically, we cannot conclude from a confidence interval that we are 95% certain that the interval contains the true population value. Rather we must conclude that 95% of such intervals in replicated studies would contain the true population value. Many researchers are strongly tempted toward the first interpretation, which can lead to incorrect inferences (see Smithson 2003). Something else that researchers often forget is that calculating confidence intervals does not get around the data having to meet parametric statistical assumptions like normality and homogeneity of variance. This is true simply because not meeting parametric assumptions influences both the size of standard errors (important since standard errors are used to compute confidence intervals) and the choice of reference distribution for constructing the interval.

Researchers who employ Bayesian statistical inference logic (to be discussed in *Fundamental concept X*) depend quite heavily on confidence intervals (or 'credibility' intervals as they are sometimes called). However, Bayesian confidence intervals can only be established upon clear specification of prior probabilities. This means that, in many cases, the confidence intervals produced by popular software packages are not appropriate for Bayesian interpretation.

Where Is This Procedure Useful?

Confidence intervals can be used almost anywhere that standard hypothesis tests are used. At a minimum, confidence intervals are important to use if a primary goal of the research is to display the precision of one's findings. It is also possible to compute confidence intervals around measures of effect size (recall *Procedure 7.7*; also see Thompson 2002). The main limiting issue is whether software packages have the capability to compute the desired confidence intervals. SPSS, for example, can compute confidence intervals for *t*-tests (means and mean differences), posthoc multiple comparison tests for ANOVA and MANOVA and for regression coefficients. However, it cannot compute confidence intervals for correlations or proportions. NCSS, on the other hand, will report confidence intervals for nearly every statistic it computes. Most good intermediate-level statistics texts will show how to compute confidence intervals by hand for a variety of statistics, if a software package cannot compute it.

Software Procedures

Application	Procedures
SPSS	SPSS offers selective capabilities for computing confidence intervals as highlighted above. For some procedures like *Independent Sample t-Tests*, SPSS will automatically produce 95% confidence intervals; for other procedures, computing confidence intervals will be an option you need to select. Many procedures (e.g. *Independent Samples t-Test, Paired Samples t-Test, One-Way ANOVA, General Linear Model* → *Univariate and* → *Multivariate, Regression* → *Linear*) offer the capability to change the probability level of the confidence intervals computed.
NCSS	Nearly every statistical procedure offered by NCSS can produce confidence intervals, either by default or as an option. Additionally, NCSS is flexible in providing the user with a choice of probabilities for confidence intervals.

(continued)

Application	Procedures
SYSTAT	SYSTAT offers selective capabilities for computing confidence intervals. For some procedures like *Independent Groups t-tests*, SYSTAT will automatically produce 95% confidence intervals; for other procedures like *Basic Statistics*, computing confidence intervals will be an option you will have to select. Most procedures that offer the option of computing confidence intervals also offer the capability to change the probability level of the confidence intervals computed. Finally, SYSTAT can, as an option, superimpose confidence interval information on many of the graphs it produces – a distinct advantage for SYSTAT.
STATGRAPHICS	STATGRAPHICS will report confidence intervals for many procedures, sometimes automatically and sometimes as an option the user must select. For all of these procedures, 95% confidence intervals are produced as the default. However, once produced, if you then move the mouse into the output pane where confidence intervals are reported and right-click, you can choose 'Pane Options...', one of which will typically be to select a probability level for the confidence interval other than 95%.
R Commander	R Commander produces 95% confidence intervals by default for many procedures. For some procedures, such as the Independent Groups t-Test, you can choose the desired probability level for the confidence interval. Field et al. (2012) illustrate how to use **R** functionalities and packages to compute confidence intervals for other statistics such as correlations and regression coefficients.

Procedure 8.4: Bootstrapping & Jackknifing

Classification   Univariate/bivariate/multivariate; inferential; nonparametric.

Purpose   To provide a general empirical system within which to establish the standard errors associated with a variety of statistics. The techniques permit hypotheses to be tested under conditions where no parametric technique exists or where parametric assumptions are strongly violated, and no nonparametric alternative exists. Bootstrapping relies on the re-creation of statistical sampling distributions using repeated random samples rather than computational formulas for standard error measurement. Jackknifing relies on successive retesting of samples where one case or observation has been deleted each time.

Measurement level   Bootstrapping and jackknifing techniques may be applied to numerical data obtained at any level of measurement. When the bootstrapping approach is applied within a standard statistical analysis framework (e.g. ANOVA, correlation, or multiple regression), the measurement requirements within those frameworks become appropriate.

Bootstrapping

Statistical *bootstrapping* (e.g. Mooney and Duval 1993) is a relatively recent innovation in statistical analysis which is slowly refining the way behavioural and social scientists approach the analysis of complex and problematic sets of data. Bootstrapping is a computer-intensive statistical procedure that could not be implemented prior to the advent of high-speed computers. Because the technique is relatively new (the concept/process was originally developed by Efron 1979), it has taken some time to be implemented in some of the widely-available statistical packages (for example, SPSS, SYSTAT, NCSS and **R** have all implemented bootstrapping capabilities for a range of procedures whereas STATGRAPHICS offers limited capabilities).

Every statistic that can be computed using sample data has an associated measure of error called a *standard error* (recall *Fundamental concept VIII*). There is, for instance, a standard error for a proportion, a standard error for the sample mean, a standard error for a correlation coefficient, a standard error for a regression weight and a standard error for the difference between two means. Each standard error can be used to establish a probabilistic band or confidence interval around the population value estimated by a statistic (recall *Procedure 8.3*).

> ***For example*** *... A polling organisation might report that 49% plus or minus 2% of the general public would vote for the Australian Labour party in the next election. The 'plus or minus 2%' expresses the standard error associated with the sample estimate of 49%.*

For many statistics, their associated standard error is computed using a simple computational formula available in almost any good statistics text. These computational formulas emerge from statistical theory and operate only on the information present in data from the original sample. These theoretical formulas work by imagining that the researcher's study has been executed an infinite number of times on random samples, each time calculating the value for the statistic of interest, and that the resulting distribution of different possible values for that statistic (called a *sampling distribution*) approximates a normal distribution with certain known characteristics. All parametric statistical hypothesis tests operate by utilising information about the standard error associated with the statistic being tested.

The goal of bootstrapping is to provide estimates of the standard error associated with a statistical measure (e.g. a proportion, mean, standard deviation, correlation, regression coefficient or mean difference) without resorting to an explicit computational formula and without requiring assumptions (e.g. normality) about the distributional form of the data. Bootstrapping bypasses the need for explicit computational formulas and their associated restrictive assumptions by treating the sample of empirical data gathered by the researcher as a quasi-population from which repeated (usually 1000 or more) random samples of the same size are taken. Bootstrap sampling generally occurs with replacement, which means that every time a case is sampled it is returned to the sampling pool to potentially be selected again. The statistic(s) of interest is calculated for each bootstrap random sample. In this way, many estimates are amassed, and the position of the original sample statistic value can be easily assessed relative to this sampling distribution of estimate values built up over the 1000 or so random samples.

The standard error associated with many statistics can therefore be empirically established rather than analytically computed. In effect, bootstrapping implements the imagine-repeated-executions-of-a-study process implicit in the concept of the standard error for a statistic rather than merely assuming that the process has occurred and relying on statistical theory to tell us the nature of the outcomes from that process. Interestingly, there are some statistics for which no theoretical formula for computing a standard error exists. Bootstrapping can provide a solution in these cases by empirically establishing that standard error using its repeated random samples approach. By not depending upon any parametric or distributional assumptions, bootstrapping offers an alternative nonparametric pathway for a wide range of statistical analyses.

> To illustrate the applicability of the bootstrapping process, consider a multiple regression analysis within the context of the QCI database in which Maree attempts to predict the **speed** scores of quality control inspectors from knowledge of their general work attitudes and conditions: **quality, training, orgsatis, jobsat** and **workcond**. Maree knows from previous analyses (e.g. see *Procedure 5.7*) that **speed** is a highly positively skewed, hence, non-normally distributed variable. Knowing this, she might surmise that some of the assumptions of the normal parametric hypothesis tests available in multiple regression (recall *Procedure 7.13*) would be violated. Maree could therefore consider an alternative analytical approach using bootstrapping methods to help her test the significance of the regression coefficients, free from the usual distributional assumptions.

The bootstrapping process applied to the above multiple regression problem in the QCI database can be described in the following series of steps.

1. Take the original data from the QCI database and conduct the usual multiple regression analysis. However, ignore any computed significance tests from this analysis and merely record the actual values for the standardised regression weights (i.e. beta weights, see *Procedure 6.4*). These values become the 'original sample estimates'.
2. Take a random sample of 112 inspectors from the QCI database and repeat the multiple regression analysis from Step 1 on this new sample of data. Note that this random sampling process must be conducted with replacement which means that, once an inspector has been sampled, his/her data are thrown back into the sampling 'pot' and can potentially be sampled again. Thus, in this process, the

new random sample, called a 'bootstrap sample', may have several replications of the same inspector's data appearing. The values for the standardised regression weights from this analysis are recorded. These values constitute 'bootstrap sample estimates'.
3. The process described in Step 2 is repeated many times and the bootstrap sample estimates are recorded each time. Researchers differ as to how large "many" should be, but generally it is in the neighbourhood of a minimum of 1000 bootstrap samples.
4. When Step 3 has been completed, the frequency distribution (recall *Procedure 5.1*) for the bootstrap estimates for each standardised regression coefficient is constructed using the recorded information from all bootstrap samples. The shape of each 'bootstrap frequency distribution' gives a good idea of the empirical shape of the population of possible values for each statistic (the original sample estimate will be somewhere near the mean of this distribution; if it differs a bit from that mean, this indicates a bias that can be corrected for). This fact allows the researcher to juxtapose the hypothesised value of the statistic with this distribution to make a judgment regarding the significance of its difference from zero (or any other hypothesised value).
5. The decision regarding significance depends upon the alpha error level set by the researcher. Assuming an alpha level of .05, it would be divided by 2 to obtain .025 or 2.5% (thereby providing for a two-tailed test). The researcher would then look for the values of the bootstrap sample estimates which fall at the 2.5 percentile and 97.5 ($= 100\% - 2.5\%$) percentile points in the bootstrap frequency distribution. These two bootstrap estimate values set the boundaries (i.e. define the error limits) of an interval within which 95% of sample estimates of the population value would be assumed to fall. If the value proposed in the null hypothesis (usually 0.0) is not contained within this interval, then the researcher is justified in rejecting the null hypothesis. This process is called the *percentile* method of bootstrap inference. However, there are situations in which this method can produce biased outcomes, so researchers have developed the *bias-corrected accelerated (BCa)* percentile method to counter this problem. SYSTAT reports outcomes using both methods; NCSS reports outcomes using the BCa method and SPSS allows you to choose which method is reported.

> Figure 8.4 presents the five bootstrap frequency distributions that result from Maree undertaking Steps 1 to 5 above where 1000 bootstrap samples (each with $n = 112$) are obtained, using SYSTAT to predict **speed** from **jobsat, workcond, quality, training** and **orgsatis**, using a simultaneous regression model. In each graph, the location and value of the original sample estimate is indicated by a thin solid vertical line. Also indicated, using vertical dashed lines, are the locations of the 2.5 percentile and 97.5 percentile bootstrap estimate values.

(continued)

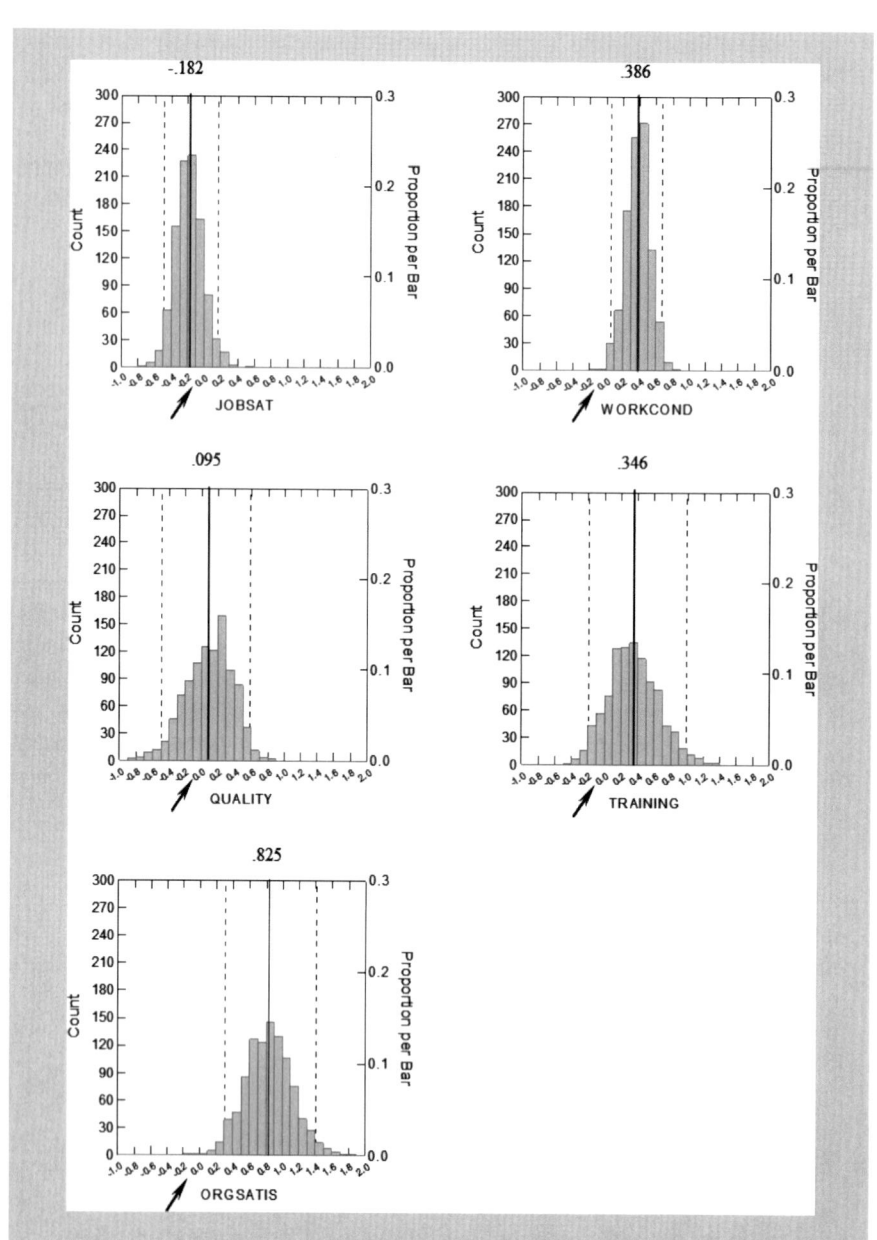

Fig. 8.4 SYSTAT bootstrap distributions of regression coefficient estimates for the five predictors of **speed: jobsat, workcond, quality, training** and **orgsatis**
Note: This is based on 1,000 random samples, each of size 112, from the QCI database. Also note that the above logic reflects the percentile method of bootstrap inference

(continued)

Using this information, it is clear that the standardised regression weights for **jobsat**, **quality** and **training** do not make statistically significant contributions to the prediction of inspection decision **speed**, because the null hypothesis value of 0 for the population parameter (arrows) falls *in between the two dotted lines*. However, **orgsatis** and **workcond** do make statistically significant contributions to the prediction of inspection decision **speed** because the null hypothesis value of 0 for the population parameter (arrows) falls *outside one of the two dotted lines*.

Thus, Maree can predict the untransformed inspection decision **speed** of quality control inspectors, but only on the basis of the inspectors' satisfaction with aspects of the organisation and its management and their perceptions of working conditions. To double-check her findings, Maree ran the same bootstrapped simultaneous regression analysis using SPSS, with BCa bootstrapped confidence intervals requested and Table 8.11 shows a summary table from that analysis. The table reports not only the 'BCa 95% Confidence Interval', but also the bootstrap-estimated two-tailed p-value ('Sig. (2-tailed)'). The conclusion is the same as that from the SYSTAT analysis: **orgsatis** ($p = .003$) and **workcond** ($p = .013$) are the only significant predictors of inspector's decision **speed**. To get a feel for whether the non-normality of **speed** scores biased the overall regression model, Maree could compare the bootstrap-estimated p-values with the p-values from the original parametric multiple regression analysis. If she did this, she would find the parametric p-values for **orgsatis** and **workcond** were .014 and .036, respectively. All other p-values were non-significant, just as in the bootstrap analysis. Thus, the non-normality of **speed** scores did not appreciably influence the inferences Maree would draw from her model.

Table 8.11 SPSS bootstrap BCa confidence intervals for regression coefficient estimates for the five predictors of **speed**: **jobsat, workcond, quality, training** and **orgsatis**

Bootstrap for Coefficients

Model		B	Bootstrap[a]			BCa 95% Confidence Interval	
			Bias	Std. Error	Sig. (2-tailed)	Lower	Upper
1	(Constant)	-1.221	-.100	1.713	.475	-4.752	1.777
	jobsat	-.182	.001	.163	.271	-.488	.140
	workcond	.386	-.002	.150	.013	.072	.681
	quality	.095	-.012	.293	.736	-.486	.618
	training	.346	.011	.312	.270	-.163	1.000
	orgsatis	.825	.029	.284	.003	.273	1.472

a. Unless otherwise noted, bootstrap results are based on 1000 bootstrap samples

Jackknifing

Bootstrapping is part of a more general class of statistical methods called *resampling methods*. Another important, but somewhat less frequently used, resampling method is called *jackknifing* (see Rodgers 2005 and Mooney and Duval 1993 p. 22; the name 'jackknife' is meant to convey a tool that is generally useful as a substitute for specialised statistical tools in a manner analogous to the trusty Boy Scout jackknife). Jackknifing was developed in 1949, well before bootstrapping methods were developed (in the late 1970s), and was designed to provide an alternative procedure for computing confidence intervals around estimates of population parameters in situations where parametric assumptions were not met.

One popular form of jackknifing works by successively dropping one observation from the original sample and recomputing the statistics(s) of interest. The process is repeated until all possible samples (each omitting one original observation) have been evaluated. In this way, the repeated samples allow you to build up a sampling distribution against which the original value of the statistic can be compared. There are several other types of jackknifing processes, some of which are described by Rodgers (2005). SPSS, for example, has an option to use a form of jackknifing called *leave-out-one classification*, which is conceptually similar to the process described above, in discriminant analysis (recall *Procedure 7.17*). Jackknifing is considered to be a form of cross-validation for statistical prediction and classification models.

Advantages

Bootstrapping provides a methodology that can be used as a nonparametric alternative to many of the more conventional parametric statistical techniques. Using bootstrapping procedures, the precision of statistical estimates can be established empirically using computer-intensive simulation methods rather than analytical computation formulas which emerge from statistical theory where certain assumptions about the data must be met. Bootstrapping can potentially be applied to a wide variety of statistical procedures (see Canty and Davison 2005 and Mooney and Duval 1993, for examples). It is now becoming routine practice to employ bootstrap estimates of standard errors to judge the significance of path coefficients in confirmatory factor analysis and structural equation models. This is an option offered by the AMOS component of SPSS (Byrne 2016), for example; see *Procedure 8.6* and *8.7*. Jackknifing is less widely used by social and behavioural scientists but does offer an alternative to bootstrapping for nonparametric estimation.

Disadvantages

Bootstrapping is a computationally-intensive methodology which requires sophisticated software and fast computer hardware. Computing speed is now a far less critical disadvantage nowadays with high-speed desktop and laptop computers but finding suitable software can still be a limiting factor in the social and behavioural sciences. Additionally, there are cases where bootstrap estimation procedures breakdown and perform less well than other alternative methodologies. In some instances, bootstrapping yields biased estimates that must be further corrected before they can be used (Mooney and Duval 1993, discuss some of these corrections, of which the BCa is one). Because the method is relatively new, research still needs to be done to test the potential for and the limitations to the applicability of bootstrapping. Its use is still far from routine in behavioural and social science research and its presentation as an appropriate approach in standard statistics texts is still relatively rare. The same story generally applies for jackknifing.

Where Is This Procedure Useful?

In practice thus far, bootstrapping appears to enjoy its greatest application in multiple regression analysis and in confirmatory factor analysis and structural equation modelling (see *Procedure 7.13, 8.6,* and *8.7*). However, its potential is much wider than this. Bootstrapping can be applied in correlational research as well as in research involving the comparisons of two or more groups (ANOVA designs).

Generally, bootstrapping may be considered as a potentially viable alternative to parametric testing wherever the satisfaction of assumptions is problematic or where the researcher wishes to test certain types of hypotheses for which no parametric or nonparametric test or standard error measures currently exist. For example, Mooney and Duval (1993) show how bootstrapping could be applied to conduct a test for the difference between two sample medians.

Jackknifing is most commonly used in procedures like discriminant analysis where classifications and predictions are being made and where a more unbiased (by assumption violation) prediction rule is desired.

Software Procedures

Application	Procedures
SPSS	SPSS offers bootstrapping capability (either the *percentile* or *BCa* method) for a substantial number of its statistical procedures, including descriptive, explore and frequency analyses, crosstab analysis, *t*-tests, ANOVA and MANOVA, general linear models, correlation and regression analyses and logistic regression analysis. Jackknifing is only available as an option for discriminant analysis classifications. When you conduct one of these procedures, you will find a 'Bootstrap' button in the main dialog box; if you hit that button, you will open another dialog box that will offer you options for controlling the bootstrapping process. AMOS, an add-on package for SPSS, offers powerful bootstrapping capabilities for confirmatory factor analysis and structural equation modelling (discussed in *Procedures 8.6* and *8.7*).
NCSS	NCSS offers a more limited set of bootstrapping capabilities (not jackknifing; and only reports the *BCa* method), compared to SPSS or SYSTAT. Those capabilities include *t*-tests, correlation and regression analyses and curve fitting analyses as well as testing the difference between two proportions.
SYSTAT	SYSTAT offers bootstrapping (reports both the *percentile* and *BCa* methods) and jackknifing capabilities for a substantial number of the statistical procedures it offers, including ANOVA, linear and multiple regression, general linear models, discriminant analysis, cluster analysis, factor analysis, MANOVA, correlation and set correlation analyses and log-linear models. It is an important software program of choice for social and behavioural scientists if bootstrap or jackknife estimation is required. These methods are available via the *Resampling* tab offered in the dialog window for many procedures.
STATGRAPHICS	STATGRAPHICS only offers bootstrapped confidence intervals as an option for its *Describe → Numerical → One Variable Analysis ...*, which is quite a limitation, compared to other major statistical packages. Although it is not clear from the software documentation, it appears that STATGRAPHICS only reports the *percentile* method of bootstrap inference.
R Commander	R Commander does not offer bootstrapping capabilities, but there are functions and packages available in the main **R** program to perform bootstrapped analyses. Field et al. (2012) illustrates such procedures in the context of correlational analysis (pp. 226–227) and multiple regression analysis (pp. 298–301).

Procedure: 8.5 Time Series Analysis

Classification   Multivariate; inferential; parametric.
Purpose   Assesses trends in the patterns of observations of a dependent measure over time, in some cases with respect to

Measurement level	an intervention, treatment or event, occurring somewhere in the series of observations. The dependent measure should be measured at the interval- or ratio-level. The main independent variable in time series analysis is a numerical measure of time such as days, weeks, months or years. If there is an intervention, this is treated as a categorical independent variable. The observations on the time-continuum are ideally assumed to occur at approximately equal intervals (e.g. successive days, weeks, months, or years).

Time series analysis is a statistical method specifically applied to a class of quasi-experimental designs called the time series designs (see, for example, McCleary and Hay 1980; Glass et al. 2008). The key feature of any time series design is multiple observations of the same phenomenon over a lengthy period of time. Time series designs differ with respect to whether the impact of a specific behavioural or social intervention, event or treatment is to be assessed.

Irrespective of the specific time series design being employed, a feature common to all time series is the potential presence of correlations between successive observations. This is a phenomenon called *autocorrelation*, which implies that observations that directly follow each other have a greater chance of being of similar magnitude than would be the case if the observations had been obtained from independent sources. The presence of substantial autocorrelations invalidates normal ordinary least squares regression analysis procedures (recall *Procedure 6.4* and *7.13*) because the standard errors associated with the regression weights (which are used in hypothesis tests about those weights) tend to be severely underestimated. A quick test for the presence of substantive autocorrelations between observations can be obtained using the Durbin-Watson test from ordinary least squares regression analysis (discussed in *Procedure 7.13*). If the value of the Durbin-Watson test is close to 2, then there is no substantive autocorrelation evident in the data and, if the residual patterns also satisfy regression assumptions, one can then justify using ordinary least squares multiple regression analysis to analyse a time series (see Ostrom 1978). If, however, there are substantive autocorrelations (signalled by value of the Durbin-Watson test approaching 1) or other influential factors such as the observations exhibiting seasonal, cyclic or periodic behaviour (such as linear trends in the data or cyclical variations over a day, a week, month or year), these must be statistically controlled before accurate tests of hypotheses about aspects of a series may be tested. Often this means undertaking a systematic search for the correct way to specify the statistical model of the time series.

In the social and behavioural sciences, when ordinary least squares regression is not appropriate, the most commonly employed time series analysis model is called the *AutoRegressive Integrated Moving Averages (ARIMA)* model (see McDowell et al. 1980, and Tabachnick and Fidell 2019, ch. 17 for discussions). ARIMA models have different parameters, *p, d* and *q*, which can be estimated for different types of time series. The *p* component reflects the autoregressive aspect of a time series; the *d* component reflects the integrated or trend aspect of a time series and the *q* component reflects the moving average component. ARIMA models are generally identified as ARIMA (*p, d, q*) where the values for *p, d* and *q* indicate how many terms must be estimated to model each component. The *d* component is especially critical as it reflects the extent to which a time series may be considered to be 'stationary' (exhibiting a flat trend, with a constant mean and variance, over time $\rightarrow d = 0$) or 'non-stationary' (trend is increasing or decreasing over time $\rightarrow d > 0$). In order for *p* and *q* to be properly identified, a non-stationary time series must first be transformed, using a process called 'differencing' so that it is stationary (see Tabachnick and Fidell 2019, section 17.4.1.1). Selecting the time series model appropriate for one's data is not an easy task; it requires practice as well as statistical insight. Some software programs, such as SPSS, offer an automated model selection process, which can simplify things a bit. However, modern graphical methods also provide an important approach for helping us understand a time series. These methods employ a variety of *smoothing* technologies to highlight specific patterns in the series data.

Time Series Analysis Without an Intervention

If no specific intervention is of interest, then a *non-interrupted time series analysis* may be conducted (see, for example, Bowerman et al. 2005, Parts III and IV; McCleary and Hay 1980; Ostrom 1978). In these designs, what is of interest is the estimation of the average level of the series, the slope (increasing or decreasing 'trend') of the series, and the periodicity (regular repeating pattern) of the observations. Non-interrupted time series designs are common in the large-scale social behaviour, marketing, finance and economics/econometrics domains. Some examples include: analysis of stock exchange prices for particular commodities over a 50-year period; analysis of the number of motor vehicle accidents, drink driving arrests or murders occurring over the past 30 years; analysis of monthly sales of a particular product over a 5-year period; or analysis of the turnover rates for a large organisation over a 60-month period.

> We can get a descriptive and visual feel for how to make sense of time series data by supposing that Maree had conducted an in-depth study of one newly hired quality control inspector with a view to understanding how that inspector's inspection decision **accuracy** changed over time, without any additional training or other intervention. She measured this inspector's inspection **accuracy** score on each of five working days, starting on a Monday, over a period of 10 successive work weeks. Figure 8.5 shows the time series plot, produced using SYSTAT, of the resulting accuracy scores on the 50 observation occasions (labelled 'Case' on the X-axis).

(continued)

Fig. 8.5 Time series plot of **accuracy** scores over 50 days for one inspector

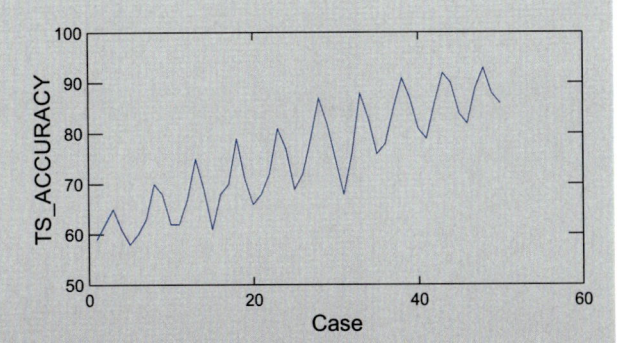

Inspection of this graph suggests several things:

- the inspector's accuracy is highly variable and seems to stay that way through the entire series (roughly constant variance over time);
- the inspector seems to be slowly but generally improving over time (plot line has an upward tilt or *trend* indicating that the series is *non-stationary*); and
- there seems to be a regular cyclical alternation between lower and higher accuracy scores embedded in the series (signalling a *seasonal* or *periodicity* effect).

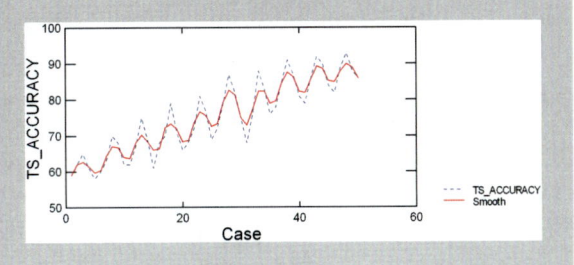

Figure 8.6, also produced using SYSTAT, shows the time series (now a dotted line) after it has been 'smoothed' by taking 3-observation moving averages of successive scores (the solid line). This type of smoothing reveals the cyclical nature of the series much more clearly. Although it is difficult to tell, given the way the X-axis is labelled (something that cannot be controlled in SYSTAT for this specific type of graph), the period repeats itself every five days. What this suggests is that the inspector starts the week performing rather poorly, gets better through the middle of the week, then starts to decline again toward the end of the week.

Fig. 8.6 Time series plot overlaid with a 3-observation moving average smoother

(continued)

Fig. 8.7 Time series plot overlaid with a LOWESS smoother

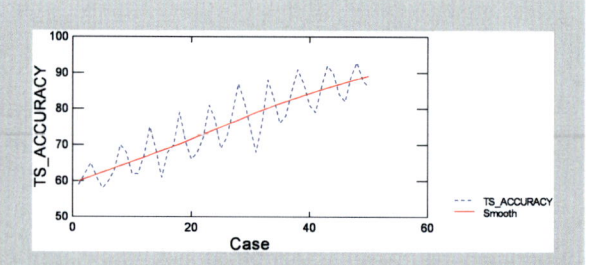

Figure 8.7, also produced using SYSTAT, shows the time series after it has been 'smoothed' by using what is called the *LOWESS* (LOcally-WEighted Scatterplot Smoothing – a type of robust regression procedure) smoother (the solid line). This smoother is designed to reveal the functional relationship between time and **accuracy** scores. In this case, the smoother clearly shows the inspector's slow but steady improvement in **accuracy** over time. Using the SPSS *Forecasting* → *Create Models...* → 'Method:' → 'Expert Modeler' procedure, this time series can best be described as an ARIMA (4, 1, 0) (four-term autoregressive, single differenced or linear trend) model. However, an ordinary least squares regression analysis shows a non-substantive value for the Durbin-Watson test (D-W = 1.959) which means that there is neglible autocorrelation between successive errors and it is therefore unlikely that the tests for significant regression coefficients in the model were distorted. For the time series in Fig. 8.5, the ordinary least squares regression model includes two predictors, the day on which an observation was made (1, 2, 3, ... 50) and the day of the work week (1, 2, 3, 4, or 5) that the observation fell on (which captures the seasonal 'period' pattern of the observations. The overall model R^2 is .337, which is significant (F(2, 47) = 13.458, $p < .001$). However, the only significant predictor in the model was observation day (B = .274, $t(47)$ = 5.182, $p < .001$). As the regression coefficient for observation day is positive, the pattern suggests that the overall trend is increasing linearly by .274 units of accuracy each day – a pattern confirmed by Fig. 8.5 and captured visually by the LOWESS smoother in Fig. 8.7.

There is a range of other more advanced approaches to non-intervention time series analysis, many of which have evolved in the finance and econometric areas. For most of these approaches, more specialised econometric software packages such as Stata, Shazam or eViews are required as the major statistical packages (recall Chap. 3), such as SPSS, SYSTAT and NCSS do not have the requisite advanced capabilities. These more advanced approaches include:

- *Autoregressive Conditionally Heteroscedastic (ARCH)* models (estimated using an ordinary least squares regression approach) and *Generalised Autoregressive Conditionally Heteroscedastic (GARCH)* models (estimated using an ARIMA approach): These time series models permit the modelling of the volatility (variances) of errors in different portions of a time series, as distinct from modelling changes in mean as ARIMA models do (Bauwens et al. 2006; Cromwell et al. 1994a). There are univariate and multivariate versions of ARCH and GARCH models. A common application of ARCH and GARCH models is the modelling of financial market volatility over time. Multivariate GARCH models permit the modelling and comparison of several markets at once (see Bauwens et al. 2006, for examples).

- *Vector AutoRegression (VAR)* and *Vector Moving Average (VMA)* models: These models permit the researcher to create and evaluate multivariate time series models (Cromwell et al. 1994b). VAR models build upon a univariate autoregression approach (the AR aspect of ARIMA models) and VMA models build on the univariate moving average approach (the MA aspect of ARIMA models). The 'vector' in each case refers to the set of scores or measurements (e.g. stock prices, productivity measures, interest rates, utility prices) available at each point in time for the time series. In some models, one or more of the variables measured on each occasion may be causally related to each other and VAR models are especially good at helping the researcher tease out those relationships (see Allen and Fildes 2001). Additionally, VAR models focus more on the dynamics of the relationships between various time series through time. [Note that there are Bayesian versions of VAR models which researchers have recently begun to use (see *Fundamental concept X* and Allen and Fildes 2001).]
- *Vector Error Correction Models (VECM)*: These models extend the logic of VAR and VMA models and allow the researcher to improve on those models through better specification of the relationships between different variables in the series (a process called 'cointegration'; see, for example, Cromwell et al. 1994b).
- *Pooled time series analysis*: This approach to time series analysis involves building a hybrid model combining time series analysis with cross-sectional data (Sayrs 1989). For example, companies in different industries or countries might be tracked over a period of 20 years in terms of their share prices or employee turnover rates where industries or countries define a cross-sectional variable for sampling.

One final important component of non-intervention time series modelling that researchers are frequently interested in is using the time series model to forecast future values of the series (see discussions in Bowerman et al. 2005). Correct identification of the time series model allows for the estimation of meaningful confidence intervals around forecasts. However, as Allen and Fildes (2001) indicate, a number of methodological details have to be attended to in order for time series forecasting to be sensible and to exhibit a desirable level of accuracy.

Time Series Analysis with an Intervention

When the impact of a specific intervention is to be assessed, the context changes to one concerning *interrupted time series analysis* (see, for example, Glass et al. 2008; McDowell et al. 1980). In interrupted time series designs, (1) a number of observations are taken on an individual (or a group or some other aggregate level, such as a city, state or nation) at numerous (ideally equally-spaced) points in time, (2) an intervention is introduced (e.g. participant(s) are exposed to some experimental manipulation or undertake some program) or experience a naturalistic event (outside of the control of the researcher; e.g. a new social policy or regulation is put in place), and (3) a number of additional observations are taken after the intervention. Each observation provides a measurement of the same dependent variable. In interrupted time series designs, the focus is on stability and change over time with respect to the dependent variable of interest and assessing the effect of the intervention. Interrupted time series designs are clearly intended to support causal inferences in that we are interested in showing that the intervention is the cause of any changes observed in the time series post-intervention.

An important feature of interrupted time series designs is that each participant serves as his/her own control (that is, a baseline pattern is established without any intervention being in place), the 'intervention' or 'treatment' is then introduced and we observe the impact of the intervention on subsequent observations relative to the baseline pattern. Figure 8.8 shows, in idealised form, some potential impacts on a dependent variable that may be observed and evaluated in an interrupted time series design. Effects may occur in three basic forms: pulse change (no duration), changes in level and changes in slope, post-intervention. Figure 8.8a shows what is called a 'step' effect (Tabachnick and Fidell 2019, section 17.2.4) for obvious reasons – only a level change occurs here. It depicts an immediate and long-lasting intervention effect such as a training program or a new social policy might produce. Figure 8.8b

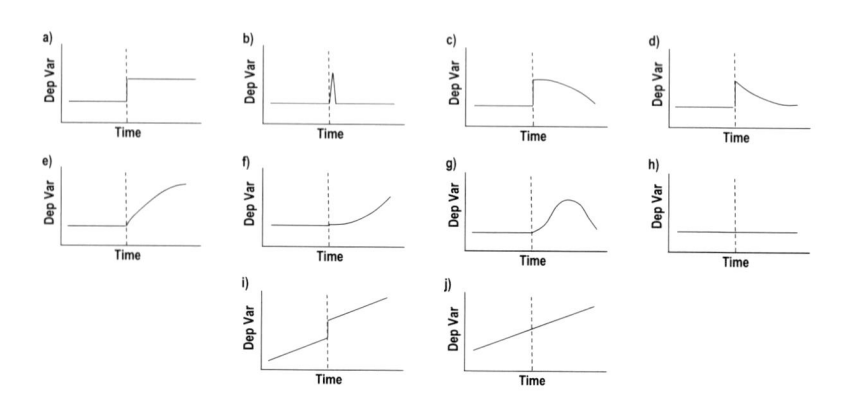

Fig. 8.8 Some illustrative intervention effects in an interrupted time series design

shows an 'impulse' effect (Tabachnick and Fidell 2019, section 17.2.4), reflecting an immediate but extremely short-lived effect, as might be produced by an event such as a flood or assassination. Figs. 8.8c and d show decay effects, where an intervention has an immediate effect which dies out over time – both level and slope changes occur here. Motivational programs or traumatic events can often have this type of effect. Figure 8.8e shows a nonlinear progressive intervention effect that levels off over time – level and slope changes occur here. Figure 8.8f shows an accelerating nonlinear progressive intervention effect– level and slope changes occur here as well. This is characteristic of a learning curve, for example. Figure 8.8g depicts an intervention effect that takes a while to build up then decays out over time – both level and slope changes occur here. This might reflect the impact of a social policy, like random breath testing, that takes time to implement fully, is effective for a while, and then gradually becomes ineffective. Figure 8.8h shows what a non-significant intervention effect might look like. Figures 8.8i and j show patterns where the series exhibits a non-zero slope over time, pre and post-intervention – reflecting a non-stationary time series. Figure 8.8i shows what a significant level change effect in such a series might look like and Fig. 8.8j shows what a non-significant intervention might look like.

In the behavioural and social sciences, the interrupted time series design is a commonly employed research design. In the more clinically-oriented behavioural sciences, such a focus has been captured in the term 'single subject designs' because time series analyses can be conducted on the data resulting from repeated observations of a single participant/patient (see Sharpley 1997). Interrupted time series designs may involve a single intervention or several interventions through time.

The typical logic of the analysis of interrupted time series data unfolds as follows:

1. Assess the ARIMA model of the time series before the intervention (establishing the baseline model), after accounting/adjusting for any seasonality evident in the data.
2. Re-estimate the ARIMA model using the entire time series sequence of observations.
3. Test for significant changes in the two models that can be attributable to the intervention; such changes may be in level and/or in slope or trend.

What is interesting is that different software packages have rather different approaches to handling interrupted time series data. SPSS, SYSTAT and NCSS can all estimate ARIMA models. However, only SPSS can test for the effects of one or more interventions but only for changes in level, not slope. The only widely-available software package that can test for both level and slope changes in ARIMA models is SAS (Statistical Analysis System)—a program that requires a fair bit of statistical sophistication to successfully operate.

A recent development in the analysis of interrupted time series is a process called segmented regression analysis (see, for example, Wagner et al. 2002). Segmented regression analysis uses ordinary least squares regression to assess the impact of an intervention and is a popular approach in medical and social/health policy research, for example. In segmented regression, both level changes and slope changes can be evaluated; period effects may also be assessed. The technique can also assess the impacts of multiple interventions through time (what are called 'change points', see Wagner et al. 2002, for more detail). The technique is easiest to implement if there are no substantive autocorrelations between model errors (assessed using the Durbin-Watson test, for example). If there are substantive autocorrelations, then the time series data need to be adjusted to remove those autocorrelation influences before the segmented regression analysis is conducted.

Segmented regression requires a minimum of three predictors in the model for the dependent variable (the time series observations): one predictor reflects the 'occasion' of measurement (e.g. day, week, year ... from 1 up to the total number of time periods measured, counting across both pre- and post-intervention occasions); the second predictor is a dummy variable coding the 'intervention' treatment/program itself, where any observation pre-intervention is coded '0' and any observation post-intervention is coded '1'; and the third predictor reflects the 'time after intervention' occasion (i.e. time after the change point; numbered 0 for all pre-intervention occasions and from 1 up to the number of post-intervention occasions for post-intervention occasions). [Note that, if desired, a period predictor can also be included, reflecting the occasion for an observation within a specific period cycle (e.g. 1 up to 7 for a 7-day cycle).] Using simultaneous regression analysis (recall *Procedure 7.13*), the 'occasion', 'intervention' and 'time after intervention' predictors are entered and assessed together (hierarchical regression could also be employed, depending upon the researcher's needs). We can illustrate the use of interrupted time series analysis, via segmented regression analysis, using another hypothetical extension of the QCI scenario.

> Suppose Maree conducted a more detailed longitudinal investigation of three newly-hired quality control inspectors (inspectors were volunteers anonymously identified as Inspectors A, B and C). She was interested in the accuracy of the inspection decisions made by each of these inspectors on a weekly basis both before and after each inspector undertook an intensive 2-day quality control training and enhancement program. She calculated the inspection decision **accuracy** of each inspector at the end of each work week as a rounded average of that inspector's accuracy performance on each of the five working days within that week. The baseline (Pre-Training Phase) period of observation was 20 weeks long. The inspectors then participated in the intensive training program over the two weekend days at the start of week 21 and were observed for a period of 20 more weeks (Post-Training Phase).
>
> Table 8.12 shows a listing of the time series data for the three QCI inspectors for the 40 observation weeks as well as the scores for the three predictors needed to run the segmented regression analyses. The nature of the three required predictors, **observation_week**, **training_program** and **week_after_program**, across the occasions of measurement can clearly be seen.

(continued)

Procedure: 8.5 Time Series Analysis

Table 8.12 Time series data for the three QCI inspectors participating in the quality control training program investigation

		Case Summaries			
observation_week	training_program	week_after_program	QCI_Insp_A	QCI_Insp_B	QCI_Insp_C
1	.00 Pre-training	.00	52.00	60.00	56.00
2	.00 Pre-training	.00	48.00	65.00	60.00
3	.00 Pre-training	.00	55.00	64.00	59.00
4	.00 Pre-training	.00	46.00	61.00	59.00
5	.00 Pre-training	.00	59.00	58.00	53.00
6	.00 Pre-training	.00	59.00	62.00	53.00
7	.00 Pre-training	.00	57.00	63.00	59.00
8	.00 Pre-training	.00	53.00	60.00	57.00
9	.00 Pre-training	.00	61.00	64.00	59.00
10	.00 Pre-training	.00	65.00	63.00	59.00
11	.00 Pre-training	.00	55.00	63.00	59.00
12	.00 Pre-training	.00	54.00	59.00	55.00
13	.00 Pre-training	.00	57.00	63.00	56.00
14	.00 Pre-training	.00	59.00	61.00	57.00
15	.00 Pre-training	.00	55.00	63.00	54.00
16	.00 Pre-training	.00	57.00	63.00	58.00
17	.00 Pre-training	.00	61.00	62.00	54.00
18	.00 Pre-training	.00	63.00	66.00	59.00
19	.00 Pre-training	.00	60.00	69.00	58.00
20	.00 Pre-training	.00	62.00	68.00	62.00
21	1.00 Post-training	1.00	87.00	70.00	61.00
22	1.00 Post-training	2.00	92.00	69.00	57.00
23	1.00 Post-training	3.00	88.00	72.00	59.00
24	1.00 Post-training	4.00	92.00	73.00	60.00
25	1.00 Post-training	5.00	93.00	70.00	61.00
26	1.00 Post-training	6.00	91.00	74.00	55.00
27	1.00 Post-training	7.00	92.00	69.00	60.00
28	1.00 Post-training	8.00	90.00	71.00	60.00
29	1.00 Post-training	9.00	88.00	73.00	64.00
30	1.00 Post-training	10.00	89.00	69.00	57.00
31	1.00 Post-training	11.00	87.00	70.00	61.00
32	1.00 Post-training	12.00	91.00	75.00	63.00
33	1.00 Post-training	13.00	87.00	72.00	55.00
34	1.00 Post-training	14.00	85.00	69.00	60.00
35	1.00 Post-training	15.00	86.00	72.00	58.00
36	1.00 Post-training	16.00	84.00	69.00	58.00
37	1.00 Post-training	17.00	88.00	70.00	56.00
38	1.00 Post-training	18.00	87.00	69.00	62.00
39	1.00 Post-training	19.00	86.00	72.00	63.00
40	1.00 Post-training	20.00	84.00	74.00	56.00

Figure 8.9 shows the graphs of the three time series resulting from Maree's investigation. The point of intervention completion (training at the start of Week 21) is indicated by the solid vertical line. The graphs were produced using SPSS, but dashed trend lines have been added to sketch the overall pattern of level and trend evident in the data. Note the variability of the

(continued)

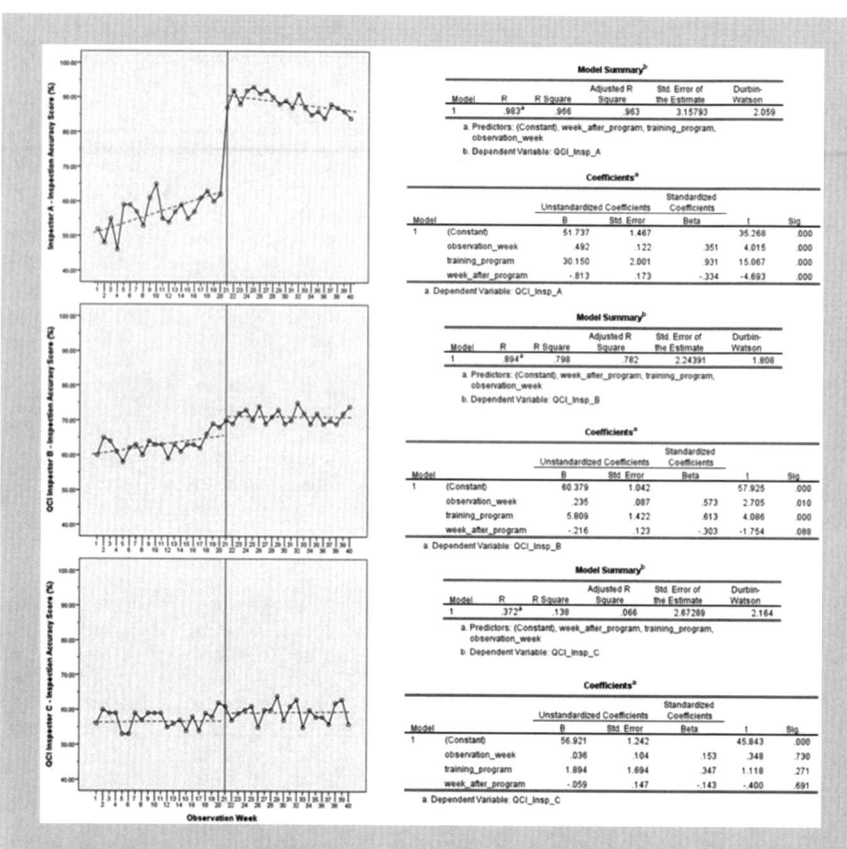

Fig. 8.9 Time series plots, produced using SPSS, for the three QCI inspectors participating in the quality control training program investigation

observations on a weekly basis for each inspector, yet it seems clear that something systematic is happening with each one, except for Inspector C. Each graph is accompanied by the basic results of the simultaneous regression analysis for the segmented regression analysis for the time series data shown in the graph (the omnibus F-test results for R^2 are not shown but are summarised below). Note that each regression analysis reports a value for the Durbin-Watson test near to or exceeding 2.0, signalling the absence of major autocorrelations between successive errors in the model. This means that the ordinary least squares approach to the segmented regression analysis is defensible for each inspector.

Looking closely at the time series plots for Inspectors A, B and C in Fig. 8.9 and their accompanying analytical tables, Maree can draw some preliminary conclusions.

Inspector A (top graph and results tables) showed a slowly increasing level of **accuracy** in the Pre-Training Phase, followed by a very dramatic jump in

(continued)

accuracy scores in the Post-Training Phase, which gently declined over time post-training. For Inspector A, the training program appears to have had a sharp impact that had lasting effects, but those effects degraded somewhat over time—a very substantial relatively permanent 'step' change. Whether or not the series would eventually return to its former baseline level for Inspector A (meaning the program had lost all potency for that inspector over time) or level off at some point higher than the baseline (meaning that the program had some substantive longer term impact) could only be addressed by obtaining many more post-training observations. The segmented regression model for Inspector A's time series was highly significant ($R^2 = .966$, $F(3, 36) = 338.276$, $p < .001$, $Durbin\text{-}Watson = 2.059$). All three predictors were significant in the segmented regression model, indicating a significant positive slope in the pre-intervention series (assessed by **observation_week**), a significant change in level of the series after the intervention (assessed by **training_program** with a dramatic jump of over 30 percentage points in **accuracy** once the intervention had occurred) and a significant change in the slope of the series (assessed by **week_after_program** with a significant negative slope post-intervention).

Inspector B (middle graph and results tables in Fig. 8.9) showed a slightly increasing trend in **accuracy** scores in the Pre-Training Phase, followed by a noticeable jump in **accuracy** performance which remained relatively steady, but variable over time. This could be described as an immediate and permanent change – a 'step' effect. The segmented regression model for Inspector B's time series was highly significant ($R^2 = .782$, $F(3, 36) = 47.513$, $p < .001$, $Durbin\text{-}Watson = 1.808$). Only two of the three predictors were significant in the segmented regression model: a significant positive slope in the pre-intervention series (assessed by **observation_week**) and a significant change in level of the series after the intervention (assessed by **training_program** with an increase of nearly 6 percentage points in **accuracy** once the intervention had occurred). There was no significant change in the slope of the series post-intervention (assessed by **week_after_program**).

Inspector C (bottom graph and results tables in Fig. 8.9) showed minimal apparent change in inspection decision **accuracy** from before to after the intervention program. This pattern likely illustrates a null intervention effect. The segmented regression model for Inspector C's time series was non-significant ($R^2 = .138$, $F(3, 36) = 1.922$, $p = .143$, $Durbin\text{-}Watson = 2.164$). Since the omnibus F-test was nonsignificant, Maree should not look any further to assess the individual predictors. Maree could thus conclude that the intensive training program was highly effective for Inspectors A and B (more so for A than B, but with the knowledge that the effects for Inspector A may not last) and not at all effective for Inspector C.

Advantages

Time series analysis is a very powerful technique for analysing time series quasi-experimental designs, with or without an intervention. Time series analysis can be done as effectively on a single participant or source of social/economic data (provided by organisations such as the Australian Bureau of Statistics) as on a sample of them. Time series analysis controls for any existing autocorrelations between successive observations—in fact, the patterns in these correlations are used to help determine the correct statistical (ARIMA) model for representing the data. Time series analysis is sensitive to many sorts of trends and effects that can appear in a specific series of observations or measurements. Many non-interrupted time series analyses have an explicit goal of generating forecasts into the future beyond the end period of observations. Economic and stock market forecasts, for example, are produced in this way. Interrupted time series analysis permits the drawing of defensible inferences regarding the impacts of interventions. This makes the interrupted time series design quite a powerful quasi-experimental causal inference approach. Glass et al (2008) provide a comprehensive discussion and illustration of techniques for assessing intervention effects in the context of ARIMA time series analysis; Wagner et al. (2002) provide an excellent discussion and illustration of the segmented regression analysis approach to assessing intervention effects.

Disadvantages

Time series analysis is a rather difficult method to understand as it is a sophisticated statistical technique, with a diverse range of potential approaches, each with its own advantages and disadvantages. Time series analysis does rely on an assumption of data normality to derive some of its parameters, which may or may not be appropriate for a specific set of observations. In addition, there is some contention as to number of observations required to produce stable estimates of time series trends. Many authors say that 50 or more observations in each phase are required whereas others (e.g. Crosbie 1993), particularly with respect to interrupted time series designs, argued that as few as 10 to 20 observations in each phase (pre- and post-intervention) were needed.

In many cases, particularly where non-interrupted time series models are being analysed, there is an element of subjectivity involved in selecting the right number of parameters for the statistical model. Much of the available software is primarily geared toward the detailed analysis of non-interrupted time series designs; analysis of interrupted time series designs using these programs is a somewhat tedious exercise. This is especially true where adjustments must be made for seasonality or other periodic cycles. Most programs assume that time or periodicity is measured in standard ways (weeks, days, months, years) and therefore lack some flexibility in handling systems where time is defined differently (for example, weeks are assumed to be comprised of seven days; if you want to deal only with working days, the standard definition of a week does not apply and is therefore difficult to handle).

One problem that can arise in the use of time series designs for interventions at the level of the single individual is the problem of multiple testing effects where the participant learns to produce better or different responses over time. Behavioural researchers must take care to ensure that such artefacts do not influence the measurements. Often this is done by using observations and physical measures where possible instead of self-report measures.

Where Is This Procedure Useful?

Time series analysis is specifically suited to time series-type quasi-experimental designs whether they employ one group (or individual), multiple groups, or even focus on societal-level indices. Designs with or without interventions can be accommodated. For econometric and financial analyses, intervention effects are typically of less interest than understanding the dynamics of the stochastic system (meaning change over time) and generating forecasts. Interrupted time series analysis is particularly well-suited to in-depth clinical investigations of single participants. However, large-scale system impacts of interventions or major events can also be assessed using such models.

> *For example* ... It would be possible to assess the impact of a major advertising media campaign on drink driving or domestic violence by tracking the number of police-reported occurrences (weekly, monthly or yearly) before and after the introduction of the media campaign.

Software Procedures

Application	Procedures
SPSS	*Analyze → Forecasting → Create Traditional Models*... and select the 'dependent Variables:' and 'Independent Variables:'; under 'Method:', choose either the 'Expert Modeler', 'Exponential Smoother' or 'ARIMA' option ('Expert Modeler' can help you decide which ARIMA model best describes your data for a specified dependent variable; the other two choices require you to explicitly specify your model); pressing the *Criteria* button gives you access to controllable aspects of the method you select; under the 'Statistics' tab, choose the statistics you wish to see; under the 'Plots' tab, choose the plots you wish to see. Tabachnick and Fidell (2019, section 17.5.2) show how an intervention can be assessed (for level change only) using 'Transfer Functions' within the ARIMA method in SPSS. Note that non-standard seasonal adjustment is difficult to do in SPSS. *Graphs→ Chart Builder*... and choose the single 'Line' chart template under

(continued)

Application	Procedures
	'Choose from:'; set your time series variable as the Y variable of the graph and the occasion variable as the X variable (can produce a graph similar to Fig. 8.8 or Fig. 8.9; the dashed lines were added later by the author). SPSS can only analyse univariate time series models and cannot deal with the more advanced types of econometric time series analyses. Interrupted time series analysis can also be done using segmented regression, where it may be possible to assess both level and slope/trend changes (as demonstrated in Table 8.12 and Fig. 8.9).
NCSS	*Analysis → Forecasting/Time Series → ARIMA (Box-Jenkins)* and specify your time series variable and choose the specifics of the model you wish to fit in terms of 'Forecasting Options', 'Data Adjustment Options', 'ARIMA Model Options', and 'Seasonality Options'. A range of plots and statistics can be selected via the 'Reports' and 'Plots' tabs. For this procedure, you must know what ARIMA model you want to fit to your data. Non-standard seasonal adjustment is difficult to do in NCSS. *Analysis → Forecasting/Time Series → Automatic ARMA* and specify your time series variable and choose the specifics of the model you wish to fit in terms of 'Forecasting Options', 'Data Adjustment Options', 'ARIMA Model Options', and 'Seasonality Options'. A range of plots and statistics can be selected via the 'Reports' and 'Plots' tabs. This procedure will attempt to identify the ARMA model that best fits your data but does not deal with the 'integrated' (i.e. differencing) component associated with the full ARIMA model. NCSS also offers a small set of smoothing options. NCSS cannot fit multivariate time series models or more sophisticated econometric time series models. NCSS can only carry out an interrupted time series analysis if you use the segmented regression approach.
SYSTAT	*Analyze → Time Series → ARIMA...* and specify your desired time series variable. You must know what ARIMA model you want to fit to your data as SYSTAT does not have an automated model testing procedure. SYSTAT also cannot assess the impacts of interventions. *Analyze → Time Series → Seasonal Adjustment...* allows you to specify your time series variable and to specify the seasonal or cyclical component (in the example in Fig. 8.5, the seasonal component would be set to 5). The seasonally adjusted series can be saved, and this new series becomes the one to use for ARIMA model fitting. *Analyze→ Time Series → Time Series Plot...* allows you to produce a time series plot like Fig. 8.5 for the variable you choose. *Analyze → Time Series → Moving Average Smoothing...* allows you to produce a time series plot like Fig. 8.6. *Analyze → Time Series → LOWESS Smoothing...* allows you to produce a time series plot like Fig. 8.7. *Analyze → Time Series → GARCH → ARCH Tests...* and choose the time series variable to analyse, the ARCH test to be applied and the lag order to be used. SYSTAT can only analyse univariate time series models. Interrupted time series analysis can only be done using segmented regression.
STATGRAPHICS	*Forecast → Automatic Model Selection...* and select the 'Data:' (variable holding the time series scores), the 'Time Indices' (variable holding the occasion index; optionally STATGRAPHICS can generate this variable for you, but the options for defining the time period are somewhat limited) and,

(continued)

Procedure 8.6: Confirmatory Factor Analysis   515

Application	Procedures
	optionally, indicate the seasonality in the model and how many forecasts are to be made using the identified model, if any; hit *OK* and in the 'Automatic Forecasting Options' dialog box, choose the modelling options to be implemented; hit *OK* and in the 'Tables and Graphs' dialog box, choose the tables and graphs you want to see. Only univariate time series models can be analysed by STATGRAPHICS. *Forecast → User Specified Model...* and select the 'Data:' (variable holding the time series scores), the 'Time Indices' (variable holding the occasion index; optionally STATGRAPHICS can generate this variable for you, but the options for defining the time period are somewhat limited) and, optionally, indicate the seasonality in the model and how many forecasts are to be made using the identified model, if any; hit *OK* and in the 'Model Specification Options' dialog box, choose the modelling specifications to be used; hit *OK* and in the 'Tables and Graphs' dialog box, choose the tables and graphs you want to see. Only univariate time series models can be specified for analysis by STATGRAPHICS. Interrupted time series analysis can only be done using the segmented regression within STATGRAPHICS.
R Commander	R Commander does not offer any time series analysis capabilities. However, for ARIMA analyses, seasonal decomposition and smoothing analyses, as well as forecasting analyses, the *Quick R* website (www.statmethods.net/advstats/timeseries.html) does offer some advice and illustrations for how to conduct various elements of a time series analysis using functionalities and packages within the main **R** program environment. Also see https://buildmedia.readthedocs.org/media/pdf/a-little-book-of-r-for-time-series/latest/a-little-book-of-r-for-time-series.pdf (Coghlan 2018) for some guidance.
Stata	Stata (StataCorp LP 2011) provides comprehensive time series analysis capabilities for the serious econometrician and social scientist. These capabilities primarily focus on non-intervention time series. Univariate (e.g. ARCH, GARCH) and multivariate time series models (e.g. VAR, ECM) can be analysed within Stata. However, using this software requires a bit more statistical sophistication to use effectively, compared to other statistical packages. Interrupted time series analysis can generally be done using segmented regression within Stata.

Procedure 8.6: Confirmatory Factor Analysis

Classification   Multivariate; inferential; parametric.
Purpose   To provide tests for the hypothesised structure of a set of latent constructs ('factors') that are theorised to contribute to the response patterns on a set of items or measures. Rather than exploring the factor structure of a set of items (recall *Procedure 6.5*), the factor structure is specified *a priori*, tested for its goodness of fit, and estimates for factor loadings and measurement errors for each item are

	produced. Confirmatory factor models constitute one important class of structural equation models (see *Procedure 8.7*).
Measurement level	Typically, the items analysed in confirmatory factor analysis are measured at the interval- or ratio-level, although this need not be the case if the items are appropriately coded for the analysis.

Confirmatory factor analysis is one important class of structural equation models where the researcher specifies the factor structure that is theoretically expected to exist among a set of items or measures (see, for example, Bryant and Yarnold 1995; Byrne 2016; Schumacker and Lomax 2016). Instead of searching for the factor structure in an exploratory fashion, it is specified prior to the analysis and the goal of the analysis is to test the fit of the structural pattern to the empirical data. If the fit is sufficiently good, then valid estimates of the hypothesised factor loadings and measurement errors for each item can be produced. Generally speaking, the parameters for a confirmatory factor analysis are estimated using specialised statistical software that can execute sophisticated maximum likelihood or generalised least squares procedures, such as AMOS (Analysis of MOment Structures, an integrated add-on program for SPSS), MPlus, RAMONA (a procedure within SYSTAT), LISREL or EQS.

> Suppose that, in the context of the QCI study, Maree designed her survey items for measuring work attitudes so that they reflected three distinct and important dimensions of work attitude: the importance of product quality to the person and to the organisation, the adequacy of training and opportunities for training in the organisation, and satisfaction with aspects of management and its policies in the organisation. She could establish such theoretical expectations on the basis of past published research or her own past research using the same types of items. These theoretically important attitude dimensions would constitute 'latent constructs' because they are not directly observed—their influence is indirectly reflected in response patterns for specifically designed survey items.

Hypothesised 'factors' in a confirmatory factor analysis are proposed to be the primary causes for the responses observed on specified items. However, we know that survey items of any sort are imperfect measures of behaviour which means that extraneous factors other than a specific latent construct may also be influencing response patterns. Since these other influences are also unobserved (e.g. Maree could never know what all of them are; many of them would be random and related to idiosyncratic aspects of individual inspector behaviour such as fatigue, boredom, hangovers, bad days effects, etc.), they must also be hypothesised as possible causes for item responses. In confirmatory factor analysis, these other extraneous influences are lumped together into a single error component referred to as *uniqueness*. Every

Procedure 8.6: Confirmatory Factor Analysis

item in a confirmatory factor analysis has one uniqueness term associated with it. It may also be the case that each latent construct shares something in common with the other latent constructs which gives rise to the possibility that the latent constructs may be correlated with each other. Thus, it is possible to hypothesise correlated factors.

We can represent a hypothesised factor structure in one of two major ways: draw a diagram of the proposed causal pathways and uniqueness influences or specify them mathematically in equations. Since it is far easier to visualise what confirmatory factor analysis accomplishes using a causal path diagram than to try to understand the mathematical representation, we will focus on the path diagram approach, which AMOS is particularly well-designed to facilitate.

> Figure 8.10 shows the causal path diagram which Maree could use to depict her hypothesised factor structure for the nine work attitude items in the QCI database. We can observe several things about her diagram which was drawn using AMOS. Ellipses are used to represent the three hypothesised latent (i.e. unobserved) constructs, *Quality*, *Training*, and *OrgSatis*. Rectangles are used to represent the observed survey items and circles are used to represent the uniqueness (*u*) components. Causal paths are depicted using single headed (directional) arrows pointing from the hypothesised cause to the item that it affects. Correlations are depicted using double-headed (bi-directional) arrows to make it clear that association, not causation, is being hypothesised.
>
> In Fig. 8.10 it is easy to see that the *Quality* latent construct as well as the three uniqueness components, $u1$, $u2$, and $u3$, are hypothesised to cause the response patterns observed on the attitude items **cultqual, qualmktg**, and **qualsat**. Similarly, the *Training* latent construct as well as the three uniqueness components, $u4$, $u5$, and $u6$, are hypothesised to cause the response patterns observed on the attitude items **acctrain, trainqua**, and **trainapp**. Finally, the *OrgSatis* latent construct as well as the three uniqueness components, $u7$, $u8$, and $u9$, are hypothesised to cause the response patterns observed on the attitude items **mgmtcomm, satsuprv**, and **polsatis**. Confirmatory factor procedures will attempt to estimate one parameter value for each proposed path (one- or two-headed arrow) in Maree's model. [Note that this logic is exactly parallel to the general linear model building logic discussed earlier in *Fundamental concept VI*: **DATA = MODEL +** ERROR, which in this case translates to **SURVEY ITEM SCORE = LATENT CONSTRUCT + UNIQUENESS**.]
>
> However, note that there are 1's superimposed on specific paths within Maree's diagram in Fig. 8.10. These 1's indicate parameters that are fixed in value (not allowed to vary in the estimation process). These 1's are used to facilitate the statistical estimation process and fix the measurement scaling of the final standardised path coefficient results. Without the 1's, the proposed

(continued)

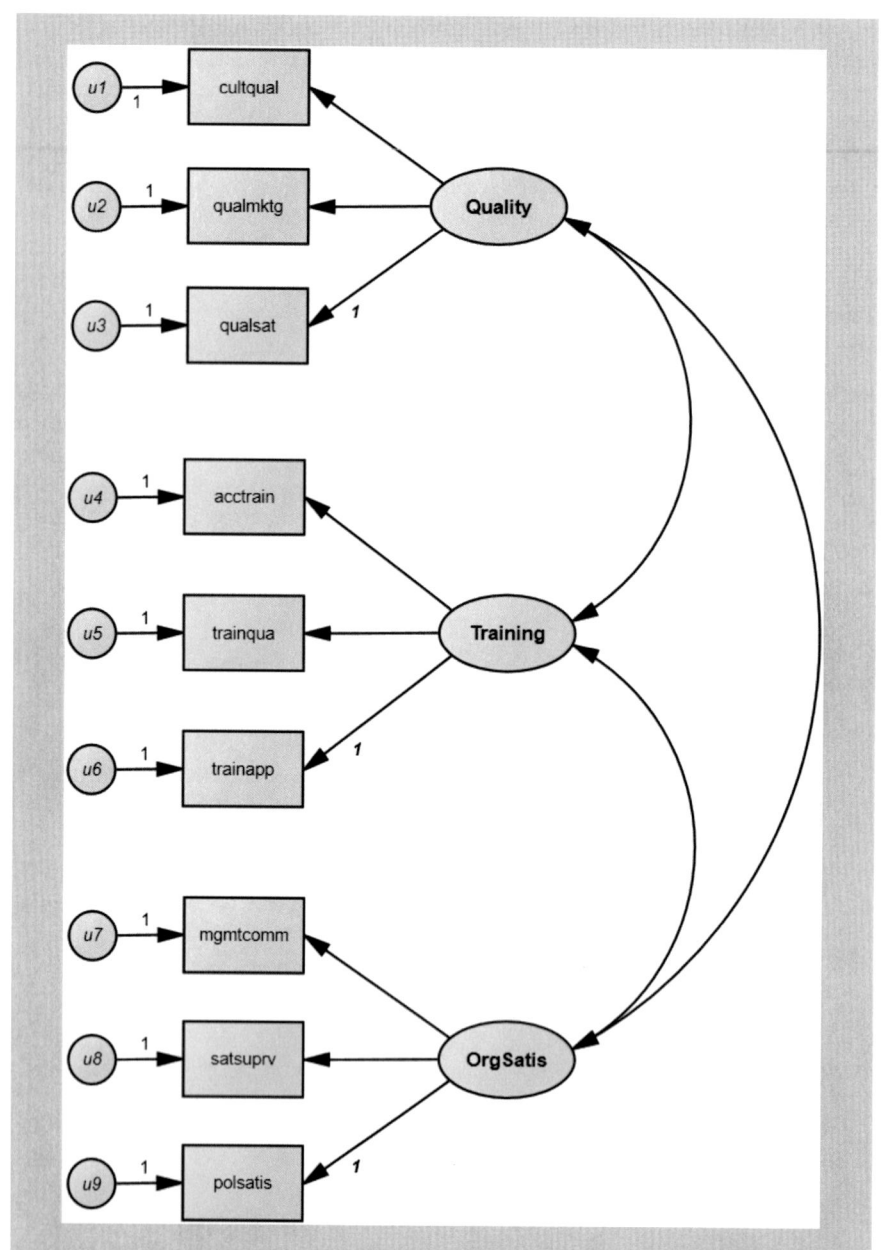

Fig. 8.10 Causal path diagram, drawn using AMOS, linking the 3 hypothesised latent work attitude constructs to the original nine QCI work attitude items

(continued)

Procedure 8.6: Confirmatory Factor Analysis

> model could not be statistically estimated. As a general rule in confirmatory factor analysis, every uniqueness or error term must be fixed as must one of the paths (the choice is usually arbitrary) from each latent construct (model components whose paths are fixed at a value of 1 for the estimation process are called 'marker indicators').
>
> [Note that Brown (2015, pp. 53–54) provides a very clear explanation as to why and where fixed parameters must be used in the estimation of confirmatory factor models.]

The outcomes from the analysis of a confirmatory factor model, analysed using a statistical program like AMOS, include among other things:

- a chi-square goodness-of-fit test for the presence of significant residuals in the model (tests the null hypothesis that the proposed model provides an adequate fit to the data; if rejected, this implies that the model is not sufficiently adequate to explain the relationships embedded in the data);
- estimates of the raw and standardised path coefficients (these are essentially regression weights) including computed standard errors where possible or bootstrap estimated standard errors (see *Procedure 8.4*) if requested;
- estimates (and associated standard errors) for the uniqueness and/or error variance components of the model;
- estimates of the correlations between the latent factors (if hypothesised); and
- a set of summary measures assessing the overall goodness-of-fit of the model in various ways.

Optional output may include what are termed 'modification indices' which can indicate places in the model where a statistically important path has perhaps been omitted by showing how fit would improve if the path were included. Behavioural researchers should be very wary of over-relying on such modification indices because the logical development of a causal model is driven primarily on *a priori* theoretical, not statistical, grounds.

> Figure 8.11 redisplays the path diagram shown in Fig. 8.10, but with the estimated (by AMOS) standardised path coefficients and latent construct correlations superimposed (significant coefficients, where the estimate is more than two bootstrap estimated standard errors away from 0.0, are shown in bold italics). The chi-square goodness-of-fit test for this model is significant ($\chi^2(24) = 48.378$, $p = .002$). This suggests that the hypothesised model does not provide a completely adequate fit to the data from the nine attitude items. However, the chi-square test is known to be biased by sample size and
>
> (continued)

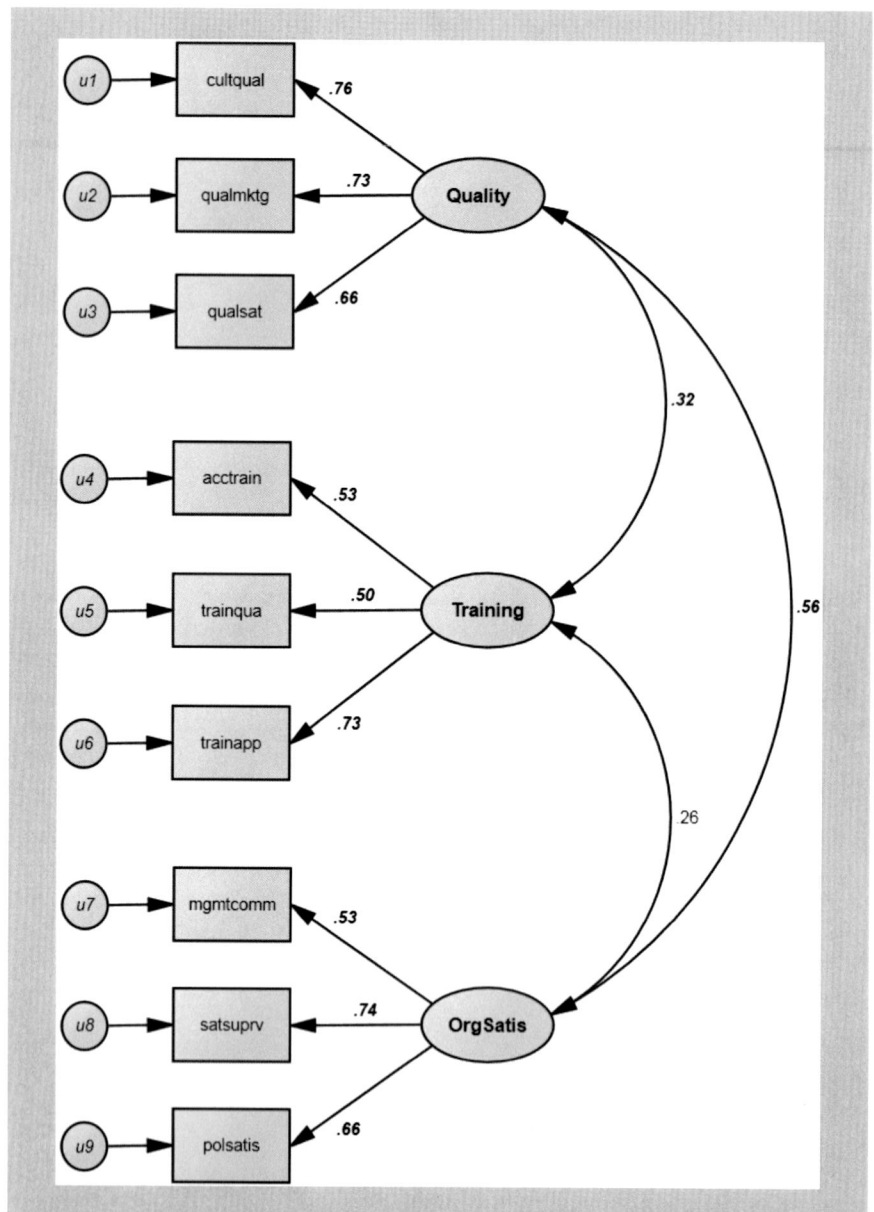

Fig. 8.11 The fitted causal path diagram from Fig. 8.10 redrawn using AMOS showing all estimated standardised path coefficients (coefficients (= parameter estimates) in *bold italics* are significant at $p < .05$)

(continued)

Procedure 8.6: Confirmatory Factor Analysis

indicates significance more often than it should (see Byrne 2016, p. 93). A somewhat better measure is chi-square divided by degrees of freedom (also called 'normed chi-square'), which for the model in Fig. 8.11, is an acceptable 2.016 (generally speaking, an acceptable value is considered to be less than 3, see Hair et al. 2010, p. 668).

It is clear that Maree's hypothesised causal paths from the latent constructs to their response items were all significant and substantial in value. Thus, the hypothesised factor structure appears to be valid (i.e. confirmed). In addition, the *Quality* and *OrgSatis* latent constructs and the *Quality* and *Training* constructs were significantly positively correlated with each other, but *OrgSatis* was not significantly correlated with the *Training* construct. The *GFI* (Goodness-of Fit Index) is one of many possible measures of the goodness-of-fit of the hypothesised model; it ranges from 0 (very poor fit) to 1.0 (perfect fit) and the value for the model in Fig. 8.11 was .921, indicating very good fit for the proposed model. Brown (2015, pp. 67–75), Byrne (2016, pp. 90–100) and Kline (2015, ch. 12) all have very good discussions of the many different measures of model goodness-of-fit. There is no consensus amongst behavioural and social scientists as to the most preferable measure (s); choices will depend upon who you read.

Advantages

Confirmatory factor analysis provides a coherent and statistically defensible method for testing hypotheses about the structure of latent constructs that may be causing the response patterns observed among a set of items or measures. It is also possible to test several different theoretically competing models within the same statistical framework in order to make a defensible choice as to the most preferred model. One common set of competing models involves a comparison of a simpler model which proposes uncorrelated latent constructs with a more complex model which proposes that the constructs are correlated (which can signal a 'second-order' factor structure in the data). We can use the difference between the two chi-square statistics (one for each model) and their respective degrees of freedom as a way of testing whether allowing the latent constructs to be correlated results in significantly improved fit to the data.

Disadvantages

Confirmatory factor analysis is very dependent upon having a strong theoretical basis for the path model(s) being tested. If the theory is wrong or the model fit poor, there may be a strong temptation to modify the model, based on statistical 'advice' (i.e. modification indices) provided by programs like AMOS. Such temptations should be resisted if one is to avoid being accused of 'fishing around' in the data for a structure. Furthermore, it is *not* 'cricket' to use exploratory factor analysis on data from a sample to see what factor structure might exist, then use confirmatory factor analysis to test the fit of this model *using the same sample of data*. The proper approach would be a cross-validation strategy (see Appendix B for one such strategy) where exploratory factor analysis is done using data from one sample to 'build the model' and the resulting confirmatory factor model is then evaluated using data from a second sample to 'test the model'. Such a strategy is more resource-and time-intensive to carry out but avoids the dangers of capitalising on chance relationships inherent in building **and** testing a model in a single sample.

Confirmatory factor analysis can run into some rather troublesome statistical problems if the data being analysed are not 'well-behaved'. The maximum likelihood algorithms that are typically used are somewhat sensitive to distortions in the data such as non-normality and excessively high correlations (i.e. multicollinearity) among the items and also do not work well if the data are categorical in nature. Researchers can be very disconcerted to find that their data yield improper estimates, due to one or more of these problems, and learning precisely where the problem lies can be difficult to diagnose. Note: AMOS offers other estimation procedures such as weighted least squares and generalised least squares, which can help circumvent some of these problems. Good data screening (recall *Procedure 8.2*; AMOS can assist here as well) can also help to counter data instability problems. Missing or incomplete data also need to be handled and the standard listwise or pairwise deletion options may not be offered by the statistical package being used. For example, AMOS does not offer listwise or pairwise deletion; instead AMOS directly estimates (imputes) missing data values before fitting a model (see the discussion in Byrne 2016, ch. 13).

The fact that no uniformly agreed goodness-of-fit measure can be put forward is also concerning and one finds different authors favouring very different indices and even different values for what counts as a 'good fit'. This means that judgments about model fit have an element of subjectivity associated with them. The chi-square goodness-of-fit test is highly sensitive to large sample sizes in that very tiny discrepancies in the model residuals can be shown to be significant such that even well-fitting models based on a large sample size might be associated with a highly significant chi-square statistic. This tendency reduces the utility of the chi-square measure itself as a measure of goodness of fit. Normal practice now seems to be to report the chi-square test, the normed chi-square value (chi-square divided by its degrees of freedom) and at least one or two of the major goodness-of-fit indices.

On the other hand, the maximum likelihood estimation procedures work best with large samples, particularly if many variables and causal paths are being analysed. Maximum likelihood estimation also works best if, in a confirmatory factor analysis context, there are at least three observable items available for each latent construct being tested. Hair et al (2010, pp. 661–662), for example, noted that the minimum sample size for structural equation modelling (which includes confirmatory factor analysis) varied with a range of considerations. They advocated ranges from a minimum $n = 100$ for simple models ("containing five or fewer constructs, each with more than three items (observed variables), and with high item communalities (.6 or higher)") to $n = 500$ or more for very complex models ("with large numbers of constructs, some with lower communalities, and/or having fewer than three measured items") (Hair et al. 2010, p. 662).

Finally, confirmatory factor models, if overly complex, may result in what are termed *identifiability* problems where the proposed model cannot be estimated (i.e. all proposed parameters cannot be identified). This usually occurs when there aren't enough fixed parameters (remember the 1's in Fig. 8.10) set up in the model or when the number of parameters to be estimated exceeds the available degrees of freedom. However, the AMOS program is especially good at detecting this specific type of problem and at proposing some possible solutions for it.

Where Is This Procedure Useful?

Confirmatory factor analysis is useful whenever a researcher proposes, up front, that the observable measures or items upon which data are to be collected in a study have a specific underlying latent structure. This hypothesised model may emerge from theoretical arguments, integration of findings of past research, or both. One analytical approach that is commonly used in behavioural research is a cross-validation strategy where exploratory factor analysis (*Procedure 6.5*) is employed to identify the underlying structure of a set of items within one sample of participants, then confirmatory factoring methods are employed to verify this underlying structure *in another sample* (perhaps from a different culture, organisation, or context of application). [Note that if the researcher has access to a very large sample, say double that required for building and testing a model, then that sample could be randomly split into two sub-samples of equal size; the model could be built using the data from one sub-sample and could be tested using the data from the other sub-sample (what is called 'split-sample', 'holdout group' or '2-fold' cross-validation).

Many researchers use confirmatory factor analysis as a way of verifying the factor structure of a previously constructed measure (often created by other researchers, using exploratory factor analysis) in a new or different research context. This process ensures that evidence for the validity of a specific theoretical construct is cumulative across studies and contexts. Failure to confirm structures across contexts (e.g. different cultures) can provide useful information for further refinement or elaboration of the theoretical constructs.

Software Procedures

Application	Procedures
AMOS (as part of SPSS)	*Analyze → IBM SPSS Amos* and the AMOS program will start up. A new interface will appear with a blank page for drawing your path diagram in a fashion similar to Fig. 8.10 and for controlling and executing your confirmatory factor model fitting analyses. The analyses will be conducted using the data file you have opened in SPSS. The analytical flow is (1) use the graphics window to diagram and set up your confirmatory factor model (AMOS uses this diagram to figure out what needs to be estimated; (2) set up the desired 'Analysis Properties' (controls all aspects of how AMOS should analyse the model, what it should output and whether or not bootstrap tests are desired; maximum likelihood estimation is the default method); (3) 'Calculate estimates' to run the computations; and (4) interpret the output (output comes in two forms, text output with all of the numbers and graphical output where AMOS superimposes the unstandardised or standardised estimates (your choice) on your model diagram). The diagrams produced in and by AMOS can be cut-and-pasted, in bitmap format, to another program for saving). An extensive manual for using the AMOS program can be accessed from the *Help* menu and Byrne (2016) provides a very good and practical guide. AMOS is probably the easiest structural equation program to learn how to use and its clear and flexible graphical interface is exceptionally useful. AMOS also offers an extensive range of data screening procedures so that researchers can detect, early on, if there are anomalies in the data that need addressing. AMOS offers bootstrapping capability for all model testing needs.
NCSS	NCSS does not offer the capability to conduct confirmatory factor analyses.
RAMONA (as part of SYSTAT)	*Analyze → Advanced → Path Analysis (RAMONA)...* and build the specifications, line by line, for your confirmatory factor model. RAMONA uses 'manifest' variables (= observed measures in the data file) and 'latent' variables (= hypothesised constructs). RAMONA does not have a graphical interface, so each path in the diagram must be specified using a line of text. For example, for the path model in Fig. 8.10, one line that would be needed is: *cultqual <- Quality* (to symbolise a correlation rather than a causal path, you would type *Training <-> Quality*). If a parameter is to be fixed, you would signal this to RAMONA by *qualsat <- Quality (0, 1.0)*, for example. Uniquenesses (or errors) are considered to be 'latent' variables and can be named on the same specification line for an observed variable as the latent construct variable. So, for example, a complete path specification line for RAMONA, that includes the uniqueness as well as the latent construct path contributions, would look like: *cultqual <- Quality E1 (0, 1.0)*. Maximum likelihood estimation is the default method. SYSTAT does not offer bootstrapping capability within RAMONA.
STATGRAPHICS	STATGRAPHICS does not offer the capability to conduct confirmatory factor analyses.

(continued)

Application	Procedures
R Commander	R Commander does not offer the capability to conduct confirmatory factor analyses. However, the *Quick R* website illustrates how one can accomplish confirmatory factor analysis using functionalities and packages within the main R program (see www.statmethods.net/advstats/factor.html).
MPlus	Geiser (2013, pp. 51–58) shows how to get MPlus to conduct a confirmatory factor analysis. Note that MPlus is a command line-driven software program, rather than a 'dialog box' program like SPSS or SYSTAT, but it does have a 'Language Generator' function that helps in setting up the necessary input file. MPlus also has a 'Diagrammer' procedure where you can draw your path diagram for estimation, similar to AMOS (but without as many 'bells and whistles'). MPlus can handle missing data by listwise deletion or by imputation. MPlus can also produce bootstrap standard errors for all components of a structural model.

Procedure 8.7: Structural Equation (Causal Path) Models (SEM)

Classification   Multivariate; inferential; covariance-based SEM approaches (e.g. maximum likelihood) are parametric; variance-based SEM approaches (e.g. partial least squares) are nonparametric.

Purpose   To estimate and test causal path or 'structural equation models' that represent complex hypotheses regarding the causal linkages theoretically thought to exist amongst several independent and/or dependent measures, some of which may be latent constructs (of the sort established by *Procedure 8.6*). This task is accomplished using correlational information.

Measurement level   Preferably, all observed variables in the causal model should be measured at the interval- or ratio-level, but this need not be strictly adhered to. Categorical variables can be included if they are correctly coded for use. All latent constructs proposed in structural equation models are considered to be measured at least at the interval-level.

Structural equation models (SEM) describe a systematic process for statistically estimating and testing specific theoretical models of the causal relationships among variables of interest (e.g. see Klem 2000; Kline 2015; Schumacker and Lomax 2016). Causal models are typically represented in diagrammatic form with arrows (paths) showing the directions of presumed causality. The *path diagram* is a pictorial representation of a specific theory that ties together the related variables in a specific domain. The actual mathematical specification of any structural equation model is in terms of a series of interconnected multiple regression equations. *Procedure 8.6* described a specific type of causal model for confirmatory factor analysis. However, more general structural equation models potentially involve relationships between observed independent variables, observed dependent variables, and latent constructs.

There are two important distinguishable components of structural equation models. The first component, which may or may not be present, concerns how specific observed variables relate to theoretical latent constructs. This is precisely what confirmatory factor analysis is concerned with and it is often the case that more general structural equation models have confirmatory factor structures embedded within them. This component represents the *measurement model* aspect of the causal modelling process. In large-scale surveys where causal models are to be tested, it is frequently the case that survey items are first linked to their underlying hypothesised latent constructs before relationships between latent constructs or between latent constructs and other variables are explored.

The second major component of a structural equation model concerns the specification of causal relationships between latent constructs or between different observed variables acting as independent or dependent variables. This is termed the *structural relations* component because it links together different constructs and/or variables and identifies all sources of error relevant to the model. Any structural equation model will always have a structural relations component but may or may not have a measurement model component.

As with a confirmatory factor model, a general structural equation model must be worked out and set up prior to the collection of data. The logic of causal modelling only works if the relationships being tested are precisely stated and defended or justified before the research commences. As before, theoretical arguments and integration of past research findings can provide the basis for specifying hypothesised causal models. If the fit of the hypothesised model is sufficiently good, then valid estimates of the path coefficient parameters and measurement errors for each variable in the causal system, either observed or latent (unobserved), can be produced.

The Covariance-Based Approach to SEM

Generally speaking, the parameters for a structural equation model are estimated using sophisticated maximum likelihood procedures or some other estimation system, such as unweighted least squares, generalised least squares or asymptotically distribution-free (ADF), which can accomplish the same purpose. Specialised statistical software, such as AMOS (add-on procedure for SPSS), LISREL, RAMONA (a procedure within SYSTAT), EQS or MPlus, needs to be used to compute such

Procedure 8.7: Structural Equation (Causal Path) Models (SEM)

models. These estimation systems implement what is called the *covariance-based approach to structural equation modelling* (Hair et al. 2014, pp. 14–19), which is most useful in research situations where (1) "the goal is theory testing, theory confirmation, or the comparison of alternative theories"; (2) errors in the model and the relationships between them need to be carefully specified; (3) the model has reciprocal feedback loops; and/or (4) the researcher wants an omnibus test of goodness-of-fit to interpret (p. 19). An alternative 'variance-based' approach, called 'partial least squares' will be discussed later in this procedure.

> Suppose that, in the context of the QCI study, Maree decided to test a causal model that attempts to explain both the **speed** and the **accuracy** of a quality control inspector's inspection decisions. She could establish such a theoretical causal model on the basis of past published research and previously existing theory or on her own past research using the same types of variables. Since her budget was limited, she could only obtain single measures of the various constructs involved in her theory; thus, she could not propose or estimate any measurement models for latent constructs—she had to work at the level of examining structural relations between observed variables.
>
> One aspect of her theory concerned whether people worked in the computer manufacturing (high technology) industries or in other lower technology industries. To this end, she recoded the **company** variable into a new variable, **computer_ind** by combining the PC and Large Business Computer manufacturing groups together (coded 1) and the remaining groups together (coded 0). Thus, **computer_ind** was a dichotomous variable. The normalised transformed version of **speed**, namely **log_speed**, was also used in her model.
>
> Figure 8.12 shows her hypothesised causal model (a structural relations model) which attempted to explain the causal chains of events that contributed to decision **speed** and **accuracy**. Note that the two demographic variables, **mentabil** and **gender**, were considered to exist prior to the causal system of interest which meant that Maree could not speculate on their causation (no arrows lead into them—these are referred to as *exogenous* variables).
>
> Both variables were thought to partially determine inspectors' **educlev** which in turn helped determine what industry they worked in, **computer_ind**. **Mentabil** and **gender** were also hypothesised to directly influence **computer_ind**. The **educlev** and **computer_ind** variables were hypothesised to causally influence both **jobsat** and **workcond** perceptions. Additionally, **workcond** perceptions were hypothesised to contribute to **jobsat** perceptions. Finally, **computer_ind**, **jobsat**, **workcond**, and **mentabil** were all hypothesised to directly influence **log_speed** and **computer_ind**, **workcond** and **log_speed** were all hypothesised to be directly causally linked to decision **accuracy**.

(continued)

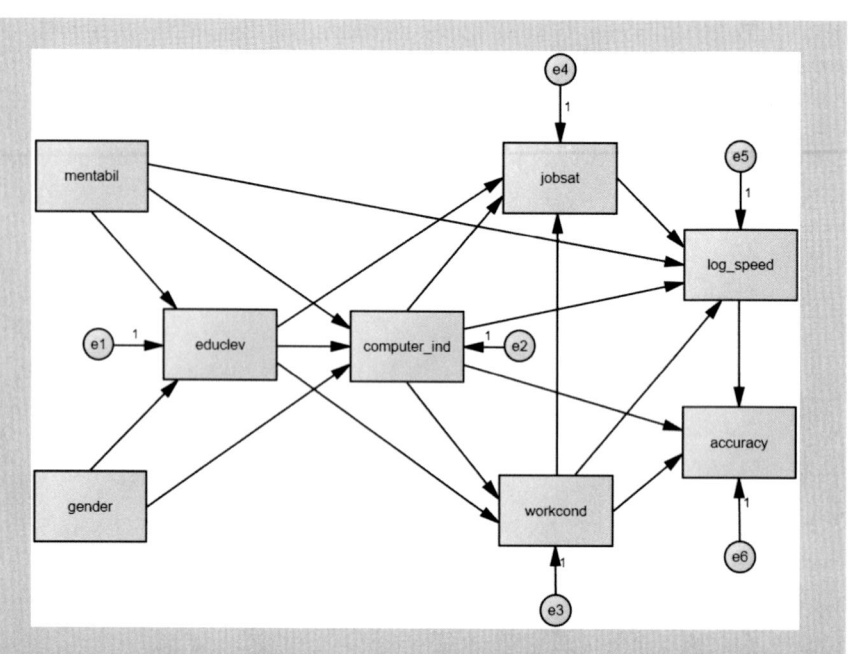

Fig. 8.12 The causal path diagram, drawn using AMOS, showing hypothesised causal links between relevant variables in the QCI database

Note that not all possible paths were represented in the model. Only those paths that Maree hypothesised to be theoretically relevant were shown; these paths were thus the only ones statistically estimated. All other potential paths were assumed to have a fixed parameter value of zero. Some causal chains were direct (where one variable was directly connected to another via a single-headed arrow). For example, **educlev** was hypothesised to directly influence **computer_ind**. Other causal chains were indirect (where one variable influenced another via a series of paths that travelled through intermediate causes, i.e., mediation). For example, **mentabil** was hypothesised to indirectly influence **jobsat** via **educlev** and **computer_ind**.

We know that survey items of any sort are imperfect measures of behaviour which means that extraneous factors, other than the specified causal variables in the hypothesised model, may be influencing response patterns. Since these other influences are also unobserved (e.g. Maree could never know what all of them were; many of them would be random and related to idiosyncratic aspects of individual inspector behaviour such as fatigue, boredom, hangovers, bad days, family situation, etc.), they must also be hypothesised as possible causes for variable responses. In structural equation models, these unobserved 'extraneous factors' are lumped together into a single *error* component. This simply reflects another variation on the **DATA = MODEL** + ERROR representation of the general linear model (recall *Fundamental concept VI*).

Every variable in a structural equation model that is the proposed effect of some known cause (i.e. any observed variable or latent construct in the model to which an arrowhead is pointing—these are referred to as *endogenous* variables) has an error term associated with it. In order to make the model statistically estimable and to facilitate the scaling of the standardised solution, every error component parameter must have its value fixed at 1 as shown in Fig. 8.12.

Note that it is possible to hypothesise that two variables (or latent constructs) are merely correlated rather than being causally linked in which case we would connect the two entities concerned using a double-headed arrow as was done in Fig. 8.11 with the hypothesised correlated factors. The model in Fig. 8.12 does not hypothesise any correlations, just causal paths. When the model is estimated, the correlation between **gender** and **mentabil** will be assumed to be zero, since there is no double-headed arrow connecting them. That is, there is no theoretical reason to expect **gender** and **mentabil** to be correlated.

The goal in structural equation modelling is to statistically evaluate each path as to its degree of importance (weight) in the overall model. If a specific path receives negligible weight, then it should be deleted from the model due to lack of statistical evidence for its existence. In the past, multiple regression (see *Procedure 7.13*) had been used to estimate these path weightings or path coefficients. But in recent years, more sophisticated statistical techniques have been developed that can better handle the simultaneous evaluation of all proposed paths in a causal model using correlations and, in addition, permit the overall evaluation of the 'goodness-of-fit' of a particular theoretical model to the observed data.

The outcomes from the analysis of a structural equation model, conducted using a covariance-based statistical program like AMOS, include among other things:

- a chi-square goodness-of-fit test for the presence of significant residuals in the model (tests the null hypothesis that the proposed model provides an adequate fit to the data; if rejected, this implies that the model is not sufficiently adequate to summarise the relationships embedded in the data);
- estimates of the raw and standardised path coefficients (these are essentially regression weights) including computed standard errors where possible or bootstrap estimated standard errors (recall *Procedure 8.4*) if requested;
- estimates (and associated standard errors) for the uniqueness and/or error components of the model;
- estimates of the correlations between variables and/or latent constructs (if hypothesised); and
- a set of summary measures assessing the goodness-of-fit of the model in various ways (the same indices as used to assess confirmatory factor models, recall *Procedure 8.6*).

Optional output may include what are termed 'modification indices' which can indicate places in the model where a statistically important path or error relationship has perhaps been omitted by showing how fit would improve if that path were included. Behavioural researchers should be very wary of over-relying on such modification indices because the logical development of a causal model is driven primarily on *a priori* theoretical, not statistical, grounds.

Figure 8.13 redisplays the causal path diagram shown in Fig. 8.12 but with the estimated standardised path coefficients (significant coefficients, where the estimate is more than two standard errors away from 0.0, are in bold italics) superimposed. The chi-square goodness-of-fit test for this model was not significant ($\chi^2(11) = 11.648, p = .391$). This indicated that the hypothesised causal model provided adequate fit to the data from the observed variables. The ratio of chi-square to degrees of freedom was 1.059, which was excellent. The 'GFI' (Goodness-of Fit Index) was .974, indicating an excellent degree of fit for the proposed model.

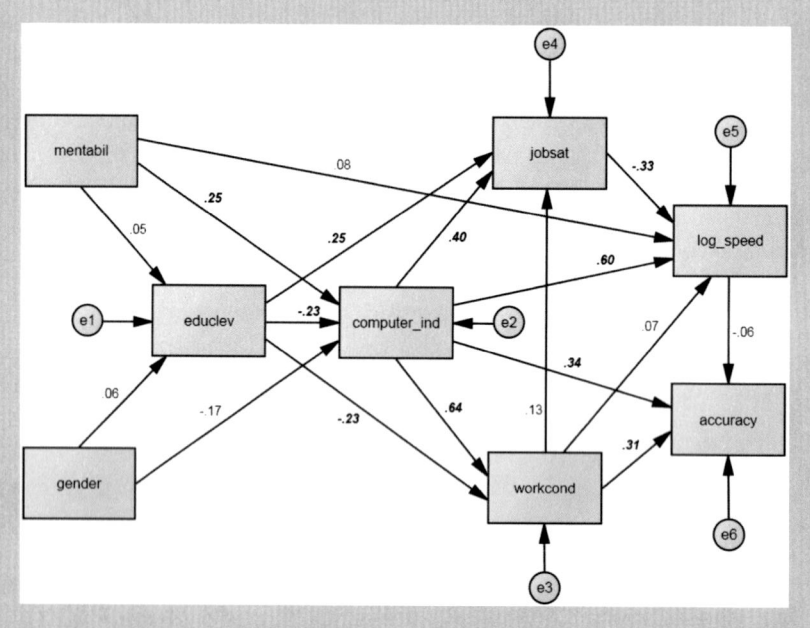

Fig. 8.13 The fitted causal path diagram from Fig. 8.12, redrawn using AMOS showing all estimated standardised path coefficients (coefficients (= parameter estimates) in *bold italics* are significant at $p < .05$)

(continued)

> Eleven of the 17 hypothesised causal paths were significant and substantial in value. Neither **mentabil** or **gender** contributed causally to **educlev**, but both **mentabil** and **educlev** were significantly influential on **computer_ind** (the negative path coefficient between **educlev** and **computer_ind** suggests that inspectors who were Tertiary-qualified tended not to work in the computer manufacturing industry—a surprising finding perhaps, given the technical nature of the computer industry). Note that **mentabil** was not significantly causally related to the **log_speed** of inspectors' decision making. Both **educlev** and **computer_ind** were significantly causally linked to both **jobsat** and **workcond** perceptions (**educlev** was negatively linked to **workcond**, suggesting that Tertiary-qualified inspectors tended to report poorer working conditions; inspectors working in the computer industry tended to report better working conditions and higher job satisfaction).
>
> **Computer_ind** was directly causally related to both inspection decision **log_speed** and **accuracy** (both path coefficients were positive, suggesting that inspectors working in the computer industry tended to make slower and more accurate inspection decisions). **Workcond** was not found to be causally related to **jobsat**. **Jobsat**, but not **workcond** perceptions, were significantly causally linked to inspection decision **log_speed** (the link between **jobsat** and **log_speed** was negative, suggesting that less satisfied inspectors tend to take longer to make their decisions). **Workcond** was significantly linked to inspection **accuracy** (better working conditions led to higher decision accuracy). Inspection **log_speed** did not causally influence inspection decision **accuracy**.
>
> If Maree were to consider refining her model to make it a more parsimonious representation of the empirical linkages between the various variables, she could do so by deleting those paths which were found to be non-significant and retesting the fit of the refined model, preferably using a new sample of inspectors (a cross-validation strategy).

Two things should be noted with respect to implementing the covariance-based approach to structural equation models. First, to implement a causal path analysis, data must be obtained on all relevant variables from the same sample of individuals and the sample must be sizable. Second, structural equation modelling is one instance where correlational information can be used to support causal inferences, but extreme caution and diligence are needed when gathering the data, using the procedure and interpreting outcomes. It is not appropriate to collect data and then try to find a set of paths which seems to explain the pattern of relationships in the data. Rather, prior to any data collection, the path models must be specified and theoretically justified. The data are then gathered to test whether the specific model proposed in theory is empirically supported by observations. In other words, covariance-based structural equation modelling is a 'confirmatory' method, not an exploratory one (Hair et al. 2014).

The Partial Least Squares (PLS) Approach to SEM

There is an alternative *variance-based approach* to estimating structural equation models called *partial least squares* (PLS; see, for example, Hair et al. 2014) that is becoming much more widely used. The difference between the covariance-based maximum likelihood approach to SEM and the PLS variance-based approach is, in some ways, analogous to the difference between common factor analysis and principal components analysis (recall *Procedure 6.5*; see also Ringle et al. 2012):

- PLS analyses the variances of variables, maximum likelihood SEM analyses the covariances between variables (what is shared in common between variables, distinguished from what is unique to each variable);
- PLS can model *formative* measurement of constructs (where response patterns on variables provide causal input to constructs – i.e. the variables *form* the construct); maximum likelihood SEM relies most commonly on *reflective* measurement of constructs (where theorised latent constructs cause patterns of response on specific variables – i.e. the variables *reflect* the construct);
- PLS is not underpinned by and does not require a strong theory of error; maximum likelihood SEM is underpinned by and requires a strong error theory (which then informs omnibus and comparative goodness-of-fit tests for models);
- PLS does not require a strong theoretical motivation or basis (i.e. it is more amenable to exploration and can tolerate greater model complexity as well as single-item measures of constructs); maximum likelihood SEM requires a strong theoretical basis (i.e. it is confirmatory in purpose and less tolerant of model complexity and of constructs measured by fewer than three items); and
- PLS is robust to small samples sizes and to data anomalies (like non-normality) whereas maximum likelihood SEM is highly sensitive to sample size and data anomalies (like non-normality).

PLS-SEM works in a similar fashion to covariance-based SEM in that the researcher sets up a path diagram to represent all the causal and correlational linkages between variables (called indicators) and constructs to be estimated. Whereas in covariance-based SEM, it is possible to separate estimation of measurement and structural relations models; in PLS-SEM, the measurement model (which links indicators to constructs) and the structural relations model (which links constructs to each other) are estimated and evaluated simultaneously (using bootstrapped *t*-values for each estimated coefficient).

Procedure 8.7: Structural Equation (Causal Path) Models (SEM)

Suppose that, in the context of the QCI study, Maree had some preliminary ideas about some important constructs, theoretically implied by her QCI variables, that might be related to each other. The central constructs she thought were operating were:

- the context of the inspector (*InspectContext*) with key formative indicators being **computer_ind** (1 = computer industry; 0 = all other industries), **educlev** (1 = tertiary educated; 0 = high school only educated), **gender** (1 = female; 0 = male) and **mentabil;**
- sentiments about training opportunities in their company (*Training*) with key reflective indicators being **acctrain, trainapp** and **trainqua**;
- sentiments about their organisation and satisfaction with its policies and practices (*OrgSatis*) with key reflective indicators being **mgmtcomm, polsatis** and **satsuprv**;
- sentiments about the focus on quality within their company (*Quality*) with reflective indicators being **cultqual, qualmktg** and **qualsat**;
- sentiments about their work in the context of their workplace (*Workplace*) with key reflective indicators being **jobsat** and **workcond**; and
- inspector's inspection performance (*InspectPerf*) with key formative indicators being **accuracy** and **speed**.

Figure 8.14, drawn using SmartPLS, shows her complete preliminary theoretical model which attempted to predict the causal chains of events that contributed to *InspectPerf*. SmartPLS depicts indicators as rectangles and constructs as circles. The structural relations model thus connected the circles in the ways that Maree hypothesised. The measurement model connected the rectangles to circles (if the measurement model was formative, as was the case for the *InspectContext* and *InspectPerf* constructs) and circles to rectangles (if the measurement model was reflective, as was the case for the *Training*, *OrgSatis*, *Quality* and *Workplace* constructs). Specifying a model for PLS analysis is much simpler than for the covariance-based SEM approach because errors and uniquenesses do not need to be specified (no 'e's or 'u's in the model) and parameters do not have to be fixed in order to properly scale the estimates (no '1's attached to any paths).

Note that not all possible paths were represented in her preliminary model. Only those paths that Maree thought would be relevant were shown; these paths thus being the only ones statistically estimated. Some causal chains were direct (where one construct was directly connected to another via a single-headed arrow). For example, Maree thought *InspectContext* would directly influence. *InspectPerf*. Other causal chains were indirect (where one variable influenced another via a series of paths that travelled through intermediate or mediating causes). For example, Maree thought that *InspectContext* would

(continued)

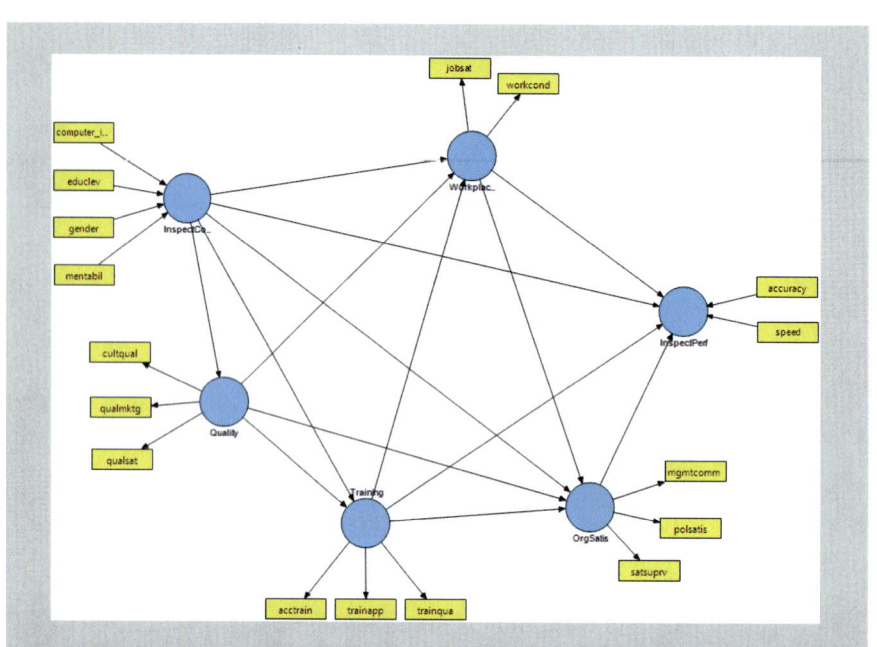

Fig. 8.14 The path diagram, drawn using SmartPLS, showing measurement model and structural relations model linkages between indicator variables and constructs in the QCI database

indirectly influence *InspectPerf* via *Workplace* perceptions. [Note that any construct may be involved in both direct and indirect causal chains, but reciprocal feedback causal loops are not permitted in PLS.]

Figure 8.15 shows the estimated standardised regression coefficients for all prescribed paths superimposed on the model. The number inside each construct circle is R^2. The R^2 value shows how much variance in a specific construct can be explained using information from all incoming constructs. Thus, for example, *InspectContext* explains 3.4% of the variance in the *Quality* construct and *InspectContext*, *Quality*, T*raining* and *Workplace* explain 43.9% of the variance in the *OrgSatis* construct. SmartPLS uses bootstrapping to provide t-tests for every estimated coefficient in both the measurement and structural relations components of the model. This means that PLS is essentially a nonparametric approach, which can easily handle data abnormalities such the severe positive skewness associated with the **speed** indicator variable.

All significant (at $p < .05$) estimated standardised regression coefficients are starred in Fig. 8.15. For the four constructs with reflective measurement models, all indicator paths are significant. For the formative construct *InspectPerf*, both **speed** and **accuracy** contribute significantly and positively.

(continued)

Procedure 8.7: Structural Equation (Causal Path) Models (SEM)

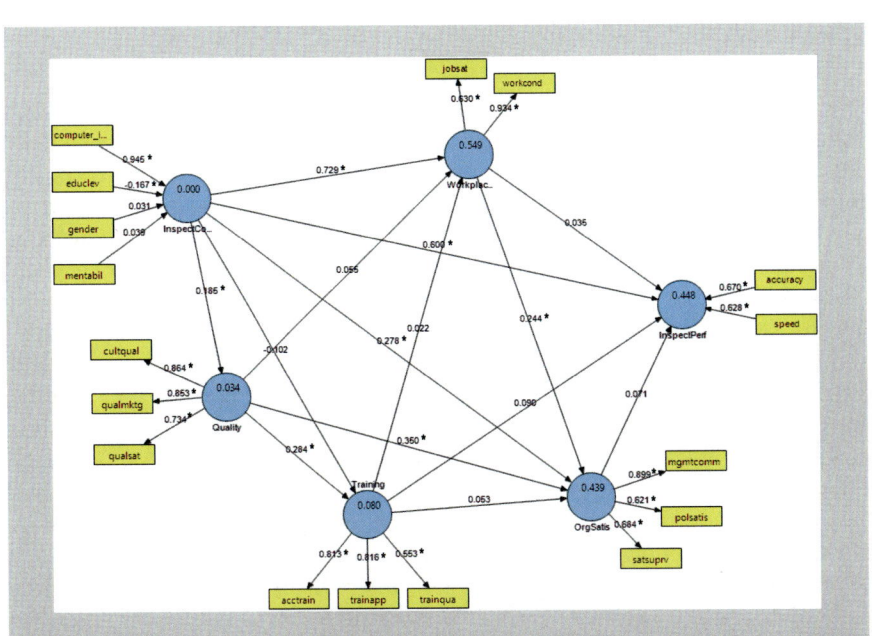

Fig. 8.15 The causal path diagram from Fig. 8.14, redrawn by SmartPLS, showing all estimated standardised regression coefficients (starred coefficients (= parameter estimates) are significant at $p < .05$)

For the *InspectContext* construct, only **computer_ind** (positively) and **educlev** (negatively) contribute significantly. Each estimated standardised coefficient is interpreted in exactly the same way as in ordinary least squares multiple regression analysis (recall *Procedure 7.13*). Thus, for example, Maree could interpret the coefficient of .350 on the path between the *Quality* and *OrgSatis* constructs as indicating that if *Quality* score is incremented by 1 standardised unit, the value for *OrgSatis* would be predicted to increase by .35 standardised units. Similarly, the coefficient of .600 on the path between *InspectContext* and *InspectPerf* indicating that if *InspectContext* score is incremented by 1 standardised unit, the value for *InspectPerf* would be predicted to increase by .60 standardised units. Maree could further unpack what this means by looking at the significant formative paths leading into each construct. For *InspectContext*, **computer_ind** has a positive path (inspectors in the computer industry perform better) and **educlev** has a negative path (high school only educated inspectors perform better) and on the *InspectPerf* side, both **accuracy** and **speed** have positive paths (higher performance is signalled by higher accuracy scores but slower speed scores).

Advantages

Structural equation models provide a coherent and statistically defensible method for testing hypotheses about the theoretical network of causal relationships which links observed variables and/or latent constructs. This is generally the only statistically appropriate procedure in which correlational information may be used to make causal statements and even then, this must be done cautiously using sound methodologies to implement the data gathering strategies. Theory revision may be undertaken through systematic assessment of various paths and their empirical importance. Unsubstantiated paths may be deleted from the model to simplify it.

It is possible, within a covariance-based SEM approach, to test several different theoretically competing models within the same statistical framework in order to make a defensible choice as to the most preferred model; however, this is only a legitimate activity when all models to be tested are set out in their entirety before the collection of the data. Frequently, two or more theories within a specific domain (e.g. theories of work motivation, job satisfaction, or self-concept) give rise to causal models that make competing predictions about hypothesised causal paths. In cases where one theoretical model is a simpler or less complex (proposing fewer causal pathways; i.e. nested) version of another theoretical model, we can use the difference between the two chi-square statistics (one from the test of the simpler model and one from the test of the more complex model) as a way of testing whether or not the more complex model results in significantly improved fit to the data.

The PLS-SEM approach offers a viable and, in many cases, more robust alternative to covariance-based SEM approaches, which means it may be suitable for a wider variety of research contexts and purposes. Hair et al. (2014) argue that PLS-SEM strongly emphasises prediction and variance explanation and provides a useful way to establish latent construct scores for further research.

Disadvantages

Covariance-based structural equation models are highly dependent upon having a strong theoretical basis for the path model(s) being tested. If the theory is wrong or the model fit poor, there may be a strong temptation to modify the model to improve statistical fit, based on statistical 'advice' (i.e. modification indices) provided by programs like AMOS. Many behavioural researchers generally condone this practice (and actually employ it) whereas others condemn the practice as engaging in a fishing expedition that capitalises on chance results. The safest and most defensible practice appears to be to resist this temptation, if at all possible, and to test any model modifications (such as deleting non-significant causal paths or adding or changing variables) that do emerge from a particular testing exercise using data from a separate sample of individuals. Furthermore, it is not 'cricket' to keep trying out different structural models *using the same sample data*, without good theoretical reasons for doing so.

The PLS-SEM approach does not suffer as strongly from a dependency on a strong theoretical background or a strong statistical error theory. However, this advantage is a two-edged sword. If the researcher wants to test strong theoretical propositions, then PLS-SEM is the wrong choice of modelling approach.

Covariance-based structural equation models can run into some rather troublesome statistical problems if the data being analysed are not 'well-behaved'. The maximum likelihood algorithms that are typically used are sensitive to distortions in the data such as non-normality and excessively high correlations (i.e. multicollinearity) among the model components. Researchers can be very disconcerted to find that their data yield improper estimates, due to one or more of these problems, and learning precisely where the problem lies can be difficult to diagnose. AMOS offers other estimation procedures such as weighted least squares and generalised least squares, which can help to circumvent some of these problems. The PLS-SEM approach is even better suited to such difficult data contexts, where small sample sizes, non-normal data and complex models can be handled quite easily.

The fact that no generally agreed goodness-of-fit measure can be put forward for the covariance-based SEM approach is also concerning and one finds different authors favouring very different indices. The chi-square goodness-of-fit test is sensitive to large sample sizes in that very tiny discrepancies in the model residuals can be shown to be significant (even well-fitting models based on a large sample size might be associated with a highly significant chi-square statistic). This tendency reduces the utility of the chi-square measure itself as a measure of goodness of fit. On the other hand, the maximum likelihood estimation procedures work best with large samples, particularly if many variables and causal paths are being analysed.

Finally, covariance-based structural equation models, if overly complex, may result in what are termed *identifiability* problems where the proposed model cannot be estimated (i.e. all proposed parameters cannot be identified). This usually occurs when there aren't enough fixed parameters (remember the 1's in Fig. 8.12) set up in the model or the error structure has been mis-specified. The AMOS program is especially good at detecting this specific type of problem and at proposing some candidate solutions for it. Identifiability issues are not a concern in the PLS-SEM approach.

The choice between using the covariance-based SEM approach and the PLS variance-based SEM approach is not always straightforward as shown by Hair et al. (2012). There are nuances of differences between the two approaches which can and should influence the choice. Sample size, type of measurement model (reflective vs formative) and data abnormalities are surface indicators for the choice. However, at a deeper level, whether or not goodness-of-fit tests for the entire model are desired and/or model comparisons are desired become critical choice factors (both of which by implication means that a strong error theory is also required, pushing the choice toward the covariance-based approach). Furthermore, whether the measurement models are formative or reflective is also a critical choice factor. Covariance-based SEM cannot easily handle formative measurement models, whereas PLS-SEM can easily do this.

Where Is This Procedure Useful?

Structural equation models are useful whenever a researcher proposes, up front, that the measures or variables for which data are to be collected in a study have a specific underlying causal structure. This hypothesised model may emerge from theoretical arguments, integration of findings of past research, or both. These grounds generally need to be very strong if a credible causal model is to be proposed that has a chance of surviving a covariance-based testing phase mostly intact. In short, SEM can be a powerful theory testing tool, but must be implemented carefully.

Causal models may have embedded measurement model sub-components (confirmatory factor models) that define the existence of latent constructs. These latent constructs, in turn, form part of the structural relations aspects of the model that causally links different variables or latent constructs to each other. This type of flexibility makes structural equation models a very powerful tool for testing behavioural and social theories if appropriate care is taken at all stages of the research. Such models should not be tested lightly. It should be noted here that more traditional parametric hypothesis testing frameworks such as multiple regression, ANOVA, ANCOVA, and MANOVA all can be represented as particular types of structural equation models, reflecting a more unified data analytic system for researchers (see Cohen et al. 2003 for a more in-depth discussion).

In some circles, PLS-SEM is considered a 'softer' form of structural equation modelling because it imposes fewer restrictions and is nonparametric (meaning a strong error theory is not needed). If the researcher is primarily dealing with formative measurement models, then PLS-SEM is a more viable approach. PLS-SEM has become especially popular in marketing and management information systems research (see Hair et al. 2012; Ringle et al. 2012); its use in management and educational research is increasing.

Software Procedures

Application	Procedures
SPSS	*Analyze → IBM SPSS Amos* and the AMOS program will start up. A new interface will appear with a blank page for drawing your path diagram in a fashion similar to Fig. 8.12 and for controlling and executing your structural equation model fitting analyses. The analyses will be conducted using the data file you have opened in SPSS. The analytical flow is (1) use the graphics window to diagram and set up your confirmatory factor model

(continued)

Procedure 8.7: Structural Equation (Causal Path) Models (SEM)

Application	Procedures
	(AMOS uses this diagram to figure out what has to be estimated; (2) set up the desired 'Analysis Properties' (controls all aspects of how AMOS should analyse the model, what it should output and whether or not bootstrap tests are desired; maximum likelihood estimation is the default method); (3) 'Calculate estimates' to run the computations; and (4) interpret the output (output comes in two forms, text output with all of the numbers and graphical output where AMOS superimposes the unstandardised or standardised estimates (your choice) on your model diagram). The diagrams produced in and by AMOS can be cut-and-pasted, in bitmap format, to another program for saving). An extensive manual for using the AMOS program can be accessed from the *Help* menu and Byrne (2016) provides a very good and practical guide. AMOS is probably the easiest structural equation program to learn how to use and its clear and flexible graphical interface is exceptionally useful. AMOS also offers an extensive range of data screening procedures so that researchers can detect, early on, if there are anomalies in the data that need addressing. AMOS offers bootstrapping capability for all model testing needs. AMOS only implements covariance-based structural equation modelling. There is also a PLS Extension module available for SPSS. However, the procedure offered in SPSS technically uses an exploratory (principal components-based approach) rather than a confirmatory approach to the model building process.
NCSS	NCSS does not offer the capability to estimate structural equation models or partial least squares models.
SYSTAT	*Analyze → Advanced → Path Analysis (RAMONA)...* and build the specifications, line by line, for your structural equation model. RAMONA uses 'manifest' variables (= observed measures in the data file) and 'latent' variables (= hypothesised constructs). RAMONA does not have a graphical interface, so each path in the diagram must be specified using a line of text. If a parameter, such as an error term, is to be fixed, this is signalled to RAMONA by appending '(0, 1.0)' after the relevant variable name in that line of text. For example, in the causal model in Fig. 8.12, one line that would be needed is: *educlev <- mentabil gender e1 (0, 1.0)* (read as 'educlev is caused by mentabil, gender and an unobserved fixed parameter error term'). To symbolise a correlation rather than a causal path, you would use '<->' between the two variables or latent constructs to be correlated. Maximum likelihood estimation is the default method. SYSTAT does not offer bootstrapping capability within RAMONA. RAMONA only implements covariance-based structural equation modelling. SYSTAT also offers a version of Partial Least Squares regression analysis which can do cross-validation tests as well as evaluate both univariate and multivariate regression models. However, the procedure offered in SYSTAT technically uses an exploratory (principal components-based approach) rather than a confirmatory approach to the model building process.
STATGRAPHICS	STATGRAPHICS does not offer the capability to estimate structural equation models. However, it does offer partial least squares analysis. *Relate → Multiple Factors → Partial Least Squares...* and select your desired 'Dependent Variables' and 'Independent Variables' for the analysis; hit *OK* and choose your desired 'PLS Options' (which includes cross-validation options); hit *OK* and choose your desired 'Tables' and 'Graphs'.

(continued)

Application	Procedures
R Commander	**R** Commander does not offer the capability to estimate structural equation models. However, you can estimate structural equation models using the lavaan package (see Rosseel 2012; see also http://www.jstatsoft.org/v48/i02/paper, accessed 19 Sept 2019) for the main **R** program. The pls package facilitates partial least squares analysis in **R** (see https://cran.r-project.org/web/packages/pls/pls.pdf, accessed 19 Sept 2019).
MPlus	Geiser (2013, pp. 62–80) shows how to get MPlus to estimate a structural equation model. Note that MPlus is a command line-driven software program, rather than a 'dialog box' program like SPSS or SYSTAT, but it does have a 'Language Generator' function that helps in setting up the necessary input file. MPlus also has a 'Diagrammer' procedure where you can draw your path diagram for estimation, similar to AMOS (but without as many 'bells and whistles'). MPlus can handle missing data by listwise deletion or by imputation. MPlus can also produce bootstrap standard errors for all components of a structural equation model. MPlus only implements covariance-based structural equation modelling.
SmartPLS	SmartPLS (Ringle et al. 2005 and see https://www.smartpls.com/) is a Java-based application for analysing PLS-SEM problems. The interface is straightforward, and it is relatively easy to diagram your model and run the analyses. There is a demonstration YouTube video at www.youtube.com/watch?v=7bqcG0GcgQ8 which shows how to set up and analyse a basic PLS-SEM analysis. SmartPLS has a basic data editor where you can input your data or, more efficiently, you could set up your indicator datafile in Excel or SPSS, then save the datafile in .csv format (with variable names saved in the first row of the datafile). SmartPLS can then import the .csv file into the PLS-SEM project you create. The graphical interface does not offer a great deal of control over formatting (e.g. font choice, font size, colours are all pre-determined) and printing the path diagrams is a bit tedious as you have to take a screenshot and paste it into a program like PowerPoint or Paint for printing. SmartPLS uses bootstrapping exclusively for evaluating the significant of individual path coefficient estimates and has some capabilities for handling missing data.

Procedure 8.8: Meta-analysis

Classification   Univariate/bivariate/multivariate; inferential; parametric.
Purpose   To provide systematic methods for statistically integrating findings within a specific behavioural or social science domain across a set of quantitative research investigations, typically those that have been peer-reviewed and published. Meta-analysis operates at the level of effect sizes in research and facilitates the computation and comparison of effect sizes within specific types of studies on specific types of participant samples.

Measurement level Since the dependent variable employed in meta-analyses is primarily an effect size measure of some variety, a ratio scale of measurement scale is implicit. The independent variables, coding different features of the studies in a meta-analysis, are typically categorical in nature, implying ordinal or nominal measurement, but some interval-level variables are also possible.

Meta-analysis is a general term for any of several systems for statistically aggregating and comparing the findings of research investigations (see, for example, Borenstein et al. 2009; Cooper et al. 2009; Wolf 1986). Meta-analysis is a methodology designed to introduce statistical rigour into the traditional review of literature that is expected in most research investigations. Additionally, in many behavioural and social science domains, a sufficient number of quantitative studies have been published over a period of years to warrant a tool that can aid in extracting and estimating generalisable trends across the studies. Instead of sampling human participants, the meta-analyst samples research studies which become the 'participants' in the research. Meta-analysis is a statistical form of document analysis.

In a typical meta-analysis, peer-reviewed published research articles and reports are sampling using some type of systematic scheme. The researcher then devises a coding system for categorising or otherwise measuring different observable aspects of each study reviewed. Each sampled study is then read by the meta-analyst and its features and outcomes are coded and recorded. Some meta-analyses may also incorporate unpublished research such as PhD and master's theses and consulting and other types of grey literature reports. Each study reviewed has its relevant aspects (e.g. year, journal, sample sizes, methodology, research context, sample composition) coded using the coding system. As well as coding relevant aspects of the research, each study's statistical research findings (whether they be means, standard deviations, χ^2 tests, *t*-tests, *F*-tests, correlations, regression coefficients, multiple correlations, or some other index of relationship or group comparison) are also recorded. These findings may be further broken down by sub-samples (e.g. for males and females, managers and non-managers, industry types, age groups) if such a breakdown is available in the written account and relates to the research questions of interest.

A standard aspect of meta-analysis methodology is to check the reliability and validity of the coding scheme that is to be employed. Content validity is especially important here, ensuring that the coding scheme captures all the relevant features of each study. As a check on coding reliability, the meta-analyst should enlist the aid of a colleague or research assistant to read and code a proportion of the sampled studies. The meta-analyst, working independently, would code the same studies. Their coding results for the same studies would then be compared and inter-observer reliabilities computed (recall *Procedure 8.1*). Any coding areas lacking in reliability (where the two coders disagree on how to code something represented in the study) should be refined and re-tested until a high level of coding agreement is reached. All remaining studies in the study would then be fully coded.

Prior to the actual conduct of the statistical aspect of meta-analysis, each statistical finding for a study must first be transformed into a common measure of *effect size*. There are a number of different measures of effect size depending upon which system of meta-analysis is being employed, but most of the measures give a standardised indication of the magnitude of relationship or group difference identified in the study. In the system advocated by Glass et al. (1981), effect size is measured by a standardised difference which looks much like a z-score (and is interpreted in much the same way, recall *Procedure 5.7*). In the system advocated by Hunter and Schmidt (2004), a standard correlational measure of effect size (such as r, eta-squared or r^2, recall *Procedure 7.7*) is computed. In one of the systems advocated by Rosenthal (1984), the p-values associated with individual research findings are aggregated.

Irrespective of the precise type of effect size measure employed, the overall goal in a meta-analysis is generally two-fold:

- provide an indication of the average effect size over the range of studies reviewed in the sample; and
- provide comparisons, using standard inferential statistical procedures, of effect sizes across different groups of studies that vary according to some coded characteristic/independent variables of interest.

We can conceptually illustrate the idea of a meta-analysis by considering the context of Maree's QCI research context.

> Suppose as a preliminary exercise prior to her research, Maree decided to do a meta-analysis of the published research on how stressful working conditions influence the accuracy of human decision making. Using appropriate resources such as library electronic literature database searches, Google Scholar and other internet-based searches, reviews of print-based journals and reviews of the reference lists from a number of key studies, Maree identified a large sample of published research investigations which related directly to her domain of interest.
>
> Suppose Maree found that a total of 900 such studies had been published over the past 20 years. She decided to employ a 50% simple random sample giving her 450 studies for her meta-analysis. Her first task was to demonstrate the content validity and reliability of her coding system. She and a research assistant independently coded a random sample of 25 studies from the pool of 450, inter-observer reliabilities were computed and codes where substantial disagreement existed were refined and retested. Once that process was completed and a refined coding system had been finalised, Maree and/or her research assistant then read each study, codified the relevant study characteristics and computed the relevant common measure of effect size (which transformed the results from studies comparing two or more groups and studies reporting simple or multiple correlations into a common standard and comparable index).

(continued)

Procedure 8.8: Meta-analysis

Maree decided to employ correlation as the measure of meta-analysis effect size (Borenstein et al. 2009, pp. 41–43) to index the strength and direction of association between the extent of stress experienced or reported in the decision context and the dependent variable of decision accuracy reported by a study. Prior to analysis, Maree transformed each originally published correlation (r) into what is called a Fisher's z (r_z) score, in line with recommendations by Borenstein et al. 2009, p. 42). Also as recommended by Borenstein et al. 2009, p. 42), Maree computed the standard error associated with each Fisher's z score (se(r_z)); this standard error was purely a function of the sample size on which the original correlation was based.

Maree was specifically interested in the following four ways of classifying the published research on which further effect size comparisons could be based:

- **Time sensitivity of the decision**: whether the decision was *time-sensitive* (i.e. had to be made within a specific time constraint) or *time-independent* (i.e. had no time constraint);
- **Measurement of stress level in the decision-making context**: whether the measurement of workplace stress was *self-reported* (using some kind of inventory or questionnaire) or *directly observed* (by a researcher or associate);
- **Impact risk associated with the decision**: whether the research focused on low impact-risk decisions which carried a low risk of impacting on or creating adverse consequences for others or high impact-risk decisions which carried a high risk of impacting on or creating adverse consequences for others; and
- **Consequences of being wrong**: whether, in the decision context, explicit penalties for making decision errors were *imposed* or *not imposed* on the research participants.

Maree's 4-dimensional classification system for coding the studies is used to identify the various rows and columns of Table 8.13. The table shows the mean original correlation value (transformed back from Fisher z-score form) between decision context stress and decision accuracy, the mean and average standard error of the Fisher z-scores computed across all the research studies falling within a specific cell of the design. Table 8.13 also shows the mean of the reported sample sizes in the studies classified into that cell as well as the total number of studies classified into that cell. Thus, for example, in the upper left corner cell of Table 8.13, the average correlation between decision context stress and accuracy was .11, obtained under *Time-Independent* conditions where a *Low Impact-Risk Decision* was made with *Penalties for Errors Not Imposed* and decision context stress measured via *Self-Report*. The average

(continued)

Table 8.13 Hypothetical statistical outcomes from Maree's meta-analysis of 450 published studies of the impact of decision context stress on decision accuracy

			Consequences of Being Wrong			
			Penalties for Errors Not Imposed		Penalties for Errors Imposed	
			Measurement of Stress in Decision Context		Measurement of Stress in Decision Context	
Time Sensitivity of Decision	Impact Risk Associated with Decision		Self-Report	Direct Observation	Self-Report	Direct Observation
Time-Independent	Low Impact-Risk					
		Mean original r	0.11	0.21	0.23	0.39
		Mean r_z	0.110	0.213	0.234	0.412
		se(r_z)	0.080	0.098	0.089	0.098
		Mean sample size	158.2	107.3	128.6	106.4
		# of studies	47	39	47	37
	High Impact-Risk					
		Mean original r	-0.06	-0.37	-0.52	-0.71
		Mean r_z	-0.060	-0.388	-0.576	-0.887
		se(r_z)	0.137	0.171	0.103	0.185
		Mean sample size	56.3	37.1	97.8	32.1
		# of studies	40	19	37	20
Time-Sensitive	Low Impact-Risk					
		Mean original r	-0.28	0.07	-0.33	-0.66
		Mean r_z	-0.288	0.070	-0.343	-0.793
		se(r_z)	0.115	0.120	0.220	0.165
		Mean sample size	78.5	72.6	23.6	39.7
		# of studies	41	24	27	32
	High Impact-Risk					
		Mean original r	-0.29	-0.73	-0.52	-0.86
		Mean r_z	-0.299	-0.929	-0.576	-1.293
		se(r_z)	0.173	0.112	0.149	0.234
		Mean sample size	36.4	82.9	47.8	21.3
		# of studies	9	8	15	8

Fisher z-score was 0.11 with an average standard error of 0.080, based on an average sample size of 158.2 participants, across n = 47 coded studies.

Using simple descriptive statistics (see *Procedure 5.4* and *5.5*), Maree computed the average Fisher z-score effect size across the entire sample of 450 studies to be $-.337$ (equates to a back-transformed value for r of $-.325$) with an average standard error of 0.141. This means that across all 450 published studies that Maree sampled, the average correlation between decision context stress and decision accuracy was $-.33$. The correlation is negative and moderately weak, indicating that as decision context stress increases, there is some tendency for decision accuracy to decrease, and vice-versa.

(continued)

Procedure 8.8: Meta-analysis

Note that the number of studies summarised in each cell of Table 8.13 suggested that there were very few studies classified as time-sensitive/high impact-risk/no error penalty imposed/self-report stress (n = 9) and relatively many studies classified as time independent/low impact-risk/self-report stress with either no error penalty imposed (n = 47) or error penalties imposed (n = 47). Furthermore, Table 8.13 made it clear to Maree that she had sampled relatively few studies focusing on time-sensitive high-impact risk decisions, irrespective of stress measurement method or error penalty condition. Maree also noted that there were interesting patterns in the mean sample sizes of studies in different cells of the table. For example, she noted that any study classified as focusing on a time-independent low-impact-risk decision had much a larger average sample size relative to studies classified elsewhere. The corollary to this observation was that effect size estimates were much more precise in these studies (i.e. had much lower standard errors for the Fisher z-scores). Thus, even reflecting on the prevalence of and sample sizes associated with different types of studies in a meta-analysis provided information about the overall landscape of the research domain Maree was investigating.

To further explore correlational patterns in the published research findings, Maree employed a 4-way factorial ANOVA procedure (recall *Procedure 7.10*) to test the average Fisher z-score effect sizes for the various main effects and interactions implicated by the design shown in Table 8.13. She found the following patterns to be particularly of interest.

- There was a significant 3-way interaction between Time Sensitivity of the Decision, Impact Risk Associated with the Decision and Consequences of Being Wrong: this interaction is shown in Fig. 8.16. For Low Impact-Risk

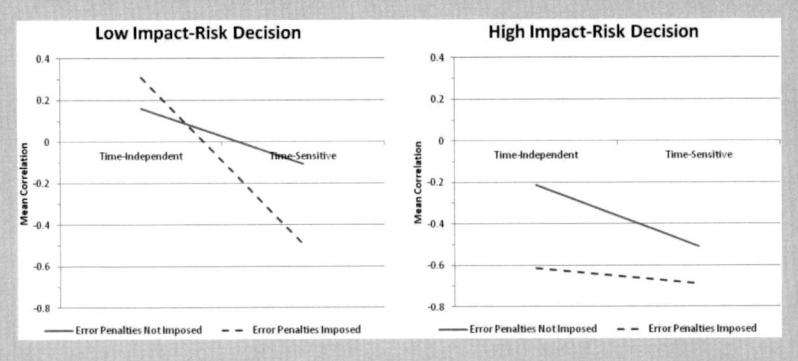

Fig. 8.16 Significant 3-way interaction from an ANOVA of Maree's meta-analysis of 450 published studies of the impact of decision context stress on decision accuracy

(continued)

Decisions (left-hand graph), the correlation between stress and decision accuracy was actually positive where the decision was Time-Independent and Error Penalties were Not Imposed (solid line), but even more positive where Error Penalties were Imposed (dashed line) – both patterns suggested that some level of stress was actually associated with enhanced decision accuracy. The pattern changed dramatically (cross-over interaction) when the decision task was Time-Sensitive – stress became moderately negatively correlated with decision accuracy where Error Penalties were Imposed (dashed line) and stayed near zero if Error Penalties were Not Imposed (solid line). The pattern for High Impact-Risk Decisions was rather different. The stress-accuracy correlation became more negative as the decision changed from Time-Independent to Time-Sensitive and from Error Penalties Not Imposed to Error Penalties Imposed. Stress was most strongly negatively correlated with accuracy under conditions involving a High Impact-Risk Decision that was Time-Sensitive and had Error Penalties Imposed. Maree concluded that this context reflected the most stressful decision-making context across all the studies she had classified; in other words, she had identified those conditions under which decision performance would be most degraded as a consequence of contextual stressors.

- There was a significant main effect for Measurement of Stress in Decision Context. The mean stress – accuracy correlation, when stress was measured via self-report, was $-.221$ (across 263 studies) and when stress was measured via direct observation, was a significantly stronger $-.421$ (across 187 studies). One explanation offered by Maree was that the stronger relationship may have emerged because of inherent biases and measurement problems associated with self-reports, compared to independent observations. Average sample size for the studies was not a likely candidate for explaining this pattern since the average sample size for self-report studies was 78.4 and, for direct observation studies, was not too dissimilar at 62.4. Differences in standard errors were thus not great across the two conditions ($se(r_z) = 0;.133$ for self-report and 0.148 for direct observation).

Advantages

Meta-analysis is an important methodology for the statistical summarisation of findings across a large number of quantitative positivist research investigations. The method treats instances of written research as the actual 'participants' in an investigation where the statistical outcomes from each study are re-expressed in terms of standardised measures of effect size. Various features of the written research can be codified for use as independent variables in comparisons of effect size differences across the different category codes. Standard inferential statistical techniques can be utilised to conduct such comparisons making it relatively easy to accomplish complex meta-analytic comparisons—specialised methods are not necessarily required.

Meta-analysis provides a rigorous way for researchers to extract the generalisable trends in findings from a large number of studies without excessive reliance on potentially biased or intuitive perceptions of research outcomes. The fact that patterns in research findings can be systematically explored in some detail by statistically comparing results as a function of different features of studies is an additional benefit. Such systematic insights are very difficult to realise and defend using more traditional narrative-oriented summarisation methods for reviewing literature.

Disadvantages

There are five very important problems associated with the conduct of a meta-analysis. First, the devising of the coding scheme for representing key features of the research studies being reviewed is not always straightforward. The researcher must decide which aspects of the research study are important to codify; if key aspects are missed then a great deal of effort must be expended in re-reading the research to generate the omitted codes. Furthermore, it is imperative that the inter-rater reliability of the coding scheme be assessed, and this adds to the time and effort required to undertake the research.

The second problem with meta-analysis concerns whether unpublished research or other grey literature should be included. Very often, unpublished research (including Ph.D. and master's dissertations and theses, papers or reports not submitted to journals, government reports) is ignored in meta-analyses. This is problematic because of what Rosenthal and Rosnow 1991, pp. 508–512) have termed the 'file drawer problem'. The file drawer problem recognises the fact that much of what goes unpublished are the insignificant results of investigations which researchers decide are not worth publishing—these studies thus stay hidden in some file drawer and forgotten. If we were to somehow include such research, our findings regarding the general effect size of some finding might be dramatically altered. The file drawer problem therefore leads to sampling bias in the studies that are included in the meta-analysis. There are some meta-analysis procedures available (see Rosenthal and Rosnow 1991) that attempt to compensate for this problem in the effect size estimation procedure.

The third problem in meta-analysis is related to the 'file drawer problem' and concerns how research studies are sampled for inclusion in the meta-analysis. The research sampling scheme is of critical importance in meta-analysis if one is to avoid claims of basing conclusions on a biased or unrepresentative sampling of studies. Bias or non-representativeness can emerge in a variety of ways, such as focusing solely on published research, focusing only on research published in specific journals, types of journals or disciplines, or focusing only on studies that use specific methodologies. In meta-analyses, stratified random or quota sampling schemes can be especially useful for ensuring representative coverage of sampled research. Of course, the best solution is to read every study published in an area, but this has obvious resource and cost implications and, given the diversity of available research outlets (both print-based and electronic), it is virtually impossible to ensure that every relevant study will be identified. Using electronic databases and search procedures does not overcome this problem because every electronic database will have its own built-in selectivity in what is included and what is excluded.

The fourth problem with meta-analysis concerns which system of meta-analysis to employ. Several systems are available, and each has its own advantages and disadvantages. The choice between them is often not obvious.

Finally, meta-analysis can only summarise the results of research that produces quantitative measures of outcomes using statistical procedures. It cannot be used as a tool for summarising the outcomes from qualitative research (meta-synthesis is one such approach). This too has an impact on the judgments one might make about what has been learned in a specific domain of investigation. Furthermore, it will strongly bias the meta-analysis toward research generated within a single social and behavioural science research paradigm—the positivist or normative paradigm. In this modern-day era of valuing multidisciplinary and mixed methods research and diversity in research perspectives, meta-analysis may provide an overly narrow and selective view of what research has learned and tend to reify or privilege quantitative data over the qualitative data. Cooksey and McDonald 2019, pp. 611–616) discuss meta-analysis (quantitative) vis-à-vis meta-synthesis (qualitative) as approaches to learning from the literature.

Where Is This Procedure Useful?

Meta-analysis can be useful in summarising any domain of research where numerous studies reporting quantitative outcomes for group comparisons or variable relationships are available. Generally speaking, meta-analysis becomes a useful tool to consider when the number of such studies exceeds at least 25 or 30 but preferably many more than 30. Meta-analyses may be conducted as the sole purpose for a research investigation (many of these types of studies are published each year in behavioural and social science journals) or it may be used as a supplemental tool in support of a more traditional review of the literature (a strategy which would certainly have to be considered if the research domain being reviewed had both quantitative and qualitative investigations available). In some areas, such as student achievement in education, there have been so many meta-analyses published that it is possible to do a meta-analysis of meta-analyses (see Hattie 2008 for an example)!

Software Procedures

Application	Procedures
SPSS	There are no specialised procedures in SPSS for conducting a meta-analysis. You can use any analytical approaches in SPSS, once your meta-analysis database has been set up. SPSS compute transformations may help you to convert specific statistical outcomes from each study into a common measure of effect size prior to conducting comparative analyses.
NCSS	*Analysis → Meta-Analysis → Meta-Analysis of Means* allows you to conduct a meta-analysis of the means and standard deviations from published studies that use two groups (experimental & control). *Analysis → Meta-Analysis → Meta-Analysis of Proportions* allows you to conduct a meta-analysis of count data from published studies that use two groups (experimental & control). *Analysis → Meta-Analysis → Meta-Analysis of Correlated Proportions* allows you to conduct a meta-analysis of count data from published studies that use two repeated measures groups (pretest & posttest or two treatment conditions). You can also use any other analytical approaches in NCSS, once your meta-analysis database has been set up. Each of the above procedures offers the option of producing a 'Forest Plot', which stacks interval comparisons on top of each other in a single plot.
SYSTAT	There are no specialised procedures in SYSTAT for conducting a meta-analysis. You can use any analytical approaches in SYSTAT, once your meta-analysis database has been set up.
STATGRAPHICS	There are no specialised procedures in STATGRAPHICS for conducting a meta-analysis. You can use any analytical approaches in STATGRAPHICS, once your meta-analysis database has been set up.
R Commander	There are no specialised procedures in **R** Commander for conducting a meta-analysis. You can use any analytical approaches in **R** Commander, once your meta-analysis database has been set up. Chen and Peace (2013) provide a thorough discussion of how functionalities and packages in **R** can be used to conduct meta-analyses. Viechtbauer (2010) describes and illustrates a specialised package for **R** called *metafor*. Everitt and Hothorn (2006, ch. 12) also illustrate how meta-analyses can be conducted using functionalities and packages in **R**.

References

References for Procedure 8.1

Athanasou, J. A. (1997). *Introduction to educational testing*. Wentworth Falls: Social Science Press, ch. 11 and 14.

Carmines, E. G., & Zeller, R. A. (1979). *Reliability and validity assessment*. Beverly Hills: Sage Publications.

Cooksey, R. W., & McDonald, G. (2019). *Surviving and thriving in postgraduate research* (2nd ed.). Singapore: Springer, ch. 18.

Gwet, K. L. (2012). *Handbook of inter-rater reliability* (3rd ed.). Gaithersburg: Advanced Analytics, LLC.

SPSS. (1998). *Using SPSS for item analysis: More reliable test assessment using statistics*, SPSS White Paper. http://www.spsstools.net/Syntax/ItemAnalysis/UsingSPSSforItemAnalysis.pdf. Accessed 16 Sept 2019.

Thorndike, R. L., & Thorndike, R. M. (1997). Reliability. In J. P. Keeves (Ed.), *Educational research, methodology, and measurement: An international handbook* (2nd ed., pp. 755–790). Oxford: Pergamon Press.

Waltz, C. F., Strickland, O. L., & Lenz, E. R. (2010). *Measurement in nursing and health research* (4th ed.). New York: Springer Publishing.

Useful Additional Reading for Procedure 8.1

Allen, P., Bennett, K., & Heritage, B. (2019). *SPSS Statistics: A practical guide* (4th ed.). Cengage Learning Australia Pty: South Melbourne, ch. 16.

George, D., & Mallery, P. (2019). *IBM SPSS statistics 25 step by step: A simple guide and reference* (15th ed.). New York: Routledge, ch. 18.

Everitt, B. S. (1995). *Making sense of statistics in psychology: A second level course*. Oxford: Oxford University Press, ch. 13.

Murphy, K. R., & Davidshofer, C. O. (2004). *Psychological testing: Principles and applications* (6th ed.). Englewood Cliffs: Prentice-Hall, ch. 6, 7, and 10.

Nunnally, J. C. (1978). *Psychometric theory* (2nd ed.). New York: McGraw-Hill, ch. 7. [a classic text in measurement theory].

Rosenthal, R., & Rosnow, R. L. (1991). *Essentials of behavioral research: Methods and data analysis* (2nd ed.). New York: McGraw-Hill, ch. 3.

Thorndike, T. L. (1982). *Applied psychometrics*. Boston: Houghton-Mifflin, ch. 6. [another classic text in measurement theory].

Walsh, W. B., & Betz, N. E. (2000). *Tests and assessment* (4th ed.). Englewood Cliffs: Prentice-Hall, ch. 3.

References for Procedure 8.2

Allison, P. D. (2001). *Missing data*. Thousand Oaks: Sage Publications.

Field, A., Miles, J., & Field, Z. (2012). *Discovering statistics using R*. London: Sage Publications, ch. 6.

Hair, J. F., Black, B., Babin, B., & Anderson, R. E. (2010). *Multivariate data analysis: A global perspective* (7th ed.). Upper Saddle River: Pearson Education, ch. 2.

Tabachnick, B. G., & Fidell, L. S. (2019). *Using multivariate statistics* (7th ed.). New York: Pearson Education, ch. 4.

Useful Additional Reading for Procedure 8.2

Beaton, A. F. (1997). Missing scores in survey research. In J. P. Keeves (Ed.), *Educational research, methodology, and measurement: An international handbook* (2nd ed., pp. 763–766). Oxford: Pergamon Press.

Cohen, J., Cohen, P., West, S. G., & Aiken, L. S. (2003). *Applied multiple regression/correlation analysis for the behavioral sciences* (3rd ed.). Mahwah: Lawrence Erlbaum Associates, ch. 4, 10 & 11.

Field, A. (2018). *Discovering statistics using SPSS for Windows* (5th ed.). Los Angeles: Sage Publications, ch. 6.
Meyers, L. S., Gamst, G. C., & Guarino, A. (2017). *Applied multivariate research: Design and interpretation* (3rd ed.). Thousand Oaks: Sage Publications, ch. 3A, 3B.

References for Procedure 8.3

Cumming, G., & Finch, S. (2001). A primer on the understanding, use, and calculation of confidence intervals that are based on central and noncentral distributions. *Educational and Psychological Measurement, 61*(4), 532–574.
Field, A. (2018). *Discovering statistics using SPSS for Windows* (5th ed.). Los Angeles: Sage Publications, section 2.8 and 2.99.
Field, A., Miles, J., & Field, Z. (2012). *Discovering statistics using R*. London: Sage Publications, ch. 2, 6, 7, 9.
Smithson, M. J. (2000). *Statistics with confidence*. London: Sage Publications, ch. 5.
Smithson, M. J. (2003). *Confidence intervals*. Thousand Oaks: Sage Publications.
Thompson, B. (2002). What future quantitative social science research could look like: Confidence intervals for effect sizes. *Educational Researcher, 31*(3), 25–32.

Useful Additional Reading for Procedure 8.3

Agresti, A. (2018). *Statistical methods for the social sciences* (5th ed.). Boston: Pearson, Ch. 5.
Dracup, C. (2005). Confidence intervals. In B. S. Everitt & D. C. Howell (Eds.), *Encyclopedia of statistics in behavioral science* (Vol. 1, pp. 366–375). Chichester: Wiley.
Thompson, B. (2006). *Foundations of behavioral statistics: An insight-based approach*. New York: The Guilford Press, ch. 7.

References for Procedure 8.4

Byrne, B. M. (2016). *Structural equation modelling with AMOS: Basic concepts, applications, and programming* (3rd ed.). New York: Routledge.
Canty, A. J., & Davison, A. C. (2005). Bootstrap inference. In B. S. Everitt & D. C. Howell (Eds.), *Encyclopedia of statistics in behavioral science* (Vol. 1, pp. 169–176). Chichester: Wiley.
Efron, B. (1979). Bootstrap methods: Another look at the jackknife. *The Annals of Statistics, 7*(1), 1–26.
Field, A., Miles, J., & Field, Z. (2012). *Discovering statistics using R* (pp. 226–227, 298–301). London: Sage Publications.
Mooney, C. Z., & Duval, R. D. (1993). *Bootstrapping: A nonparametric approach to statistical inference*. Newbury Park: Sage Publications.
Rodgers, J. L. (2005). Jackknife. In B. S. Everitt & D. C. Howell (Eds.), *Encyclopedia of statistics in behavioral science* (Vol. 2, pp. 1005–1007). Chichester: Wiley.

Useful Additional Reading for Procedure 8.4

Dracup, C. (2005). Confidence intervals. In B. S. Everitt & D. C. Howell (Eds.), *Encyclopedia of statistics in behavioral science* (Vol. 1, pp. 366–375). Chichester: Wiley.
Field, A. (2018). *Discovering statistics using SPSS for Windows* (5th cd., pp. 265–268). Los Angeles: Sage Publications.
Mooney, C. (2004). Bootstrapping. In M. S. Lewisbeck, A. Bryman, & T. F. Liao (Eds.), *The SAGE encyclopedia of social science research methods* (Vol. 1, pp. 75–78). Thousand Oaks: Sage Publications.
Westfall, P. H., & Young, S. (1993). *Resampling-based multiple testing: Examples and methods for p-value adjustment.* New York: Wiley.

References for Procedure 8.5

Allen, P. G., & Fildes, R. (2001). Econometric forecasting. In J. S. Armstrong (Ed.), *Principles of forecasting: A handbook for practitioners and researchers* (pp. 303–362). Boston: Kluwer Academic Publishers.
Bauwens, L., Laurent, S., & Rombouts, J. V. K. (2006). Multivariate GARCH models: A survey. *Journal of Applied Econometrics, 21*(1), 79–109.
Bowerman, BL O'Connell, RT & Koehler, AB 2005, Forecasting, time series, and regression: An applied approach, 4, Brooks/Cole, Belmont, Parts III and IV.
Coghlan, A. (2018). *A little book of R for time series (Release 0.2).* https://buildmedia.readthedocs.org/media/pdf/a-little-book-of-r-for-time-series/latest/a-little-book-of-r-for-time-series.pdf. Accessed 18 Sept 2019.
Cromwell, J. B., Labys, W. C., & Terraza, M. (1994a). *Univariate tests for time series models.* Thousand Oaks: Sage Publications.
Cromwell, J. B., Hannan, M. J., Labys, W. C., & Terraza, M. (1994b). *Multivariate tests for time series models.* Thousand Oaks: Sage Publications.
Crosbie, J. (1993). Interrupted time-series analysis with brief single-subject data. *Journal of Consulting and Clinical Psychology, 61*(6), 966–974.
Glass, G. V., Willson, V. L., & Gottman, J. M. (2008). *Design and analysis of time-series experiments.* Charlotte: Information Age Publishing.
McCleary, R., & Hay, R. A., Jr. (1980). *Applied time series analysis for the social sciences.* Beverly Hills: Sage Publications.
McDowell, D., McCleary, R., Meidinger, E. E., & Hay, R. A., Jr. (1980). *Interrupted time series analysis.* Beverly Hills: Sage Publications.
Ostrom, C. W., Jr. (1978). *Time series analysis: Regression techniques.* Beverly Hills: Sage Publications.
Sayrs, L. W. (1989). *Pooled time series analysis.* Newbury Park: Sage Publications.
Sharpley, C. F. (1997). Single case research: Measuring change. In J. P. Keeves (Ed.), *Educational research, methodology, and measurement: An international handbook* (2nd ed., pp. 451–456). Oxford: Pergamon Press.
StataCorp, L. P. (2011). *Stata time-series reference manual: Release 12.* College Station: Stata Press.
Tabachnick, B. G., & Fidell, L. S. (2019). *Using multivariate statistics* (7th ed.). New York: Pearson Education, ch. 17.
Wagner, A. K., Soumerai, S. B., Zhang, F., & Ross-Degnan, D. (2002). Segmented regression analysis of interrupted time series studies in medication use research. *Journal of Clinical Pharmacy and Therapeutics, 27*, 299–309.

Useful Additional Reading for Procedure 8.5

Gottman, J. M. (1981). *Time series analysis: A comprehensive introduction for social scientists*. London: Cambridge University Press.
Hamilton, L. C. (2013). *Statistics with Stata: Version 12*. Boston: Brooks/Cole, ch. 12.
Huitema, B. (2004). Analysis of interrupted time-series experiments using ITSE: A critique. *Understanding Statistics, 3*(1), 27–46.
IBM Corporation. (2017). *IBM SPSS Forecasting 25*, available as downloadable user's guide. ftp://public.dhe.ibm.com/software/analytics/spss/documentation/statistics/25.0/en/client/Manuals/IBM_SPSS_Forecasting.pdf. Downloaded 19 Sept 2019.

References for Procedure 8.6

Brown, T. A. (2015). *Confirmatory factor analysis for applied research* (2nd ed.). New York: The Guilford Press.
Bryant, F. B., & Yarnold, P. R. (1995). Principal components analysis and exploratory and confirmatory factor analysis. In L. G. Grimm & P. R. Yarnold (Eds.), *Reading and understanding multivariate statistics* (pp. 99–136). Washington, DC: American Psychological Association.
Byrne, B. M. (2016). *Structural equation modeling with AMOS: Basic concepts, applications, and programming* (3rd ed.). New York: Routledge.
Geiser, C. (2013). *Data analysis with MPlus*. New York: The Guilford Press, section 3.4.
Kline, R. B. (2015). *Principles and practice of structural equation modelling* (4th ed.). New York: The Guilford Press, ch. 9, 12, 13.
Schumacker, R. E., & Lomax, R. G. (2016). *A beginner's guide to structural equation modeling* (4th ed.). New York: Routledge, ch. 6.

Useful Additional Reading for Procedure 8.6

Hair, J. F., Black, B., Babin, B., & Anderson, R. E. (2010). *Multivariate data analysis: A global perspective* (7th ed.). Upper Saddle River: Pearson Education, ch. 12.
Long, J. S. (1983). *Confirmatory factor analysis*. Beverley Hills: Sage Publications.
Meyers, L. S., Gamst, G. C., & Guarino, A. (2017). *Applied multivariate research: Design and interpretation* (3rd ed.). Thousand Oaks: Sage Publications, ch. 11A, 11B.

References for Procedure 8.7

Byrne, B. M. (2016). *Structural equation modeling with AMOS: Basic concepts, applications, and programming* (3rd ed.). New York: Routledge.
Cohen, J., Cohen, P., West, S. G., & Aiken, L. S. (2003). *Applied multiple regression/correlation analysis for the behavioral sciences* (3rd ed.). Mahwah: Lawrence Erlbaum Associates, ch. 12.
Geiser, C. (2013). *Data analysis with MPlus*. New York: The Guilford Press.

Hair, J. F., Hult, G. T. M., Ringle, C. M., & Sarstedt, M. (2014). *A primer on partial least squares structural equation modelling (SEM)*. Los Angeles: Sage Publications.
Hair, J. F., Sarstedt, M., Ringle, C. M., & Mena, J. A. (2012). An assessment of the use of partial least squares structural equation modeling in marketing research. *Journal of the Academy of Marketing Science, 40*, 414–433.
Klem, L. (2000). Structural equation modeling. In L. G. Grimm & P. R. Yarnold (Eds.), *Reading and understanding more multivariate statistics* (pp. 227–260). Washington, DC: American Psychological Association (APA).
Kline, R. B. (2015). *Principles and practice of structural equation modelling* (4th ed.). New York: The Guilford Press, ch. 6, 7, 8, 10.
Ringle, C. M., Wende, S., & Will, S. (2005). *SmartPLS 2.0 (M3) Beta*. Hamburg: University of Hamburg.
Ringle, C. M., Sarstedt, M., & Straub, D. W. (2012). A critical look at the use of PLS-SEM in *MIS Quarterly*. *MIS Quarterly, 36*(1), iii–xiv and S3–S8.
Rosseel, Y. (2012). lavaan: An R package for structural equation modelling. *Journal of Statistical Software, 48*(2), 1–36.
Schumacker, R. E., & Lomax, R. G. (2016). *A beginner's guide to structural equation modeling* (4th ed.). New York: Routledge.

Useful Additional Reading for Procedure 8.7

Hair, J. F., Black, B., Babin, B., & Anderson, R. E. (2010). *Multivariate data analysis: A global perspective* (7th ed.). Upper Saddle River: Pearson Education, ch. 11, 12.
Hair, J. F., Ringle, C. M., & Sarstedt, M. (2011). PLS-SEM: Indeed a silver bullet. *Journal of Marketing Theory and Practice, 19*(2), 139–152.
Hair, J. F., Ringle, C. M., & Sarstedt, M. (2013). Partial least squares structural equation modeling: Rigorous applications, better results and higher acceptance. *Long Range Planning, 46*(1-2), 1–12.
Hoyle, R. H. (Ed.). (2012). *Handbook of structural equation modeling*. New York: Guilford Press.
Long, J. S. (1983). *Covariance structure analysis*. Newbury Park: Sage Publications.
Meyers, L. S., Gamst, G. C., & Guarino, A. (2017). *Applied multivariate research: Design and interpretation* (3rd ed.). Thousand Oaks: Sage Publications, ch. 12A to 14B.
Pedhazur, E. J. (1997). *Multiple regression in behavioral research: Explanation and prediction* (3rd ed.). South Melbourne: Wadsworth Thomson Learning, ch. 18 and 19.
Sellin, N., & Keeves, J. P. (1997). Path analysis with latent variables. In J. P. Keeves (Ed.), *Educational research, methodology, and measurement: An international handbook* (2nd ed., pp. 632–640). Oxford: Pergamon Press.
Tabachnick, B. G., & Fidell, L. S. (2019). *Using multivariate statistics* (7th ed.). New York: Pearson Education, ch. 14.
Thompson, B. (2000). Ten commandments of structural equation modeling. In L. G. Grimm & P. R. Yarnold (Eds.), *Reading and understanding more multivariate statistics* (pp. 261–284). Washington, DC: American Psychological Association (APA).
Tuijnman, A. C., & Keeves, J. P. (1997). Path analysis and linear structural relations analysis. In J. P. Keeves (Ed.), *Educational research, methodology, and measurement: An international handbook* (2nd ed., pp. 621–632). Oxford: Pergamon Press.
Ullman, J. B., & Bentler, P. M. (2004). Structural equation modeling. In M. Hardy & A. Bryman (Eds.), *Handbook of data analysis* (pp. 431–458). London: Sage Publications.

References for Procedure 8.8

Borenstein, M., Hedges, L. V., Higgins, J. P. T., & Rothstein, H. R. (2009). *Introduction to meta-analysis*. Chichester: Wiley.
Chen, D.-G., & Peace, K. E. (2013). *Applied meta-analysis using R*. London: Chapman & Hall/CRC.
Cooksey, R. W., & McDonald, G. (2019). *Surviving and thriving in postgraduate research* (2nd ed.). Singapore: Springer.
Cooper, H., Hedges, L. V., & Valentine, J. V. (2009). *The handbook of research synthesis and meta-analysis* (2nd ed.). New York: Russell Sage Foundation.
Everitt, B. S., & Hothorn, T. (2006). *A handbook of statistical analyses using **R***. Boca Raton: Chapman & Hall/CRC.
Hattie, J. (2008). *Visible learning: A synthesis of over 800 meta-analyses relating to achievement*. New York: Routledge.
Hedges, L., & Olkin, I. (1985). *Statistical methods for meta-analysis*. Orlando: Academic, ch. 1 provides a good foundation.
Viechtbauer, W. (2010). Conducting meta-analyses in **R** with the metafor package. *Journal of Statistical Software, 36*(3), 1–48.
Wolf, F. (1986). *Meta-analysis*. Beverly Hills: Sage Publications.

Useful Additional Reading for Procedure 8.8

Durlak, J. A. (1995). Understanding meta-analysis. In L. G. Grimm & P. R. Yarnold (Eds.), *Reading and understanding multivariate statistics* (pp. 319–352). Washington, DC: American Psychological Association.
Glass, G. V., McGaw, B., & Smith, M. L. (1981). *Meta-analysis in social research*. Beverly Hills: Sage Publications.
Hunter, J. E., & Schmidt, F. L. (2004). *Methods of meta-analysis: Correcting error and bias in research findings* (2nd ed.). Thousand Oaks: Sage Publications.
Konstantopoulos, S., & Hedges, L. (2004). Meta-analysis. In D. Kaplan (Ed.), *The Sage handbook of quantitative methodology for the social sciences* (pp. 281–300). Thousand Oaks: Sage Publications.
Lipsey, M. W., & Wilson, D. B. (2001). *Practical meta-analysis*. Thousand Oaks: Sage Publications.
Rosenthal, R. (1984). *Meta-analytic procedures for social research*. Beverly Hills: Sage Publications.
Rosenthal, R., & Rosnow, R. L. (1991). *Essentials of behavioral research: Methods and data analysis* (2nd ed.). New York: McGraw-Hill, ch. 22.

Chapter 9
Specialised Statistical Procedures

In this chapter, we explore some rather more specialised, but nonetheless important statistical procedures. These techniques can accomplish a variety of tasks that are specific to certain types of research questions, often within particular disciplines (e.g. marketing, education, sociology, quality control and assurance). More and more social and behavioural research makes use of these specialised procedures, yet it remains unusual to encounter explanations of what these procedures are intended to do in standard statistics and research methods texts. In this chapter, you will encounter the last fundamental concept, namely *Bayesian statistical inference*. Here we explore an alternative logic for statistical hypothesis testing (i.e. an alternative to the 'classical' statistical hypothesis testing logic discussed in *Fundamental concept V*).

Rasch models and *item response theory* provide alternative approaches to item analysis (compared to *Procedure 8.1*) in the creation and evaluation of tests and scales; approaches especially of interest in the education and training disciplines. *Survival/failure analysis* provides techniques for assessing the factors that predict how long a particular outcome event (e.g. success or failure) will take to be realised; approaches especially of interest in the medical, nursing, human resource management and quality assurance disciplines. *Quality control charts* provide methods for tracing, measuring and analysing the quality of products, services and processes, as well as for monitoring quality before, during and after quality improvement interventions; approaches especially of interest in the quality assurance, management and marketing disciplines. *Conjoint measurement* and *choice modelling* provide a sophisticated statistical methodology for assessing consumer decision making, often through survey-based experimental designs; an approach found to be especially useful in the marketing and psychological disciplines. *Multi-level (hierarchical linear) models* provide a set of techniques designed to facilitate the estimation of multiple regression models at different levels of analysis (e.g. students, teachers and schools; employees, managers and departments or companies); approaches that have proven especially useful in the education, management, organisational behaviour and industrial/organisational psychology disciplines. *Classification and regression trees* describe a range of techniques designed to facilitate detection of prediction patterns, especially interaction patterns, in data in ways that ordinary multiple regression models (recall *Procedure 6.4*) cannot; this approach that has been found useful in the medical, sociological and social policy disciplines.

Social network analysis is a quantitative approach to understanding and displaying the nodes, interactions and relational connections in a social network (e.g. between friends, between internet (e.g. Facebook) connections, between members of a group or team, between workers in a department or between multiple departments, between multiple agencies and organisations ...). We will also introduce you to a range of other more *specialised forms of regression analysis*, including robust, multinomial, fuzzy, nonlinear, ridge, generalised least squares and generalised linear models, 2-stage, tobit and probit regression and nonparametric regression methods such as ordinal regression. We will also look at how mediator and moderator effects can be tested in a regression modelling context. *Data mining* encompasses a range of methods (including classification and regression trees and cluster analysis) for detecting and learning patterns in data, particularly in large data sets. Here you will encounter such methods as neural network analysis, latent class and latent profile analysis, data visualisation and machine learning techniques. *Text mining* is a more specialised form of data mining which focuses on the detection and analysis of concepts and patterns in qualitative data. In effect, text mining is a quantitative approach to extracting meaning out of qualitative textual data (such as responses to open ended survey questions, transcripts of interviews or reports, texts, books and other types of documents). Finally, we will briefly explore quantitative approaches to *simulation and computational modelling* which allow the researcher to build statistical, mathematical or virtual models of the world and then use those models to generate data for testing research conceptualisations, theories and hypotheses.

Consider the following potential research objectives or questions for which it would be appropriate for Maree to utilise one of the techniques illustrated in this chapter (note that most of these questions require conceptual extensions and expansions of the QCI research context to incorporate different research designs and variables).

- Can Maree show how items on a knowledge test for quality control inspection procedures behave in terms of difficulty and, at the same, show how quality control inspectors differ in their ability? [See *Procedure 9.1* – Rasch models and item response theory.]
- Can Maree test how long quality control inspectors will stay in their jobs (in months) given their initial level of decision accuracy and whether they participate in a quality control training program? [See *Procedure 9.2* – Survival/failure analysis.]

(continued)

9 Specialised Statistical Procedures   559

- Can Maree evaluate quality inspectors' daily decision accuracy over a month against expectations that accuracy will not stray below a specific lower boundary? [See *Procedure 9.3* – Quality control charts.]
- Can Maree assess quality control inspectors' preferences for potential reward and benefit packages, comprising different mixtures of pay, holiday, bonus and profit share outcomes, conditional on their performance? [See *Procedure 9.4* – Conjoint measurement and choice modelling.]
- Can Maree show how predictions of inspectors' decision accuracy, based on job satisfaction and perceptions of working conditions, will vary for different types of companies? [See *Procedure 9.5* – Multi-level (hierarchical linear) models.]
- Can Maree predict what industry an inspector works for based on different combinations of cutting points on inspectors' inspection performance (accuracy and speed), satisfaction and working condition measures? [See *Procedure 9.6* – Classification & regression trees.]
- If QCI inspectors are observed at their monthly gatherings in a company over a span of 6 months, who communicates with whom, how frequently and is one person more central than any other? Can informal leaders, for example, be identified through the interaction patterns? [See *Procedure 9.7* – Social network analysis.]
- In the QCI database, is there evidence to show that the amount of job satisfaction reported by QCI inspectors moderates the relationship between workplace stress and the speed and accuracy of their inspection decisions? Is decision speed an effective mediator variable between workplace stress and decision accuracy? What if quality inspection decisions were classified, not as binary 'yes/no' decisions, but as trinary 'yes/maybe/no' decisions – how could Maree develop a prediction model for those decision outcomes? [See *Procedure 9.8* – Specialised forms of regression analysis.]
- How, in a company that has a history of keeping detailed spreadsheet records of every customer complaint, product fault and product return, along with demographic and other details associated with each event, might useful patterns be discovered that could help to improve product and service quality? [See *Procedure 9.9* – Data mining & text mining.]
- Suppose in an effort to see just how strongly product quality figures in the culture and mindset of companies, Maree downloads a strategic plans and annual reports for a large sample of companies; she might then be interested in statistically analysing the texts of those reports to see if quality-related concepts can be found and, if so, what their connections are to other key concepts and ideas. [See *Procedure 9.9* – Data mining & text mining.]
- Suppose Maree was interested in experimenting with changes to quality control inspection procedures (e.g. product assembly workflow rates, design of work shifts, location of decision points within the workflow of

(continued)

> product assembly, and so on) to see what impact these might have on inspection decision speed and accuracy – how might she do this without actually having to intervene in and disrupt a company's work context? [See *Procedure 9.10* – Simulation and modelling.]

Fundamental Concept X: Bayesian Statistical Inference – An Alternative Logic

In *Fundamental concept V*, we explored the standard or 'classical' logic of statistical inference as employed by most social and behavioural science researchers. There are, however, other possible systems of logic for statistical inference that lay claim to being more appropriate than the classical approach for hypothesis testing (for example, see the debated issues discussed in Gigerenzer et al. 2004). The most prominent of these alternative logical systems is *Bayesian statistical inference*, named after the Reverend Thomas Bayes, an English clergyman who, in the 1700s, derived what has since been called 'Bayes theorem'. For illustration purposes, we will discuss Bayesian statistical inference in the context of testing for group mean differences. However, you should know that the potential applications are much wider.

In classical statistical inference, the researcher gathers a set of data in order to test two competing hypotheses: a null hypothesis (assuming no relationship, no group difference and the like) and an alternative hypothesis (proposing the existence of a relationship or difference). Using statistical evidence from the data, the researcher decides to reject or fail to reject the null hypothesis based on the probability that the observed result would have been found in a world where the null hypothesis was true. If this probability is sufficiently small (by convention, p less than or equal to .05), the researcher is justified in rejecting the null hypothesis in favour of the alternate hypothesis. In essence, the evidence sought in classical statistical inference is the probability that the statistical estimate or effect observed in the data sample would have arisen *given* that we make the assumption that the null hypothesis is true. In shorthand form, we could represent this as p(sample data evidence|null hypothesis is true) where the '|' symbol means 'given' (formally, this is called a 'conditional probability', recall *Fundamental concept I*).

Fundamental Concept X: Bayesian Statistical Inference – An Alternative Logic

In the QCI database, Maree may wish to test for differences between tertiary qualified and high school educated inspectors in terms of their inspection decision accuracy in their company. The data in the QCI database thus provide observed outcomes from measuring **accuracy** on a sample of inspectors classified at the different educational levels (codified in the **educlev** variable). To test for such differences using classical statistical inference logic, Maree sets up two competing hypotheses: the null stating there is no **educlev** difference on **accuracy** scores and the alternate stating there is an **educlev** difference on **accuracy** scores.

Maree must assume the null hypothesis is true until and unless she discovers sufficient evidence in her data to reject that assumption. Using an independent groups *t*-test statistic (recall *Procedure 7.2*), Maree then tests the hypothesis using statistics computed from her data set (e.g. means and standard deviations for tertiary qualified and high school educated inspectors, respectively). She evaluates the test outcome by finding the probability that the study would have produced a test result as large as or larger than the one she observed in a world where the null hypothesis of no **educlev** difference was true. She compares this probability to her preset level of significance (say, .05) and decides to reject the null hypothesis only if her test result has a probability of occurring less than or equal to $p = .05$.

The results from such a *t*-test would show: mean **accuracy** scores for tertiary qualified and high school educated inspectors of 79.94 and 84.48, respectively; a *t*-test statistic value of 2.683 (with 106 degrees of freedom); and a *p*-value for this result of .008. She would therefore reject the null hypothesis because her observed *p*-value was much less than the $p = .05$ criterion. By conducting her test in this manner, Maree essentially quantifies how likely it is (i.e. $p = .008$) that she would have observed the **educlev** difference in **accuracy** scores computed from her data (i.e. 4.54), given her prior assumption that the null hypothesis was true.

Bayesian statistical inference takes the position that the classical approach in fact tests the wrong hypothesis. The argument is that researchers are not really interested in *p*(sample data evidence|null hypothesis is true); instead they are (or should be) interested in finding *p*(null hypothesis is true|sample data evidence). That is, the Bayesian perspective says that we need to find the likelihood or probability of the null hypothesis being true *given* the statistical evidence available in the data. This is the reverse conditional probability from that sought in classical statistical inference.

On the surface, it may look like these are really asking the same question in different ways, but there is one important difference: to find *p*(null hypothesis is true| sample data evidence), the researcher needs to know or make a guess as to how likely it was that the null hypothesis was true in the world before the researcher conducted his or her study (i.e. independent of any statistical evidence from a research sample). In other words, Bayesian statistical inference requires knowledge of a second probability, *p(null hypothesis is true)*; something that is called a *base rate probability* or, more formally, a *prior probability*. This prior probability effectively modifies the standard *p*-value so that it correctly reflects the desired (i.e. reversed) conditional probability, which is formally called the *posterior probability*. Technically, what this means is that the standard *p*-value from classical statistical inference testing is multiplied by the prior probability to give the desired posterior probability (this is a highly simplified version of the process but serves to illustrate the logic).

Knowledge of a prior probability can come from past research or, more commonly, can be calculated with reference to a specifically assumed prior probability distribution, such as the normal distribution (in either case, this would yield an 'objective' prior probability). Alternatively, Bayesian statistical inference works just as well if the researcher simply makes an educated guess as to the prior probability (in which case, this would be a 'subjective' prior probability). Iversen (1984) describes this process much more fully, for several hypothesis testing scenarios, including testing for mean and group proportion differences and for significance of correlations and regression coefficients.

It may help your understanding to describe the Bayesian perspective in the context of medical diagnosis. A doctor sees a patient showing a range of symptoms (the data). Her goal is to diagnose the illness based on the symptoms she observes. Medical research provides ample information on the likelihood of symptoms being observed given that one possible diagnosis might be cancer (*p(symptoms|cancer)*). However, the decision the doctor really confronts is judging the likelihood (posterior probability) of the patient having cancer given the symptoms observed, *p(cancer| symptoms)*, so that appropriate treatment can be prescribed. Bayesian inference requires the doctor to know (perhaps from previous epidemiological research) or guess at the prior probability of the specific type of cancer being considered occurring in the population in general (i.e. *p(cancer)*) in order to establish the desired posterior probability.

Suppose Maree wishes to employ Bayesian statistical inference for the *t*-test scenario described above. In order to do this, she would need to specify the prior probability for the null hypothesis being true in the population before her data were collected. Suppose she manages to find a published meta-analysis (recall *Procedure 8.8*) indicating a 40% chance of finding judgment accuracy differences between tertiary educated and high school educated people in a variety of situations. She decides that this result from the research literature provides for a reasonable estimate of the prior probability in her situation. Thus, for a 'no difference' null hypothesis, her prior probability becomes .6 (= 1.0 − .4). If the standard *p*-value from her *t*-test was .008, this would mean that

(continued)

> her posterior probability would be .6 times .008 or .0048. This means that, from a Bayesian perspective, the probability that there are no **educlev** differences, given the data she has collected, is .0048. Since this is a small probability, Maree would be justified in concluding that *there was only a very small chance that tertiary educated and high school educated inspectors did not differ in their inspection* **accuracy**.

One major reason why Bayesian statistical inference logic is not more widely applied in the social and behavioural sciences (it is somewhat more commonly applied in the business and economics disciplines and in some areas of psychology) is that many researchers object to having to specify the prior probability. To many, this smacks of excessive subjectivity in the hypothesis testing process since researchers often lack a very sound basis on which to establish the prior probability and must therefore resort to making a subjective estimate (see Lindman 1997 and Sohn 1998 for further discussion of this issue as well as of other potential problems associated with Bayesian statistical inference logic).

Bayesian statisticians argue that this is no more or less subjective than many other aspects of the statistical inference process (including the setting of the level of significance by 'convention') and results in a more appropriate test of hypotheses because the correct probability is being targeted. Another problem is that Bayesian methods are not uniformly accessible in the statistical packages typically used by social and behavioural science researchers, except for selected types of analyses, For example, SYSTAT offers Bayesian regression (Panik 2009); SPSS offers Bayesian options for posthoc multiple comparison tests in ANOVA and MANOVA, multinomial regression analysis, certain types of cluster analysis and time series analysis as well as a *Bayesian Stastics* analysis menu pathway; AMOS and MPlus offer the Bayesian Information Criterion (BIC) index as a measure of goodness-of-fit for structural equation models; AMOS offers Bayesian estimation and imputation methods for structural equation modelling; and **R** offers some approaches to Bayesian estimation of parameters (see the website https://a-little-book-of-r-for-bayesian-statistics.readthedocs.org/en/latest/src/bayesianstats.html for some examples).

The debate still rages on, with classical statistical inference still having a dominant hold in most disciplines, as reflected by the volume of research being published that relies on this logic. The spin-off effect of this is that students and fledgling researchers tend not to get exposed to Bayesian statistical inference logic in their undergraduate (or even postgraduate) training because it is not typically covered in the standard textbooks. Until this situation changes, it is unlikely that Bayesian statistical inference will achieve prominence in most social and behavioural sciences.

Procedure 9.1: Rasch Models & Item Response Theory

Classification	Multivariate; inferential; parametric.
Purpose	Rasch models provide an alternative (many researchers say more defensible) approach to item analysis that simultaneously estimates item difficulties and person abilities along the same interval-level measurement scale. Item response theory encompasses a more general set of analytical models for analysing items, including Rasch models as special cases, which permit the estimation of a range of item parameters such as difficulty, discrimination and guessing.
Measurement level	Items for Rasch model or item response theory analysis are typically measured at least at the ordinal-level (dichotomous – right/wrong; polytomous – several Likert-type rating categories).

Rasch Models

Rasch models are named after Georg Rasch, the Danish mathematician who derived the original model, and are designed specifically for the analysis of test and scale items (Bond and Fox 2015). They emerged from a measurement perspective that sought to construct scientifically valid interval-level measurement systems (where intervals between numbers on the measurement scale were exactly, not approximately equal) from data that were generally ordinal in structure. The theory behind Rasch models takes a very different approach to measurement processes from that taken in classical item analysis (recall *Procedure 8.1*). The theory builds on the idea that item difficulty and person ability are intimately linked: people of higher ability should have a higher probability of getting more difficult items correct.

Rasch models provide one of the few ways that a measurement scale can be defensibly transformed to a higher level of measurement from a lower level of measurement. This is especially important in tests of ability and achievement where the researcher wants to move from items scored as right or wrong to a more general scale that reflects a person's underlying ability or achievement level (sometimes called a 'latent trait'). Thus, an interesting benefit that emerges with Rasch models is the capacity to not only analyse items on a test but also each person's performance on the test in terms of their 'ability'. The Rasch model fitting process positions both items and persons along the same scale so that they can be compared. For test items, the comparisons between items are made based on item difficulty; for persons, the comparisons between people are made on the basis of person ability. Part of the Rasch model fitting process involves assessing the degree of fit (or misfit) between items and the expectations of the model and between persons and the expectations of the model. Misfitting items or persons are typically deleted from the model fitting process during scale refinement.

The most basic form of Rasch model is used for analysing tests where items (such as multiple choice, true/false) are scored on a correct/incorrect basis. This means that there are only two categories for response: the person gets an item right or gets it wrong. Note that Rasch modelling requires sophisticated statistical estimation procedures and complete Rasch analyses (where both items and persons are scaled and displayed) require specialised software, because the more commonly available packages (except for SYSTAT) do not have Rasch modelling capability.

Suppose Maree wishes to test the quality control knowledge and abilities of a random sample of 75 quality control inspectors. She designs a ten-item multiple choice test, each item offering five answer options, to accomplish this goal. Figure 9.1, produced using SYSTAT, shows what are called 'item characteristic curves' (SYSTAT calls them 'Latent Trait Model Item Plots') for each of the ten questions. Item characteristic curves are estimated from the observed responses to the questions and plot the proportion of people getting the item correct as a function of person ability.

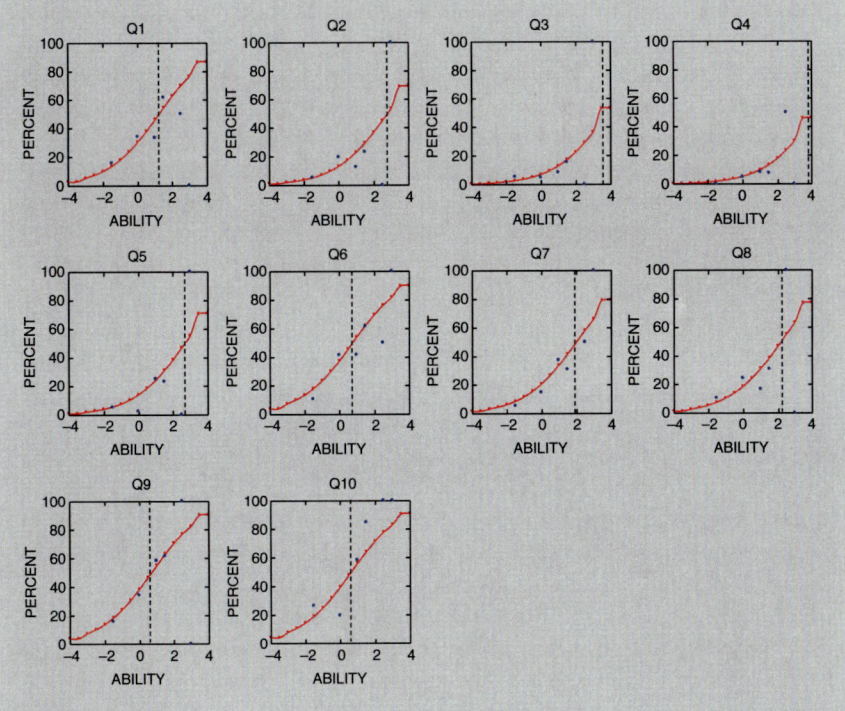

Fig. 9.1 Item characteristic curves, fitted by the Rasch model using SYSTAT (Latent Trait Model Item Plots)

(continued)

The Rasch model fitting process determines the shape of these curves; they all have a characteristic 'S-shape' to them (the full 'S-shape' may not show on the graph, depending upon the position of the curve). The position of each curve along the ability axis (X-axis) reflects the difficulty of the item; curves shifted to the right reflect more difficult items than curves shifted to the left. In a Rasch model, the numerical value for item difficulty is defined as the ability point on the item characteristic curve where 50% of people in the sample get the item right (dashed vertical line superimposed on each graph).

In Fig. 9.1, questions 3 and 4 (Q3 and Q4) are the most difficult items, requiring a much higher level of ability to have a 50% chance of getting them correct, whereas questions 6, 9 and 10 (Q6, Q9 and Q10) are the easiest items. It is important to note that the Rasch model requires the researcher to assume that all items have an equal level of discrimination or ability to differentiate between people of lower and higher ability. What this means is that all item characteristic curves, like those shown in Fig. 9.1, have the same slope, which, for each of these ten items, is .395. As we will see later on, it turns out that item Q2 actually does not fit the Rasch model very well because people are almost equally attracted to every answer option for the item, irrespective of their latent ability; this item would probably need to be deleted from the test upon subsequent revision. Thus, item Q2, in particular, does not meet the assumption of equal discrimination capacity, compared to other items.

If we use a more specialised software package, called MINISTEP (a evaluation version of the package called WINSTEPS, downloadable from the website www.winsteps.com/ministep.htm), we can produce a very clear comparison between items and persons along a common measurement scale.

Figure 9.2 shows this comparative item-person *map*. The 75 people in the testing sample are labelled according to unique initials, so that each person can be precisely located on the map. The map is vertically-ordered, using the measurement scale (in units called 'logits') on the left-hand side; from easier items/lower abilities toward the bottom to harder items/higher abilities toward the top. On the left side of the vertical axis on the map (the '+' signs), the items are listed in descending order of difficulty: Q4 is the hardest item (takes a person of very high ability to have a 50% chance of getting it correct) and Q10 and Q6 are equally the easiest items. On the right side of the map, people are listed, in descending order, at the level of their estimated latent ability, given their performance on the 10 questions. Here, it is clear that inspector EM shows the highest estimated ability on the test and inspectors AA and AH joined a group of 19 inspectors having the lowest estimated ability levels.

(continued)

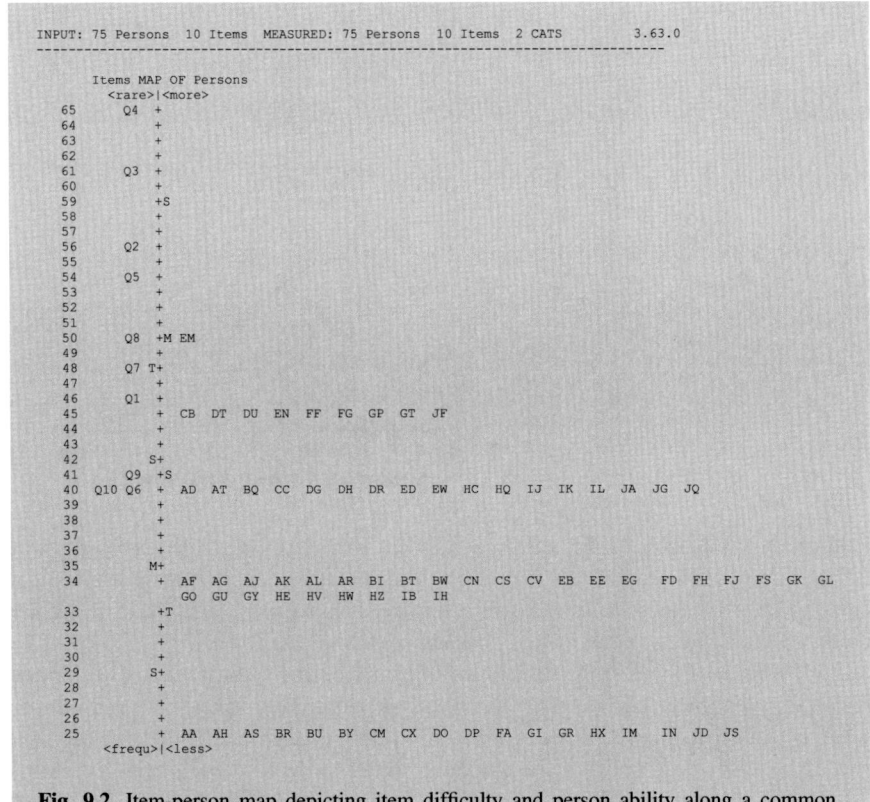

Procedure 9.1: Rasch Models & Item Response Theory

```
INPUT: 75 Persons  10 Items   MEASURED: 75 Persons  10 Items  2 CATS      3.63.0
-------------------------------------------------------------------------------
        Items MAP OF Persons
              <rare>|<more>
   65       Q4  +
   64           +
   63           +
   62           +
   61       Q3  +
   60           +
   59           +S
   58           +
   57           +
   56       Q2  +
   55           +
   54       Q5  +
   53           +
   52           +
   51           +
   50       Q8  +M EM
   49           +
   48       Q7  T+
   47           +
   46       Q1  +
   45           +   CB  DT  DU  EN  FF  FG  GP  GT  JF
   44           +
   43           +
   42           S+
   41       Q9  +S
   40   Q10 Q6  +   AD  AT  BQ  CC  DG  DH  DR  ED  EW  HC  HQ  IJ  IK  IL  JA  JG  JQ
   39           +
   38           +
   37           +
   36           +
   35           M+
   34           +   AF  AG  AJ  AK  AL  AR  BI  BT  BW  CN  CS  CV  EB  EE  EG  FD  FH  FJ  FS  GK  GL
                    GO  GU  GY  HE  HV  HW  HZ  IB  IH
   33           +T
   32           +
   31           +
   30           +
   29           S+
   28           +
   27           +
   26           +
   25           +   AA  AH  AS  BR  BU  BY  CM  CX  DO  DP  FA  GI  GR  HX  IM  IN  JD  JS
              <frequ>|<less>
```

Fig. 9.2 Item-person map depicting item difficulty and person ability along a common measurement scale, fitted by the Rasch model using MINISTEP

A more complex multiple category (polytomous) version of Rasch model is also available (see Andrich 2005; Ostini and Nering 2006), which can address one of two types of problems:

- analysis of rating scale items where several scale points are available (as with Likert-type scales, where there is no 'correct' answer); and
- analysis of items where partial credit for answers is possible (thus offering more than two categories for correctness/incorrectness).

Figure 9.3, also produced using MINISTEP, shows what is called a 'category characteristic curve' for a 5-point agree/disagree Likert-type attitude item. Category characteristic curves are estimated from the observed proportions of choices for each category on an item's scale. Here, what is being estimated by the Rasch modelling process is not latent ability or achievement but latent attitude (note that the X-axis label in Fig. 9.3 is a bit misleading in this regard – the axis scales attitude intensity and direction, rather than item difficulty). People who have a greater latent positive

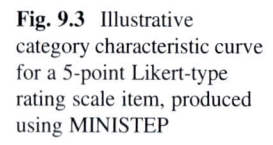

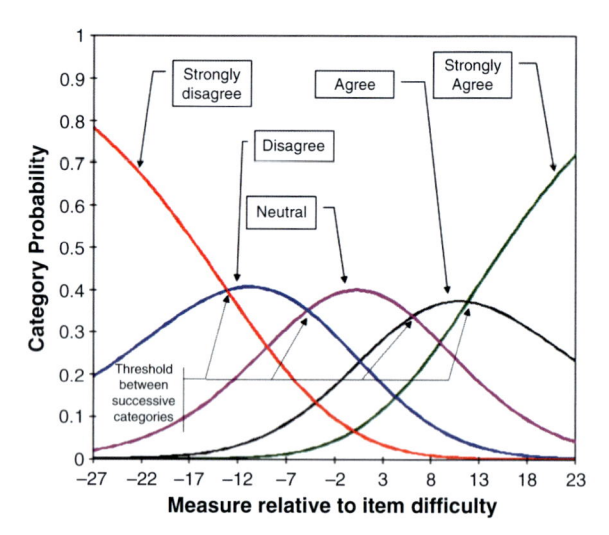

Fig. 9.3 Illustrative category characteristic curve for a 5-point Likert-type rating scale item, produced using MINISTEP

attitude on the item, as reflected by agreement, will have a greater probability of falling in one of the right-hand curves on the scale and people with a greater latent negative attitude on the item, as reflected by disagreement, will have a greater probability of falling in one of the left-hand curves on the scale.

Note that there are five curves plotted in Fig. 9.3, one curve for each of the five response categories (solid labelled arrows). The intersection of two curves produces what is called a *threshold*, which identifies the conceptual boundary between successive rating categories (superimposed dotted arrows). A well-behaved multi-categorical rating scale will have thresholds that increase in orderly fashion from left to right. If the thresholds are equally spaced, then the item demonstrates a true interval scale. Figure 9.3 shows only an approximate interval scale, because the interval is wider between the two thresholds around the 'Neutral' scale point.

More General Item Response Theory Models

It turns out that Rasch models comprise a subset of a more general theoretical perspective called *item response theory* (IRT; sometimes called *latent trait models*, see, for example, Baker 2001; de Ayala 2009; Henard 2000). In item response theory, we are still interested in analysing items (via item characteristic curves) and estimating person abilities but using a more general set of logistic equation models. Logistic item response theory models can involve the estimation of up to three parameters per item: *item difficulty, item discrimination* and *guessing*, respectively, in order of theoretical importance. Rasch models focus on estimating only the first parameter, item difficulty. For this reason, Rasch models are often called *one-parameter models*.

Procedure 9.1: Rasch Models & Item Response Theory

The item discrimination parameter refers to an item's sensitivity to reflecting differences in ability. An item with a low discrimination parameter will have a hard time clearly differentiating between people of low and high ability. In an item characteristic curve, the item discrimination parameter influences the slope of the middle of the S-shaped curve—curves with shallower slopes are less discriminating; curves with steeper slopes are more discriminating. If both item difficulty and item discrimination parameters are being estimated in an item response model, we have what is called a *two-parameter model*. It is important to know that Rasch models assume that all items have the same item discrimination parameter (thus, it is a constant in the model). If such an assumption is untenable (may be reflected, for example, by large numbers of misfitting items), then a Rasch model is not the proper model to fit the data, and the two-parameter model will be more appropriate. In an item characteristic curve estimated using a two-parameter model, difficulty will be reflected in the horizontal positioning of the curve and discrimination will be reflected in the vertical steepness of the curve.

Figure 9.4, produced using SYSTAT, shows the outcomes of fitting a two-parameter logistic model to Maree's ten-question quality control test. Note that item difficulty is still defined the same way (the ability level where 50% of people get the question right) and is reflected in the same way on these

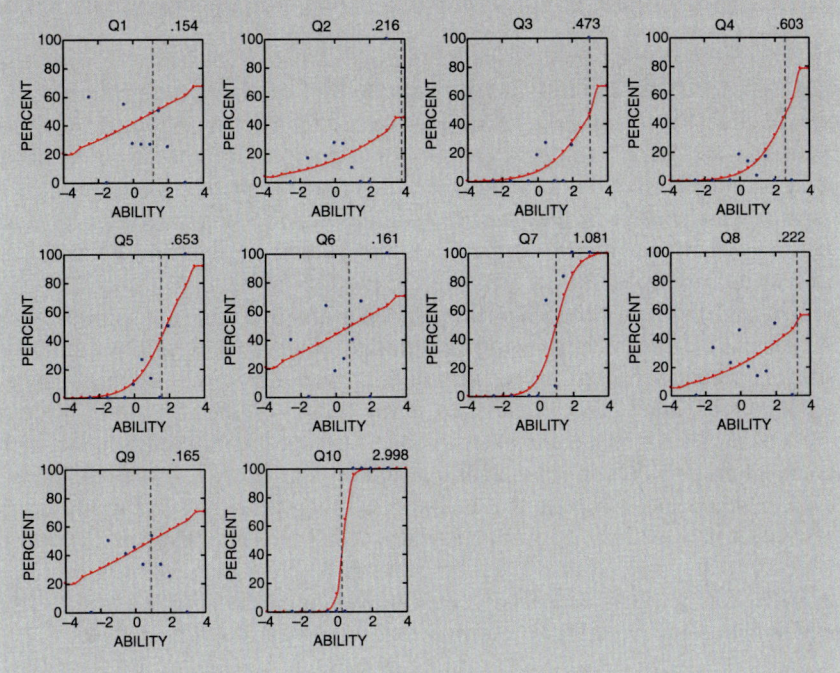

Fig. 9.4 Item characteristic curves, fitted by the two-parameter logistic model using SYSTAT

(continued)

graphs as the position of the curve along the ability axis (see the superimposed vertical dashed lines). However, each curve now has a different slope or tilt to it, reflecting the influence of the estimated item discrimination parameter (the estimated discrimination value for each item is superimposed on the top right of each plot in Fig. 9.4). Items Q1, Q2, Q6, Q8 and Q9 all had rather flattish slopes, indicating these items did not discriminate well between inspectors of lesser and greater ability (item Q1 least of all with a discrimination parameter estimate of .154). Item Q10 had the highest discrimination parameter estimate at 2.998, reflecting an extremely sharp capacity to differentiate between inspectors of lower and higher ability, followed by item Q7.

It should be noted that because the two-parameter model simultaneously estimates the discrimination parameter for each item, item difficulty may differ, sometimes dramatically, from what a Rasch model would have estimated, particularly if the discrimination parameter for the item is quite different from other items. Item Q2, for example, shows a much higher level of estimated item difficulty compared to its Rasch model estimated value in Fig. 9.1. The low discrimination level for item Q2 is thus one reason why the item did not fit the Rasch model very well. The fact that these 10 questions differed so dramatically in discrimination suggests that the Rasch model may not have been the best model to fit to the data.

There is also a three-parameter model that one can estimate (see the discussion in Baker 2001), but this requires very sophisticated software. A three-parameter item response model involves the simultaneous estimation of item difficulty, item discrimination and guessing parameters. Guessing refers to the chances of someone getting an item correct by guessing, i.e. without having the knowledge needed to reason out an answer to the question. Multiple choice and true-false tests are notorious for having known probabilities for guessing (for true-false tests, the guessing probability is .50 (or 50%) = a coin toss; for multiple choice tests, the guessing probability is 1 divided by the number of answer options—so for a 5-option multiple choice item, the guessing probability would be 1/5 or .20 (or 20%)). Through judicious use of sound item writing and test composition strategies, guessing probabilities can be influenced downward, but if they are still considered to be a factor, then the three-parameter model is technically the correct model to fit. Note that Rasch models do not include a parameter for guessing. Therefore, fitting a Rasch model means accepting the assumption that guessing does not play a substantive role in overall performance on the test. Bond and Fox (2015, ch. 13) present an interesting discussion of the relative benefits of Rasch models and two- and three-parameter item response models and they provide a compendium of useful references to help people more fully inform themselves about Rasch modelling.

Advantages

Rasch models offer a powerful alternative to classical item analysis. Its advantages include the fact that Rasch models yield item difficulties and ability measures indexed along on the same scale and produce genuine interval-level scales as a result. This capacity is why Bond and Fox (2015) refer to the Rasch model as a fundamental measurement model. By ordering items according to difficulty and people according to ability, decisions about test composition (is the test too easy or too hard?) and ability ranking (which people demonstrate higher or lower levels of the latent ability or achievement learning trait?) become easier to make and more publicly defensible at the same time. Rasch models are useful for analysing not only dichotomously scored items (right/wrong) but also for analysing items where partial credit is allowed, and attitude items measured using Likert-type scales. The advantage of using Rasch models to analyse Likert-type scales is that the researcher need not simply assume that the measurement scale is interval-level; the Rasch model ensures they are interval-level.

Two-parameter models, while less commonly utilised, offer some advantages over Rasch models in that both item difficulty and item discrimination are estimated. Such models would be most useful in circumstances where the assumption of equal item discrimination, required by Rasch models, is untenable. If guessing is a problem, then three-parameters can correctly incorporate this into the statistical estimation of the model.

Finally, the development of the Rasch model in the context of item response theory has facilitated the evolution of a testing technology called *computer-adaptive testing* (*CAT*; see, for example, Magis et al. 2017). CAT refers to a type of testing where the items for presentation to the test taker are automatically and sequentially sampled from a large item bank containing Rasch-scaled items, conditional upon the test taker's response to the current item. If the test taker gets an item correct, a more difficult item is then sampled and presented; if or she gets an item wrong, a less difficult item is then sampled and presented. The process continues in this iterative fashion until a stable estimate of latent ability or achievement is produced. This process usually results in a shorter test (in terms of item numbers) than a standard fixed-format test.

Disadvantages

Rasch models are accompanied by some important disadvantages and limitations. First and foremost, Rasch models assume that the collection of items being analysed is unidimensional; that is, that a single underlying ability factor (recall *Procedure 6.5*) characterises the latent (unobservable but intended) structure of the items. Item response theory models, of whatever complexity (Rasch, two-parameter; three-parameter), use a model fitting logic that is the reverse of the normal statistical model fitting process. Normally, statistical models are fitted to the data and are then evaluated for completeness as well as for deficiencies. Item response modelling fits the data to the statistical model. Many researchers have a problem with this logic which discards items and people from a statistical model on the basis that they don't fit the model (rather than modifying the model to better fit the data). For many researchers, item response theory is also much harder to grasp conceptually than is classical item analysis; a fact that has probably inhibited wide-spread adoption outside educational contexts.

To properly fit data to a Rasch model, a relatively large item pool is needed (20 or more is the normal minimum prescribed, but some say an item bank containing upwards of 100 items or more is desirable) as well as a large number of participants (perhaps up to 1000 or more). This is to provide stability in the statistical estimation process as well as to allow for misfitting items and persons to be identified and deleted from the modelling process. CAT is especially dependent upon a stable and large item bank. The advantages of estimating a two-parameter or three-parameter model come with some considerable computational complexity and costs and sensible solutions may not always be achieved. Furthermore, such models generally require highly specialised and more complex (from a user perspective) software to produce sound statistical estimates, more complete diagnostics and useful outcome displays.

Where Is This Procedure Useful?

Rasch models and other item response theory models are most commonly used for the construction and analysis of ability, knowledge and achievement tests in virtually any discipline or context. Some authors (e.g. Bond and Fox 2015; Ostini and Nering 2006) have argued that item response theory models, including Rasch models, should replace classical item analysis for Likert-type scales. Their reasoning here is two-fold:

- it is better to explicitly test and rescale items on a test to have interval properties rather than to simply assume that the scale has interval properties as is common practice now; and
- item response theory models analyse both items and persons simultaneously, yielding much more information, compared to classical item analysis, which tends to focus solely on items. Item response theory models can also be useful where the researcher wishes to obtain a very clear and defensible positioning of persons along a latent ability dimension.

Software Procedures

Application	Procedures
SPSS	SPSS does not currently have built-in capacities to conduct either Rasch model analyses or item response theory analyses. [Note, however, that there are some SPSS extension commands (for version 17 of SPSS and higher) that can be downloaded, which can permit SPSS to conduct various types of item response analyses; see https://www.ibm.com/support/pages/item-response-theoryrasch-models-spss-statistics. Most of these extensions require **R** and/or Python plug-ins and you need to have access to the SPSS Programmability module – not feasible for most users.]
NCSS	*Analysis* → *Descriptive Statistics* → *Item Response Analysis* and select your desired 'Item Variables'; you must also have a variable containing the correct responses for the selected items (NCSS can only fit a dichotomous two-parameter model for ability or achievement tests; it cannot fit polytomous, categorical or partial credit models); under the *Reports* tab, select the output reports you want to see (e.g. select 'Abilities report' if you want the list of all people in the sample along with their estimated ability score) and under the *Plots* tab, indicate that you want to see the item response plots. NCSS cannot fit a one-parameter Rasch model or a three-parameter model. Also, NCSS cannot produce the side-by-side item-person map shown in Fig. 9.2.
SYSTAT	*Analyze* → *Advanced* → *Test Item Analysis* → *Logistic...* and select the desired 'Test Item(s):' for analysis. For multiple choice items, you must specify a correct response for each item selected; for a Rasch model analysis, choose the 'One-parameter logistic' *Model Option*; to fit a two-parameter model, choose the 'Two-parameter logistic' *Model Option*. By default, SYSTAT will produce a list of all test respondents along with their estimated ability score. SYSTAT cannot fit polytomous Rasch models, so it is only useful for ability and achievement test item analysis. Also, SYSTAT does not provide the range of statistical and display outcomes available using more specialised packages. For example, SYSTAT cannot produce the side-by-side item-person map shown in Fig. 9.2. SYSTAT cannot fit three-parameter models.
STATGRAPHICS	STATGRAPHICS does not currently have capacities to conduct either Rasch model analyses or two- or three-parameter IRT model analyses.
R Commander	**R** Commander does not currently have capacities to conduct either Rasch model analyses or two- or three-parameter IRT model analyses. However, there are packages available for the main **R** program which can carry out item response analyses (see, for example, the *ltm* package, described by Rizopoulos (2006), which fits all types of IRT dichotomous and polytomous models, and the *psych* package, described by Revelle (2019), which fits only the two-parameter dichotomous IRT model and links IRT with factor analysis).
MINISTEP/ WINSTEPS	A self-contained Rasch modelling software package. It can read SPSS data files, but also has its own data entry capabilities. Only Rasch (one-parameter) models can be fitted by this package; both dichotomous and polytomous models can be fit. The package produces a very extensive range of statistics, diagnostic indices and plots and graphical displays. MINISTEP is the evaluation version, relatively easily learned, which can be

(continued)

Application	Procedures
	downloaded from the website: www.winsteps.com/ministep.htm. WINSTEPS is the full version of the software, which can be purchased.
RUMM	RUMM (**R**asch **U**nidimensional **M**easurement **M**odels) is another self-contained package for conducting Rasch analysis and fitting both dichotomous and polytomous models. A trial version can be downloaded from the website: www.rummlab.com.au. RUMM takes a bit more technical knowledge to conduct a Rasch analysis than does MINISTEP.

Procedure 9.2: Survival/Failure Analysis

Classification   Bivariate/multivariate; descriptive/inferential; parametric.
Purpose   To describe, analyse and possibly predict 'time to event' data based on group comparisons or continuous predictors or covariates.
Measurement level   The 'time to event' data are generally initially measured at the interval- or ratio-level; group membership is recorded at the nominal- or ordinal-level; continuous predictors and covariates are typically measured at the interval/ratio-level.

Survival/failure analysis is a set of procedures designed to help researchers understand what predicts 'time to event' or survival data (see, for example, Wright 2000; these methods also go by the name *event history analysis*, see Allison 2004). 'Time to event' can be defined in any number of ways. Medical researchers may define 'time to event' as survival time from diagnosis of a disease until patient death. Quality control researchers may define 'time to event' as the time it takes for specific parts to fail in a machine or process (thus, constituting 'failure time'). Educational researchers may define 'time to event' as time spent in enrolled in school until dropping out (before completion of year 12). Occupational health and safety researchers may be interested in how long it takes for an accident to occur in a specific context. In each case, the key feature of the dependent variable is elapsed time until some specified event occurs.

What is different about the analysis of survival/failure data is the fact that some observations may not actually reach the 'event' during the course of an investigation (a patient may not die; a part may not fail; a student may not drop out; an accident may not occur). Such observations are termed 'censored' (technically, 'right-censored' because the right end-point on a horizontal timeline has not been observed; if the event occurred before the data were collected, the observation would be 'left-censored'). Right-censored observations still constitute viable data, but only in a more general sense since we cannot know when the event would actually be reached by the observed entity; we only know they did not reach the event during the time the data were collected.

Procedure 9.2: Survival/Failure Analysis

What we are often interested in with survival/failure analysis is which independent variables predict survival or failure rates. We may also have covariates of interest and some of the independent variables may code for different groups or for a treatment intervention. Thus, using survival/failure analysis, we can evaluate experimental or quasi-experimental interventions (such as training programs, medical treatments, advertising campaigns) designed to enhance survival rates or reduce failure rates.

While there are several different types of survival/failure analysis (e.g. Life Tables; Kaplan-Meier), we will focus on a version that is conducted using a variation of multiple regression analysis (see Tabachnick and Fidell, ch. 11). The specific procedure we will explore is called *Cox regression* and it permits the evaluation of covariates as well as treatment interventions and other independent variables. Life tables display the number of people surviving to specific successive points in time. Such data can then be used to produce a descriptive graph, the 'survival function', for displaying the probability of survival as time progresses and the survival rates for different groups or treatments can also be displayed.

> Suppose Maree received an additional grant for her study of quality control inspectors from industry unions. The unions were interested in turnover rates (which were extremely high amongst quality control inspectors) and how they might be influenced by a new mentoring program as well as by training offered by the companies and inspectors' previous experience in the industry. Maree conducts an intervention study where 67 randomly selected inspectors are tracked, from the point at which they are hired, over a subsequent 24-month period.
>
> Of the 67 inspectors, 32 were initially randomly assigned to a mentoring program where they were paired with a senior quality control inspector in the company (a volunteer) for the first four weeks of their employment. The remainder of the inspectors were not assigned to a mentor. Maree gathered data on each inspector in terms of the number of years of experience they had in their specific industry and the number of training days they experienced while working with their current company. Maree also recorded the number of weeks that each inspector remained with the company—this measure served as the survival dependent variable for her analysis, combined with a dummy variable indicating whether an inspector had left the employment of their company (= 1) during the time period of the study or remained employed by their company at the time the study concluded (= 0). By the conclusion of her study, four inspectors remained employed with their respective companies, thus giving four right-censored observations.
>
> Maree employs the *SPSS Cox Regression* procedure to analyse the number of weeks remaining with the company. Maree first has a descriptive look at the survival data for the 67 inspectors, plotting a separate survival function for

(continued)

Fig. 9.5 SPSS Survival function plot for mentored and unmentored quality control inspectors

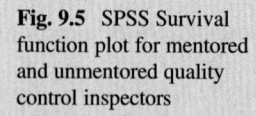

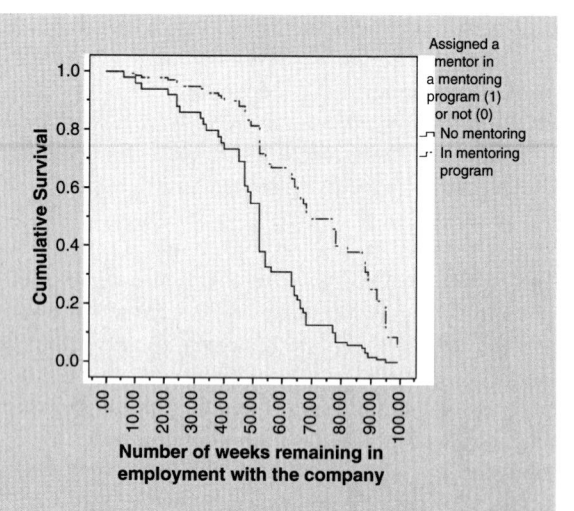

unmentored (solid line) and mentored (patterned line) inspectors. Figure 9.5, produced using the *SPSS Cox Regression* procedure, shows the survival function plot that compares the two groups of inspectors directly. The figure clearly depicts a difference in survival rates for the two groups of inspectors. The plot looks like a pair of raggedly descending staircases because what is being plotted is the cumulative proportion of people still left (i.e. 'surviving') in employment with their company as weeks pass. It appears as if the mentoring program has had an effect in terms of increasing the length of time that inspectors stay with their respective companies. The slopes of the 'staircases' are somewhat different suggesting to Maree that mentoring probably not only influenced overall survival rate, but also the rapidity with which survival rate declined (the curve for mentored inspectors was less steep, signalling a slowing of the rate at which inspectors left employment with their current company). Maree then needed to evaluate the mentoring effect, taking into account the other information she had to hand about each inspector.

The Cox regression procedure allows for a hierarchical regression analysis of the survival rates data. Maree entered the number of training days (**N-training_days**) and years of previous experience in the industry (**past_ind_exp**) as covariates in the first step in her analysis, to control for any effects they might have. Then she tested for the **mentoring** effect itself in the second step of the analysis. Table 9.1 ('Block 1') and Table 9.2 ('Block 2') show selected outcomes from the two stages in this hierarchical analysis. [Note that these outcomes resemble those from logistic regression analysis (recall *Procedure 7.14*). This is because survival analysis works by transforming the original time-based variable into proportions, then analysing any predictive patterns that emerge.]

(continued)

Procedure 9.2: Survival/Failure Analysis

Table 9.1 SPSS Cox regression analysis for mentored and unmentored quality control inspectors – Block 1

Block 1: Method = Enter [the covariates: **N_training_days**; **past_ind_exp**]

Omnibus Tests of Model Coefficients[a]

-2 Log Likelihood	Overall (score)			Change From Previous Step			Change From Previous Block		
	Chi-square	df	Sig.	Chi-square	df	Sig.	Chi-square	df	Sig.
423.591	9.763	2	.008	10.015	2	.007	**10.015**	**2**	**.007**

a. Beginning Block Number 1. Method = Enter

Variables in the Equation

	B	SE	Wald	df	Sig.	Exp(B)
N_training_days	-.150	.050	8.904	1	.003	.861
past_ind_exp	.051	.062	.680	1	.410	1.052

Table 9.2 SPSS Cox regression analysis for mentored and unmentored quality control inspectors – Block 2

Block 2: Method = Enter [the independent/treatment variable: **mentoring**]

Omnibus Tests of Model Coefficients[a]

-2 Log Likelihood	Overall (score)			Change From Previous Step			Change From Previous Block		
	Chi-square	df	Sig.	Chi-square	df	Sig.	Chi-square	df	Sig.
408.659	23.736	3	.000	14.933	1	.000	**14.933**	**1**	**.000**

a. Beginning Block Number 2. Method = Enter

Variables in the Equation

	B	SE	Wald	df	Sig.	Exp(B)
N_training_days	-.222	.056	15.514	1	.000	.801
past_ind_exp	.039	.062	.392	1	.531	1.040
mentoring	-1.073	.279	14.847	1	.000	.342

'Block 1' (Table 9.1) evaluates the covariates entered together on the first step. The 'Omnibus' table (focusing on the numbers in boldface) showed that the covariates explained a significant amount of variance in survival rates over and above no predictors at all ($\chi^2(2) = 10.015, p = .007$). The 'Variables in the Equation' table provides a closer look at which of the covariates was doing most of the explanatory work and it was clear that only the number of training days (statistics in boldface) significantly predicted survival rates (B = -.150, Exp(B) = .861, Wald statistic (1) = 8.904, $p = .003$). The Exp(B) value of .861 can be interpreted as showing that the odds were slightly more than 1:1 of an inspector not leaving employment with their current company (surviving) with an increasing number of training days (because B is negative, to get this interpretation, Maree had to take the reciprocal of .861 which is 1/.861 = 1.16).

'Block 2' (Table 9.2) tests for the effect of **mentoring** over and above what the covariates could account for. The 'Omnibus' table revealed that adding the **mentoring** variable into the model added a significant amount of explanatory

(continued)

power ($\chi^2(1) = 14.933$, $p < .001$) over and above what the covariates could explain. The 'Variables in the Equation' table provides a closer look at the **mentoring** effect; **mentoring** significantly predicted survival rates (B = −1.073, Exp(B) = .342, Wald statistic (1) = 14.847, $p < .001$). The Exp (B) value of .342 can be interpreted as showing that the odds were almost 3:1 of an inspector not leaving employment with their current company (surviving) if that inspector participated in the mentoring program (again, the negative B coefficient meant that Maree had to take the reciprocal of Exp(B) to get to this interpretation: 1/.342 = 2.92). Thus, Maree concluded that, when past experience and training days were controlled for, the mentoring of quality control inspectors in their first four weeks of employment significantly increased the chances that inspectors would remain longer with their company.

Advantages

The major advantage of survival/failure analysis is that the impact of independent variables on survival or failure rates can be directly tested, even in cases where some of the observations are censored. The effect of covariates and treatment variables can be separately assessed and interpreted in a manner very similar to logistic regression. Survival function plots provide a useful visual display of treatment group differences in survival rates. Horizontal displacement of survival curves indicates probable treatment effect; non-parallel slopes in the survival functions for different groups indicate differential declines in survival rates over time.

Disadvantages

Survival/failure analysis, via the Cox regression method, is subject to the same assumptions that apply for logistic regression. Multicollinearity amongst the covariates can be a real problem for survival/failure analysis. Tabachnick and Fidell (section 11.3.2.7) suggest that survival/failure analysis is extremely sensitive to small sample sizes, because of the statistical estimation algorithm used. The 10 to 1 ratio of cases to covariates is a useful guide to requisite sample size. Finally, the nature of the research design itself can influence the outcomes from a survival/failure analysis. If the study is undertaken over a lengthy period of time, one needs to be careful that the conditions that may be influencing survival or failure are not changing as the study progresses. This becomes more of a problem the longer the time horizon is for the study. Finally, if there is something systematically different about what causes the censored observations compared to the uncensored observations, and this systematicity is not captured in the explanatory variables, then the applicability of the survival/failure analysis can be severely compromised.

Where Is This Procedure Useful?

Survival/failure analysis is useful in any context where 'time to event' data are available. The procedure is especially useful in disciplines such as medicine, nursing and health, psychology, education, public policy and management and quality control where events such as deaths and injuries, bankruptcy, accidents, resignations/sackings, compliance, relapses after therapy or part/process failures/faults are recorded. Sometimes all that is of interest is actual survival rates; sometimes what is really of interest is what helps us to understand and predict those survival rates. Survival/failure analysis can be of great assistance in both respects. Treatment effects can be explicitly evaluated in experimental and quasi-experimental designs.

Software Procedures

Application	Procedures
SPSS	*Analyze → Survival → Cox Regression...* and select the 'Time:' variable, the 'Status:' variable (a variable that indicates whether or not an observation is censored – you also have to tell SPSS which code is used to 'Define Event...') and one or more 'Covariates' (covariates may include a categorical variable such as an indicator of group membership and may be entered in hierarchical blocks); select the *Categorical* button and choose which of your covariates is a categorical variable (the coding method is also selectable – 'Indicator' means dummy-coded); it is also possible to produce a survival function plot by selecting the *Plots* button and ticking the 'Survival' box, then in the 'Covariate Values Plotted at:' box, select the categorical variable that defines the groups to be compared and move that variable across into the 'Separate Lines for:' box. This will produce a plot similar to Fig. 9.5. SPSS offers three other types of survival analysis as well: *Life Tables...*, *Kaplan-Meier...* and *Cox w/ Time Dep Cov...* (Cox regression with time-dependent covariates).
NCSS	*Analysis → Survival/Reliability → Cox Regression* and select the 'Time Variable:', the 'Censor Variable:' (and the code indicating the failure event), the 'Xs: Numeric Independent Variables:', and any 'Categorical Independent Variables:'; under the *Model* tab, choose your desired approach to building the model ('Hierarchical Forward' builds the model adding one variable at a time, in the order you list them under the *Variables* tab, but starting with the categorical variables; 'None' does simultaneous entry of all predictors); under the *Reports* and *Plots* tabs, choose what you want to see produced.

(continued)

Application	Procedures
	Analysis → Survival/Reliability → Life Tables Analysis and select the 'Time Variable:', the 'Censor Variable:' (and the code indicating the failure event), and a 'Group Variable:'; under the *Reports* and *Plots* tabs, choose what you want to see produced — the survival plot is produced by default in this analysis. NCSS also offers *Kaplan-Meier Curves (Logrank Tests)* as a survival analysis method.
SYSTAT	*Analyze → Advanced → Survival Analysis... → Parametric and Cox* and select the 'Time:' variable, any 'Covariate(s):' of interest, the 'Censor Status:' variable and the 'Strata:' variable (used to identify a categorical grouping variable); under the *Tables and Graphs* tab, choose the tables and graphs you want to see (the 'Survivor Function' graph is default, similar to Fig. 9.5). Hierarchical entry of covariates is not possible in SYSTAT (only Stepwise and Complete (= simultaneous) predictor entry methods are offered). SYSTAT also offers a nonparametric version of survival analysis.
STATGRAPHICS	*Relate → Life Data → Parametric Models...* and choose your 'Dependent Variable:', the '(Censored:)' variable, the 'Quantitative Factors:' you want to test and the 'Categorical Factors:' you want to test. STATGRAPHICS can only build simultaneous entry Cox regression models. STATGRAPHICS can also do Life Table analysis, which can produce a survival function plot.
R Commander	**R** Commander does not currently have any survival analysis capability. However, there are packages available for use with the main **R** program, which can carry out survival analyses of various sorts. For example, Deitz (2013) describes how such analyses can be accomplished within **R**.

Procedure 9.3: Quality Control Charts

Classification	Univariate/bivariate/multivariate; descriptive/inferential; parametric.
Purpose	To monitor quality control measurements over time and conditions and provide guides for judging whether product or service quality is meeting standards or expectations.
Measurement level	Quality control measurements and observations are generally obtained using interval- or ratio-level scales.

Quality control is a concern in many manufacturing and service-oriented organisations. Quality control charts and the statistics and relationships they display have been developed to help researchers and decision makers track measurements relevant to the quality control of products and services over time to identify where, how often and under what conditions processes get 'out of control' (i.e. fall below quality standards or customer expectations). A wide variety of quality control statistics and charts have been developed (see, for example, Adhikari et al. 2009a; Benneyan 1998; Black et al. 2007), each designed to serve a different purpose and to inform different types of decisions. For illustration purposes, we will examine three specific types of quality control charts: Pareto charts, Control X-bar charts and R (or Range) charts.

Pareto Charts

Pareto charts (named after a well-known economist) are designed to show the proportion or percentage of outcomes of various types that have been judged, using some type of quality control process, to be defective, below standard or rejected/discarded. Pareto charts are simply a variation on the standard bar graph (recall *Procedure 5.2*) that allows one to visually identify where most of the quality assurance problems appear to be occurring. There is a general maxim, often reflected in Pareto charts, that about 80% of problems with quality come from about 20% of the components in the system. Note that economists have even argued that this maxim applies to wealth in certain types of economies: 80% of wealth tends to be concentrated in the hands of 20% of the population – what has come to be known as the 'Pareto's principle' or the '80-20 Rule'.

> Suppose Maree was interested in studying the quality control processes in more depth within the PC manufacturing industry. She conducts a study where she obtains counts, over a 16-week period, of the number of computer part rejection decisions made by quality control inspectors. Maree records the number of rejections for each of seven categories of PC parts (these part-type categories coincide with seven distinct divisions within the industry). Figure 9.6, produced using NCSS, shows what a combined Pareto chart of the outcomes from the study might look like. The bar graph in Fig. 9.6 shows a standard or absolute version of the Pareto chart. Each bar indicates the total count of part rejection decisions over the 16 observations periods plotted against the category of computer part concerned.
>
> Note that the part categories are arranged along the X-axis in descending order of total number of rejections (this ordering is a standard feature of Pareto charts). This means that the most problematic areas automatically appear toward the left of the graph. Absolute Pareto graphs typically have a form like a descending staircase; Fig. 9.6 clearly shows that the bulk of parts rejection decisions concerned Peripheral Devices (like keyboards, speakers and mouse), Power & Cooling components or Monitors. Cases & Covers was the least problematic parts category.
>
> The line graph superimposed on Fig. 9.6 presents the cumulative version of the Pareto chart; it shows the cumulative percentage of part rejection decisions across the part categories. The line graph clearly shows that 78% of rejection

(continued)

Fig. 9.6 Combined absolute/cumulative Pareto chart for number of rejected parts by part type

Pareto Chart of Total of n_rejects

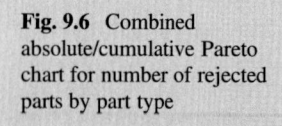

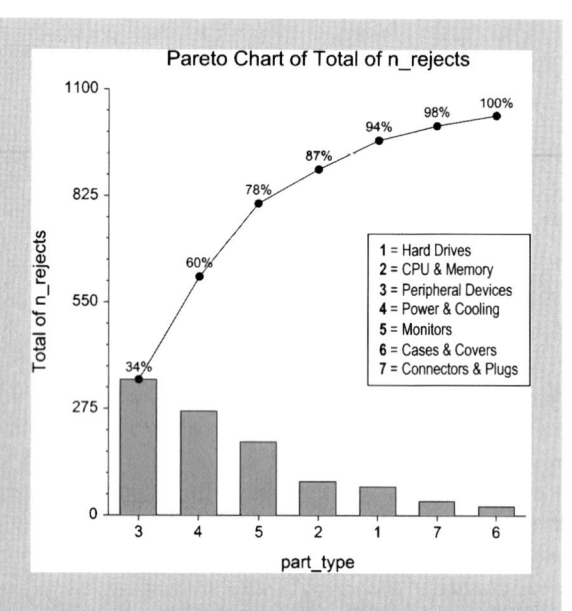

1 = Hard Drives
2 = CPU & Memory
3 = Peripheral Devices
4 = Power & Cooling
5 = Monitors
6 = Cases & Covers
7 = Connectors & Plugs

decisions concerned just three of the seven parts categories: Peripheral Devices (like keyboards, speakers and mouse), Power & Cooling components or Monitors. Consistent with the general Pareto principle (80/20 rule), the majority of quality problems were concentrated in a small proportion of the total manufacturing system. Using this combined chart, Maree could identify key target areas for quality improvement interventions.

Control X-Bar Charts

We have seen that Pareto charts can help localise quality process problems through counts of the number of problems observed in different areas. In contrast, the *Control X-bar chart* is designed to display a different kind of data, namely quality measurements obtained over time. An X-bar chart is very similar in structure to a time series plot (recall *Procedure 8.5*) but with additional information superimposed on it. An X-bar chart, as its name suggests, also displays the mean of the series of quality measurements over time as well as the upper control limit (UCL) and lower control limit (LCL) for the process. Where the LCL and UCL are set is entirely under the control of the researcher or manager; the limits might reflect objectively established physical tolerances, standard error bands (similar to confidence intervals; recall *Procedure 8.3*) or an industry-level agreed maximum and minimum standards for quality. In many cases, the LCL is the critical limit for quality control managers to attend to – where quality drops below some minimum acceptable level.

Control X-Bar Charts

With such information superimposed on the quality measurement time series, it becomes very easy to see how quality varies around the average and isolate where specific observations wander outside the UCL or LCL. Such 'wandering' observations signal situations or observations where the process is 'out-of-control'. In other words, observations falling outside the LCL or UCL indicate where quality of outcome has failed to meet the required standard. The researcher or manager can then dig more deeply into the 'out-of-control' observations to try and identify why they were out of control as the first step toward planning quality improvement interventions. Additionally, the observed mean of the series can be compared to an established standard, preset level or expectation, to give an overall indication of whether the performance of the quality control system is up to par.

> In the context of Maree's study of quality control processes in PC manufacturers, suppose she also collected data on the accuracy of quality control inspection decisions, made by a quality control inspector, 'Joe', each day over a 90-day period, within one PC manufacturing company. The PC industry was moving to adopt a new standard of 90% accuracy in quality control decisions and 3-sigma (standard deviation) limits, where sigma is set to 5 percentage points, around this standard. One goal of the investigation was to see how close the standard was to being achieved in this one company. Obviously, there is an upper limit ceiling of 100% imposed by the nature of the accuracy measure itself which means that the desired quality standards are an average level of accuracy of 90% and all variations from this level must fall between an LCL of 75% (90 – 3 times a sigma of 5) and a theoretical UCL of 100% (the maximum possible score).
>
> Figure 9.7, produced using NCSS, shows the resulting X-Bar chart for the observations Maree gathered in this research context. Joe's performance is variable as one might expect given that human judgment and behaviour is involved. This indicates that his accuracy judgments are not consistent each day. Joe's mean accuracy level (the X-Bar for the chart) is 87.6%, 2.4% points lower than the desired average standard of 90%. The LCL (3 x sigma) for his performance is 71.6% (4.4% points lower than the desired LCL) and the UCL is computationally 103.6%, but actually 100% (this UCL meets the desired UCL standard). The overall impression is that Joe has a bit of work to do to achieve the desired quality control judgment standard. However, a closer look at Fig. 9.7 reveals that on six days (days 7 and 8 and days 33, 34, 35 and 36), Joe's inspection judgment performance fell just at or below his own LCL. Such extremely low performance levels would of course influence both the overall mean and the standard deviation of the performance series. Management could look more closely into Joe's performance on these specific days to try and tease out why his performance declined so markedly on those occasions.

(continued)

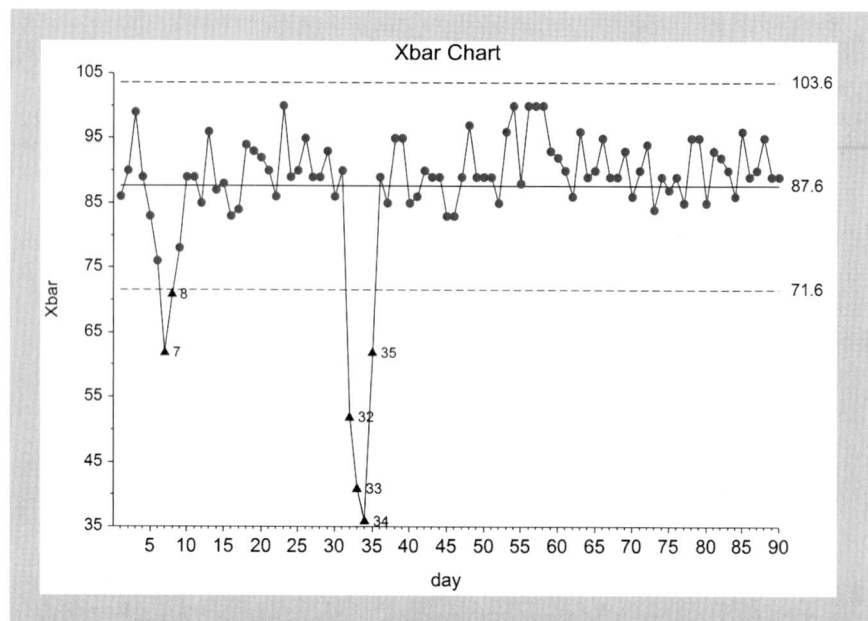

Fig. 9.7 Control X-bar chart for Joe's daily inspection accuracy measures

Joe's problems are obviously acute, not chronic, and his performance always lifts back into its normal range—indicating that the problem is not one of declining capacity or need for more training. Maree conducted a follow-up interview with Joe which revealed that, for the four-day period in question, he had a cold and was taking cold medication that made him drowsy. The two-day period coincided with a period when his mother went into hospital for an operation and he was worried about the outcome. What this shows is that, while an X-Bar chart might reveal when quality problems may have occurred, further data gathering is necessary to tease out causal factors prior to planning interventions—the data in the chart do not tell the whole story.

R charts

The R chart is similar in structure to the X-Bar chart in that it displays a series of data over time. However, for an R chart, the data displayed are not raw quality measurements, but ranges, defined in the case of single observations per unit of time, as the difference between successive measurements. An R chart also displays the mean of the ranges of quality measurements over time and the upper control limit (UCL) and lower control limit (LCL) for those ranges. An R chart is useful for tracking extreme swings in successive observations, something that may signal the rapid onset of a process failure. For quality control measurement designs where random samples of products are selected and measured at specific points in time, the R chart plots the range of measurements observed within each sample.

> Consider the data displayed in the X-Bar chart in Fig. 9.7. Maree could easily construct an R chart for those data as well. Figure 9.8, produced using NCSS, shows what such an R chart would look like. Since the data are based on samples of size 1 over a run of 90 days, each range plotted is the difference between the measurement obtained on the day of interest and the measurement obtained the day before. The R chart therefore shows successive changes over time and can be most useful in helping to identify dramatic changes in quality system stability.
>
> The average range over Joe's entire series is 6.0% points, indicating that on average, from day to day, Joe varied from his previous day's performance by plus or minus 6.0% points in accuracy. The UCL is 19.6% and the LCL is 0.0%. However, it is obvious in the graph that there are three very extreme ranges that have influenced these statistics. The range point labelled '32' shows the dramatic and immediate decline of 38 percentage points in Joe's judgment performance from day 31 to day 32; this coincides with day 32 on the X-Bar graph.
>
> The range points labelled '35' and '36' shows the dramatic and immediate improvement noted from day 34 to day 35 to day 36 on the X-Bar chart (perhaps signalling the period in which Joe was recovering from his cold
>
> **Fig. 9.8** R(ange) chart for Joe's daily inspection accuracy measures

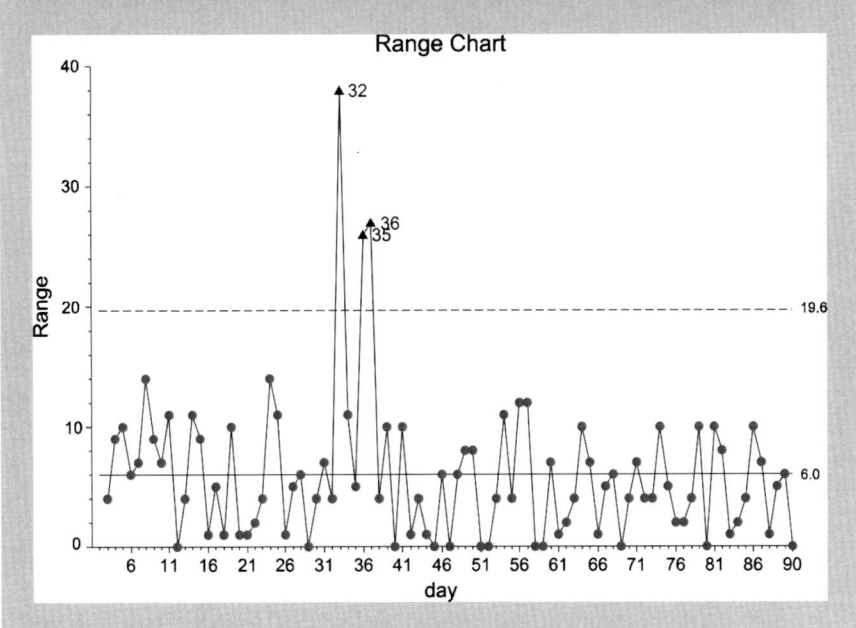

(continued)

> and/or quit taking his medication?). Treating the cause(s) of such extreme variations would help in stabilising the series within the UCL and LCL boundaries, as well as reducing the overall range of accuracy in quality inspection judgments.
>
> Given that an interview with Joe revealed the problem to be illness compounded by medication, management can now take appropriate steps to keep the problem from recurring (new policy on taking sick leave; helping Joe find alternative non-drowsy medication, etc.).

Advantages

Quality control charts provide very flexible and visual tools for monitoring quality over time and conditions. The quality measurements and observations are often made according to strict statistical sampling plans, ensuring maximum precision in statistical estimates and maximum generalisability. Such charts are extremely useful in comparing quality process outcomes to objective standards and identifying observations which signal that the process is out-of-control. This then provides diagnostic indicators for how out-of-control a process is and where to intervene to improve quality. After an intervention, control charts can assist in monitoring and evaluating quality improvements. Many different quality control management/ improvement systems (e.g. Six Sigma, Total Quality Management, Benchmarking, Just-in-Time Inventory Systems, Business Process Re-Engineering) can be informed by the information provided by quality control charts of various types.

Disadvantages

Quality control charts can only be accurate and effective tools for decision making if the quality control measurement process is valid and reliable. They can provide information about where, how serious and under what conditions quality has gone out-of-control, but they cannot tell decision makers precisely why the process has gone out-of-control (a causal conclusion) nor how to intervene to achieve quality improvement. This is an especially important limitation where human systems are involved because it is one thing to know where to intervene to improve quality and quite another to know how best to intervene and to plan for/cope with natural human resistance to change that normally accompanies such interventions. In such contexts, quality control charts should not be the only tool relied upon to inform managerial decision making.

Control charts function best where physical objective (i.e 'hard') statistical measures of quality are available (number of rejected parts, number of parts outside tolerance limits, number of errors made, number of accidents). In such contexts, quality standards are often much more easily and objectively set.

> *For example ... Six Sigma (see Breyfogle 2003) is one quality management system that operates on such principles—processes must produce outcome measures that stay within a 6-standard deviation window to be considered 'in-control'. This means that 99.74% (this standard comes straight from the normal distribution; recall* Procedure 5.7) *of all products produced must be within 3 standard deviations either side of the set tolerance/quality level.*

The 'softer' measures that often arise in human systems are generally of lesser measurement quality than physical 'hard' measures and this means that gaining control over an out-of-control process creates a more complex decision-making task for judging how best to intervene to improve quality. Quality standards are much harder to objectively establish in such contexts. This ultimately means that quality control charts will function best in product-oriented industries where physical measures abound, such as manufacturing, food preparation, transport and packaging. Much more care must be taken in service-oriented industries such as hospitality, human resource management, banking, health and education, to create and validate the appropriate quality measurement systems necessary to realise full value from control chart information.

Where Is This Procedure Useful?

Quality control charts are useful in any discipline or industry where issues of quality control and quality improvement are important to address. Quality control charts provide important sources of statistical information to guide managers in making decisions related to quality improvement. Quality control charts also provide ways of publicly demonstrating the quality of a product or service over time. This can be especially useful where specific experimental or quasi-experimental designs have been used to implement systematic interventions designed to improve quality.

The most obvious industry for application of quality control charts is the manufacturing industry. However, they would prove useful in other industries including health, medicine and nursing, food preparation, hospitality, education (including universities), banking and other financial institutions and any other industry where measurement of quality against standards is important. Finally, quality control charts could prove useful in areas such as pharmaceuticals, mechanical repairs and servicing, occupational health and safety and ergonomics, where compliance with strict policy and legislative regulations is required.

Software Procedures

Application	Procedures
SPSS	*Analyze → Quality Control → Pareto Charts...* and select 'Simple' when prompted to *Define* the chart; choose the variable to define the 'Category Axis:', choose to have the 'Bars Represent:' counts or a 'Sums of variable' and choose the desired variable. *Analyze → Quality Control → Control Charts...* and select your desired 'Variables Chart' (e.g. 'X-bar, R, s') and, for 'Data Organization', choose either 'Cases are units' or 'Cases are subgroups' when prompted to *Define* the chart; choose the 'Process Measurement:' variable, the 'Subgroups defined by:' variable and the 'Identify points by:' variable and under 'Charts', choose 'X-bar using standard deviation'; under the *Options...* button, select your sigma value; under the *Control Rules...* button select one or more control rules to govern the quality process; under the *Statistics...* button, set the 'Upper:' and 'Lower:' and 'Target:' 'Specification Limits and select the statistics you want to see reported. This procedure produces both X-bar and R charts. SPSS is not a particularly powerful package for quality control charts and analyses; it offers a limited range of approaches.
NCSS	*Analysis → Quality Control → Pareto Charts* and select the 'Data Variables:', the 'Category Variable:' and appropriate 'Chart Arrangement' definition (e.g. 'Total By Category'); under the *Reports* and *Plots* tabs, select what you want to see. You may have to tinker with the settings for the plot to get it looking exactly how you want it. *Analysis → Quality Control → XBar-R (Variables) Charts* and select the 'Data Variables:' and the 'Label Variable:'; select the *Options* tab to set desired control limits for the process. This procedure can produce a number of quality control charts, including X-bar and R charts (selected under the *XBar & R Charts* tab). You may have to tinker with the settings for the plots to get them looking exactly how you want it. NCSS offers a wide range of quality control functions and analyses; it is a powerful statistical package in this regard.
SYSTAT	*Advanced → Quality Analysis → Pareto Chart...* and select the 'Y-variable' (the quality control measure) and the 'X-variable' (the different categories of product or service); tick 'Cumulative frequencies' and 'Aggregate by subgroup'. *Advanced → Quality Analysis → Control Chart → Shewhart...* and select the 'Y-variable' (the quality control measure), the 'X-variable' (e.g. observation times like days) and 'Chart Type' (e.g. 'X-bar' or 'Range (r)' amongst others); if you have preset control limits, tick 'Preset' and enter the desired lower and upper limit values. SYSTAT (see Adhikari et al. 2009b) offers a wide range of quality control functions and analyses; it is another powerful statistical package in this regard.
STATGRAPHICS	*SPC → Quality Assessment → Pareto Analysis...* and for 'Data', choose a variable reflecting either 'Untabulated – Observations' or 'Tabulated – Counts' as appropriate; choose a categorical variable to serve as '(Labels:)' for the horizontal axis of the chart (e.g. **part_type**); hit *OK* and choose your desired 'Pareto Analysis Options'; hit *OK* and choose the graphs you want. Note that for Pareto analysis, STATGRAPHICS expects the data to either be

(continued)

Application	Procedures
	organised as aggregated counts by category (= Tabulated – Counts, where each row of the data matrix represents the total observational count for one category of defect or problem) or listed by type of problem observed on each occasion (= Untabulated – Observations, where each row of the data matrix reflects the specific category of defect or problem observed on that specific occasion, e.g. day). SPC → Control Charts → Basic Variables Charts → X-Bar and R... and choose the variable for the 'Observations' 'Data'; choose a variable to assess 'Date:Time/Labels or Size' and enter the desired upper ('USL') and lower ('LSL') specification limits; hit OK and select your desired 'X-Bar and R Chart' options; hit OK and choose 'X-Bar Chart' and/or 'Range Chart', as desired. STATGRAPHICS offers a wide range of quality control functions and analyses; it is another powerful statistical package in this regard.
R Commander	R Commander does not have quality control chart or analysis capabilities. However, there is a package for R called *qcc* which offers a fairly wide range of quality control charting and analysis functions (see Scrucca 2004 for a complete description of the *qcc* package).

Procedure 9.4: Conjoint Measurement & Choice Modelling

Classification   Multivariate; inferential; parametric.
Purpose   To provide a coherent methodology for gathering and analysing decision-making using choice-based or preference-based data, given specific stimuli or sets of stimuli, described by a set of attributes. Conjoint measurement and choice modelling typically result in the assessment of attribute and choice utilities as well as overall attribute importance weights.
Measurement level   The dependent variable in a conjoint analysis may be interval or ratio (or metric) in scale, but may also be ordinal (or nonmetric) for some forms of conjoint analysis and choice models. The independent variables are typically categorical (nominal or ordinal) variables, designed to represent the array of attributes available for each choice.

Conjoint measurement and *choice modelling* are two approaches in the same general category for data gathering and analysis methodology (see, for example, Bakken and Frazier 2006; Hair et al. 2010, ch. 6; Louviere 1988; Louviere et al. 2000). The goal in conjoint measurement and choice modelling is to employ experimental designs to present configurations of object attributes to participants for judgment. Thus, these methods are extremely useful for the analysis of human judgment and preferences in a variety of circumstances. Conjoint measurement and choice modelling tasks are typically administered using paper-based or online surveys; these methods thus provide two of the very few ways we have of building a formal experimental design and structure into a survey instrument.

Conjoint Measurement or Preference Modelling

Conjoint measurement models are most useful for research designs where stated preferences (in the form of ratings or rankings) are obtained from participants. Conjoint measurement and analysis are typically used to fit preference models to data from individual participants or judges. This has the benefit of controlling for anything that is unusual in the participant's thinking about the task, by holding these factors constant through repeated measurements.

Conjoint measurement preference tasks have a very systematic structure and execution process. A set of relevant attributes (or characteristics or properties) of the objects or alternatives to be judged is established by the researcher and two or more levels or values for each attribute are concretely defined. An alternative, therefore, is defined as a bundle of attributes. Then, using a specific experimental design, alternatives are systematically constructed using combinations of attribute levels/values.

In conjoint measurement experimental designs, the typical procedure involves participants rating or ranking each bundle of attributes (= an alternative) separately—a methodology referred to as the 'full profile approach'. The specific experimental designs employed in conjoint measurement research can range from very simple to very complex. A simple experimental design involves a small set of attributes (say 3 or 4), each having only two levels. A *complete factorial design* (recall *Procedure 7.10*), representing all possible combinations of levels of the attributes, generates the alternative configurations for preference judgment. For 3 attributes, each with 2 levels, this gives $2 \times 2 \times 2$ or 8 possible attribute configurations; for 4 attributes, each with 2 levels, this gives $2 \times 2 \times 2 \times 2$ or 16 possible attribute configurations. You can immediately see that increasing either the number of attributes beyond four or the number of levels for each attribute beyond two will generate a large number of possible configurations for judgment. The number would soon exceed the participant's capacity to cope (e.g. 5 attributes, each with 2 levels, would produce 32 different configurations; 6 attributes, each with 2 levels would generate 64, and so on).

Thus, the problem with full factorial conjoint designs is that the number of potential configurations for judgment can virtually explode. For this reason, conjoint measurement researchers have often relied on what are called *fractional factorial designs*. Sometimes another design variation has been found to be more useful, namely a *balanced incomplete block design*. These are designs where only a subset of all possible configurations needs to be tested. The trick is that it must be the right subset in order to estimate the desired effects in the preference model.

Fractional factorial designs and balanced incomplete block designs employ a specific proportion (e.g. ½ or ¼) of the overall set of possible configurations and the choice of the subset to use requires the researcher to deliberately sacrifice his or her ability to estimate certain model effects. Most commonly what conjoint researchers do is to sacrifice some or all information about interactions between attributes (recall *Fundamental concept IX*) so that a specific subset of configurations can be used that will allow all model main effects (one associated with each attribute) to be estimated. Sacrificing interaction information means deliberately assuming that certain interactions are not important and therefore need not be estimated. Louviere (1988) provides a very good and clear discussion of the issues associated with using fractional factorials and other partial design strategies to fit preference models.

> A researcher may wish to conduct a conjoint measurement experiment, using the full profile approach, for a situation where alternatives (say, possible tourist destinations) are described by 7 attributes (key features associated with tourist destinations), each having two possible levels.
>
> The full factorial form of this design would have $2 \times 2 \times 2 \times 2 \times 2 \times 2 \times 2$ or 128 possible attribute configurations (each constituting one possible alternative for the participant to provide a preference rating for), which is a very large judgment task. Alternatively, by sacrificing interaction information, it is feasible to obtain statistical estimates for all 7 attribute main effects using a smaller set of alternatives for judgment. For example, if the researcher uses a ½ fractional factorial design, only some interactions (the researcher would have to choose which) must be sacrificed; however, this would still leave 64 possible configurations—still a large number for a participant to plow through.
>
> The researcher could instead opt to use a ¼ fractional factorial design, yielding 32 attribute configurations to rate—a design that just might be feasible with good will on the part of participants. However, running a ¼ fractional factorial design means sacrificing even more interaction information in order to estimate the 7 attribute main effects. Clearly, conjoint researchers have a great deal to think about and decide upon when designing their experiments—each choice sacrifices some information to gain other more desirable information.

Conjoint analysis employs multiple regression procedures (different types of regression models may be fitted, depending upon the research design and preference recording format) to provide two types of information (see discussions in Hair et al. 2010, ch. 6 and Louviere et al. 2000):

- a set of part-worths or utilities, which are related to regression coefficients, that show how each level of each attribute influences preference ratings. Part-worths can be aggregated across all levels of an attribute to form a measure of overall attribute importance to preference.

Different judges will most likely have different attribute weights informing their preferences—something that would be overlooked if only a group-level (aggregate) model had been constructed.
- an index of model goodness-of-fit, usually measured by either Pearson Correlation (for a metric conjoint model) or Kendall's tau correlation for a non-metric conjoint model. This indexes how well the conjoint model has actually captured the judgment process, as reflected in the part-worths, employed by the participant.

Conjoint preference models can range in complexity from simple models that test only for the main effects of the attributes to complex models where interactions between different attributes can be tested. Model complexity is limited by the nature of the experimental design that guides the number and construction of the alternatives to be judged; often a main effects model is all that can be tested.

> Suppose Maree was interested in conducting a very preliminary pilot study of inspector preferences for different working conditions, as a prelude to a much larger scale study for which she intends to apply for grant funds. She designs a simple 3-attribute conjoint measurement experiment to test inspector's preferences for different sets of possible working conditions.
>
> The three attributes, each of which has two possible values, were described as follows:
>
> - **Office Plan**: open-plan office (partitions, not offices) or traditional plan (separate offices for each inspector)
> - **Breaks**: paid 10-minute breaks every 2 hours or paid 20-minute breaks every 4 hours
> - **Pay/Penalty**: bonus payments for high (90% or above) accuracy but penalty payments for low (less than 75%) accuracy or no bonus or penalty payments
>
> Given the attributes as described above, there are eight possible experimental configurations of working conditions, namely a $2 \times 2 \times 2$ factorial design where each level of an attribute is paired with each of the levels of the other two attributes. Preferences for potential work conditions were rated using a 7-point Likert-type rating scale ranging from $1 =$ not at all preferred to $7 =$ very strongly preferred – yielding a metric conjoint task.
>
> Maree's survey instrument for this stated preference experiment is shown in Fig. 9.9. Maree administered this pilot experimental survey to two quality control inspectors, A and B (the ratings from inspectors A and B are shown in the dashed box in Fig. 9.9). She used SYSTAT, fitting a metric linear model, to analyse the conjoint experimental design and construct a separate preference model for each inspector. The edited results are shown in Table 9.3 and Fig. 9.10 (for inspector A) and Table 9.4 and Fig. 9.11 (for inspector B).
>
> (continued)

Procedure 9.4: Conjoint Measurement & Choice Modelling

Please rate the following packages of possible working conditions according to your degree of preference as a set of conditions you would like to work under.

To make your rating, please use the 7-point scale provided below and write your rating in the space provided.

The characteristics of each working conditions package are defined as follows:

Office Plan	Traditional	= an office plan where every inspector has his/her own office
	Open	= an open office plan with 1.5 meter high partitions between cubicles for inspectors
Breaks	10 min/2 hrs	= paid 10-minutes work breaks every 2 hours
	20 min/4 hrs	= paid 20-minute work breaks every 4 hours
Reward/Penalty	No Bonus/Penalty	= no bonus payments for high accuracy; no penalty payments for low accuracy
	Bonus/Penalty	= bonus payments for high (90% or more) accuracy; penalty payments for low (less than 75%) accuracy

Rating Scale: Not at all Preferred   Moderately Preferred   Very Strongly Preferred
                      1   2   3   4   5   6   7

Working Conditions Package	Characteristics of the Package				A	B
A	Traditional	20 min/4 hrs	Bonus/Penalty	Preference = ____	5	4
B	Open	10 min/2 hrs	No Bonus/Penalty	Preference = ____	4	5
C	Open	20 min/4 hrs	No Bonus/Penalty	Preference = ____	3	3
D	Traditional	10 min/2 hrs	Bonus/Penalty	Preference = ____	4	6
E	Traditional	20 min/4hrs	No Bonus/Penalty	Preference = ____	7	1
F	Open	20 min/4 hrs	Bonus/Penalty	Preference = ____	2	2
G	Traditional	10 min/2hrs	No Bonus/Penalty	Preference = ____	6	2
H	Open	10 min/2hrs	Bonus/Penalty	Preference = ____	1	7

Fig. 9.9 Maree's conjoint experiment survey instrument

Table 9.3 SYSTAT parameter estimates (part-worths) expressed by inspector A

Parameter Estimates (Part Worths)					
Open	Traditional	10 min/2 hrs	20 min/4 hrs	Bonus/Penalty	No Bonus/Penalty
−0.330	0.754	−0.205	−0.024	−0.459	0.264
1.084		0.229		0.723	
53.2%		11.2%		35.5%	

Goodness of Fit (Pearson Correlation)

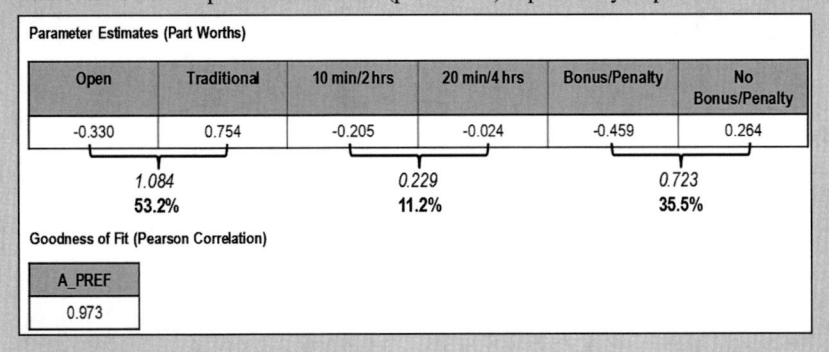

A_PREF
0.973

Table 9.3 reports the preference modelling outcomes for inspector A. The overall goodness-of-fit of the preference model to inspector A's ratings was assessed by a Pearson correlation value of .973 meaning that inspector A's observed ratings and ratings predicted by the preference model were very strongly correlated.

(continued)

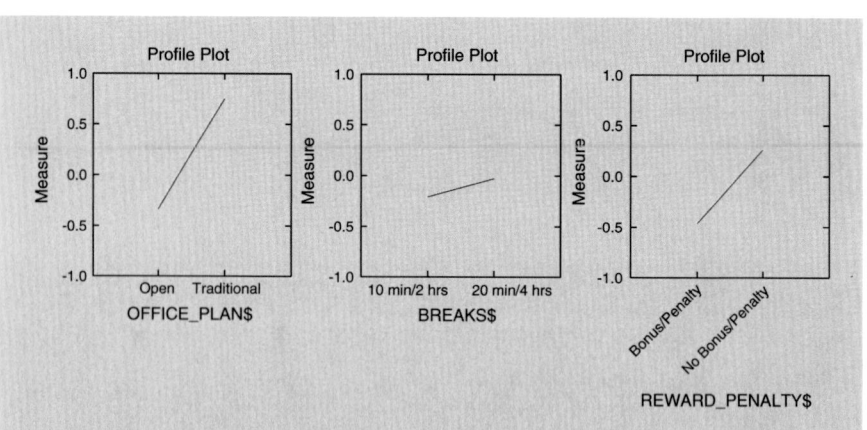

Fig. 9.10 Graphical display of attribute part-worths for inspector A

Table 9.4 SYSTAT parameter estimates (part-worths) expressed by inspector B

Parameter Estimates (Part Worths)					
Open	Traditional	10 min/2 hrs	20 min/4 hrs	Bonus/Penalty	No Bonus/Penalty
0.190	-0.228	0.558	-0.459	0.373	-0.464
0.418 / 18.4%		1.017 / 44.8%		0.837 / 36.9%	

Goodness of Fit (Pearson Correlation)

B_PREF
0.845

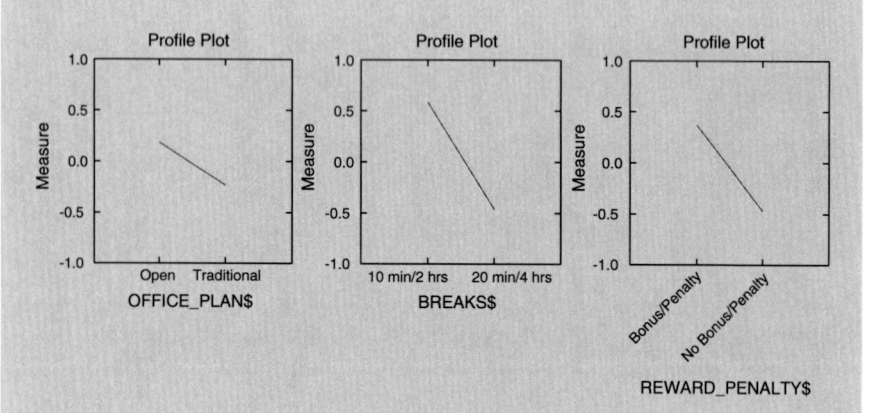

Fig. 9.11 Graphical display of attribute part-worths for inspector B

(continued)

The part-worths or utilities show the impact of each attribute level on preference ratings. One part-worth is calculated for each level of each attribute. Thus, 'Open' and 'Traditional' report the part-worths for the open office design and traditional plan office design as $-.330$ and $.754$, respectively. So, when inspector A saw a working conditions package offering a traditional office design, his preference had a very strong tendency to increase (hence the positive sign on the part worth); when he saw a working conditions package offering an open plan office design, his preference decreased.

The left-most graph in Fig. 9.10 shows this trend quite clearly by plotting the part-worths. Maree would make a similar interpretation for the other two attributes, 'BREAKS$' and 'REWARD_PENALTY$, by inspecting their respective part-worths and graphs. In summary, inspector A's preferences are oriented toward a traditional plan office design and toward having a No Bonus/Penalty system in place, whereas the length and frequency of paid breaks was a relatively irrelevant attribute.

In support of these conclusions, Maree computed overall importance weights for the three attributes, either as proportions or as percentages, using the part-worths reported by SYSTAT (a process discussed in Hair et al. 2010, pp. 302–303). For the 'OFFICE_PLAN$' attribute, Maree found the range of the two part-worths by subtracting the minimum value from the maximum value (gave the quantity pointed to by the brace). She repeated this process for the other two attributes, added up all the ranges and divided each attribute's range by this total. If she multiplied each result (which indicates a proportional importance weight) by 100%, she would obtain a percentage of weight or importance given to each attribute when inspector A made his judgments. Table 9.3 shows that, for inspector A, Office Plans (53.2%) was the most important consideration, followed by the Reward/Penalty scheme (35.5%); the configuration of Breaks was only minimally important (11.2%) to inspector A.

Table 9.4 shows the preference modelling outcomes for inspector B. The overall goodness-of-fit of the preference model to inspector B's ratings was a Pearson correlation of .845, indicating that inspector B's observed ratings and ratings predicted by the preference model were also very strongly correlated, but not quite at the same level as for inspector A. The part-worths for inspector B, shown in Fig. 9.11, revealed a preference orientation for an open office plan, for paid 10-minute breaks every 2 hours, and for a Bonus/Penalty system. In terms of overall importance for inspector B, the Breaks attribute was most important (44.8%) in her rating of preferences with the Bonus/Penalty attribute being next most important (36.9%) and the Office Plan attribute being least important (18.4%).

(continued)

> The preference models for inspectors A and B were thus rather different and this would lead Maree to predict that each might feel more comfortable working under somewhat different working conditions. What is available for the Bonus/Penalty attribute is important for both inspectors, but their preferences run in opposite directions.

Choice Modelling

Choice models are most useful for research designs where stated choices (choosing the most preferred option from specific choice sets of alternatives) are obtained from participants. Choice modelling employs binary logistic or multinomial logistic regression analysis (recall *Procedure 7.14*) for the modelling fitting exercise. Choice models are an interesting innovation for a number of reasons. Firstly, choice models can employ all the experimental design innovations and strategies used in conjoint measurement preference modelling. Thus, fully factorial, fractional factorial or balanced incomplete block designs can be implemented in choice modelling.

Secondly, choice models can be estimated that can accurately predict choice behaviour, but through having participants actually make choices, rather than through having them rate full profiles. This constitutes a choice-based, rather than a full profile, approach. Choice modelling introduced an innovation into conjoint research methodology, namely the *choice set*. By presenting the participant with a small number of statistically optimised choice sets, where each set may contain two or three simultaneously presented alternatives (such as three alternative tourist destinations), described by potentially a large number of attributes, and by having the participant choose their most preferred alternative in each set, the desired choice model effects (those permitted by the experimental design used) can be estimated. It is a very efficient (and less tiring) way to obtain data from participants.

Thirdly, choice models can be estimated at the individual level or at the group or aggregate level, depending upon the needs of the researcher. This allows a researcher to begin building a case for the generalisation of choice models for a specific type of choice problem. Some choice models can only be estimated at an aggregate level.

Best-Worst Scaling Models

Best-worst scaling methodology (see Augur et al. 2007; Lee et al. 2008; Louviere et al. 2015; the method is also sometimes called the 'max-diff' approach to scaling) is a relatively recently developed variant of choice modelling. Best-worst scaling works like choice modelling in that:

- specific experimental designs are implemented;
- choice sets are created and presented to participants; and
- choice data form the basis for estimating the choice models.

What best-worst scaling does is, for each choice set, require the participant to indicate both the best choice in the set and the worst choice in the set. The choice sets are constructed so that specific statistical features of the choice model, determined by the experimental design, can be estimated. However, because the participants have been asked to indicate both the best and the worst choice in each set, it becomes possible not only to estimate choice models, but also to create a genuine interval or ratio scale that orders and precisely positions each and every alternative along the scale. Most preferred alternatives then become easy to identify, they simply 'bubble up' to the top of the interval scale. Best-worst scaling is also applicable in research contexts where Likert-type scales would otherwise be used, such as in the measurement of values (see Lee et al. 2008, for a concrete example). In these cases, best-worst scaling avoids some of the major pitfalls and social desirability biases inherent in the use of self-rating Likert-type scales, producing a stronger less biased measurement framework overall. The computations behind best-worst scaling are specialised but not all that intensive. However, the major software packages like SPSS, NCSS and SYSTAT do not typically have the capability to handle best-worst scaling computations. Instead, Lee et al. (2008) indicated that a program for performing best-worst scaling could be obtained from them. Furthermore, Augur et al. (2007) employed a simple approximation for best-worst scaling that involved, for each alternative, subtracting the total number of times an alternative was chosen as worst from the total number of times it was chosen as best.

> To illustrate the simplest approximation approach to best-worst scaling, suppose Maree wanted to understand what experts felt were the most important aspects of the performance of a quality control inspector. In interviews with a sample of experts she learned that there were five key aspects of performance that were of interest in the role of quality control inspector: (1) accuracy of decisions; (2) speed of decisions; (3) Cost implications of decisions; (4) Consistency of decisions; and (5) agreement with other inspectors. Maree decides that a simple best-worst scaling exercise, similar to what Augur et al. (2007) employed, would give her the data she needs. The left side of Fig. 9.12, configured using Excel, shows the structure of the experimental task. She gets three quality control experts to run through the task. The task involves 5 possible performance aspects (relabelled as A, B, C, D and E, for simplicity of representation) and she decides that a choice set size of three will give her the right size of task to do the job. Five aspects in sets of 3 gave Maree a total of 10 distinct choice sets, listed in the left-hand table of Fig. 9.12.

(continued)

A = Accuracy of decisions
B = Speed of decisions
C = Cost implications of decisions
D = Consistency of decisions over time
E = Agreement with other inspectors

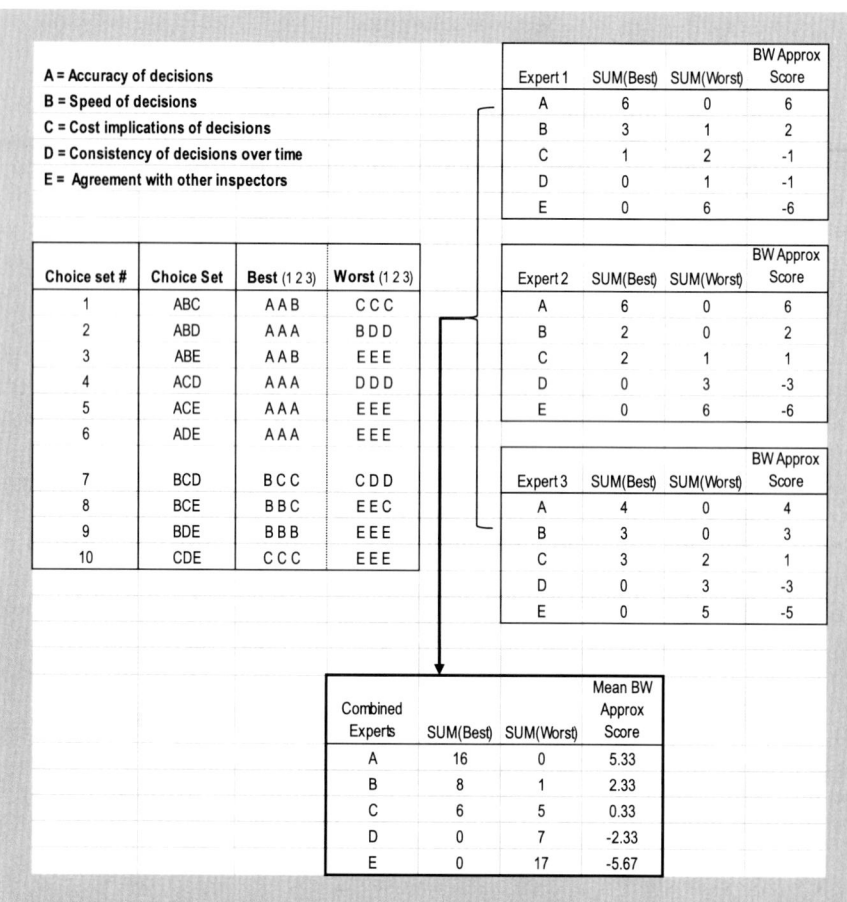

Expert 1	SUM(Best)	SUM(Worst)	BW Approx Score
A	6	0	6
B	3	1	2
C	1	2	-1
D	0	1	-1
E	0	6	-6

Choice set #	Choice Set	Best (1 2 3)	Worst (1 2 3)
1	ABC	A A B	C C C
2	ABD	A A A	B D D
3	ABE	A A B	E E E
4	ACD	A A A	D D D
5	ACE	A A A	E E E
6	ADE	A A A	E E E
7	BCD	B C C	C D D
8	BCE	B B C	E E C
9	BDE	B B B	E E E
10	CDE	C C C	E E E

Expert 2	SUM(Best)	SUM(Worst)	BW Approx Score
A	6	0	6
B	2	0	2
C	2	1	1
D	0	3	-3
E	0	6	-6

Expert 3	SUM(Best)	SUM(Worst)	BW Approx Score
A	4	0	4
B	3	0	3
C	3	2	1
D	0	3	-3
E	0	5	-5

Combined Experts	SUM(Best)	SUM(Worst)	Mean BW Approx Score
A	16	0	5.33
B	8	1	2.33
C	6	5	0.33
D	0	7	-2.33
E	0	17	-5.67

Fig. 9.12 Illustration of the simplified best-worst scaling approach using three expert judges

Each expert examined each choice set and indicated which aspect of performance they considered to be the most important with respect to the performance of quality control inspectors in that set of 3 and which aspect was the least important. The best and worst choices indicated by each of the three experts are shown in the columns labelled 'Best' and 'Worst'. The three smaller tables on the right side of Fig. 9.12 show the best-worst (BW) scaling outcomes for each individual quality expert. For Expert 1, for example, the 'BW approx score' for aspect A (accuracy of decisions) was found by counting the number of times A was chosen as best by that expert and the number of times it was chosen as worst, then subtracting the worst total from the best total.

(continued)

Maree could easily see that the experts varied slightly in how they scaled each aspect of performance, but across the three experts, the rank ordering of the aspects was nearly identical. The small table at the bottom of Fig. 9.12 shows the final outcome from combining the assessments of all three experts, giving Maree a final interval-level BW score for each aspect.

It is important to realise that the best-worst scaling process illustrated in Fig. 9.12 produces only an approximate BW scaling of the choice alternatives. To obtain the statistically optimal estimates, a more formal regression-based analytical process must be undertaken.

Advantages

Conjoint measurement, choice modelling and best-worst scaling all employ explicitly controlled experimental designs as a vehicle for gaining insights into judgment and decision processes. Judgments are analysed to produce estimated part-worths for detailed evaluation of choices and preferences with a high degree of statistical precision. Such methods afford a high degree of contextual control over the data gathering process, but as research has shown (see Louviere et al. 2000 for some discussion), laboratory-based choice experiments can and do yield externally valid and generalisable predictions about people's choice behaviour in the real world. Conjoint measurement, choice modelling and best-worst scaling are all very well-suited to online data gathering (where data recording, preparation and analysis can all be undertaken automatically) as well as to more traditional paper-based survey methods. Best worst scaling methodology is a very useful innovation; it is a simple task for participants to cope with and it can yield genuine interval- or ratio-level scaling of alternatives.

The analysis and modelling of individual judgments or choices is a real benefit in that individualised conjoint or choice models can provide the first step in a market segmentation study (where, for example, part-worths can be cluster-analysed (recall *Procedure 6.6*) to identify market segments having similar preference structures). The fact that choice models can be constructed at the individual level as well as at an aggregate level means that choice modelling and best-worst scaling provide very flexible and widely applicable methodologies for analysing decision making in a range of contexts.

Disadvantages

Conjoint measurement, choice modelling and best-worst scaling are limited, primarily, by what participants can reasonably be asked to cope with. This means that there are inherent limitations in the number of alternatives, attributes and levels that can be used in a full factorial design. If more attributes or attributes with more levels (or both) are required, then more sophisticated experimental designs are required in order to keep the number of attribute configurations or choice sets that a participant has to judge at a feasible level.

Complete factorial designs must be 'cut down to a feasible size', while maintaining the desired statistical properties and this requires some fairly sophisticated thinking about what information to sacrifice in order to make the design feasible. Some types of designs, notably those involving mixtures of different levels for attributes, cannot be optimally cut down in size, which often means that the research design has to be modified before the desired choice model can be estimated. Choice modelling is thus a statistically sophisticated methodology, which can render wide-spread usage difficult to achieve. For this reason, many companies will employ a choice modelling consultant to conduct and analyse choice experiments on behalf of the company.

The overall decision task or some of its associated choices and attribute configurations employed in conjoint measurement, choice modelling or best-worst scaling research, may strike some participants as unrealistic, irrelevant and artificial. This may impact on their motivation or capacity to give an accurate depiction of their preferences and on their willingness to work through a large design (considerations collectively known as 'incentive compatibility'). This means that a critical issue in the use of conjoint measurement, choice modelling or best-worst scaling is the correct and meaningful specification of relevant attributes and of their levels. Realism is often difficult to achieve in such methodologies. Decision researchers who employ these methodologies must always remain aware of the fact that decision or choice experiments are simulation tasks. Making such tasks as realistic and meaningful as possible is an important step in generalising what is learned beyond the simulated context. If a researcher focuses solely upon the experimental design parameters of the research, thereby ignoring or minimising contextual issues, then what is learned in the experiment will have little meaning outside the simulation environment and may, in fact, have little meaning to the participants, which of course will yield lower quality data.

Where Is This Procedure Useful?

Conjoint measurement and choice modelling enjoy widespread use in the marketing disciplines as avenues for evaluating preferences and choices amongst products and services. For example, the first step in deciding whether or not to develop and launch a new product or service may be to conduct a choice experiment to see where the new product/service fits with respect to those already available in the marketplace.

Conjoint measurement and choice modelling methods can also useful in:

- medical contexts (e.g. evaluating alternative treatment regimes);
- public policy (e.g. choosing best location for planned developments, services);
- tourism and hospitality (e.g. choosing hotels or tourist destinations);
- education (e.g. preferences for careers);
- business and management (e.g. analysing strategic choices, options, preferences);
- human resource management (e.g. choosing between different work contract conditions);
- psychology (e.g. preferences for different values people can live by—best-worst scaling would be especially useful here; see Lee et al. 2008, for an example); and
- in any other disciplines where an understanding of people's judgments, choices or decisions is desired.

If an understanding of individual-level decision making is desired, conjoint measurement models and certain types of choice models can be of great assistance. This could, for instance, be an especially important step to undertake to provide the foundation for identifying meaningful aggregates of similar preference or choice models (e.g. market segmentation; forming policy typologies and so on).

Software Procedures

Application	Procedures
SPSS	To analyse data from a conjoint analysis investigation, you need to use an SPSS command procedure called *Conjoint* as there is no menu-driven procedure available. The SPSS Help menu can assist in configuring this command in the SPSS *Syntax* window. *Data → Orthogonal Design → Generate...* is a procedure for generating orthogonal experimental designs, which can be useful in conjoint/choice experiments. To generate a design, you name each desired factor ('Factor Name:'), *Add* the factor to the list and define the next factor; then, via *Define Values...*, indicate the number of levels each factor is to have. SPSS will then produce a smaller set of stimulus profiles for conjoint analysis that have been properly configured to produce unconfounded main effect estimates. *Data → Orthogonal Design → Display...* can be used to have SPSS display the stimulus profiles generated, either for experimenter or research participant use.
NCSS	NCSS does not currently have the capability to properly analyse conjoint measurement and choice modelling experiments. To use this program for such research designs, you would have to manage every step of the analytical process manually using different procedures within NCSS. NCSS does have extensive capabilities to generate a wide variety of experimental designs for use in conjoint research: *Analysis → Design of*

(continued)

Application	Procedures
	Experiments and choose the desired experimental design plan generator, such a 'D-Optimal Designs' 'Fractional Factorial Designs' or 'Balanced Incomplete Block Designs'.
SYSTAT	*Advanced → Conjoint Analysis* ... and select the dependent variable (s) ('Dependent(s)') of interest (preference ratings, rankings or choices) and independent variables ('(Independent(s):'; the attribute variables); if the preference rating scale or ranking system is oriented so that higher numbers indicate more preferred options, choose *Polarity* to be 'Positive'; if the system is oriented so that lower numbers indicate more preferred options, choose *Polarity* to be 'Negative'; for a metric conjoint analysis, choose the 'Loss' function as 'Stress' and the 'Regression' approach as 'linear'; for a nonmetric conjoint analysis, choose the 'Loss' function as 'Tau' and the 'Regression' approach as 'Monotonic'. For more details, see Wilkinson (2009). SYSTAT also offers extensive capabilities to generate a wide variety of experimental design plans, including fractional factorials, via the following pathway: *Utilities → Design of Experiments → Wizard...* and select the desired configuration for the experimental plan. For more details, see Stenson (2009).
STATGRAPHICS	STATGRAPHICS does not currently have the capability to properly analyse conjoint measurement and choice modelling experiments. To use this program for such research designs, you would have to manage every step of the analytical process manually using different procedures within STATGRAPHICS. *DOE → Experimental Design Wizard...* offers the capability to generate experimental designs of various types; the wizard has a series of twelve steps (implemented via control buttons, which progressively become active for the next step in the sequence) to guide you through the experimental design process.
R Commander	**R** Commander currently does not offer capabilities to either create experimental designs for conjoint analysis investigations or to conduct conjoint analysis itself. However, there is a package available for the main **R** program, called *ChoiceModelR*, which can conduct various sorts of conjoint analyses (see Sermas and Colias 2013 for details). Aizaki and Nishikura (2008) describe and illustrate a set of approaches for creating and analysing conjoint experiments using **R**.
LIMDEP	This is a well-known and extensively featured econometric statistics program that is explicitly designed to estimate linear and nonlinear regression models of all types. It seems to be the preferred software package for researchers who do choice modelling, because analysing choice models, as distinct from conjoint measurement models, using almost any other statistical package (including SPSS, NCSS and STATGRAPHICS) is very tedious, if not impossible. More details about LIMDEP and its multinomial extension NLOGIT can be found at its website: www.limdep.com.
LatentGOLD w/Choice add-on	This software package offers a wide range of latent class analysis models. It also has an add-on module called *Choice* for analysing conjoint measurement and discrete choice data; see https://www.statisticalinnovations.com/latent-gold-5-1/.

Procedure 9.5: Multilevel (Hierarchical Linear/Mixed) Models

Classification	Multivariate; inferential; parametric.
Purpose	To fit and evaluate a multiple regression model that predicts a dependent variable using predictors or independent variables at two or more different levels of analysis.
Measurement level	The dependent variable must be measured at the interval or ratio-level and the independent variables, for any level of analysis, may be obtained using any measurement scale as long as any categorical (nominal or ordinal) measures are appropriately coded for the analyses.

Multilevel models (or hierarchical linear models – HLM) (e.g. Bickel 2007; Cohen et al. 2003, ch. 14; Field 2018, ch. 21; Raudenbush and Bryk 1997, 2002; Robson and Pevalin 2016; Tabachnick and Fidell 2019, ch. 15) embodies a sophisticated multiple regression approach to constructing and testing statistical models. Multilevel modelling is a more recent evolution in statistical methodology and is enjoying increased exposure in the literature. Other names for hierarchical linear models include mixed models, mixed regression models and random coefficient models, although there are some slight nuances in differences between these types of models, depending upon who you read. The key distinguishing feature of a multilevel model is that it has predictors obtained from two or more levels of analysis in the same investigation. A 'level of analysis' implicates measurement at different levels of aggregation within a sample. Thus, individual people may comprise one level of analysis where measurements may focus. Groups may comprise a second level of analysis where measurements may focus (where groups are aggregates of level one individuals). The dependent variable in a multilevel model is always obtained at the *lowest* level of analysis to be considered.

Multilevel models can only be constructed in contexts where the data are obtained at multiple levels of analysis and where the physical linkages between the lowest level and the highest level of analysis are tightly maintained. Thus, if individuals and groups are the two levels of analysis of interest, we would first draw and measure a sample of groups, then sample and measure individuals from inside those groups—groups and individuals must therefore have a 'nested' relationship where individuals are 'nested' inside groups. The best sampling strategies for multilevel models involve some sort of multistage random sampling plan (recall *Fundamental concept VIII*), starting with choosing sampling units at the highest level of analysis first, then working progressively down into the lower levels of analysis within those units sampled at the higher level.

> In the business arena, a multilevel model might be constructed to predict organisational commitment from measures of performance and turnover at the department level within a large organisation as well as from measures of performance and job satisfaction for individual employees. This would yield a two-level HLM: individuals and departments.
>
> A multilevel model in the educational discipline might comprise predictors linked to individual students, to their classrooms and to their schools. Thus, we might wish to build a model that predicts children's mathematical achievement from measures of individual student classroom performance, mathematical and logical intelligence as well as from measures of classroom size, teacher style and composition and school socioeconomic status. The sample in such a study might comprise 25 randomly sampled schools within a metropolitan city, 50 randomly sampled classrooms within those 25 schools, and 1000 individual students randomly sampled from within those 50 classrooms. This would yield a three-level HLM: students, classrooms and schools.
>
> In a social psychology context, we might be interested in building a multilevel model predicting team member performance as a function of measures of ability and commitment from individual team members as well as measures of leadership and communication styles at the team level. In a nursing context, we might be interested in constructing a multilevel model predicting patient recovery time from individual measures of patients' treatment responsiveness as well as from measures of nursing care quality at the ICU level and measures of resourcing adequacy at the hospital level.
>
> In each example, the levels of analysis involve different extents of aggregation across individuals. It is unusual to find a multilevel model involving more than three levels, simply because of the inherent complexity and sheer sample sizes such modelling would require. Many published multilevel models tend to involve two levels of analysis.

While the mathematical process for estimating a multilevel or hierarchical linear model is rather complex, the basic concept of how such a model is conceived is relatively easy to grasp (see also Robson and Pevalin 2016, for a plain language discussion). In a two-level hierarchical linear model, we might have a sample comprising a number of groups from within which we draw further samples of individuals. We take measurements both at the individual level and at the group level (such measurements might be obtained using surveys, interviews, observations, official records or some combination of these or other methodologies). Measurements at the group level might also be aggregated upward from the individual level of measurement.

In constructing a multilevel or mixed model, we need to understand the distinction between a 'fixed' effect and a 'random' effect in such models. A 'fixed' effect is a coefficient estimate that does not vary across different contexts (e.g. groups or schools) – the coefficient is therefore assumed to remain constant across contexts. A

Procedure 9.5: Multilevel (Hierarchical Linear/Mixed) Models

'random' effect emerges if we allow for regression coefficient estimates to vary as a function of the different contexts in which the coefficient is estimated. Bickel (2007, pp. 126–127) argues that a 'random' effect actually has two components: a 'fixed' component, which represents the average coefficient value across contexts and a 'random' component which represents the variance of the coefficients across contexts. A key benefit of multilevel modelling is that it handles the problems of non-independent (i.e. correlated) model errors when research participants are 'nested' within a larger grouping (e.g. students within a classroom, taught by the same teacher, would be more alike than students from different classrooms because of their identical contextual circumstances).

To visualise how a multilevel model might be constructed, imagine conducting a series of simple linear regression analyses (recall *Procedure 7.13*), predicting some dependent variable of interest using a single individual-level predictor; each analysis being conducted using individuals inside one of a number of groups (this would be the *Level 1 analysis*). The effects that would be assessed at this level are called fixed effects (effects are 'fixed' within groups). The result would be a set of regression slopes and intercepts where the set would contain as many slope/intercept pairs as there were groups for analysis.

Then imagine conducting a second set of regression analyses, using predictors that index group characteristics or performance to separately predict the slopes and the intercepts from the first set of regressions (this would be called the *Level 2 analysis*). The effects that would be assessed at this level are called 'random effects' to signal that regression model statistics from the Level 1 analyses may vary significantly (i.e. have a variance) as a function of the Level 2 variables (the amount of variation in the dependent variable attributable to 'random' effects can be assessed using an *intraclass correlation*). The resulting statistical model would then comprise a mixture of 'fixed' effects and 'random' effects, each of which could be assessed for significance. This process is the reason why another label that is often applied to such models is *mixed models*. In such models, it is also possible to evaluate interactions between Level 1 and Level 2 predictors, giving what are called *cross-level interactions*. Finally, it is also possible to define and test 'random' effects and 'fixed' effects at the same level of a model.

> Suppose Maree received a large grant to continue her research in the QCI context. She decides to conduct a multilevel investigation into the accuracy of product and process quality decisions. Within the five basic types of manufacturing company she was interested in (PC, large electrical appliances, small electrical appliances, large business computer and automobile), she randomly samples 3 specific manufacturing organisations, giving a total of 15 organisations. Within each of these 15 companies, she targets her survey to all employees who have responsibilities for some aspect of decision making about product or process quality. She obtains an average of about 67 survey

(continued)

respondents from each manufacturing organisation; her total sample size amounted to 1008 employees across all 15 organisations. For each employee, she measured quality decision accuracy (**accuracy**), the log of quality decision speed (**log_speed**), perception of working conditions (**workcond**) and job satisfaction (**jobsat**). She also obtained and dummy-coded information about each employee's gender and education level (all measures were defined and measured in the same way as for her original investigation that gave rise to the QCI database).

Maree's research design could be described as a three-level investigation: individual employees ($n = 1008$), organisations ($n = 15$) and industry type ($n = 5$). Her chief interest, however, was in testing a two-level hierarchical linear model. To get a feel for the potential for such a model, Maree examined 'fixed effect' linear regression analyses predicting employee's quality decision **accuracy** scores using employee's **workcond** perceptions within each of three 15 organisations in her sample. In essence, these were simple exploratory Level 1 analyses for Maree. She wishes to see if the intercepts and/or slopes from these 15 Level 1 regressions differed across the 15 organisations – an exploratory Level 2 analysis.

Figure 9.13, produced using SYSTAT, plots the outcomes from this Level 1 analysis, compared across the 15 Level 2 organisations, in terms of the fitted regression lines. While the graph appears very busy showing data points from all organisations as well as the 15 fitted regression lines, it is the pattern of the fitted lines that is key. It is clear that the fitted regression lines are separated from each other and not all are parallel to each other. The lines also show different intercept points along the Y-axis. Thus, this pattern made it visually

Fig. 9.13 Graph of fitted regression lines, produced using SYSTAT, predicting **accuracy** from **workcond** within each of fifteen manufacturing organisations

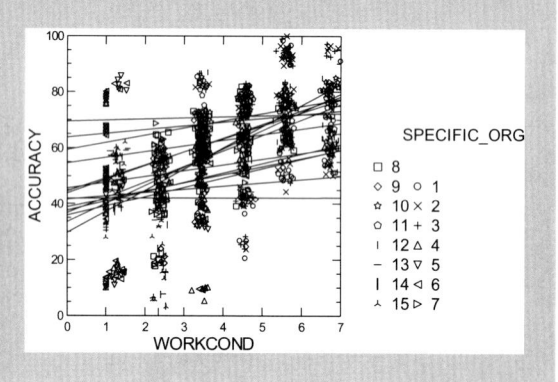

(continued)

obvious to Maree that **workcond** predicted **accuracy** more strongly in some organisations than in others and from different starting points in terms of accuracy. This pattern suggested to Maree that there might be merit in properly testing a formal two-level hierarchical linear model on her data. A Level 2 analysis could test whether or not the slopes for the 15 regression lines were significantly different from each other and could test whether or not the intercepts (the point where each line crosses the Y-axis) from the 15 regression lines were significantly different from each other. However, Maree also wanted to use the other predictor information she had to hand as well.

To this end, Maree evaluated a two-level HLM where **jobsat**, **log_speed**, dummy-coded **gender** and dummy-coded **educlev** were defined as Level 1 'fixed' main effect predictors. The **workcond** variable was defined as a Level 2 'random' main effect, with the groups for that effect being defined by a variable called **specific_org** which carried the information about which one of the 15 organisations an employee worked for. Finally, she was also interested in examining four two-way 'cross-level interactions' to see if the relationship between working conditions and other specific predictors varied across the 15 organisations: **workcond** by **jobsat**, **workcond** by **log_speed**, **workcond** by **gender** and **workcond** by **educlev**. She used SPSS to obtain the output shown in Table 9.5 (output has been edited to remove unnecessary detail; note that SPSS refers to continuous predictors like **jobsat** and **log_speed** as *covariates* for this type of analysis). [Note that SPSS does not report the *intraclass correlation*, but for Maree's new data set, the value is .335 which means that 33.5% of the variance in **accuracy** scores is between groups (i.e. organisations) variance, suggesting that fitting a two-level HLM would be worth doing. However, to find this value, an 'intercepts-only' model must be fitted – see Tabachnick and Fidell (2019, section 15.6.1 and 15.4.1). A formula for calculating the intraclass correlation coefficient appears in Appendix A.]

The top subtable in Table 9.5 reports the outcomes for the fixed effects Level 1 portion of the analysis. Both **jobsat** and **educlev** (the line of statistics aligns with the category automatically dummy-coded '1' by the procedure) were significant Level 1 predictors (in boldface). The regression coefficient for **jobsat** suggested that a higher level of job satisfaction predicted a lower level of decision **accuracy**. The regression coefficient for **educlev** was negative which suggested that high school-only educated employees averaged significantly lower **accuracy** compared to tertiary-educated employees in this sample (confirmed by the bottom subtable of covariate-adjusted group means for **educlev**.

The subtable labelled 'Estimates of Covariance Parameters' reports the outcomes for Level 2 of the HLM analysis, where **workcond** and its cross-level interactions with the fixed effect predictors in Level 1 were tested as 'random'

(continued)

Table 9.5 Mixed model analysis results, produced by SPSS, for Maree's two-level HLM

Estimates of Fixed Effects[a]

Parameter	Estimate	Std. Error	df	t	Sig.	95% Confidence Interval Lower Bound	95% Confidence Interval Upper Bound
Intercept	88.652433	2.915479	20.949	30.408	.000	82.588463	94.716403
jobsat	-1.143773	.314461	83.490	-3.637	.000	-1.769170	-.518377
log_speed	-1.720678	2.089763	43.502	-.823	.415	-5.933680	2.492325
[gender=1.00]	-1.953087	1.078583	37.389	-1.811	.078	-4.137738	.231563
[gender=2.00]	0[b]	0					
[educlev=1.00]	**-2.657170**	**.840508**	**23.743**	**-3.161**	**.004**	**-4.392887**	**-.921453**
[educlev=2.00]	0[b]	0					

a. Dependent Variable: accuracy.
b. This parameter is set to zero because it is redundant.

Estimates of Covariance Parameters[a]

Parameter		Estimate	Std. Error	Wald Z	Sig.	95% Confidence Interval Lower Bound	95% Confidence Interval Upper Bound
Residual		41.226518	2.013763	20.472	.000	37.462660	45.368529
Intercept [subject = specific_org]	Variance	51.847915	28.410716	1.825	.068	17.713514	151.760194
workcond [subject = specific_org]	Variance	2.074524	2.387175	.869	.385	.217487	19.788113
jobsat * workcond [subject = specific_org]	Variance	.083385	.043410	1.921	.055	.030058	.231322
workcond * log_speed [subject = specific_org]	**Variance**	**3.964270**	**1.778975**	**2.228**	**.026**	**1.645071**	**9.553046**
gender * workcond [subject = specific_org]	**Variance**	**.618178**	**.285500**	**2.165**	**.030**	**.250031**	**1.528389**
educlev * workcond [subject = specific_org]	Variance	.189758	.108732	1.745	.081	.061725	.583369

a. Dependent Variable: accuracy.

1. gender[a]

gender	Mean	Std. Error	df	95% Confidence Interval Lower Bound	95% Confidence Interval Upper Bound
1.00 Male	77.991[b]	2.185	12.008	73.232	82.750
2.00 Female	79.944[b]	2.175	11.707	75.193	84.696

a. Dependent Variable: accuracy.
b. Covariates appearing in the model are evaluated at the following values: jobsat = 5.4409, log_speed = .6722.

2. educlev[a]

educlev	Mean	Std. Error	df	95% Confidence Interval Lower Bound	95% Confidence Interval Upper Bound
1.00 High School only	77.639[b]	2.176	11.841	72.892	82.386
2.00 Tertiary qualified	80.296[b]	2.131	10.808	75.596	84.996

a. Dependent Variable: accuracy.
b. Covariates appearing in the model are evaluated at the following values: jobsat = 5.4409, log_speed = .6722.

effects. [Note that the column labelled 'Estimates' report what are called the 'variance components' of the multilevel model. Variance components reflect the most commonly employed modelling approach for understanding how much variance 'random' effects account for in a multilevel model (see Bickel 2007, pp. 92–93 and Cohen et al. 2003, p. 560.]

There were two significant (highlighted in boldface in the 'Estimates of Covariance Parameters' subtable in Table 9.5) cross-level interactions: **workcond * log_speed** and **workcond * gender**. This means that the two

(continued)

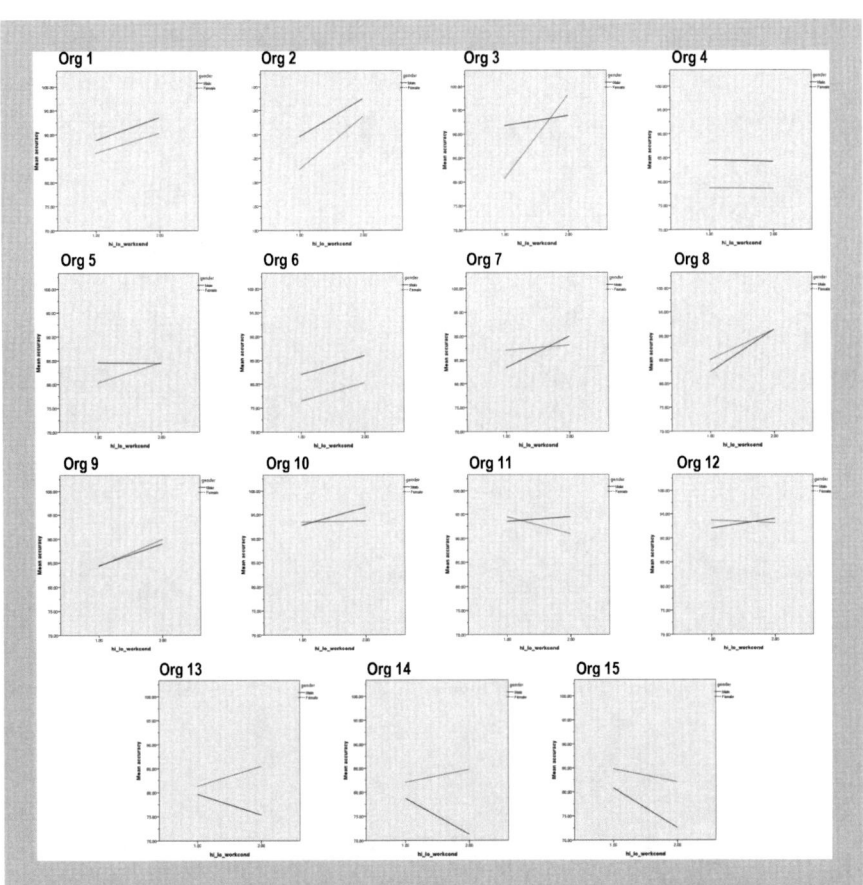

Fig. 9.14 Two-way workcond by gender interactions for the significant 'random' effect for the 15 organisations in Maree's new study

interaction regression coefficients had a significant variance across the various organisations at Level 2 of the analysis. The relationships between **workcond** and **gender** and between **workcond** and **log_speed** therefore varies from one organisation in the sample to the next. To get a visual feel for these trends, Figs. 9.14 and 9.15, produced using SPSS, display these interactions for each of the 15 organisations.

Figure 9.14 shows the **workcond** by **gender** interactions and it is clear that the interaction was strongly evident in some organisations (e.g. Org 3, Org 7, Org 13, Org 14) and non-existent in others (e.g. Org 1, Org 4, Org 6). Take Org 3, for example: the interaction showed that female employees were much less accurate under high stress conditions (left-end of X-axis) compared to male employees, but under low stress working conditions (right-end of X-axis), the situation was reversed. On the other hand, in Org 1 for instance,

(continued)

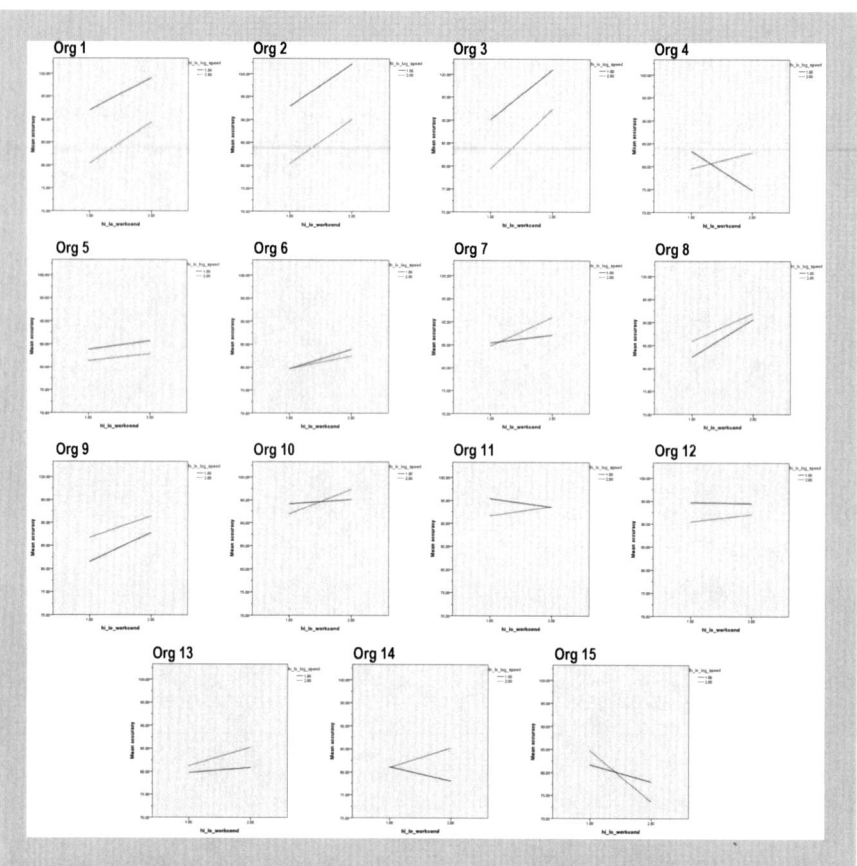

Fig. 9.15 Two-way workcond by log_speed interactions for the significant 'random' effect for the 15 organisations in Maree's new study

female and male employees were both more accurate in their decisions under low stress working conditions compared to high stress working conditions, with males always being more accurate than females.

Figure 9.15 shows the **workcond** by **log_speed** interactions where it is again clear that the interaction was strongly evident in some organisations (e.g. Org 4, Org 10, Org 15) and non-existent in others (e.g. Org 1, Org 2, Org 5, Org 9). Take Org 4, for example: the interaction showed that employees who made slower decisions were more accurate under high stress conditions (left-end of X-axis) compared to employees who made faster decisions, but under low stress working conditions (right-end of X-axis), the situation was dramatically reversed. On the other hand, in Org 2 for instance, slow and fast decision makers were both more accurate in their decisions under low stress working conditions compared to high stress working conditions, with slower decision makers always being more accurate than faster decision makers.

Procedure 9.5: Multilevel (Hierarchical Linear/Mixed) Models

There are other types of mixed models possible, namely models that involve repeated measures components (e.g. longitudinal studies and growth models; see Bickel 2007, ch. 11; Garson 2013; Luke 2004 and Tabachnick and Fidell 2019, ch. 15, for some discussion and examples). Technically, models from such research designs are not considered to be multilevel, but just mixed (the models have both 'fixed' and 'random' effects represented in them at the same level).

Advantages

The big advantage of multilevel models is that they permit the assessment of context effects, thereby automatically handling the correlated error problems that context effects create. By this is meant that not only are individual-level variables being examined for their impact on a dependent variable, higher level contextual effects are also being examined. Student achievement in a classroom, for example, may be a product, not only of their own abilities and motivation, but also of such classroom-based contextual influences as the teacher's style and ways of dealing with students, resources available in the classroom and so on. Team member effectiveness may be a product, not only of the team member's own abilities and personality, but also of team-based contextual influences such as the team leader's style, communication preferences and ways of exerting power. Here, we could even envisage that team member effectiveness might also depend upon how well the organisation resources its teams to do their work (a 'Level 3' analysis).

Multilevel models open up the possibilities for social and behavioural science researchers to expand their theoretical thinking and conceptualising beyond one level of analysis. The statistical technology has now evolved to a state where such theories and the models they hypothesise are explicitly testable. Producing a visual graph of Level 1 relationships broken down by values of a Level 2 variable (as was done in Figs. 9.13, 9.14 and 9.15) provides additional assistance in achieving meaningful interpretation and understanding of trends. An additional advantage of multilevel modelling is that it is a very general approach that can accommodate a wide range of models including HLM, repeated measures and growth/change models.

Disadvantages

Multilevel models are complex statistical processes to understand (visual representations aside) and implement. To achieve stable estimates when fitting a multilevel model, adequate sample sizes, *at each level of intended analysis*, are required. For example, Bickel (2007, pp. 272) suggests that "[t]he most commonly offered rule of thumb with regard to sample size for multilevel models is at least 20 groups and at least 30 observations per group". However, both Bickel (2007) and Tabachnick and Fidell (2019, section 15.3.2.1) are quick to remind researchers that these are only very rough guidelines, influenced, sometimes rather strongly, by contextual factors such as the extent to which key assumptions are met and the size of the intraclass correlation, and/or the number of predictors in the model at each level (including cross-level interactions). If multiple regression models are to be tested in a multilevel context, the sample size requirements begin to balloon very quickly. Many researchers lack the necessary resources to amass a sufficiently large database to permit multilevel model testing.

Fitting multilevel models requires highly sophisticated maximum likelihood statistical procedures that may be unstable at particular levels of analysis and therefore may not yield a viable or interpretable solution. This can especially be a problem if there is an insufficient number of groups for the Level 2 analysis. Even if a stable solution is achieved, the results can often be complex to interpret. Another problem with multilevel modelling is that it is extremely sensitive to multicollinearity amongst the predictors. If highly correlated predictors are employed, this will likely introduce further instabilities into the model fitting process. Another issue with multilevel models is that different types of models may need to be tested before settling upon the final configuration. For example, in order to estimate the intraclass correlation coefficient, you must test a very simple model, the 'intercepts-only' model, that involves only the intercept predictors at each level of the model (fixed effect intercept at Level 1; random effects intercept (defined by the Level 2 group identifier variable) at Level 2; see the discussion in Tabachnick and Fidell 2019, section 5.6.1).

Multilevel models are subject to the usual assumptions required of any regression analysis (e.g. normality, homogeneity of variance, absence of multicollinearity amongst predictors) but with the added complexity that these assumptions should be met at each level of the analysis—a much tougher task. Multicollinearity is especially tricky to handle in multilevel models and may involve centring predictors at one level of analysis for use at another level of analysis (see Tabachnick and Fidell 2019, section 15.7.1.3). Finally, the software approaches to testing multilevel models for the most part are rather tricky and non-intuitive to implement. This is one key drawback to a very general analytical approach like multilevel modelling in that the approach is so general that setting out the specifications for analysing a particular model is not easy (to say nothing of the fact that different software packages often use different nomenclature to address specific aspects of a model, adding to the confusion).

Where Is This Procedure Useful?

Multilevel models may be useful in any research context where theoretical or practical expectations involve considerations at more than one level of analysis,

Procedure 9.5: Multilevel (Hierarchical Linear/Mixed) Models

assuming that adequate sample sizes at each level can be achieved. Such models are thus potentially useful wherever higher-order context effects are thought to be operating through to the Level 1 variables. Multilevel models may also be useful in some accounting, finance and economics areas where large secondary databases, containing measurements at multiple levels and perhaps at multiple times, may be available. Finally, multilevel models can provide an important alternative analysis pathway for handling longitudinal data, compared to what standard repeated measures ANOVA offers (recall *Procedure 7.11*), while providing ways to test context effects that might be operating over time. Such models, however, are very complex to set up, estimate and interpret. Tabachnick and Fidell (2019, section 15.2) discuss a range of types of research questions that multilevel models may be suitable for addressing.

Software Procedures

Application	Procedures
SPSS	SPSS offers a procedure called *Mixed Models* (*Analyze* → *Mixed Models* → *Linear…*) which is a very general multilevel modelling procedure (for details, see Norušis 2012). The procedure is so general that it can be difficult to figure out exactly how to fit a specific type of multilevel model. A fair degree of statistical knowledge is needed to competently execute and interpret the outcomes from this modelling process. Field (2018, ch. 21), Garson (2020) and Tabachnick and Fidell (2019, ch. 15) provide some coherent guidance to using SPSS for conducting a range of multilevel analyses.
NCSS	NCSS offers a procedure called *Mixed Models* (*Analysis* → *Regression* → *Mixed Models*). However, the range of models appears to be limited to single level linear models rather than hierarchical linear models, where there is a higher-order nesting structure for research participants. Hintze (2007, ch. 220) describes and illustrates how to use NCSS to estimate a variety of mixed regression models. These capabilities are somewhat restricted compared to the capabilities offered by other software systems such as SPSS, SYSTAT, HLM, MPlus and Stata.
SYSTAT	SYSTAT offers several procedures called *Mixed Models* (*Analyze* → *Mixed Models* → *Hierarchical Linear Mixed Models…* or *Linear Mixed Models…* or *Mixed Regression…* or *Variance Components…*) which are all very general multilevel modelling procedures (for details, see Hedeker et al. 2009). They are also so general that it can be difficult to figure out exactly how to fit a specific type of multilevel model, although they are a bit easier to use than the SPSS procedure. Still, a fair degree of statistical knowledge is needed to competently execute and interpret the outcomes from this modelling process. One big advantage in using SYSTAT as a multilevel modelling environment is that it can produce the group-based regression line overlay plots like Fig. 9.13, which can greatly facilitate interpretation.
STATGRAPHICS	STATGRAPHICS offers a limited range of multilevel modelling capabilities; namely *Compare* → *Analysis of Variance* → *Variance Components…* analysis (in the context of an analysis of variance model) and *Relate* →

(continued)

Application	Procedures
	Multiple factors → General Linear Models... analysis (where certain random effects models can be set up). These capabilities are somewhat restricted compared to the capabilities offered by other software systems such as SPSS, SYSTAT, HLM, MPlus and Stata.
R Commander	**R** Commander does not have multilevel modelling capabilities. However, Field et al. (2012), Finch et al. (2017) and Garson (2020) describe and illustrate a range of multilevel analysis procedures available through functionalities and packages within the main **R** software environment. A fair degree of statistical knowledge is needed to competently execute and interpret the outcomes from **R**-based approaches.
MPlus	Geiser (2013, ch. 5) and Finch and Bolin (2017) describe and illustrate the multilevel modelling capabilities available in the MPlus software package. A demo version of the MPlus software can be downloaded from http://www.statmodel.com/demo.shtml. A fair amount of statistical knowledge is needed to competently execute and interpret the outcomes from this package as well as sound familiarity with the MPlus data analysis environment.
HLM	Raudenbush et al. (2011) and Garson (2020) describe and illustrate the extensive multilevel modelling procedures available in the HLM 7 software package. A student version of the HLM 7 software can be downloaded from http://www.ssicentral.com/index.php/2-uncategorised/87-hml-student (note that this version has some restrictions attached to the size and configuration of models tested). A wide variety of linear and nonlinear as well as multivariate mixed models can be tested using HLM 7. HLMs with up to four levels of analysis can be configured and analysed. A substantial amount of statistical knowledge is needed to competently execute and interpret the outcomes from this package. [Note that a newer version, HLM 8, is available, but without an accompanying student version.]
Stata	Hamilton (2013, ch. 13) and Garson (2020) discuss and illustrate how to evaluate a wide range of multilevel models within the Stata package. Stata offers some nice graphical options for displaying multilevel model outcomes as well and Hamilton (2013) reviews and illustrates some of these capabilities. A substantial amount of statistical knowledge is needed to competently execute and interpret the outcomes from this package as well as sound familiarity with the Stata data analysis environment.

Procedure 9.6: Classification & Regression Trees

Classification Purpose   Multivariate; inferential; parametric/nonparametric.
Classification trees classify participants into subgroups based on information from one or more independent or predictor variables.

Regression trees form subgroups of participants who have similar patterns of response on some dependent variable of interest based on information from one or more independent or predictor variables.

Procedure 9.6: Classification & Regression Trees   615

Measurement level   For classification trees, the dependent variable is categorical (nominal or ordinal) in nature; the predictors may be obtained using any measurement scale as long as any categorical (nominal or ordinal) measures are appropriately coded for the analyses.

For regression trees, the dependent variable is continuous (interval or ratio) in nature; the predictors may be obtained using any measurement scale as long as any categorical (nominal or ordinal) measures are appropriately coded for the analyses.

Classification and regression trees (CART) are techniques that have emerged from the machine learning literature as more general approaches to analysing predictive systems (see, for example, Breiman et al. 1984; Everitt 2005: Ma 2018; Steinberg and Colla 1997; see also www.statsoft.com/textbook/classification-and-regression-trees/). Classification trees provide an alternative methodology to discriminant analysis (recall *Procedure 7.17*), logistic regression (recall *Procedure 7.14*) and cluster analysis (recall *Procedure 6.6*). The goal of classification trees is to work out how best to predict group membership (a categorical dependent variable) using information from several predictors. Regression trees provide an alternative methodology to multiple regression (recall *Procedures 6.4* and *7.13*) and ANOVA (recall *Procedure 7.10*). The goal of regression trees is to work out how best to predict a continuous dependent variable using information from several predictors).

You may have guessed that a central feature of classification and regression trees is the 'trees' aspect. Instead of trying to estimate statistically optimal prediction equations, classification and regression trees search for patterns in the relationships between a dependent variable and one or more independent variables or predictors. This is accomplished by successively splitting the sample of observations into subgroups that have relatively homogeneous response patterns on the dependent variable. Splitting is done using information from one predictor at a time in a way that attempts to maximise the explanatory power of the split. Traditional *F*-tests (for regression trees) or chi-square tests (for classification trees) are used to determine which variables and which splits are optimal. Splitting stops when no more significant predictors on which to base a split can be identified. The history of subgroup splits is then displayed in a tree diagram. The tree diagram can then be used as a predictive tool, rather like a decision tree.

One of the more interesting benefits of using classification and regression trees for predictive analysis is that they are far more sensitive to identifying potential interactions between independent or predictor variables that more traditional parametric techniques (such as multiple regression or discriminant analysis) have difficulty detecting, especially in small samples. In this sense, classification and regression trees can be considered as quasi-nonparametric methods. The overall quality of a classification or regression tree solution is indexed by a *proportional reduction in error (PRE) measure*. The PRE measure reflects how much we can reduce our prediction errors by using the tree structure produced by the analysis. The PRE measure ranges in value from 0.0 (no reduction in prediction errors) to 1.0 (prediction errors are effectively zero, i.e. perfect predictions are made by the tree). Thus, a high PRE value (toward 1.0) indicates a good prediction tree; a small value (towards 0) indicates a poorly performing prediction tree.

> Suppose Maree was interested in trying to predict whether quality control inspectors work within the computer manufacturing industry using the predictors available in the QCI database. She conducted a classification tree analysis using the **educlev, gender, mentabil, workcond, jobsat, speed, accuracy, quality, training** and **orgsatis** variables to predict a new binary variable, **computer**, which categorised inspectors as being either from the computer manufacturing industry (either PC or Large Business Computers) or as inspectors from all other industries. Her goal was to see what variables predicted employment in the computer manufacturing industry and how. Figure 9.16 (produced using SYSTAT with the phi coefficient defining the loss function) shows the resulting classification tree. The PRE measure for the entire tree was .642, indicating that errors in predicting which industry an inspector worked in were reduced by over 64%, given the subgroup splitting structure embodied in the tree. The top node in the tree showed the modal group in the entire sample, prior to any predictive split (in this case, there were many more inspectors from other industries; 24% of entire sample of inspectors were not from other industries, but from the computer industry)
>
> For this classification tree, only three predictors resulted in significant subgroup splits: **speed** (reduced errors most by a PRE of .370), **accuracy** (added an additional .068 to reduction of prediction errors) and **workcond** (added a further .204 reduction in prediction errors). For each tree 'node' (i.e. box) in Fig. 9.16, three statistics were shown: mode (which category of inspectors occurred most frequently at that node), impurity (proportion of inspectors that were misclassified into the wrong group) and N (the number of inspectors classified at that node).
>
> The first split broke the inspectors into two groups: one with **speed** scores less than 3.83 (44 inspectors predominantly from other industries who were faster decision makers, 6.4% of whom were misclassified as being from the computer industry) and the other with **speed** scores greater than or equal to 3.83 (54 inspectors predominantly from the computer industries who were slower decision makers, 22.2% of whom were misclassified as being from other industries).
>
> The left **speed**-split subgroup was then further split into two groups based on **accuracy** scores less than 93% (39 less accurate inspectors, all of whom were correctly classified as working in other industries) and greater than or equal to 93% (5 highly accurate inspectors predominantly from the computer industry, 24% of whom were misclassified as working for other industries).

(continued)

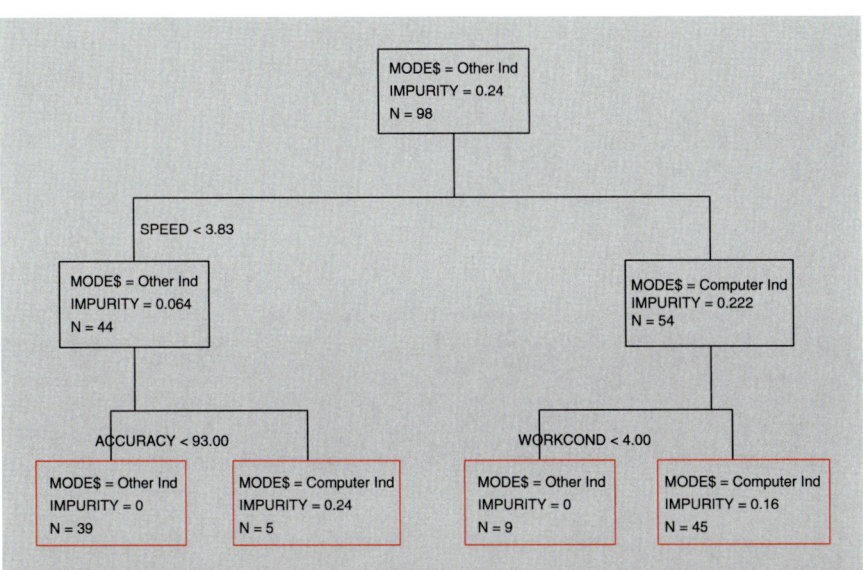

Procedure 9.6: Classification & Regression Trees

```
                    ┌─────────────────────┐
                    │ MODE$ = Other Ind   │
                    │ IMPURITY = 0.24     │
                    │ N = 98              │
                    └─────────────────────┘
                    SPEED < 3.83
        ┌───────────────────────┐       ┌───────────────────────┐
        │ MODE$ = Other Ind     │       │ MODE$ = Computer Ind  │
        │ IMPURITY = 0.064      │       │ IMPURITY = 0.222      │
        │ N = 44                │       │ N = 54                │
        └───────────────────────┘       └───────────────────────┘
        ACCURACY < 93.00                WORKCOND < 4.00
  ┌──────────────┐ ┌──────────────┐  ┌──────────────┐ ┌──────────────┐
  │MODE$=Other Ind│ │MODE$=Comp Ind│  │MODE$=Other Ind│ │MODE$=Comp Ind│
  │IMPURITY = 0   │ │IMPURITY=0.24 │  │IMPURITY = 0   │ │IMPURITY=0.16 │
  │N = 39         │ │N = 5         │  │N = 9          │ │N = 45        │
  └──────────────┘ └──────────────┘  └──────────────┘ └──────────────┘
```

Fig. 9.16 Classification tree predicting whether quality control inspectors work within the computer manufacturing industry

Finally, the right **speed**-split subgroup was further split into two groups based on **workcond** scores less than 4 (9 inspectors working in perceived more stressful working conditions, all of whom were correctly classified as working in industries other than computer) and greater than or equal to 4 (45 inspectors predominantly from the computer industry working in perceived less stressful working conditions, 16% of whom were misclassified as working for other industries.)

The fact that the initial **speed**-split groups were subsequently split by *different* variables indicates that there were important interactions at play in the data set; interactions that could not be picked up using discriminant analysis or logistic regression analysis. For the left **speed**-split subgroup, prediction was improved by acknowledging an interaction between **speed** and **accuracy** and this interaction related primarily to predicting non-computer industry inspectors. For the right **speed**-split subgroup, prediction was improved by acknowledging an interaction between **speed** and **workcond** and this interaction related primarily to predicting computer industry inspectors.

Basically, what the classification tree analysis conveyed was that inspectors who worked in non-computer industries tended to be faster decision makers but with lower accuracy whereas inspectors who worked in the computer industry tended to be slower decision makers who perceived that they had less stressful working conditions. The tree model was better at predicting inspectors who didn't work in the computer industry than it was at predicting those who do.

(continued)

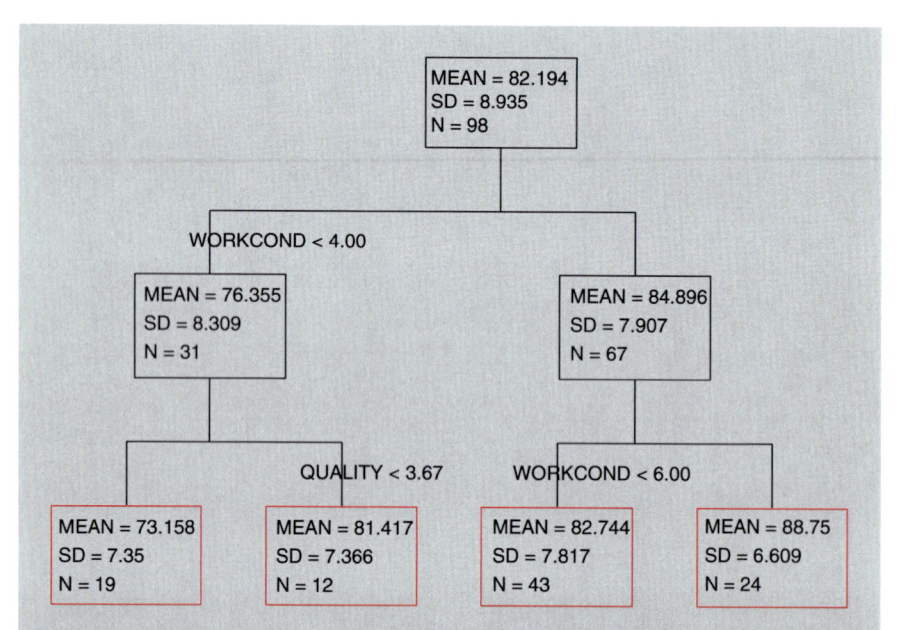

```
                        ┌──────────────────┐
                        │ MEAN = 82.194    │
                        │ SD = 8.935       │
                        │ N = 98           │
                        └──────────────────┘
              WORKCOND < 4.00
   ┌──────────────────┐              ┌──────────────────┐
   │ MEAN = 76.355    │              │ MEAN = 84.896    │
   │ SD = 8.309       │              │ SD = 7.907       │
   │ N = 31           │              │ N = 67           │
   └──────────────────┘              └──────────────────┘
         QUALITY < 3.67                 WORKCOND < 6.00
┌────────────┐ ┌────────────┐   ┌────────────┐ ┌────────────┐
│MEAN=73.158 │ │MEAN=81.417 │   │MEAN=82.744 │ │MEAN=88.75  │
│SD = 7.35   │ │SD = 7.366  │   │SD = 7.817  │ │SD = 6.609  │
│N = 19      │ │N = 12      │   │N = 43      │ │N = 24      │
└────────────┘ └────────────┘   └────────────┘ └────────────┘
```

Fig. 9.17 Regression tree predicting inspection decision **accuracy** of quality control inspectors

Suppose now that Maree also wanted to try to predict inspector decision **accuracy** using the **educlev, gender, mentabil, workcond, jobsat, speed, accuracy, quality, training** and **orgsatis** variables. Since **accuracy** was a continuous measure, this analysis resulted in a regression tree. Figure 9.17 (produced using SYSTAT employing a least squares loss function) shows the resulting tree. The PRE measure for the entire tree was .336, indicating that errors in predicting the decision **accuracy** of inspectors were reduced by nearly 34%, given the subgroup splitting structure embodied in the tree.

For this regression tree, only two distinct predictors resulted in significant subgroup splits: **workcond** (reduced errors most by a PRE of .200), **quality** (added an additional .065 to reduction of prediction errors) and **workcond** (added a further .072 reduction in prediction errors). For each tree 'node' (i.e. box) in Fig. 9.17, three statistics were shown: mean (average **accuracy** scores for the identified subgroup), SD (standard deviation of **accuracy** scores for the subgroup) and N (the number of inspectors contained in the subgroup at that node). The node at the top of the regression tree displayed the accuracy statistics for the entire sample, prior to any predictive split.

The first split broke the inspectors into two groups: one with **workcond** scores less than 4.00 (31 inspectors who worked in perceived more stressful working conditions and who had an lower average **accuracy** score of 76.355%) and the other with **workcond** scores greater than or equal to 4.00

(continued)

(67 inspectors who worked in perceived less stressful working conditions and who had a higher average **accuracy** score of 84.946%).

The left **workcond**-split subgroup was then further split into two groups based on **quality** scores less than 3.667 (12 inspectors who rated their company less highly with respect to **quality** focus but who showed a higher average **accuracy** score of 81.417%) and greater than or equal to 3.667 (19 inspectors who rated their company more highly with respect to **quality** focus but who showed a lower average **accuracy** score of 73.158%).

Finally, the right **workcond**-split subgroup was further split into two groups based on **workcond** scores less than 6.00 (43 inspectors working in perceived more stressful conditions who showed an average **accuracy** score of 82.744%) and greater than or equal to 6.00 (24 inspectors working in perceived very much less stressful conditions who showed a very high average **accuracy** score of 88.75%).

The fact that the initial **workcond** groups were subsequently split by *different* variables indicates that there were important interactions at play in the data set; interactions that could not be picked up using discriminant analysis or logistic regression analysis. For the left **workcond**-split subgroup, prediction was improved by acknowledging an interaction between **workcond** and **quality** and this interaction related primarily to predicting less accurate inspectors who felt that they worked in more stressful working conditions but who had more positive perceptions of quality in their companies. For the right **workcond**-split subgroup, prediction was improved by acknowledging a further split based on **workcond** and this finer split related to predicting highly accurate inspectors who perceived that they enjoyed extremely positive working conditions.

Basically, the regression tree analysis conveyed that inspectors who felt that they worked in more stressful working conditions tended to be less accurate decision makers, especially if they perceived that their company had a better focus on quality, whereas inspectors who perceived that they had extremely good working conditions were more accurate than those inspectors who perceived that they had less than optimal working conditions. **Workcond** was thus a very potent predictor for accuracy and worked in different ways depending upon where, along the **workcond** rating continuum, inspectors tended to fall.

Advantages

Classification and regression trees provide distinctly different ways of sifting through predictive relationships in a database. In particular, they focus on identifying relatively homogeneous (with respect to the dependent variable) subgroups using statistically computed cut-off points. Interactions amongst predictors are almost always detected using classification or regression trees, whereas as the researcher would be hard-pressed to do so using their more traditional statistical parallels, such as discriminant analysis and multiple regression. In particular, classification and regression trees automate the search for effective predictor relationships and interactions, a very useful strategy for exploring large databases having many potential predictor variables. In fact, another name for the classification tree method is CHAID (CHi-square Automatic Interaction Detection) and for the regression tree method is AID (Automatic Interaction Detection). Classification and regression trees can easily be cross-validated and doing so would enhance the utility of the resulting tree structure.

Classification and regression trees are hybrid technologies in the sense that they are parametric, in terms of relying on statistical tests to flag significant predictors, but also nonparametric, in that they do not estimate statistical parameters, they identify cut-points for the most effective placement of subgroup boundaries. If a dependent variable does not satisfy parametric assumptions, all is not lost as regression trees will tend to isolate extreme observations as distinct subgroups, so that predictive relationships can be more precisely understood. In addition, different loss functions (e.g. least squares, trimmed means) can be used to drive the construction of the tree, and some loss functions, such as trimmed means and least absolute deviations, work quite well in circumstances where the data have extreme outliers (see Wilkinson 2009 for more discussion of this issue).

Disadvantages

Classification and regression trees provide non-standard outcomes in terms of the usual sorts of statistical model fitting that is done in social and behavioural research. This can make it difficult to compare results from a classification or regression tree analysis to results from more traditional statistical analyses used in other studies of the same variables. Another difficulty with classification and regression trees is that the software for computing them is only now becoming widely available and this has contributed to a slow spread of use.

Furthermore, there are some problematic issues that have emerged with respect to the use of classification and regression trees. A sound classification or regression tree analysis requires a clear rule for when to stop growing the tree. Traditional F-tests or chi-square tests have been used to provide a stopping rule, namely stop splitting out subgroups when no predictor will produce a significant split as gauged by one of these statistical tests at the researcher's established level of significance. However, the use of such a rule can snare the unwary researcher in the trap of conducting multiple significance tests on a single data sample, possibly leading to more significant predictors (meaning more identified splits and subgroups) than is really justified. This is the same trap that snares researchers who use stepwise regression or discriminant analysis methods. For this reason, cross-validation of a tree structure should be considered a mandatory exercise before seeking to publish tree-based outcomes. This would normally be done by holding out a randomly selected group of participants from the process of building the initial tree, then applying the resulting tree to classification/prediction of these holdout participants. This will give a clearer sense of the actual predictive utility of the tree.

Where Is This Procedure Useful?

Classification and regression trees may be useful in contexts where the assumptions of more traditional statistical procedures are not strictly met by one's data or where the researcher explicitly wants a decision-tree like outcome to facilitate decision making. Researchers in the medical disciplines as well as market researchers have found classification and regression trees to be of practical value in guiding diagnosis, decision making and market segmentation. In fact, it is not a big jump to move from classification and regression trees to the creation of an expert system that can query data and sift them through a series of nodes to arrive at a conclusion. Classification and regression trees realise their maximal potential in the 'mining' of large databases where predictive trends and patterns are being sought (see also *Procedure 9.9*; in some quarters, classification and regression trees are considered to be examples of machine learning algorithms for data mining, see, for example, Loh 2011). In this sense, classification and regression trees are exploratory rather than confirmatory or hypothesis testing procedures.

Software Procedures

Application	Procedures
SPSS	SPSS offers classification and regression tree analysis only as part of an add-on procedure called *Decision Tree*. Access to this package would depend upon local licensing arrangements at your institution. More details on this procedure can be found at ftp://public.dhe.ibm.com/software/analytics/spss/documentation/ statistics/25.0/en/client/Manuals/IBM_SPSS_Decision_Trees.pdf.
NCSS	NCSS currently does not have the capability to compute classification or regression trees.
SYSTAT	*Advanced → Trees (C&RT)...* and select the desired 'Dependent:' variable, independent variable(s) ('Independent(s):') and 'Loss:' Function. If the dependent variable is categorical, choose either the 'Phi coefficient', Gini index' or 'Twoing' loss function and SYSTAT will automatically produce a classification tree analysis; if the dependent variable is continuous, choose the 'Least squares', 'Trimmed mean' or 'Least absolute deviations' loss function and a regression tree analysis will be reported. SYSTAT *Trees*

(continued)

Application	Procedures
	(C&RT) also offers a range of options for stopping criteria and how to display the nodes in the tree. For more details, see Wilkinson (2009).
STATGRAPHICS	STATGRAPHICS recently added classification and regression trees analysis capabilities via the *Classification and Regression Trees* procedure in version 18 (see http://blog.statgraphics.com/classificationregressiontrees).
R Commander	**R** Commander currently does not have the capability to compute classification or regression trees. However, there are functionalities and packages available within the main **R** software environment for conducting classification and regression tree analyses; see, for example, http://www.statmethods.net/advstats/cart.html.

Procedure 9.7: Social Network Analysis

Classification	Multivariate; descriptive; correlational.; inferential
Purpose	To describe, summarise and visually display nodes (e.g. individuals, groups, organisations) and relational links between those nodes (e.g. social ties, kinship, strength, communications, interactions, liking, trust), connected in a social network.
Measurement level	Social network connections may be quantified simply using binary indicators (e.g. a connection is 'present/absent') or counts (e.g. number of contacts or communication interactions) or on the basis of constructs measured at ordinal-, interval- or ratio-level (e.g. degree of connection strength, liking or trust). Links may be uni-directional, bi-directional or reciprocal and may reflect symmetric or asymmetric relationships.

Social network analysis (e.g. see Borgatti et al. 2013; Brieger 2004; Scott 2017) is commonly employed in sociological, educational and organisational research to explore the nature and relationships between and among a set of entities. Borgatti et al. (2013) discuss the unique methodological and data collection issues associated with social network analysis. The focus of data gathering is rather different from standard approaches, simply because the focus is often on asking (typically via questionnaire or interview) sampled people (often called 'actors' or 'agents') to identify other actors within a network of interest (which may be a group, a team, a committee, a department, an organisation, a collection of social groups or organisations and so on) and answer questions/provide ratings about their relationships with those other actors. Social networks may also be observed (e.g. at meetings or other gatherings), where the researcher records certain aspects of behaviour relative to the

Procedure 9.7: Social Network Analysis

relationships between actors. Finally, some aspects of knowledge about networks may be accumulated through documents relevant to the network. In some cases, the data gathered may be qualitative in nature, which then creates the need for a preliminary step in the analytical process to convert the qualitative data into meaningful quantitative measures, if formal mathematical analyses of the network are to be pursued (consistent with positivist guiding assumptions). Complexity is added if the research focus shifts to the relationships between actors and different groups/networks or between networks themselves. In many cases, the boundary and even the membership of a network may not be fully known or accessible, which can create interesting missing data problems for the social network analyst.

The data in a social network may comprise information attached to each node (attributes) and relational information attached to each link. The types of *attribute data* that can potentially be gathered with respect to nodes or actors include:

- demographic information (e.g. gender, ethnic background, job role, role within the group or network, tenure in the network)
- background and contextual data (e.g. reasons why the actor is in the network or the group, roles, expectations and concerns, motivations; Scott (2017) refers to this type of data as *ideational data*)

The types of *relational data* that can potentially be gathered with respect to network links may include:

- indications of the nature of the role-based connections with other actors in the network (e.g. supervisor, subordinate, friend, relative, mentor)
- simple binary indicators (e.g. indicating a connection to or relationship with another actor or not; present/absent-type data);
- counts (e.g. number of times actor A initiates an interaction or communicates with actor B);
- rankings (e.g. rankings of other actors according to degree of liking or closeness);
- ratings (e.g. ratings of level of trust, liking or strength of relationship).

> To illustrate social network analysis in action, suppose Maree has an opportunity to observe the management group responsible for overseeing all quality control activities in a large computer manufacturing company. The group comprises seven members and the following provides some preliminary attribute data about each member:
>
> - *Jean* is the formally appointed group leader; she is the most senior manager in the group and has been group leader for two years. Jean is close friends with Jill.
> - *Bill* is a middle manager from the Finance department. He was nominated by his boss to join the group to provide advice on financial and other resourcing matters related to quality control.
>
> (continued)

> - *Jill* is a manager from the Human Resources department and is close friends with Jean. Jean asked for her to be a member of the group to provide advice about recruiting, employing and managing the performance of quality control inspectors.
> - *Sue* is a quality control inspector and is a union representative in the group, nominated by the union president to be a member of the group to provide a union perspective on workload, workflow and management issues.
> - *Joe* is a quality control inspector, who is not a union member, nominated by his boss to be a member of the group to provide a non-union perspective on workload, workflow and management issues.
> - *John* is a manager from the Manufacturing department. He is in the group to provide process-related input to quality control procedures and to take feedback back to his department about quality problems that need addressing.
> - *Mary* is the newest member of the group, having joined 6 months ago. She is the manager of the Organisational Development & Training department and is in the group to provide input and guidance on quality control training and development issues.
>
> Maree gathered data on this management group, as a network, through participant observation of a number of meetings of the group over a 6-month time interval. During those observations, she would record data such as who initiated communication with whom.

A social network comprises two key components: a set of nodes (the entities of interest in the network) and relational links (connections between any two nodes in the network). The number of actors in a network determines the size of the network. Relationships in a social network may either be *directed* (where relationships are asymmetric; e.g. Jean might like Bill but Bill might not like Jean) or *undirected* (where relationships are symmetric; e.g. where Jean and Jill are friends).

Breiger (2004) identified at least three different types of networks that a researcher could observe or gain insights into:

- *One-mode network*, also described as a 'who to whom' network, where the focus is on the membership of a specific network and relationships between those members;
- *Two-mode network*, also described as a 'who to which' network or an 'affiliation network', where the focus is on which networks actors belong to and have relationships with; and
- *'Ego-centred' network*, which looks at a network from the perspective of a focal actor.

Descriptively, a social network may be displayed in one of two ways: *matrix* (which may be symmetric or asymmetric in form) or *graph/network diagram*

Procedure 9.7: Social Network Analysis

(sometimes referred to as a 'sociogram' and may be directed or undirected in configuration). Analytically, social network analysis has strong roots in mathematical graph theory (giving rise to an analytical approach known as *sociometrics*, see Scott 2017) well as to multivariate statistics. These roots provide the foundation for social network analysis as a general analytical framework for relational data.

> For Maree's targeted management group, described above, Table 9.6 shows a matrix representation of simple 'actively talked to' data where a "1" signalled that Maree observed the actor identified in a row initiating at least one conversation with the actor identified in a column and "0" indicated that no such conversations were initiated. Everywhere there is a "1" in the matrix, there is a corresponding arrow in the diagram running from the row actor who started the conversation to the column actor who was the target of the conversation. Note that between a number of pairs of actors, arrows run in both directions, indicating a reciprocal relationship where each actor was willing to initiate at least one conversation with the other. The matrix in Table 9.6 is asymmetric in form, meaning that the numbers above the main diagonal (where each actor intersects him or herself) do not match the numbers below the main diagonal.
>
> Figure 9.18 represents the information in the matrix from Table 9.6 in a network diagram. Jean is highlighted as the formal group leader in the darker circle. This social network has seven nodes or actors and is a directed network where relationships may be asymmetric. In this case, for example, Jean has initiated at least one conversation/communication with Mary, but Mary has not initiated a conversation with Jean. Jean and Bill, however, have each initiated conversations with each other. At a glance, it appears that Jean, the group leader has had reciprocal conversations with every actor in the group except for Mary. Mary appears to be somewhat isolated within the group.
>
> **Table 9.6** Social network matrix showing 'actively talks to' connections (row person talking to column person)
>
> | | Bill | Jill | Sue | Jean | Joe | John | Mary |
> |------|------|------|-----|------|-----|------|------|
> | Bill | 0 | 1 | 1 | 1 | 1 | 1 | 0 |
> | Jill | 1 | 0 | 0 | 1 | 1 | 0 | 0 |
> | Sue | 1 | 0 | 0 | 1 | 0 | 1 | 0 |
> | Jean | 1 | 1 | 1 | 0 | 1 | 1 | 1 |
> | Joe | 0 | 1 | 0 | 1 | 0 | 1 | 0 |
> | John | 1 | 0 | 1 | 1 | 1 | 0 | 1 |
> | Mary | 0 | 0 | 0 | 0 | 1 | 0 | 0 |

(continued)

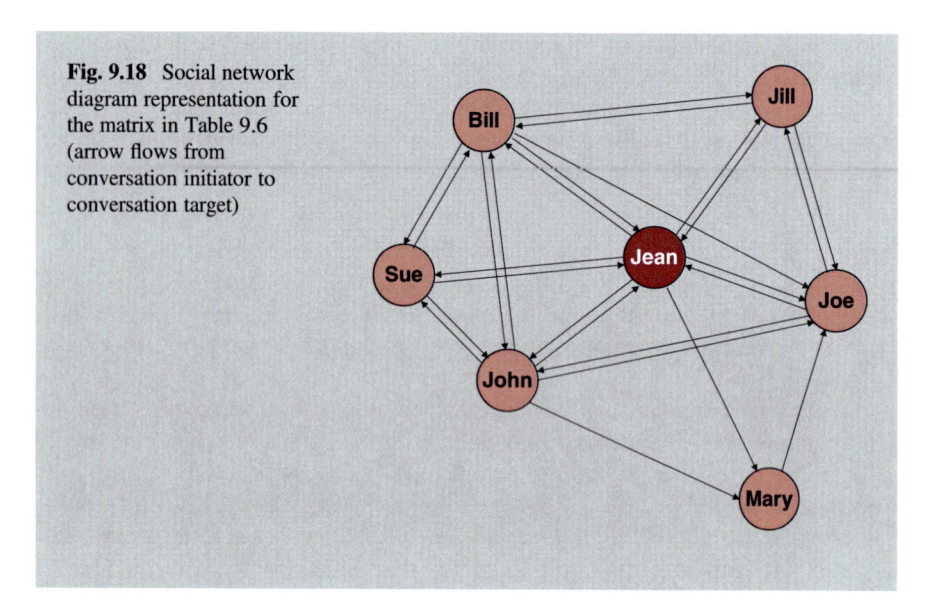

Fig. 9.18 Social network diagram representation for the matrix in Table 9.6 (arrow flows from conversation initiator to conversation target)

There are a number of key analytical concepts, relevant to social network analysis, which capture different structural and relational features of a network. These concepts generally have one or more quantitative indices associated with them (see, for example, Borgatti et al. 2013 and Knoke and Yang 2008). In the discussion below, we will refer to 'actors' to make things a bit more concrete, but you should understand that these concepts apply to nodes in general, regardless of what they represent. The concepts discussed here are just a sampling of possible ways to assess meaningful aspects of a social network. In most cases, the indices can take on values at the individual actor level, but may also be aggregated across other actors, other groups or even the entire network.

- *Clique*: refers to a sub-group within a network where all of the actors are directly connected to each other. One clique cannot be completely embedded within another clique; however, cliques may overlap each other. For example, in Maree's management group example, two readily identifiable cliques are: Jean, Joe and Jill and Jean, Bill, Sue and John. Jean is thus a member of both cliques.
- *Component*: a sub-group within a network comprising actors who are connected with each other, but not connected to actors in other parts of the network. There are no distinctly identifiable components in the management group.
- *Geodesic distance*: refers to the shortest possible path from one actor to another in the network; measured by counting the number of links to get from one to the other. For example, in Fig. 9.18 the geodesic distance from Mary to Bill is 3 (Mary → Joe → Jill → Bill), from Bill to Jean, it is 1; from Joe to Sue, it is 2 (Joe → John → Sue). The largest value for the geodesic distance in the network defines the *diameter* of the network, which has a value of 3 for the management group.

Procedure 9.7: Social Network Analysis

- *Centrality*: examines the positioning of each actor with respect to the relational links they have with others in the network.
    - *Degree centrality*: reflects the number of links an actor has. In an asymmetric or directed network, we can distinguish between *in-degree* (number of links inward toward an actor = count arrowheads) and *out-degree* (number of links outward away from the actor = count tails of arrows). For example, in the management group, Bill has an *in-degree* value of 4 and an *out-degree* value of 5; Jean, the leader, has an *in-degree* value of 5 and an *out-degree* value of 6; Mary has an *in-degree* value of 2 and an *out-degree* value of 1. Generally speaking, out-degree links reflect influence; in-degree links indicate prestige, popularity or respect (e.g. as for a mentor).
    - *Betweenness centrality*: reflects the extent to which an actor is part of the geodesic paths between all other pairs of actors in the network. The higher the level of betweenness centrality for an actor (i.e. the more geodesic pathways they are part of), the more power that actor has within the network. For example, Mary's betweenness centrality value is 0; John's value is 3 (Bill → John → Mary; Sue → John → Mary; Joe → John → Mary).
    - *Closeness centrality*: focuses on the distance between an actor and all other actors in the network. For any actor, closeness is found as the reciprocal of the sum of the geodesic distances to every other actor in the network. The closer an actor is to all others (= the larger the value for closeness), the more favourable their position in the network is. As an example, the closeness measure for Mary would be 1/13 or .077; for Jean, it would be 1/7 or .143.
- *Density*: the number of observed or identified links in a network expressed as a proportion or percentage of all the links possible in a network of that size. In the management group, the total number of links possible in an asymmetric network with seven nodes was 42 (= # nodes x (# nodes −1)) and the number of observed links was 26, giving a density of 26/42 or .619 (i.e. 61.9% of the possible links were present in the network). [Note that in a symmetric network, the number of possible links would be half of the number possible if the network was asymmetric.] Higher density in a network reflects a higher level of 'social capital' and a faster speed for diffusion of information amongst the actors.
- *Reachability*: reflects whether any two actors in the network are connected either through a direct pathway or through one or more indirect pathways. In the management group, every member is reachable via some pathway (e.g. Mary is reachable by Bill indirectly via Jean; John is reachable by Mary through Joe).
- *Structural equivalence*: refers to different actors in a network displaying the same kinds of relational links with other actors (e.g. two managers in a network having similar types/patterns of relationships with their own subordinates). Structural equivalence often equates to social position or roles within a network.
- *Structural hole*: refers to a gap between two components in a network where very few or no cross-component links exist. An actor who has or can build a link spanning such a hole can develop an advantage in power between the two components.

- *Isolate*: a single actor that is not connected to any other actors in the network or a component that is not connected to any other components in the network. If a component is also an isolate, this may represent a faction within a network. There is no isolate in the management group, but Mary is very close to being an isolate.

> Table 9.7 shows a network matrix summarising the frequency of conversations initiated by a row actor with a column actor over Maree's 6-month meeting observation period. The numbers in the main diagonal are the total number of times each actor was observed. The numbers off the diagonal are the number of time Maree saw a row actor initiate a conversation with a column actor. Figure 9.19 displays the network diagram for these data where the thickness of the link between any two actors is a function of the frequency of conversation initiation.
>
> To analyse the social network matrix in Table 9.6, Maree uses a comprehensive social network analysis software package known as UCINET (Borgatti et al. 2002; see also https://sites.google.com/site/ucinetsoftware/home). Table 9.8 reports a very small sampling of illustrative summary tables from the analysis of this binary matrix, namely a matrix of centrality measures and analyses of cliques.
>
> The top subtable in Table 9.8 reports 'MULTIPLE CENTRALITY MEASURES' in what is called a normalised form (meaning that the indices are scaled relative to the largest value for the index. Columns 1, 2 and 9 are the key columns to focus on here. Column 1 reports the normalised out-degree centrality index for each actor where Jean has the maximum value (enjoying the most influence over the network) and Mary the minimum (having the least influence over the network). Column 2 reports the in-degree centrality index for each actor where Jean and Joe are both at maximum (both are equally popular) and Mary is at minimum (least popular). The second subtable, labelled 'CLIQUES', shows that the network has four cliques and the membership of each clique is listed. Note that Mary is not part of any clique; she is on her own. The 'Clique Participation Scores' table reports the proportion of each clique that each actor is adjacent to in the network (e.g. Bill is adjacent to two of the three members of clique 3, Jill and Jean, but is not adjacent to Joe). The 'Actor-by-Actor Clique Co-Membership Matrix' shows the number of cliques that each actor is a member of along the main diagonal (e.g. Bill is in two cliques; Jean is in all four cliques) and the number of cliques that any two actors are members of are shown in the off-diagonal cells (e.g. Bill is in two cliques with Jean and one clique with Jill).
>
> Figure 9.20 shows a network diagram produced by a program called NetDraw, which is part of the UCINET software system. The frequency matrix in Table 9.7 provided the data for the diagram as did a separate file containing some coded attributes of the actors (their gender and role within the group-- Representative (union or non-union), Manager or Leader)). The diagram was

(continued)

Procedure 9.7: Social Network Analysis

Table 9.7 Social network diagram showing frequency of communication initiation connections (row person talking to column person)

	Bill	Jill	Sue	Jean	Joe	John	Mary
Bill	53	12	9	5	11	12	0
Jill	13	41	0	15	9	0	0
Sue	7	0	52	12	0	12	0
Jean	17	14	9	75	15	9	5
Joe	0	13	0	6	66	9	0
John	14	0	5	9	7	58	4
Mary	0	0	0	0	9	0	21

Fig. 9.19 Social network diagram, from the data in Table 9.7, showing the directional relationships scaled by the frequency of communication initiation amongst the 7 group members (arrow goes from communication initiator to communication target)

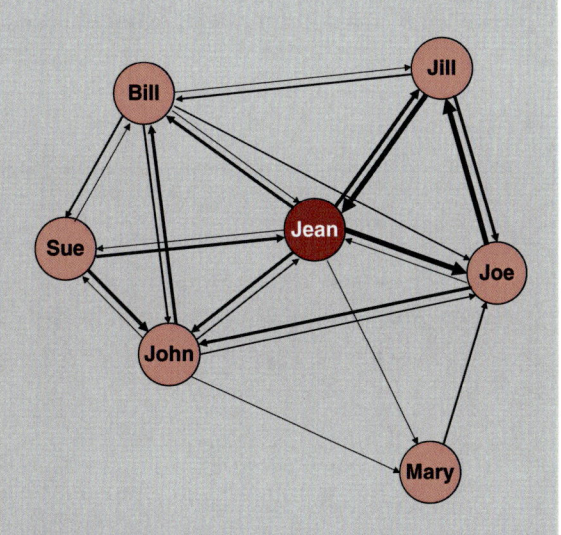

produced using a metric multidimensional scaling analysis (recall *Procedure 6.7*) and shows that Jean does occupy a central position in the network (as one would expect for a leader). The resulting spatial configuration of the network, based on frequency of communication initiation, is displayed in Fig. 9.20, with the size of the links and arrowheads between actors scaled according to frequency of communication initiation (the actual frequency is shown as well with the number toward the head of each arrow in most cases). Note that the nodes or actors have also been scaled: symbol shape and size codes group role (large triangle = group Leader; medium squares = Managers and small circles = Representatives) and symbol colour codes gender (light = females; dark = males). Mary's near isolation is quite evident (Maree might

(continued)

Table 9.8 Social network statistics, produced by UCINET, for the network matrix in Table 9.6

```
MULTIPLE CENTRALITY MEASURES
-------------------------------------------------------------------

Treat data as:                          Directed
Type of scores to output:               Normalized

Centrality Measures

              1     2     3     4     5     6     7     8     9
           OutDe Indeg OutBo InBon Out2S In2St OutAR InARD Betwe
           ----- ----- ----- ----- ----- ----- ----- ----- -----
   1 Bill  0.833 0.667 3.368 2.677 1.000 0.833 0.917 0.806 0.067
   2 Jill  0.500 0.500 2.266 2.298 1.000 1.000 0.750 0.750 0.022
   3  Sue  0.500 0.500 2.472 2.216 1.000 0.833 0.750 0.722 0.000
   4 Jean  1.000 0.833 3.481 3.315 1.000 1.000 1.000 0.917 0.206
   5  Joe  0.500 0.833 2.197 3.181 1.000 1.000 0.750 0.917 0.189
   6 John  0.833 0.667 3.026 2.854 1.000 1.000 0.917 0.833 0.117
   7 Mary  0.167 0.333 0.551 1.545 0.667 1.000 0.528 0.667 0.000

UCINET 6.485 Copyright (c) 1992-2012 Analytic Technologies
```

```
CLIQUES
-------------------------------------------------------------------

Minimum Set Size:                       3

NOTE: Directed graph. You may prefer to symmetrize first.
4 cliques found.

    1:  Bill Sue Jean John
    2:  Bill Jill Jean
    3:  Jill Jean Joe
    4:  Jean Joe John

Clique Participation Scores: Prop. of clique members that each node is adjacent to

              1     2     3     4
           ----- ----- ----- -----
   1 Bill  1.000 1.000 0.667 0.667
   2 Jill  0.500 1.000 1.000 0.667
   3  Sue  1.000 0.667 0.333 0.667
   4 Jean  1.000 1.000 1.000 1.000
   5  Joe  0.500 0.667 1.000 1.000
   6 John  1.000 0.667 0.667 1.000
   7 Mary  0.000 0.000 0.000 0.000


Actor-by-Actor Clique Co-Membership Matrix

           1 2 3 4 5 6 7
           B J S J J J M
           - - - - - - -
   1 Bill  2 1 1 2 0 1 0
   2 Jill  1 2 0 2 1 0 0
   3  Sue  1 0 1 1 0 1 0
   4 Jean  2 2 1 4 2 2 0
   5  Joe  0 1 0 2 2 1 0
   6 John  1 0 1 2 1 2 0
   7 Mary  0 0 0 0 0 0 0

UCINET 6.485 Copyright (c) 2002-12 Analytic Technologies
```

(continued)

Procedure 9.7: Social Network Analysis

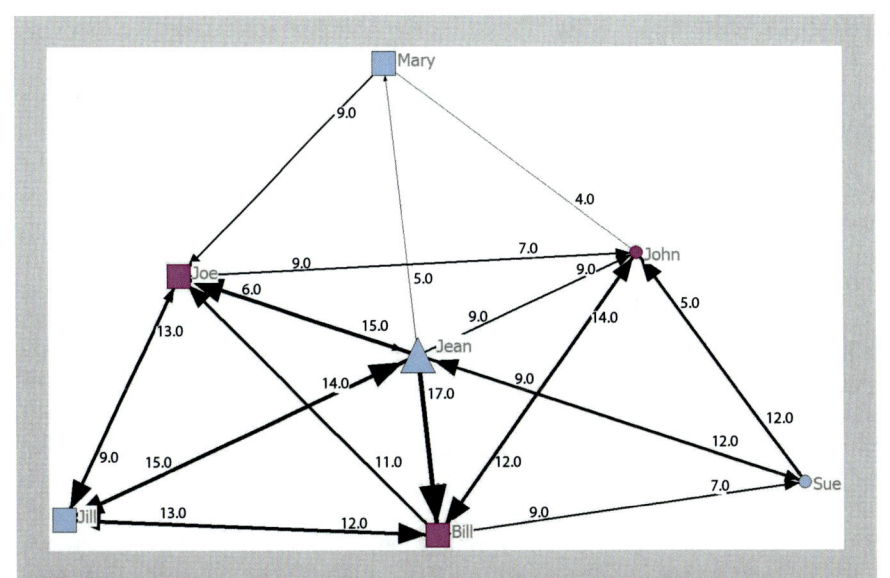

Fig. 9.20 Social network diagram, produced by NetDraw, for the matrix in Table 9.7

conclude that Mary has not been accepted by the group) and there appear to be slightly weaker connections between Jean, John and Sue than there are between Jean, Joe, Bill and Jill. Jean's strong connection to Jill (they are friends) is also evident.

Advantages

Social network analysis offers a highly coherent and mathematically defensible approach to analysing relationships in networks and as well as the associations between actors and their relational links. Social network analysis is adaptable to multiple levels of analysis, from individual actors in a network to large scale multi-organisational, even multinational, networks. The metrics of social network analysis work at any level of analysis, setting the stage for defensible cross-level comparisons, if desired. The stories that emerge can provide insights into issues such as the structural composition and relationships of networks, who is central and who is isolated, in- and out-group relationships, how power is distributed and perhaps even enacted, how and which cliques and factions form and under which circumstances and many other issues. Stories between as well as within networks may also emerge from social network analyses.

Disadvantages

Social network analysis has a few important limitations. Most social network analyses tend to be static in nature, focusing on actors, links and behaviours at a specific point in time. However, in many cases, changes in the network through time may be quite critical to understand. Examining and evaluating dynamics over time can be tricky with standard social network analysis. Even with a static focus, understanding the complete composition of a network and its boundaries, especially if it is large or virtual, may be difficult. Furthermore, individual actors in a network may not be able to self-report on their relationships with complete candour or with infallible memories. Assessing larger-scale networks, where relationships between groups or organisations are of interest, creates some interesting methodological choices that need to be made. These choices can include: who to obtain data from within each group in a network (sampling), how to best understand the contextual influences that may shape network connections and interactions (contextualising the interactions) and what to actually measure about the relationships (measurement issues). Finally, social network analysis functionality has been slow to make it into some of the major statistical packages. UCINET appears to be the most widely used of the stand-alone packages available.

Where Is This Procedure Useful?

Social network analysis is useful in any investigation where the researcher plans to or has explicitly gathered node/link type of data on a network or networks of interest, at virtually any level of analysis. Many studies of human and social capital in organisations, for example, rely on social network analysis (e.g. Kilduff and Tsai 2003). Social network analysis is gaining wider acceptance in a range of social and behavioural science disciplines and is growing in sophistication. The capacity to weight and measure aspects of each link in a network as well as to track key attributes of nodes provides opportunities for a more complete exploration of a social network. It is also possible to use social network analysis to explore possibilities for change through intervention in a network (e.g. to cover/exploit structural holes, improve communication and distribution of resources, leverage specific power and influence relationships or study factional behaviour).

Software Procedures

Application	Procedures
SPSS	There is an add-on module for SPSS called *IBM SPSS Modeler Social Network Analysis* (see https://www.ibm.com/support/knowledgecenter/en/SS3RA7_15.0.0/com.ibm.spss.sna.doc/product_overview.htm). This

(continued)

Procedure 9.8: Specialised Forms of Regression Analysis 633

Application	Procedures
	module can facilitate the conduct of a range of group-based (e.g. density, centrality, clique identification) and diffusion-oriented (locating the actors most influenced by other actors) social network analyses. If you prepare your network matrix in an Excel spreadsheet, that matrix can be imported into SPSS where certain analyses, such as multidimensional scaling and cluster analysis, can be conducted. [Note that access to social network analysis procedures presumes your institution has included *IBM SPSS Modeler* in its licensing arrangements.]
R Commander	**R** Commander currently offers no social network analysis capabilities. However, there is a package called *sna* (see https://cran.r-project.org/web/packages/ sna/sna.pdf) available for use with the main **R** program, which can carry out a wide range of social network analyses, including numerical and graphical analyses.
UCINET and NetDraw	*UCINET* and *NetDraw* (see https://sites.google.com/site/ucinetsoftware/home) are social network analysis software programs that are very widely used in research. NetDraw is a free package that works with UCINET (trial version is available for download at the above website; the full version is not that expensive). Both Borgatti et al. (2002, 2013) have discussed and illustrated the use of UCINET and NetDraw for a wide range of social network analyses. UCINET is one of the more comprehensive packages available. The user interface is a bit clunky and sterile and output modes are a bit inflexible, but the software package works well as a whole. UCINET comes with built-in data editors and facilities to create both network matrices (one and two-mode) and node attribute matrices.
InFlow	*InFlow* is another package offering a fairly standard suite of social network analysis capabilities (see http://orgnet.com/inflow3.html). Pricing information can only by obtained by contacting the company directly.
Sentinel Visualizer	*Sentinel Visualizer* (see http://fmsasg.com/Products/SentinelVisualizer/) is a data visualisation software package with excellent social network analysis capabilities. In addition, this package has functionalities designed to enhance the analysis of networks over time. The package is rather expensive. A trial version can be obtained by contacting the company directly.
NodeXL	*NodeXL* is a free open source software template for Excel (see https://www.smrfoundation.org/nodexl/). This template is very useful for graphically displaying and analysing social networks. It takes a bit of effort to learn how to use but is fairly flexible in what it can do. Hansen et al. (2011) provides a useful tutorial learning guide for NodeXL as well as case studies illustrating its use in understanding various types of social media networks (such as Facebook, Flickr and YouTube).

Procedure 9.8: Specialised Forms of Regression Analysis

Classification Multivariate; correlational.; inferential; parametric/nonparametric.

Purpose To predict and/or explain different types of dependent variables, using alternative, perhaps less assumption-sensitive, regression model estimation methods, evaluate nonlinear predictive relationships and/or assess mediator or moderator effects in regression models.

Measurement level The nature of the dependent variable for each specialised type of regression analysis changes depending upon the specific regression modelling approach adopted. As for the ordinary least squares form of multiple regression analysis (recall *Procedure 7.13*), predictors may be at any level of measurement so long as they are correctly coded for use in a regression model (recall *Fundamental concept VI*). For some specialised analyses (e.g. nonlinear regression, moderated regression), specific additional predictors may need to be created by the researcher prior to commencing the model building process.

Multiple regression analysis, and by logical and statistical extension, the general linear model, has evolved into a number of more diverse forms, each having a more specialised purpose or function. For some of these more specialised forms of regression analysis, the evolution has occurred so that nominal or ordinal dependent variables can be predicted or explained (we have already reviewed one such procedure, logistic regression (recall *Procedure 7.14*), for predicting a dichotomous categorical dependent variable). In the case of ordinal dependent variables, the term *generalised linear models* is typically used to refer to a class of regression-type approaches to modelling such variables (see Cohen et al. 2003, ch. 13). For other forms of regression, the evolution has focused on improving the model estimation procedures, perhaps making that procedure more robust to violations of the assumptions of ordinary least squares regression analysis. Finally, some adaptations and extensions have evolved for ordinary least squares regression analysis so that more sophisticated models, involving nonlinear or interactive relationships, can be assessed.

Other Forms Of Regression Analysis

Probit Regression Analysis for a Dichotomous Dependent Variable

Probit regression analysis (Tabachnick and Fidell 2019, section 10.5.4) is a type of generalised linear model and is very closely related to logistic regression analysis (recall *Procedure 7.14*) in that a dichotomous ("0" or "1" coded) dependent variable is modelled using information from one or more predictors (measured at any level if appropriately coded for use in a regression model). Technically, the dependent variable analysed is the proportion of cases in the categories of the dependent variable and the goal of the model is to estimate the proportion of cases that will fall into the dependent variable category coded "1" for different patterns of the predictors. Where probit regression differs from logistic regression is in two major areas: (1) probit regression uses a *probit* transformation of the category proportions, based on z-score positions in the cumulative normal distribution (in contrast, logistic regression uses a *logit* transformation), and (2) interpretation of the probit model focuses on "effective values of predictors for various rates of response" (Tabachnick and Fidell 2019, section 10.5.4; logistic regression focuses on interpretation of odds ratios). Probit regression analysis is often used in health and medical research to assess effectiveness of drug dosages (e.g. warfarin, beta blockers) and other treatments (e.g. exercise, diet control) on the probability of experiencing a specific health event (like a heart attack or stroke). Probit regression may also be useful in predicting loan defaulters in financial lending contexts. The key limitation of probit regression is its dependence upon an underlying normal distribution, which means that it is a more restrictive form of analysis, relative to logistic regression.

Tobit Regression Analysis for Censored or Truncated Data

Tobit regression analysis is another type of generalised linear model; one of several approaches to building regression models using incomplete information, involving either censored or truncated data (Breen 1996; Fu et al. 2004). *Truncated* data, as described by DeMaris (2004, pp. 315–316), result from deliberately not sampling observations occurring either below a specific cut-off score on the dependent variable (i.e. truncation from below) or above a specific cut-off score on that dependent variable (i.e. truncation from above). For example, in a study of high-achieving children in mathematics, our sampling intention might be to only focus on those children who scored above a certain level on some test of math achievement, say 80% (meaning that the truncation value is effectively 79%). Any child scoring at or less than the truncation value of 79% would not be sampled, meaning that we would have no information in our sample about such children. Truncation thus imposes restrictions on the representativeness of the resulting sample.

Censored data refer to data that have had their measurement continuum modified such that scores falling either at or below (left-censoring) or at and above (right-censoring) a specific value are recorded as simply scoring at that value (DeMaris 2004, pp. 318–319). In our earlier example of high-achieving children in mathematics, instead of not sampling any child scoring less than 80%, we might simply record any child scoring at or less than 80% as having scored 80%. Censoring also imposes restrictions on the representativeness of the resulting sample, but the restriction applies to the effective range of the measurement scale, not the composition of the sample. In either case, with truncated or censored data, ordinary least squares regression is not appropriate for the analysis.

Where censored data are present, tobit regression is commonly employed to allow a researcher to properly estimate the parameters of a regression model. Where truncated data are present, truncated regression analysis is employed; however, this is a much less common data scenario (see DeMaris 2004, pp. 321–324 for a discussion of truncated regression analysis). Tobit regression assumes the normal distribution is the appropriate underlying reference distribution for the dependent variable but that this distribution is missing information from one or the other tail. Where a tail is missing information in a normal distribution, the area under the curve is no longer 1.0, as it is in a full normal distribution (recall *Fundamental concept II*). The tobit transformation compensates for this missing tail by effectively rescaling the distribution of the dependent variable so that its area equals 1.0. The appropriate regression model parameters can then be estimated, usually using a maximum likelihood (as distinct from a least squares) estimation procedure. The resulting regression coefficient estimates have the same interpretation as for ordinary least squares regression, once the tobit estimation process has been completed. Tobit regression also produces a pseudo-R^2 measure that reflects the variance in the censored dependent variable explained by the regression model.

Tobit regression is sensitive to the assumption of normality and will therefore be inappropriate if that reference distribution is incorrect or untenable. Tobit regression may be useful for building models in research contexts such as predicting the severity of reported child abuse from family history predictors (where severity is left-censored at 0, with families with no recorded instances of child abuse being scored as "0") or predicting the magnitude of individual income tax refund for a financial year as a function of the number of charitable donations made in that year and the number of dependents who qualified as living at home in that year (where magnitude of refund is recorded as "0" for anyone not receiving a refund or not filing a tax return).

Ordinal Regression Analysis

Ordinal regression (sometimes called *ordinal logistic regression*; Cohen et al. 2003, pp. 522–525; Long and Cheng 2004) is yet another type of generalised linear model and is an approach to regression model building for predicting or explaining an ordinal multi-category dependent variable. It is a variation of multinomial logistic regression analysis (recall *Procedure 7.14*) that can recognise the explicit ordering of the categories. An ordinal dependent variable with k categories will generally be coded so that the "1" category reflects the least amount of characteristic being measured and "k" reflects the most. No assumptions about the relationships between adjacent categories are required, other than a higher number reflects more of what is being assessed relative to a lower number. In ordinal regression analysis, predictors may be at any level of measurement, if they are properly coded for the analysis. However, ordinal regression typically refers to categorical predictors (nominal- or ordinal-scale) as 'factors' and continuous predictors (interval or ratio scale) as 'covariates'.

Ordinal regression (see Norušis 2012, pp. 69–83), which focuses on the prediction of probabilities in the ordered categories of the dependent variable, can be conducted using a logit or probit transformation (typically called a *link function*). The significance of the resulting model is evaluated using a chi-square goodness-of-fit test based on maximum likelihood estimation and the overall quality of the model is reflected in Nagelkerke's pseudo-R^2 measure of the strength of

Procedure 9.8: Specialised Forms of Regression Analysis

association (Cox and Snell and McFadden pseudo-R^2 measures are also typically reported). The parameter estimates for an ordinal regression model are typically referred to as coefficients of *location* and have analogous interpretation to standard logistic regression coefficients: if positive, the coefficient signals a higher predicted probability of membership in higher numbered categories of the dependent variable; if negative, the coefficient signals a higher predicted probability of membership in lower numbered categories of the dependent variable.

> Suppose in the QCI research context, Maree was interested in assessing the impact of a QCI inspector's job satisfaction and perception of their working conditions on their overall quality control decision making performance. To measure the latter construct, Maree uses a QCI performance index, **QCI_perf_index**, created by efficiency experts, which supposedly reflects both an inspector's decision accuracy and their decision speed (where accuracy is always indicative of better performance but slower decision speed is only indicative of better performance up to a point, after which speed is so slow that performance actually declines and may cost the company money due to a reduction in efficiency). Maree then classifies each QCI inspector into one of three ordered groups (a top third group ($= 3$), a middle third group ($= 2$) and a lower third group ($= 1$)), based on their score on **QCI_perf_index**. This gives her a new variable she called **QCI_perf_class**. She then conducts an ordinal regression analysis with SPSS using **jobsat** and **workcond** as covariates and **QCI_perf_class** as the dependent variable. Table 9.9 shows an edited report of this analysis. [It should be noted that the **Goodness-of-fit** table from this analysis has not been shown as the figures reported in the table are extremely sensitive to the presence of large numbers of cells with no observations, something that is likely to occur in models involving several continuous predictors. This is because ordinal regression basically treats every value of a predictor as another way to define a group of participants and when combined with the categories of other predictors as well as with the categories of the dependent variable, the number of cells can be enormous. In the analysis reported in Table 9.9, there were 56 (47.9%) empty cells.]
>
> The **Case Processing Summary** table shows that there were 34 QCI inspectors in each of the lower third (worst performers), middle third and top third (best performers) performance groups, as expected. The **Model Fitting Information** table reports two log likelihood chi-square ratios, one for a model with no predictors (the *Intercept Only* model; constitutes the null hypothesis model) and one where all predictors have been entered (the *Final* model). The difference between these two log likelihood ratios yields the value of 6.888 reported in the 'Chi-Square' column, which has 2 degrees of freedom and is significant at $p = .032$. This tells Maree that the model with the predictors does significantly better than the model with no predictors. The

(continued)

Table 9.9 Outcomes from an SPSS ordinal regression analysis using **jobsat** and **workcond** to predict **QCI_perf_class**

Case Processing Summary

		N	Marginal Percentage
QCI_perf_class	1.00	34	33.3%
	2.00	34	33.3%
	3.00	34	33.3%
Valid		102	100.0%
Missing		10	
Total		112	

Model Fitting Information

Model	-2 Log Likelihood	Chi-Square	df	Sig.
Intercept Only	151.984			
Final	145.096	6.888	2	.032

Link function: Logit.

Pseudo R-Square

Nagelkerke	.073

Link function: Logit.

Parameter Estimates

		Estimate	Std. Error	Wald	df	Sig.	95% Confidence Interval	
							Lower Bound	Upper Bound
Location	jobsat	.129	.118	1.201	1	.273	-.102	.361
	workcond	.228	.116	3.900	1	.048	.002	.455

Link function: Logit.

Pseudo R-Square table reports a Nagelkerke value of .073, indicating that about 7.3% of the variance in ordinal category probabilities can be explained using knowledge of the two predictors, **jobsat** and **workcond**.

Using the hurdle step logic of significance testing, Maree can now move on to interpret the individual predictor contributions to the model. The **Parameter Estimates** table shows that only **workcond** contributes significantly to prediction of QCI performance classification. The location estimate coefficient for **workcond** is positive suggesting that less stressful working conditions are associated with a higher level of QCI performance classification.

Latent Class Regression Analysis

Latent class regression analysis (also called *mixture regression analysis*; see Wedel and DeSarbo 2002) is a hybrid analytical technique combining a type of cluster analysis methodology (to identify latent classes or segments of a sample) with regression analysis to assess predictive models within those latent classes. A 'latent class' refers to an unobserved sub-grouping structure or mixture within a larger sample. For example, for marketing researchers, latent classes may correspond to market segments; for a health psychologist, latent classes may correspond to a typology of psychological characteristics or disorder symptoms. The latent class aspect of the analysis focuses on estimating parameters of the latent class mixture and trying to identify the optimal number of classes to work with. The regression aspect of the analysis tries to fit predictive models (these may be multiple regression or logistic regression models) within each of the latent classes identified, which then permits comparisons of model characteristics across the different classes. Latent class regression analysis is often used in conjoint measurement studies (recall *Procedure 9.4*) and in consumer-oriented marketing research. You generally need more specialised computer programs, such as MPlus, to carry out a latent class regression analysis.

Fuzzy Regression Analysis

Fuzzy regression analysis (see Panik 2009, pp. 311–345) is an extension of fuzzy set theory and depends upon a view of measurement that is imprecise or *fuzzy*, embodying ambiguity or degree of membership in categories. In ordinary logistic or multinomial regression, for example, membership in a category of the variable is considered to be 'crisp'; that is, every person is unambiguously a member of one and only one category (e.g. male or female). In contrast, in fuzzy regression, membership in a category of a dependent variable is considered to be fuzzy, i.e. degree of membership in a category varies according to some distribution (usually triangular, giving rise to what fuzzy set theorists called 'triangular' numbers). For example, the categories of 'rich' and 'poor' people may be thought of as fuzzy categories because 'richness' and 'poorness' are vague in definition and may be subjectively inferred. Thus, any one person, depending upon their income, could have some degree of membership in both the 'rich' and the 'poor' category. The same could be said of many other types of category systems that social and behavioural researchers might employ: sick vs healthy, ethical vs unethical, good vs bad, etc.

Fuzzy regression analysis is useful in circumstances where the relationships between the predictors and the dependent variable are vague and imprecise, perhaps where subjective judgment and intuition are involved. These conditions often emerge in the study of complex systems, for example. Fuzzy regression analysis may also be useful in circumstances where the sample size is very small (but where no substantive outliers are present) and/or the model to be constructed is poorly specified (perhaps where some relationships or influences remain unknown or uncertain). In fuzzy regression analysis, errors tied to the model, its specification and precision of measurement are distinguished from data error (which is captured by ordinary least squares and is reflected in the residuals). Fuzzy regression analysis is a rather more esoteric form of regression analysis, which means that software support for the approach tends to be limited to statistical packages like SAS.

Alternative Estimation Approaches for Regression Modelling

There are a number of different approaches to estimating the parameters of a regression model, besides ordinary least squares. Some of these approaches would be considered nonparametric and others are alternative parametric approaches. Table 9.10 lists and summarises several of these alternative estimation approaches. Each approach discussed in Table 9.10 is designed to address a problem with one or more specific assumptions associated with ordinary least squares regression analysis.

More Sophisticated Regression Modelling Approaches

Nonlinear and Power Polynomial Regression Analysis

Nonlinear regression and *power polynomial regression* are two forms of regression analysis that are designed to assist the researcher in handling nonlinear relationships in their statistical models. Nonlinear regression analysis involves a rather complicated set of considerations (see Panik 2009, pp. 667–668). In fact, we have already explored some nonlinear regression approaches in the form of generalised linear models (such as logistic, probit, tobit and ordinal regression). These models are nonlinear in their original form but linear in their estimated model form. Such models are called *intrinsically linear in the variables.* For example, in logistic regression, while the original model for estimating probabilities is a nonlinear logistic model, the application of the logit transformation effectively linearises the nonlinear model into a log odds model, which looks very much like a standard general linear model. In contrast, there are models where, no matter what transformation is applied, the model cannot be reduced to a linear form. Such models are called *intrinsically nonlinear* and require much more sophisticated nonlinear least squares or maximum likelihood statistical methods to estimate. One type of an intrinsically nonlinear model is any model where the errors are multiplicative rather than additive (i.e. **DATA** = **MODEL**∗ERROR as opposed to **DATA** = **MODEL** + ERROR as for the general linear model).

Power polynomial regression is an estimation approach that uses linear multiple regression to model nonlinear relationships between a predictor and a dependent variable (Cohen et al. 2003, pp. 193–214 and 221–254). Power polynomial regression works by including higher powers of a predictor (e.g. the square of the predictor, the cube of the predictor, etc.) in the model along with the original predictor. Such models are considered to be *intrinsically linear in the parameters.* The logic is relatively straightforward; if the regression model has:

- only the original predictor included in it, this model tests for a *linear* relationship (i.e. a straight-line relationship) between the predictor and the dependent variable;
- the original predictor as well as the square of the predictor included in the model, this model tests for a *quadratic* relationship (i.e. a relationship curve with one bend in it, like a U-shaped function);

Procedure 9.8: Specialised Forms of Regression Analysis

Table 9.10 Alternative estimation approaches for regression modelling

Alternative estimation approach	Major problem(s) handled relative to OLS[a]	Description of the approach
Nonparametric regression analysis	Handling normality or heterogeneity of variance problems	*Nonparametric regression analysis* (Panik 2009, pp. 189–201) refers to a class of regression approaches designed to estimate regression models in situations where the normality and/or homogeneity of variance assumptions are severely violated. Two different approaches are available: • *Ordinary Median Rank regression* –applicable in situations where the regression assumptions such as normality and homogeneity of variance have been severely violated and/or where substantive outliers or influential observations are present. It uses ordinary medians to produce nonparametric estimates of regression coefficients. • *Weighted Median Rank regression* –applicable in situations where the regression assumptions such as normality and homogeneity of variance have been severely violated and/or where substantive outliers or influential observations are present. It uses weighted medians to produce nonparametric estimates of regression coefficients. Here the weight applied to each case reflects the rank of the absolute value of its model residual.
Robust regression analysis	Handling outliers	*Robust regression analysis* (Das et al. 2009; Panik 2009, pp. 285–310) refers to a class of regression approaches designed to estimate regression models in situations where strong outliers are evident in the data. These approaches occupy a sort of middle ground between nonparametric regression analysis and OLS regression analysis. An approach is considered 'robust' if it produces estimates that are resistant to the influence of outliers or extreme observations, either on the side of the predictors (X-space) or the side of the dependent variable (Y-space) or both. In all cases, the goal of robust regression is to produce influence-resistant estimates of regression coefficients, which are then interpreted in the same way as coefficients produced by OLS regression. Several approaches to robust regression are possible (see the discussion in Das et al. 2009, for an overview): • *Least Absolute Deviations (LAD) regression* – applicable in situations where Y-space outliers are present (but where no X-space outliers are present). The method minimises the sum of the absolute values of model residuals, rather than the sum of squared model residuals (as in OLS regression). • *M regression* (see also Panik 2009, pp. 289–291) – is also best applied in situations where the outliers are in the Y-space, not the X-space. M regression is a Maximum likelihood procedure and is more effective under conditions where the homogeneity of variance assumption is also violated.

(continued)

Table 9.10 (continued)

Alternative estimation approach	Major problem(s) handled relative to OLS[a]	Description of the approach
		• *Least Median Squares (LMS) regression* – applicable in situations where both X-space and Y-space outliers are present. It focuses on minimising the median of the squared residuals.
		• *Least Trimmed Squares (LTS) regression* – also applicable in situations where both X-space and Y-space outliers are present. This approach minimises the sum of squared residuals, but for an optimally selected subset of the original sample (where the influential outlier observations have been 'trimmed' from the sample).
		• *Scale (S) regression* – also applicable in situations where both X-space and Y-space outliers are present. This approach is very effective against the influence of outliers in the X-space and works by minimising a scale function of the residuals.
Weighted least squares analysis	Handling heterogeneity of variance problems	*Weighted least squares* (Panik 2009, pp. 109–112; Norušis 2012, pp. 351–359; Stolzenberg 2004, pp. 204–205) refers to an approach to estimating regression models in situations where the homogeneity of variance assumption has been severely violated. This approach works by weighting observations by the reciprocal of their error variance when estimating the regression coefficients and R^2. This has the effect of equalising dependent variable variances across the range of predictor values.
2-stage least squares analysis	Handling correlated errors	*Two-stage least squares* (Norušis 2012, pp. 341–349) refers to an approach to estimating regression models under conditions where the independence of model prediction errors assumption has been severely violated. Technically, the issue addressed by 2-stage least squares estimation is where one or more predictors are correlated with the residual term in the regression model. This is a phenomenon that can easily happen if there is a dynamic feedback loop between a predictor and the dependent variable (as there is, for example, between price and demand in econometric models or between job satisfaction and employee productivity). When this happens, the OLS estimated regression coefficients are biased. In order for 2-stage least squares estimation to work, two classes of variables need to be identified: (1) variables in the model that are causally dependent upon each other (called *endogenous variables*) and (2) variables that are not endogenous and therefore may influence a specific endogenous variable in the model but are not themselves influenced by such variables (called *instrumental variables*). Consider, for example, a regression model intended to predict employee productivity from employee job satisfaction and number of years with the company. We know that job satisfaction and productivity are

Procedure 9.8: Specialised Forms of Regression Analysis 643

reciprocally related in that each can influence the other (more satisfied → better productivity; better productivity → higher job satisfaction). In this model, job satisfaction and productivity would be considered as *endogenous* variables. Suppose we also had measures of work ethic and personality orientation (Type A or B personality). These latter two variables could be *instrumental* in predicting job satisfaction, while not being influenced by either productivity or number of years with the company (see the small diagram of the model below; the goal is to negate the link covered by the X by finding a way to replace the job satisfaction variable in the model). Instrumental variables should correlate highly with an endogenous variable and correlate zero with model error terms.

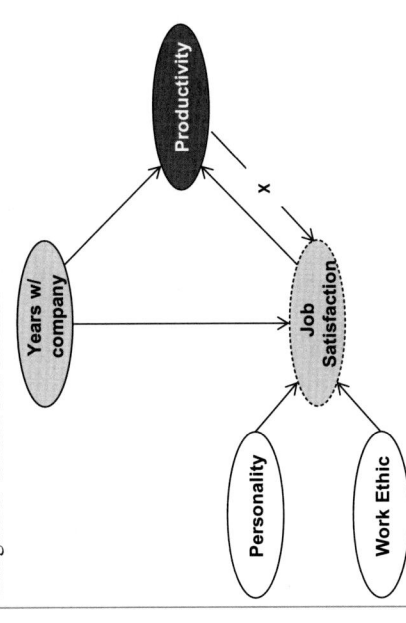

Two-stage least squares estimation, for this scenario, works as follows:

- *Stage 1*: the *instrumental* variables, including years w/ company are used to predict job satisfaction and to generate predicted values for job satisfaction – these predicted values are then substituted for the observed job satisfaction values (symbolised by the dashed line surrounding the job satisfaction ellipse); then
- *Stage 2*: the complete regression model is estimated using number of years with the company and instrumentally-predicted job satisfaction to predict productivity.

(continued)

Table 9.10 (continued)

Alternative estimation approach	Major problem(s) handled relative to OLS[a]	Description of the approach
Generalised least squares analysis	Handling heterogeneity of variance and correlated error problems	*Generalised least squares* (Panik 2009, pp. 109–112) refers to a more general approach to estimating regression models in situations where both the homogeneity of variance and independence of errors assumptions have been severely violated. This approach works by weighting observations by a function of the reciprocal of their error variances combined with their error correlations when estimating the regression coefficients and R^2.
Ridge regression analysis	Handling multicollinearity amongst predictors	*Ridge regression analysis* (Cohen et al. 2003, pp. 427–428; DeMaris 2004, pp. 231–239; Panik 2009, pp. 285–310) refers to a class of regression approaches designed to estimate regression models in conditions where predictors are very strongly correlated with each other, thereby violating the assumption of absence of multicollinearity. Multicollinearity severely distorts the standard error for each regression coefficient, thereby providing a very misleading picture of its precision (as well as a biased test of significance). The ridge regression procedure works by adding a constant amount (called the 'ridge factor' or 'ridge constant') to the variance of each predictor prior to estimating the regression coefficients. This will automatically bias somewhat the regression coefficient estimates, but with the potential to dramatically improve the precision of the estimates (therefore leading to a more correct test of significance). The trick is to choose the right value for the constant that will produce regression coefficient estimates that are not too substantively biased but are much more precisely estimated. In other words, ridge regression trades a bit of bias for gaining a lot more precision.

[a]*OLS* Ordinary Least Squares multiple regression estimation

Procedure 9.8: Specialised Forms of Regression Analysis

- the original predictor as well as the square of the predictor and the cube of the predictor included in the model, this model tests for a *cubic* relationship (i.e. a relationship curve with two bends in it, like an S-shaped function).

Each additional term included in a power polynomial model has its own estimated regression coefficient associated with it. In general, in order to obtain an interpretable power polynomial model as well as to greatly reduce the high multicollinearity between the predictor, its square and its cube, the recommended practice is to *centre* the predictor first (centring simply involves subtracting the predictor mean from each observed value of the predictor), then use the centred predictor and its higher powers to fit the model (Cohen et al. 2003, pp. 201–204). Power polynomial models are normally constructed and evaluated in hierarchical fashion, assessing the linear relationship first, then the quadratic relationship over and above the linear relationship, then the cubic relationship over and above the linear and quadratic relationships, and so on. Each additional increment in R^2 represents the unique contribution of each functional form to the model over and above all previously entered forms and can be easily tested using a partial *F*-test (recall *Procedure 7.13*).

> To illustrate the power polynomial procedure, consider more closely the QCI decision performance index (called **QCI_perf_index**) created by efficiency experts in the earlier ordinal regression example. This index was intended to reflect both an inspector's decision accuracy and their decision speed. The best scores on this index would theoretically reflect an inspector who made very accurate decisions with moderate speed; the index would report poorer performance if decision accuracy was lower and/or decision speed was either too fast or too slow. Thus the relationship between the new index and **accuracy** was intended to be linear and the relationship between the new index and **log_speed** (the normalised transformation of decision speed) was intended to be quadratic (i.e. an inverted U-shaped curve). Maree wished to verify whether or not this new QCI performance index in fact captured the nonlinear relationship with **log_speed**. Using the SPSS *Curve Estimation* procedure and requesting that a quadratic trend be tested, she obtained the graph shown in Fig. 9.21. This graph shows both the fitted linear (solid line) and quadratic (dot-dash line) curves. The R^2 for the linear model was .006, which was not significant at $p = .441$ (this is why the line is nearly flat in Fig. 9.21). The R^2 for the linear plus quadratic model was .058, which was just significant at $p = .048$. This result indicated that the linear plus quadratic model predicted nearly 6% of the variability in **QCI_perf_index** scores. The change in R^2 from the linear model stage to the linear plus quadratic model stage was $.058 - .006 = .052$. The partial *F*-test for this R^2 change value was significant ($F(1, 102) = 5.63, p = .020$). This says that the quadratic trend was the only significant trend in the relationship. Maree could conclude that the new index does reflect

(continued)

Fig. 9.21 Fitted linear and quadratic curves for the relationship between **QCI_perf_index** and **log_speed**

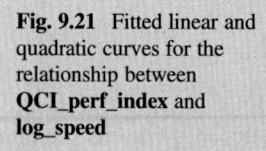

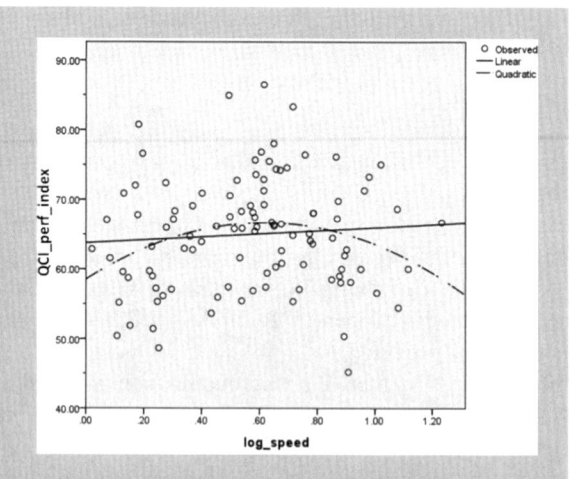

the intended inverted U-shaped relationship with **log_speed**; however, the strength of this relationship was rather weak (explaining only 5.2% of the variability in index values). The unstandardised regression coefficient for the quadratic trend was −21.349; the fact that it was negative signalled that the relationship was an inverted U-shape rather than normal U-shape--a pattern confirmed by the fitted curve in the graph.

Latent Growth Models in Multilevel Regression Analysis

Latent growth models (Cohen et al. 2003, pp. 588–595; Preacher et al. 2008) is a multilevel regression approach to evaluating growth or change over time. Such models are often estimated using structural equation modelling (recall *Procedure 8.7*) as well as by more traditional hierarchical linear model procedures (recall *Procedure 9.5*). Latent growth models can address a range of research hypotheses, including:

- What is the shape of the mean trend over time?
- Does the initial level predict rate of change?
- Do two or more groups differ in their trajectories?
- Does rate of change or degree of curvature in the mean trend predict key outcomes?
- What variables are systematically associated with change over time?
- Are theoretical hypotheses about the trajectory tenable given observed data?
- Does significant between-person variability exist in the shape of the trajectory?

- Is change over time in one variable related to change over time in another variable? (Preacher et al. 2008, pp. 2–3)

If a growth model involves hypotheses about heterogeneous mixtures of subgroups within a sample, each potentially demonstrating a different growth trajectory over time, part of the analytical exercise will be to identify the latent classes comprising that mixture. If this is the case, then latent growth models describe the modelling context. If, however, the researcher can or is willing to assume that his or her sample is relatively homogeneous in composition, then whole-of-sample growth modelling looking at change over time becomes the focus, meaning that the growth trajectory hypotheses are not multilevel in nature.

Mediated and Moderated Regression Analysis

Mediated and *moderated regression analyses* represent two different types of theory testing logic within the multiple regression/general linear model framework. Each type of analysis hypothesises a different type of theoretical causal relationship between predictor variables and a dependent variable. Both mediation and moderation concern the potential effects of predictor variables on or with each other and how those effects impact on their relationships with the dependent variable of interest. Thus, testing a mediation or a moderation hypothesis involves elaborating the **MODEL** in the **DATA** = **MODEL** + ERROR general linear model system. It is also possible to test more complicated models such as mediated moderation and moderated mediation, usually in the context of more sophisticated structural equation modelling (see, for example, Jose 2013, ch. 7).

Mediated regression analysis (Iacobucci 2008; Hayes 2018, Part II; Jose 2013, ch. 3 & 4; Miles and Shevlin 2001, ch. 7) uses a sequence of multiple regression models or a structural equation model to evaluate whether a specific variable mediates the relationship between a predictor and the dependent variable. If a variable is a mediator, then it is operating as a type of conduit through which the mediated predictor transmits its influence on a dependent variable. Thus, the relationship between a predictor and a dependent variable may be partially or totally attributable to the intervention of the mediating variable. For example, in the left diagram below, the relationship between the design of an organisation (e.g. along a mechanistic – organic continuum; white ellipse = the hypothesised mediated variable) and the behaviour of a leader (e.g. along a transactional – transformational continuum; dark ellipse) may be mediated by the extent to which that design and the leader's leadership style (e.g. along a task-focused – relationship-focused continuum) are congruent with each other (reflecting organisational 'fit' – grey ellipse = the hypothesised mediator variable). The mediated relationship is reflected in the pathway depicted by the two solid arrows. There is also the possibility that there is a direct relationship (dashed arrow) as well. Part of testing a mediation hypothesis involves assessing the relative strength/viability of the indirect and direct causal pathways. The right-hand diagram below shows another example. The relationship between children's writing ability and their writing achievement in school may be mediated by the extent to which their classroom environment supports effective writing.

In the discussion to follow, we will focus on the design – 'fit' – leader model illustration provide context. Basic mediated regression analysis for this model would proceed through the following four stages (Miles and Shevlin 2001, p. 190):

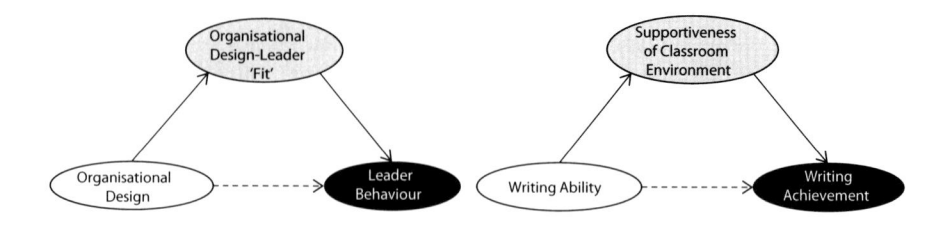

1. *Assess the predictive relationship between the mediated variable and the dependent variable*: Predict the Leader Behaviour (LB) measure using the Organisational Design (OD) measure and show that the resulting standardised regression coefficient (i.e. beta) is significant;
2. *Assess the predictive relationship between the mediated variable and the mediator variable*: Predict the Organisational Design-Leader 'Fit' (OD-L Fit) measure using the OD measure and show that the resulting standardised regression coefficient is significant;
3. *Assess the predictive relationship between the mediator variable and the dependent variable, controlling for the influence of the mediated variable*: Predict the LB measure using both the Organisational Design (OD) measure and the OD-L Fit measure (a simultaneous regression analysis) and show that the standardised regression coefficient for the OD-L Fit measure is significant (this effectively shows the relationship between OD-L Fit and LB, *controlling* for the influence of OD); and finally
4. *Compare the coefficients for the mediated variable from Step 1 and Step 3*: Examine the standardised regression coefficient for OD from the regression analysis done in Step 3. If OD-L Fit is a complete mediator between OD and LB, then the standardised OD regression coefficient should effectively be zero (i.e. no longer significant) at Step 3. If OD-L Fit is a partial mediator between OD and LB, then the standardised OD regression coefficient will be reduced in magnitude compared to what it was in the Step 1 regression, but still significant. The difference between the standardised OD regression coefficient from the Step 1 regression and the standardised OD regression coefficient from the Step 3 will reflect the degree of mediation that OD-L Fit exerts. [There is a statistical test called *Sobel's z-test* which can evaluate the significance of the reduction between the two regression coefficients (in their *unstandardised* forms), see Jose 2013,

pp. 51–55); this effectively tests for the presence of a significant mediation effect (see Appendix A for the formula for Sobel's z-test).]

While the above describes the basic approach to testing a mediator, in many cases the mediation hypotheses are more complicated, perhaps involving multiple mediators and multiple predictors. In these more complex modelling circumstances, a structural equation modelling approach becomes the only viable way to proceed (see the discussions in Iacobucci 2008).

Moderated regression analysis (Hayes 2018, Part III; Jose 2013, ch. 5 & 6; Miles and Shevlin 2001, ch. 7) uses hierarchical regression analysis to assess the joint impact of two or more predictors on predictions of the dependent variable. Information about the joint impact of two or more predictors is carried in interaction terms. Thus, in moderated regression analysis, we need to clearly distinguish between main effects (the impact of single predictors, ignoring all others) and interactions (the joint impact of two or more predictors). Furthermore, we need to distinguish between categorical predictors and continuous (or quantitative) predictors, as we deal with those predictors somewhat differently in a moderated regression analysis. The two diagrams below display two different moderated regression modelling contexts. The left diagram shows a scenario where the relationship between an employee's job skills (e.g. score on a job skills test; white ellipse) and their work performance (e.g. supervisor's rating; dark ellipse) is moderated by their stress level (self-report stress instrument; grey ellipse = the moderator). In this scenario, the interaction between a measure of employee job skills and a measure of employee stress levels carries this moderating effect. The right diagram shows an educational scenario where the relationship between the quality of a child's home environment (e.g. score on a home quality inventory) and their attendance rate at school (e.g. number of days absent over a 6-month period) is moderated by the extent to which the child is experiencing cyberbullying (e.g. number of reported incidents from teachers and/or parents; grey ellipse). In this scenario, the interaction between a measure of quality of home environment and a measure of cyberbullying experienced carries this moderating effect.

In the discussion to follow, we will focus on the cyberbullying model presented above to provide context. Moderated regression analysis proceeds through the following four stages (Cohen et al. 2003, ch. 7–9; Miles and Shevlin 2001, pp. 165–187):

1. *Centre any quantitative/continuous predictors and properly code any categorical predictors*: As both Quality of Home Environment (QHE) and Extent of Cyberbullying Experienced (ECE) are quantitative/continuous predictors, we *centre* each predictor first by subtracting the predictor mean from each individual's observed score (see Cohen et al. 2003, pp. 261–267 for a discussion of the process, interpretation and merits of centring for continuous predictors). Categorical predictors need not be centred, they just need to be properly coded (e.g. effect coding) for use as regression predictors. Centring of quantitative predictors accomplishes two goals: (1) the resulting regression coefficients are more interpretable and (2) the influence of multicollinearity between interaction terms and their constituent main effect predictors is greatly reduced.
2. *Create the relevant interaction terms*: We then create a new predictor, called QHE∗ECE, which is simply the product of the two centred predictors. This product method works for any combination of categorical and/or quantitative predictors. However, if a categorical predictor is involved, you need to create as many new product variables as there are effect-coded variables for that predictor. So, for example, if one predictor is a three-category variable, two new effect-coded variables will be needed to represent this predictor information in the model (recall *Fundamental concept VI*). If the other predictor is quantitative, two new interactions terms would then need to be created, each by multiplying one of the effect-coded variables by the quantitative predictor. If the other variable is also categorical (also say with three categories → two effect-coded variables), then you would need to create 2 × 2 new interaction terms, each multiplying one specific effect-coded variable for each predictor.
3. *Enter all predictor main effects into the regression model first*: Using hierarchical regression, enter all main effect predictors (use the centred ones from Step 1 for quantitative predictors) into the regression model. The main effects are entered into the model first so that any overlap they share with the interactions they are involved with are controlled for.
4. *Enter the interaction terms as predictors into the regression model*: For stage 2 of the hierarchical regression analysis, enter all the interaction terms created in Step 2 into the regression model, ensuring that you obtain R square change and partial F-tests. This means that interaction effects are always assessed and interpreted over and above what their constituent main effect predictors can account for. This provides for clean, uncontaminated and unambiguous interpretations, unconfounded by any main effects.

Cohen et al. (2003), Hayes (2018) and Miles and Shevlin (2001) each provide excellent discussions of how to interpret the outcomes from a moderated regression analysis as well as how to construct effective graphs of the interactions themselves. Also recall that the role and impact of an interaction as a moderating influence was graphically illustrated in *Fundamental concept IX*. [It should be noted that Hayes 2018, pp. 304–312 and Jose 2013, pp. 158–159, convincingly dispute the need for always centring predictors when forming interaction terms. They show how uncentred interaction terms can be effectively graphed and interpreted. Thus, the practice of centring is not uniformly advocated amongst regression analysis experts.]

Software Procedures

Application	Procedures
SPSS	Norušis (2012) describes a range of SPSS capabilities for implementing alternative forms of regression models and regression estimation approaches. These capabilities include: *Curve Estimation* regression (for power polynomials analysis), *Ordinal* regression, *Probit* regression, *Multinomial Logistic* regression, *Weight Estimation* (for weighted least squares), *Nonlinear* regression, *2-Stage Least Squares* regression and *Generalized Linear Models*. Moderated and mediated regression analyses can be accomplished via the *Linear Regression* procedure, if the variables are properly set up first. Hayes (2018) has created a freely available SPSS macro called PROCESS which can be used to conduct and display outcomes from a wide variety of mediated and moderated regression analyses: see https://processmacro.org/index.html. AMOS can evaluate mediation and moderation models as well as various types of growth models (but not latent class growth models or regression models).
NCSS	NCSS has a limited but useful range of alternative regression analysis capabilities including: *Nonlinear Regression, Probit Analysis, Ridge Regression, Logistic Regression* (including multinomial logistic regression), general *Curve Fitting* and robust regression via the *Multiple Regression* procedure. Moderated and mediated regression can also be handled via the *Multiple Regression* procedure, if the variables are properly set up first.
SYSTAT	SYSTAT has extensive capabilities in terms of specialised types of regression analysis and a range of estimation approaches. These capabilities are well-described by Das et al. (2009) and Wilkinson and Coward (2009) and include: a range of types of robust and nonparametric regression (including *Rank, Scale, LAD, LTS, LMS* and *M* estimation), *Two-Stage Least Squares* regression, *Ridge* regression, *Polynomial* regression, *Bayesian* regression, *Probit* regression and *Nonlinear* regression. Growth models (but not latent growth models) can be set up and evaluated using *Path Analysis (RAMONA)*. Moderator and mediator analyses can be done using the *Linear → Least Squares* procedure, if the variables are properly set up first.
STATGRAPHICS	STATGRAPHICS offers a somewhat limited but useful range of alternative regression analysis capabilities including: *Logistic Regression, Probit Regression, Nonlinear Regression, Ridge Regression, Polynomial Regression* and for several procedures, there is an option for specifying weights for a weighted least squares solution. Moderated and mediated regression can also be handled via the *Multiple Regression* procedure, if the variables are properly set up first.
R Commander	R Commander can be used to conduct generalised linear model analyses (with the probit transformation as one option, the logit transformation as another), multinomial logit analysis and ordinal regression analysis, all accessible via *Statistics → Fit Models*. Mediated and moderated regression analyses can be conducted using the *Linear Regression...* procedure accessible via *Statistics → Fit Models*. Other types of regression analyses are available via specific packages for the main **R** program. • The package *censReg* offers options for tobit regression; see https://cran.r-project.org/web/packages/censReg/vignettes/censReg.pdf. • The *ridge* package can be used to conduct ridge regression analyses; see https://cran.r-project.org/web/packages/ridge/ridge.pdf.

(continued)

Application	Procedures
	• Linzer and Lewis (2011) describe a package for **R** called *poLCA*, which is designed to facilitate latent class analysis, including latent class regression; see https://cran.r-project.org/web/packages/poLCA/poLCA.pdf. • The *QuantPsy*c package facilitates moderated and mediated regression; see https://cran.r-project.org/web/packages/QuantPsyc/QuantPsyc.pdf. • There is also a tutorial series for **R** that explores some basic functions within R to carry out different types of regression analyses; see http://rtutorialseries.blogspot.com/search/label/R%20Tutorial%20Series. The Quick-**R** website also offers some guidance and links to other resources; see https://www.statmethods.net/index.html.
MPlus	Geiser (2013, ch. 4, 5 and 6) describes how to use MPlus to accomplish a range of latent class analyses and latent growth model analyses. Geiser (2013, ch. 3) also describes how to use structural equation models to evaluate mediation and moderation hypotheses. MPlus is one of the best packages for doing any type of latent class analysis.
Stata	Hamilton (2013, ch. 8 and 9) shows how to use Stata to conduct a variety of types of regression analysis including robust regression, nonlinear regression models, tobit, probit and other types of logistic and multinomial logistic models.
LatentGOLD	This software package offers a wide range of latent class analysis clustering and regression models and offers capacities for estimating growth models; see https://www.statisticalinnovations.com/latent-gold-5-1/.
GOLDMineR	This software package offers a range of dichotomous and ordinal regression analyses and graphical procedures; see https://www.statisticalinnovations.com/shop/goldminer/.

Procedure 9.9: Data Mining & Text Mining

Classification Multivariate; correlational.; inferential; parametric/nonparametric.
Purpose To discover hidden or unknown patterns and relationships in large quantitative or qualitative (e.g. text) databases.
Measurement level Databases may comprise complex arrays of quantitative data, in the form of categorical (nominal/ordinal) and/or continuous (interval/ratio) variables. Data in the form of texts, web content, social media and/or multimedia content may also be mined, using quantitative approaches for analysing qualitative data.

Data Mining Approaches

Data mining is a collective term for any of the myriad quantitative analysis approaches that can be employed to discover unknown or hidden patterns and relationships amongst variables in large databases (Han et al. 2012). For this reason, data mining is commonly referred to as Knowledge Discovery in Databases (KDD – see Sharda et al. 2018, ch. 3); however, more recently, data mining has become associated with *data science* and the processes for managing and analysing *big data* (see, e.g. O'Neil and Schutt 2014). These large databases may be assembled by researchers, marketers, retailers, organisations, businesses, agencies, professions or governments using manual or automatic data gathering methods. In many cases, these databases emerge from a systematic process called *data warehousing* (Han et al. 2012, section 1.3.2; Sharda et al. 2018, ch. 3), where data are gathered, cleaned, coded, pooled, integrated and stored for use in supporting future research and decision making. The definition of data has gradually widened to encompass not only quantitative variables, but also documents and other textual materials, web content, social media content, audio and video content. In short, virtually anything that can be gathered from data sources of interest and measured or stored in some way can be warehoused in a database for use by others. In the twenty-first century, the amount of data potentially available for gathering and storage by some agency has exploded exponentially in concert with technological advances in data recording and methods for controlling access to those data (which may involve some ethical implications regarding privacy and data usage concerns). The internet has contributed greatly to this trend. There are now consumer/customer relationship databases, criminological and geographical databases, linguistic and literature databases, doctor, patient and health databases, legal and historical databases, astronomical databases, educational databases, business management, economic and financial databases, population (census) databases and so on.

It is one thing to gather masses of data from people, companies, governments and so on, and quite another to know what can be learned from those data that is useful for some purpose. The volume of data available has become so large and databases so massive that humans can no longer see the patterns and relationships they might contain. This is where *data mining* (and, where textual materials are concerned, *text mining*) has come into play, becoming virtually an industry in its own right, and providing algorithmic, analytical and data management support to humans in search of those patterns and relationships. The term *knowledge discovery* is an apt phrase to use in this regard – data mining is used to build new knowledge from the masses of data we collect. The pathways to knowledge discovery and learning via data mining may occur using a *supervised* or *unsupervised* learning approach. Supervised learning involves pattern and relationship prediction amongst known classes or categories of objects or entities. Unsupervised learning involves pattern and relationship discovery where the classes or categories of objects and entities are initially

Table 9.11 Some specialised algorithmic approaches to data mining

Approach or algorithm	Description
Decision Tree Induction	This procedure is a logical extension of classification and regression trees (recall *Procedure 9.6*). It is a classification and prediction approach to data mining and involves using categorical and/or continuous predictors (for some algorithms, continuous predictors may be categorised into classes) to classify entities or observations into categories on a dependent variable of interest (see Han et al. 2012, pp. 330–350). The outcome of decision tree induction is a decision tree (of a form similar to Figs. 9.14 or 9.15), where the emergent rules for classification are embedded in the structure of the tree. Several different induction algorithms are available including classification and regression tree (CART), Chi-squared Automatic Interaction Detection (CHAID, which uses categorised continuous predictors) and exhaustive CHAID. Because decision tree induction can result in very extensive, possibly overfitted structures, they are often 'pruned', either forward by halting tree growth based on some criterion value such as information gain, or backward (post-pruning) by working backward through a completed tree and converting less productive branches into end nodes or 'leaves'. There are variations of decision tree induction that use either *naive Bayesian classifiers* or *Bayesian belief networks,* utilising conditional and prior probability information, to help build the decision tree (see Han et al. 2012, pp. 350–355 and 393–398).
Genetic algorithms	Genetic algorithms are a specialised class of classification-oriented data mining procedures, loosely based on the metaphor of 'genetic natural selection', where a random population of classification rules is selectively pared down and fine-tuned using a 'survival of the fittest' logic (the rules that work better are retained longer) (see Han et al. 2012, pp. 426–427). In a genetic algorithm, the natural selection metaphor is further reflected by mutations (random variations within a single rule) and crossovers (combining aspects of two different rules) may be introduced to enhance the robustness of the emerging rule-based system. Learning and discarding of rules progresses until a relatively stable system of effective classification rules is arrived at. Han et al. (2012, pp. 415–429) also discuss some other types of algorithms for developing classification and prediction systems in a data mining context including *K nearest neighbour classifiers, association rule analysis, case-based reasoning, rough sets classification* and *fuzzy sets classification.*
(Artificial) Neural Network (ANN) Analysis	Neural network analysis (Garson 1998) has its roots in brain physiology and learning theories. A neural network embodies a biological metaphor as well as a mathematical representation of learning patterns based on neurons and their connections in the brain (as these neurons are models, they are often referred to as 'artificial'). Neural network analysis is most useful for classification and prediction forms of data mining (supervised learning; however, they can be used for unsupervised learning as well) and looks to build classification and prediction rules through a training/learning process where connections between predictors and

(continued)

Procedure 9.9: Data Mining & Text Mining 655

Table 9.11 (continued)

Approach or algorithm	Description
	one or more dependent variables are repeatedly reinforced. There are a number of approaches to building a neural network model including *multi-layer perceptrons, radial basis function, backpropagation* and *Kohonen self-organising maps* to name a few (see Garson 1998 for a more comprehensive introduction). Most algorithms grow a common 3-layer structure for a neural network: an *input layer* comprising all of the desired categorical and/or continuous predictors in the database; a *hidden layer* which comprises a specific set of learned relationships between the various predictors; and an *output layer* where the hidden layer builds connections to the dependent variables. [It is important to note, however, that a neural network need not be restricted to just a single *hidden* layer. Depending upon the nature of the database and the problem at hand, several hidden layers of 'neurons' may be required to adequately learn the predictive relationships.] The importance of the predictors to classifying or predicting a dependent variable is carried in the relative strength of the connections between the predictor and dependent variable working through the hidden layer. Neural network models are essentially nonparametric (requiring no distributional assumptions about any variables), can capture nonlinear and complex interactive relationships, which cannot be anticipated in advance, can adapt to changing contextual circumstances and can be constructed and evaluated in very large databases.
Support Vector Machines (SVM) algorithm	The SVM algorithm is a classification/prediction approach that can handle both linear and nonlinear data (Han et al. 2012, pp. 408–415). For a dichotomous dependent variable, SVM works by trying to identify a best-fitting higher-order plane (called a 'hyperplane' classifier), combining predictor information in such a way as to maximally distinguish (= predict membership in) one dependent variable category from the other category. When data are 'linearly separable', the best-fitting hyperplane separates categories using straight line/linear hypersurface decision boundaries. When data are nonlinearly separable, the best-fitting hyperplane separates categories using nonlinear hypersurface decision boundaries (Han et al. 2012, pp. 413–414). When a dependent variable has more than two categories, SVM generates a separate hyperplane to maximally distinguish each category from all remaining categories. This means that for a dependent variable with p categories, SVM builds p hyperplane classifiers, which, when combined, permit classification predictions across the entire dependent variable category system.
Apriori algorithm	The Apriori algorithm is a data mining approach for discovering association rules and patterns (see Han et al. 2012, pp. 248–256). This algorithm focuses on finding what are called *frequent itemsets* (dichotomous presence/absence data for a number of items that tend to occur or take on the value of 1 (= present) together; e.g. counting the number of times that items A, B and C are observed together across cases in the database – larger counts

(continued)

Table 9.11 (continued)

Approach or algorithm	Description
	point to more frequent itemsets). The algorithm is called Apriori because it defines frequent itemset properties using prior knowledge, or in this case, a principle where all possible subsets (having at least one item in them) of a frequent itemset must also be frequent. This principle constrains and focuses the amount of analytical effort needed to identify frequent itemsets. Once the frequent itemsets have been identified, a set of strong association rules can be defined, which can then be used for classification purposes.
FP-Growth algorithm	FP-Growth or *Frequent Pattern Growth* is a data mining approach for finding frequent sequential patterns (see Han et al. 2012, pp. 256–259). A *frequent sequential pattern* is an extension of the concept of frequent itemset, where we count the occurrences that unfold in a sequence (first A, then B, then C is observed). The FP-Growth algorithm grows an *FP-Tree*, which shows the frequent sequences that have been identified. The FP-Tree as a representation then embodies the association rules than have been learned.
Latent Class Analysis	*Latent Class Analysis* (sometimes called *latent structure analysis* or *finite mixture modelling*) is a more general approach for unsupervised learning where patterns in data are unknown, but assumed to be latent (i.e. unobserved, but discoverable) and therefore detectable with appropriate statistical methods (see Geiser 2013, ch. 6; Madigan and Vermunt 2004). The approach has a similar goal to cluster analysis but takes a rather different and more model-driven approach to finding the latent classes. McCutcheon (1987, p. 7) characterised latent class analysis as a "qualitative data analog to factor analysis". Latent class analysis therefore focuses on the relationships amongst categorical variables. However, as a data mining approach, it is most useful in identifying homogenous classes or subgroups in a large database and exploring the relationships between those latent classes and other variables available in the database. Latent class analysis uses conditional probabilities of membership in categories of relevant predictors to build up a mixture model that estimates the probability of membership in a specific latent class. Latent class analysis can also facilitate comparisons of latent class structures across different samples or sub-samples.

unknown. Data mining procedures may focus on classification and prediction (supervised learning), clustering (unsupervised learning) or discovery of association rules (unsupervised learning).

Procedures for data and text mining have emerged from statistics, computer science and business/management information and decision support systems. The fields of statistical analysis (e.g. model building and pattern detection) and artificial intelligence (machine learning and pattern recognition), in particular, have contributed to developments in data mining technology. From statistics, a range of procedures we have already explored in this book, including factor and cluster analysis (recall *Procedures 6.5* and *6.6*, as well as a variant of cluster analysis called *nearest neighbour analysis*), multidimensional scaling and correspondence analysis (recall *Procedure 6.7*), classification and regression trees (recall *Procedure 9.6*), regression, logistic regression and discriminant analysis (recall *Procedures 7.13, 7.14, 7.17* and *9.8*), time series analysis (recall *Procedure 8.5*) and latent class analysis (to be discussed below), can be used for data mining purposes. From computer science and artificial intelligence, we get neural network analysis, genetic algorithms, Support Vector Machines (SVM) and a range of specific algorithms for machine learning of association rules (such as Apriori and FP-Growth) . Cluster analysis, multidimensional scaling, correspondence analysis, latent class analysis and association rule algorithms are all examples of unsupervised learning approaches to data mining. Classification and regression trees, regression analysis, discriminant analysis, genetic algorithms, SVM and neural network analysis are all examples of supervised learning approaches to data mining. Table 9.11 provides additional details on specialised algorithmic approaches to data mining which have not been previously explored in this book.

There are also holistic integrated processes for data mining which attempt to standardise and logically sequence the data mining activities. Two such holistic models are: *Cross-Industry Standard Process for Data Mining* (CRISP-DM) and *Sample, Explore, Modify, Model, Assess* (SEMMA) (see Shafique and Qaiser 2014). For example, the CRISP-DM process has six distinct stages that may unfold in a linear or nonlinear (involving iterative alternations between stages) fashion: (1) Business Understanding; (2) Data Understanding; (3) Data Preparation; (4) Model Building; (5) Testing & Evaluation; and (6) Deployment. This type of process contextualises the data mining activity so that it can produce viable outcomes for deployment and application by end users. Most of the analytical procedures mentioned earlier as potential data mining techniques could come into play in Stage 4 (Model Building) of the CRISP-DM process. Stage 5 (Testing & Evaluation) could involve some sort of validation or cross-validation process using new cases, particularly for supervised classification and prediction models. One approach that can be taken in Stage 5 is *split sample cross-validation* where a very large data sample is randomly split into three sub-samples. One sample, called the *training sample*, is used to develop the classification/prediction rules and estimate the parameters of the network. The second sample, called the *test* sample, is used to monitor learning errors made during the training phase for building the neural network and is designed to prevent what is called 'overtraining' or 'overfitting' of the resulting neural network model. The final sample, called the *holdout* sample, is used to provide 'new' cases for evaluating the newly developed classification/prediction rules. This type of validation approach is most useful for classification and prediction-oriented data mining. However, it is important to note that there are other types of validation/

model quality evaluation approaches available including bootstrapping and jackknifing (recall *Procedure 8.4*), leave-one-out validation, Receiver-Operator Curve (ROC) analysis and gains and lift charts.

Neural Network Analysis

To illustrate the neural network (also referred to as *artificial neural network* or ANN*)* analysis approach to data mining more concretely, we will extend the QCI data context.

> Assume that, in the QCI data context, an international consortium of manufacturing organisations, coordinated by Maree, had gotten together and agreed to assemble comprehensive data about all their quality control inspectors. Organisations from six nations (Australia, USA, UK, China, Japan and Germany) cooperated to assemble QCI statistics, similar to what Maree had gathered in her survey, across the five types of manufacturing companies (PC, large electrical appliance, small electrical appliance, large business computer and automotive). Out of this cooperative venture emerged a comprehensive database of 1456 QCI inspectors, where the following variables were recorded for each inspector: **country**, **company**, **gender** and educational level (**educlev**), inspector's salary at the time of survey (**inspector_salary**; measured in $1000 US-equivalent), numbers of years inspector had been with their current company (**yrs_w_company**; measured in years and decimal fraction of a year), a measure of the accuracy of the inspector's quality control decisions (**accuracy**; measured as a percentage score), a measure of the speed of their quality control inspection decisions (**log_speed**; measured in log seconds), a measure of the inspector's overall job satisfaction (**jobsat**; 1 to 7 Likert-type scale like Maree's survey) and a measure of the stressfulness of their working conditions (**workcond**; 1 to 7 Likert-type scale like Maree's survey).
>
> The consortium was interested how the variables in this database might be used to predict both the accuracy and the speed of quality control inspector decision making. This required mining the large database from a classification and prediction perspective. Maree decided to employ a multi-layer perceptron neural network analysis (via the SPSS *Neural Networks* procedure in this instance), using as **country**, **company**, **gender** and **educlev** as categorical predictors or 'factors', **inspector_salary**, **yrs_w_company**, **jobsat** and **workcond** as 'covariates' (= continuous predictors) and **accuracy** and **log_speed** as the dependent variables. She decided to randomly allocate

(continued)

50% of the database into the *training* sample, 30% into the *test* sample and the remaining 20% into the *holdout* sample.

The overall prediction errors for the network model were reflected in the following statistics: for the *Training* sample, average overall relative error was .478 (.498 for **accuracy**; .458 for **log_speed**); for the *test* sample, average overall relative error was .583 (.574 for **accuracy**; .591 for **log_speed**) and for the *holdout* sample (predicting 'new' cases), average overall relative error was .611 (.589 for **accuracy**; .630 for **log_speed**). Note that the error rates climb as one looks from the *training* sample (where the optimised model is constructed) to the *test* sample (where over-training and dependence on spurious data patterns are corrected) to the *holdout* sample (where the model is applied to predicting new cases). It is in the *holdout* sample where the final applied model reveals how it will perform as a predictive tool and the model will always make more errors when predicting new cases. In the end, the neural network model does slightly better at predicting **accuracy** than at predicting **log_speed**. Figure 9.22 shows the complete final neural network model for this analysis.

The full neural network model in Fig. 9.22 revealed several things: (1) the three layers of the model appeared left to right: *input* layer with all of the predictors, the *hidden* neuron layer showing how the predictors weighted together in a particular pattern, and the *output* layer where the nine emergent neuron patterns were weighted to predict the two dependent variables; (2) the categorical predictors were handled by coding each category of the variable as a separate input into the model (the labels for each category and variable have been superimposed on the left side of the figure); (3) weight strength was signalled by connector line thickness; and (4) overall predictor weight was signalled by the size of the box in the *input* layer.

Figure 9.23 shows the aggregate weight in the neural network for each predictor, reported by SPSS in both unnormalised and normalised (= scaled relative to the largest value, which is set at 100%) forms. The sideways bar graph displays the normalised aggregate weights in descending order. It is clear that **company** was the most important predictor in the network, followed by **workcond, inspector_salary** and **jobsat. Country** differences did not weigh very heavily in the predictor system. However, a closer look at specific connector weights, in a table not reported here for space reasons, revealed that Australia, the USA and the UK tended to be most heavily connected to specific neurons in the *hidden* layer of the network).

(continued)

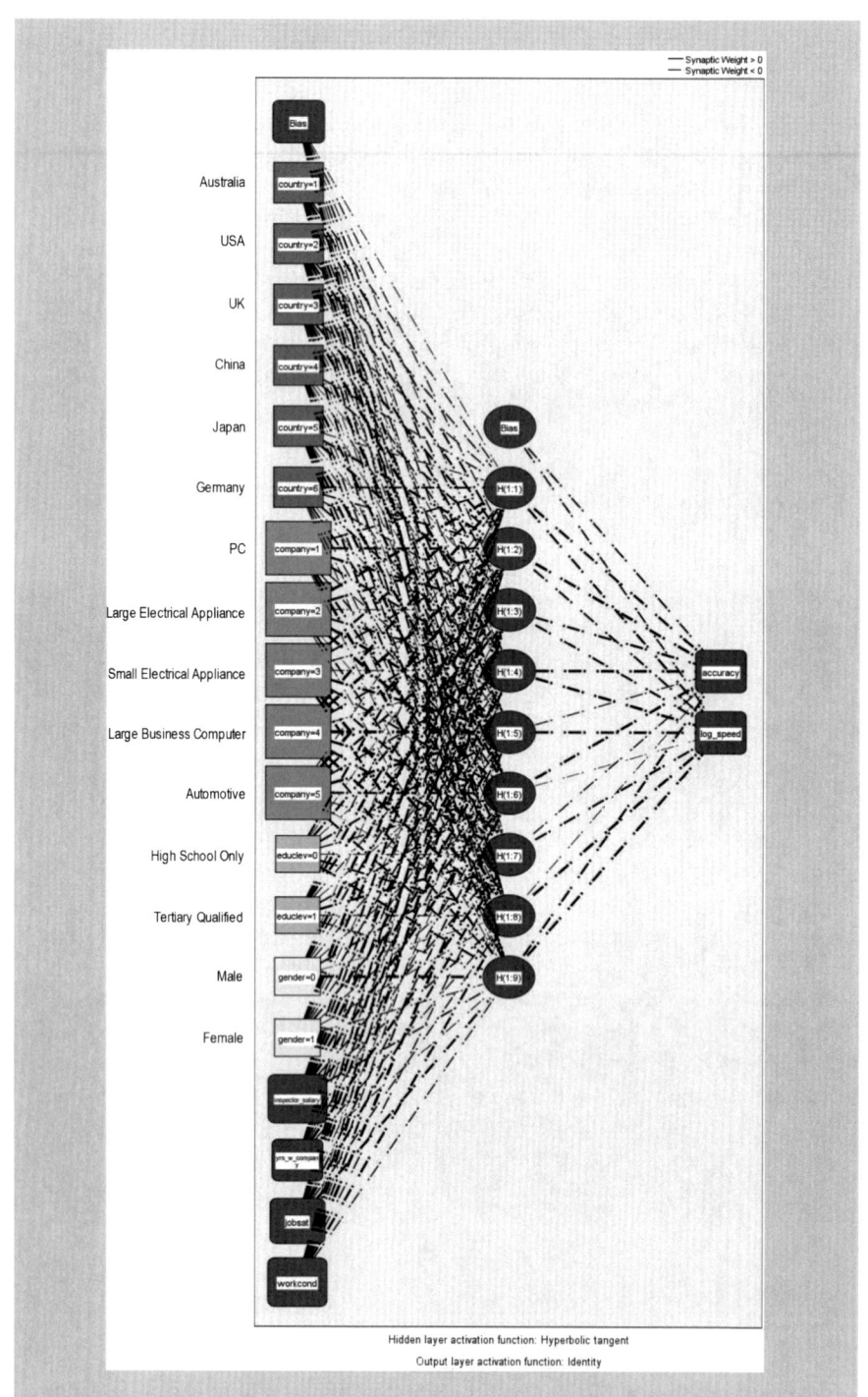

Fig. 9.22 Artificial neural network constructed using the multi-layer perceptron algorithm

(continued)

Independent Variable Importance

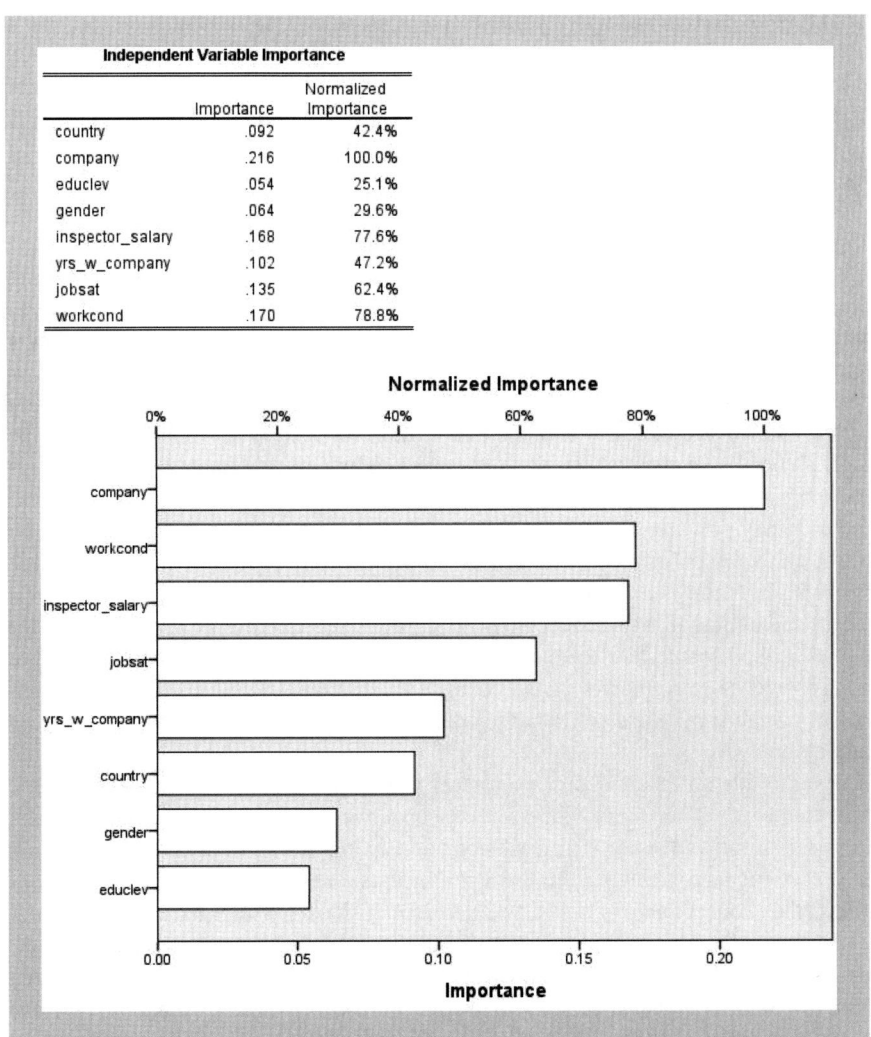

	Importance	Normalized Importance
country	.092	42.4%
company	.216	100.0%
educlev	.054	25.1%
gender	.064	29.6%
inspector_salary	.168	77.6%
yrs_w_company	.102	47.2%
jobsat	.135	62.4%
workcond	.170	78.8%

Fig. 9.23 Estimated aggregate predictor importance (unnormalised and normalised) in the artificial neural network in Fig. 9.22

Text Mining Approaches

Data mining of qualitative texts, called *text mining*, has now become a commonplace form of quantitative analysis. Databases comprising various types of assembled texts (such as interview transcripts, documents, books, reports, articles, media stories, language corpuses, answers to open-ended survey questions, ethnographic field notes) can now be content-analysed using a range of text mining procedures.

Quantitative content analysis of texts is an approach consistent with positivist paradigm guiding assumptions (as distinct from qualitative analysis of texts which is much more consistent with interpretivist/constructivist paradigm guiding assumptions). Quantitative content analyses typically focus on word counts and counting co-occurrences of words and phrases with the dual goals of identifying the most frequently occurring concepts and exploring the relationships between concepts.

We will illustrate the Leximancer quantitative text mining/content analysis approach here. *Leximancer* (Smith and Humphreys 2006; Leximancer Pty Ltd 2011) takes a combined statistical/machine learning approach to text mining, emphasising not only the automated identification of key words and concepts in text but also the co-occurrences of such concepts in the text. Leximancer takes a textual database (comprising, for example, pdf files, Word document files, web content) and looks first to identify key 'word-like concepts' in the text, building up an extensive dictionary and thesaurus, then to identify relationships amongst those 'word-like concepts'. These steps can be completely automated (in which case, the analysis is essentially a machine learning exercise) or partially automated (where the user may provide some or all of the concept seed words to start the learning part of the analysis and/or edit the thesaurus of identified concepts, deleting undesirable concepts, merging similar concepts (such as the singular and plural versions of a word) and adding in additional concepts). Leximancer can also keep track of proper names (called 'name-like' concepts) as well. Leximancer offers the user complete control over the parameters of the concept learning and thesaurus generating processes or can simply use the default settings to obtain a completely automated text mining analysis.

For any identified concept, Leximancer can show you how frequently it occurs, how frequently it connects with other concepts, where in specific texts that concept has been found and so on. The final step in a Leximancer analysis is the production of a concept map, showing the learned concepts and their co-occurrence relationships. This concept map is basically a multidimensional scaling (recall *Procedure 6.7*) diagram of the co-occurrence data. Leximancer also cluster analyses the coordinates of the concept map to identify theme clusters – clusters of similar concepts that tend to co-occur in the same texts. The concept map is dynamic in that the user can interact and augment the map in various ways to highlight specific relationships and pathways.

To illustrate the Leximancer approach, we will use the text of the previous edition of this entire textbook (up through *Procedure 9.8*) as the text database. The Word-version of the book was submitted to Leximancer for analysis (the table of contents and index were not included in the text file that was mined). An automated analysis of the text was conducted first, then the learned concept list was edited to remove uninteresting words such as 'the', 'via' and 'before', merge different forms of certain words (e.g. 'inspector' and 'inspectors' and 'shows' and 'showing') and add new concepts such as 'hierarchical', 'parametric' and 'causal'. Figure 9.24 shows the concept map resulting from this refined Leximancer analysis where all the learned concepts are shown in relation to each other. The sizes of the concept points on the map are proportional to the frequency of occurrence of that concept in the text being

Procedure 9.9: Data Mining & Text Mining

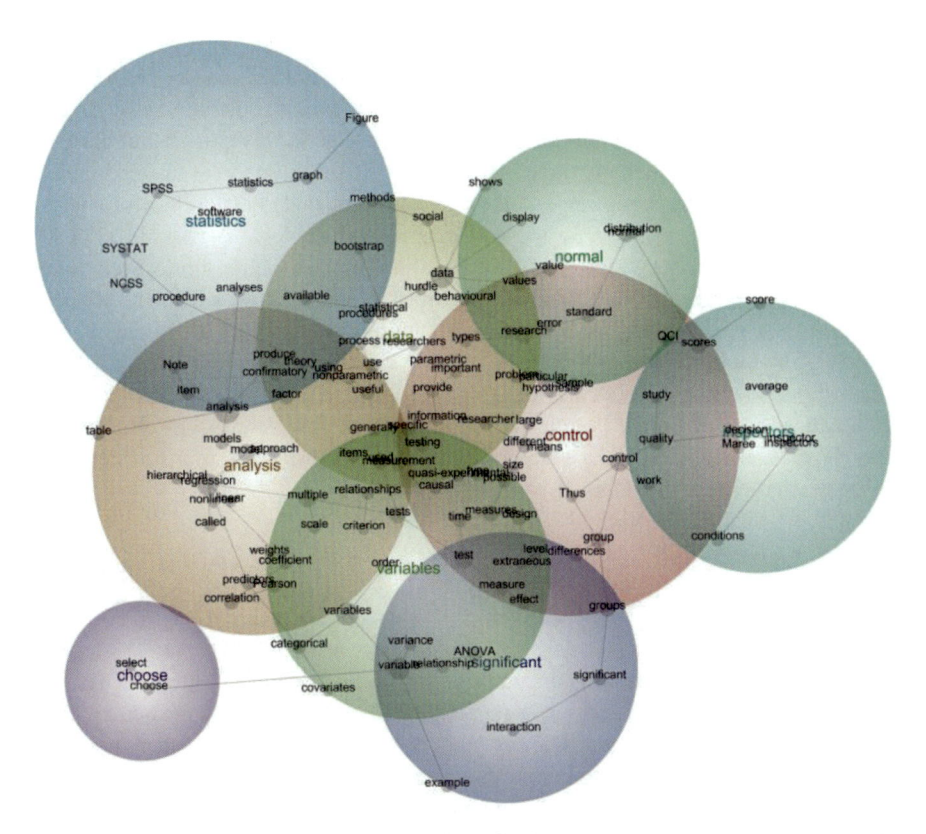

Fig. 9.24 Leximancer concept map showing thematic clusters

mined (larger = more frequent). The grey lines connecting the various concept points comprise the 'minimum spanning tree' pathway between all the points on the map. The concept map also shows thematic clusters that have been identified. On a computer monitor, the theme clusters are 'heat-map' colour-coded so that the most important thematic cluster is red, the next most important is orange and so on around the colour wheel (the progression is thus from the warmest colours of red and orange to the coolest colours of blue and green). Figure 9.24 shows that the most important thematic cluster is the 'control' cluster, followed by the 'analysis' cluster, then the 'data' cluster.

There are some interesting observations one can make about the concept map in Fig. 9.24. The concepts that appear in the 'control' cluster all relate to research design issues. The concepts that appear in the 'analysis' cluster all relate to model building and theory/relationship testing. The 'statistics' cluster contains the main software packages used to generate most of the illustrations in this textbook. QCI-related concepts tend to appear in the 'inspectors' cluster. The 'variables' cluster encompasses word-like concepts related to different types of variables employed in statistical analyses. The 'significant' cluster covers concepts related to assessing group differences. The 'data' cluster appears to encompass concepts related to the handling of data and the researcher's role in managing data and testing relationships more generally in statistical analyses. The 'normal' cluster ties together concepts related to the normality assumption underpinning parametric statistical analyses (note that the concept 'parametric' does not lie in this cluster, but close to it in the 'data' cluster, perhaps signalling the key role the researcher plays in making a decision about the assumption and the consequences that decision has for analytical pathways choices made by the researcher downstream). The small 'choose' cluster is most closely connected to the 'variables' cluster via its shortest relational pathway to the concept point 'variable'. The most efficient pathway from the concept 'SPSS' to the concept 'Figure' runs through the related concepts of 'statistics' and 'graph'. The cluster spheres overlap, signalling relationships across cluster boundaries (this is actually a reflection of the hierarchical clustering algorithm used by Leximancer and the number of clusters shown can be completely controlled by the user; fewer clusters will produce larger more overlapping spheres; more clusters will produce smaller less overlapping spheres).

Advantages

Data mining and text mining encompass important systems of approaches for discovering unknown patterns and relationships in large datasets. Data mining approaches can handle variables at any level of measurement, are nonparametric in nature (requiring no distributional or other statistical assumptions) and can focus on classification and prediction, clustering and/or patterns of association as required by the researcher. The utility of these tools can only increase as society continues to indulge its fascination with gathering and recording data in all walks of life. The internet generates prodigious amounts of data each day and both data mining and text mining approaches can help us to make sense of those data. Data/text mining embeds learning as a key functionality of analytical approaches, meaning that the analytical systems can be trained to detect patterns and relationships (and to adapt to contextual changes in these) and what is learned needs to be evaluated and validated. By situating data/text mining in a larger business context, for example, using CRISP-DM logic, meaning can be extracted from masses of data with the express purposes of informing executive and institutional decision making and directing action.

Disadvantages

The chief drawbacks associated with data mining and text mining approaches are their sheer diversity and complexity coupled with the speed with which the field evolves. There are a very large number of approaches and algorithms for the researcher to choose from and it may not be easy to figure out what the best approach to take should be. Each algorithm or approach requires certain assumptions and decisions to be made by the researcher with respect to data management, data formatting and how missing data should be handled. Furthermore, every approach is controlled by a specific set of parameters and it is not always clear what the best choices for those parameter values ought to be for a specific problem (which may create a heavy dependence on default settings in software for data mining/text mining procedures). Adding to this complexity is the fact that data and text mining, because of their strong roots in disciplines other than the social sciences, tend to use a much more specialised jargon, which makes understanding algorithms and choices somewhat less than transparent. For example, where a social scientist might easily resonate to the concepts in the left column of the table below, rather fewer would realise how these concepts are referred to in data mining jargon (see Garson 1998; Han et al. 2012, ch. 1).

Meaningful concept to a social scientist	Data mining equivalence
Survey respondent or participant	'tuple' or 'pattern'
Survey or measurement item	'attribute'
Statistical estimation	'training', 'learning', 'self-organisation'
Independent or predictor variable	'input'
Dependent or criterion variable	'output'
Prediction or classification	'supervised learning'
Clustering	'unsupervised learning'
Decision rule	'classifier'

Garson (1998, pp. 16–17) discussed some of the reasons why neural network models were not in more widespread use by social scientists and highlighted that jargon was a significant barrier to adoption of the neural network approach. The data mining concepts of supervised and unsupervised learning are foreign to a social scientist with more formal training in statistical analysis (e.g. few social scientists would realise that the general linear model is basically a rudimentary supervised learning system). One final but important drawback to data mining approaches is that while they are very powerful methods for detecting and predicting patterns and relationships, they are very poor at helping the researcher infer causality in those connections. In this sense, data mining and text mining approaches are generally not strong approaches for testing theory. While software providers may try to sell their text mining technologies as broad approaches to qualitative analysis, the researcher needs to be careful in how far to accept such claims. Knowing one's own guiding paradigm assumptions is essential to evaluating these claims (see the discussion in Cooksey and McDonald 2019, ch. 9, for example). Leximancer, for example, is a very useful approach to carrying out quantitative content analysis of qualitative texts, but it is not a qualitative data analysis tool in the sense that an interpretivist or constructivist would view such a tool. Leximancer only focuses on what is mechanistically countable in texts (entirely consistent with positivist guiding assumptions); it cannot read or code for deeper meaning and it cannot adopt the hermeneutical perspective of the author of the text. Leximancer cannot do what a Computer-Aided

Qualitative Software Analysis System (CAQDAS; see Lewins and Silver 2014) such as NVivo, MAXQDA, dedoose or Atlas-ti can help the researcher do.

Where Is This Procedure Useful?

Data mining procedures can be useful wherever the researcher has access to a large database or data warehouse and wants to discover what patterns and relationships might exist between variables and measurements in the database. Data mining is useful for quantitative databases; text mining is useful for qualitative texts of any sort. The larger the database or corpus of text(s), the more useful data and text mining procedures become. In many cases, data mining or text mining may be the only viable systematic ways to extract meaning from such databases, under positivist guiding assumptions. The fact that most data mining procedures have built-in logic for validating the rules and patterns that are learned is an added bonus. The rules and patterns are not just discovered; they can immediately be applied and tested so that the researcher can see how well they function.

Software Procedures

Application	Procedures
SPSS	The base SPSS package offers some limited specialised data mining capabilities in the form of the *Tree...* procedure (for building decision trees) and *Nearest Neighbor...* procedure (both accessible via the *Analyze → Classify* menu sequence). This is in addition to the plethora of standard data analysis and model building capabilities as described elsewhere in this book. There is an SPSS *Neural Networks* add-on module (SPSS Inc 2017) which provides capabilities for building and evaluating both multi-layer perceptron and radial basis function types of neural networks. The *Multilayer Perceptron* procedure was used to generate the analyses reported in Figs. 9.22 and 9.23. SPSS *Modeler Text Analytics* (SPSS Inc 2018) add-on module offers a range of text mining capabilities, including mining for concepts and categories, mining for text links, categorising text data, clustering concepts and text link analysis. It also offers a range of analysis visualisation capabilities. Using text analytics in SPSS can also help build links between text mining and data mining, such that concepts and links can be predictively related to other variables from other data sources.
R Commander	R Commander offers no data mining capabilities, but there is a plug-in for R Commander for text mining called *RcmdrPlugin.temis* (see https://journal.r-project.org/archive/2013-1/bouchetvalat-bastin.pdf). There is, however, a wide range of open-source data mining packages and functionalities associated with the main R program itself. Torgo (2010) describes a range of data mining capabilities with R, illustrating their use via specific case studies. Williams (2011) introduces a package for R called *Rattle*, which provides a graphical interface for data mining with R (not unlike the logic behind R Commander). *Rattle* can be obtained from https://rattle.togaware.com and offers

(continued)

Procedure 9.9: Data Mining & Text Mining

Application	Procedures
	access to both supervised and unsupervised learning approaches to data mining. *Rattle* can build neural network models, boosted decision trees, support vector machines and generalised linear models as well as offering procedures for cluster analysis, association rules and data visualisation. There is also a package called *tm* which provides some text mining capabilities, in the form of word frequency analyses (see http://www.rdatamining.com/examples/text-mining for some examples; *tm* can be obtained from http://tm.r-forge.r-project.org/).
RapidMiner	RapidMiner is a free open source software package for data mining. It also has an add-on text processing/mining extension to enhance its capabilities. RapidMiner can implement decision trees, neural network analysis, cluster analysis, discriminant and regression analysis and association rule analysis. It offers an innovative and intuitive workspace interface within which to design your data mining processes. The text mining extension facilitates word frequency analysis and some word association analyses, although the text mining capabilities fall well short of what can be achieved using Leximancer, for example. North (2012) provides a general introduction to data mining, using RapidMiner as its platform and CRISP-DM as its guiding framework. The website for RapidMiner is https://rapidminer.com/ and it provides access to the software as well as to various supporting materials including tutorials and the user's manual.
XLMiner	XLMiner is an Excel add-on software package, offering a wide range of data mining capabilities, in terms of both supervised and unsupervised learning. XLMiner can carry out neural network analyses, a range of classification and regression tree analyses as well as other statistical analyses such as cluster analysis and can build Bayesian classifiers and association rules. It also offers a range of data visualisation techniques as well as techniques for 'cleaning' data. The XLMiner website has additional information on this add-on package for Excel, including access to a 15-day trial version: https://www.solver.com/xlminer-data-mining.
MPlus	MPlus has extensive capabilities for conducting latent class analyses. Geiser 2013, ch. 6) shows how to configure and carry out a range of such analyses. The MPlus website (https://www.statmodel.com/) provides access to a demo version of the software and provides a range of papers and documents that discuss and illustrate MPlus capabilities.
Latent GOLD	Latent GOLD is a stand-alone software package for conducting a wide range of latent class analyses, including latent class clustering, latent class regression and discrete factor models. Multi-level latent class models can also be estimated. The website for Latent GOLD is https://www.statisticalinnovations.com/latent-gold-5-1/ and provides access to a demo version of the program as well as a range of supporting resources.
Leximancer	Leximancer Pty Ltd (2011) and https://info.leximancer.com/ describe how to use Leximancer to carry out quantitative content analysis on a wide variety of types of text. The Leximancer interface offers control options for all aspects of the concept learning → thesaurus assembly → concept map creation and configuring process. There are a range of options available for tagging text content (including, for example, speakers in interview transcripts), seeding concepts for learning, editing, adding and deleting concepts from the thesaurus, distinguishing between word-like and name-like concepts, configuring the concept map as a map or as a concept cloud and tracing pathways through the concept map. The concept map is accompanied by a series of tabs offering statistical, analytical and textual quote

(continued)

Application	Procedures
	data pertinent to the map and the concepts it contains. Queries can also be conducted on concepts and concept relationships in the map. Leximancer is not a cheap package to acquire. It runs through an internet browser and makes heavy use of Java functionality.
WordStat	WordStat offers capabilities that encompass word frequency and co-occurrence forms of quantitative text mining and content analysis. In this regard, WordStat shares some capabilities in common with Leximancer, including some automated machine learning algorithms, but also offers a greater range of statistical analyses that can relate text concepts and co-occurrences to categorical or numerical variables. See http://provalisresearch.com/products/content-analysis-software/ for more information on this package and its capabilities.

Procedure 9.10: Simulation & Computational Modelling

Classification   Univariate; Multivariate; descriptive; inferential; nonparametric

Purpose   To build, test and potentially validate models of the world using Monte Carlo simulation or computational simulation methods.

Measurement level   Variables at any level of measurement can be generated using simulation and modelling methods, although continuous (interval/ratio) variables tend to be used more frequently than categorical (nominal/ordinal) variables, especially in computational modelling.

In this, the final *Procedure* in the book, we discuss methods that social and behavioural scientists use to explore data scenarios and complex dynamic problems using generated data rather than empirically measured observations. In short, we are focusing on methods that create data, using specific parameters, for testing models about the social world. You might wonder why a researcher would want to create data rather than empirically gather data; it might even strike you that we are venturing into and advocating the ethically objectionable practice of faking data. However, there is a key difference between openly generating data for model exploration and testing purposes and covertly generating data to ensure research results come out in a specific way. The open approach is entirely ethically defensible because the researcher is completely open about the fact that they have generated data, clearly communicates how and why the data are generated and then reports any results in that light – avoiding deception and potentially adverse ethical impacts and maximising learning via simulation are the goals here (see, for example, Axelrod 2007; Gilbert and Troitzsch 2005). The covert approach is ethically reprehensible because the researcher is hiding the fact that they have created data for the purposes of supporting a specific hypothesis – deception for personal gain is the only goal here.

Now to the question of 'why would a researcher want to create data'? Researchers may wish to or be constrained to create data in order to test specific types of hypotheses. In terms of constraints, in certain nonparametric situations, there may be no available statistical test for evaluating a specific hypothesis. In this situation, the researcher may create many samples of random data, generated using specific parameters and distributional characteristics, in order to evaluate the hypothesis. This is one purpose for what are called *Monte Carlo methods* (after the famous casino in Monte Carlo, Monaco). Note that, technically, bootstrapping (recall *Procedure 8.4*) is a type of Monte Carlo approach. A second purpose for generating random data involves conducting experimental research explicitly to study the behaviour of data under different assumptions and/or generating algorithms and seeing how those differences impact on the conclusions one might draw, using different analytical approaches. For example, Glenn Milligan achieved an international reputation in the field of cluster analysis in the 1980s through his Monte Carlo investigations of different clustering methods, clustering criteria and stopping rules (see, for example, Milligan 1981a, b). In this situation, it was not feasible for the researcher to gather all of the data needed to implement the desired experiment and test the hypotheses, so the required large volumes of data were artificially created with known characteristics. A third purpose for generating random data is for teaching purposes, a purpose that is reflected throughout this book. For this book, I created an artificial multivariate data set, embedded within a hypothetical QCI scenario (associated with the fictitious researcher, Maree Lakota), entirely for pedagogical purposes. The data were constructed using a random sample from a multivariate normal distribution, where the distribution was configured in such a way that specific kinds and patterns of relationships were generated between different variables and sets of variables, in a manner consistent with the structure and intent of the hypothetical scenario.

Monte Carlo methods and associated procedures comprise one class of approaches for generating data, primarily using random sampling algorithms. There is another class of approaches where data are effectively generated for model testing purposes, not via random sampling, but via the structure and parameters of the model itself; a class known as *computational modelling*. Computational models encompass many different approaches including dynamic systems and systems thinking models, agent-based models, network models and other types of models, represented and implemented using computer programming. Computational modelling allows the researcher to set up and test models (even theories) about the social world without needing to gather empirical observations. In many cases, the models or theories are so complex that empirical data gathering is infeasible or at least far too costly to obtain. The only option then is to simulate data within a controllable computational environment. Computational modelling can also allow the researcher to test certain types of hypotheses that are infeasible, perhaps even unethical, to test in the real world.

For example, if a social scientist wishes to explore the potential impacts of different types of government policies for dealing with immigrants and refugees, it is unreasonable (and probably unethical as well) for the researcher to expect to be able to conduct a social field experiment where they put a specific policy in place and gather empirical data in order to evaluate the impact(s) of that policy. However, if that researcher can construct a convincing computational model that appropriately represents the problem and the important variables and relationships associated with it, then the social experiment can effectively be run within that computational or virtual world. These types of computational models can also be used for teaching purposes – allowing students to vary specific parameters of a social or business system model and observe what happens as a consequence of their decisions (without the actual social or business system ever being at risk of damage or unintended side effects!). In dynamic systems modelling, for example, the 'beer game management flight simulator', a role-play environment developed by John Sterman at MIT (see http://web.mit.edu/jsterman/www/SDG/beergame.html), was designed to help managers and management teams learn how to manage beer production, inventory and stock movements, supply chain relationships and distribution through the making of simulated management decisions within one particular role/sector (e.g. Factory) and see the consequences of those decisions for other roles/sectors (e.g. Retailer, Distributor or Wholesaler) in the game. The game was also designed to capture the complexities and messiness of the decision environment within which the managers have to work (thereby simulating a real-life business context).

Monte Carlo Simulation Methods

Monte Carlo methods (sampling via random data generation; see, for example, Carsey and Harden 2013; Karandikar et al. 2009; Mooney 1997) encompass a wide range of approaches. For any Monte Carlo method, the researcher's goal is to decide which data distribution(s) and associated parameters are relevant for defining the population(s) to be randomly sampled. The researcher also designs the sampling plan (how many observations are generated, under which conditions). All Monte Carlo methods depend upon random number generators, which most statistical packages have access to. The most commonly employed random number generator is the Mersenne-Twister algorithm. This algorithm technically generates pseudorandom rather than genuine random numbers (because the numbers emerge from a calculation), but the patterns of numbers that emerge are effectively considered to be random. Data from a wide variety of univariate distributions may be randomly generated (e.g. binomial or multinomial distribution, uniform distribution, normal distribution, chi-square distribution, Weibull distribution); categorical as well as continuous data can be created in this way. If continuous data are generated in z-score form, they can easily be rescaled to any other measurement scale using a simple formula (see Appendix A for this formula). Once the artificial dataset(s) have been created, the researcher can then use those data in the same ways they would use empirically gathered data. Some statistical packages, such as SYSTAT, have the capability to generate multivariate data where there is a specific correlational structure amongst all the variables (this is how I generated the data for the QCI scenario).

Procedure 9.10: Simulation & Computational Modelling

	Population Means		
	S1V1	S1V2	S1V3
	0.000	0.000	0.000

	Population Correlation Matrix		
	S1V1	S1V2	S1V3
S1V1	1.000		
S1V2	0.200	1.000	
S1V3	0.500	0.300	1.000

1000 simulated data profiles; no systematic outliers

Number of Non-Missing Cases: 1,000

Means
S1V1	S1V2	S1V3
-0.025	0.005	0.045

Pearson Correlation Matrix
	S1V1	S1V2	S1V3
S1V1	1.000		
S1V2	0.191	1.000	
S1V3	0.523	0.272	1.000

Bartlett Chi-Square Statistic	:	399.072
df	:	3
p-Value	:	0.000

1015 simulated data profiles; 15 systematic outliers

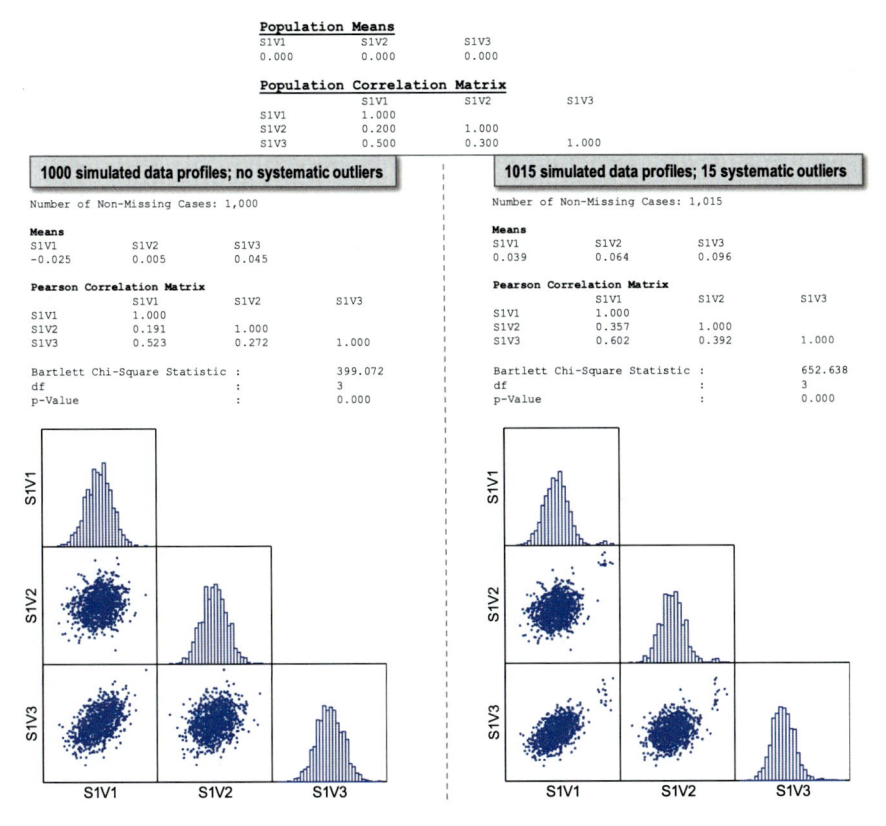

Number of Non-Missing Cases: 1,015

Means
S1V1	S1V2	S1V3
0.039	0.064	0.096

Pearson Correlation Matrix
	S1V1	S1V2	S1V3
S1V1	1.000		
S1V2	0.357	1.000	
S1V3	0.602	0.392	1.000

Bartlett Chi-Square Statistic	:	652.638
df	:	3
p-Value	:	0.000

Fig. 9.25 Monte Carlo simulation, using SYSTAT, of a 3-variable multivariate data set, showing starting population values, 1000 cases with no systematic outliers added and 1015 cases with 15 systematic outliers added

Figure 9.25 provides a simple illustration of a Monte Carlo simulation approach for generating a multivariate data sample. In this illustration, suppose we want to see what impact a small but systematic set of outliers might have in a multivariate dataset. Using SYSTAT, a multivariate dataset comprising just three variables was generated. The two subtables at the top of Fig. 9.25 shows the initial (i.e. known) population parameters that drove the random data generation process; for the multivariate sample, the population mean for each variable (I used 0.0 here for each variable, which was what the mean of a population of z-scores on a variable would be) and the population correlation matrix amongst all three variables had to be specified. I then asked SYSTAT to generate 1000 cases, effectively obtaining a sample of $n = 1000$ from the known population.

The statistics and scatterplot matrix on the left-side of the vertical dashed line in Fig. 9.25 shows what the correlation analysis of these 1000 sampled data points looked like. Note that because we effectively performed random sampling, the sample means and correlations did not exactly equal the population parameter values – this will always be the case with Monte Carlo methods. The Bartlett's chi-square test evaluates whether or not there are significant correlations anywhere in the correlation matrix. In this case, the test was significant which confirmed the presence of at least one significant correlation in the matrix.

The means, correlations and the Bartlett chi-square test are all known to be highly susceptible to influence by outliers and non-normal distributions. To see what impact a set of systematic outliers would have on this dataset, I did a mini-experiment by adding just 15 cases to the original sample of 1000 cases, where the 15 added cases were set to extreme values (z-scores in excess of +3.5) for all three variables. Thus, the augmented dataset had $n = 1015$. The statistics and scatterplot matrix on the right-side of the vertical dashed line in Fig. 9.25 shows what the correlation analysis of this augmented dataset looked like. The means for the three variables were all higher, which reflected the pull of the extreme positive outliers. The correlations were all similarly influenced by the addition of the 15 extreme cases – the extreme positive scores artificially inflated the correlation estimates. The Bartlett test was very strongly inflated by the addition of the 15 outlier cases (jumping in value from 399 to over 650). While the outlier set was very small relative to the overall sample size (constituting just 1.548% of the sample of 1015), its impacts on the statistical estimates and test were quite dramatic. This demonstrates the potential learning value of a Monte Carlo study: generate data under known assumptions, systematically vary the conditions under which the data are generated and evaluate what happens.

Other types of Monte Carlo procedures include *Monte Carlo Exact tests* and *Monte Carlo Markov Chains*. Exact tests (see, for example, the discussions in Pesarin 2001; SYSTAT Inc 2009a, b) are an alternative strategy to parametric and nonparametric tests of hypotheses in situations where small sizes and/or other data anomalies exist. An exact test (also called a *permutation* test) finds the exact p-value for a statistic by computing all possible permutations of the data in a sample. In situations where this is not possible, a Monte Carlo sampling approach is used, hence a Monte Carlo Exact test. Exact tests are computationally expensive to conduct and the Monte Carlo approach, where a large number of random samples of data are examined, is typically more efficient (Mooney 1997, pp. 72–88). In SYSTAT, for example, there are Monte Carlo exact tests available for continuous data, rates and proportions data and correlational data.

A Markov chain is a discrete model that represents a set of possible states for an observation or event along with the probability that any particular state will change to another state or a particular event will lead to another event on the next or subsequent observation (see Jackman 2009, ch. 4). This permits the construction and testing of models where previous outcomes can influence current outcomes. Each observational occurrence is called a 'step' and the probabilities are called 'transitional probabilities' (probabilities of transitions between states or events). Since Markov chains can model probabilistic (or stochastic) transitions between states or events, they have a natural alignment with Bayesian inference (recall *Fundamental concept X*). More complex models can be evaluated by combining Markov chains with Monte Carlo simulation, leading to an analytical approach

Procedure 9.10: Simulation & Computational Modelling

called *Monte Carlo Markov chains* (see Jackman 2009, ch. 5 and 6). Areas where Markov chain and Monte Carlo Markov chain models have been used include weather forecasting, genetic analysis between generations, medical and epidemiological analysis, analysis of board games that use dice, analysis of stock market behaviour and modelling of the economic and political development of a nation.

A simple example of a two-step Markov chain would be a social worker visiting a household unannounced on two separate occasions. On occasion 1, the social worker observes evidence of child abuse or not (the two possible states would be a child was abused or a child was not abused); on occasion 2, the social worker returns unannounced to the household and makes another assessment, with the same two possible outcomes. The diagram below shows what this situation might look like if $n = 167$ social workers recorded their observations on two occasions (note that a Bayesian would say that occasion 1 observations provide prior probability estimates for occasion 2 observations).

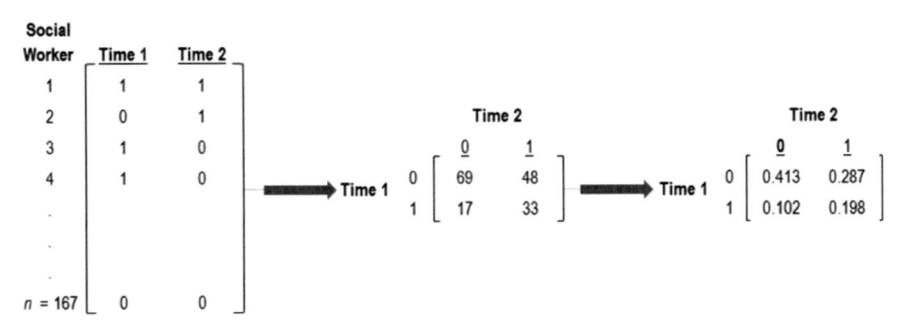

Social Worker	Time 1	Time 2
1	1	1
2	0	1
3	1	0
4	1	0
.		
$n = 167$	0	0

$$\text{Time 1} \begin{array}{c} \\ 0 \\ 1 \end{array} \begin{bmatrix} \text{Time 2} \\ 0 \quad 1 \\ 69 \quad 48 \\ 17 \quad 33 \end{bmatrix} \quad\longrightarrow\quad \text{Time 1} \begin{array}{c} \\ 0 \\ 1 \end{array} \begin{bmatrix} \text{Time 2} \\ 0 \quad\quad 1 \\ 0.413 \quad 0.287 \\ 0.102 \quad 0.198 \end{bmatrix}$$

0 = no evidence of child abuse observed on unannounced visit
1 = evidence of child abuse observed on unannounced visit

The original observation matrix (left-hand matrix) is condensed into a square matrix of counts (middle matrix) which is then converted into observed probabilities (right-hand matrix). Markov chain analysis could allow us to model the transitional probabilities between the states on the two occasions, using any number of potential theoretical distributions. If we want to model a much larger number of steps, then Monte Carlo Markov chain analysis, using random samples from the original observation matrix, would provide the best approach to evaluating the model.

Computational Modelling

Computational modelling (see Taber and Timpone 1996, for an overview) takes advantage of computing power coupled with computer programming to (1) facilitate the construction of a theoretical model, (2) generate data needed to monitor the behaviour of the model and (3) provide avenues for validating both the model processes and its outcomes. Taber and Timpone (1996) observed that there were several categories of computational models, including dynamic systems simulation models (which subsume Monte Carlo simulation), knowledge-based models (e.g. semantic networks and expert rule-based systems) and machine learning models. We have already explored several approaches associated with the latter two categories in *Procedure 9.9*: knowledge-based models in the form of the Leximancer concept map (a semantic network model of sorts) and association rules (in data mining) and machine learning systems (e.g. neural network models and clustering models for supervised and unsupervised learning). Here we want to explore dynamic systems simulation as used in the social and behavioural sciences. We will look at two broad approaches: *dynamic systems modelling* and *agent-based modelling*. Both approaches follow a similar trajectory but have rather different purposes.

The basic stages involved in dynamic systems simulation are: (1) development of a theory or conceptual framework for the model and its associated processes and anticipated outcomes; (2) using that framework to guide the development of the model itself (usually in conjunction with and support of an integrated computer programming environment such as *Insight Maker, Stella, iThink, Vensim* or *NetLogo*); (3) evaluating the behaviour of the model by running simulations and experiments; and (4) refining the model based on learning from Stage 3 as well as perhaps from external validation research (comparing model outcomes to observable outcomes in real systems, where and when such data can be sourced). Computer software systems can facilitate the navigation of Stages 1, 2 and 3 as well as providing a platform for incorporating and evaluating further refinements.

Dynamic Systems Modelling

System dynamics as an applied research discipline had its genesis in the work of Jay Forrester at MIT in the 1950s. From there, *dynamic systems modelling* began to grow as a concrete pathway for developing and testing simulation models with developments keeping pace with advances in computer programming and technology. Moving into the twenty-first century, dynamic systems modelling melded with systems thinking principles (from researchers such as Peter Senge and Russell Ackoff) as well as complexity theory and new mathematic approaches for modelling complex nonlinear systems. Such models may be *deterministic* in emphasis (where a system of equations calculates exact quantities for each step of the simulation process) or *stochastic* (where simulated random data are used as part of the modelling process; this is where Monte Carlo methods can play a role). Software environments to support researchers and problem solvers in the conceptualisation, construction and testing of dynamic systems models have become a necessity. For example, *Insight Maker* (http://insightmaker.com/) now provides a free-to-access online platform for building and sharing dynamic systems models as well as system conceptualisations. *Stella*, and its industrial strength older 'brother' *iThink* (https://iseesystems.com/store/products/stella-architect.aspx), as well as *Vensim (*https://

vensim.com/) and *Powersim* (https://www.powersim.com/) are all integrated packages for mapping and modelling dynamic systems (each of these packages offer a trial version for user evaluation).

It is easier to understand the dynamic systems approach using a concrete illustration. Richmond (2004) provides a nice conceptual introduction to systems thinking and dynamic systems modelling (using *Stella* as the computer platform). In designing and constructing a dynamic systems model, the researcher first constructs a representation or story that conveys the essence of the process being modelled. A common representation device for such thinking is the *causal loop diagram* and Kim and Anderson (1998) have shown that, for many human and organisational systems, there are certain archetypal representations, such as the 'Fixes the Fail' archetype, the 'Limits to Growth or Success' archetype, the 'Tragedy of the Commons 'archetype, the 'Escalation' archetype and the 'Shifting the Burden' or 'Addiction' archetype.

To illustrate an archetype in action and to follow its development into a dynamic systems model, consider the following scenario. We are interested in modelling the dynamics between training and job performance for quality control inspectors, noting that in many cases, it appears that the more we train someone the worse they eventually perform. Training is often the 'fix' of choice by management when workers are performing below expectations and one would expect training to enhance performance. However, while it may do this in the short term, in the longer-term, this fix might begin to fail (i.e. backfire), creating more of a problem, rather than less. These anecdotal observations suggest that the system dynamics here are congruent with the 'Fixes that Fail' system archetype.

Figure 9.26 shows what a 'Fixes that Fail' archetype causal link diagram of this problem might look like (building upon a causal loop template produced by Bellinger 2013). The systems diagram has several key concepts: undergoing training (the action or 'fix'), the current level of QCI inspection performance, the desired level for that performance, the performance gap (desired level minus current level) and a concept referred to as training fatigue. These concepts are linked to each other via causal loops. There are two kinds of causal loops and both are represented in Fig. 9.26: a *balancing loop* (represented by the 'seesaw' icon), where the system tries to stabilise itself, seeking an equilibrium point (rather like a thermostat controlling the temperature of a room) and a *reinforcing loop*, (represented by the 'snowball rolling down a hill' icon) where the system gets progressively more destabilised, moving away from equilibrium (rather like positive feedback through a microphone). The arrows show the theorised direction of causality and the plus or minus sign attached to each arrow signals the direction of the influence (plus → more begets more or increase in cause (tail of the arrow) leads to increase in effect (head of the arrow); minus → more begets less or increase in cause leads to decrease in effect). Note that in Fig. 9.26, the hallmark of a balancing loop is that it has all plus signs except one. Thus, undergoing training improves performance which reduces (i.e. shrinks) the performance gap which reduces the need for more training. The hallmark of a reinforcing loop is that all signs point one direction (either all plus or all minus in the loop). Thus, undergoing more training stimulates training fatigue

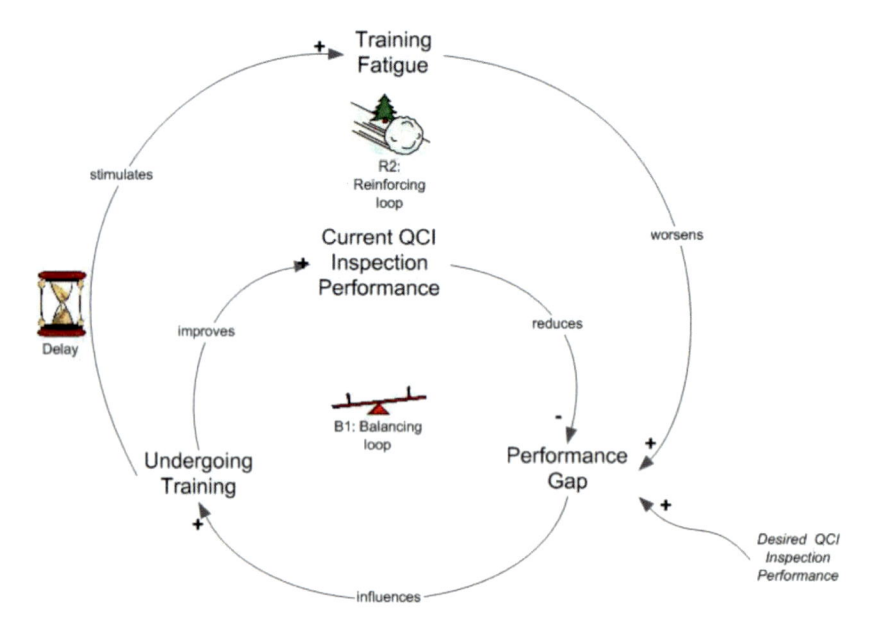

Fig. 9.26 The 'Fixes that Fail' systems archetype. (Loosely adapted from Bellinger's 2013 'causal loop' template, see https://insightmaker.com/insight/24847/Fixes-that-Fail-Archetype)

which worsens (i.e. widens) the performance gap which creates the need for more training. The 'hourglass' icon symbolises that the causal effect between undergoing training and training fatigue is a delayed effect.

Insight Maker provides a simple interface for translating the causal loop diagram in Fig. 9.26 into the dynamic systems model shown in Fig. 9.27 (sometimes called a 'stocks & flows' model; this one builds on a 'simulation' template uploaded by Bellinger 2013 to the *Insight Maker* website community). Figure 9.27 is an actual screen shot of *Insight Maker*, where the dynamic systems model appears on the left-hand side of the screen surrounding by the dashed line. You can see that in order to make the model functional and more realistic, some new concepts have to be represented in the model: 'boredom' to capture what actually contributes to training fatigue, which is influenced by the 'volume' of training undergone; 'training effectiveness' to signal that no training will completely close a performance gap, only part of it; 'demoralisation' as a way of capturing the feeling that would emerge from the confluence of current performance level and training fatigue and 'psychological resilience' to provide a controlling influence on just how strongly an inspector might develop a sense of demoralisation (more resilience means more robustness against feelings of demoralisation). For each concept and major arrow in the model, a constant or equation is assigned to capture the meaning and/or relationship (including rates that reflect how things would be predicted to change over time). The window on the right-hand side of Fig. 9.27 is the interactive panel, where sliders allow the researcher to control inputs into the simulation. When control inputs have

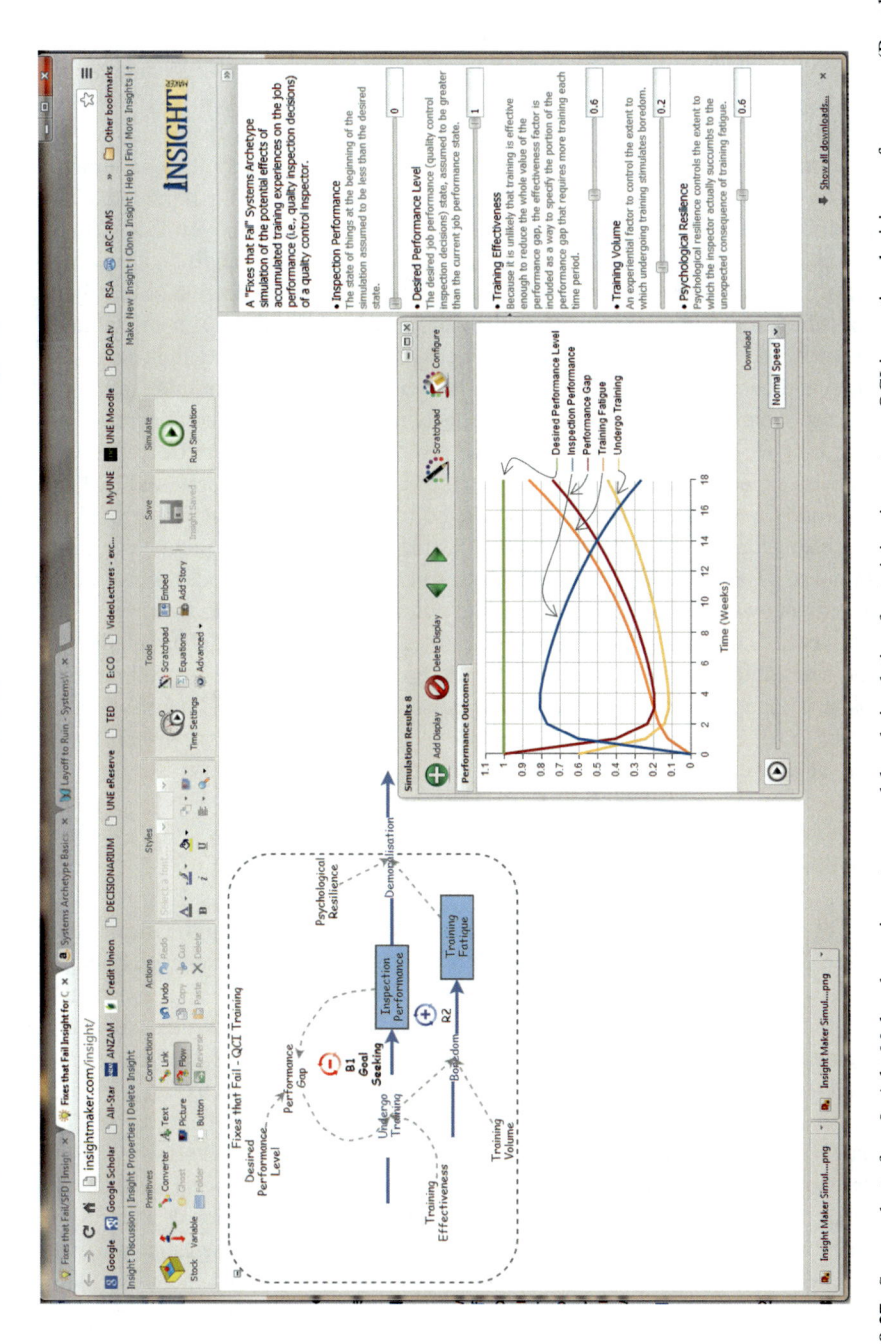

Procedure 9.10: Simulation & Computational Modelling

Fig. 9.27 Screen shot of an *Insight Maker* dynamic systems model and simulation for training impacts on QCI inspection decision performance. (Based on Bellinger's 2013 stocks and flows 'simulation' template, see https://insightmaker.com/insight/24847/Fixes-that-Fail-Archetype)

been set, the researcher simply presses the 'Run Simulation' icon. The outcomes from the simulation, given the control parameters and relationships built into the dynamic model are then graphed over time (in this case a simulated time period of 18 weeks). It is this type of control panel that gives rise to the term 'flight simulator' when describing the simulation environment of a dynamic systems model.

The graph in the middle of the screen shot in Fig. 9.27 plots the behaviour of several components in the model (differentiated by colour on a computer monitor and flagged by arrows for additional clarity) shows several trends. The top horizontal line represents the desired level of performance which is a constant. As you look across time, you can see that (1) inspection performance increases with additional training earlier in the time period (as one would expect if training is effective at all) then reverses direction and gets progressively worse (reflecting the impact of training fatigue); (2) the performance gap at first gets dramatically smaller, than reverses direction and becomes progressively larger, which feeds a demand for more training, which also begins to increase over time, accompanied by a steady increase in training fatigue. Thus, the behaviour of the model does appear to mirror what we would have expected given our anecdotal observations. The step that would remain for externally validating this model would be to actively seek data from quality control inspectors, monitor them over time as they undergo training and have their performance assessed, obtaining key measures of the system (the model shows what measurements would be needed) at multiple points in time, mapping their resulting behaviours and comparing the observed outcomes with what the model has predicted via the simulation. The model can also undergo *sensitivity testing* as well, which means systematically varying some of the inputs (e.g. psychological resilience, for example, to see how different levels might affect the behaviour of the system) and looking at how changes in inputs produce changes (or not) in outputs.

Agent-Based Modelling

Agent-based modelling (ABM; see, for example, Gilbert 2008 and Gilbert and Troitzsch 2005, ch. 8 and 9) is a more recent development in computational modelling. ABM focuses on social science problems that surround how agents (e.g. individual people, teams, departments, organisations, nations) interact with each other dynamically through time. It shares some similar interests to social network analysis (recall *Procedure 9.7*) but from a dynamic computational modelling perspective. Agents are modelled and interact with each other in a virtual world, controlled by the parameters of the model and the simulation conditions. Agent-to-agent interaction represents information flow (verbal, nonverbal, observational, decisions and choices) as well as relational connection and quality (formal, informal, cognitive, emotional, attitudinal, cross-cultural, trust, prior history). Gilbert (2008) highlighted some interesting areas in social science where ABM has proven useful: social influences on consumer behaviour, collaboration/competition in industrial networks, team interactions (e.g. interactions of interdependent teams), opinion dynamics (how opinions spread through a population), models of urban behaviour (e.g. neighbourhood growth; segregation dynamics) and supply chain dynamics (inventory management and movements between various organisations in a supply chain).

ABM requires very precise theoretical conceptualisation and mathematical modelling which, similarly to dynamic systems modelling, requires a computer support environment to facilitate. For ABM, a very popular software platform, that is free and open source, is *NetLogo* (Wilensky 1999). *NetLogo* offers a model development interface (note that dynamic systems models can also be constructed within *NetLogo*; a screen shot of the NetLogo interface appears in Fig. 9.28). Again, we will illustrate what agent-based modelling can accomplish using a concrete example. The example, in this case, is the 'team assembly' agent-based model included in the *NetLogo* library (see Bakshy and Wilensky 2007, building upon Guimera et al. 2005, for the more formal story about the model). The 'team assembly' model was designed to dynamically simulate the collaborative interactions between new and existing small work teams in creative enterprise networks. At each step (called a 'tick') of the simulation, a new team is created (on a probability basis, controllable by the researcher using a slider bar); the team may be comprised of 'newcomers' (never participated in a team before) or 'incumbents' (have previous experience on teams). If an 'incumbent' is selected, then the probability that an incumbent will be chosen for the team based on having previously collaborated with another team member is set at .65 (65%). Also controllable with slider bars are the length of time an agent is idle (i.e. has not participated in a new team = max_downtime) before being deleted from the map and the size of the teams that are created.

The screen shot in Fig. 9.28 shows the simulation interface for the 'team assembly' model 152 'ticks' into a simulation run, where the probability of a new team member being an 'incumbent' was set at .40 (40%), team size was set at 4 and max_downtime was set at 40 ticks. The simulation field (centre graphical area in Fig. 9.28) shows the evolution of the network of teams, where lines show the most recent collaborative links (while it is not apparent in greyscale print, some of the links are red, which connects incumbents who have collaborated with each other a number of times). There are several graphs that also dynamically update as the simulation progresses, to provide insights into the mixtures of newcomers and incumbents (the 'Link counts' stacked histogram in the lower left-hand corner), the percentage of team members in the largest connected network in the simulation field ('% of agents in the giant component' line graph, top right-hand side) and the average size of isolated collaborative networks ('average component size' graph, bottom right-hand side). Notice that the collaborative network is densest and most connected in the centre of the simulation field and isolated collaborative networks gravitate toward the periphery. By experimenting with different parameter settings, the behaviour of the collaborative networks can be studied. For example, if the only parameter that is changed is team size, from 4 to 6, then at tick 152, the large collaborative network in the centre will tend to be much more densely connected, but with more isolated teams on the periphery.

There is a critical observation to make about agent-based models which is that, at some point, the model needs to be connected or related to empirical observations in order to verify/validate the behaviour of the model (Boero and Squazzoni 2005).

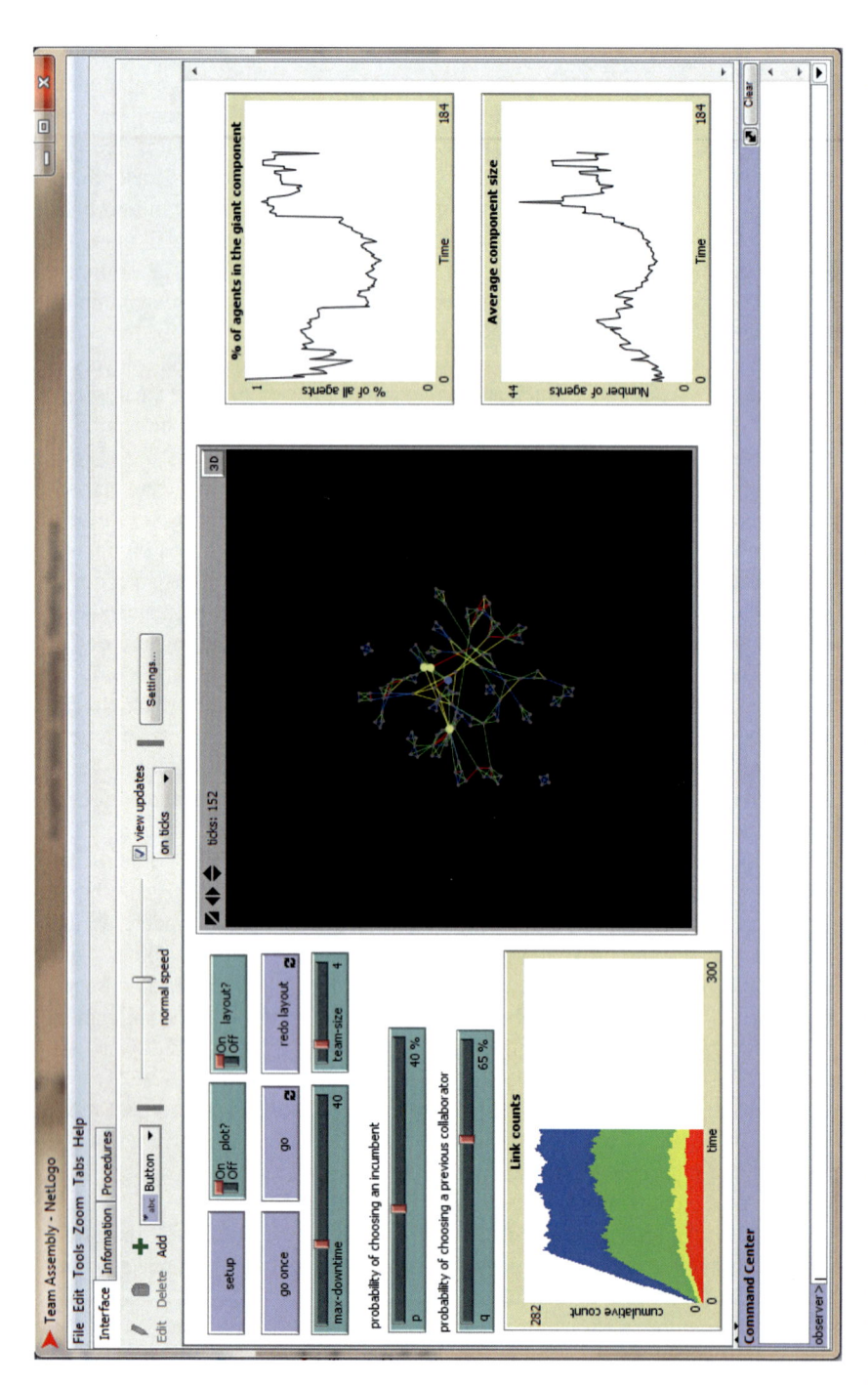

Fig. 9.28 Screen shot of NetLogo simulation of an agency-based model for team assembly and collaboration. (Based on Bakshy and Wilensky 2007)

Verification involves double-checking the conceptualisation of the model itself with experts and other data sources and validation involves comparing model outcomes to empirical and observable outcomes in real contexts. One useful validation process is called *docking*, which involves attempting to replicate the outcomes from one model using another model (see, for example, Olaru et al. 2009). Docking is most useful in contexts where empirical data for validating a model is either infeasible or too costly to gather. Verification and validation are essential processes in ABM research as they move agent-based models from the realm of abstract computer gaming to a more meaningful playing field where a model's usefulness can be more convincingly demonstrated.

Advantages

Simulation and computational modelling are becoming increasingly important tools in the toolkit for social and behavioural researchers. They allow the researcher to experimentally explore, without risk, models and theories using generated data rather than empirical data. Such models can incorporate complexities and nonlinearities that may render testing using empirical data or possibly existing statistical procedures infeasible. In fact, a computational model may help the researcher overcome, at least partially, the limits to linear models discussed in *Fundamental concept VI*. Dynamic simulation permits model testing and evaluation of the effects of intervention and changes in a social system, without having to physically intervene in the social system. Computer support environments for constructing and testing simulation and computational models are now at a stage where dynamic model building is within reach of the average social scientist. Monte Carlo methods facilitate the testing of stochastic (i.e. probabilistic) models and may afford the only defensible approach for certain types of hypothesis tests. Computational modelling allows the researcher to incorporate and accommodate more realistic social scenarios, relative to pure mathematical or statistical modelling, including the specification and testing of dynamic feedback loops, something that linear statistical models have great difficulty with (even in the context of structural equation modelling).

Disadvantages

It is important to acknowledge that there are critical limitations and caveats attached to simulation and computational modelling. Conceptualising and building these types of models is not easy; it may require a great deal of background work, including studying the literature and gathering qualitative and quantitative data from experts and potential users in order to verify the model conceptualisation and the processes embedded within it. Every computational model and every simulation model requires the researcher to make specific and simplifying assumptions about what concepts and relationships to include and which must be excluded, what parameters are needed (including what values and ranges they should reflect), how concepts should be measured and what kinds of outputs one should expect the model to be able to produce. These assumptions then shape and delimit the specifications and characterisation of the model. This means that any simulation or computational model is only as good as its underlying assumptions and model specifications will allow. This imposes an extra burden of transparency on the researcher, to be very clear about the entire conceptualisation → construction → testing → validation process for a model. In addition, model performance may be critically dependent upon the anchoring assumptions and the choices for controlling parameters. This means that thorough experimentation, sensitivity testing and validation (wherever possible) of simulation and computational models is essential, if the model is to be accepted as being of any use beyond an intellectual exercise. In some cases, external validation of model outcomes may be difficult or impossible to obtain because the data necessary to do so are not accessible. Docking has emerged as one way of trying to cope with this potential limitation, but to be successful, docking requires an appropriate alternative model to compare outcomes with – something that may not exist.

Where Is This Procedure Useful?

Simulation and computational modelling methods are most useful in research contexts where reasonably strong theory exists or can be developed to underpin the model. If complexity needs to be modelled instead of assumed away, then a simulation or computational model may be the only feasible research pathway to pursue. Simulation and computational modelling methods are also very useful in circumstances where it is not possible to physically test or perhaps even measure the sorts of behaviours that the model is focusing on. Working within a simulated environment may allow for difficult constructs to be dealt with, especially where the constructs do not easily translate into direct measurement systems. Computational modelling methods, especially in the form of dynamic systems models, may also be useful for pedagogical and training purposes, allowing students to play with the parameters of a system and to see the consequences of their decisions without risk of harming themselves or anyone else.

Procedure 9.10: Simulation & Computational Modelling

Software Procedures

Application	Procedures
SPSS	SPSS has Monte Carlo simulation capabilities (accessed via *Analyze→ Simulation*; see IBM SPSS Inc 2013 and https://www.ibm.com/support/knowledgecenter/en/SSLVMB_22.0.0/com.ibm.spss.statistics.help/spss/base/simulation.htm). Additionally, SPSS can generate univariate random variables from various distributions (e.g. normal, binomial, uniform) via the *Transform → Compute Variable* menu pathway. Generated data can also be saved as new variables for use in other analytical procedures. SPSS cannot, however, do computational modelling. There is also an add-on module for SPSS that can conduct a range of *Exact Tests* (see http://www.spss.com.hk/software/statistics/exact-tests/).
NCSS	NCSS has some useful Monte Carlo simulation capabilities but cannot do computational modelling and can perform only a very limited set of exact tests. Monte Carlo simulation via NCSS is accessible via the *Analysis → Descriptive Statistics → Data simulation* menu pathway. A wide variety of discrete and continuous distributions may be randomly sampled as well as different combinations and mixtures of distributions and NCSS can provide a range of statistics and graphical displays of the generated data. Generated data can also be saved as new variables for use in other analytical procedures.
SYSTAT	SYSTAT has a wide range of Monte Carlo simulation capabilities but cannot do computational modelling. Monte Carlo simulation procedures can be accessed via the *Addons → Monte Carlo...* menu pathway (SYSTAT 2009a). Univariate discrete and continuous distributions can be sampled as well as five different multivariate distributions having a specifiable correlation structure amongst the variables. SYSTAT can also perform Monte Carlo Markov Chain (*MCMC*) analyses as well as something called Integration and Importance Distribution Monte Carlo (*IIDMC*) sampling. Generated data can also be saved as new variables for use in other analytical procedures. SYSTAT has an add-on module for *Exact tests* as well (SYSTAT 2009b), which can carry out a wide range of exact tests for statistical procedures.
STATGRAPHICS	STATGRAPHICS has some useful Monte Carlo simulation capabilities but cannot do computational modelling. Monte Carlo procedures are accessible via the *Tools → Monte Carlo Simulation...* menu pathway. *Data Simulation Models and Random Number Generation* can both be done as well as *ARIMA Time Series Models* and each procedure will provide some descriptive analyses of the generated data. Several discrete and continuous distributions can be sampled, but technically STATGRAPHICS cannot generate a true multivariate dataset with a specific correlation structure. Each approach will generate the specified dataset which can be saved and used in other analytical procedures.
R Commander	**R Commander** offers some limited Monte Carlo capability in terms of being able to randomly sample a wide variety of distributions and store the created dataset. These capabilities are accessed via the *Distributions* menu and you can choose whether you want to sample a discrete or a continuous distribution. More generally, there are several packages and functionalities available for use within the main **R** program environment for carrying out various types of Monte Carlo and Monte Carlo Markov Chain simulations

(continued)

Application	Procedures
	(see Roberts and Casella 2009). The *MCMCpack* package (https://cran.r-project.org/web/packages/MCMCpack/MCMCpack.pdf) also provides Monte Carlo Markov Chain analysis capabilities. With respect to computational modelling, the package *RNetLogo* (https://cran.r-project.org/web/packages/RNetLogo/RNetLogo.pdf) provides an interface for working with the *NetLogo* agent-based modelling system program within **R**. The 3D version of *NetLogo* can also be managed through this interface.
Insight Maker	*Insight Maker* (https://insightmaker.com/) is a free online dynamic systems modelling environment for the layperson, with all of the necessary model development tools, tutorials/webinars and manuals accessible from the website. Using the online interface, you can sketch your own causal loop diagram and build your own stock & flow model (or you can clone and adapt those of others). You can construct a 'flight simulator' with slider control bars for key parameters and run experimental simulations with the model you design. *Insight Maker* can display graphical outcomes from each simulation you run. The website also gives you access to an extensive online library of models that others have built and uploaded (such as the Bellinger 2013 templates used for Figs. 9.26 and 9.27). *Insight Maker* is best suited for creating and testing dynamic systems models, but you can use it to build and test certain types of agent-based models as well.
Stella	*Stella* (https://iseesystems.com/store/products/stella-architect.aspx) is a well-known software package for creating and testing dynamic systems models. Richmond (2004) provides a very straightforward introduction to systems thinking and modelling in the *Stella* environment. *Stella* provides an integrated interface for developing and testing models and displaying the outcomes from simulations. The user can develop a diagram of their system in the 'Map' workspace of the interface and develop their model from that diagram in the 'Model' workspace. An entire 'flight simulator' or 'learning laboratory' surrounding a model can be developed in *Stella*, providing control over all desirable parameters of the model and over the outputs that the user wishes to see displayed, via the 'Interface' workspace. Using the interface, systematic experiments can be run. Dynamic experiments can also be run, where the simulation is periodically halted as a function of some emergent pattern or issue and the user is asked to make a decision or choice. Then the simulation can be resumed to track what follows from that choice. A trial version of *Stella* can be downloaded from its website.
NetLogo	*NetLogo* (Wilensky 1999; http://ccl.northwestern.edu/netlogo/index.shtml) is a free open-source programming and simulation environment for constructing and testing agent-based models (some dynamic systems models can also be constructed using *NetLogo*). *NetLogo* comes with an extensive user's manual and it takes some effort to learn how to use effectively to program, build and test ABMs. The interface allows the user to build and customise a graphics-based simulation control platform. Fortunately, *NetLogo* comes with a diverse range of publicly accessible, pre-built, pre-tested and documented ABMs (one of which, by Bakshky and Wilensky 2007, was used to produce Fig. 9.28) and system dynamics models from different disciplines in its 'Models Library', including the social sciences, earth sciences (including climate change), art and general networks. There is also a 'User Community' that contributes publicly accessible *NetLogo* models.

References

References for Fundamental Concept X

Gigerenzer, G., Krauss, S., & Vitouch, O. (2004). The null ritual: What you always wanted to know about significance testing but were afraid to ask. In D. Kaplan (Ed.), *The SAGE handbook of quantitative methodology for the social sciences* (pp. 391–408). Thousand Oaks: Sage.

Iversen, G. R. (1984). *Bayesian statistical inference*. Thousand Oaks: Sage.

Lindman, H. R. (1997). Bayesian statistics. In J. P. Keeves (Ed.), *Educational research, methodology and measurement: An international handbook* (2nd ed., pp. 456–461). New York: Pergamon.

Panik, M. (2009). *Regression modelling: methods, theory, and computation with SAS* (pp. 265–283). Boca Raton: CRC Press.

Sohn, D. (1998). Statistical significance and replicability: Why the former does not presage the latter. *Theory & Psychology, 8*(3), 291–311.

Useful Additional Reading for Fundamental Concept X

Lambert, B. (2018). *A student's guide to Bayesian statistics*. Los Angeles: Sage.

Lee, P. M. (2012). *Bayesian statistics: An introduction* (4th ed.). Chichester: Wiley.

Phillips, L. D. (2005). Bayesian statistics. In B. S. Everitt & D. C. Howell (Eds.), *Encyclopedia of statistics in behavioral science* (Vol. 1, pp. 146–150). Chichester: Wiley.

References for Procedure 9.1

Andrich, D. (2005). Rasch models for ordered response categories. In B. S. Everitt & D. C. Howell (Eds.), *Encyclopedia of statistics in behavioral science* (Vol. 4, pp. 1698–1707). Chichester: Wiley.

Baker, F. B. (2001). *The basics of item response theory* (2nd ed.). College Park: ERIC Clearing House on Assessment and Evaluation.

Bond, T. G., & Fox, C. M. (2015). *Applying the Rasch Model: Fundamental measurement in the human sciences* (3rd ed.). New York: Routledge.

Henard, D. H. (2000). Item response theory. In L. G. Grimm & P. R. Yarnold (Eds.), *Reading and understanding more multivariate statistics* (pp. 67–98). Washington, DC: American Psychological Association.

Magis, D., Yan, D., & von Davier, A. A. (2017). *Computerized adaptive and multistage testing with R: Using packages catR and mst*. Cham: Springer Nature.

Ostini, R., & Nering, M. L. (2006). *Polytomous item response theory models*. Thousand Oaks: Sage.

Revelle, W. (2019). *An overview of the psych package*. Downloaded from https://cran.r-project.org/web/packages/psych/vignettes/overview.pdf. Accessed 20 Sept 2019.

Rizopoulos, D. (2006). ltm: An R package for latent variable modelling and item response theory analyses. *Journal of Statistical Software, 17*(5), 1–25.

Useful Additional Reading for Procedure 9.1

Allerup, P. (1997). Rasch measurement theory. In J. P. Keeves (Ed.), *Educational research, methodology and measurement: An international handbook* (2nd ed., pp. 836–874). New York: Pergamon.
Andrich, D. (1988). *Rasch models for measurement*. Thousand Oaks: Sage.
Baker, F. B., & Kim, S.-H. (2004). *Item response theory: Parameter estimation techniques* (2nd ed.). New York: Marcel Dekker.
de Ayala, R. J. (2009). *The theory and practice of item response theory*. New York: The Guilford Press.
Fischer, G. H. (2005). Rasch modeling. In B. S. Everitt & D. C. Howell (Eds.), *Encyclopedia of statistics in behavioral science* (Vol. 4, pp. 1691–1698). Chichester: Wiley.
Stocking, M. L. (1997). Item response theory. In J. P. Keeves (Ed.), *Educational research, methodology and measurement: An international handbook* (2nd ed., pp. 836–840). New York: Pergamon.

References for Procedure 9.2

Allison, P. (2004). Event history analysis. In M. Hardy & A. Bryman (Eds.), *Handbook of data analysis* (pp. 369–385). London: Sage.
Deitz, D. M. (2013). *Survival analysis in R*. Unpublished paper, downloaded from https://www.openintro.org/download.php?file=survival_analysis_in_R&referrer=/stat/surv.php. Accessed 20 Sept 2019.
Tabachnick, B. G., & Fidell, L. S. (2019). *Using multivariate statistics* (7th ed.). New York: Pearson Education. ch. 11.
Wright, R. E. (2000). Survival analysis. In L. G. Grimm & P. R. Yarnold (Eds.), *Reading and understanding more multivariate statistics* (pp. 363–407). Washington, DC: American Psychological Association (APA).

Useful Additional Reading for Procedure 9.2

Cohen, J., Cohen, P., West, S. G., & Aiken, L. S. (2003). *Applied multiple regression/correlation analysis for the behavioral sciences* (3rd ed., pp. 596–600). Mahwah: Lawrence Erlbaum Associates. ch. 15.
Klein, J. P., van Houwelingen, H. C., Ibrahim, J. G., & Scheike, T. H. (Eds.). (2014). *Handbook of survival analysis*. Boca Raton: Chapman & Hall/CRC.
Kleinbaum, D. G., & Klein, M. (2012). *Survival analysis: A self-learning text* (3rd ed.). New York: Springer.
Landau, S. (2005). Survival analysis. In B. S. Everitt & D. C. Howell (Eds.), *Encyclopedia of statistics in behavioral science* (Vol. 4, pp. 1980–1987). Chichester: Wiley.
Meyers, L. S., Gamst, G. C., & Guarino, A. (2017). *Applied multivariate research: Design and interpretation* (3rd ed.). Thousand Oaks: Sage. ch. 20A, 20B.
Mills, M. (2011). *Introducing survival and event history analysis*. London: Sage.

References for Procedure 9.3

Adhikari, S. K., Manna, B. K., & Pillai, P. R. P. (2009a). Quality analysis. In SYSTAT Software, Inc (Ed.), *SYSTAT 13 Statistics VI* (pp. VI-89–VI-199). Chicago: SYSTAT Software, Inc.
Adhikari, S. K., Chaudhuri, A. K., Dirghangi, A. K., Manna, B. K., & Pillai, P. R. P. (2009b). *SYSTAT 13 Quality Analysis – 1*. Chicago: SYSTAT Software Inc.
Benneyan, J. C. (1998). Use and interpretation of statistical quality control charts. *International Journal for Quality in Health Care, 10*(1), 69–73.
Black, K., Asafu-Adjaye, J., Khan, N., Perera, N., Edwards, P., & Harris, M. (2007). *Australiasian business statistics*. Milton: Wiley. ch. 18.
Breyfogle, F. W., III. (2003). *Implementing six sigma: Smarter solutions using statistical methods* (2nd ed.). New York: Wiley. ch. 5, 10.
Scrucca, L. (2004). qcc: An R package for quality control charting and statistical process control. *R News, 4*(1), 11–17.

Useful Additional Reading for Procedure 9.3

Gerald, MS 2004, *Statistical process control and quality improvement*, 5th edn, Prentice Hall, Englewood Cliffs . ch. 4, 6, 7 and 9.
Montgomery, D. C. (2019). *Introduction to statistical quality control* (8th ed.). New York: Wiley. ch. 5, 6 and 7.

References for Procedure 9.4

Aizaki, H., & Nishikura, K. (2008). Design and analysis of choice experiments using R: A brief introduction. *Agricultural Information Journal, 17*(2), 86–94.
Augur, P., Devinney, T. M., & Louviere, J. J. (2007). Using best-worst scaling methodology to investigate consumer ethical beliefs across countries. *Journal of Business Ethics, 70*, 299–326.
Bakken, D., & Frazier, C. L. (2006). Conjoint analysis: Understanding consumer decision making. In R. Grover & M. Vriens (Eds.), *Handbook of MARKETING RESEARCH* (pp. 288–312). Thousand Oaks: Sage.
Hair, J. F., Black, B., Babin, B., & Anderson, R. E. (2010). *Multivariate data analysis: A global perspective* (7th ed.). Upper Saddle River: Pearson Education. ch. 6.
Lee, J. A., Soutar, G. N., & Louviere, J. (2008). The best-worst scaling approach: An alternative approach to measuring Schwartz's values. *Journal of Personality Assessment, 90*(4), 335–347.
Louviere, J. J. (1988). *Analyzing decision making: Metric conjoint analysis*. Newbury Park: Sage.
Louviere, J. J., Flynn, T. N., & Marley, A. A. J. (2015). *Best-worst scaling: Theory, methods and applications*. New York: Cambridge University Press.
Louviere, J. J., Henscher, D. A., & Swait, J. D. (2000). *Stated choice methods: Analysis and application*. Cambridge: Cambridge University Press.
Sermas, R., & Colias, J. V. (2013). *Package 'ChoiceModelR'*. https://cran.r-project.org/web/packages/ChoiceModelR/ChoiceModelR.pdf. Accessed 21 Sept 2019.
Stenson, H. (2009). Design of experiments. In SYSTAT Software, Inc (Ed.), *SYSTAT 13 Statistics I* (pp. I-341–I-385). Chicago: SYSTAT Software, Inc.
Wilkinson, L. (2009). Conjoint analysis. In SYSTAT Software, Inc (Ed.), *SYSTAT 13 Statistics I* (pp. I-125–I-155). Chicago: SYSTAT Software, Inc.

Useful Additional Reading for Procedure 9.4

Henscher, D. A., Rose, J. M., & Greene, W. H. (2005). *Applied choice analysis: A primer*. Cambridge: Cambridge University Press.

Hess, S., & Daly, A. (Eds.). (2014). *Handbook of choice modelling*. Cheltenham: Edward Elgar.

Louviere, J. J. (1991). Experimental choice analysis: introduction and overview. *Journal of Business Research, 23*(4), 291–297.

Louviere, J. J. (1994). Conjoint analysis. In R. Bagozzi (Ed.), *Handbook of Marketing Research* (pp. 223–259). Oxford: Blackwell Publishers.

References for Procedure 9.5

Bickel, R. (2007). *Multilevel analysis for applied research*. New York: The Guilford Press.

Cohen, J., Cohen, P., West, S. G., & Aiken, L. S. (2003). *Applied multiple regression/correlation analysis for the behavioral sciences* (3rd ed.). Mahwah: Lawrence Erlbaum Associates. ch. 14.

Field, A. (2018). *Discovering statistics using SPSS for Windows* (5th ed.). Los Angeles: Sage. ch. 21.

Field, A., Miles, J., & Field, Z. (2012). *Discovering statistics using R*. London: Sage. ch. 19.

Finch, W. H., & Bolin, J. E. (2017). *Multilevel modeling using MPlus*. Boca Raton: CRC Press.

Finch, W. H., Bolin, J. E., & Kelley, K. (2017). *Multilevel modeling using R*. Boca Raton: CRC Press.

Garson, G. D. (Ed.). (2013). *Hierarchical linear modelling: Guide and applications*. Los Angeles: Sage.

Garson, G. D. (2020). *Multilevel modelling: Applications in Stata, IBM SPSS, SAS, R & HLM*. Los Angeles: Sage.

Geiser, C. (2013). *Data analysis with MPlus*. New York: The Guilford Press.

Hamilton, L. C. (2013). *Statistics with Stata: Version 12*. Boston: Brooks/Cole.

Hedeker, D., Marcantonio, R., & Pechnyo, M. (2009). Mixed regression. In SYSTAT Software Inc (Ed.), *SYSTAT 13 Statistics II* (pp. 439–502). Richmond: SYSTAT Software, Inc.

Hintze, J. L. (2007). *NCSS 8 help system: Mixed Models*. Kaysville: Number Cruncher Statistical Systems. ch. 220.

Luke, D. A. (2004). *Multilevel modeling*. Thousand Oaks: Sage.

Norušis, M. J. (2012). *IBM SPSS statistics 19 advanced statistical procedures companion*. Upper Saddle River: Prentice Hall. ch.10.

Raudenbush, S. W., & Bryk, A. S. (2002). *Hierarchical linear models: Applications and data analysis methods* (2nd ed.). Thousand Oaks: Sage.

Raudenbush, S. W., & Bryk, A. S. (1997). Hierarchical linear modeling. In J. P. Keeves (Ed.), *Educational research, methodology and measurement: An international handbook* (2nd ed., pp. 549–556). New York: Pergamon.

Raudenbush, S. W., Bryk, A. S., Cheong, Y. F., Congdon, R. T., Jr., & du Toit, M. (2011). *HLM 7: Hierarchical linear and nonlinear modelling*. Lincolnwood: Scientific Software International.

Robson, K., & Pevalin, D. (2016). *Multilevel modeling in plain language*. Los Angeles: Sage.

Tabachnick, B. G., & Fidell, L. S. (2019). *Using multivariate statistics* (7th ed.). New York: Pearson Education. ch. 15.

Useful Additional Reading for Procedure 9.5

Browne, W., & Rabash, J. (2004). Multilevel modelling. In M. Hardy & A. Bryman (Eds.), *Handbook of data analysis* (pp. 459–479). London: Sage.
Hox, J. (2002). *Multilevel analysis*. Mahwah: Lawrence Erlbaum Associates.
Keeves, J. P., & Sellin, N. (1997). Multilevel analysis. In J. P. Keeves (Ed.), *Educational research, methodology and measurement: An international handbook* (2nd ed., pp. 394–403). New York: Pergamon.
Meyers, L. S., Gamst, G. C., & Guarino, A. (2017). *Applied multivariate research: Design and interpretation* (3rd ed.). Thousand Oaks: Sage. ch. 8A, 8B.
Miles, J., & Shevlin, M. (2001). *Applying regression & correlation: A guide for students and researchers*. London: Sage. ch. 8.
Ployhart, R. E. (2005). Hierarchical models. In B. S. Everitt & D. C. Howell (Eds.), *Encyclopedia of statistics in behavioral science* (Vol. 2, pp. 810–816). Chichester: Wiley.
Scott, M. A., Simonoff, J. S., & Marx, B. D. (2013). *The Sage handbook of multilevel analysis*. Los Angeles: Sage.

References for Procedure 9.6

Breiman, L., Friedman, J., Stone, C. J., & Olshen, R. A. (1984). *Classification and regression trees*. New York: Chapman & Hall.
Everitt, B. S. (2005). Classification and regression trees. In B. S. Everitt & D. C. Howell (Eds.), *Encyclopedia of statistics in behavioral science* (Vol. 1, pp. 287–290). Chichester: Wiley.
Loh, W.-Y. (2011). Classification and regression trees. *Wiley Interdisciplinary Reviews: Data Mining and Knowledge Discovery, 1*(1), 14–23. http://pages.stat.wisc.edu/~loh/treeprogs/guide/wires11.pdf. Accessed 22 Sept 2019.
Ma, X. (2018). *Using classification and regression trees: A practical primer*. Charlotte: Information Age Publishing.
Steinberg, D., & Colla, P. (1997). *CART—Classification and regression trees*. San Diego: Salford Systems.
Wilkinson, L. (2009). Classification and regression trees. In SYSTAT Software, Inc (Ed.), *SYSTAT 13 Statistics I* (pp. I-49–I-70). Chicago: SYSTAT Software, Inc.

Useful Additional Reading for Procedure 9.6

Haughton, D., & Oulabi, S. (2001). Direct marketing modeling with CART and CHAID. *Journal of Direct Marketing, 11*(4), 42–52.

References for Procedure 9.7

Borgatti, S. P., Everett, M. G., & Freeman, L. C. (2002). *UCINET for Windows: Software for social network analysis*. Harvard: Analytic Technologies.
Borgatti, S. P., Everett, M. G., & Johnson, J. C. (2013). *Analyzing social networks*. London: Sage.

Breiger, R. L. (2004). The analysis of social networks. In M. Hardy & A. Bryman (Eds.), *Handbook of data analysis* (pp. 505–526). London: Sage.
Hansen, D., Scheiderman, B., & Smith, M. A. (2011). *Analysing social media networks with NodeXL: Insights from a connected world*. Burlington: Morgan Kaufman/Elsevier.
Kilduff, M., & Tsai, W. (2003). *Social networks and organizations*. London: Sage.
Knoke, D., & Yang, S. (2008). *Social network analysis* (2nd ed.). Los Angeles: Sage.
Scott, J. G. (2017). *Social network analysis: A handbook* (4th ed.). Los Angeles: Sage.

Useful Additional Reading for Procedure 9.7

Klovdahl, A. S. (1997). Social network analysis. In J. P. Keeves (Ed.), *Educational research, methodology, and measurement: An international handbook* (2nd ed., pp. 684–690). Oxford: Pergamon Press.
Scott, J. G., & Carrington, P. J. (Eds.). (2011). *The Sage handbook of social network analysis*. London: Sage.
Yang, S., Keller, F. B., & Zheng, L. (2016). *Social network analysis: Methods and examples*. Los Angeles: Sage.

References for Procedure 9.8

Breen, R. (1996). *Regression models: Censored, sample selected or truncated data*. Thousand Oaks: Sage.
Cohen, J., Cohen, P., West, S. G., & Aiken, L. S. (2003). *Applied multiple regression/correlation analysis for the behavioral sciences* (3rd ed.). Mahwah: Lawrence Erlbaum Associates. ch. 13, 15.
Das, S., Ghosh, S., Jore, R., & Kulkarni, K. R. (rev Badashah, S. N. & Kulkarni, M.) (2009). Robust regression. In SYSTAT Software, Inc (Ed.), *SYSTAT 13 Statistics IV* (pp. IV-235–IV-269). Chicago: SYSTAT Software, Inc.
DeMaris, A. (2004). *Regression with social data: Modeling continuous and limited response variables*. Hoboken: Wiley.
Fu, V. K., Winship, C., & Mare, R. D. (2004). Sample selection bias models. In M. Hardy & A. Bryman (Eds.), *Handbook of data analysis* (pp. 409–430). London: Sage.
Geiser, C. (2013). *Data analysis with MPlus*. New York: The Guilford Press.
Hamilton, L. C. (2013). *Statistics with Stata: Version 12*. Boston: Brooks/Cole.
Hayes, A. F. (2018). *Introduction to mediation, moderation and conditional process analysis: A regression-based approach* (3rd ed.). New York: The Guilford Press.
Iacobucci, D. (2008). *Mediation analysis*. Los Angeles: Sage.
Jose, P. E. (2013). *Doing statistical mediation and moderation*. New York: The Guilford Press.
Linzer, D. A., & Lewis, J. B. (2011). poLCA: An R package for polytomous variable latent class analysis. *Journal of Statistical Software, 42*(10), 1–29.
Long, J. S., & Cheng, S. (2004). Regression models for categorical outcomes. In M. Hardy & A. Bryman (Eds.), *Handbook of data analysis* (pp. 259–284). London: Sage.
Miles, J., & Shevlin, M. (2001). *Applying regression & correlation: A guide for students and researchers*. London: Sage. ch. 6 and 7.
Norušis, M. J. (2012). *IBM SPSS statistics 19 advanced statistical procedures companion*. Upper Saddle River: Prentice Hall. ch. 10.

Panik, M. (2009). *Regression modeling: methods, theory, and computation with SAS*. Boca Raton: CRC Press.

Preacher, K. J., Wichman, A. L., MacCallum, R. C., & Briggs, N. E. (2008). *Latent growth curve modelling*. Los Angeles: Sage.

Stolzenberg, R. M. (2004). Multiple regression analysis. In M. Hardy & A. Bryman (Eds.), *Handbook of data analysis* (pp. 165–207). London: Sage.

Tabachnick, B. G., & Fidell, L. S. (2019). *Using multivariate statistics* (7th ed.). New York: Pearson Education. ch. 10.

Wedel, M., & DeSarbo, W. S. (2002). Mixture regression models. In J. A. Hagenaars & A. L. McCutcheon (Eds.), *Applied latent class analysis* (pp. 366–382). New York: Cambridge University Press.

Wilkinson, L., & Coward, M. (rev Ghosh, S. & Kulkarni, S. R.). (2009). Linear models I: Linear regression. In SYSTAT Software, Inc (Ed.), *SYSTAT 13 Statistics II* (pp. II-37–II-119). Chicago: SYSTAT Software, Inc.

Useful Additional Reading for Procedure 9.8

Aguinis, H. (2004). *Regression analysis for categorical moderators*. New York: The Guilford Press.

Bickel, R. (2007). *Multilevel analysis for applied research*. New York: The Guilford Press. ch. 11.

Darlington, R. B., & Hayes, A. F. (2017). *Regression analysis and linear models: Concepts, applications, and implementation*. New York: The Guilford Press. chs. 12–15, 18.

Field, A. (2018). *Discovering statistics using SPSS for Windows* (5th ed.). Los Angeles: Sage. ch. 11.

Jaccard, J., & Turrisi, R. (2003). *Interaction effects in multiple regression* (2nd ed.). Thousand Oaks: Sage.

Muthén, B. (2004). Latent variable analysis: Growth mixture modelling and related techniques for longitudinal data. In D. Kaplan (Ed.), *The Sage handbook of quantitative methodology for the social sciences* (pp. 345–368). Thousand Oaks: Sage.

References for Procedure 9.9

Cooksey, R. W., & McDonald, G. (2019). *Surviving and thriving in postgraduate research* (2nd ed.). Singapore: Springer Nature.

Garson, G. D. (1998). *Neural networks: An introductory guide for social scientists*. London: Sage.

Geiser, C. (2013). *Data analysis with MPlus*. New York: The Guilford Press.

Han, J., Kamber, M., & Pei, J. (2012). *Data mining: Concepts and techniques* (3rd ed.). Waltham: Morgan Kaufmann Publishers.

Lewins, A., & Silver, C. (2014). *Using software in qualitative research: A step-by-step guide* (2nd ed.). Los Angeles: Sage.

Leximancer Pty Ltd. (2011). *Leximancer manual Version 4*. Brisbane: Leximancer Pty Ltd. Downloaded from https://www.leximancer.com/site-media/lm/science/Leximancer_Manual_Version_4_0.pdf. Accessed 23 Sept 2019.

Madigan, J., & Vermunt, J. K. (2004). Latent class models. In D. Kaplan (Ed.), *The Sage handbook of quantitative methodology for the social sciences* (pp. 175–198). Thousand Oaks: Sage.

McCutcheon, A. L. (1987). *Latent class analysis*. Newbury Park: Sage.

North, M. (2012). *Data mining for the masses*. Athens: Global Text Project. Free eBook available from https://docs.rapidminer.com/downloads/DataMiningForTheMasses.pdf. Accessed 23 Sept 2019. [A companion website offering additional resources for this eBook is available at https://sites.google.com/site/dataminingforthemasses/]

O'Neil, C. & Schutt, R. (2014). *Doing data science: Straight talk from the frontline*. Sebastopol, CA: O'Reilly Media Inc.

Shafique, U., & Qaiser, H. (2014). A comparative study of data mining process models (KDD, CRISP-DM and SEMMA). *International Journal of Innovation and Scientific Research, 12*(1), 217–222.

Sharda, R., Delen, D., & Turban, E. (2018). *Business intelligence, analytics, data science: A managerial perspective* (4th ed.). Boston: Pearson Education. ch. 3, 4, 5 & 6.

Smith, A. E., & Humphreys, M. S. (2006). Evaluation of unsupervised semantic mapping of natural language with Leximancer concept mapping. *Behavior Research Methods, 38*(2), 262–279.

SPSS Inc. (2017). *IBM SPSS neural networks 25*. Chicago: SPSS Inc. ftp://public.dhe.ibm.com/software/analytics/spss/documentation/statistics/25.0/en/client/Manuals/IBM_SPSS_Neural_Network.pdf. Accessed 23 Sept 2019.

SPSS Inc. (2018). *IBM SPSS modeler text analytics 18.2 user's guide*. Chicago: SPSS Inc. ftp://public.dhe.ibm.com/software/analytics/spss/documentation/modeler/18.2/en/ModelerTextAnalytics.pdf. Accessed 23 Sept 2019.

Torgo, L. (2010). *Data mining with R: Learning with case studies*. London: Chapman & Hall/CRC.

Williams, G. (2011). *Data mining with Rattle and R: The art of excavating data for knowledge discovery*. New York: Springer Science.

Useful Additional Reading for Procedure 9.9

Abdi, H., Valentin, D., & Edelman, B. (1999). *Neural networks*. Thousand Oaks: Sage.

Hagenaars, J. A., & McCutcheon, A. L. (Eds.). (2002). *Applied latent class analysis*. New York: Cambridge University Press.

References for Procedure 9.10

Axelrod, R. (2007). Simulation in the social sciences. In *Handbook of research on nature inspired computing for economy and management* (pp. 90–100).

Bakshy, E., & Wilensky, U. (2007). *NetLogo Team Assembly model*. Evanston: Center for Connected Learning and Computer-Based Modeling, Northwestern University. http://ccl.northwestern.edu/netlogo/models/TeamAssembly. Accessed 24 Sept 2019.

Bellinger, G. (2013). *Fixes that fail archetype.*. https://insightmaker.com/insight/24847/Fixes-that-Fail-Archetype. Accessed 24 Sept 2019.

Boero, R., & Squazzoni, F. (2005). Does empirical embeddedness matter? Methodological issues on agent-based models for analytical social science. *Journal of Artificial Societies and Social Simulation, 8*(4), 1–31.

Carsey, T. M., & Harden, J. J. (2013). *Monte Carlo simulation and resampling methods for the social sciences*. Los Angeles: Sage.

Gilbert, N. (2008). *Agent-based models*. Thousand Oaks: Sage.

Gilbert, N., & Troitzsch, K. G. (2005). *Simulation for the social scientist* (2nd ed.). New York: McGraw-Hill International.

Guimera, R., Uzzi, B., Spiro, J., & Amaral, L. (2005). Team assembly mechanisms determine collaboration network structure and team performance. *Science, 308*(5722), 697–702.
IBM SPSS Inc. (2013). *Better decision making under uncertain conditions using Monte Carlo simulation*. New York: IBM Corporation.
Jackman, S. (2009). *Bayesian analysis for the social sciences*. Chichester: Wiley.
Karandikar, R. L., Krishnan, T., & Panchanana, M. R. L. N. (2009). *SYSTAT 13 Monte Carlo – 1*. Chicago: SYSTAT Software Inc.
Kim, D. H., & Anderson, V. (1998). *System archetype basics: From story to structure*. Waltham: Pegasus Communications Inc.
Milligan, G. W. (1981a). A Monte Carlo study of thirty internal criterion measures for cluster analysis. *Psychometrika, 46*(2), 187–199.
Milligan, G. W. (1981b). A review of Monte Carlo tests of cluster analysis. *Multivariate Behavioral Research, 16*(3), 379–407.
Mooney, C. Z. (1997). *Monte Carlo simulation*. Thousand Oaks: Sage.
Olaru, D., Purchase, S., & Denize, S. (2009). Using docking/replication to verify and validate computational models. In *Proceedings of the 18th World IMACS/MODSIM Congress* (pp. 4432–4438).
Pesarin, F. (2001). *Multivariate permutation tests: With applications in biostatistics* (Vol. 240). Chichester: Wiley.
Richmond, B. (2004). *An introduction to systems thinking: STELLA software*. Lebanon: isee Systems.
Roberts, C. P., & Casella, G. (2009). *Introducing Monte Carlo methods with R*. New York: Springer.
SYSTAT. (2009a). *SYSTAT 13 Monte Carlo – 1*. Chicago: SYSTAT Software Inc.
SYSTAT. (2009b). *SYSTAT 13 Exact Tests – 1*. Chicago: SYSTAT Software Inc.
Taber, C. S., & Timpone, R. J. (1996). *Computational modeling*. Thousand Oaks: Sage.
Wilensky, U. (1999). *NetLogo*. Evanston: Center for Connected Learning and Computer-Based Modeling, Northwestern University. http://ccl.northwestern.edu/netlogo/. Accessed 24 Sept 2019.

Useful Additional Reading for Procedure 9.10

Bonabeau, E. (2002). Agent-based modelling: Methods and techniques for simulating human systems. *Proceedings of the National Academy of Sciences*, 7280–7287.
Dunn, W. L., & Shultis, J. K. (2011). *Exploring Monte Carlo methods*. Amsterdam: Elsevier Science.
Fisher, D. M. (2011). *Modeling dynamic systems: Lessons for a first course* (3rd ed.). Lebanon: ISEE Systems.
Gilbert, N., & Terna, P. (2000). How to build and use agent-based models in social science. *Mind & Society, 1*, 57–72.
Kroese, D., Taimre, T., & Botev, Z. I. (2011). *Handbook of Monte Carlo methods*. Wiley.
Meadows, D. H. (2008). *Thinking in systems: A primer*. White River Junction: Chelsea Green Publishing.
Sterman, J. (1994). Learning in and about complex systems. *Systems Dynamics Review, 10*(2/3), 291–330.
Sterman, J. S. (2000). *Business dynamics: Systems thinking and modelling for a complex world*. New York: McGraw-Hill/Irwin.

Appendices

Appendix A: A Primer on Basic Mathematics, Statistical Notation & Simple Formulas for Measures

When you embark on a more in-depth study of statistical methods with a view to understanding how the various procedures discussed in this book are computed, you will have to come to grips with some basic mathematical and notational principles. The purpose of the next section is to introduce you to some of these principles with a view to de-mystifying them before you move into more intensive study.

Notation Used in Statistical Texts and Formulas

The following list provides guidance as to the types of symbols typically used in statistical texts to represent certain quantities and characteristics of data as well as representing certain mathematical operations (see also Howell 2011, Appendix B). Generally, statistics texts employ lower case letters of the Greek alphabet to represent specific population parameters or probabilities, upper case Greek letters to represent mathematical operations or formal theoretical distributions, and Arabic letters to represent sample statistics.

Variables

Symbol	Meaning
X	Symbol used to represent the possible values taken by a single variable or measurement. In bivariate or multivariate statistics, X is used to represent a variable which defines a 'predictor' or an 'independent' variable. Thus, in simple linear regression (see Procedure 6.3), X would be used to represent the predictor in the regression equation.
Y	Symbol used to represent the possible values taken by a single variable or measurement. In bivariate or multivariate statistics, Y is used to represent a variable which defines the 'criterion' or 'dependent' variable. Thus, in simple linear regression (see Procedure 6.3), Y would be used to represent the criterion in the regression equation.
X_i or Y_i	Variables with a subscript ($_i$) are often used so that the value of a specific variable for a specific participant (the ith participant) can be isolated. Typically, i can take on any value from 1 (participant 1) to n (participant n where n represents the number of participants in the sample being considered). Thus, we could refer to the score on the X variable obtained by participant number 10 as X_{10}.
X_{ij} or Y_{ij}	In more complex statistical techniques, a double subscript is used so that the participant number can be isolated (i) and the group of which the participant is a member (j) can also be denoted. Typically, i can take on any value from 1 to n_k where n_k represents the number of participants in the group k); j can take on any value from 1 to k where k represents the number of groups of participants in the sample. For example, the Procedure 7.6 (One-way ANOVA) could use this notation to depict the score obtained by participant 5 in group 3 as X_{53}.

Population Parameters

Symbol	Meaning
α	Greek lowercase Alpha. Symbol used to represent the probability of making a Type I error in statistical inference—that is, it represents the probability of the researcher rejecting the null hypothesis when it is true (and therefore should not be rejected). This is the so-called 'level of significance' value which is established by the researcher (see *Fundamental concept V*).
β	Greek lowercase Beta. Symbol used to represent the probability of making a Type II error in statistical inference—that is, it represents the probability of the researcher failing to reject the null hypothesis when it is false (and therefore should be rejected). This is related to the level of 'power' in a statistical test which is defined as $1 - \beta$ (see *Fundamental concept V*).
	Confusingly, β is also used to represent the parameter value for the population regression coefficient (see Procedures 6.3, 6.4 and 7.13). Even more confusingly, β is also used to represent the sample value for a standardised regression coefficient (the so-called 'beta weight'). It may be subscripted by an j to denote its being the beta weight for the jth predictor in a multiple regression equation (e.g. β_j; see Procedure 7.13).
μ	Greek lowercase Mu. Symbol for the parameter value of the population mean. This symbol is often subscripted by an X or a Y to indicate the variable for which it is the population parameter (e.g. μ_X or μ_Y)

(continued)

Symbol	Meaning
σ^2	Greek lowercase Sigma squared. Symbol for the parameter value of the population variance. This symbol is often subscripted by an X or a Y to indicate the variable for which it is the population parameter (e.g. σ^2_X or σ^2_Y).
σ	Greek lowercase Sigma. Symbol for the parameter value of the population standard deviation. This symbol is often subscripted by an X or a Y to indicate the variable for which it is the population parameter (e.g. σ_X or σ_Y).
ρ	Greek lowercase Rho. Symbol for the parameter value of the population correlation coefficient (see Procedure 6.1). This parameter may be subscripted (e.g. ρ_{ij}) when used in the context of multiple regression to denote the parameter value of the population correlation coefficient between variable i and variable j.
χ^2	Greek lowercase Chi squared. Symbol used to represent a statistic from the chi-squared distribution. This symbol may refer to a computed value for the statistic (in contingency table analysis—see Procedure 7.1) or to a theoretical value obtained from a statistical table of the χ^2 distribution.

Sample Statistics

Symbol	Meaning
$\overline{X}$ or $\overline{Y}$	Symbol for the statistic value of the sample mean for variable X or variable Y.
s^2	Symbol for the statistic value of the sample variance. This statistic may be subscripted by an X or a Y to indicate the variable for which it is the sample statistic (e.g. s^2_X or s^2_Y).
s	Symbol for the statistic value of the sample standard deviation. This statistic may be subscripted by an X or a Y to indicate the variable for which it is the sample statistic (e.g. s_X or s_Y).
r	Symbol for the statistic value of the sample correlation coefficient. This statistic may be subscripted (e.g. r_{ij}) when used in the context of multiple regression to denote the value of the sample correlation coefficient between variable i and variable j.
b	Symbol for the statistic value for the sample raw or unstandardised regression coefficient. It may be subscripted by a j to denote its being the sample unstandardised regression coefficient for the jth predictor in a multiple regression equation (e.g. b_j).
z	Symbol used to represent a standardised score for a specific data value in a sample (see Procedure 5.7). This symbol may also be used to refer to a theoretical value obtained from a statistical calculation from the standard normal distribution.
t	Symbol used to represent a statistic from the t-distribution. This symbol may refer to a computed value for the statistic (e.g. computed for the related or independent groups t-test—see Procedures 7.2 or 7.4) or to a theoretical value obtained from a statistical calculation from the t-distribution. Every t-statistic has an associated 'degrees of freedom' which is a direct function of the sizes of the groups being compared.
F	Symbol used to represent a statistic from the F-distribution. This symbol may refer to a computed value for the statistic (e.g. computed for ANOVA, ANCOVA, or multiple regression—see Procedures 7.6, 7.10, 7.10, 7.13, or 7.15) or to a theoretical value obtained from a statistical calculation from the F-distribution. Every F-statistic has two associated 'degrees of freedom': one of which is a direct function of the sizes of the groups being compared (called the 'denominator' degrees of freedom) and the other of which is a direct function of the number of groups being compared or number of predictors being tested (called the 'numerator' degrees of freedom).

(continued)

Symbol	Meaning
η^2	Greek lowercase Eta squared. Here we encounter an exception to the 'Greek-alphabet-for-population-parameters' rule. Eta-squared represents the proportion of variance in a dependent variable that can be explained by an independent variable which defines different groups of participants (as in ANOVA and ANCOVA). Eta-squared is a descriptive sample-based measure of effect size (see Procedure 7.7).
ω^2	Greek lowercase Omega squared. Omega-squared represents the proportion of variance in a dependent variable that can be explained by an independent variable which defines different groups of participants (as in ANOVA and ANCOVA). Omega-squared estimates the population proportion of variance explained, another measure of effect size.

Mathematical

Symbol	Meaning
$+, -, \times, \div$ (or /)	Symbols for the mathematical operations of addition, subtraction, multiplication, and division.
a/b	A fraction read as "a divided by b"; a is called the 'numerator' and b is called the 'denominator' of the fraction.
$a \approx b$ or $a \cong b$	Read as "a is approximately equal to b".
$a < b$	Read as "a is less than b". Mnemonic: the symbol points at the smaller number.
$a > b$	Read as "a is greater than b".
$a \leq b$	Read as "a is less than or equal to b".
$a \geq b$	Read as "a is greater than or equal to b".
$a < b < c$	Read as "a is less than b is less than c". That is, b lies between the values of a and c.
$a \pm b$	Read as "a plus or minus b".
$\lvert a \rvert$	Read as "the absolute value of a" which means that the sign attached to a is ignored. Thus, the absolute value of -3 is 3 and the absolute value of 17 is 17.
$\frac{1}{a}$	Read as "the reciprocal of a" which the result of dividing 1 by a.
a^n	Read as "a raised to the power of n" which indicates that a is multiplied by itself n times. For example, a^3 is equal to $a \times a \times a$.
$\sqrt{a} = a^{1/2}$	Read as "the square root of a" which indicates a number which when multiplied by itself gives a as the result. For example, the square root of 9 is 3 since $3 \times 3 = 9$.
Σ	This is called the 'summation sign' (Greek uppercase Sigma) which is read as "find the sum of". This is a shorthand representation for addition of the values of a variable in statistical formulas. For example, ΣX means add up or find the sum of all the values for the X variable in the sample. Frequently, subscripts will be used in conjunction with the Σ sign so that the precise values over which the addition is supposed to occur is clear.
	Consider the formula $\sum_{i=1}^{n} X_i$. Translated into ordinary English, this says "find the sum of all of the values for the X variable, starting with the value for participant 1 and ending with participant n" which basically says we should add up all the values for X in the sample.

Appendices

Some Basic Mathematical Principles

In this section, we review some basic mathematical principles which are useful in the understanding of computational formulas for procedures (Howell 2011, Appendices A and C and Levine and Stephan 2009, Appendices A and B, also review some important principles). Knowing these principles will help you decode what is intended by a specific statistical formula. The most important thing to learn is the order in which mathematical operations are expected to occur.

1. Any calculation which is surrounded by brackets is always computed before any calculation outside the brackets. Different types of brackets may be used, such as ( ), [ ], or { }, but all of them work in pairs and will surround a calculation which is to be done before working outside of them. For example, in the calculation $(3 + 5) - (4-1)$, we first find $(3 + 5) = 8$ and $(4 - 1) = 3$, then we complete the calculation by subtracting 3 from 8 to get a final answer of 5.
2. Anything inside brackets is treated as a single symbol or result. For example, in the formula $(\Sigma X)/n$, we must find the sum of all the values of the X variable first, and then divide that sum by the value of n. If we have $n = 3$ observations of X which have the values of 5, 7, and 9, then $\Sigma X = 5 + 7 + 9$ or 21 and $21/3 = 7$. [Note: We have just computed the average or the mean of the three observations for X.]
3. If computations are represented in a fraction, then compute the top part of the fraction and the bottom part of the fraction separately (as if each part were contained in brackets) and then divide the top result by the bottom result. For example, in the formula $X^2/(n - 1)$, we must first find the value for X^2 and the value for $n - 1$, then we complete the division using the two results. If $X = 5$ and $n = 6$, our answer should be found as $X^2 = 25$, $n - 1 = 5$, and $25/5 = 5$.
4. If two symbols or results are written immediately next to each other, they are to be multiplied. For example, $X_i Y_i$ means "take the ith value of the X variable and multiply it by the ith value of the Y variable". If $X_i = 17$ and $Y_i = 3$, then $X_i Y_i = 17 \times 3$ or 51.
5. Except where determined by brackets, the order of performing mathematical operations is:

   - do any squaring (or raising to any other power);
   - do any multiplication or division;
   - do any operations denoted with the Σ sign; and finally
   - do any addition or subtraction

To illustrate many of the above principles, consider the following formula, which may seem very complex, but we can work through it logically by breaking it into its component parts.

$$\sqrt{\dfrac{\sum_{i=1}^{n} X_i^2 - \dfrac{\left(\sum_{i=1}^{n} X_i\right)^2}{n}}{n-1}}$$

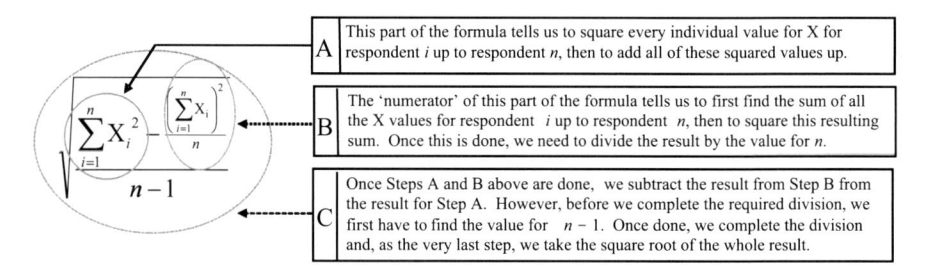

A — This part of the formula tells us to square every individual value for X for respondent i up to respondent n, then to add all of these squared values up.

B — The 'numerator' of this part of the formula tells us to first find the sum of all the X values for respondent i up to respondent n, then to square this resulting sum. Once this is done, we need to divide the result by the value for n.

C — Once Steps A and B above are done, we subtract the result from Step B from the result for Step A. However, before we complete the required division, we first have to find the value for $n - 1$. Once done, we complete the division and, as the very last step, we take the square root of the whole result.

Note that this is the formula for computing a standard deviation where n is the number of participants in the sample. The formula under the square root sign is the variance.

Simple Formulas for Computing Certain Quantities by Hand

In this section, we show some simple but essential formulas to calculate certain statistical quantities by hand, using specific numbers produced by SPSS, for example. These formulas may prove useful in circumstances where your statistical package doesn't produce the quantities you need for interpretation. For example, SPSS does not report the η^2 associated with an independent groups or related groups t-test (recall Procedure 7.2) nor will SPSS or any other major package produce Model 2 F-tests for a hierarchical regression analysis (recall Procedure 7.13 and *Fundamental concept IV*). Some programs will not report the interquartile range (or semi-interquartile range) measure of variability associated with the median statistic (recall Procedure 5.5). If you are doing a mediated regression analysis (recall Procedure 9.8), then you may need to compute Sobel's z-test by hand (using information you can get SPSS to provide) to assess the significance of the indirect mediation effect.

- **Interquartile Range** (IQR; also called the 'Hinge-spread' or 'H-spread', recall Procedure 5.6 where stem-and-leaf plots and boxplots were discussed). To find the IQR, all scores in a distribution need to be ranked in order from lowest to highest. Then identify that score that falls at the 25th percentile point (called the first quartile score Q_1) and the score that falls at the 75th percentile point (called the third quartile score Q_3) [note that the median is the score falling at the 50th percentile or second quartile and is sometimes labelled as Q_2]. The IQR is then found as:

$$IQR = Q_3 - Q_1$$

Semi-Interquartile Range (S-IQR; also called the 'midhinge', see Glass and Hopkins 1996, ch. 5):

$$S - IQR = \frac{IQR}{2}$$

- **Measure of effect size *(eta-squared)* for an independent or related-groups *t*-test** (Rosenthal and Rosnow 1991, p. 292):

$$\eta^2 = \frac{t^2}{(t^2 + df_{error})}$$

- **Measures of effect size for F-tests** (see Tabachnick and Fidell 2019, section 3.4; where SS means 'Sum of Squares' and MS means 'Mean Square', from ANOVA or MANOVA computer output produced by SPSS, for example):
Omega-squared:

$$\omega^2_{effect} = \frac{(SS_{effect} - (df_{effect} \times MS_{error}))}{(SS_{total} + MS_{error})}$$

Eta-squared:

$$\eta^2_{effect} = \frac{SS_{effect}}{SS_{total}}$$

Partial eta-squared:

$$\text{Partial } \eta^2_{effect} = \frac{SS_{effect}}{(SS_{effect} + SS_{error})}$$

Rosenthal and Rosnow (1991, p. 323) also showed a way of computing η^2 from an *F*-value and its associated degrees of freedom:

$$\eta^2_{effect} = \frac{(F \times df_{effect})}{((F \times df_{effect}) + df_{error})}$$

- **Partial F-test using Model 2 error** in hierarchical regression for a predictor set:

$$\text{Model 2 } F - \text{test} = \frac{R^2_{Change}/df_{change}}{((1 - R^2_{Full\ Model})/df_{Residual\ Full\ Model})}$$

where R^2_{change} and df_{change} are the values from the step at which the predictor set entered the regression model and $R^2_{Full\ Model}$ and $df_{Residual\ Full\ Model}$ are from the final step in the analysis, when all predictor sets have been entered.

For assessing an individual predictor, the formula is:

$$\text{Model 2 } F-\text{test} = \frac{sr^2_{Change}}{\left((1-R^2_{Full\ Model})/df_{Residual\ Full\ Model}\right)}$$

where sr^2 is the squared semi-partial correlation for the predictor being assessed (what SPSS calls the 'Part Correlation').

- **Recomputing the p-value for a Model 2 F-test** using the Microsoft Excel FDIST function (accessed via *Formulas* → *More Functions* → *Statistical* → *FDIST*). The formula is:

$$FDIST\left(F-value, df_{effect}, df_{error}\right)$$

You simply select an empty cell in the Excel spreadsheet, then select the function. A dialog box will open where you can enter the three required quantities. Hit *OK* and the *p*-value will be shown in the cell you selected. Alternatively, you can cut and paste a table of 'Model Statistics' from an SPSS hierarchical regression analysis into an Excel spreadsheet, then use cell references to complete the *FDIST* function specification in the cell editing line (e.g. "= FDIST(A1, B1, C1)" where the *F*-value, df_{effect} and df_{error} are in the three cells A1, B1 and C1, respectively). There are other functions in Excel that work similarly to compute the *p*-value for a *t*-test (*TDIST*) or a chi-square test (*CHIDIST*).

- **Changing a z-score to another unit of measurement** (defined by a specific *mean* and standard deviation (*sd*), see Howell 2013, ch. 3):

$$(\text{desired } sd \times z) + \text{desired } mean$$

For example, to change a *z*-score of +1.25 to a T-score (which has a mean of 50 and a standard deviation of 10 (recall *Fundamental concept II*), the T-score would be computed as (10 × +1.25) + 50 = 62.5.

- **Intraclass correlation in a multilevel model** (which is a squared correlation, see Tabachnick and Fidell 2019, section 15.6.1). This statistic is computed using specific variances from an '*intercept-only*' multilevel model fitted using SPSS (recall Procedure 9.5; the formula below applies only to a 2-level model):

$$r^2_{intraclass} = \frac{s^2_{Intercept\ [Subject=???]}}{\left(s^2_{Intercept\ [Subject=???]} + s^2_{Residual}\right)}$$

where ??? refers to the name of the variable defining the groups at Level 2 of the model.

- **Sobel's z-test for an indirect mediating effect in a mediated regression analysis** (see Jose 2013, pp. 51–55): This statistic is computed using specific unstandardised regression coefficients and their associated standard errors from specific steps in the mediated regression analysis conducted using SPSS, for example. To run this test, you need to run the regressions set out in steps 1, 2 and 3 outlined in Procedure 9.8. The diagram below shows the information required for the test.

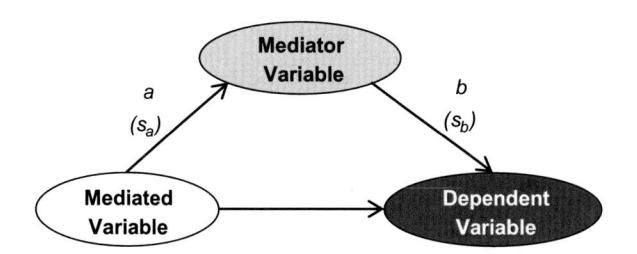

The numbers for a and s_a are the unstandardised regression coefficient and its associated standard error from the regression predicting the mediator variable using the mediated variable. The numbers for b and s_b are the unstandardised regression coefficient and its associated standard error *for the mediator variable* from the simultaneous regression predicting the dependent variable using both the mediator variable and the mediated variable. The formula for the test then combines this information in the following way:

$$\text{Sobel's z-test} = \frac{a \times b}{\sqrt{(b^2 \times s_a^2 + a^2 \times s_b^2)}}$$

As this is effectively a z-score, it can be evaluated for significance by reference to a standard normal distribution (recall *Fundamental concept II*; for an alpha level of .05, the test value should exceed 2.0). Note that there is there is also a website available where this calculation can be carried out by plugging in the relevant quantities from SPSS output from the two regression analyses described above: http://quantpsy.org/sobel/sobel.htm.

References

Glass, G. V., & Hopkins, K. D. (1996). *Statistical methods in education and psychology* (3rd ed.). Upper Saddle River: Pearson.
Howell, D. C. (2011). *Fundamental statistics for the behavioral sciences* (7th ed.). Belmont: Cengage Wadsworth, Appendices A, B and C.
Howell, D. C. (2013). *Statistical methods for psychology* (8th ed.). Belmont: Cengage Wadsworth, ch. 3.

Jose, P. E. (2013). *Doing statistical mediation and moderation.* New York: The Guilford Press.

Levine, D. M., & Stephan, D. F. (2009). *Even you can learn statistics: A guide for everyone who has been afraid of statistics* (2nd ed.). Upper Saddle River: Pearson Prentice Hall, Appendices A and B.

Rosenthal, R., & Rosnow, R. L. (1991). *Essentials of behavioral research: Methods and data analysis* (2nd ed.). New York: McGraw-Hill Inc, ch. 15.

Tabachnick, B. G., & Fidell, L. S. (2019). *Using multivariate statistic* (7th ed.). New York: Pearson Education, ch. 3.

Appendix B: General Analytical Sequences for Multivariate Investigations

There are several analytical sequences that can prove useful to behavioural and social researchers, depending upon the research goals being pursued. These analytical pathways rely on different blends of a number of procedures described in this book. Four specific pathways are outlined here. However, these four should not be taken as an exhaustive range of possible pathways. You will notice that some of the very early steps are common to all the pathways. You should also notice that the hurdle logic for significance testing (recall *Fundamental concept V*) is reflected in these analytical pathways. I developed these pathways building upon discussions in Hair et al. (2010, ch. 1), Harlow (2005, ch. 1 and 2) and Tabachnick and Fidell (2019, ch. 2). Hardy and Bryman (2004) reinforced the common threads amongst different techniques of data analysis (and they included qualitative data analysis in their thinking) in terms of the goals they are intended to achieve. These goals are reflected in the various pathways.

An Exploratory Pathway

A fair proportion of multivariate research occurs without a strong *a priori* idea of the internal structure of a questionnaire or other instruments. Such research can be deemed exploratory because at least part of the analytical process is devoted to finding structure within the multivariate data set in order to reduce the dimensionality of the data (i.e. variable *condensation*), simplify the conduct of statistical tests and reduce the chances of making false claims of significance (i.e. reduce the chances of making a Type I error). This is often the pathway that researchers must employ if they design their questionnaire from scratch. For steps 6, 7 and 8, the researcher may need to consider nonparametric alternatives (shown in [ ]) if required assumptions are not satisfied. The general sequence of steps for an exploratory multivariate study are:

Step	Suggested relevant procedures & fundamental concepts
1. Undertake data preparation: to minimise errors in transferring the raw data to the spreadsheet database required for analysis using a specific statistical package.	Procedure 5.1
2. Conduct data exploration and screening, assumption checking & missing value analysis: use descriptive statistics, graphs and other appropriate procedures to examine the nature of variables and build up a story about participants/cases in the sample, check for satisfaction of necessary assumptions for later hypothesis tests, check for any patterns attributable to missing data and make any necessary data transformations and missing value adjustments.	Procedures 5.1, 5.2, 5.3, 5.4, 5.5, 5.6, and 8.2; possibly Procedure 5.7 or 8.3 *Fundamental concept II*
3. Conduct exploratory factor analysis: to reduce the dimensionality of key sets of variables (components) or discover the latent structure of key sets of variables (factors). Note that this step helps to establish construct validity and may be iterative in that structures may be further refined depending upon the outcomes of initial analyses.	Procedure 6.5; possibly Procedures 6.6 or 6.7 *Fundamental concept III*
4. Establish reliability of scales: conduct relevant item analyses and compute reliability coefficients to demonstrate the overall consistency of the components or factors (scales) identified in step 3.	Procedure 8.1; possibly Procedure 9.1
5. Create scale scores: compute component or factor scores for each participant in the sample, preferably by summing (only if there are no missing data) or averaging the non-missing defining items for each component or factor (generically called 'scales').	Procedure 6.5; possibly Procedure 5.7 or 9.1; use specific transformation functions in your statistical package
6. (a) Conduct multivariate tests of hypotheses: using the new component or factor (scale) scores, other relevant dependent variables and relevant independent variables, conduct relevant multivariate group comparison tests and compute effect sizes; and/or	Procedures 7.16, 7.17, 7.7; possibly Procedure [8.4] *Fundamental concepts V, VII, IX*
(b) Build & test correlation/regression/general linear model: using the new component or factor (scale) scores, other relevant dependent variables and relevant predictors (with suitably coded categorical predictors, where appropriate) to build predictive or explanatory models using simultaneous or hierarchical entry and compute effect sizes.	Procedures, 6.3, 6.4, 6.8, 7.13, 7.14, 7.17, 7.18, 8.5 or 9.5; possibly Procedure [8.4] or 9.8 *Fundamental concept IV, V, VI*

(continued)

Step	Suggested relevant procedures & fundamental concepts
7. (a) Conduct univariate tests: where appropriate, follow up significant multivariate group comparison tests with univariate group comparison tests on the component or factor (scale) scores or other relevant dependent variables and compute effect sizes; and/or	Procedures 7.2, 7.4, 7.6, 7.7, 7.10, 7.11, 7.15; possibly Procedure 9.4 or [8.4, 7.3, 7.5, 7.9, or 7.12] *Fundamental concepts V, IX*
(b) Evaluate individual predictors or bivariate relationships: test for the contributions of individual predictors in a model and compute effect sizes or conduct correlation tests of hypotheses.	Procedures 6.1, 6.2, 7.1; 7.13, 7.15, 7.18; possibly Procedure [8.4] or 9.8 *Fundamental concepts III, V, VI*
8. Conduct posthoc tests to isolate group differences: where appropriate, follow up significant univariate group comparison tests (involving three or more groups) with posthoc multiple comparison tests (this step only required if MANOVA/ANOVA analyses are conducted at steps 6 and/or 7).	Procedure 7.8

A Confirmatory Model Testing Pathway

Some multivariate research is undertaken where the researcher does have a strong *a priori* idea of the internal structure of a questionnaire or other instruments, based on theoretical grounds and/or on previous research. Such research can be deemed confirmatory because the main goal is to verify the theorised or intended structure (measurements → constructs → linkages) within a specific research context. Sometimes, confirmatory research may be undertaken to test an instrument's theoretical structure in different sampling contexts (e.g. cross-cultural settings) to test the generality of the constructs embedded in the instrument. Any research that proposes a causal model to be tested, using a structural equation modelling approach, is essentially a confirmatory study. The general sequence of steps for a confirmatory model testing multivariate study are:

Step	Suggested relevant procedures & fundamental concepts
1. Build the conceptual/theoretical model: set out and justify all hypothesised structural linkages prior to data collection; decide upon the specific approach to structural equation modelling to be adopted: covariance-based SEM or variance-based partial least-squares SEM.	Procedure 8.7

(continued)

Appendices

Step	Suggested relevant procedures & fundamental concepts
2. Undertake data preparation: to minimise errors in transferring the raw data to the spreadsheet database required for analysis using a specific statistical package.	Procedure 5.1
3. Conduct data exploration and screening, assumption checking & missing value analysis: use descriptive statistics, graphs and other appropriate procedures to examine the nature of variables and build up a story about participants/cases in the sample, check for satisfaction of necessary assumptions for later hypothesis tests, check for any patterns attributable to missing data and make any necessary data transformations and decide on how best to deal with/impute missing values.	Procedures 5.1, 5.2, 5.3, 5.4, 5.5, 5.6, 8.2; possibly Procedure 5.7 or 8.3 *Fundamental concept II*
4. Set up the conceptual/theoretical model: use the appropriate software interface to build the specifications for the model to be tested/confirmed, making sure all paths and error terms are correctly represented, where appropriate.	Procedures 8.6 and/or 8.7
5. Test the conceptual/theoretical model: use the appropriate software interface and analysis options to execute the required model fitting computations (if bootstrap errors for all parameter estimates are possible, they should also be computed).	Procedures 8.6 and/or 8.7; [8.4]
6. Make necessary model modifications: using the test results from step 5, revise the model specifications as appropriate.	Procedures 8.6 and/or 8.7
7. Obtain a new sample: in order to test the refined model, a new sample should be drawn, and a new set of data collected (this cross-validation strategy avoids over-optimising results through using a single sample to both fit and refine a model).	Procedures 8.6 and/or 8.7
8. Confirm final model refinements on new sample: test the refined model using the data from the new sample.	Procedures 8.6 and/or 8.7; [8.4]

A Hybrid Cross-Validation Pathway

Some multivariate research is undertaken with a strong applied intent. In this type of research, the goal is to construct a model or a mathematical decision rule that can be used to make predictions or decisions about new cases not used in the creation of the model. Since virtually all model building statistical methods are optimised for the sample on which they are constructed, this means that the models will inevitably lose some predictive power when applied in new samples. The hybrid cross-validation pathway permits the explicit testing of this predictive loss as a way of assessing the final utility of the model. The pathway is called 'hybrid' because it blends the exploratory pathway with a cross-validation strategy. The general sequence of steps for a hybrid cross-validation multivariate study are:

Step	Suggested relevant procedures & fundamental concepts
1. Undertake data preparation: to minimise errors in transferring the raw data to the spreadsheet database required for analysis via a specific statistical package.	Procedure 5.1
2. Create development and holdout groups: randomly split sample into *development* and *holdout* groups (needs double the usual sample size).	Use specific functions in your statistical package
3. Conduct data exploration and screening, assumption checking & missing value analysis: for the *entire sample* use descriptive statistics, graphs and other appropriate procedures to examine the nature of variables and build up a story about participants/cases in the sample, check for satisfaction of necessary assumptions for later hypothesis tests, check for any patterns attributable to missing data and make any necessary data transformations and missing value adjustments.	Procedures 5.1, 5.2, 5.3, 5.4, 5.5, 5.6, 8.2; possibly Procedure 5.7 or 8.3 *Fundamental concept II*
4. Conduct exploratory factor analysis: to reduce the dimensionality of key sets of variables (components) or discover the latent structure of key sets of variables (factors) in the *development* group. Note that this step helps to establish construct validity and may be iterative in that structures may be further refined depending upon the outcomes of initial analyses.	Procedure 6.5; possibly Procedures 6.6 or 6.7 *Fundamental concept III*
5. Establish reliability of scales: conduct relevant item analyses and compute reliability coefficients in the *development* group to demonstrate the overall consistency of the components or factors (scales) identified in step 3.	Procedure 8.1; possibly Procedure 9.1
6. Create scale scores: compute component or factor scores for each participant in the *development* group, preferably by summing (only if there are no missing data) or averaging the non-missing defining items for each component or factor (generically called 'scales').	Procedure 6.5; possibly Procedure 5.7 or 9.1; use specific transformation functions in your statistical package

(continued)

Step	Suggested relevant procedures & fundamental concepts
7. Build the predictive model: using the new component or factor (scale) scores and other relevant variables from the *development* group, estimate the desired predictive model using multiple regression; discriminant analysis; logistic regression, multilevel or other type of regression model, time series analysis or classification/regression tree, as appropriate.	Procedures 7.13, 7.17, 7.14, 8.5, 9.5, 9.6 and/or 9.8; possibly Procedure [8.4]
8. Use the predictive model to predict/classify the cases in the *holdout* sample: compute component or factor (scale) scores for each participant in the *holdout* group and use the model resulting from step 7 to predict or classify cases in the *holdout* group (thus, the *holdout* group provides the 'new' cases for testing the predictive model).	Procedures 7.13, 7.17, 7.14, 8.5, 9.5, 9.6 and/or 9.8
9. Check the accuracy of predictions/classifications: compare the predictions/classifications resulting from step 8 to the actual scores or group memberships for participants in the *holdout* group.	Procedures 7.13, 7.17, 7.14, 8.5, 9.5, 9.6 and/or 9.8; possibly, Procedures 7.1, 7.18

Note that these steps describe what is called a **single cross-validation** study. However, the process can actually be run in both 'directions' using each sample as both a *development* and a *holdout* group (this will give two predictive models) and averaging the results. This would yield what is called a **double cross-validation study**.

Also note that having a very large sample can allow the researcher to use this hybrid cross-validation strategy as part of a confirmatory model testing approach. This would avoid the necessity for collecting a new sample to test model refinements.

A Typology Discovery and Testing Pathway

Some multivariate research is undertaken with the goal of discovering a previously unknown typology of cases or participants and testing for key differences amongst different classification groups within the typology. Typologies are classifications of cases based on patterns of responses to a range of variables. A typology can then be 'validated' by comparing the resulting different classification groups with respect to variables not used to construct the typology. The sequence of steps for a typology construction and testing multivariate study are:

Step	Suggested relevant procedures & fundamental concepts
1. Undertake data preparation: to minimise errors in transferring the raw data to the spreadsheet database required for analysis using a specific statistical package.	Procedure 5.1
2. Conduct data exploration and screening, assumption checking & missing value analysis: use descriptive statistics, graphs and other appropriate procedures to examine the nature of variables and build up a story about participants/cases in the sample, check for satisfaction of necessary assumptions for later hypothesis tests, check for any patterns attributable to missing data and make any necessary data transformations and missing value adjustments.	Procedures 5.1, 5.2, 5.3, 5.4, 5.5, 5.6, 8.2; possibly Procedure 5.7 or 8.3 *Fundamental concept II*
3. Conduct exploratory factor analysis: to reduce the dimensionality of key sets of variables (components) or discover the latent structure of key sets of variables (factors). Note that this step helps to establish construct validity and may be iterative in that structures may be further refined depending upon the outcomes of initial analyses.	Procedure 6.5; possibly Procedures 6.6 or 6.7 *Fundamental concept III*
4. Establish reliability of scales: conduct relevant item analyses and compute reliability coefficients to demonstrate the overall consistency of the components or factors (scales) identified in step 3.	Procedure 8.1; possibly Procedure 9.1
5. Create scale scores: compute component or factor (scale) scores for each participant in the sample, preferably by summing (only if there are no missing data) or averaging the non-missing defining items for each component or factor (generically called 'scales').	Procedure 6.5; possibly Procedure 5.7 or 9.1; use specific transformation functions in your statistical package
6. Generate and refine the typology: use cluster analysis methods to generate and refine the typology of participants, based on the new component or factor (scale) scores and/or other relevant variables (this step may require the new component or factor (scale) scores to be standardised prior to the analysis).	Procedures 6.6, 9.6 or 9.9; possibly Procedure 6.7 for certain applications
7. Assign cases to clusters: create a new variable that codes the cluster or group that each case/participant has been classified into.	Procedures 6.6, 9.6 or 9.9

(continued)

Appendices

Step	Suggested relevant procedures & fundamental concepts
8. Conduct multivariate tests of hypotheses: using dependent variables not used to construct the typology (external validation of clusters), conduct relevant multivariate cluster comparison tests (using the new variable from step 7 to define the groups) for differences between groups/clusters in the typology. Multivariate procedures and/or graphs can also be used to provide a descriptive picture of the clusters emerging from step 6 (internal validation of clusters).	Procedures 5.1, 5.2, 5.3, 6.6, 7.16, 7.17, 7.7; possibly Procedure 8.4 *Fundamental concepts V, VII, IX*
9. Conduct univariate tests: where appropriate, follow-up significant multivariate cluster comparison tests with univariate cluster comparison tests on the component or factor (scale) scores or other dependent variables and/or univariate tests comparing clusters with other categorical demographic variables (follow-up step for external validation of clusters).	Procedures 6.2, 6.6, 7.1, 7.6, 7.7, 7.10, 7.15, 7.17, 7.18; possibly Procedure 9.4 or [8.4, 7.3 or 7.9] *Fundamental concepts V, IX*
10. Conduct posthoc tests to isolate cluster differences: where appropriate, follow-up significant univariate cluster comparison tests (involving three or more clusters) with posthoc multiple comparison tests or cell-by-cell statistics for contingency table analyses (final follow-up step for external validation of clusters).	Procedure 7.1 and/or 7.8

Mindmap of SPSS Capabilities to Support Analytical Pathways

Figure B.1 shows a detailed mindmap of the capabilities within SPSS to support different activities in the analytical pathways. Different SPSS functions and procedures can be utilised to facilitate data preparation, data screening, getting to descriptive analytical stories, getting to relational analytical stories and getting to hypothesis testing inferential stories. There are some important cross-linkages between SPSS procedures used to help build up descriptive, relational and hypothesis testing stories. Often, for example, relational and hypothesis testing stories emerge concurrently, such that a relationship (e.g. a correlation or a predictive model) is not only described by some statistic (e.g. r, b weight) but also evaluated for significance in the population. When groups are compared (hypothesis testing), a descriptive story about the groups themselves is also assembled.

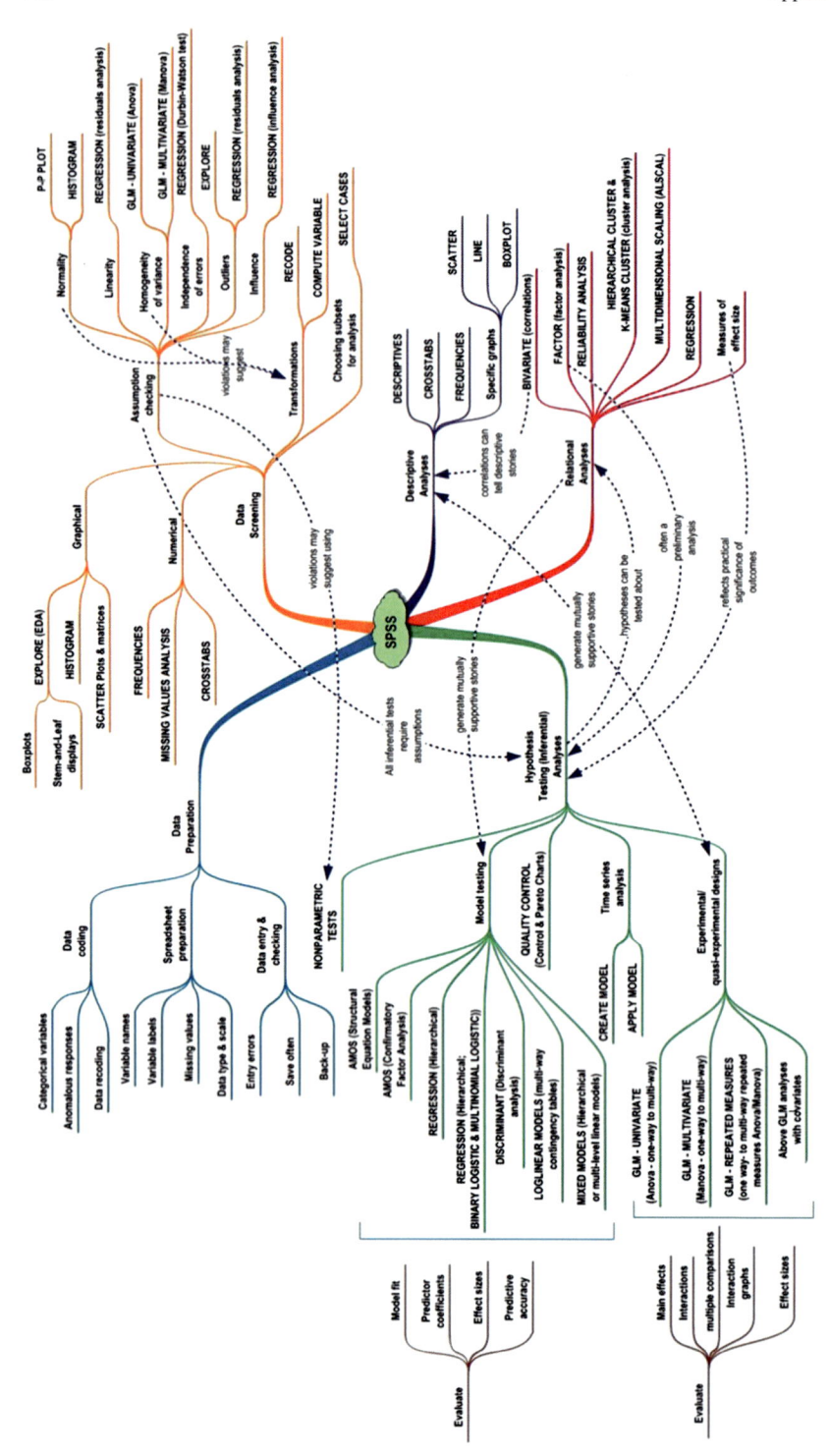

Fig. B.1 Mindmap of SPSS capabilities to support different activities in the various analytical pathways

References

Hair, J. F., Black, B., Babin, B., & Anderson, R. E. (2010). *Multivariate data analysis: A global perspective* (7th ed.). Upper Saddle River: Prentice Hall, ch. 1.

Hardy, M., & Bryman, A. (2004). Introduction: Common threads among techniques of data analysis. In M. Hardy & A. Bryman (Eds.), *Handbook of data analysis* (pp. 1–13), London: Sage Publications.

Harlow, L. L. (2005). *The essence of multivariate thinking: Basic themes and methods*. Mahwah: Lawrence Erlbaum Associates, ch. 1 and 2.

Tabachnick, B. G., & Fidell, L. S. (2019). *Using multivariate statistics* (7th ed.). New York: Pearson Education, ch. 2.

Appendix C: Writing Up Statistical Analyses

This appendix provides some guidance for the writing up of specific types of statistical analyses, in terms of the language used, the level of detail required and a desirable balance between the use of text, numbers, tables and figures/graphs. The bigger pictures associated with carrying out and reporting on quantitative research in general and with writing up statistical stories are also discussed and displayed using mindmaps.

The Finer Points of Writing Up

What are illustrated here are some schemata (i.e. skeletal sketches) for reporting statistical outcomes. You would still need to make substantive interpretations of what the results meant with respect to your research questions/hypotheses, theory and so on. The schemata assume the analyses were conducted using SPSS. Cooksey and McDonald (2019, pp. 946–959) review some additional considerations associated with the writing up and presentation of statistical results. You will also find that Miller (2005) provides extensive advice on how to write about multivariate analyses, including how to construct effective tables and figures. One important rule to remember is to studiously avoid simply cutting and pasting whole analytical outputs, tables and/or figures from SPSS (or any other computer package) into your document draft. Instead be judicious and selective (but not biased) in reporting only the evidence needed to convey your story and interpretations clearly and concisely.

Statistical procedure	Schemata [comments on what is being said in a sentence]
Contingency table/ crosstabulation analysis	A contingency table analysis relating X and Y [two categorical variables] showed a significant association (chi-squared [χ^2](df) = ???, p - .???, Cramer's V = .???). Inspection of adjusted standardised residuals revealed that, for category x1 of variable X, significantly more than expected observations fell into category y1 of variable Y whereas significantly fewer than expected observations fell into category y2 ... [adjust interpretation as needed given the crosstabulation table patterns; you may wish to comment on row % and column % patterns as well].
Pearson correlation	Variables X and Y were found to be significantly positively correlated (r = +.???, p = .???) showing that as scores in X increased, scores in Y also tended to be higher [adjust interpretation as needed given the Pearson correlation pattern, in terms of sign and magnitude; the negative correlation pattern is: as scores in X increased, scores in Y tended to be lower].
Principal components analysis (with reliability analysis and component scoring process)	A principal components analysis was conducted on the set of attitudinal measures [specify as appropriate for the pool of measures being analysed]. Promax rotation was employed to allow for correlated components and the eigenvalue greater than one rule, coupled with inspection of the scree plot, was used to determine the number of components to interpret. Inspection of Bartlett's test of sphericity and the Kaiser-Meyer-Olkin Measure of Sampling Adequacy showed that the correlation matrix was factorable [check SPSS output for acceptable values for each index]. ?? interpretable components were identified. Table ?? [build a table that displays the pattern coefficients and component correlations – see example shown left which shows a 2-component solution] shows the pattern coefficients for the items on each component (only coefficients greater than an absolute value of .40 are shown). Table ?? also shows the correlations amongst the components. Inspection of the item content for component 1 suggested that the component was measuring ??? [provide a description of the construct or theme the component appeared to be measuring and provide a short mnemonic name to use as an identifier]. This component showed a Cronbach's alpha internal consistency reliability value of .??. The item content for component 2 signalled that the component was measuring ??? [provide a description of the construct or theme the component appeared to be measuring and provide a short mnemonic name to use as an identifier]. This component showed a Cronbach's alpha

(continued)

Statistical procedure	Schemata [comments on what is being said in a sentence]
Pattern Matrix 		Component	
	1	2
X4	.876	
X5	.847	
X6	.748	
X2		.818
X3		.803
X1		.780 | **Component Correlation Matrix** | Component	1	2
1	1.000	.198
2	.198	1.000		internal consistency reliability value of ??.... [adjust interpretation as needed given the number of components identified]. The two components were only weakly positively correlated. Participants [cases] were given a unit-weighted score on each component by averaging the non-missing values for those items that defined each component.
Simultaneous regression analysis	A simultaneous regression analysis was conducted predicting Y [dependent variable/ criterion] from X1, X2, X3 ... [predictors]. Inspection of relevant indicators showed that all regression assumptions were satisfied [inspect Durbin-Watson test, distribution of residuals, normal probability plot of residuals and plot of ZPRED vs ZRESID to determine if this statement applies; if it does not, you need to explain which assumption was not satisfied and what you did about it]. The predictors, considered together, explained a significant amount of the variance in Y (R^2 = .???, adjusted R^2 = .???, $F(df1, df2)$ = ???, p = .???). Predictor X2 was found to make a significant contribution to prediction of Y (B = +???, beta = .???, t(df residual) = ???, p = .???, squared part correlation [sr^2] = .???). An increase in X2 by 1 unit predicted an increase in Y of B units [if B is negative, the predicted value for Y will decrease instead of increase] ... [adjust interpretation as needed given the regression analysis

(continued)

Statistical procedure	Schemata [comments on what is being said in a sentence]
	outcome patterns for the various predictors; optionally you could compare sr^2 values to inform a judgment about which of the significant predictors was relatively more important in the prediction model].
Hierarchical regression analysis	A hierarchical regression analysis was conducted predicting Y [dependent variable/ criterion] from X1, X2, X3 … [predictors]. The predictors were entered into the regression model in the following order, based on a quasi-temporal ordering [or based on prior theory; in either case, need to unpack the logic clearly for the reader]: X1 and X2 were entered first, followed by X3 and X4… Model I F-tests (see Cohen et al. 1983, pp. 171–179) were used to evaluate all tests of significance. Inspection of relevant indicators showed that all regression assumptions were satisfied [inspect Durbin-Watson test, distribution of residuals, normal probability plot of residuals and plot of ZPRED vs ZRESID to determine if this statement applies; if it does not, you need to explain which assumption was not satisfied and what you did about it]. Predictors X1 and X2, when entered first, explained a significant amount of variability in Y scores ($R^2 = .???$, Partial F(df1, df2) [F Change]= ???, $p = .???$). Of the two predictors, only X2 contributed significantly to explaining variance in Y scores (b = ???, beta = .???, t(df residual) = ???, $p = .???$, squared part correlation [sr^2] = .???). The prediction pattern was positive such that an increase in X2 scores predicted an increase in Y scores. [adjust interpretation as needed given the regression analysis outcome pattern for the first set entered (look at SPSS Model 1 Summary stats); adjust directionality of the prediction pattern according to the sign of beta or the part correlation; the individual predictor statistics are from the Model 1 block of the Coefficients table produced by SPSS]. Predictors X3 and X4 were entered next, they explained a significant proportion of the variability in Y scores over and above what the first set of predictors could explain ($sr^2 = .???$ [R Square Change], Partial F(df1, df2) [F Change]= ???, $p = .???$). Both X3 (b = ???, beta = .???, t(df residual) = ???, $p = .???$, squared part correlation [sr^2] = .???) and X4 (b = ???, beta = .???, t(df residual) = ???, $p = .???$, squared part correlation [sr^2] = .???) contributed significantly to explaining variance in Y scores. The prediction pattern for X3 was positive such that an increase in X3 scores predicted an increase in Y scores. The prediction pattern for X4 was negative such that an increase in X4 scores predicted a decreased in Y scores. [adjust interpretation as needed given the

(continued)

Statistical procedure	Schemata [comments on what is being said in a sentence]
	regression analysis outcome pattern for the first set entered (look at SPSS Model 2 Summary stats); adjust directionality of the prediction pattern according to the sign of beta or the part correlation; the individual predictor statistics are from the Model 2 block of the Coefficients table produced by SPSS]. [continue with interpretations, until all predictor sets that have been entered into the model have been interpreted. If you decide to use Model II error for all partial F-tests, you need to recalculate each test appropriately and include those outcomes in the story, instead of the Model I error tests reported by SPSS - recall the discussion in Procedure 7.13.]
One-way MANOVA (with univariate follow-up interpretation pathway (recall Fig. 7.27) and posthoc multiple comparison tests)	A one-way between groups MANOVA was conducted using X as the independent variable [we will assume three groups for this example] defining the groups to be compared and Y1, Y2 and Y3 [tailor as needed] as the dependent variables [if there is only one Y, then a univariate ANOVA is being reported]. Inspection of relevant indicators, including Box's M test [only for MANOVA] and separate Levene's tests, revealed no problems conforming to the assumptions of normality [these indicators would be available from preliminary explorations of the dependent variables] and homogeneity of variance. A significant multivariate effect of X on the set of dependent variables was observed (Wilks' lambda = .???, MV F (Hypothesis df, Error df) = ???, p = .???, MV partial eta-squared (η^2) = .???) [these statistics available in the Multivariate Tests table of the SPSS output; MV signals a multivariate index; measures of effect must be requested]. Follow-up univariate ANOVA tests revealed that significant differences for X were observed on Y1 (F(X df, Error df) = ???, p = .???, partial eta-squared (η^2) = .???) and Y3 (F(X df, Error df) = ???, p = .???, partial eta-squared (η^2) = .???) [these statistics are available in the Tests of Between Subjects Effects table of the SPSS output, from the relevant row of the table]. For Y1, posthoc Tukey's HSD tests revealed that group x1 showed a significantly higher mean score compared to group x2, but neither group differed significantly from x3 (x1 mean = ??? (se = ???); x2 mean = ??? (se = ???); x3 mean = ??? (se = ???)). For Y3, posthoc Tukey's HSD tests revealed that group x2 showed a significantly higher mean score compared to group x3, but neither group differed significantly from x1 (x1 mean = ??? (se = ???); x2 mean = ??? (se = ???); x3 mean = ??? (se = ???)) [the Tukey's HSD results are taken from the Multiple Comparisons table of the SPSS output; the means and standard errors (se) are available from the Estimated

(continued)

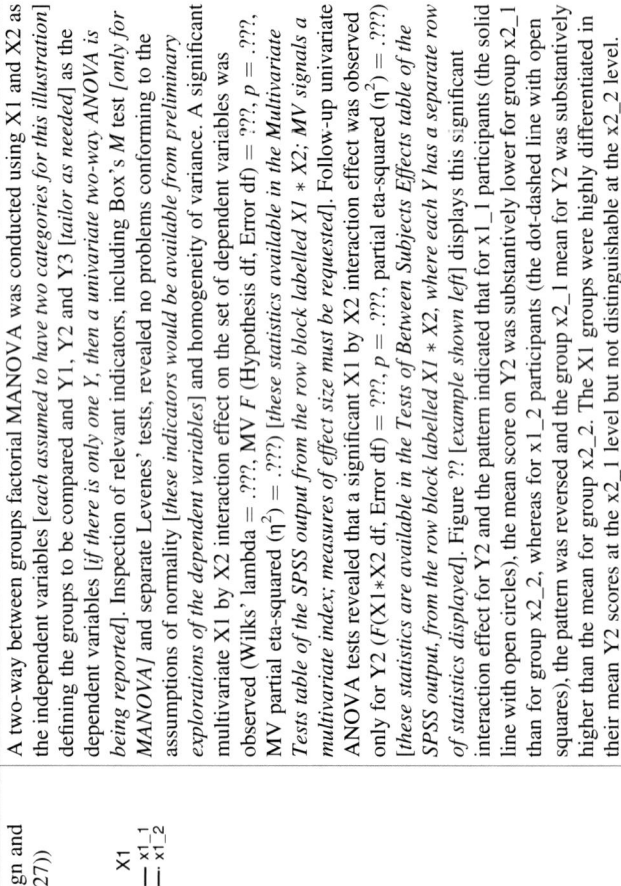

Statistical procedure	Schemata [comments on what is being said in a sentence]
	Marginal Means table of the SPSS output]. [tailor all multivariate, univariate and multiple comparison effects reported to the pattern of results that were seen in the tables].
Two-way MANOVA (for 2 × 2 factorial MANOVA design and univariate follow-up interpretation pathway (recall Fig. 7.27)) **Estimated Marginal Means of Y2** *[Line graph showing Estimated Marginal Means (y-axis, 4.50 to 5.75) vs X2 (x-axis, x2_1 to x2_2), with two lines: X1 solid line (x1_1) decreasing from ~5.75 to ~4.50, and X1 dashed line (x1_2) increasing from ~4.75 to ~5.25]*	A two-way between groups factorial MANOVA was conducted using X1 and X2 as the independent variables [each assumed to have two categories for this illustration] defining the groups to be compared and Y1, Y2 and Y3 [tailor as needed] as the dependent variables [if there is only one Y, then a univariate two-way ANOVA is being reported]. Inspection of relevant indicators, including Box's M test [only for MANOVA] and separate Levenes' tests, revealed no problems conforming to the assumptions of normality [these indicators would be available from preliminary explorations of the dependent variables] and homogeneity of variance. A significant multivariate X1 by X2 interaction effect on the set of dependent variables was observed (Wilks' lambda = .???, MV F (Hypothesis df, Error df) = ???, p = .???, MV partial eta-squared (η^2) = .???) [these statistics available in the Multivariate Tests table of the SPSS output from the row block labelled X1 * X2; MV signals a multivariate index; measures of effect size must be requested]. Follow-up univariate ANOVA tests revealed that a significant X1 by X2 interaction effect was observed only for Y2 (F(X1*X2 df, Error df) = ???, p = .???, partial eta-squared (η^2) = .???) [these statistics are available in the Tests of Between Subjects Effects table of the SPSS output, from the row block labelled X1 * X2, where each Y has a separate row of statistics displayed]. Figure ?? [example shown left] displays this significant interaction effect for Y2 and the pattern indicated that for x1_1 participants (the solid line with open circles), the mean score on Y2 was substantively lower for group x2_1 than for group x2_2, whereas for x1_2 participants (the dot-dashed line with open squares), the pattern was reversed and the group x2_1 mean for Y2 was substantively higher than the mean for group x2_2. The X1 groups were highly differentiated in their mean Y2 scores at the x2_1 level but not distinguishable at the x2_2 level. The multivariate main effect for X2 on the set of dependent variables was also significant (Wilks' lambda = .???, MV F (Hypothesis df, Error df) = ???, p = .???, MV partial eta-squared (η^2) = .???) [these statistics are available in the Multivariate Tests table of the SPSS output from the row block labelled X2; MV signals a

(continued)

Statistical procedure	Schemata [comments on what is being said in a sentence]
	multivariate index]. Follow-up univariate ANOVA tests revealed that a significant main effect for X2 was observed for Y1 (F(X2 df, Error df) = ???, p = .???, partial eta-squared (η^2) = .???) and Y3 (F(X2 df, Error df) = ???, p = ???, partial eta-squared (η^2) = .???) [these statistics are available in the Tests of Between Subjects Effects table of the SPSS output, from the row block labelled X2, where each Y has a separate row of statistics displayed]. For Y1, group x2_1 showed a significantly higher mean score compared to group x2_2 (x1 mean = ??? (se = ???); x2 mean = ??? (se = ???)). For Y3, group x2_2 showed a significantly higher mean score compared to group x2_1 (x1 mean = ??? (se = ???); x2 mean = ??? (se = ???)) [the means and standard errors (se) are available in the Estimated Marginal Means table of the SPSS output]. The multivariate test for the main effect of X1 on the set of dependent variables was not significant (Wilks' lambda = .???, MV F (Hypothesis df, Error df) = ???, p = .???, MV partial eta-squared (η^2) = .???). [tailor all multivariate, univariate and multiple comparison effects reported to the pattern of results that were seen in the tables].
Cluster analysis (with internal and external validation analyses (recall Procedure 6.6))	A hierarchical cluster analysis was conducted using scores on Y1, Y2, Y4 and Y5 [tailor as needed] to form a typology of participants [cases]. Ward's minimum variance method was employed using squared Euclidean distance as the measure of dissimilarity. All variables were standardised to z-scores prior to the clustering process. Inspection of the resulting dendrogram and examination of changes in distance coefficient for the final few cluster merges revealed that 3 [tailor this number to the number observed; we will assume a 3-cluster solution for this example] clusters appeared to be the appropriate number of clusters to interpret. The clusters were internally validated through an exploration of their differences on the clustering variables. Figure ?? [example shown at left] shows the profiles for the 3 clusters using means in their original scale of measurement. Cluster 1 (n = ???; solid line with open circles) was characterised by relatively higher average responses on Y1, Y2 and Y3 and a relatively lower average response on Y4 relative to Clusters 2 and 3, leading to this cluster being labelled ???. Cluster 2 (n = ???; dot-dashed line with open squares) was characterised by mid-level average responses on all 5 clustering variables; this cluster was thus labelled ???. Cluster 3 (n = ???; dashed line with xs) was

(continued)

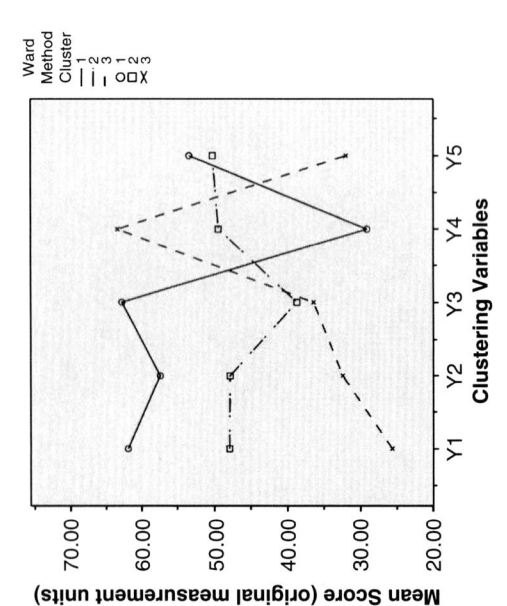

Statistical procedure	Schemata [*comments on what is being said in a sentence*]
[Chart: Line plot titled "Clustering Variables" with x-axis labels Y1, Y2, Y3, Y4, Y5 and y-axis "Mean Score (original measurement units)" ranging 20.00 to 70.00. Legend: Ward Method Cluster 1, 2, 3.]	characterised by much lower average responses on Y1, Y2 and Y5 compared to cluster 1 and a relatively higher average response on Y4 compared to Clusters 1 and 2; cluster was thus labelled *???* [*you want to give a meaningful label to each cluster to summarise its pattern, based on the actual measures used to build the clusters*]. The three clusters were externally validated by assessing their relationship with other demographic variables and other measurements that were not used to produce the clusters. Contingency table analysis relating the clusters with Y6 [*a categorical variable like gender or ethnic background; here we will assume Y6 has three categories*] showed a significant association (chi-squared $[\chi^2]$(df) = ???, p - ???, Cramer's V = ???). Inspection of adjusted standardised residuals revealed that, for Cluster 1, significantly more than expected observations fell into category y6_1 of variable Y6 whereas significantly fewer than expected observations fell into category y6_2 and y6_3. The reverse trend was noted for both Cluster 2 and Cluster 3 ... [*adjust interpretation as needed given the crosstabulation table patterns and, where appropriate, use the cluster labels, rather than the generic 'Cluster?' labels; you may wish to comment on row % and column % patterns as well*]. A one-way between groups MANOVA was conducted using the clusters to define the independent variable to be compared and Y7, Y8 and Y9 [*tailor as needed*] as the dependent variables [*if there is only one Y, then a univariate ANOVA is being reported*]. Inspection of relevant indicators, including Box's M test [*only for MANOVA*] and separate Levenes' tests, revealed no problems conforming to the assumptions of normality [*these indicators would be available from preliminary explorations of the dependent variables*] and homogeneity of variance. A significant multivariate effect of cluster membership on the set of dependent variables was observed (Wilks' lambda = ???, MV F (Hypothesis df, Error df) = ???, p = .???, MV partial eta-squared (η^2) = .???) [*these statistics available in the Multivariate Tests table of the SPSS output; MV signals a multivariate index; measures of effects size must be requested*]. Follow-up univariate ANOVA tests revealed that significant differences for clusters were observed on Y7 (F(X df, Error df) = ???, p = ???, partial eta-squared (η^2) = .???) and Y9 (F(X df, Error df) = ???, p = .???, partial eta-squared (η^2) = .???) [*these statistics available in the Tests of Between Subjects*

(continued)

Appendices

Statistical procedure	Schemata [comments on what is being said in a sentence]
	Effects table of the SPSS output, from the relevant row of the table]. For Y7, posthoc Tukey's HSD tests revealed that Cluster 1 showed a significantly higher mean score compared to Cluster 2 (Cluster 1 mean = ??? (se = ???); Cluster 2 mean = ??? (se = ???). For Y9, Cluster 2 showed a significantly higher mean score compared to Cluster 1 and Cluster 3 (Cluster 1 mean = ??? (se = ???); Cluster 2 mean = ??? (se = ???); Cluster 3 mean = ??? (se = ???)) [the means and standard errors (se) are available from the Estimated Marginal Means table of the SPSS output; if there are three or more clusters, you need to request and interpret Tukey's HSD multiple comparison tests for the dependent variables]. [tailor all multivariate, univariate and multiple comparison effects reported to the pattern of results that were seen in the tables].

The Bigger Picture of Writing up

Figure C.1 presents a big picture mindmap of some important issues and considerations associated with the conduct and reporting of quantitative research. Before any writing commences, you should review closely the considerations that underpinned your research design and the choices you made in order to obtain your data. Your thinking needs to be reflected when you write up the methodology, analysis and conclusion stories. Your research questions and hypotheses emerge from your overall research framework and should guide the bulk of analyses undertaken. Part of the research story will concern the character and quality of your quantification systems (i.e. your measurement strategies). There will likely be specific statistical analyses you can draw upon to flesh out the measurement stories you will need to write. You need to consider carefully how to manage your data analyses as well as how you write them up. You need to remain fully aware of what research quality criteria demand of your research as well as what readers/reviewers will be looking for in your stories. Keeping these issues in mind along with your research questions and hypotheses as you conduct your analyses may help you to decide which analytical procedures to use, what types of tables and graphs you might need, how to effectively bring descriptive, relational and inferential stories together and so on. Hardy and Bryman (2004) reinforced the common threads amongst different techniques of data analysis (and they included qualitative data analysis in their thinking) in terms of the goals they are intended to achieve. These goals need to be kept in mind as you write up your analysis stories so that you can maximise the power of your stories to convince a reader of the meaningfulness of what you have learned.

Figure C.2 presents an extensive mindmap that displays more general questions, issues and considerations associated with the production of statistical stories as well as highlighting some of the key linkages between them. The preliminaries help to establish the foundations upon which convincing statistical stories will be based in the write-up. You would want to unfold a narrative about some of these preliminary activities in order to set the stage for the reader for the analytical activities and stories to come. Descriptive, relational and hypothesis testing stories all focus on addressing key questions for the reader, only some of which may be formally set out in terms of research questions, hypotheses or propositions. For example, a demographic story about your sample is usually expected in any report of social or behavioural research, but this story is seldom the target of a formal research question – it is often just part of setting the stage. Some aspects of data exploration may never actually be reported to a reader – they merely signal analytical processes you might use to gain an early first-blush understanding of what your data look like and how they generally behave. Note that part of any hypothesis testing story focuses on statistical significance, but that story needs to be backed up by indicators of and reflections on practical significance in terms of effect size.

Appendices

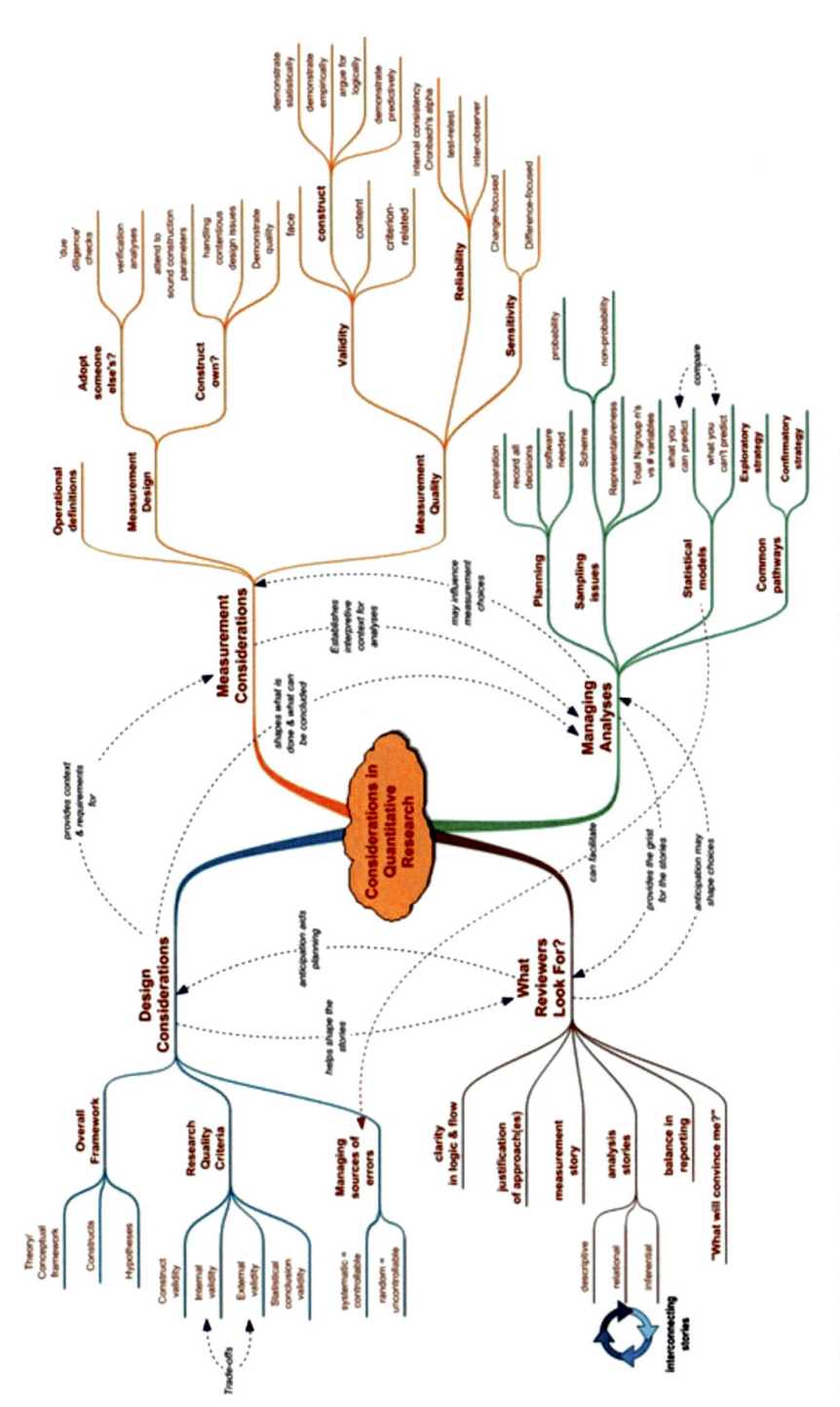

Fig. C.1 Mindmap of issues and considerations associated with quantitative research at the big picture level

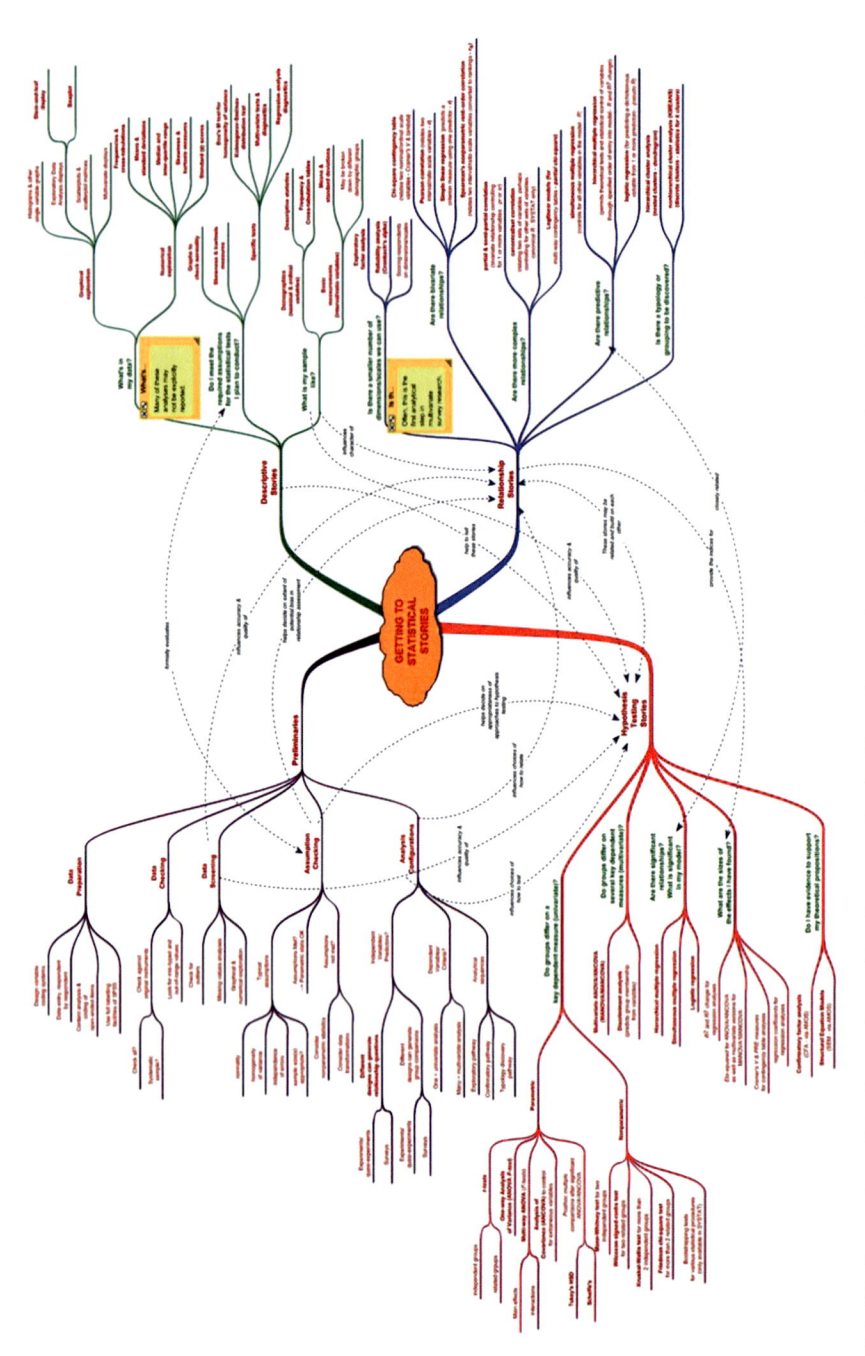

Fig. C.2 Mindmap of questions, issues & considerations associated with getting to relevant stories in reporting statistical analyses

Sequencing of Stories

There is a definite linear sequence to unfolding statistical stories in the larger context of research writing and this sequence needs to be properly configured (balancing narrative with strategically chosen numbers, tables and graphs) and written to maximise the chances of convincing the reader about what you have found and about the meaningfulness/importance/usefulness of your conclusions. Writing up of methodological stories and statistical stories is generally done in the past tense; any implications of those stories for the here and now can be written in present tense.

1. **Contextual and methodological stories:**
    - Summarise the essence of the research questions/hypotheses (*narrative*).
    - Contextual story about the research endeavour as a whole, to place the data and participants in their proper larger contexts (*narrative*).
    - Methodological story about how the data were gathered, design and quality of measurements, how participants were sampled, and ethical protections provided and a foreshadowing of the analytical approach to be taken to connect the data stories with the research questions (*narrative*).

2. **Demographic story about the research participants**:
    - Story about who (or what, in the case of documents or secondary data) provided the measurements to be analysed, in terms of key demographic characteristics and patterns of response and non-response (*narrative backed up by relevant numbers, one or two tables and/or graphs*).
    - Comment, where appropriate, about representativeness of the sample relative to the intended population – helps to shape the reader's expectations about the extensional reach of your generalisations (*narrative backed up by key numbers and/or a table*).

3. **Stories summarising preliminary explorations and screening of the data:**
    - Story about what preliminary explorations of the data revealed about distributional patterns and meeting of key assumptions (e.g. normality, homogeneity of variance, independence of errors, presence of outliers, etc.) for the planned analyses (*generally narrative backed up with numbers and perhaps a strategically chosen graph*).
    - Story about patterns of non-response/missing data and comments on possible biases in your sample (*narrative with numbers and perhaps a key table or graph*).

4. **Where relevant, stories about data condensation and/or construct validation**:
    - Short story about the approach to data condensation/construct validation and the analytical choices made and decision criteria used (*narrative*)

- Story about how measurement items were condensed into smaller number of composites/components/factors including interpretation and clear labelling of each emerging composite/component/factor and incorporating any outcomes of refining analyses (*narrative backed up with numbers and a key table*)
- Story about the reliability of each composite/component/factor and an indication of how participants were scored on each composite/component/factor (*narrative backed up with numbers*).

5. **Stories focusing on the research questions/hypotheses**:

- Story that summarises and interprets the analyses for targeting specific research questions/hypotheses using relevant descriptive statistics and/or relational statistics, accompanied by hypothesis test results to signal statistical significance and using effect sizes to indicate practical significance; focus should be on a basic interpretation of what the analysis results meant, what the important patterns were and what they indicate with respect to the research question/hypothesis (*narrative backed up with numbers and, where strategic to do so, key tables or graphs*).

6. **Stories emerging from any supplementary analyses**:

- Story that summarises and interprets the analyses for targeting specific research questions using relevant descriptive statistics and/or relational statistics, accompanied by hypothesis test results to signal statistical significance and using effect sizes to indicate practical significance; focus should be on a basic interpretation of what the analysis results meant, what the important patterns were and why the supplemental analyses were important to conduct (*narrative backed up with numbers and, where strategic to do so, key tables or graphs (perhaps reported in an appendix), but with less emphasis than for the research question/hypothesis analyses*).

7. **Overall concluding stories**:

- Story summarising what has been learned from the analyses relative to the research questions (*narrative*).
- Story logically arguing some justifiable (based on your research evidence and in recognition of any limitations to/caveats on your inferences) conclusions and generalisations about the patterns identified through those analyses, including the implications of any data or sample anomalies for those conclusions (*narrative*).

References

Cohen, J., Cohen, P., Aiken, S. G., & West, L. S. (1983). *Applied multiple regression/correlation analysis for the behavioral sciences* (3rd Ed.). Mahwah, NJ: Lawrence Erlbaum Associates.

Cohen, J., Cohen, P., West, S. G. & Aiken, L. S. (2003). *Applied multiple regression/correlation analysis for the behavioral sciences* (3rd ed). Mahwah: Lawrence Erlbaum Associates.

Cooksey, R. W., & McDonald, G. (2019). *Surviving and thriving in postgraduate research* (2nd ed.). Singapore: Springer Nature, ch. 21, 22.

Hardy, M., & Bryman, A. (2004). Introduction: Common threads among techniques of data analysis. In M. Hardy & A. Bryman (Eds.), *Handbook of data analysis* (pp. 1–13), London: Sage Publications.

Miller, J. E. (2005). *The Chicago guide to writing about multivariate analyses*. Chicago: University of Chicago Press.

Index

A
Agent-based modelling, 678
Alternative hypothesis, 243
Analysis
  cluster, 203
  common factor, 192, 193
  component, 193
  confirmatory factor, 516
  conjoint, 349
  correspondence, 219
  of covariance, 403
  factor, 192
    confirmatory, 192
    exploratory, 192
  interrupted time series, 506
  item, 457
  level 1, 605
  level 2, 605
  MANCOVA, 409
  MANOVA, 409
  missing values, 470, 476
  non-interrupted time series, 502
  principal component, 193
  Rasch model, 564
  reliability, 456
  survival/failure, 574
  time series, 501
    with intervention, 506
Analysis of Moment Structures (AMOS), 45
ANCOVA, 403
  'homogeneity of regression lines'
      assumption, 407
ANOVA, 312, 342
  between-groups, 342
  factorial, 312
  factorial design, 343
  interaction, 344
  main effect, 344
  repeated measures, 357
  within-groups, 342
Apriori algorithm, 655
A priori contrasts, 324
Artificial neural network, 658
Assessing
  central tendency, 94
  variability, 100
Association, 160–168
Asymmetric, 112
Autocorrelation, 501
Autoregressive Integrated Moving Averages
      (ARIMA), 501

B
Bar chart, 69
Bayesian statistical inference, 560
Bell-shaped distribution, 128
Best-worst scaling models, 596
Beta (β), 245
  error, 245
  population parameter, 696
  weights, 179
Biserial correlation, 150, 155
Bivariate, 62, 68, 150, 160, 168, 280, 318
      372, 393, 429, 455, 487, 492, 540,
      574, 580
  statistics, 4
Bonferroni correction, 253
Bootstrapping, 15, 493
Box-and-whisker plot, 114

Box's *M* test, 417
Boxplot, 102, 114

C

Canonical
  correlation, 224
  variate, 225
Causal models, 525
Cause and effect, 4
Censored data, 574, 635
Central tendency, 94
Chart
  bar, 69
  Control X-bar, 582
  frequency polygon, 70
  histogram, 70
  line graph, 73
  multivariate, 80
  Pareto, 581
  pie, 69
  quality control, 580
  R, 584
  scatterplot, 75
  scatterplot matrix, 80
  spider, 82
Chernoff's faces plot, 88
Chi-squared, 251, 697
Choice modelling, 596
Cluster analysis, 203
  external validation, 208
  hierarchical, 204
  non-hierarchical, 203
Cluster sample, 276
Coefficient
  phi, 161
Cohen's kappa coefficient of agreement, 464
Common factor analysis, 193
Communalities, 195
Component analysis, 193
Computational modelling, 668, 669, 673
Computer-adaptive testing, 571
Conditional probability, 110
Confidence
  intervals, 130, 487
Confirmatory factor analysis, 516
Confounding, 267
  variable, 11
Conjoint
  analysis, 349
  measurement, 349
Construct validity, 25
Contingency tables, 66

test of significance, 280
Contrasts
  a priori, 324
  orthogonal, 324
  Scheffé, 330
  uncorrelated, 324
Control
  by design, 266, 347
  procedural, 266
  through statistical analysis, 266, 403, 409
Control X-bar charts, 582
Convenience sample, 277
Correlation, 6, 142
  biserial, 150, 155
  canonical, 224
  factor, 196
  matrix, 150
  part, 186, 383
  partial, 189, 403
  Pearson, 145, 150
  point-biserial, 150, 154
  semi-partial, 186
  set, 191, 224, 228
  Spearman's rho, 150, 152
  spurious, 13
  tetrachoric, 150, 156
Correlational, 150, 160, 168, 175, 191, 203,
       215, 224, 318, 455, 622, 633, 652
  statistics, 7 (*see also* Chapter 6)
Correspondence analysis, 219
Covariate, 267, 402, 403
Covariation, 5
Cox regression, 575
Cramer's V statistic, 162
Criterion, 373
  measure, 8
Cronbach's alpha, 456
Crosstabulation, 64
Cross-validation, 183, 421, 707
Cumulative distribution, 63

D

Data mining, 205, 653
  Apriori algorithm, 655
  decision tree induction, 654
  FP-Growth algorithm, 656, 657
  genetic algorithms, 654, 657
  latent class analysis, 656
  neural network analysis, 654, 657, 658
  Support Vector Machine (SVM) algorithm,
       655, 657
Data screening, 470–476

Index 731

Degrees of freedom, 251
Dendrogram, 204, 205
Density trace, 118
Dependent
　measure, 8
　variable, 4, 8
Dependent samples *t*-test, 297
Descriptive, 62, 68, 79, 94, 100, 112, 121, 469, 574, 580, 622, 668
　statistics, 7 (*see also* Chapter 5)
Deviation, 103
Discriminant analysis, 420
　canonical correlation, 422
　descriptive, 420
　discriminant functions, 420
　multivariate pathway for MANOVA, 411, 415
　predictive, 420
Distractor analysis, 460
Distrator analysis, 460
Distribution, 8, 62
　bell-shaped, 128
　cumulative, 63
　frequency, 63
　Gaussian, 128
　normal, 15, 128
　sampling, 246, 493
Dummy coding, 182, 259
Durbin-Watson test, 381
Dynamic systems modelling, 674

E

Effect size, 245, 318
　eta-squared, 320
　logic, 257
　in a meta-analysis, 542
　omega-squared, 320
Eigenvalues, 195
Endogenous variables, 529
Error
　beta, 245
　model, 256
　Model I error (hierarchical regression), 387
　Model II error (hierarchical regression), 387
　Type I (alpha), 244
　Type I (beta)
　　Bonferroni correction, 253
　　inflation problem, 253
　Type II, 245
　Type II (beta), 245
Eta-squared, 320
　partial, 321

eViews, 43
Exact tests, 672
Excel, 44
Exogenous variables, 527
Experimental design, 264, 270
　balanced incomplete block, 590
　conjoint measurement, 590
　factorial posttest only control, 272
　fractional factorial design, 349
　interrupted time series, 273
　non-equivalent control group pretest-posttest, 273
　non-equivalent static group, 273
　posttest only control group design, 271
　pretest-posttest control groups, 271
　quasi-experimental, 272
　repeated measures, 297, 357, 367
　Solomon four group design, 271
　within-groups factorial, 342
Exploratory data analysis (EDA), 112
External validation of clusters, 208
Extraneous variable, 11, 265

F

Factor
　correlations, 196
　pattern matrix, 196
　rotation
　　oblique, 194
　　orthogonal, 194
　　promax, 194, 196
　　varimax, 194, 196
　structure matrix, 196
Factor analysis, 192
　common, 192
　confirmatory, 192
　exploratory, 192
　extraction methods, 193
　principal component analysis, 193
Factorial design
　between-groups ANOVA, 342
　complete, 590
　fractional, 590
FP-growth algorithm, 657
Frequency
　crosstabulation, 64
　distribution, 63
　polygon, 70
　relative, 63
　tabulation, 62
Friedman rank test, 367
F-test, 414

G

Gamma statistic, 164
Gaussian distribution, 128
General linear model, 256
Goodman & Kruskal lambda statistics, 163
Group design
    non-equivalent static, 273
    posttest only control, 271
    pretest-posttest control, 271

H

Heteroscedasticity, 146, 147, 474
Hierarchical
    cluster analysis, 204
    linear models, 603
    regression, 185, 373
Hierarchical linear models (HLM), 603
Hinge-spread, 102
Histogram, 8, 70
    plot, 88
Homogeneity of variance, 254, 287, 312, 380
H-spread, 102
Hypothetical constructs, 23

I

Icon plot, 87
Imputation, 483
Independent groups t-test, 251, 286
    variable, 11
Inferential, 280, 286, 292, 297, 305, 312, 323, 335, 341, 357, 366, 372, 393, 402, 409, 420, 429, 487, 492, 500, 515, 525, 540, 564, 574, 580, 589, 603, 614, 622, 633, 652, 668
    statistics, 12 (*see also* Chapter 7)
Interaction, 14, 350
Internal consistency reliability, 456
Inter-observer reliability, 464
Interquartile range (IQR), 101
Interrupted time series design, 273
Interval scale, 27
Intraclass correlation, 605, 702
IQ scores, 126
Item
    analysis, 457
    difficulty, 459
    discrimination, 459
    response theory, 568

J

Jackknifing, 498

K

Kaiser-Meyer-Olkin (KMO) measure of smapling accuracy, 194
Kendall's Tau statistic, 164, 592
Kolmogorov-Smirnov test, 295
Kruskal-Wallis
    Rank test, 316, 335
Kurtosis, 8

L

Latent class analysis, 656
Latent trait models, 568
Levene's F-test, 417
Levene's test, 717, 720
Leximancer, 662
Likert-type scale, 27
Line graph, 73
Linear transformation, 126
Listwise deletion, 482
Logistic regression, 393
Log-linear models, 429
    general, 430
    hierarchical, 430
    logit, 432

M

Main effect, 14, 350
MANCOVA, 409
Mann-Whitney U-test, 292
MANOVA, 409
    multivariate F-test, 409
    stepdown F-tests, 414
Mardia's multivariate tests for skewness & kurtosis, 476
Marginal distributions, 65
Mathematical
    principles, 699–700
    symbols, 698
Mean, 8, 13, 94
    sample, 19
    substitution, 483
Measure, 21
Measurement scales, 26
Median, 96
Mediating variable, 13
Mesokurtic, 15
Meta-analysis, 541
Midspread, 103
Mindmap
    considerations in quantitative research, 724
    getting to statistical stories, 726
    limits to general linear models, 262
    SPSS, 711

Index   733

Minimum variance method, 204
Missing values analysis, 470, 476
    missing completely at random (MCAR), 477
    missing not at random (MNAR), 477
Mode, 97
Models
    best-worst scaling, 596
    causal, 525
    general linear, 256
    hierarchical linear, 603
    latent trait, 568
    log-linear, 429
    mixed, 605
    multilevel, 603
    one-parameter, 568
    Rasch, 564
    structural equation, 525
    three-parameter, 570
    two-parameter, 569
Moderating variable, 14
Modification indices, 519
Monte Carlo simulation methods, 668
    Markov Chain, 672
MPlus, 44
Multicollinearity, 522
Multidimensional scaling, 216
Multilevel
    latent growth model, 646
Multilevel models, 603
Multiple
    correlation coefficient, 180
    correspondence analysis, 219
    regression, 176, 373
Multiple comparison test
    Tukey's HSD, 328
Multiple regression
    coefficients, 376
    hierarchical model, 373
    normal Q-Q (or P-P) plot, 381
    residuals histogram plot, 381
    residuals-other indices, 382
    residuals plot (predicted *vs.* residual values), 380
    simultaneous model, 373
    standardised (beta) regression coefficients, 377
    stepwise model, 373
    unstandardised (B) regression coefficients, 377
Multiplot, 84
Multistage sample, 276
Multivariate, 79, 175, 191, 203, 215, 224, 357, 372, 393, 409, 420, 429, 455, 469, 487, 492, 500, 515, 525, 540, 564, 574, 580, 589, 603, 614, 622, 633, 652, 668
    analysis of
        covariance, 409
        variance, 409
    graphs, 80
    statistics, 14

N
η^2, 320
Nagelkerke R Square, 395
Neural network analysis, 654, 658
Nominal scale, 26
Nonhierarchical cluster analysis, 203
Nonlinearity, 147
Nonparametric, 150, 160, 203, 215, 280, 292, 305, 335, 366, 492, 614, 633, 652, 668
    statistics, 15
Normal distribution, 15, 62, 128
Nuisance variable, 11
Null hypothesis, 243
Number Cruncher Statistical system (NCSS), 37
n-way ANOVA, 342

O
Omega-squared, 320
Omnibus test of model coefficients, 395
One-parameter models, 568
One-way analysis of variance (ANOVA), 312
Operational definition, 24
Ordinal scale, 26
Orthogonal
    contrasts, 324
    polynomial tests, 327
Orthogonal (uncorrelated) contrasts, 324

P
p, 252
Paired t-test, 297
Pairwise deletion, 482
Parallel coordinate plot, 86
Parameters, 16
    population, 696–697
Parametric, 150, 168, 175, 191, 224, 286, 297, 312, 323, 341, 357, 372, 393, 402, 409, 420, 429, 487, 500, 515, 525, 540, 564, 574, 580, 589, 603, 614, 633, 652
    statistics, 16
Pareto charts, 581

Partial correlation, 189
Partial least squares (PLS-SEM), 532
Partialling, 185
Path diagram, 526
Pathway
   confirmatory model testing, 706
   exploratory, 704
   hybrid cross-validation, 707
   typology discovery and testing, 709
Pearson
   chi-square, 280
   correlation coefficient, 145, 150
   product-moment correlation coefficient, 145, 150
Percentiles, 63
Phi coefficient, 161
Pie chart, 69
Planned comparison test, 324
Plot
   boxplot, 102, 114
   Chernoff's faces, 88
   dendrogram, 204
   density trace, 118
   histogram, 88
   icon, 87
   interaction, 350
   multiplot, 84
   parallel coordinate, 86
   profile, 88
   radar, 82
   spider graph, 82
   sunray, 91
   violin, 118
Point-biserial correlation, 150, 154
Population, 17
Posthoc multiple comparison tests, 328
Power polynomial regression, 389, 640
PRE statistics, 163, 615
Predictors, 12, 373
   coding categorical
      dummy coding, 182, 259
      effect coding, 259
Preference modelling, 590
Preparing data, 46
Pretest-posttest control group design, 271
Principal components analysis, 193
Probability, 17, 62
   base rate, 562
   conditional, 110
   posterior, 562
   prior, 562
   simple, 108
Profile plot, 88

Proportional reduction in error
   measure, 615
   statistics, 163
Pseudo R^2, 395
Purposive sample, 277
p-value, 252, 370

Q
QCI database, 51
Quality control charts, 580
Quantitative stories, 56, 722
Quasi-experimental design, 264, 270
   interrupted time series (single participant) design, 273
   non-equivalent control group pretest-posttest design, 273
   non-equivalent static group design, 273
   single group pretest-posttest design, 272
   single-shot posttest only design (many surveys), 272
Quota sample, 276

R
r, 145
R, 39
R^2, 180, 187, 373
   pseudo, 395
   testing, 375
R charts, 584
R Commander, 40
Radar plot, 82
Random sampling, 18, 274
Range, 100
Rasch model
   analysis, 459, 564
Ratio scale, 28
Regression
   centring, 645
   coefficients, 376
   Cox, 575
   fuzzy, 639
   generalised least squares, 644
   hierarchical, 185, 373
   latent class, 638
   line, 169
   logistic, 393
   mediated, 647
   moderated, 647
   multiple, 176, 373
   nonlinear, 640
   nonparametric, 641

Index 735

  ordinal, 636
  power polynomial, 640
  probit, 634
  ridge, 644
  robust, 641
  segmented, 507
  simple linear, 169
  tobit, 635
  two-stage least squares, 642
  weighted least squares, 642
  weights–standardised, 179, 379
Related groups t-test, 297
Relative frequency, 63
Reliability
  internal consistency (Cronbach's alpha), 456
  inter-observer, 464
  inter-rater, 464
Reliability analysis, 456
Repeated measures
  ANOVA, 357
  design, 342
Resampling methods, 498
Residuals
  diagnostic patterns, 382

S

Sample, 19
  cluster, 276
  convenience, 277
  mean, 19
  multistage, 276
  purposive, 277
  quota, 276
  simple random, 275
  snowball, 277
  stratified random, 275
  systematic, 276
  volunteer, 277
Sampling, 274
  distribution, 246, 493
  non-probability scheme, 274
  probability scheme, 274
  random, 18, 274
Scale
  interval, 27
  Likert-type, 27
  measurement, 26
  nominal, 26
  ordinal, 26
  ratio, 28
Scatterplot matrix, 80

Scatterplots, 75
Scheffé
  contrasts, 330
Score
  factor, 200, 705
  IQ, 126
  standard, 121
  stanines, 126
  T, 126
  z, 121
Semi-interquartile range (S-IQR), 102
Semi-partial correlation, 186, 383
Set correlation, 191, 224, 228
Sign test, 310
Simple
  linear regression, 169
  probability, 108
  random sample, 275
Simulation, 668
Single group pretest-posttest design, 272
Single-shot posttest only design, 272
Six sigma, 587
Skewness, 8
Snowball sample, 277
Sobel's z-test, 648, 702
Social network analysis, 622
Software
  AMOS, 45
  eViews, 43
  Excel, 44
  HLM, 614
  ICLUST, 205, 215
  InFlow, 633
  InsightMaker, 674, 684
  Latent GOLD, 667
  Leximancer, 662, 667
  LIMDEP, 602
  MINISTEP/WINSTEPS, 566, 573
  MPlus, 44
  NCSS, 37
  NetDraw, 628, 633
  NetLogo, 679, 684
  NodeXL, 633
  R, 39
  RapidMiner, 667
  R Commander, 40
  RUMM, 574
  SAS, 43
  Sentinel Visualizer, 633
  SmartPLS, 540
  SPSS, 33
  Stata, 43
  STATGRAPHICS Centurion, 38

Software (*cont.*)
   STATISTICA, 43
   Stella/iThink, 675, 684
   SYSTAT, 35
   UCINET, 628, 633
   UNISTAT, 44
   XLMiner, 667
   XLSTAT, 44
Somer's D statistic, 164
Spearman's rho correlation, 150, 152
Spider graph, 82
SPSS, 33, 713
Spurious correlation, 13
Standard
   deviation, 105
   error, 278, 493
   score, 121
Standardised regression weights, 179
Stanines, 126
STATGRAPHICS Centurion, 38
Statistic, 19
   Cramer's V, 162
   gamma, 164
   Goodman & Kruskal lambda, 163
   Kendall's Tau, 164
   Proportional Reduction in Error (PRE), 163
   Somer's D, 164
STATISTICA, 43
Statistical analysis system (SAS), 43
Statistical inference, 19, 243
   courtroom analogy, 244
   hurdle logic for testing, 253
Statistics
   bivariate, 4
   correlational, 7 (*see also* Chapter 6)
   descriptive, 7 (*see also* Chapter 5)
   inferential, 12 (*see also* Chapter 7)
   multivariate, 14
   nonparametric, 15
   parametric, 16
   sample, 697–698
   univariate, 21
Stem & leaf displays, 113
Stratified random sample, 275
Structural equation models, 525
   covariance-based, 526
   partial least squares (PLS-SEM), 532
Sunray plot, 91
Supervised learning, 653
Support vector machine (SVM) algorithm, 655
Survival/failure analysis, 574
Symmetry, 15
SYSTAT, 35
Systematic sample, 276

T
Temporal ordering, 5
Test
   Anderson-Darling, 472
   Bartlett's chi-square test of sphericity, 194, 672
   dependent samples t, 297
   F, 414
   Friedman rank, 367
   independent groups t-test, 251, 286
   Kolmogorov-Smirnov, 295, 472
   Kruskal-Wallis Rank, 316, 335
   Little's MCAR, 482
   Mann-Whitney U-test, 292
   omnibus test of model coefficients, 395
   orthogonal polynomial, 327
   paired t, 297
   Pearson's chi-square, 280
   planned comparison, 324
   posthoc multiple comparison, 328
   related groups t, 297
   Scheffé, 315
   Shapiro-Wilk, 472
   Sign, 310
   of significance, 20, 280
   statistic, 251
   Tukey's HSD, 315
   Wilcoxon
      Rank Sum, 295
      Signed Ranks, 306
Tetrachoric correlation, 150, 156
Text mining, 661
Three-parameter models, 570
Time series analysis, 501
   advanced modelling approaches, 504
   ARIMA, 501
   interrupted, 506
   non-interrupted, 502
   segmented regression, 507
Transformation, 131, 148
   log, 130, 473
Treatment, 12
T-scores, 126
t-test, 286, 297
Tukey's HSD test, 315, 328
Two-parameter models, 569
Type II error, 245

U
Uniqueness, 516
UNISTAT, 44
Univariate, 62, 68, 94, 100, 112, 121, 286, 292, 297, 305, 312, 323, 335, 341, 357,

366, 402, 469, 487, 492, 540, 580, 668
statistics, 21
Unsupervised learning, 653
U-test, 292

V
Validity
  construct, 25
  external, 264
  internal, 264
Variability, 8, 21, 100
  explained, 22
Variables, 6, 21, 696
  dependent, 8
  extraneous, 11, 265
  independent, 11
  mediating, 13
  moderating, 14
  two dependent, 4
Variablewise deletion, 482
Variance, 103
  accounted for, 22
  explained, 22
Variance partitioning, 374

hierarchical Venn diagram, 374
simultaneous Venn diagram, 374
Venn diagram
  partial correlation, 189
  r^2, 170
  semi-partial correlation, 186
Violin plot, 118
Volunteer sample, 277

W
ω^2, 320
Ward's Method, 204
Wilcoxon
  Rank Sum test, 295
  Signed Ranks test, 306
Wilks' lambda, 409

X
χ^2, 280
XLSTAT, 44

Z
z-score, 121

Printed in the United States
by Baker & Taylor Publisher Services